Lecture Notes in Electrical Engineering

Volume 139

David Jin and Sally Lin (Eds.)

Advances in Electronic Engineering, Communication and Management Vol. 1

Proceedings of 2011 International Conference on Electronic Engineering, Communication and Management (EECM 2011), held on December 24–25, 2011, Beijing, China

Springer

Prof. David Jin
International Science & Education Researcher Association
Wuhan
China
E-mail: dayang1818@163.com

Prof. Sally Lin
International Science & Education Researcher Association
Guang Zhou
China
E-mail: julia499780828@163.com

ISBN 978-3-642-27286-8 e-ISBN 978-3-642-27287-5

DOI 10.1007/978-3-642-27287-5

Lecture Notes in Electrical Engineering ISSN 1876-1100

Library of Congress Control Number: 2011943324

Typeset by Scientific Publishing Services Pvt. Ltd., Chennai, India.

Printed on acid-free paper

9 8 7 6 5 4 3 2 1

springer.com

Preface

International Science & Education Researcher Association (ISER) puts her focus on studying and exchanging academic achievements of international teaching and scientific research, and she also promotes education reform in the world. In addition, she serves herself on academic discussion and communication too, which is beneficial for education and scientific research. Thus it will stimulate the research interests of all researchers to stir up academic resonance.

EECM2011 is an integrated conference concentrating its focus upon Electronic Engineering, Communication and Management. In the proceeding, you can learn much more knowledge about Electronic Engineering, Communication and Management of researchers all around the world. The main role of the proceeding is to be used as an exchange pillar for researchers who are working in the mentioned field. In order to meet high standard of Springer, LNEE series, the organization committee has made their efforts to do the following things. Firstly, poor quality paper has been refused after reviewing course by anonymous referee experts. Secondly, periodically review meetings have been held around the reviewers about five times for exchanging reviewing suggestions. Finally, the conference organization had several preliminary sessions before the conference. Through efforts of different people and departments, the conference will be successful and fruitful.

EECM2011 is co-sponsored by International Science & Education Researcher Association, Beijing Gireida Education Co. Ltd and Wuhan University of Science and Technology, China. The goal of the conference is to provide researchers from Electronic Engineering, Communication and Management based on modern information technology with a free exchanging forum to share the new ideas, new innovation and solutions with each other. In addition, the conference organizer will invite some famous keynote speaker to deliver their speech in the conference. All participants will have chance to discuss with the speakers face to face, which is very helpful for participants. During the organization course, we have got help from different people, different departments, different institutions. Here, we would like to show our first sincere thanks to publishers of Springer, LNEE series for their kind and enthusiastic help and best support for our conference, especially to Mr. Ditzinger Thomas for his great help. Secondly, the authors should be thanked too for their enthusiastic writing attitudes toward their papers. Thirdly, all members of program chairs, reviewers and program committees should also be appreciated for their hard work.

In a word, it is the different team efforts that they make our conference be successful on December 24–25, 2011, Beijing, China. We hope that all of participants can give us good suggestions to improve our working efficiency and service in the future. And we also hope to get your supporting all the way. Next year, in 2012, we look forward to seeing all of you at EECM2012.

October 2011 ISER Association

Committee

Honor Chairs

Prof. Chen Bin	Beijing Normal University, China
Prof. Hu Chen	Peking University, China
Chunhua Tan	Beijing Normal University, China
Helen Zhang	University of Munich, China

Program Committee Chairs

Xiong Huang	International Science & Education Researcher Association, China
LiDing	International Science & Education Researcher Association, China
Zhihua Xu	International Science & Education Researcher Association, China

Organizing Chair

ZongMing Tu	Beijing Gireida Education Co. Ltd, China
Jijun Wang	Beijing Spon Technology Research Institution, China
Quanxiang	Beijing Prophet Science and Education Research Center, China

Publication Chair

Song Lin	International Science & Education Researcher Association, China
Xionghuang	International Science & Education Researcher Association, China

International Committees

Sally Wang	Beijing Normal University, China
LiLi	Dongguan University of Technology, China
BingXiao	Anhui University, China
Z.L. Wang	Wuhan University, China
Moon Seho	Hoseo University, Korea
Kongel Arearak	Suranaree University of Technology, Thailand
Zhihua Xu	International Science & Education Researcher Association, China

Co-sponsored by

International Science & Education Researcher Association, China
VIP Information Conference Center, China

Reviewers of EECM2011

Chunlin Xie	Wuhan University of Science and Technology, China
LinQi	Hubei University of Technology, China
Xiong Huang	International Science & Education Researcher Association, China
Gangshen	International Science & Education Researcher Association, China
Xiangrong Jiang	Wuhan University of Technology, China
LiHu	Linguistic and Linguidtic Education Association, China
Moon Hyan	Sungkyunkwan University, Korea
Guangwen	South China University of Technology, China
Jack H. Li	George Mason University, USA
Marry Y. Feng	University of Technology Sydney, Australia
Feng Quan	Zhongnan University of Finance and Economics, China
PengDing	Hubei University, China
Songlin	International Science & Education Researcher Association, China
XiaoLie Nan	International Science & Education Researcher Association, China
ZhiYu	International Science & Education Researcher Association, China
XueJin	International Science & Education Researcher Association, China
Zhihua Xu	International Science& Education Researcher Association, China
WuYang	International Science & Education Researcher Association, China
QinXiao	International Science & Education Researcher Association, China
Weifeng Guo	International Science & Education Researcher Association, China
Li Hu	Wuhan University of Science and Technology, China
ZhongYan	Wuhan University of Science and Technology, China
Haiquan Huang	Hubei University of Technology, China
Xiao Bing	Wuhan University, China
Brown Wu	Sun Yat-Sen University, China

Contents

Depth Distributions of Some Linear Codes

Zhi-min Li, Xin Xu, and Cun-hua Li

School of Computer Engineering, Huaihai Institute of Technology, Lianyungang, China
{lizhimin1981,xuxin,cunhua-li}@gmail.com

Abstract. The depth distribution of a linear code is a new characterization. In this paper, we study the depth distributions of parity-check codes, $[p^m + r, 2r]$ MDS codes and some direct and outer product codes. Furthermore, we derive four depth-equivalence classes for the binary [24, 12, 8] Golay codes.

Keywords: Depth, depth distribution, depth-equivalence classes, derivative.

1 Introduction

The depth distribution was investigated by Etzion [4] as a new characterization for linear codes. Etizon [4] showed that the depth distribution of the nonzero codewords of an $[n, k]$ linear code consists of exactly k nonzero values, and its generator matrix can be constructed from any k nonzero codewords with distinct depths. He discussed the depth distributions of some well-known binary linear codes, such as the Hamming codes, the extended Hamming codes, and the first-order Reed-Muller codes. He also showed how the set of equivalent codes can be partitioned into the depth-equivalences classes. Mitchell [5] gave a characterization for infinite sequences of finite depth in terms of their periodicity. He determined the depth spectrums for all linear cyclic codes. Yuan Luo.*al* [6] derived some enumeration formulas concerning the depth distributions of linear subcodes of a linear code. He determined the depth distributions of the Reed-Muller codes and showed that there are exactly nine depth-equivalence classes for the ternary [11, 6, 5] Golay codes. In this short paper, we get the depth spectrums of parity-check codes, $[p^m + r, 2r]$ MDS codes, some direct and outer product codes, moreover, we derive some depth-equivalence classes of the binary [24, 12, 8] Golay codes.

The paper is organized as follows. In Section 2, we give the results that including the depth distributions of parity-check codes, $[p^m + r, 2r]$ MDS codes, some direct and outer product codes. We study the depth-equivalence classes of the binary [24, 12, 8] Golay codes in Section 3.

2 Depth Spectrums of Some Linear Codes

Let $V_n(q)$ be the n-dimensional vector space over the Galois field GF(q). The operator $D: V_n(q) \to V_{n-1}(q)$, which is called *derivative*, is defined as follows:

$$D(x_1, \cdots x_n) = (x_2 - x_1, x_3 - x_2, \cdots, x_n - x_{n-1}).$$

D. Jin and S. Lin (Eds.): Advances in EECM Vol. 1, LNEE 139, pp. 1–6.
springerlink.com

It is easy to see that D is a linear operator, i.e., for $x, y \in V_n(q)$, $\lambda \in GF(q)$,

$$D(x+y) = D(x) + D(y) \qquad D(\lambda x) = \lambda D(x).$$

For $\lambda \in GF(q)$, we denoted by $[\lambda^m] = (\lambda, \cdots, \lambda)$ the all λ's row vector of length m.

Definition 1: The depth of x, depth(x), is the smallest integer i such that $D^i x = [0^{n-i}]$. If no such i exists, the depth of x is defined to be n.

It is obvious that the depth of vector $x \in V_n(q)$ is i if and only if $D^{i-1}x = [\lambda^{n-i+1}]$ for a nonzero element $\lambda \in GF(q)$, and the depth of x is at most n.

Definition 2: For an $[n, k]$ linear code C over GF(q), let D_i be the number of codewords of depth i. The numbers $D_0, D_1, \cdots, D_n$ are called the depth distribution of C. Note that $D_0 = 1$.

Definition 3: For an $[n, k]$ linear code C over GF(q), the index set $\{i | 1 \le i \le n, D_i \ge 0\}$ is called the depth spectrum of C.

Lemma 1[6]: Let $v = (v_1, \cdots, v_n) \in V_n(q)$ and p be the characteristic of GF(q). If $p^m \le n$ then $D^{p^m-1}v = (v_1 + v_2 + \cdots + v_{p^m}, v_2 + v_3 + \cdots + v_{p^m+1}, \cdots, v_{n-p^m+1} + v_{n-p^m+2} + \cdots + v_n)$

If $p^m < n$, then we have the result that $D^{p^m}v = (v_{p^m+1} - v_1, v_{p^m+2} - v_2, \cdots, v_n - v_{n-p^m})$.

Let $v = (v_1, \cdots, v_n) \in V_n(q)$ and r be the largest integer such that $q^r < n$. Let $v' = (v_1, \cdots, v_{q^r})$, $u = D^{q^r}v$. Then by Lemma 1, $u = (-v_1 + v_{q^r+1}, -v_2 + v_{q^r+2}, \cdots, -v_{n-q^r} + v_n)$.

Algorithm 1 [6]**:** The depth of $v \in V_n(q)$ can be recursively as follows:

(1) If $v = [0^n]$, then depth (v) =0.

(2) If $v = [\lambda^n]$, for a nonzero element $\lambda \in GF(q)$, then depth (v) =1.

(3) If $u = [0^{n-q^r}]$, then depth (v) = depth (v').

(4) If $u \ne [0^{n-q^r}]$, then depth (v) = q^r +depth (u).

Lemma 2 [6]**:** Let C be an $[n, k]$ linear code over GF(q). Then for all nonzero codewords of C, there are exactly k distinct depths $l_1 < l_2 < \cdots < l_k$. Furthermore, $D_{l_i} = (q-1)q^{i-1}, (1 \le i \le k)$.

Lemma 3 [5]**:** Suppose C is an $[n, k]$ linear cyclic code, and let $g(x)$, of degree n-k, be the generator polynomial for C. Then $(x-1)^s \| (x^n - 1)/g(x)$ if and only if C has a depth spectrum as $\{1, 2, \cdots, s\} \cup \{n, n-1, \cdots, n-k+s+1\}$.

Theorem 1: Let P be an $[n, n\text{-}1]$ parity-check code over $F_2^n, n = 2^s t, (2, t) = 1$ then P has depth spectrum $\{1, 2, \cdots, 2^s - 1\} \cup \{2^s + 1, \cdots, n\}$, for $s \ge 0$.

Proof. Since P is an $[n, n\text{-}1]$ parity-check code, i.e., P is the set of the even weight words in F_2^n. As the following matrix A is equivalent to the standard generated matrix G of P.

$$A=\begin{pmatrix}1&1&0&\cdots&0\\0&1&1&\cdots&0\\\vdots&\vdots&\vdots&\ddots&\vdots\\0&0&0&\cdots&1\end{pmatrix}\qquad G=\begin{pmatrix}1&0&0&\cdots&1\\0&1&0&\cdots&1\\\vdots&\vdots&\vdots&\ddots&\vdots\\0&0&0&\cdots&1\end{pmatrix}$$

It is obvious that any row vector of A is a codeword of P. Thus if $C=<g(x)>$, where g(x) $=x$-1, A is its generated matrix, then we have P=C. By Lemma 3, we can easily have the desired result.

Lemma 4: An $[n, k]$ code over GF(q) is MDS if and only if any set of k columns of its code matrix contains all q^k possible row vectors exactly one time.

Proof. Let A be the code matrix of C. Then if C is MDS, we know that each k coordinates of C can be taken as message symbols, that is to say, any $q^k \times k$ submatrix of A contains each k-tuple exactly once.

If each set of k columns of A contains all q^k possible row vectors exactly one time, we have that any k coordinates of C can be taken as message symbols, then we have the desired result.

Theorem 2: Let C be an $[p^m+r,2r]$ MDS code over GF(q), p be the characteristic of GF(q), $r<p^m, l_1,l_2,\cdots,l_{2r}$ is the depth spectrum of C. Then $l_1<l_2<\cdots<l_r\le p^m$ and $l_{r+i}=p^m+i, 1\le i\le r$.

Proof. Let $c=(c_1,\cdots,c_{p^m+r})\in C$, then $D^{p^m}c=(-c_1+c_{p^m+1},\cdots,-c_r+c_{p^m+r})$.

Since C is MDS, by Lemma 4, we can calculate that there are exactly q^r-1 nonzero codewords satisfying depth $\le p^m$ in C, in another words, the number of nonzero codewords in C satisfying $c_i=c_{p^m+i}$ $1\le i\le r$, is exactly q^r-1. Moreover,

$$D_{l_1}+D_{l_2}+\cdots+D_{l_r}=q-1+(q-1)q+\cdots+(q-1)q^{r-1}=q^r-1\,.$$

Then by Lemma 2, the depth spectrum of C satisfies $0<l_1<l_2<\cdots<l_r\le p^m$ and $l_{r+i}=p^m+i$, $1\le i\le r$, since the length of C is p^m+r .

We now determine the depth spectrum of a direct product code in terms of the depth spectrums of its component codes. Let A, B be $[n_1,k_1],[n_2,k_2]$ linear codes respectively, $A\otimes B$ is an $[n_1n_2,k_1k_2]$ code. For two vectors $X=(x_1,x_2,\cdots,x_{n_1})$ and $Y=(y_1,y_2,\cdots,y_{n_2})$, denote their product by $x\otimes y=(x_1y_1,x_1y_2,\cdots,x_{n_1}y_{n_2})$.

Theorem 3: A is a $[n_1,k_1]$ code, B is a $[n_2,k_2]$ code, $n_2=p^m, m\in N$, over GF(q), p is the characteristic of GF(q). The depth spectrum of A is $\{l_1^{(1)},l_2^{(1)},\cdots l_{k_1}^{(1)}\}$, and that of B is $\{l_1^{(2)},l_2^{(2)},\cdots l_{k_{21}}^{(2)}\}$, where $l_1^{(i)}<l_2^{(i)}<\cdots<l_{k_i}^{(i)}, i=1,2$. Then the depth spectrum of $A\otimes B$ is

$\{(l_i^{(1)}-1)p^m+l_j^{(2)} | 1\le i\le k_1, 1\le j\le k_2\}$. Furthermore, the corresponding depth distribution is $D_{ls}=(q-1)q^{(i-1)k_2+j-1}, 1\le s\le k_1k_2$.

Proof. Let $x=(x_1,x_2,\cdots,x_{n_1})\in A$, depth $(x)=s$, $y=(y_1,y_2,\cdots,y_{n_2})\in B$, depth(y) = t, and $x\otimes y=(x_1y_1,\cdots,x_1y_{n_2},\cdots,x_{n_1}y_1,\cdots,x_{n_1}y_{n_2})=(x_1y,\cdots,x_{n_1}y)$. Since the length of y is p^m, by Lemma 3, we have $D^{p^m}(x\otimes y)=((x_2-x_1)y,\cdots,(x_{n_1}-x_{n_1-1})y)$. If $x_i\ne x_j, i\ne j$ (i.e. s>1), then repeat the procedure. We repeat this s-1 times and obtain that $D^{(s-1)p^m}(x\otimes y)=(y,\cdots,y)$. Hence, by the depth Algorithm, depth $(x\otimes y)=(s-1)p^m+t$ .If $x_i=x_j$, (i.e., s=1), then $x\otimes y=(y_1,\cdots,y_{n_2},\cdots,y_1,\cdots,y_{n_2})$. Since the length of y is p^m, by Lemma 2, we have depth $(1\otimes y)=$ depth(y)=t.

Thus the depth spectrum of $A\otimes B$ is $\{(l_i^{(1)}-1)p^m+l_j^{(2)} | 1\le i\le k_1, 1\le j\le k_2\}$, and its corresponding depth distribution is $D_{ls}=(q-1)q^{(i-1)k_2+j-1}$, where $1\le s\le k_1k_2, 1\le i\le k_1, 1\le j\le k_2$, according to Lemma 2.

We can also determine the depth spectrum of an outer product code in terms of the depth spectrums of its component codes.

Let $1_n=(1,\cdots,1)$ denote the all-one vector of length n. Let n_1 and n_2 be two positive integers. For x= $(x_1,x_2,\cdots,x_{n_1})$ and y= $(y_1,y_2,\cdots,y_{n_2})$, $x_i, y_i\in F_q, 1\le i\le n_1, 1\le j\le n_2$, the outer product $x\circ y$ is defined as $x\circ y=(x_1+y_1,x_1+y_2,\cdots,x_{n_1}+y_1,\cdots x_{n_1}+y_{n_2})$.

Let C_1 and C_2 be [n_1,k_1] and [n_2,k_2] linear codes respectively. The outer product $C_1\circ C_2$ is defined to be the set $C_1\circ C_2=\{x\circ y | x\in C_1, y\in C_2\}$.

Then the outer product $C_1\circ C_2$ is an $[n_1n_2,k]$ linear code, where

$$k=\begin{cases} k_1+k_2-1 \, if & 1_{n_1}\in C_1 \, and \quad 1_{n_2}\in C_2; \\ k_1+k_2 & otherwise. \end{cases}$$

Let $\{x_1,x_2,\cdots x_{k_1}\}$ and $\{y_1,y_2,\cdots y_{k_2}\}$ be a basis for C_1 and C_2 respectively. Then $\{x_i\circ 0_{n_2}, 0_{n_1}\circ y_j | 1\le i\le k_1, 1\le j\le k_2\}$ is a basis for $C_1\circ C_2$ except for $1_{n_1}\in C_1$, $1_{n_2}\in C_2$.

If $1_{n_1}\in C_1$ and $1_{n_2}\in C_2$, let $\{1,x_2',\cdots,x_{k_1}'\}$ and $\{1,y_2',\cdots,y_{k_2}'\}$ be a basis for C_1 and C_2 respectively. Then $\{1_{n_1n_2}, x_i'\circ 0_{n_2}, 0_{n_1}\circ y_j' | 2\le i\le k_1, 2\le j\le k_2\}$ is a basis for $C_1\circ C_2$.

Theorem 4: Let C_1 be an [n_1,k_1] code, C_2 be an [n_2,k_2] code, $n_2=p^m, m\in N$, over F_q, p be the characteristic of F_q. The depth spectrum of C_1 is $\{l_1^{(1)},l_2^{(1)},\cdots,l_{k_1}^{(1)}\}$ and that of C_2 is $\{l_1^{(2)},l_2^{(2)},\cdots,l_{k_2}^{(2)}\}$ where $l_1^{(i)}<l_2^{(i)}<\cdots<l_{k_i}^{(i)}, i=1,2.$ Then the depth spectrum of $C_1\circ C_2$ is $\{l_j^{(2)},(l_i^{(1)}-1)p^m+1 | 1\le i\le k_1, 1\le j\le k_2\}$.

Proof. Let $0\ne x=(x_1,x_2,\cdots,x_{n_1})\in C_1$, depth(x) = s; y= $(y_1,y_2,\cdots,y_{n_2})\in C_2$, depth(y)=$t$, and $x\circ y=(x_1+y_1,\cdots,x_1+y_{n_2},\cdots,x_{n_1}+y_1,x_{n_1}+y_{n_2})=(x_1\cdot 1_{n_2}+y,\cdots x_{n_1}\cdot 1_{n_2}+y)$.

Since the length of y is p^m, by Lemma 1, we have the following equation that $D^{p^m}(x \circ y) = ((x_2 - x_1)1_{n_2}, \cdots, (x_{n_1} - x_{n_1-1})1_{n_2})$. If s=1, then $D^{p^m}(x \circ y) = (0 \cdot 1_{n_2}, \cdots 0 \cdot 1_{n_2})$. Then by the depth Algorithm, depth ($x \circ y$) = depth ($1 + y_1, \cdots, 1 + y_{n_2}$). Hence, depth ($x \circ y$) = t, depth (y) = t. If $x_i \neq x_j, i \neq j (s > 1)$, then repeat the procedure. We repeat this s-1 times and obtain that $D^{(s-1)p^m}(x \circ y) = (1_{n_2}, \cdots, 1_{n_2})$. Hence, by the depth algorithm, depth $(x \circ y) = (s-1)p^m + 1$. If x = 0, then $x \circ y = (y_1, \cdots, y_{n_2}, \cdots, y_1, \cdots, y_{n_2})$. Thus, depth ($x \circ y$) = t since depth (y) = t. Hence, we have that the depth spectrum of $C_1 \circ C_2$ is

$$\{l_j^{(2)}, (l_i^{(1)} - 1)p^m + 1 \| 1 \leq i \leq k_1, 1 \leq j \leq k_2\}.$$

Remark: If $1_{n_1} \in C_1$, $1_{n_2} \in C_2$, then $l_1^{(1)} = l_1^{(2)} = 1$ and $(l_1^{(1)} - 1)p^m + 1 = 1$. Thus

$$\left|\{(l_j^{(2)}, (l_i^{(1)} - 1)p^m + 1 : 1 \leq i \leq k_1, 1 \leq j \leq k_2)\}\right| = k_1 + k_2 - 1.$$

3 The Depth-Equivalence Classes of Binary Golay Code

Two [n, k] linear codes C_1 and C_2 over GF(q) are called equivalent if C_1 can be obtained from C_2 by a permutation on the n coordinates. Two [n, k] linear codes are defined to be depth-equivalent if they are equivalent and have the same depth distribution. This definition gives us a new way to classify and partition the equivalent codes into the depth-equivalence classes. Etzion [4] showed that there are exactly four depth-equivalence classes for the binary [8, 4, 4] extended Hamming code. Yuan Luo *et.al* [6] derived that there are exactly nine depth-equivalence classes for the ternary [11, 6, 5] Golay code. In this section, we use the method in Section 3 to illustrate how to get some depth-equivalence classes for the binary [24, 12, 8] Golay codes.

We know an [24, 12, 8] Golay code can be formed by concatenating as follows: $G = \{(a + x, b + x, a + b + x) | a \in \overline{H}, b \in \overline{H}, x \in \overline{H^*}\}$ .H is the [7, 4] Hamming code, H^* is obtained by reversing the order of the symbols in the words of H, $\overline{H} \cap \overline{H^*} = \{0,1\}$. The depth spectrums of the four depth-equivalence classes for the [8, 4, 4] extended Hamming code $\overline{H}$ showed in Etizon [4] are listed as follows:

$$F_1 = \{1,2,3,5\}, F_2 = \{1,2,5,6\}, \ F_3 = \{1,3,5,7\}, F_4 = \{1,5,6,7\}.$$

We can verify $\overline{H}$ and $\overline{H^*}$ have the same depth distribution. Let $h = (h_1, h_2, \cdots, h_7, h_8) \in \overline{H}$, then $h^* = (h_7, h_6, \cdots, h_1, h_8) \in \overline{H^*}$.

Case 1 depth (h) = 1
That is to say $h = (1,1,\cdots,1)$, then $h^* = (1,1,\cdots,1) = h.$

Case 2 depth (h) = 2, 3
Since $D^2(h) = (h_3 - h_1, h_4 - h_2, h_5 - h_3, h_6 - h_4, h_7 - h_5, h_8 - h_6)$, and

$$D^2(h^*) = (h_5 - h_7, h_4 - h_6, h_3 - h_5, h_2 - h_4, h_1 - h_3, h_8 - h_2).$$

If depth (h) =2, then $h = h_3 = h_5 = h_7, h_2 = h_4 = h_6 = h_8$, and $h_i \neq h_{i+1}, 1 \le i \le 8$. Thus $D^2(h^*) = 0$, and $D\ (h^*) \neq 0$.

It is similar to verify depth (h^*) = 3. Conversely, if depth (h^*) =2 or 3, we can verify that depth (h) =2 or 3 similarly.

Case 3 depth (h) = 5, 6, 7

Since $D^4(h) = (h_5 - h_1, h_6 - h_2, h_7 - h_3, h_8 - h_4)$, and $D^4(h^*) = (h_3 - h_7, h_2 - h_6, h_1 - h_5, h_8 - h_4)$.

If depth (h) = 5, then $D^4(h)$= (1, 1, 1, 1), thus we have $h_7 - h_3 = h_5 - h_1 = 1$. Then we have depth (h^*) = 5.

If depth (h) = 6, 7, $D^6(h) = (h_7 - h_3 - h_5 + h_1, h_8 - h_4 - h_6 + h_2)$. Since $D^6(h^*) = (h_1 - h_5 - h_3 + h_7, h_8 - h_4 - h_2 + h_6)$ and $h_i \in F_2, i = 1, \cdots, 8$. Then $D^6(h^*) = D^6(h)$.

If depth $(h^*) < 6 \le$ depth (h), then it is a contradiction to the case when depth (h) < 6. Thus depth(h^*)=6 if depth(h)= 6 and depth(h^*)= 7 if depth(h)= 7 . Then use the depth algorithm, we can calculate four depth-equivalence classes for the binary [24,12,8] Golay code, they are:

$$G_1 = \{1,2,3,5,9,10,11,13,17,18,19,21\},\ G_2 = \{1,2,5,6,9,10,13,14,17,18,21,22\},$$
$$G_3 = \{1,3,5,7,9,11,13,15,17,19,21,23\},\quad G_4 = \{1,5,6,7,9,13,14,15,17,21,22,23\}.$$

Of course, they are not all of the depth-equivalence classes for the binary [24, 12, 8] Golay codes. For calculating other depth-equivalence classes of the [24, 12, 8] Golay codes, people can refer to the method in [6].

Acknowledgement. This work is sponsored by Jiangsu Provincial Natural Science Foundation of China (No.BK20082140), Item of Scientific Research Fund for Talents of Huaihai Institute of Technology (No.KQ10121), Research Fund of Huaihai Institute of Technology (No.KX10530).

References

1. Blinovsky, V., Erez, U., Litsyn, S.: Weight distribution moments of random linear/coset codes. Designs, Codes and Cryptography 57(2), 127–138 (2010)
2. Brackx, F., De Knock, B., De Schepper, H.: On the Fourier Spectra of Distribution In Clifford Analysis. Chinese Annals of Mathematics-Series B 27(5), 495–506 (2006)
3. Chen, W., Torleiv, K.: Weight hierarchies of linear codes satisfying the almost chain condition. Science in China Series F: Information Sciences 46(3), 175–186 (2005)
4. Etzion, T.: The depth distuibution–A new characterization for linear codes. IEEE Trans. Inform. Theory 43, 1361–1363 (1997)
5. Mitchel, C.J.: On integer-valued rational polynomials and depth distributions of binary codes. IEEE Trans. Inform. Theory 44, 3146–3150 (1998)
6. Luo, Y., Fu, F.-W., Wei, V.K.: On the depth distribution of linear codes. IEEE Trans. Inform. Theory 46, 2197–2203 (2000)

Design of Dual Phase Signals Generator Based on AD9833

Jian Guo and Pingping Dong

Information School, Beijing Wuzi University, Beijing 101149, China
{guojian,dongpingping}@bwu.edu.cn

Abstract. On the basis of analyzing the working principle of AD9833 chip, taking MSP430F135 as main circuit, a novel dual tunable phase difference signal generator based on AD9833 is designed. The hardware circuits consist of two channel signal generating circuit, dual phase amplitude modulating circuit, key and microcontroller circuit, display driving circuit and power source and so on. Testing results show that the signal generator has a lot of a lot of merits such as continuous adjustable phase difference of two channel signals, wide output frequency, stable frequency output, low output error, which can efficaciously address the shortcomings of modulating the phase difference of two channel output signals by phase-shifting network.

Keywords: dual signal generator, adjustable phase difference, direct digital synthesis, AD9833.

1 Introduction

Signal generator that can be used to generate arbitrary waveforms is a common signal source and it is employed widely in fields such as communication control, navigation, radar, medical treatment and experiment of electronic information course in university and so on[1]. Dual phase signals generator has two channel signal outputs which can be ac or dc signal or signal with phase difference. The usual method to carry out dual phase signals is adopting phase-shifting network to modulate the phases of two channel signals. However, this method has a lot of shortcomings such as that phase output waveform is vulnerable by the input waveform, the degree of phase difference is related to the size and nature of load, the accuracy of phase difference is not high and the resolution is relatively low and so on[2-3].

Direct digital frequency synthesis technology (DDFS) is a new synthesizing frequency technology. Comparing to the traditional frequency synthesis method, DDFS has a lot of characteristics such as high frequency resolution, fast switching, continuous phase when switching frequency, low phase noise, whole digitization and easy integration and so on. DDFS is developed rapidly since it can satisfy many demands of modern electronic systems [4-5]. Basing on DDS chip AD9833, taking controller MSP430F135 as main circuit, a dual phase signal generator with tunable phase difference is designed. The experiment results show that this signal generator has features such as simple structure and fine performance.

D. Jin and S. Lin (Eds.): Advances in EECM Vol. 1, LNEE 139, pp. 7–13.
springerlink.com

2 The Working Principle of Chip AD9833

AD9833 is a programmable waveform generator of low power consumption, which can generate sine, triangle and square wave. It needs not external components and the output frequency and phase can be modulated by software program. The frequency register is 28-bit and the main frequency clock is 25MHz, the accuracy is 0.1Hz. When the main frequency clock is 1MHz, the accuracy can be up to 0.004Hz [7-8].

The internal circuits consists of numerically controlled oscillator (NCO), frequency and phase adjuster, sine ROM, digital analog converter(DAC) and voltage regulator. The core is 28-bit phase accumulator, which is composed of adder and phase register. When a clock pulse coming, phase accumulator is incremented by a step. The output of phase register is added to the phase control word and the result is imported into the sine query table address. The sine query table contains the information of digital amplitude of one cycle sine wave. Each address is corresponding to a phase point of 0-360 degree of a sine wave. The input phase information is mapped into the digital signal of sine wave amplitude by the query table and then DAC is driven and the analogy is output [9]. The phase register is come back to the original status after 2^{28} reference clock frequency and the corresponding sine query table is back to the initial position through a cycle and so a cycle of sine wave is output. The frequency of sine wave is $f_{OUT=}M (F_{MCLK}/2^{28})$. Here, M is the frequency control word which is set by external programming and its range is $0\sim2^{28}-1$. F_{MCLK} is reference clock frequency.

3 The Hardware Design of System

The aim of the design is to get two same signals with phase required and the phase difference is continuously adjustable with required phase resolution. So, firstly, the signals should be generated synchronously. And then, the phase difference of signals should be adjusted continuously. Signal synchronization design requires that control input port of two signal generator attach to the same control signal line. The starting command need simultaneously reach the two signal generators. The concrete hardware circuits consist of two channel signal generating circuit, dual phase amplitude modulating circuit, key and microcontroller circuit and display driving circuit and power source.

3.1 Structure Design of System

The whole system consists of main controller MCU (MSP430F135), DDS chip (AD(9833), keys, LED display unit (74LVC07), low pass filter and the power source. The signal is generated by the MSP430F135 control module and DDS chip. Keys and display unit is used to realize the function of man-machine interaction. The users input the order from keys and the data will be shown in LED and the order is transferred to MSP430F135. And then the signal output by DDS chip is processed by follow-up signal conditioning circuit and then output. The Filter module is used to signal post-processing.

Design of Two Channel Signal Generating Circuit. Two channel signal generating circuit consist of DDS chip AD9833, active crystal oscillator and the op amp adjusting circuit. The concrete circuit is shown as Fig.1.

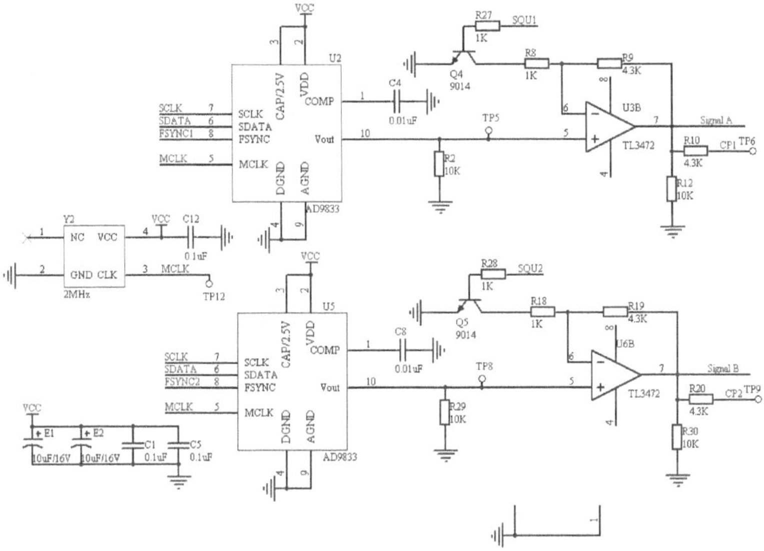

Fig. 1. Two channel signal generating circuit

Design of Dual Phase Amplitude Modulating Circuit. Dual phase amplitude modulating circuit consists of operational amplifier, analog switches HCF4051 and control signal electric level conversion circuit. The concrete circuit is shown as Fig.2.

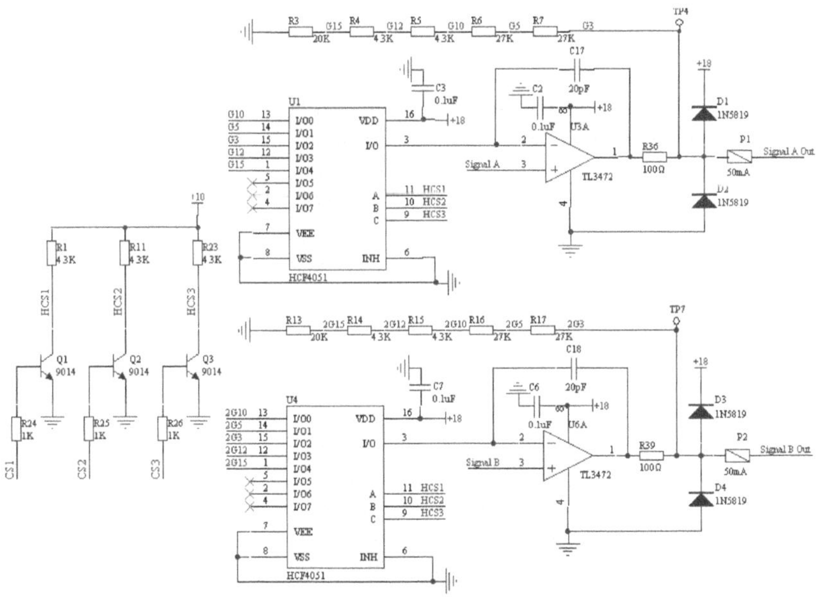

Fig. 2. Dual phase amplitude modulating circuit

Design of Key and Microcontroller Circuit. Microcontroller circuit includes microcontroller MSP430F135 and reset circuit. Key consist of function key, enter key, left key, up and down key. The concrete circuit is shown as Fig.3.

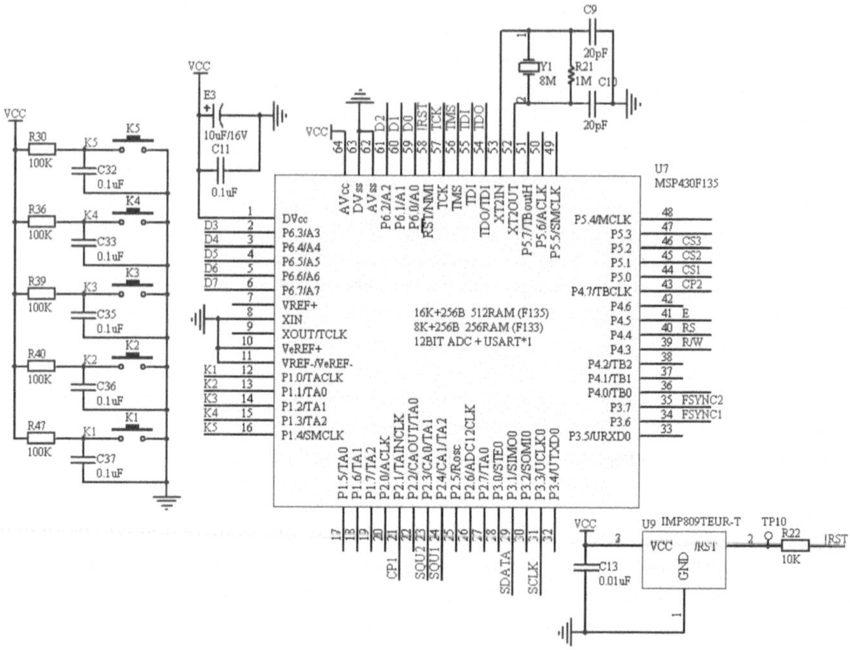

Fig. 3. Key and microcontroller circuit

Design of Display Driving Circuit. The display part consists of signal electric level conversion circuit and LCM1602D module. The concrete circuit is shown as Fig.4.

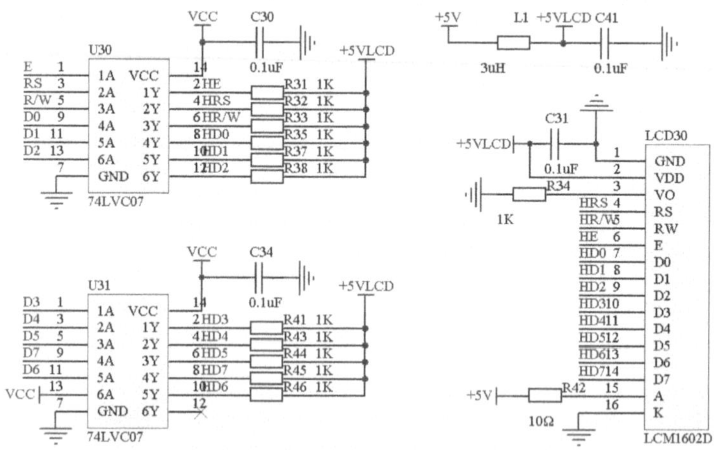

Fig. 4. Display driving circuit and power source circuit

Design of Power Source. The power source includes +18V, +5V and +3.3V powers. Three chips powered by two lithium batteries. The concrete circuit is shown as Fig.5.

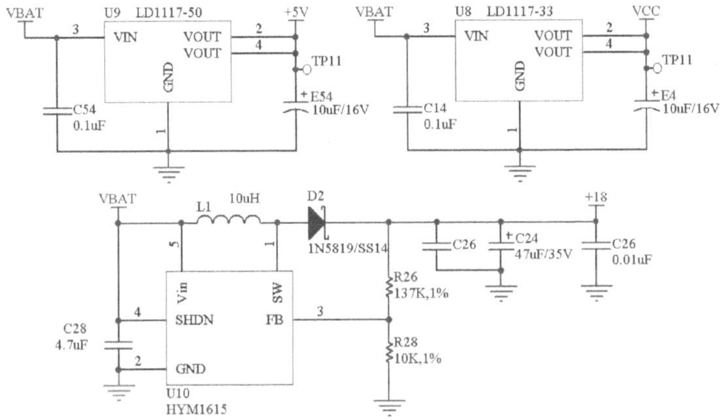

Fig. 5. Power source circuit

4 The Software Design of System

The main task of controller is to respond to keyboard input, to refresh liquid crystal display, to control waveform output. Firstly, the controller is set to the default working state. If there is a key pressed, the keys will be responded. The required frequency, amplitude and wave parameters are input by keys. The frequency word is computed by the controller based on input frequency. At last, the corresponding waveform is output according to the frequency control word, amplitude and waveform parameters.

Software Design of Generating Signal. AD9833 is a programmable signal generator and it has two 28-bit frequency register and two 12-bit phase register. a 16-bit working mode word is written into the chip when working, and then the working status is confirmed and frequency register and phase register are chosen. One or two frequency control words are written to control output frequency. At last, the phase control word is written and then DDS signal generator can output the corresponding waveform based on the mode control word. Here, the frequency control word is obtained according to a certain algorithm based on an input frequency. The amplitude of waveform is gotten by amplitude modulating circuit corresponding to the key input by microcontroller.

Software Design of Key Input. The processing procedure of key is as following. The key interrupt is generated when a key is pressed. The key pressed is identified and the corresponding key is gotten in interrupting procedure. If the key value of main function is not zero, the key process is implemented. According to the different key pressed, corresponding operation, such as parameter choice, data change and so on, will be implemented in key process program. To different parameters, the range of input data can be limited and finally, the enter key is pressed, the settings take effect. The system will run according to the new setting.

Software Design of Display Circuit. The working mode is classified into normal working mode (waveform output mode) and parameter setting mode. Under different working mode, the screen displays difference interface to facilitate users. If the working mode is that of waveform output, the cursor and character blink is cut off and the current amplitude is displayed in the first row and the current frequency is shown in the second row. It is not displayed when the high bit is zero. When the mode is that of parameter setting, the cursor is open, the data character that will be modified is blink. The parameter is revised from the minimum digit acquiescently. Each digit of parameter can be changed by the up and down key and the digit changed can be chosen by the left key. When the digit is the highest one of parameter, the minimum digit can be chosen by pressing the left button.

5 Conclusions

Dual phase signals generator, widely used in fields such as communication, navigation, radar, medical treatment and so on, has two channel signal outputs and can generate arbitrary waveform. Direct digital synthesis technology is widely applied in various fields because of characteristics of high stability, wide frequency band, and continuous phase and so on. AD9833 is a programmable chip of DDFS with virtue of low power consume easy modulation and no external components. Taking MSP430F135 as main control circuit, coordinating with peripheral I/O devices, a dual phase signal generator is designed based on AD9833 chip. The generator has a lot of advantages such as simple structure, high performance, and wide output frequency band, which can efficaciously address the shortcomings of modulating the phase difference of two channel output signals by phase-shifting network.

Acknowledgment. The paper is supported by the Funding Project for Academic Human Resources Development in Institutions of Higher Learning Under the Jurisdiction of Beijing Municipality (PHR201007145, PHR201108311), Funding Project for Base Construction of Scientific Research of Beijing Municipal Commission of Education(WYJD200902) and Funding project for Beijing excellent talents (2010D005009000002).

References

1. Yazhen, Z., Rongshan, W.: Design and Implementation a Signal Generator Based on AD9833 Chip. China Instrumentation 3, 60–63 (2010)
2. Shangsong, C., Jia, L., Qing, G.: Electronic Measurement and Instrument. Publishing House of Electronic Industry (2005)
3. Qing, G., Hailing, Y., Shangsong, C.: Design of Dual Tunable Phase Difference Signal Generator. Electronic Measurement Technology 30, 191–200 (2007)
4. Ping, Y., Danhui, W., Liangyu, Y.: Application of DDS Technology in Sine Wave. Computer Measurement & Control 6, 1738–1740 (2008)
5. Yun, X., Zhongjun, C., Kaihong, Z., Hong, Z.: Study of Double Channel Arbitrary Wave Generator Based on Direct Digital Frequency Synthesis. Chinese Journal of Scientific Instrument 27, 515–519 (2006)

6. Jiliang, Z., Sijiu, L.: Design of Signal Generator Based on AD9833 DDS Chip, http://www.paper.edu.cn
7. Guoliang, L., Liqing, L., Jinping, S.: Programmable Waveform Generator AD9833 and Its Application, vol. 6, pp. 44–48 (2006)
8. Bin, H., Yingzheng, H., Kangsheng, Z.: Design of High Precision Programmable Waveform Generator System Based on AD9833. Electronic Design Engineering 17, 6–7 (2009)
9. Shenping, X., Yingyan, D., Hongbing, Z., En, Z.: Design of Signal Generator Based on AD9833. Development & Innovation of Machinery & Electrical Products 21, 67–69 (2008)

The Design of Temperature and Humidity Control System Based on Fuzzy Control in Multi Incubators

Ruilan Wang

School of Information and Control Enginerring, Weifang University
Weifang, Sandong, China
wrl2836@163.com

Abstract. hatching systems requiring high control accuracy of temperature and humidity. Particularly temperature, small temperature change will significantly affect hatching time and hatching rate, and temperature and humidity is the controlled amount of interaction between two. Due to the coupling characteristics of temperature and humidity is difficult to establish a precise mathematical model, caused uncertainty in its control. Fuzzy control method witch proposed in this article will not only improve control precision, and to improve the robustness of control algorithms and adaptive capacity.

Keywords: Incubator, temperature and humidity, fuzzy control, coupling.

1 Introduction

Incubation equipment is an important equipment, it is a bionic device witch based on the poultry biology of incubation, using economic and reasonable technical means, creating the artificial control ecological environment for incubation and hatching. With the development of the domestic farming industry, hatchery to the major direction of development is the trend of the times. But due to the incubator's temperature and humidity control difficult, the actual control accuracy is not high, so need to study effective control strategies.

2 The Mathematical Model of the Temperature and Humidity Control Object

Hatching systems require high control accuracy of temperature and humidity. Particularly temperature, small temperature change will significantly affect hatching time and hatching rate , and temperature and humidity is the amount of interaction between two charged. Temperature and humidity control is essentially a non-linear, hysteresis, strong coupling, time-varying system control. Temperature and humidity can be approximated as the two order differential of the lag link, but scaling factors and hysteresis parameters are different for each equation, its mathematical model as shown in the figure 1.

Traditional control method of temperature and humidity in incubator is PID control, and control instruments are independent, universal humidity and temperature control instruments. Because of the different types of incubator, great changes take place in its

D. Jin and S. Lin (Eds.): Advances in EECM Vol. 1, LNEE 139, pp. 15–18.
springerlink.com

control object, these common control method of poor adaptability, and the temperature and humidity independent control, does not take into account their strong coupling effects, control accuracy is generally unsatisfactory. This paper focuses on the problem, proposed a fuzzy control method for controlling, it can not only improve control precision, and to improve the robustness of control algorithms and adaptive capacity.

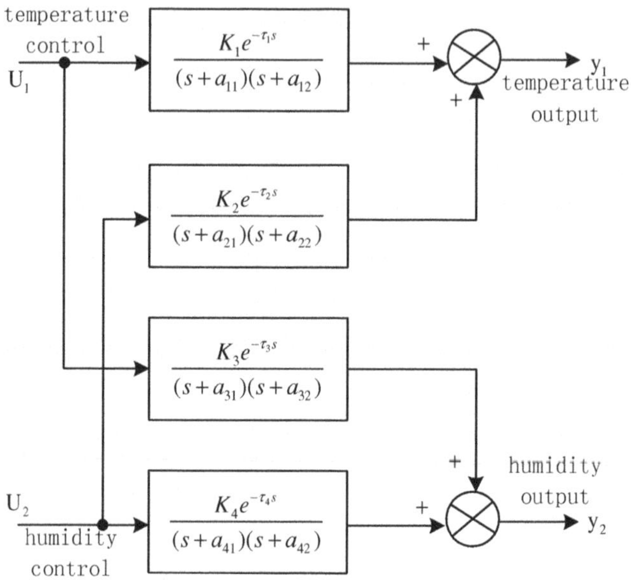

Fig. 1. The mathematical model of temperature and humidity objects

3 The Design of Fuzzy Controller

The prominent advantage of fuzzy control is adaptable to different process control, with strong robustness and simple structure, easy to implement. This control does not need to know the exact model of the object, is a language controller to achieve control of nonlinear, large-lag characteristic object; can also be a very good application on the mathematical model does not clearly or time-varying parameters of objects and general PID control algorithm is not the ideal occasion. Therefore, adopt fuzzy control to temperature and humidity on the incubation system.

3.1 Determination of Input and Output Variables

The control structure of the fuzzy controller as shown in Figure 2.The control of temperature and humidity adopt two dimensional input single-output fuzzy controller, input respectively for temperature error and the error rate of change of temperature, error of humidity and humidity changes in error rates.. After processing by fuzzy decoupling algorithm get output for temperature compensation of (U_t) and humidity compensation (U_h) and the throttle opening (U_w).

3.2 Fuzzy

According to the actual situation and the experience, setting temperature deviation of basic universe is [-2°C, + 2°C], the maximum value is T+2 °C, the minimum value is T-2°C, humidity deviations of the basic universe is [-20%,+20%], the maximum value is H+20%, the minimum value is H-20%. Fuzzy algorithm does not perform outside of this range, but with maximum output directly or zero output.

Select the temperature and humidity fuzzy sets on the domain n=4, that is[-4,-3,-2,-1,0,1,2,3,4]. The variable selection in the input language is: NL (negative large), NM (negative median), NS (negative small), ZE (zero), PS (positive small), PM (positive median), PL (positive large). The basic universe of output control u is [0,1,2,3,4].

Heating energy output fuzzy domain is [0,1,2,3,4], corresponding resistance heater 5 kinds of work status: ZE (not heated), PS (smaller heating), PM (medium heat), PL (larger heating), PF (full heating).

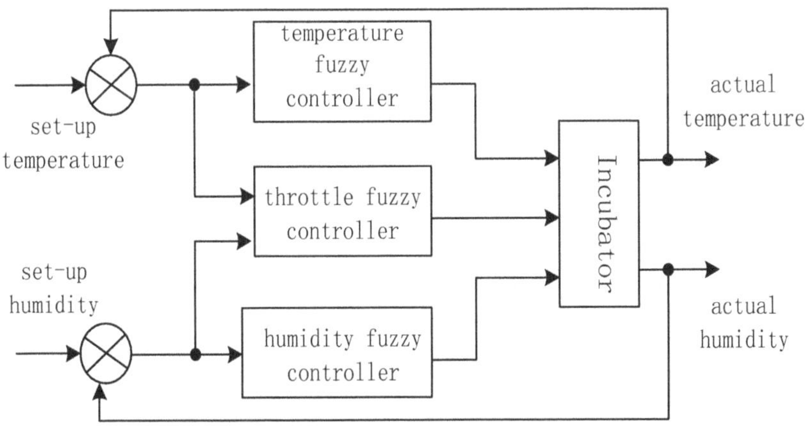

Fig. 2. The fuzzy control chart of temperature and humidity

3.3 Establishment of Fuzzy Rules

The membership function use triangular distribution taking into account the distribution of temperature and humidity inside the incubator. Effect of temperature on humidity in the incubator is much larger than the effect of humidity on the temperature, that is, when after a change of temperature and humidity will change, but little humidity temperature may change after change. Therefore, in the dynamic changes of temperature and humidity inside the incubator, be temperature driven and humidity changes, temperature is active, and changes in humidity is passive. In addition, because the major impact factor of egg hatching is temperature so select temperature as the primary control factor, temperature control, humidity control subsidiary principle of fuzzy control rules in this system. From here, get fuzzy control rule of temperature and humidity based on aggregating and summarizing experience. Temperature control rule table as shown in table 1.

Table 1. The fuzzy control rules table of control U_t

EC_t \ E_t	NL	NM	NS	ZE	PS	PM	PL
NL	PF	PF	PF	PF	PF	PF	PF
NM	PF	PF	PL	PL	PL	PM	PM
NS	PF	PL	PM	PS	PS	ZE	ZE
ZE	PM	PM	PS	ZE	ZE	ZE	ZE
PS	PM	PS	ZE	ZE	ZE	ZE	ZE
PM	ZE	ZE	ZE	ZE	ZE	ZE	ZE
PL	ZE	ZE	ZE	ZE	ZE	ZE	ZE

4 Conclusion

This article analyses the control object on the basis of the characteristics of temperature and humidity incubator, has proposed using fuzzy control method controlled temperature and humidity of incubator. The method has the following characteristics: (1) control of high accuracy, fast response and overshoot a small amount, Actual control accuracy of temperature can be up to ± 0.1°C, accuracy of humidity up to ± 3%. (2) due to the robustness and strong adaptability of fuzzy control, even if the system has certain changes in the object model structures and parameters, can also meet the requirements of actual control.

References

1. Wang, S., Meng, Q., Wu, J.: The DCS of Waste Heat Power Generation of Cement Plant Based on Fuzzy Control. In: IEEE HIS 2009, pp. 247–251 (2009)
2. Li, Z.: Automatic Thermal Control system. China Electric Rower Press, Beijing (2001)
3. Zadeh, L.A.: Fuzzy sets. Information and Control, 338–353 (1965)
4. Zhu, J.: Fuzzy Control Principle & Application. China Machine Press, Beijing (2001)
5. Sun, B., Zhang, H.: The design of constant pressure water supply system based on fuzzy control. Electrical Automation (March 2002)

Construction Data Mining Information Management System Based on FCA and Ontology

XiuYing Sun

Modern Education Technology Center, Huanghe Science and Technology College, Zhengzhou, 450063, Henan, China

Abstract. Ontology is a formal, explicit specification of a shared conceptual model and provides a way for computers to exchange, search and identify characteristics. Data mining is a drawing work from areas including database technology, machine learning, statistics, pattern recognition, neural networks, artificial intelligence, and data visualization. As a branch of applied mathematics, formal concept analysis comes of the understanding of concept in philosophical domain. Finally, data mining information management system in E- Commerce is proposed based on integrating of ontology and formal concept analysis. The experimental results indicate that this method has great promise.

Keywords: formal concept analysis(FCA), ontology, data mining, information management.

1 Introduction

An important objective of intelligent data analysis is to reveal and indicate diverse non-trivial features or views of a large amount of data. Therefore, these modal-style data operators can provide a unified way to examine, characterize and construct different types of knowledge. Formal concept analysis (FCA) proposed by Wille, which provides a theoretical framework for the design and discovery concept hierarchies from relational information system. The main contribution of the authors is that they showed a way to incorporate the idea of thresholds into the definition of extent- and intent-forming operators in such a way that some desirable properties remain available.

The mathematical model of a process can be obtained through the study of variables involved in the process. In general, many physical models have non-linear characteristics and analytical complexities that avoid its application in on-line systems. In other words, data storage, data management, data transmission and even analysis rely on computer and network technology. At the same time, as a result of the vigorous developments and accessibility of the World Wide Web (WWW), a great quantity of information was suddenly made available to people. In text-based communications, users can express their identity with their ID, nickname, and profile.

Ontology is a conceptualization of a domain into a human understandable, machine-readable format consisting of entities, attributes, relationships, and axioms[1]. As the technology advances, the deployment of information systems and technology in business and government increases rapidly. Effectively integrating the information and material flows within the demand and supply process is the main concern for Supply

D. Jin and S. Lin (Eds.): Advances in EECM Vol. 1, LNEE 139, pp. 19–24.
springerlink.com

Chain Management (SCM). A knowledge base is a special database type for representing domain expertise. The repository comprises collections of facts, rules, and procedures organized into schemas. Ontology building is a task that pertains to ontology engineers, an emerging expert profile that requires the expertise of knowledge engineers (KEs) and domain experts (DEs).

Finally, data mining information management system in E- Commerce is proposed based on integrating of ontology and formal concept analysis. This information is generated from multiple actors and shared or exchanged within the construction workflow. The paper offers a methodology for building ontology and carries on ontology merging for knowledge sharing and reusing based on concept lattice union. To achieve this goal, we first need to identify appropriate concepts that can be classified as part of the situation awareness domain.

The goal of the early software tutoring systems was to build user interfaces that provide efficient access to knowledge for the individual learners. The experimental results indicate that this method has great promise.

Efficient implementation of the method yields an incremental FCI-miner whose performances are compared to those of a known batch procedure. The advantage of the proposed model consists in the measure is convenient to be calculated from objects and attributes classes sets, especially for the large context.

2 Data Mining Information Management Based on Ontology

With the application of more and better computer techniques in education and the involvement of more adults in software tutoring systems, the learner control strategy has become more appreciated than tutor or program control[2]. Many commercial tools exist that have made substantial progress primarily in terms of text-matching and text-categorization techniques that facilitate information retrieval of metadata.

Ant colony optimization was introduced in the early 1990s as a novel technique for solving hard combinatorial optimization problems and inspired by the behavior of ants in finding paths from the colony to food. These data are classified into two categories: (1) geospatial data, and (2) model data. Geospatial data includes maps, the numerical grid, the bathymetry, and the digital elevation model, while model data includes the state variables and all coefficients and constants, which are geo-referenced to the geospatial data sets.

Ontologies are representations of the knowledge within a domain of interest, defined via the terminology (concepts) used within the domain and the properties and relationships among domain objects. The project's approach is to extend EML to support the semantic annotation of ecological data sets, such that EML data-set descriptions can use terms drawn from OWL-DL ontologies. Both intelligent agent and semantic web service technologies are able to reach remarkable achievements and in some cases have overlapping functionalities.

2.1 Information Management System Based Ontology

To computerize the cognitive processes, problem solving steps are captured and stored in knowledge base and then be retrieved based on cognitive model of problem solver. Ontology provides a formal semantic representation of the objects for case representation. Even though automatic ontology learning methods (such as text mining and knowledge extraction significantly support ontology engineers by speeding up their task, there is still the need of a significant manual effort, in the completion, consolidation, and validation of the automatically generated ontology. The triple $K=(G,M,I)$ is called a *(formal) context*. If $A \subseteq G$, $B \subseteq M$ are arbitrary subsets, then the *Galois connection* is given by the following *derivation operators*.

$$\begin{aligned} A' &:= \{m \in M | gIm \text{ for all } g \in A\} \\ B' &:= \{g \in G | gIm \text{ for all } m \in B\} \end{aligned} \quad (1)$$

When an expert system for security is developed, an important question is to decide if the normality of a domain should be analyzed or, on the contrary, its abnormality.

However, providing the appropriate keywords for a search is not easy, especially when a user is unfamiliar with the subject being searched. Knowledge-based systems (KBSs) currently provide powerful and flexible means for utilizing expertise to address problems associated with a specific domain[3]. A knowledge base can be rule-based, frame-based, or utilize scripts. Recently, markup languages, such as XML, have been applied in developing knowledge bases.

Therefore, the order in the concept lattice is given as follows: if (A,B) and (C,D) are two concepts, $(A,B) \leqslant (C,D)$ iff $A \subseteq C$ (or, equivalently, $B \supseteq D$). Consequently, ontology mapping this study conducts similarity matching for concept names and considers the similarities of essential information and relationship to precisely identify the similarity between concepts. However, for the purposes of this paper, we simply assume that an ontology is a schema agreed upon by a group of interest in order to formalize the data relevant for the domain in question.

In OWL, properties are binary relations. The class at the tail of an arrow is the domain of the relation and the class at the head of the class is the range of the relation. The ontology is a computational model of some portions of the world. It is a collection of key concepts and their inter-relationships collectively providing an abstract view of an application domain.

2.2 Data Mining System Algorithm Based on FCA and Ontology

Data mining is a young multidisciplinary field, drawing work from areas including database technology, machine learning, statistics, pattern recognition, neural networks, artificial intelligence, and data visualization etc. Nowadays, it is one of the most active and exciting areas of the database research community. RDF can be used to describe the resources of a given web page, using a meaning graph of the RDF to represent a problem. The RDF accentuates the exchange and automation processing of web resources designated by a Unified Resource Identifier (URI), a string of web resources or an element of XML.

The classes of the ontology correspond to the tables of the database: the name of a class in the ontology is consistent with the name of the table in the database, and the slots of the class correspond to the column of the table. Data analysis is therefore implemented to mine the useful relations in the database. The ontology, which is good at the management and reuse of relations, is used to save the results of data analysis. Algorithm 1 gives the detailed steps to evaluate the structural validity of ontology, Algorithm 1.

```
BuildTree(Tho);
/* Build a tree structure for thesaurus ontology, and
terms of the latter map to the nodes of the former; */
for (all l ∈L(Fgs)) do
 if ( SimBetweenNodes (r, l) > Threshold)
 then do
  numbern • the group number of l;
  list.add(n);
 endif
endif;
  if  S T (X).count[i]≥min_count  and  I(X∪i) then{
   build S T (X∪i) based on S T (X) ; }
end//S T(X∪i)count[ ]中X∪i
```

3 Using Ontology Model to Build Concept Lattices System

The contributions of our work described here may be of relevance to several communities, i.e. the information extraction community as we show how a variety of information extraction techniques on different kinds of data can be integrated into an end-to-end system which constantly monitors the web and automatically maintains a coherent knowledge base. One of the great advantages of having situations represented in a formal language is that facts that are not explicitly stated can be derived using an inference engine[4]. In this section, we consider how one derives facts within a single situation, given that some other facts are known.

In addition, the ontology is no longer a mere research topic. Nowadays, its relevance has been recognized in several practical fields. The application areas include the natural language translation, geographic information system, biology and medicine, agent systems, knowledge management systems, and e-commerce platforms, etc. Therefore, the system mixes TF-IDF with Entropy to define the weight of the key terms. The formula 2 which calculates the similarity of Agents is as follows.

$$similarity(A_i, A_j) = \alpha \times C(N_i, N_j) + \beta \times \left(\frac{\sum_{k=1}^{f}(W_{i,k} \bullet W_{j,k})}{\sqrt{\sum_{k=1}^{f} W_{i,k}^2}\sqrt{\sum_{k=1}^{f} W_{j,k}^2}}\right) \quad (2)$$

Ontology extraction and similarity calculation are the two main components of the prototype system. Therefore, the part of ontology extraction was implemented and there are 16 randomly chosen documents used to be evaluated the result. The aim of ontology is to obtain, describe and express the knowledge of related domain. Ontology provides common understanding of the domain knowledge and confirms common approbatory vocabulary in the domain, as well as gives specific definition of the relation between these vocabularies from formal model of different levels. Finally, we test the average cost time of distributed discovery system and centralized discovery system. Fig.1 describes the method of building concept lattices in e-business based on ontology model and data mining.

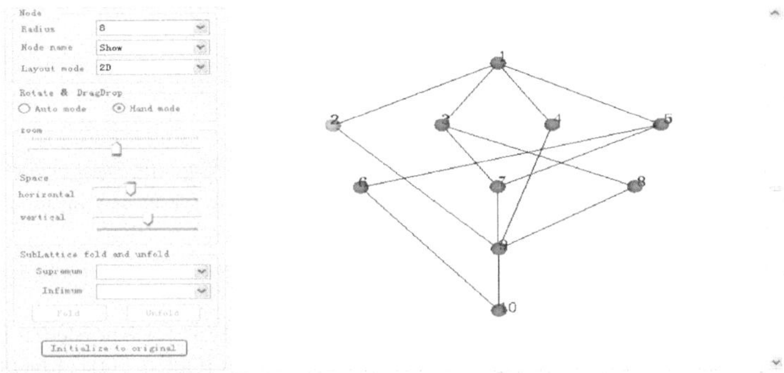

Fig. 1. Building concept lattices based on ontology and data mining

4 Construction Data Mining System Based Ontology and FCA

According to the goal of ontology, the pivotal problem of ontology constructing is to find the concepts and relationship among concepts after confirming the field, but these are connotative in the brain or store the file of this field in e-business. The OWL formal semantics provide five property characteristics that can be used to reason. In this paper, we apply the theory of concept lattices and ontology to automatically construct the concept hierarchy of mining in e-business and to match up the binary relation matrix of documents and terms to express the independence, intersection and inheritance between different concepts to form the concept relationship of ontology. Fig.2 shows the detailed comparison results of FCA and ontology. The experimental results indicate that this method has great promise.

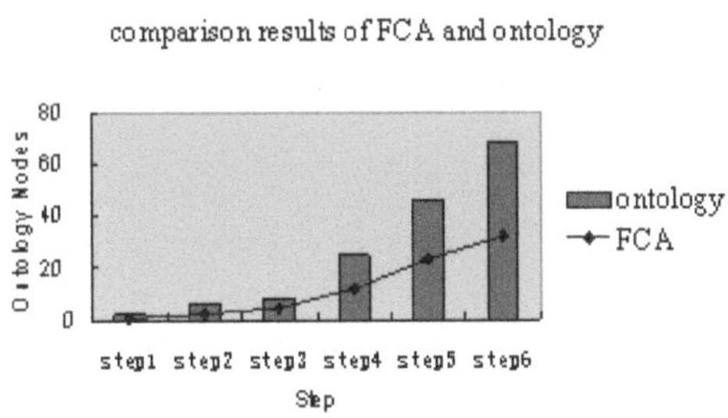

Fig. 2. Building data mining system comparison results of FCA and ontology

5 Summary

As the foundation of the semantic web, ontology is a formal, explicit specification of a shared conceptual model and provides a way for computers to exchange, search and identify characteristics. In this paper, we adopt ontology and FCA to improve the performance of data mining from documents. The paper probes into building e-business domain data mining information management system based on concept lattices and ontology in order to suffice the needs of theory and application in e-commerce recommendation system.

References

1. Berners-Lee, T., Hendler, J., Lassila, O.: The Semantic Web. Scientific American, 5 (2001)
2. Missikoff, M., Navigli, R., Velardi, P.: Integrated Approach for Web Ontology Learning and Engineering. IEEE Computer 35(11), 60–63 (2002)
3. Ontologies, F.D.: Silver Bullet for Knowledge Management and Electronic Commerce. Springer, Berlin (2000)
4. Valtchev, P., Missaoui, R., Godin, R.: Formal Concept Analysis for Knowledge Discovery and Data Mining: The New Challenges. In: Eklund, P. (ed.) ICFCA 2004. LNCS (LNAI), vol. 2961, pp. 352–371. Springer, Heidelberg (2004)

The Apply of the Adjustable Modifying Factor Fuzzy-PID Control in Constant Pressure Water Supply System

Ruilan Wang

School of Information and Control Enginerring, Weifang University
Weifang, Sandong, China
wrl2836@163.com

Abstract. The paper revolves around the design of large and medium-size water supply system in which the adjustable modifying factor fuzzy-PID control algorithm is adopted. The design offers a solution which is targeted at two difficult problems; the first one is the closed-loop control by analysis mathematical model is not efficiently achievable, because of the complicated water pipelines and numerous curves, as well as the fact that the objects to be controlled are characterized by their high order, big lagging, close coupling, non-linear and time-varying parameters, and the second one is about the over-current and over-voltage surge produced in the process in which the variable-frequency power sources are switched to frequency stabilized power sources when multiple engines of water pumps are operated in parallel. Experimental results show that the design can enhance the stability and reliability of the frequency control water supply system.

Keywords: constant pressure water supply system, adjustable modifying factorl, Fuzzy-PID control.

1 Introduction

The constant-pressure control is the output frequency control of inverter, and adjusting the speed of the pump motor to adapt to the pressure changing in the pipe network real time through the inverter based on the deviation of a given value and actual value of water pressure. Installing a pressure transmitter as a feedback component in the pipe network to maintain the pressure of the water supply network unchanged. The mathematical model of water supply system has the characteristics of high order, big lagging, close coupling, non-linear and time-varying parameters, and it is not efficiently achievable, because of medium and high power pump is driven by three-phase squirrel-cage induction motor, coupled with dead-zone and non-linear resistance characteristic of water supply network in the general water supply system. In addition, the water supply system with multi-parallel pumps cycle runs, water pressure switch the continuous state to flutter state, is equivalent to adding a serious interference signal when the frequency stabilized power sources are switched to variable-frequency power sources in the system.

The PID control is commonly used in the little change water supply system for water consumption. However, only using PID control would make the dynamic performance

D. Jin and S. Lin (Eds.): Advances in EECM Vol. 1, LNEE 139, pp. 25–28.
springerlink.com

is poor, not only adjusting time is long, but also serious oscillation may appear, even causing system instability when change of water consumption is large.

2 The Design of Adjustable Modifying Factor Fuzzy-PID Controller

The fuzzy control has good dynamic properties and PID control with excellent static properties. Using fuzzy control to speed up the response time when the pressure fluctuation is large, and using PID control to maintain static accuracy when the pressure fluctuation is small.

Although fuzzy-PID compound control for improving performance of static, but it is difficult to meet the requirements for systems in different states due to its control rules unchanging, so as to impair the control effect. Therefore we constitute a adjustable modifying factor fuzzy-PID control of constant-pressure water supply system, with a view to reach a satisfactory control effect by introduction of adjustable modifying factor fuzzy control to meet the adaptive capacity of control system. Its structure as shown in Fig. 1.

In Figure 1, the input variable is the error e and error rate $\dot{e}$ between the measured pressure value y of pipe network and the setting value r of system, the output variable is a real-time control correction variable u of system , K_e is the quantized factor of error e, as K_{ec} is the quantized factor of error rate $\dot{e}$, K_u is the proportion factor of the control variable u .

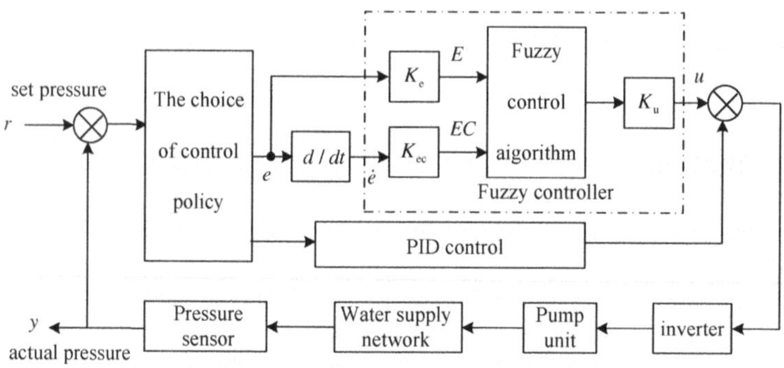

Fig. 1. The diagram of the adjustable modifying factor fuzzy-PID control system

The choice of constant pressure water supply system control policy is according to system's information, and judge between the adjustable modifying factor fuzzy control and the PID control. The basic principle for choosing control policy is: Take the choice by the hydraulic pressure actual observed value and the setting value error's size as the condition, $|E| = E_b$ is the hydraulic pressure control switching point.if the size is in a big error range($|E| > E_b$) , adopt the adjustable modifying factor fuzzy control so as to raise the dynamic response speed and strengthen the auto-adapted ability; but if in a small error range ($E_a < |E| < E_b$) , adopt the PID control so as to eliminate the static error and improve the control precision. Moreover, in order to prevent the system from switching too frequently, the system provides no switch at the erroneous switching point ($|E| = E_b$) and systems within the allowed range ($|E| \le E_a$), just maintains the previous movement.

The adjustable modifying factor fuzzy-PID controller's actual output is:

$$u = \begin{cases} -K_u[\alpha E + (1-\alpha)EC] & |E| > E_b \\ u_o & |E| = E_b, |E| \le E_a \\ K_p[e(k) - e(k-1)] + K_i e(k) + K_d[e(k) - 2e(k-1) + e(k-2)] & E_a < |E| < E_b \end{cases}$$

where $\alpha = (\alpha_s - \alpha_0)|E|/N + \alpha_0$, $0 < \alpha_0 < \alpha < \alpha_s < 1$

Define the universe of discourse of E 、 EC 、 U :{-7,…,-1,0,1,…,7}

where $|E| \le E_a$ is water supply system error target, u_o is the system's last action, K_u is the output control quantity scale factor, K_p is the proportional coefficient, K_i is the integral coefficient, K_d is the differential coefficient, $e(k)$ is the error between hydraulic pressure actual observed value and setting value in the Kth sampling.

3 The Analysis of Simulation Result

This paper simulates the water supply system, and lets the hypothesis water supply pressure 0.5Mpa. The system's approximation model is:

$$G(s) = \frac{20e^{-4s}}{(10s+1)(0.6s+1)}$$

In the PID control, K_p =0.08, K_i =0.025, K_d =0.04 ; In fuzzy control, K_e =70, K_{ec} =2.5 , K_u =0.45 ; In the fuzzy PID control, K_p =0.1 , K_i =0.022 , K_d =0.02, K_e =60, K_{ec} =2.45 , K_u =0.6. The hydraulic pressure regulating process simulation based on the fuzzy-PID control is shown in Fig. 2.

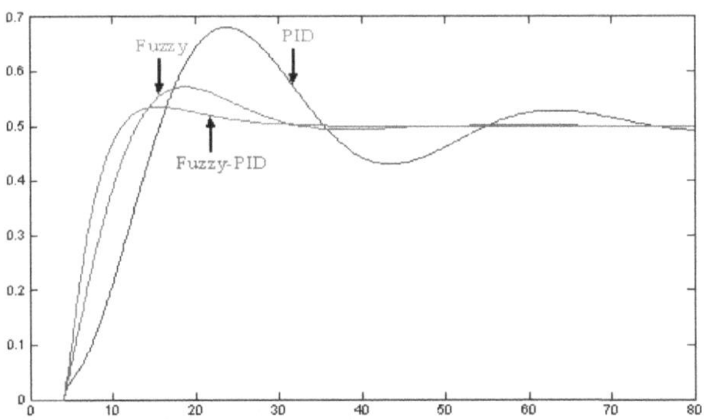

Fig. 2. The simulation diagram of hydraulic conditioning

The simulation results can be seen from Fig. 2, the system based on adjustable modifying factor fuzzy PID control has shorter settling time, smaller overshoot, and higher accuracy than the PID controlling. This control method not only has the fast dynamic response speed and the good control precision, but also the system has the good robustness and adaptability, when the parameters and the structure changes.

References

1. Jiang, X., Dai, H.: The mechanism analysis of the frequency conversion constant pressure water supply system. Electric Drive Automation (April 2002)
2. Jin, C., Mao, Z.: The application of the frequency and speed modulation technology in water pump control system. Applications of Electronic Technique (September 2000)
3. Sun, B., Zhang, H.: The design of constant pressure water supply system based on fuzzy control. Electrical Automation (March 2002)

A On-Site Monitoring System of Reservoir Sediment for Three Gorges

Yaodong Du[1,2] and Xingyuan Song[2]

[1] Bureau of Hydrology, Changjiang Water Resources Commission; Wuhan, China, 430010
[2] State Key Lab. of Water Resources and Hydropower, Wuhan University, Wuhan, China, 430072

Abstract. The On-site reservoir sediment monitoring system plays a very important role in the reservoir operation and the sediment research. The LISST-100X (Laser On-site Scattering and Transmissometer-100X) is able to obtain the information of the sediment concentration, the sediment gradation, and the water temperature at the dam's upstream and downstream rapidly and precisely, therefore it has been applied to the Three Gorges Water Project Reservoir Sediment Monitoring System. The real-time sediment data collected by the system can be immediately uploaded to the Control Center and consequently taken as the support for the realtime schedule of the project, providing a guarantee for the reservoir operation center to fully and timely receive the hydrology and sediment information. This paper studies some key technologies of the system in the building-up process and reach some valuable experience of the data analysis, the sediment concentration calculation, and the design of the system, which finally helps obtain some useful results. The system could the future of On-site reservoir sediment monitoring system in China.

Keywords: LISST-100X, Sediment monitoring, On-site monitoring.

1 Introduction

In the context of technological innovations of the Changjiang River hydrological measurement, the use and promotion of the new hydrological equipment and methods have speeded up in the Changjiang River. The LISST-100X (Laser in-situ Scattering and Transmissometer-100X), produced by the Sequoia Ltd., capable of measuring both the sediment concentration and the suspended sediment gradation, has been entering into our daily measuring work, the use of which could greatly reduce the workload of the sand measurement and enhance the timeliness of the sand measurement data. In measuring the sediment concentration and the sediment gradation, compared with those traditional measurement methods, the LISST-100X possesses the characteristics as good timeless, and small workload. It can make real-time monitoring, on-site analysis, and laboratory sample analysis of the sediment concentration in real time. What's more, it can also simultaneously measure the sediment concentration, the sediment gradation, the water depth, the water temperature, the light transmission level and the attenuation of the light. The Hydrographic Bureau of the Changjiang

D. Jin and S. Lin (Eds.): Advances in EECM Vol. 1, LNEE 139, pp. 29–38.
springerlink.com

Water Resources Commission has carried out correlation analysis studies since many years ago and has finished comparative measuring experiments in a number of hydrological stations. At present, the Three Gorges On-site monitoring system has been established by the using of the LISST-100X, which includes the following six hydrological stations: Cuntan, Yichang, Zhicheng, Shashi, Luoshan, and Datong. Since May 2010, the LISST- 100X instrument has been used in the On-site measurement of the suspended sediment gradation and the reporting of the sediment concentration to the Hydrologic Information Analysis Center of the Changjiang River and the China Three Gorges Corporation. The measurement and reporting of the suspended sediment gradation data information are being experimented on, which helps making preparation for a further increase of the suspended sediment gradation in the Three Gorges Water Project On-site Reservoir Sediment Monitoring System.

2 The Three Gorges On-Site Reservoir Sediment Monitoring System

2.1 Introduction to the LISST-100X Equipment

The LISST-100X is an On-site laser particle analyzer produced by the Sequoia, which can be used both in the suspended sediment gradation measurement, but also in the sediment concentration measurement.

Fig. 1. LISST-100X Instrument

The measurement principle of the LISST-100X is the laser diffraction, which is not affected by the color or the size of the particles, with large-size particles of small diffraction angles and small-size particles of large diffraction angles. After the light irradiates the particles, the diffracted light bypasses the particles and is focused through a convex lens to a photodiode detector composed of 32 rings, and then the laser energy received is preserved and converted to the size distribution of the particles. At the same time, the degree of light penetration measured by the system will be used to compensate for the diffraction attenuation caused by the concentration. The 32 detection rings can measure the particle distribution of 32 levels. With the particle concentration of each size being calculated by the energy received on each measure ring, the total particle concentration of 32 levels is then the total concentration of the suspended solids.

The minimum measurable concentration of the LISST-100X is 0.1mg / L and the maximum 10kg / m^3. Battery and data storage are merged in the system, making the LISST-100X self-contained and availably used in online measurement. The pressure rating of the system is 300-5000 meters, which can meet the measurement requirements of different depth levels. The LISST-100X can measure simultaneously the volume concentration, the size distribution of the graded particle, the water depth, the water temperature, the degree of light transmission, the attenuation of light and the like.

According to the size of sediment concentration, different optical path controller can be installed in the instrument, whose measuring mode is divided into the following categories: the standard mode (with no optical path controlled being installed), the 2.0 optical path mode (with 2.0 optical path controller being installed), the 4.0 optical path mode (with 4.0 optical path controller being installed), the 4.5 optical path mode (with 4.5 optical path controller being installed) and the like.

2.2 Composition of the Three Gorges On-site Reservoir Sediment Monitoring System

The Three Gorges On-site Reservoir Sediment Monitoring System is mainly composed of the six hydrological stations along the Changjiang River, the Hydrologic Information Analysis Center, and the receiving units (namely the Hydrologic Information Analysis Center of the Changjiang River and the China Three Gorges Corporation). The six hydrological stations from upstream to downstream are respectively Cuntan, Yichang, Zhicheng, Shashi, Luoshan, and Datong, among which Cuntan is taken as the Three Gorges Reservoir sediment input monitoring station and Yichang the sediment output monitoring station. Zhicheng and Shashi are two stations monitoring the effects of the Jingjiang River's being eroded by the Three Gorges Reservoir discharge to the suspended sediment. Luoshan station monitors the sediment changes of the Changjiang River below the Dongting Lake export. And Datong station monitors the effects of the Three Gorges Reservoir discharge to the sediment downstream.

The Three Gorges On-site Reservoir Sediment Monitoring System adopts the On-site monitoring sediment data obtained by the LISST-100X at daily 8:00 a.m. The sediment monitoring stations input the data to the Automatic Flood Forecast System, and the data, through VHF, telephone, satellite or other information channels, will then be transmitted to the Hydrologic Information Sub-Center, where the data will be reported to Hydrologic Information Center of the Changjiang River. After the data information being summarized there, it will finally be given to the China Three Gorges Corporation.

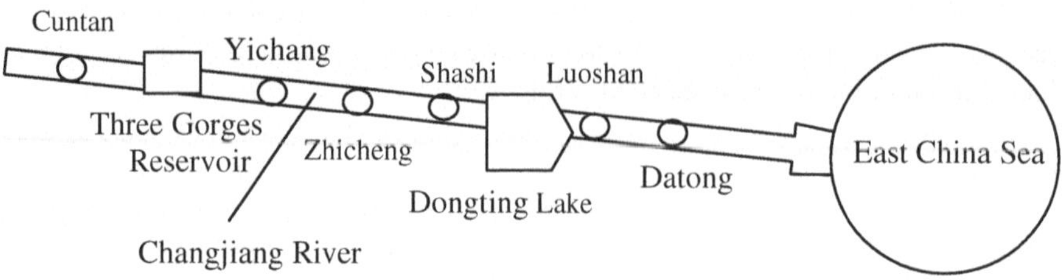

Fig. 2. Distribution of the Three Gorges On-site Reservoir Sediment Monitoring System

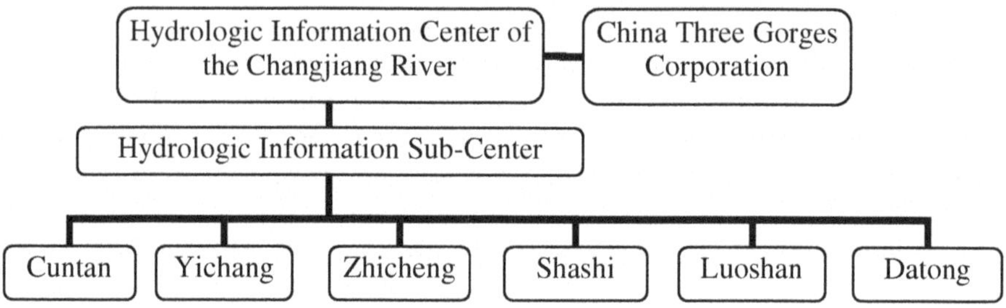

Fig. 3. Abridged general view of the Three Gorges On-site Reservoir Sediment Monitoring System

3 Application Experience of the Lisst-100x

Based on years' practical application of the LISST-100X and a series of discussions by experts in professional seminars, a complete set of application experience of the LISST-100X has been established by the Hydrographic Bureau of the Changjiang Water Resources Commission. The application experience formed includes the following:

3.1 The Design of Professional Installation Equipment

Through the comparison and experiment of a variety of ways of equipment installation, the installation of the Jingjiang River Hydrological Bureau of Changjiang Water Resources Commission is finally selected, whose installation way of the LISST-100X is shown in Fig. 4. The biggest advantage of this installation way is as follows:

1) The instrument is combined with the lead fish, which makes the system firm;
2) It can not only prevent the instrument from being badly crashed, but also won't change the natural flow pattern after entering into the water;
3) It won't produce bigger wear and tear and sand accumulation, which ensures a relative reliability of the measuring results.

Fig. 4. Equipment installation method

3.2 Operation Procedures of Measurement

In order to make further good use of the LISST-100X, related operation procedures of measurement have been made, where detailed requirements of more than ten items are carried out as the following: pre-screening of the instrument, equipment installation, set-up of the operating parameters, measurement records, data acquisition, maintenance and storage of the equipment and the like. The operation procedures of measurement ensures each station to use the LISST-100X under the guidance of unified and scientific technical requirements, which consequently ensures the quality of measurement results.

3.3 Professional Data Processing Software

In order to further improve the application of the LISST-100X, with a combination with the application requirements, a professional data processing software has been worked out by the Hydrological Techniques Institute of the Hydrographic Bureau of the Changjiang Water Resources Commission. The software includes the following modules as data preprocessing, data processing, result calculation, the result table generation and the like. The report form that finally can be derived is as follows:

1) Results table of the average vertical sediment concentration and the sediment concentration of the fixed points;

2) Results table of the average vertical sediment gradation and the time-averaged sediment gradation of the measuring points;

3) Calculation results table of the sediment transport rate;

4) Parameters table of the water depth, the temperature, the battery voltage and the like;

4 Improvements of the Sediment Calculation Method

According to the principles of the Laser method, the data of the LISST-100X relates primarily to the volume size of the water impurities, which needs to be converted while the sediment concentration is being calculated. At present, there are two conversion methods:

1) With the relationship between the output value measured by the LISST-100X and the sediment concentration, the value of the conversion coefficient VCC can be calculated, and then the sediment concentration can be derived through the VCC value and the measured output value;

2) With the relationship between the biased attitude of the sediment gradation and the conversion coefficient VCC being established, the value of the conversion coefficient VCC can be derived, and then the sediment concentration can be derived by the VCC value and the measured output value.

However, both methods have their defects. As to the former method, with a fixed VCC value being adopted, the changes of the sand density and the value of the conversion coefficient VCC, which result from the changes of the sand quality in the water, is not taken into account. And as to the latter method, though it takes into account the effect of the changes of the sand quality to the value of the conversion coefficient VCC, it doesn't take full advantage of the sediment gradation information of the LISST-100X. Therefore, in this paper, the data information of the LISST-100X is fully pursued. By using the LISST-100X, the advantages of sediment gradation can be simultaneously measured. Considering different distribution of the sand quality of each sediment gradation to the sediment concentration, the multiple linear regression method has been proposed to estimate the sediment concentration. The method takes into account the distribution of the sand quality of each sediment gradation to the sediment concentration, which can tell the effect of different sand quality to the relationship between the measured output value and the conversion of the sediment concentration.

4.1 The Existing Sediment Calculation Method

The existing sediment calculation method (the original method) of the LISST-100X (hereinafter referred to as the original method) works as the following: Through the relationship between the output value measured by the LISST-100X and sediment concentration, the value of the conversion coefficient VCC can be calculated, and then the sediment concentration can be derived through the VCC value and the measured output value. The calculation is as follows:

$$VCC = \frac{\sum_{i=1}^{k} \frac{output_i}{C_i}}{k} \tag{1}$$

Where: VCC for the conversion coefficient, C_i for the sediment concentration of the ist sample, $output_i$ for the output value measured by the LISST-100X of the ist sample, k for the total number of the samples.

$$C = \frac{output}{VCC} \tag{2}$$

Where: C for the sediment concentration of the measured point, output for the output value measured by the LISST-100X, VCC for the conversion coefficient.

With the Eq. 1 the VCC value being firstly calculated, then by using the calculated VCC value and the output value measured by the LISST-100X, the sediment concentration C can be calculated with the Eq. 2.

Another method works as the following: With the relationship between the biased attitude of the sediment gradation and the conversion coefficient VCC being established, the value of the conversion coefficient VCC can be derived, and then the sediment concentration can be derived by the VCC value and the measured output value. As the data of the relationship between the biased attitude of the sediment gradation and the conversion coefficient VCC used in this paper is not that good, the method can not be used and therefore is not further described either.

4.2 The Improved Sediment Calculation Method

In order to consider the contribution of the sand quality of all levels to the sediment concentration and to illustrate the effects of different sand qualities to the conversion relationship between the measured output value and the sediment concentration, the multiple linear regression method has been proposed in the paper to calculate the sediment concentration.

$$C = output \times \sum_{i=1}^{n} a_i p_i \tag{3}$$

Where: C for the sediment concentration of the measured point, p_i for the percentage of the sediment gradation of the ist level, a_i for the undetermined coefficient, output for the output value measured by the LISST-100X, n for the total level numbers of sediment gradation.

According to the measurement theory, the intercept of the multiple linear regressions can be eliminated by collection of background scattering data. Therefore, a multiple linear regression model without intercept is used in this paper. According to the actual situation of the section, the sediment gradation in this paper is divided into eight levels, respectively 0-4μm, 4-8μ, 8-16μm, 16-32μm, 32-62μm, 62-125μm, 125-250μm, and 250-500μm. By using the comparative measured data, and through the Least Square method described by the following, the value of a_i can be calculated. Then with the output value calculated by the LISST-100X and the sediment gradation data p_i being put to the Equation 3, the sediment concentration of measured points can be calculated directly.

$$\hat{a} = (X'X)^{-1} X'Y \tag{4}$$

where, $$X = \begin{bmatrix} p_{11} & p_{21} & \cdots & p_{n1} \\ p_{12} & p_{22} & \cdots & p_{n2} \\ \vdots & \vdots & \vdots & \vdots \\ p_{1k} & p_{2k} & \cdots & p_{nk} \end{bmatrix}, \quad \hat{a} = \begin{bmatrix} a_1 \\ a_2 \\ \vdots \\ a_n \end{bmatrix}, \quad Y = \begin{bmatrix} C_1 / output_1 \\ C_2 / output_2 \\ \vdots \\ C_k / output_k \end{bmatrix}$$

Where: p_{ij} for the percentage of the ist-level of the jst sample, $output_i$ for the output value of each sample measured by the LISST-100X, a_i for the undetermined coefficient, C for the sediment concentration of each sample, n for the total level numbers of sediment gradation, k for the total number of the samples.

4.3 Calculation Results

The data of the Nanzui hydrological station and the Hankou (Wuhan Guan) hydrological station measured by the LISST-100X (calculated by original method and multiple linear regression method) are chosen to make a comparative analysis with the corresponding data of the sediment concentration measured by those traditional measurement methods (namely: taking water samples by horizontal water samplers and weighing the sand by the weight-dried method). The two hydrological stations are the representative station of the Dongting Lake and the Changjiang River, so the analysis conclusion is of good representation. The standard mode, the 2.0 optical path mode, the 4.0 optical path mode, and the 4.5 optical path mode are respectively been adopted in the Nanzui hydrological station. The comparative measured data are mainly as the following: the data of 94 standard modes (0.012-0.099 kg/m^3), the data of 27 4.0 optical path modes (0.239-0.331 kg/m^3), and the data of 47 4.5 optical path modes(0.352-0.712 kg / m^3). Among them, 79 data of the standard modes are used in the parameter estimation with the rest 15 being used in the parameter test; 40 data of the 4.5 optical path modes are used in the parameter estimation with the rest 7 being used in the parameter test; the total 27 data of the 4.0 optical path modes are all used in the parameter estimation. The comparative measured data of the Hankou (Wuhan Guan) hydrological station are much less, with the total 36 data (0.064-0.364 kg/m^3) also all being used in parameter estimation and none being used in the parameter test.

The calculation error of part of the data being used in the parameter estimation of the Nanzui hydrological station by respectively using the original method and the multiple linear regression method is shown in Table 1.

Table 1. Error statistics of the Nanzui Hydrological Station in parameter estimation

Index	Standard mode		4.5 optical path mode		4.5 optical path mode	
	the original method	the multiple linear regression method	the original method	the multiple linear regression method	the original method	the multiple linear regression method
System error (kg/m^3)	0.000	0.000	0.000	0.000	0.000	0.000
Deterministic coefficient	0.862	0.968	0.410	0.923	0.381	0.715
95% confidence interval (kg/m^3)	0.014	0.007	0.149	0.054	0.045	0.030

The calculation error of part of the data being used in the parameter test of the Nanzui hydrological station by respectively using the original method and the multiple linear regression method is shown in Table 2.

Table 2. Error statistics of the Nanzui Hydrological Station in parameter test

Index	Standard mode		4.5 optical path mode	
	the original method	the multiple linear regression method	the original method	the multiple linear regression method
System error (kg/m^3)	0.003	0.001	0.029	-0.003
Deterministic coefficient	0.773	0.929	-1.714	0.907
95% confidence interval (kg/m^3)	0.021	0.012	0.251	0.046

The calculation error of the data used in the parameter estimation of the Hankou (Wuhan Guan) hydrological station by respectively using the original method and the multiple linear regression method is shown in Table 3.

Table 3. Error statistics of the Hankou (Wuhan Guan) hydrological station

Index	Standard mode	
	the original method	the multiple linear regression method
System error (kg/m^3)	0.000	0.001
Deterministic coefficient	0.319	0.521
95% confidence interval (kg/m^3)	0.112	0.094

As can be seen from the above three tables, the 95% confidence interval of the sediment concentration calculated by using the multiple linear regression method is significantly reduced compared with that of the original method, and the calculation error in the parameter test period is close to that of the parameter estimation period and has a better stability.

The following Figure 5 are comparison charts of the sediment concentration measured by the traditional measurement method, the sediment concentration calculated by the original method, and the sediment concentration calculated by the multiple linear regression method in the Nanzui hydrological station. Seen from the Figure 5, it can be found that the sediment concentration calculated by the multiple linear regression method is more consistent than the sediment concentration measured by the traditional measurement method.

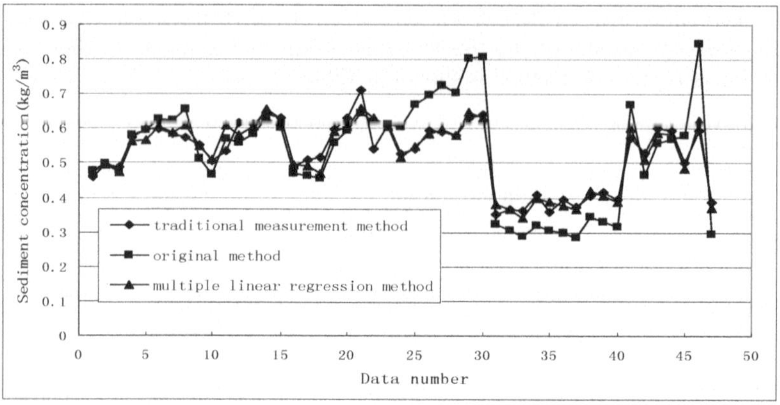

Fig. 5. Comparison of the measured sediment concentration(Nanzui Hydrological Station, 4.5 Optical Path Mode)

5 Conclusions

The successful operation of the Three Gorges Water Project On-site Reservoir Sediment Monitoring System has proved the feasibility of constructing an On-site reservoir sediment monitoring system by using the LISST-100X. The real-time sediment data collected by the system can be immediately uploaded to the Control Center and consequently taken as the basis for the schedule of the project, providing a guarantee for the reservoir operation center to fully and timely receive the hydrology and sediment information. The building-up of this system also shows a new way for the set-up of the in-time reservoir sediment monitoring system. The improved equipment installation method and calculation method of the sediment concentration proposed in this paper has been proved to effectively increase the accuracy and stability of the LISST-100X, making the accuracy of the On-site Reservoir Sediment Monitoring System meet the requirements.

References

1. Mikkelsen, O.A., Pejrup, M.: In situ particle size spectra and density of particle aggregates in a dredging plume. Marine Geology (170), 443–459 (2000)
2. Mikkelsen, O.A., Pejrup, M.: The use of a LISST_100 laser particle sizer for in_situ estimates of floc size, density and settling velocity. Geo_Marine Letters (20), 187–195 (2001)

Research on the Algorithm of Management Clustering for Wireless Sensor Network

Wenbo Zhang and Deyu Zhang*

School of Information Science and Engineering, Shenyang Ligong University
Communication and network institute, Shenyang Ligong University 110159, China
jessezwb@163.com

Abstract. In terms of the dynamic nature of the satellite nodes and the features of the management field division in satellite network, the fault diagnosis algorithm of the satellite network was established. If the satellite could respond to the network management instruction, the network management technique was used to diagnose the interfaces in the satellite. If the satellite could not respond to the network management, the intra-domain cooperation or inter-domain cooperation would be activated. The suspectable fault satellite could be, diagnosed through cooperation among the faulty diagnosis agents in the satellite. The simulation results shows that in the circumstance of the low faulty frequency, the new method put forward in this paper could be effectively used in satellite network with short cooperative time and low throughput.

Keywords: Wireless sensor networks, network management, clustering, MSCBRE.

1 Introduction

Wireless sensor network management technology puts forward new challenges for generating clusters[1]. Firstly wireless sensor network management of cluster head must be evenly distributed in order to improve the efficiency of the network management and decrease energy consumption of the intermediate nodes forwarding message[2][3]; Secondly because wireless sensor network link is vulnerable to damage and interference, in the process of generating cluster, cluster heads can still provides management services for as many member nodes as possible considering link failures; Finally cluster heads need more energy consumption than ordinary member nodes, so when choosing cluster-heads, the surplus energy of the nodes must be considered[4].

Clustering algorithm of this paper is based on the layout of the maximum expected coverage problem (MEXCLP) model ,after considering the energy issues of wireless sensor networks, to design the maximum successfully communication clustering algorithm (MSCBRE), which is based on residual energy [5] [6][7].

2 Overview of MEXCLP Model

2.1 MCLP Model

MCLP is positioning M facilities in N candidate nodes. The M facilities provide service for N request nodes as far as possible. That is to say, these M facilities cover

* Corresponding author.

N nodes to maximize the request which makes the request quantity can be maximized. MCLP model uses the locations of N candidate nodes, requests, facilities services radius D and facilities number M as input parameters. Through calculating the maximum resolving the best positions of M facilities can be got. The following is the formulation description of the model of M facilities in N nodes.

$$\text{Max.:} \sum_{k=1}^{N} h_k y_k \tag{1}$$

$$\Sigma\tau.\ y_k - \sum_{i=1}^{N} a_{ki} X_i \le 0 \qquad k = 1,2\cdots,N \tag{2}$$

$$\sum_{i=1}^{N} X_i \le M \tag{3}$$

$$\left.\begin{array}{ll} X_i = 0,1, & i = 1,\cdots,N \\ y_k = 0,1, & i = 1,\cdots,N \end{array}\right\} \tag{4}$$

Among them h_k is equal to the request produced by Node k.

$$y_k = \begin{cases} 0 & \text{node k isn't covered by facilities} \\ 1 & \text{node k is covered by facilities} \end{cases}$$

$$X_l = \begin{cases} 0 & \text{facilities aren't in the nodes} \\ 1 & \text{facilities are in the nodes} \end{cases}$$

$$a_{ki} = \begin{cases} 0 & d_{ki} > D (\text{The facilities at node i covere node k}) \\ 1 & d_{ki} \le D (\text{The facilities at node i don't covere node k}) \end{cases}$$

(Here D refers to the effective distance covered service facilities, namely in this range nodes are covered)

2.2 MEXCLP Model

In the above MCLP model, assume that all facilities can work normally. In fact, not all facilities can correctly response to each request. If facilities can react to request, it is called "work" condition, otherwise called "doesn't work" or "damage" condition. Probability p is used to say probability when facilities are not at the state of work. Assuming probability p for all facilities is known and same, and further assumes all facilities whether work is independent of each other. In these assumptions, work facilities quantity T should obey binomial distribution $P\{T = j\} = C_M^j (1-p)^j p^{M-j}$. It can also calculate the probability of the coverage of a given node k, in node k m a facility, which is covered by the working facilities.

$$P\{\text{k the probability of working facilities to be covered}\} = 1 - P\{\text{m facilities were not working}\} = H_{k,m}$$

Let random variable $H_{k,m}$ express the number of requests at node k, which are covered by m working facilities.

$$H_{k,m} = \begin{cases} h_k & \text{probability } 1 - \mathrm{p}^{\mathrm{m}} \\ 0 & \text{probability } \mathrm{p}^{\mathrm{m}} \end{cases}$$

And E ($H_{k,m}$) $= h_k (1 - p^m), \forall k, m$

When the number of covering node k, change from m-1 to m, the expected increase for coverage of Node k may be expressed as:

$$\Delta E(H_{k,m}) = E(H_{k,m}) - E(H_{k,m-1}) = h_k p^{m-1}(1-p) \qquad m = 1,2,\cdots,M$$

The number of facilities to cover nodes k is decided by $\sum_{i=1}^{N} a_{ki}x_i$ .Using the above definition, Daskin summarizes the maximum expected covering location model (MEXCLP) as following, that is model (1) to (4) improve results.

$$\begin{aligned} \text{Max} \quad & \sum_{k=1}^{N}\sum_{j=1}^{M}(1-p)p^{j-1}h_k y_{jk} = \sum_{k=1}^{N}\sum_{j=1}^{M} w_j h_k y_{jk} \\ & \sum_{j=1}^{M} y_{jk} - \sum_{i=1}^{N} a_{ki}X_i \le 0 \qquad \forall k \\ & \sum_{i=1}^{N} X_i \le M \end{aligned} \tag{5}$$

Among them

$$y_{jk} = \begin{cases} 0 & \text{If the node k is covered by at least j devices} \\ 1 & \text{If the equipment covered by the node k is less than k} \end{cases}$$

X_i = the number of facilities located in node i

$w_j = (1-p)p^{j-1}, \quad j = 1;\cdots, M$

K value of each objective function is concave on j, which shows that: $y_{jk} = 1$ then $y_{1k} = y_{2k} = \cdots = y_{jk} = 1$

$y_{lk} = 0$ then $y_{lk} = y_{l+1,k} = \cdots = y_{Mk} = 0$.

3 The Proposition of MSCBRE Clusters Algorithm

Although MEXCLP model takes into account the case of service failure, it does not take into account the problem of limited energy in the network of the wireless sensor nodes and it can not be directly used for wireless sensor networks. This article will improve MEXCLP, and put forward the management clusters algorithm MSCBR (Maximum Succeed Communication Based on Remainder Energy).

In MEXCLP model, suppose the probability p is known, and its value is the same regarding all servicing facilities. However, considering in the wireless sensor network node, surplus energy is not the same, here the probability of the service failure of the nodes has been revised. The excess energy parameter and the influence factor have been joined in the service failure rate. The failure probability p_j of the node j has been revised as following.

$$p_j = p - \alpha p + \alpha(1 - Er_j / Ei_j) \tag{6}$$

In MSCBRE, supposed that the situation if each node works is mutually independent. Therefore by the theory of probability, the probability that the node k can be covered and served by facilities is expressed as following:

$$\begin{aligned}&P\{\text{K is the probability of work facilities covered}\}\\&=1-P\{\text{the probability of M a facilities are not working}\}\\&=1-p^m\end{aligned}$$

However, taking into account the service failure rate of each facility node is not the same, the probability that m facilities cannot succeed to provide the service cannot use the expression p^m,but the expression $\prod_{j=1}^{m} p_j$. If we use the variable y_{jk} indicated whether the node k is covered by facilities j,when the node k is covered by the facility j, its value is 1; otherwise its value is $1/p_j$. Then we can put the expression $\prod_{j=1}^{m} p_j$ as $\prod_{j=1}^{M} p_j y_{jk}$. This is because when the node k is not covered by the facility j, $y_{jk} = 1/p_j$, $p_j y_{jk} = p_j * 1/p_j = 1$. Obviously when the node k is not covered by the facility j, $p_j y_{jk}$ has no effect on the value of $\prod_{j=1}^{m} p_j$, so in $\prod_{j=1}^{m} p_j$ adding M-m facilities in which Node k is not covered by does not have the influence on the result of $\prod_{j=1}^{m} p_j$.Therefore the following expression is tenable:

$$p^m = \prod_{j=1}^{m} p_j = \prod_{j=1}^{M} p_j y_{jk}$$

Therefore, the probability of the node k request success and is serviced by work facilities is expressed as:

$$P\{\text{the probability of working facilities services}\} = 1 - p^m = 1 - \prod_{j=l}^{M} p_j y_{jk} \tag{7}$$

It supposes that node k works the facility cover place the request quantity is used the expression $H_{k,m}$,it can be expressed as follows:

$$H_{k,m} = \begin{cases} h_k & \text{probability } 1-\mathrm{p}^{\mathrm{m}} \\ 0 & \text{probability } \mathrm{p}^{\mathrm{m}} \end{cases}$$

And $\mathrm{E}(H_{k,m}) = h_k(1-p^m), \forall k, m$

By the formula (7) shows:

$$E(H_{k,m}) = h_k(1-p^m) = h_k(1-\prod_{j=1}^{M} p_j y_{jk})$$

Therefore, in the network all nodes succeed the service expected value be possible to express as follows:

$$\sum_{k=1}^{N} \mathrm{E}(H_{k,m}) = \sum_{k=1}^{N} h_k (1 - p^m) = \sum_{k=1}^{N} h_k (1 - \prod_{j=1}^{M} p_j y_{jk})$$

Then the formula (6) can get MSCBRE model is as follows:

$$\text{Max.} \sum_{k=1}^{N} (1 - \prod_{j=1}^{M} p_j y_{jk}) h_k \tag{8}$$

$$\text{St.} \quad p_j = p - \alpha p + \alpha(1 - Er_j / Ei_j) \tag{9}$$

Among them, $y_{jk} = \begin{cases} 1 & \text{If the node k is covered by j a equipment} \\ \frac{1}{p_j} & \text{If the node k isn't covered by j a equipment} \end{cases}$

P is the cluster head node failure rate of the service, that is the failure rate of the network. α is the adjustment factor, It regulates energy in the influence degree clusters algorithm. Ei_j is initial energy of the node j, Er_j is residual energy of node j. M is the number of Cluster-heads nodes, N is the number of nodes.

4 MSCBRE Clustering Algorithm Simulation

In this paper, clustering algorithm for MSCBRE cluster head selection is simulated. The input parameters of MSCBRE clustering algorithm include the location of the node, the node's residual energy and node communication radius and so on. In this article simulation's experiment scene data is as follows: Nodes randomly deployed in the area $20m \times 30m$; The node correspondence radius is 8; The adjustment factor takes 0.5; In the network nodal point number is 40; cluster head node number is 5; In the wireless sensor network's service failure rate is 0.9; Each node's initial energy is 100; In the network each node's position, the request number and excess energy as shown in Table 1.

Table 1. Clusters simulation data tables

serial number	location	request number	excess energy	serial number	location	request number	excess energy	serial number	location	request number	excess energy
1	(8,27)	24	17	15	(25,10)	29	78	29	(39,33)	14	49
2	(6,2)	8	92	16	(10,24)	22	10	30	(27,19)	17	49
3	(29,28)	24	77	17	(13,2)	8	92	31	(39,29)	12	96
4	(36,28)	16	66	18	(32,8)	10	67	32	(29,28)	8	2
5	(32,16)	24	14	19	(8,20)	19	45	33	(25,12)	1	78
6	(4,23)	1	25	20	(9,7)	1	65	34	(31,3)	19	20
7	(9,5)	23	21	21	(24,30)	9	94	35	(9,21)	14	20
8	(13,24)	18	12	22	(1,16)	1	71	36	(37,5)	8	40
9	(33,34)	26	17	23	(4,37)	25	65	37	(39,7)	19	3
10	(34,36)	8	74	24	(2,36)	9	93	38	(4,18)	14	23
11	(33,21)	6	15	25	(25,10)	28	17	39	(34,3)	26	6
12	(36,10)	7	76	26	(0,5)	28	77	40	(31,21)	9	89
13	(22,39)	27	47	27	(6,33)	20	3				
14	(20,39)	3	34	28	(32,28)	11	14				

Carrying on the simulation to the MSCBRE algorithm, simulation chooses a cluster head node is : 1、 2、 3、 39、 40. Their distribution map is shown in Figure 1.

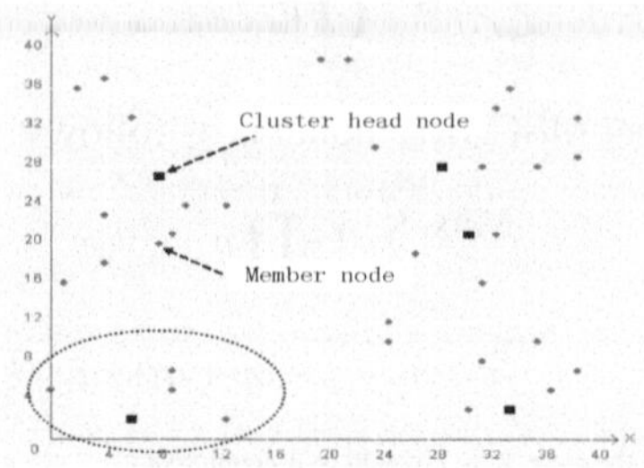

Fig. 1. Cluster head choice result diagram

From the chart, it can be seen that the cluster head nodes of the MSCBRE algorithm is distributed evener, and cluster head is in the place where there are a large number of nodes. The cluster head nodes of the MSCBRE algorithm are the nodes that have relatively much surplus energy. In the figure above left bottom dotted portion's 5 nodes respectively are 2、 7、 17、 26、 20. Their location and residual energy are 2[(6, 2),92], 7[(9,5),21], 17[(13,2),92], 20[(9,7),65], 26[(0,5),77]. By the above data it can be seen that the residual energy of node 2 and the node 17 is the same, but because node correspondence radius supposition is 8, the node 17 cannot cover the node 26 correspondences, the node 2 can actually cover the node 17, so the node 2 is more suitable than node 17 as the cluster head. It can be seen although MSCBRE is a centralized clusters algorithm, the cluster heads formed through it which are more suitable for wireless sensor networks.

Acknowledgment. This paper is sponsored by Liaoning province innovation group project.

References

1. Kansal, A., Hsu, J., Zahedi, S., et al.: Power management in energy harvesting sensor networks. ACM Transactions on Embedded Computing Systems (TECS) 6(4) (2007)
2. Ruiz, L.B., Siqueira, I.G., Oliveira, L.B., et al.: Fault management in event-driven wireless sensor networks. In: Proc. of Modeling, Analysis and Simulation of Wireless and Mobile Systems, pp. 149–156. ACM, New York (2004)
3. Lee, W.L., Datta, A., Cardell-Oliver, R.: Network Management inWireless Sensor Networks. Handbook of Mobile Ad Hoc and Pervasive Communication, pp. 234–253 (2007)
4. Schurgers, C., Tsiatsis, V., Ganeriwal, S., et al.: Optimaizing sensor networks in the energy-latency-density design space. IEEE Transactions on Mobile Computing (1), 70–80 (2002)
5. Wei, Z., Yao, L., Yu, Q., et al.: H-WSNMS: A web-based heterogeneous wireless sensor networks management system architecture. In: Proc. of NIBS 2009, pp. 155–162. IEEE, Indianapolis (2009)

Research on the Satellite Networks Management Based on Policies

Wenbo Zhang and Deyu Zhang*

School of Information Science and Engineering, Shenyang Ligong University
Communication and network institute, Shenyang Ligong University 110159, China
jessezwb@163.com

Abstract. According to structure feature of satellite network management, the global policy servers and the local policy servers have been established. Function modules of satellite networks were designed on the base of system architecture, such as policy management tools, policy library, policy decision point and policy execution point, etc. The policy description method based on XML has been implemented and policy decision point LDAP communication module and COPS communication module have been intensive studied. The policy transmission has been implemented among the policy management tools, the policy library and policy decision poinst through LDAP module and the transmission between the policy decision points and policy implementation points were also implemented by COPS.

Keywords: Satellite networks, network management, management architecture, communication module.

1 Introduction

With the rapid development of the computer network, network management become increasingly importance, and because the network scale continually expands and the network complexity constantly increases. The management difficulty also has been increasing .How to construct the efficient practical networks management system has become one of the research hotspots and important issue in the computer network domain. The networks management technology based on policy emerged in this situation. Policies are a set of rule set that are used for managing, configurating and controlling network resource access. Administrator only needs to define policy without paying attention to specific details of implementing this policy and the situation of relevant equipment through separating management behavior and specific implementation. The networks management system based on policies automatically manage network resources according to policy rules. Administrator could implement distribution management of networks and dynamically adjust networks behavior to adapt to changing business requirement and reduce the cost of networks management through customizing kinds of policies. The networks management based on policies

* Corresponding author.

D. Jin and S. Lin (Eds.): Advances in EECM Vol. 1, LNEE 139, pp. 45–50.
springerlink.com

regard network as a whole to manage and control, meanwhile management policy that have no reference to specific network equipment is management intellective abstract and ensure the reliability and consistency of the network management.

Through building policy management architecture and implementing policy management tools, policy library, policy decision points and policy implementation point and defining policy description methods, storage method in satellite networks, satellite networks management based on policies connect each management function modules together to implement independent management of the satellite network according to policies.

2 Satellite Networks Management Architecture Based on Policy

Satellite network is a kind of network that has highly complex, isomerism and dynamic network characteristic and these unique network operation characteristic repel management architecture of full centrality and complete distribution, and it is a kind of networks management architecture based on dynamic layering structure. The satellite networks management architecture based on policies has been designed, which has independent, flexible, expandability, task-oriented and other characteristics. As shown in figure 1.

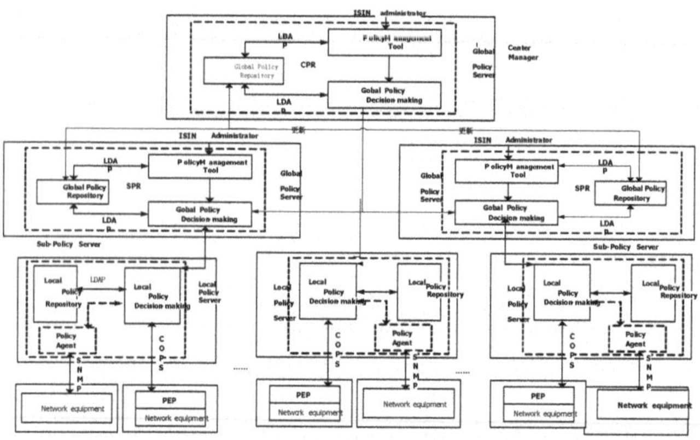

Fig. 1. Satellite networks management architecture based on policies

Policy management station uses Center Policy Server (Center Policy Server, CPS) as the center for radial structure in this architecture. Sub-Policy Server (Sub-Policy Server, SPS) locates in different regions and communicates with Center Policy Server. Structure of Sub-Policy Server is identical with structure of Center Policy Server and includes policy management tools, policy decision-making device and policy library. Each policy servers ensure consistency of the policy by communicating with other servers. Satellites constitute management clustering by clustering algorithm in satellite network, and then a local policy server is made up of cluster-heads .Local Policy Server consists of local policy decision-making server engine, local policy repository and policy agent. Local Policy Server is responsible for handling some simple policies that store in satellite without assistance of ground station, so it could improve networks management efficiency.

Administrator could create, browse and change policy that is in the Global Policy Repository; meanwhile the effectiveness of policy and policy conflict could be preliminary judged by PMT that is an interface of administrator and the whole network management system. PMT distributes in Center Policy Server and Sub-Policy Server, so administrator inputs and manages policies and monitoring network situation through policy management tools.

Policy repository store satellite network policy rule and directory services of policy, meanwhile policy repository also contain other network information such as user profile and IP infrastructure data etc. Policy library can be divided into Global Policy Repository (Global Policy Repository, GPR) and Local Policy Repository (Local Policy Repository, LPR).GPR is responsible for storing all the policy information in the policy management process, which is located in the Center Policy Server and Sub-Policy Server of ground management station that have the updated mechanism between them. LPR locates in satellite cluster-heads and is responsible for storing part of policies that use for satellite management. LPR is a subset of Global Policy Repository of satellite network and stores policy backup t.

PDP takes charge of accessing policy mode that exists in policy repository and making decision according to the policy information, and is set up in the ground sub-management station in this architecture which is built in this paper. Local Policy Decision Point (Local Policy Decision Point, LPDP) is set in the management satellite .It is in the charge of receiving PEP request of satellite agent, retrieving Local Policy Repository and making policy decision . If Local Policy Repository has no policy that PEP requests, LPDP can also transmit PEP policy request to PDP. Policy information that PEP executes LPDP or PDP to produce is stored in cache. It transmits information to LPDP and PDP to make LPDP understand the change of Network or equipment condition.When PEP of satellite agent requests policy, it transmits policy request to LPDP, and then LPDP makes policy decision and sends policy information to PEP through inquiring the local repository. If LPDP does not retrieve the required policy information in local policy repository, LPDP will transmit PEP policy request to ground PDP through the LDAP protocol. PDP retrieves global policy repository of satellite networks and gets its required policy information to send it to LPDP. LPDP makes policy decision according to policy information after obtaining required policy information, and transmits the decision information to PEP, meanwhile it will stores requested policy backup in the Local Policy Repository.

3 The Realization of the Function of System

3.1 Implementation of PMT

Policy management tool as a platform that policy edits and browses is a direct user -oriented management tool .Policy management tool is made up of three modules: user interface, logic converter and the repeat.

A function that uses XML to describe logic converter is that it will parse XML language and separate out policy content,and then transform it into policy format which PCIM support.Read of XML adopts Dom4j packet that JAVA support to analyze.

3.2 Implementation of PR

Policy repository mainly realizes the establishment of LDAP directory server and data structure of satellite network fault management that based on policy.Schema could been customized after completing configuration of Open LDAP and starting its service. LDAP refers to object classes, property type, grammar and matching rules as Schema,and it has lots of the LDAP object class, property type, grammar and matching rules,which these system Schema are seted in LDAP standard .

3.3 Implementation of PDP

PDP mainly contains five modules: LDAP communication module, policy explanation module, policy decision module, COPS communication module and policy conflict resolution module. LDAP communication module connects PMT, PR and PDP, and policy which user input should store in the directory server through the LDAP protocol, and then PDP will read policy from PR and implement. COPS communication module that realizes policy transmission between PDP and PEP connects them. So LDAP communication module and COPS communication module are a key that the whole policy network form a unified whole, and implementation policy network management that it primary completed is LDAP communication module and COPS communication module.

LDAP communication module mainly implements policy network management to connection of PR.LDAP communication module completes through LDAPPolicy and LDAPEngine, as shown in the diagram.

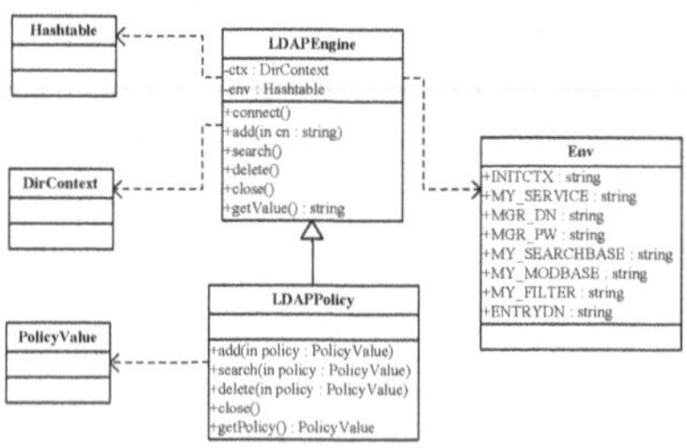

Fig. 2. LDAP communication module of UML diagrams

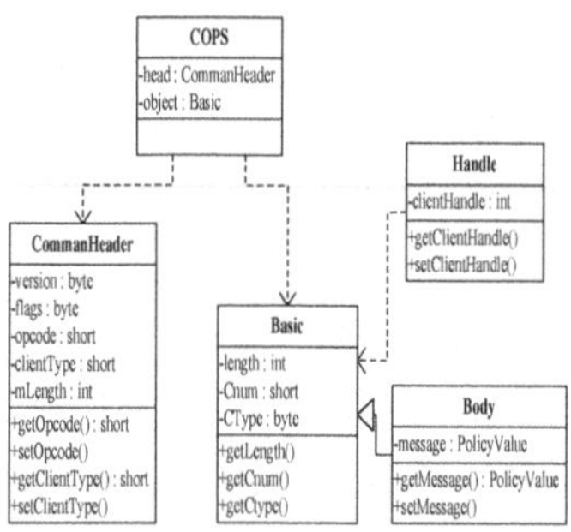

Fig. 3. COPS object static class figure

LDAPEngine implements some basic operations in terms of directory server through the DirContext class of JAVA supports (in java. naming. directory. * package).It includes: establishing connection with directory serve, adding new data, inquiring or searching data, editing or deleting data and closing connection with directory server etc. LDAP Policy class which is responsible for adding, inquiring, deleting policy information inherites LDAPEngine .Data information that query returns is encapsulated into PolicyValue by the getPolicy () method .

COPS protocol which is made up of a general header and other standard object message sets message format of COPS protocol. COPS protocol which could Load specific format subobject to satisfy management service in the existing standard object is extensible. General header and COPS object are encapsulated in this paper according to the format of the COPS protocal message, as shown in figure.

Class implements the formal description in terms of COPS Standard object through JAVA language. Firstly, COPS class creates a cops object which contains two attributes, and head represents the head of COPS message and object indicates COPS object. Head which includes version, flags, opcode, clientType and mLength that respectively represents the version of COPS protocol head, request identifier, operation codes, clientType and message length is an object of the ComanHeade class. Object is an object of Body class, which contains length, Cnum, Ctype, etc. Body is substantive section of COPS protocol, which contains configuration information and policy information, etc. COPS communication module undertake communication between PDP and PEP, this is static class figure COPS communication module.

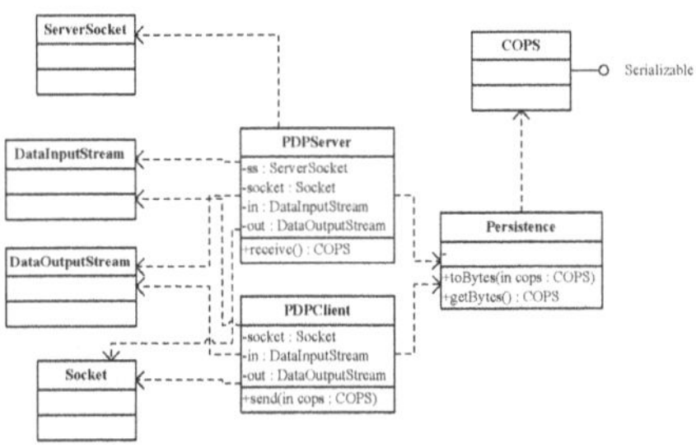

Fig. 4. COPS communication module

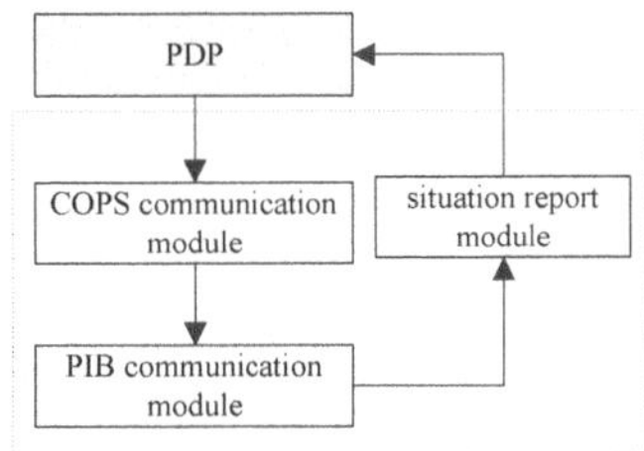

Fig. 5. PEP structure

3.4 Policy Execution Point(PEP)

Policy execution point would be a application and execution equipment of different policies, also would been a respondent of network situation and collector of information.

PEP mainly includes three modules: COPS communication module, PIB control module and situation report module, its structure as shown in figure 5. COPS communication module is responsible for the communication between PEP and the PDP. PIB control module take charge of maintaining.PIB information repository.Situation report module is in charge of regularly reporting implementation and operation of PEP policy to PDP.

Communication is carried out through COPS protocol between PEP and PDP.PEP sends request to PDP and receives policy decision that PDP sends Communication mode based on the principle of request/decision-making mode between PEP and

PDP.PDP usually sends some unsolicited decision to PEP, and compels PEP to execute.It would initiatively report implementation situation to PDP,when PEP has finished execution of policy later.

4 Conclusion

This paper not only has went into the satellite networks management base on policy, but also established the system architecture of networks management based on policy, and then designed and implemented policy management tool, policy repository, policy decision point and policy implementation point under this structure, meanwhile LDAP communication module and COPS communication module of policy decision point have been lucubrated through analyzing the feature of satellite comprehensive information network.Satellite networks management simulation demonstration prototype system based on policy has been established by the HLA/RTI simulation system structure at last. Simulation structure shows that the design of this paper meets the request of satellite networks policy management.

Acknowledgment. This paper is sponsored by Liaoning province innovation group project.

References

1. Vivero, J., Serrat, J.: MANBOP: Management of Active Networks Based on Policies. IEEE Communications Magazine 32(7), 135–139 (2002)
2. Zhao, W., Shao, J.-L.: Policy-based Network Management. Journal of PLA University of Science and Technology (Natural Science) 6(2), 61–66 (2001)
3. Caini, C., Firrincieli, R.: End-to-end TCP enhancements performance on satellite links. In: Proceedings of IEEE. Symposium on Computers and Communications (ISCC 2006), Cagliari, Italy, pp. 1031–1036 (June 2006)
4. Caini, C., Firrincieli, R.: Packet spreading techniques to avoid bursty traffic in satellite TCP connections. In: Proceedingsof IEEE 59th Vehicular Technology Conference, VTC 2004-Spring, Milan, Italy, pp. 2906–2910 (May 2004)
5. Hu, Y., Li, V.: Satellite-based internet: a tutorial. IEEE Communication Magazine 39(3), 154–162 (2001)
6. Floreani, D., Wood, L.: Internet to Orbit. Cisco Systems Packet Magazine 17(3), 19–23 (Third quarter 2005)
7. Herscovici, N., Christodoulou, C., Lappas, V., Prassinos, G., Baker, A., Magnuss, R.: Wireless Sensor Motes for Small Satellite Applications. Antennas and Propagation 8. IEEE Magazine 48(5), 175–179 (2006)
8. Wen, X.Y., Feng, Y.X., Wang, G.X.: Algorithm of Cluster Generation and Management Domain Decision in Integrated Satellite Information Network. Mini-micro Systems 25(10), 1742–1745 (2004)

The Application of Ontology in Knowledge Discovering and Data Acquisition

ZhongXia Hu*

Department of Computer Application, Qingyuan Polytechnic College, Qingyuan, 511510, Guangdong, China

Abstract. The research of knowledge discovery and data acquisition is an interdisciplinary area focusing upon methodologies for extracting useful knowledge from data. Ontologies present two main advantages for practitioners: the knowledge contained in ontology is shareable and reusable, so the same ontological content can be used in different tasks and applications. In this paper ontologies will be the knowledge discovering and data acquisition technology, and performance measurements will be based on ontology-related activities. Finally, knowledge discovering and data acquisition information management system in E- Commerce is proposed based on ontology. The experimental results indicate that this method has great effective promise.

Keywords: knowledge discovering, ontology, data mining, information management.

1 Introduction

The deployment of the semantic web depends on the rapid and efficient construction of the ontology. Ontologies also play an important role in biomedical informatics and in knowledge management. Therefore, these modal-style data operators can provide a unified way to examine, characterize and construct different types of knowledge. As the technology advances, the deployment of information systems and technology in business and government increases rapidly. The value of information to the decision maker is often measured indirectly by evaluating information systems against some surrogate criteria. For example, the value of a decision support system (DSS) such as that described has been evaluated from the perspective of developing a DSS.

Over the past decade the ability of the internet has expounded tenfold and has replaced the previous common meaning of knowledge discovery. Modern knowledge discovering no longer begins or ends with the fabled stack of encyclopedias and journal research. Ontologies and knowledge bases are often constructed from texts and attempt to model the meaning of the text[1]. The rationale of such developments is easing integration and reusability of environmental datasets, models and processing dataflows. Yet, many of the potential benefits of a semantically explicit

* Author Introduce: ZhongXia Hu (1970.6-),Male, Han, Master of computer department of JiNan University, Research area: Development of Computer Application System and Virtualization Technology.

D. Jin and S. Lin (Eds.): Advances in EECM Vol. 1, LNEE 139, pp. 51–56.
springerlink.com

environmental modelling remain unexplored. In other words, data storage, data management, data transmission and even analysis rely on computer and network technology. Traditionally, IA refers to devices designed to perform specific functions, especially electrical household devices, such as toasters, coffee machines, mixers and refrigerators, and which do mechanical functionalities so efficiently and with little conscious effort from the user. The essence of evaluation is to obtain useful data for the individuals, the teacher and the institution. In this sense, evaluation should be continuous, and individuals should get quick feedback.

Domain ontology can help users locate and learn related information more effectively. Hence, building the ontology rapidly and correctly has become an essential task for content based search on the Internet. The main part of the article is the description of a new visual language for graph-based knowledge definition. The ontology provides a convenient basis for adding detailed semantic annotations to scientific data, which crystallize the inherent "meaning" of observational data. A knowledge base is a special database type for representing domain expertise. The repository comprises collections of facts, rules, and procedures organized into schemas. Ontology building is a task that pertains to ontology engineers, an emerging expert profile that requires the expertise of knowledge engineers (KEs) and domain experts (DEs).

Finally, this paper puts forward the methodology for knowledge discovering and data acquisition information management system in E- Commerce based on ontology. This information is generated from multiple actors and shared or exchanged within the construction workflow. The paper offers a methodology for building knowledge discovering and data acquisition information management system for knowledge sharing and reusing based on ontology. Because the key tool to allow this unification is the use of structured knowledge (organized in ontologies) to inform data and metadata compilation, model conceptualization, and simulation, we start with a brief introduction to ontologies geared to environmental modelling applications.

Owing to the rapid growth of the internet and the increasing importance of consumer electronics in our daily lives, marketers actively and continuously seek to expand all sorts of 3C products and digital contents in order to profit from the business opportunities produced by this digital revolution. The experimental results indicate that this method has great promise. It is now common to identify an ontology with a web-accessible document that can be created, edited and validated using ad hoc tools. Ontologies are used as references to annotate resources with concepts in standardized ways, e.g. in RDF.

2 Ontology-Based Knowledge Discovering Information System

With the vigorously generalization of semantic web by W3C, semantics oriented web information integrating method has been the major point of the research on web information integration technology. Both intelligent agent and semantic web service technologies are able to reach remarkable achievements and in some cases have overlapping functionalities.

In this era of the knowledge economy, how enterprises create value via product knowledge has become a critical problem most enterprises face. Historically, information retrieval was derived from well-organized search engine collections of textual based data. This paper is constructed a symbolic model for representing knowledge and a tree structure for differentiating between different knowledge levels, knowledge was then inferred using a fuzzy algorithm that integrated different knowledge sources.

Ontology could efficiently express the general knowledge in specific domain, which is adaptable to be the common semantic model of semantics oriented web information integration. One major issue with traditional expert system methodologies lies in the fact that all knowledge is usually lumped together to form a "knowledge base". Both intelligent agent and semantic web service technologies are able to reach remarkable achievements and in some cases have overlapping functionalities.

2.1 Knowledge Discovering System Based on Ontology Model

Ontology construction – It is important to find the exceptions as well as the average knowledge if the ontology is to be generally applicable, and many exceptions relate to human life. As a consequence, the knowledge that a user can acquire from such systems depends predominantly on a user's skill to query the system and to assimilate the results. Moreover, traditional IR systems cannot handle heterogeneous and semi-structured document corpora[2]. Therefore, there is a need for efficient KM systems that can organize and access knowledge contained in such corpora. Effectively integrating or sharing valuable product knowledge to satisfy the knowledge demands of all participants within a product's lifecycle is impossible. First, the system calculates the TF-IDF and Entropy value of keywords, then will define the TF-IDF threshold θ_t and the Entropy threshold θ_e. If the TF-IDF of the keyword is greater than θ_t or the Entropy value of the keyword is greater than θ_e will be selected. The related formulas for essential information similarity matching are as presented in Eqs1.

$$\text{Essential information}\left(Ci^A, Cj^B\right) = \frac{\left|CiE^A \cap CjE^B\right|}{\left|CiE^A \cup CjE^B\right|} \tag{1}$$

As one example of the importance of ontology quality consider the area of business process modelling. IDSS add artificial intelligence (AI) functions to traditional DSS with the aim of guiding users through some of the decision making phases and tasks or supplying new capabilities.

Effective data discovery is particularly problematic in ecology, where traditionally small, focused studies employed largely ad hoc data management solutions, often consisting of flat files or spreadsheets with minimal formal structure and little to no metadata documentation. A knowledge base can be rule-based, frame-based, or utilize scripts. Recently, markup languages, such as XML, have been applied in developing knowledge bases.

Consequently, the next step for the development of numerical models should focus on a standardized data description, an improved functionality that permits better sharing of both codes and model data, and provide a platform for preprocessing, execution, and retrieval of simulation results in an environment that is operating system independent. However, for the purposes of this paper, we simply assume that an ontology is a schema agreed upon by a group of interest in order to formalize the data relevant for the domain in question.

The semantic web approach aims to improve the quality of information in the World Wide Web. This approach is based on RDF, a language developed especially for modeling knowledge. The ontology is a computational model of some portions of the world. It is a collection of key concepts and their inter-relationships collectively providing an abstract view of an application domain.

2.2 The Development of Knowledge Discovering System

Databases are an essential, widely used technology for biomedical research projects. In projects ranging from clinical studies to genomics research, databases are used to maintain, integrate and share data. These describe a series of modelling stages that invariably include some form of information view, which structures knowledge and information according to a pre-defined class hierarchy[3]. Consequently, modern business entities are challenged with identifying effective means of reducing production costs, improving product and service quality, reducing time-to-market delivery and accelerating response to customer requirements.

Knowledge acquisition collects information then converts it into useful knowledge, in particular, focusing on nonarticulated practitioners' experiences. *Knowledge map* can navigate critical medical resources, including valuable equipment and invaluable practitioners in and out of organizations, and support to make a decision. Individual knowledge needs to be converted to group available knowledge, and data and text need to be converted to usable knowledge, not to be lost in the piles and piles of data and text that are available. Algorithm 1 gives a number of AI-related competencies (310 in total) that formulate an *isPartOf* hierarchy.

```
   Input:Produce the binary relation matrix between
documents and terms.
   Output:knowledge discovering and data acquisition.
   if(topage!=null){
     showpage=Integer.parseInt(topage);
    if(showpage>pagecount){
    showpage=pagecount;
          }else if(showpage<=0){
           showpage=1 };
    for  i = max order  down to  1  do{
  FP[ ++ length ] = item ( i) ;
  output { item (i) } and its support count [ i ]/ n ;
  build S T ( i) based on Unid_FP-tree ;
  if  ( there is an non-root node in S T ( i ) )
  Generate the inter-relationships of concepts;
  endif
  end
```

3 Using Ontology Model to Build Data Acquisition System

Using the knowledge base as a basis for the construction method requires the prior construction of knowledge bases in related domains. The knowledge base must include basic rules and simple examples. One of the great advantages of having situations represented in a formal language is that facts that are not explicitly stated can be derived using an inference engine[4].

The Internet has become the most popular information and knowledge source. Web searching is a basic daily activity for computer users. The application areas include the natural language translation, geographic information system, biology and medicine, agent systems, knowledge management systems, and e-commerce. The formula 2 which calculates the similarity of data acquisition is as follows.

$$\mathrm{CV}(Q_i) = \{(K_1,W_{i1}),(K_2,W_{i2}),\ldots(K_j,W_{ij})\ldots(K_n,W_{in})\} \tag{2}$$

Given knowledge of the environment and the threats that it presents, decision makers are able to make informed decisions and assess the impact of those decisions more effectively.Ontology provides common understanding of the domain knowledge and confirms common approbatory vocabulary in the domain, as well as gives specific definition of the relation between these vocabularies from formal model of different levels. Finally, we test the average cost time of distributed discovery system and centralized discovery system. Data acquisition system currently provides powerful and flexible means for utilizing expertise to address problems associated with a specific domain.

4 The Application of Ontology in Knowledge Discovering and Data Acquisition

The aim of ontology is to obtain, describe and express the knowledge of related domain. In this paper, we apply the technology of ontology to automatically construct the concept hierarchy of knowledge discovering and data acquisition in e-business

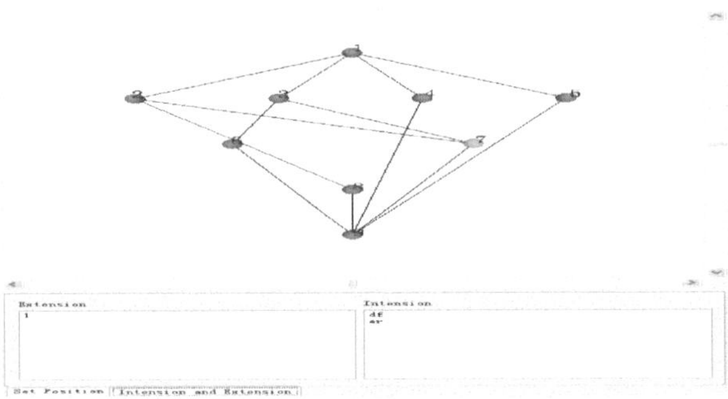

Fig. 1. The application results of ontology in knowledge discovering and data acquisition

and to match up the binary relation matrix of documents and terms to express the independence, intersection and inheritance between different concepts to form the concept relationship of ontology. Fig.1 shows the detailed application of ontology in knowledge discovering and data acquisition. The experimental results indicate that this method has great promise.

5 Summary

In the field of ontology, ontological framework is normally formed using manual or semi-automated methods requiring the expertise of developers and specialists. In this paper, we adopt ontology to improve the performance of data acquisition from documents. The paper probes into building domain knowledge discovering information management system e-business based on ontology.

References

1. Berners-Lee, T., Hendler, J., Lassila, O.: The Semantic Web. Scientific American, 5 (2001)
2. Wray, R., Lisse, S., Beard, J.: Ontology infrastructure for execution-oriented autonomous agents. Rob. Auton. Syst. 49, 113–122 (2004)
3. Chen, T.-Y., Chen, Y.-M., Su, C.-Y.T.-Y.: Designing a multiple-layer ontology-based knowledge representation model in virtual enterprises. Journal of Information Management 15(1), 239–262 (2008)
4. Zhang, H., Chen, Y.: Web information acquisition and mining towards TCM. J. Chin. Hyg. Inf. Manage. Syst. 1(2), 72–76 (2004)

Face Feature Extraction and Face Recognition

Dashen Xue*, Zhaohui Li, and Shenglan Hao

Transportation Management College,
Dalian Maritime University, Dalian, 116026, China
xds59@dlmu.edu.cn

Abstract. Face recognition is a key subject of the research on biological features. This paper describes the whole process of face recognition, and makes the relevant study on feature extractions and identifies ways of the face recognition from the angle of learning. It listed four basic methods of feature extractions and comparing.

Keywords: Face recognition, Biological features, Security.

1 Introduction

As the most direct, natural, and easily acceptable biometric technology, face recognition is devoted to explore how to make the machine can automatically identify the user's identity according to the user's face image [1]. It relates to pattern recognition, computer vision, artificial intelligence, image processing, psychology, medicine, physiology, cognitive science and many other important disciplines [2]. It is closely related to man-computer interaction field and based on biometrics personal identification. Although people can easily identify each other through facial image, however, due to various influencing factors during the image processing, there are huge changes on face image for the same person. So it is very challenging to set up an automatic system to accomplish recognition [3]. At present, there have been many practical systems exists, but only under extremely imaging conditions, we can get the satisfying effect during the image recognition [4].

2 Image Input and Preprocessing Phase of Face Recognition

This stage includes face detection and image preprocessing phase. Face detection is the given random image that determines if there is a face in this stage and gives the information of face position, size and status. The image reorganization can pick pure facial image and geometrical or grey unitary processing after face alignment and normalization. Face detection is affected by illumination noise, facial gradient and masks. So it has started to become an independent research topic and attracted attention widely in recent years. Lack of space only reaches feature extractions and recognition of the face recognition topic here.

* Corresponding author.

D. Jin and S. Lin (Eds.): Advances in EECM Vol. 1, LNEE 139, pp. 57–62.
springerlink.com

3 Face Feature Extraction

To obtain image classification in image recognition is impractical. First, image data occupied large storage space, direct identification is to be time consuming, laborious and its computation is unacceptable. Secondly, the image contains much relevant information with recognition, so we must achieve feature extractions and selections. Therefore, this is helpful for image recognition by compressing numerous identified image data.

3.1 Feature Extractions

Feature extractions means that it transforms higher dimensional characteristics into lower dimensional characteristic of the feature space by mapping or transformation method in the theory of pattern recognition. There are many methods in feature extractions, what characters can be extracted from one mode according to different modes. And it has a straight relation with the aim and method of recognition. The general principle of feature extractions is to reduce probably the processing time and error identification of the identification system.

3.2 Feature Extractions Based on Principal Components Analysis

Principal Component Analysis (PCA) method is also called K-L Transform method, because it is basically to use the K-L transform of the image. The K-L transform is a kind of common orthogonal transform, it applies to any probability distribution. And it acquires the best transform of data compression in mean squared error. The K-L transform is to be used for statistic features extraction thus formed the basis for pattern recognition.

When PCA is to be used for face recognition, we need hypothesis that face is at low dimensional linear space and different faces have separability. Due to high dimensional image space can get a new group of orthogonal basis after K-L transformation, so it can generate low dimensional face space by keeping partial orthogonal basis. But, the low dimensional space basis is obtained through analyzing statistical characteristics of face training sample set. The generator matrix of the K-L transform is total scatter matrix of the training sample set. It is presented with A (i, j) or [aij], its rows and columns marked a point of the image, and the relative element of the matrix marked the graspable value. An image with N×M can link and constitute a vector with N×M according to the column.

$$x = (a_{12}a_{21}...a_{N1}a_{12}a_{22}...a_{N2}...a_{1N}a_{2N}...a_{NM})^T \tag{1}$$

The generator matrix of K-L coordinate system is Ψ=E[xxT], owing to sample's average vector u without category labels, we should use sample's covariance matrix:

$$\psi = E[(x-u)(x-u)^T] \tag{2}$$

As the generation matrix of K-L transform, sometimes, is also called sample's total scatter matrix. So formula (2) can instead as follows:

$$\psi = \frac{1}{C}\sum_{i=0}^{c-1}(x_i - u)(x_i - u)^T \tag{3}$$

Therefore, xi is the ith training sample's image vector, u is the average image vector of training sample set, and C is the total number of training samples.

Owing to the dimension of the covariance matrix is N×M, the computer can not bear the large amount of calculation, if wanted to directly evaluate the eigenvalues and eigenvectors of the matrix. Therefore generally using SVD decomposition method to indirectly evaluate eigenvalues and eigenvectors, SVD theorem is as follows:

Let x be a n*r dimension matrix with rank r, and there are two orthogonal matrixes:

$$\begin{aligned} U &= [u_0, u_1, \ldots, u_{r-1}] \in \Re^{n\times m} \\ U &= [v_0, v_1, \ldots, v_{r-1}] \in \Re^{n\times m} \end{aligned} \tag{4}$$

and diagonal matrix:

$$\Lambda = diag[\lambda_0, \lambda_1, \ldots, \lambda_{r-1}] \in \Re^{r\times r} \tag{5}$$

and $\lambda_0 \geq \lambda_1 \geq \ldots \geq \lambda_{r-1}$ must meet the following requirements:

$$X = U\Lambda^{\frac{1}{2}}V \tag{6}$$

In these formulas, λ_i is the nonzero eigenvalue of the X^TX and XX^T matrixes, μ_i and v_i are the eigenvectors of the XX^T and XX^T matrixes, The decomposition is the singular value decomposition of matrix X, so we can get the formula from what have been discussed above:

$$U = XV\Lambda^{-\frac{1}{2}} \tag{7}$$

Ψ can be represented as follows:

$$\psi = \frac{1}{C}\sum_{i=0}^{c-1}(x_i - u)(x_i - u)^T = \frac{1}{C}XX^T \tag{8}$$

In the formulas,

$$\mathrm{X} = [x_0 - u, x_1 - u, \ldots, x_{M-1} - u]$$

Therefore, we can structure the matrix:

$$R = X^TX \in \Re^{N\times M} \tag{9}$$

The eigenvalues of the matrixes in formula (9) are evaluated, and then the eigenvectors U of the total scatter matrix can be evaluated with formula (6). If all the eigenvectors were selected, the dimension is very big, so according to the characteristics of K-L transform, we just choose the eigenvectors K corresponding to the first K and biggest eigenvalues. Each of the eigenvectors corresponds to an image, for these images look a lot like face that is called feature face, so the K-L transform is used to detect face called feature face method. These eigenvector formed a reduced order subspace, original face image can be thought as a point of the space, the projection of these images in this subspace can get a group of coordinate coefficient as follows:

$$y_i = \overline{U}_k^T (x_i - u) \tag{10}$$

This group of coefficient indicates the position of original face images in orthogonal subspace with principal component analysis method, can be used as the basis for face recognition. If you could be obtained coordinate coefficient of the projection for identifying samples in the subspace, you can determine which person in the face database by comparing the coordinate coefficient with face database.

3.3 Feature Extractions Based on Linear Differential Method

The basic idea of the Linear Method means that the original high-dimensional pattern samples is projected to the best differential vector space, in order to achieve the effect of extracting classification information and compressing characteristic space dimension, and guarantee the largest distances between classes and the smallest inner-class distance of pattern samples in the new subspace after projection.The key of this method is how to solve the best differential vector, and eigenvector set emphasizes the differences of different faces instead of lighting conditions face expressions and direction changes.

3.4 Feature Extractions Based on Discrete Cosine Transform

Discrete Cosine Transform (DCT) is a common transforming coding method of bit rate sequence coding; its compression quality is close to the optimal transform of information compression that is K-L transformation. Only cosine items is in Fourier transform for any consecutive real symmetric function, thus cosine items is as same as Fourier transform with definite physical meaning.

The characteristics of the Discrete Cosine Transform is when the frequency variation factor u and v is bigger, the coefficients Y (u, v) of high-frequency regional DCT is very small; otherwise the bigger coefficients Y (u, v) is main in the smaller top-left corner of the u, v area which is lower frequency area that means the place concentrated useful information. When you rebuild the DCT coefficient images, you should retain few low frequency component of the discrete cosine transform, but surrender most of high frequency component. Then you can obtain the recovery image that is similar with the original image by inverse transform. The new image had a certain error with the original image, but the important information has been preserved. This theory based theoretical basis of data compression has been widely used in image and other data compression field.

3.5 Feature Extractions Based on Independent Component Analysis Method

Independent Component Analysis (ICA), which is an analysis method based on higher order statistics, has better local phenetic ability. Taking non-gaussian distribution as premise in the application, the method is suitable for natural data distribution. So it is suitable for use as face feature extraction method.

There are three methods on the more popularsolution method at present. One is based on the information theory, the second is the fixed algorithm, and the third is the characteristic matrix combined with approximate diagonalizing. Independent component analysis (ICA) is a method of blind source separation that first has been used with some success. Its first application in face was proposed by Bartlett. He regards the face image as the linear superposition composed of multiple independent base images, and finishes the verification on the subset of the FERET database. At last, the experimental results demonstrate that it is superior to Principal Component Analysis.

4 Face Recognition

Face recognition belongs to one of patter recognition, so its basic recognition method has statistical recognition method and structure recognition method. And the design and realization is the two processes which correspond to the methods. The design (is also called training set and learning set) is used to design the classifier with a number of samples, and the realization is used to make classification decision for the samples to be realized with design classifiers.

The recognition system in this paper is the method based on statistical recognition. Its training process shows as Figure 1.

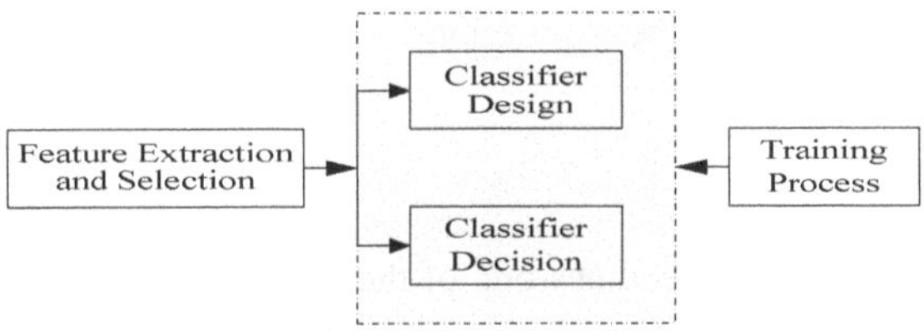

Fig. 1. The schematic diagram of training process

Classification decision is used in the feature space to identify the researched objects as belonging to a certain type with statistical methods. The basic principle is to distinguish a judgment based on prior training samples and make the judgment smallest errors and losses to the researched objects. Then it identifies the defect type by means of the nearest neighbor classifier in this paper.

4.1 Set Up the Sample Database of Face Recognition

Define the establishing of a good face database is important to statistical recognition, especially for face recognition.

The database of face recognition includes face images and face features. Original images and face features are the face feature vectors.

The process of establishing samples is as follows:

- The background of noises of the original images is removed, then finish edge detection and thresholding processing.
- Pretargeting, positioning and precisely locating face images.
- Processing feature extractions and selections to the positioning images.

Sample management:

- Addition: The addition of sample database is processing the original images and obtaining their eigenvectors, then respectively adding face images and features to the sample database.
- Insertion: Insert operation means a user-specific sample to be added to the position, if the position
- already exists sample, it will ask whether cover it. Select the answer that override the original sample, otherwise recede the sample one position on the specified location or after the location. Then specified sample is inserted into specified location.
- Deletion: Deleting samples is first to make the calibration processing of some delete parameter, if the calibration is correct, the system will delete calibration designated samples, and move forward the specified sample one position. The process is a relatively simple task.

4.2 The Nearest Neighbor Method

The initial nearest neighbor method is first put forward by Cover and Hart in 1968[2]. Owing to this method in theory is analyzed deeply, so far it is still one of the most important patter recognition in non-parametric method. The nearest neighbor method is one kind of these non-parametric methods, which is used for classification decision based on the distance between the two samples.

5 Conclusion

Face recognition technology becomes one of the most challenging research topics in the field of computer vision, pattern recognition and artificial intelligence. Large numbers of important researches are made in recent years, but it is still far from the ability of human perception system. the existing automatic face recognition business system is far from satisfied with the present state of the performance under non-ideal application conditions. The automatic face recognition technology is just researched in this paper, provide reference for others.

References

1. Pan, Z., Bolouri, H.: High speed face recognition based on discrete cosine transforms and neural networks. University of Hertfordshire, UK (1999)
2. Zhao, H., Zhong, J.: The automatic face recognition based on shape and vein. The Research and Development of Computer 40(3), 22–23 (2003)
3. Abate, A.F., Nappi, M., Riccio, D., Sabatino, G.: 2D and 3D face recognition: A survey. Pattern Recognition Letters 28, 1885–1906 (2007)
4. Bronstein, A.M., Bronstein, M.M., Kimmel, R.: Three-dimensional face recognition. International Journal of Computer Vision 64(1), 5–30 (2005)

Method of Automatic Identification of Rifling Mark Based on Similarity

Hong Wang[1], Ruiying Zhou[2], and Lihui Zhou[1]

[1] College of Science, Hebei United University, Tangshan, Hebei, China
wanghong9907@126.com, zhoulh324@126.com
[2] College of Light Industry, Hebei United University, Tangshan, Hebei, China
tianjinying102@163.com

Abstract. This paper provides an effective method for rifling marks identification based on similarity. Here, secondary edges of rifling marks are the object of study. In detail, first, we extract feature of each curve, all of which compose a cross section, by the method of quadratic curve fitting. Second, the similarity of every two secondary edges from different two bullets is computed, through the improved Euclidean distance calculation. Third, these similarities are compared by program in MATLAB. Finally, doing two experiments, it is shown that if their similarity is much lower, less than 0.4, they are from the same gun, otherwise, they are from different guns. So it is proved this method can correctly identify the rifling marks of bullets whether from the same gun.

Keywords: Similarity, rifling mark, automatic identification.

1 Introduction

It is used to determine whether the two warheads fired by the same gun according to rifling mark. There are two drawbacks of traditional approaches: First, it is inefficient, because in many cases, rifling marks are " similar but not ", so the number of rifling marks, such as tens, hundreds, almost can not be compared. Second, Samples of warheads are stored difficultly, because the bullet is damaged easily by corroded.

Automatic comparison consists of two steps, as follows: The first step, 3-D data of rifling mark is collected by optical equipment, and saved as 8 files. Measurement reference plane is set to this space xoy plane Cartesian coordinate system, which is fixed to the measuring device. Measured step, along the x-axis direction and along the y-axis, is 2.75 microns; z-axis measuring accuracy is 1 micron, the data unit is mm. Four secondary edges are sorted in one direction uniform, for example, from the bottom of the warhead to the head, four secondary edges are sorted counterclockwise. These data files can be opened in MATLAB by data import, you can also open in WordPad. The data of each file includes more than 40 million lines, each line has 3 columns. Three data in each row, correspond to the spatial coordinates of a point (x, y, z) on the surface of the warhead, the first column corresponds to X coordinate; the second column corresponds to Y coordinate; the third column corresponds to Y coordinate. Reference plane was set near the rifling marks, and when measuring the

D. Jin and S. Lin (Eds.): Advances in EECM Vol. 1, LNEE 139, pp. 63–68.
springerlink.com

location of warheads should be adjusted such that: (1) the cylinder center line of warhead parallel to the reference plane as far as possible; (2) z-axis parallel to the scratches as the direction.

In all, it is difficult to recognition accuracy by above method, so the similarity is introduced for automatic identification in this paper.

2 Similarity Analysis

2.1 Feature Extraction

The main features of rifling marks are extracted from the massive data. Feature extraction includes several aspects as follows. First, the main part of each piece of rifling marks are displayed as different sizes, different shades of lines. Second, the current data must reflect these features. Third, as far as possible less data reflect most features of rifling marks.

Extraction method of two features: With Fig. 1 and data, the method of the breadth and depth of feature extraction is provided: Different kinds of rifling produce different rifling marks; it is reflected in the depth and width. The minimum value is defined as the first minimum point, which is on the right of the maximum of the rifling marks cross-section, the absolute width w is defined as the horizontal distance between the maximum point and the first maxima point which is on the right of minimum, the absolute depth h is defined as the vertical distance between the first maximum point on the right of the minimum point and the minimum point[1]. Fig.1 is a simplified diagram of a surface section. In Fig.1, the horizontal axis is the X-axis; vertical axis is the Z axis. By observing the graphic, you can find rifling marks in XOZ projection surface has a highest point A. when the bullet through the barrel of a gun, the bullet due to the large impact force, was rifled bore extrusion, resulting in a maximum point, therefore, for each bullet; there is the highest point of such as A.

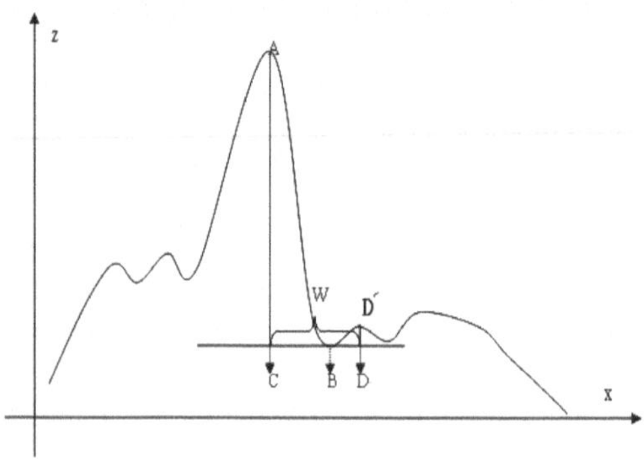

Fig. 1. Simplified diagram of rifling mark cross section surface

The position of rifling mark is on the right of the point A, so the curve on the left of point A does not make sense. Searching for only in the curve, which is on the right of point A, the first minimum point found is as the minimum, marked as point B. and then we make one vertical line from point A to X-axis, and make another vertical line from point B to Z-axis, these two vertical lines are cross at point C. Then searching for in the curve, which is on the right of point B, the first maximum point can be found, marked as D'. and then we make one vertical line from point D' to x-axis, and make another vertical line from point B to Z-axis, these two vertical lines are cross at point D. Then w = | CD |, h = |D D' |. The extraction method of rifling mark surface: according to the rifling mark surface, it can be determine which gun the bullet was fired, such as Fig.2.From Fig.2, it can be found that rifling mark data of the bullet is on the right of the highest point. Due to the wear and the speed of bullet, different rifling, and rifling marks are not the same[2]. Therefore, the curve on the right of the highest point makes sense. But the curve on the right of the highest point is much complex, it is needed to intercept part of the image. Method of interception is as follow: Interception of this part of the image, from the highest point A, to its right 0.4mm (eliminate X, Y coordinates of a translation error). According to the data of the image portrayed, it can be determined this parts of the image, from the highest point A to its right 0.4mm, can well described the characteristics of the bullet.

First, we find the minimum point B from the X coordinate of the highest point A to its right range of 0.1mm, so the X coordinate of point B is in [0, 0.1 mm], and then continue to move along the right 0.3mm, we get a point marked as D, So this curve is divided into AB, BD two parts. And the two curves can reflect a high degree of absolute depth and absolute features, which is extracted in (1). The detailed steps are as follows: First, we find X coordinates of the highest point in the map of three-dimensional image on the plane XOZ (a total of 756 surface), from the x-coordinate of the highest point to its right for a distance of 0.4mm, the corresponding X and Z coordinates can be considered. The X coordinate is got by certain step size (2.75 microns), so in the 0.4mm range, the total of data is 146. Therefore, for 40-million mass data, 146 * 756 sets of data can be extracted. Second, curve fitting. Taking one section from 756 sections, and then the 146 sets of data, obtained from this section, is divided into two parts of curves(First, we find the minimum point B from the X coordinate of the highest point A to its right range of 0.1mm, and then continue to move along the right 0.3mm, we get a point marked as D, So this curve is divided into AB, BD two parts), and then programming in MATLAB, by quadratic curve fitting for these two curves separately, we can obtained two curves Z1 and Z2[3].Finally, reset the value. As the part near point A in the curve AB does not well described rifling marks, it is necessary to re-set his value. These points are as the approximate point of curve AB, which is located in the left of B, and the distance between the point B is less than or equal to 20 steps. These points are as the approximate points of BD curve, which are located in the right side of B point, and the distance between the point B is less than or equal to 100 steps; in order to eliminate the possible errors caused by the parallel movement along the X-axis, Z-axis direction, the location of the origin is moved to point B. Take any one section, by the steps, taking 121 X coordinates, which are these X coordinates of the approximate points of curve AB and curve BD, and then they are substituted into the fitting curve Z1, Z2, a series of values can be obtained. At last, moving the origin, we can get the new sets of values $Z1'$ and $Z2'$. If we marked $Z1'$, $Z2'$ as Z' (total 121), there are 121 * 756 new sets of Z' in total.

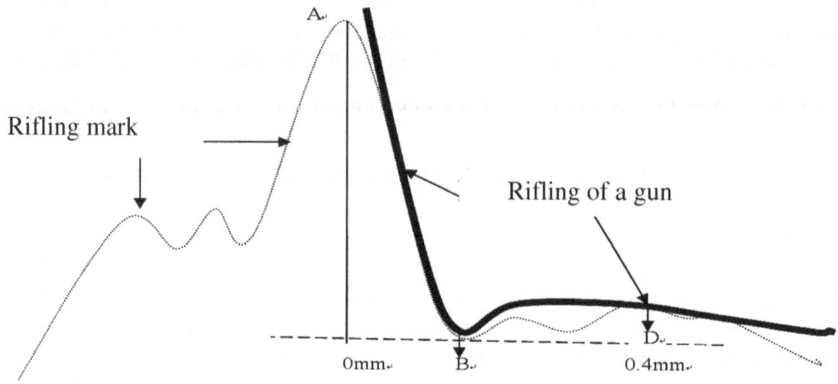

Fig. 2. Surface section of rifling mark

2.2 Calculation of Similarity

In statistics, the distance between the two can be used to describe the similarity. In fact, the distance can be transformed into similarity, If d_{ij} is a distance, the similarity is defined as $C_{ij} = \frac{1}{1+d_{ij}}$ [4]. Therefore, the distance between the two is calculated first. The most common formula of distance is Euclidean distance $d_{ij} = \left[\sum_{k=1}^{p} (x_{ik} - x_{jk})^2 \right]^{1/2}$ $(i=1,2,3,\cdots,n\ ;\ j=1,2,3,\cdots,p)$. But Euclidean distance has a drawback that the size of distance is affected by their data units, it means that dimension is different.

Therefore, standardization of data should be made before comparing, so we can get: $x'_{ij} = \frac{x_{ij} - \overline{X}_j}{S_j}$, in which $\overline{X}_j = \frac{1}{n}\sum_{i=1}^{n} x_{ij}$, $S_j = \left[\frac{1}{n-1}\sum_{i=1}^{n} (x_{ij} - \overline{X}_{ij})^2 \right]^{1/2}$. The standardized data into the formula d_{ij}, the distance can be obtained without the dimension, then by the formula C_{ij}, the similarity can be computed between the two, and it meets $C_{ij} \in [0,1]$. For the problem in this paper, we can compute the similarity of Z' $(p=1)$.

Then the similarity of the two sub-edges is solved. Rifling marks on the comparison from the two warheads, the data M1、M2,which from one secondary edge of each rifle marks, is taken, and then, 756 sets data of $\dot{Z}$ can be calculated according to the above four steps[5]. Because there may be up to 0.03mm error generated by the parallel movement in the Y axis, so when we compute the similarity of M1 and M2, it is needed to take the maximum similarity of M1and C2, which is in the interval [Y-0.03, Y +0.03], after Y coordinate of M1 is determined.

3 Comparison of Rifling Marks

3.1 Automatic Identification of Rifling Mark

First, we compare two rifling marks, which are from two guns. It is Supposed that the rifling mark of bullet issued by first gun has four secondary edges, and their number is C1, C2, C3, C4, and another rifling mark of bullet issued by the second gun has four secondary edges, and their number is C5, C6, C7, C8. Then we can obtain four values of similarity, which can constitute a row vector; So, a matrix A with 4×4 can be got and is expressed as the following chart:

$$A = \begin{pmatrix} C1C5 & C2C6 & C3C7 & C4C8 \\ C1C6 & C2C7 & C3C8 & C4C5 \\ C1C7 & C2C8 & C3C5 & C4C6 \\ C1C8 & C2C5 & C3C6 & C4C7 \end{pmatrix}$$

Second, the minimum value of each line is removed, and we compute the average of the remaining three similarity $\overline{C}_j\left(j=1,2,3,4\right)$. It is concluded that the similarity of two bullets is $C = \max \overline{C}_j$.

3.2 Experiment

In experiment, we compare two bullets every time, which are randomly selected from 22 pairs of bullets. First, the rifling marks data of two bullets, which is issued from the same gun, is provided. According to the above method, we can get matrix A.

$$A = \begin{pmatrix} C1C5 & C2C6 & C3C7 & C4C8 \\ C1C6 & C2C7 & C3C8 & C4C5 \\ C1C7 & C2C8 & C3C5 & C4C6 \\ C1C8 & C2C5 & C3C6 & C4C7 \end{pmatrix} = \begin{pmatrix} 0.3236 & 0.3875 & 0.2292 & 0.3122 \\ 0.3384 & 0.4320 & 0.2911 & 0.3159 \\ 0.3013 & 0.3868 & 0.2738 & 0.3269 \\ 0.3349 & 0.4294 & 0.2861 & 0.2409 \end{pmatrix}$$

The minimum values of each line are removed, and then we compute the average of the remaining three similarities. The list is below:

Table 1. The average similarities of each line

Number of line	average similarity
$\overline{C}_1$	0.341
$\overline{C}_2$	0.362
$\overline{C}_3$	0.338
$\overline{C}_4$	0.350

According to the above table, the similarity of two bullets is $C = \max \overline{C}_j = 0.362$. From the results of above comparison, it was found that the similarity value of two bullets, from the same gun, is greater than 0.4, and the similarity of two bullets from different guns, is lower, less than 0.4. Therefore, this experiment proved that this algorithm can correctly identify the rifling marks.

4 Conclusion

This paper provides an effective method for rifling marks identification based on similarity. We extract feature of each curve, which compose the cross section, by the method of quadratic curve fitting, and the similarity of every two secondary edges from different two bullets is computed, through the improved Euclidean distance calculation. These similarities are compared by program in MATLAB. By two experiments, it is proved that this method can correctly identify the rifling marks of bullets whether from the same gun.

References

[1] Zheng, Q., et al.: Measurement Error Analysis and Data Processing, pp. 168–170. Press of Beijing University of Aeronautics and Astronautics, Beijing (2007)

[2] He, X.: Modern Statistical methods and Application, pp. 263–267. China Renmin University Press, Beijing (2007)

[3] Wang, M.: MATLAB and Scientific Computing. Electronic Industry Press, Beijing (2003)

[4] Liu, C., et al.: Application of Numerical Analysis. Metallurgical Industry Press, Beijing (2005); Meseguer, J., Quesada, J.F.: Maude: pecification and Programming in Rewriting Logic. Theoretical Computer Science 285(2), 187–243 (2002)

[5] Zheng, X., et al.: Nonlinear Optimization, pp. 144–145. National Defense University Press, Changsha (2003)

Research on Construting of Sudoku Puzzles

Hong Wang[1], Yi-shu Zhai[2], and Shao-hong Yan[1]

[1] College of Science, Hebei United University, Tangshan, Hebei, China
wanghong9907@126.com, shaohong@heut.edu.cn
[2] College of Science, Tianjin University of Technology Education, Tianjin, China
ebook@126.com

Abstract. Sudoku is generally considered a good way of training one's thinking. Based on the first puzzle calculated randomly and satisfying metrics, this paper adopts the comparative deletion algorithm to obtain the unique Sudoku the first puzzle thus constructs Sudoku game bank. The algorithm constructs nine TG_i in accordance with the first puzzle. After finding the first key grid, the possible point is decreased in the TG_i , and another key grid is found. By doing so, the game is made very interesting and attractive.

Keywords: Sudoku Puzzles, 9×9 grid, comparative deletion.

1 Introduction

First, we make nine sheets of 9×9 grid figures to replace empty which may be digits. Selecting digits through the first puzzle and game rules, we obtain nine sheets of currently optimized 9×9 grid figures (if we find all the solutions at the final puzzles, we define the first puzzle as the entrance level and store it into the game bank.). Second, if we combine these nine sheets of 9×9 grid figures and seek those grids with only one digit possible, that is key grids, then the digit in the grid is determined. (If we find all the solutions at the final puzzles, we define the first puzzle as the primary level and store it into the game bank.). Third, if we separate and clear up the nine sheet of 9×9 grid and select digits on each of them, we obtain the currently optimized 9×9 grid figures(if we find all the solutions at the final puzzles, we define the first puzzle as the secondary level and store it into the game bank.). At last we combine, separate and calculate these sheets repeatedly and obtain the solution (if we find all the solutions at the final puzzles, we define the first puzzle as the advanced level and store it into the game bank.).

The complexity of comparative deletion algorithm is low which can be measured by the times we use metrics. The worst scenario, the difficulty level of the algorithm can be expressed as $T=O[(81-s)^{10}]$ .What's more, we can obtain the unique solution to the Sudoku puzzles, that is to say, we can obtain solution by controlling the empty grids and key grids. If the number of empty grids is zero, then we can obtain the unique solution for the first puzzle. If the number of key grids is zero, then the first puzzle doesn't have the unique solution or no solution.

D. Jin and S. Lin (Eds.): Advances in EECM Vol. 1, LNEE 139, pp. 69–74.
springerlink.com

2 Model Analysis and Solution

2.1 Model Analysis

We are required to model a Sudoku algorithm and construct a Sudoku puzzle with varying difficulty. Develop metrics to define difficulty level.

Algorithm and metrics can be used to solve Sudoku puzzles of different difficulty levels. You should illustrate the algorithm with at least 4 difficulty levels and guarantee the unique solution. Analyze the complexity of your algorithm and minimize its complexity and meet the above requirement. Here, a_{ij} : the digit at the i row and j column in the 9×9 grid figure. First puzzle: T , matrix set $G=\left\{B \,\middle|\, B=\begin{pmatrix} b_{11} & \dots & b_{19} \\ \dots & \dots & \dots \\ b_{91} & \dots & b_{99} \end{pmatrix}_{9\times9}, b_{i,j}\in\{0,1,2,\dots,9\}\right\}$ · $G_1=\begin{pmatrix} 1 & 1 & \dots & 1 \\ 1 & 1 & \dots & 1 \\ \dots & \dots & \dots & \dots \\ 1 & 1 & \dots & 1 \end{pmatrix}_{9\times9}$ is a square matrix with the digit 1, $G_2=2G_1$, $G_3=3G_1$, $\cdots$, $G_9=9G_1$. The mapping $F:U_1\to U_2$, with $U_1,U_2\in G$ is used to define the three rules of the game. TG_i is a 9×9 grid which the digit i may appear in its grids. M_n is called feasible cell which is the combined solution at the corresponding position after i times of combination and separation. s is the number of digits in first puzzles.

Model hypothesis is following:

1) The empty grid in the 9×9 grid figure is expressed as 0 in the matrix.
2) The number of grid cells in the first puzzle is equal to or larger than 17.
3) Each randomly-obtained first puzzle is different. If a unique solution is obtained, it can be stored in the game bank.

2.2 Model Solution

The puzzle is most frequently a 9×9 grid made up of 3×3 subgrids (called region). Game players should put the nine digits 1 to 9 into the empty grids in accordance with the rules of the game which are as follows:

1) Each digit may appear at each row only once.
2) Each digit may appear at each column only once.
3) Each number may appear at each region only once.

We are required to model a Sudoku algorithm and construct a Sudoku puzzle with varying difficulty. Develop metrics extensible to define a varying number of difficulty levels. The algorithm should be illustrated with at least 4 difficulty levels and guarantee the unique solution.

Based on the above requirements and considering the great difficulty of obtaining the first puzzle through final puzzle, we randomly obtain first puzzles satisfying rules, construct algorithm with comparative deletion method and create the Sudoku game bank. segmentation is one of the main problems of image processing.

$G_{i1} = F(T \cap G_i)$, $i = \{1,2,...,9\}$ is the result G_i and first puzzles T intersected under the mapping F . For example, the 9×9 grid of first puzzle(see the following Figure 1). If the corresponding matrix of first puzzle in Figure 1 is T_1 , $TG_1 = F(T_1 \cap G_1)$, with $T_1 \cap G_1$ in Figure 2, TG_1 in Figure 3.

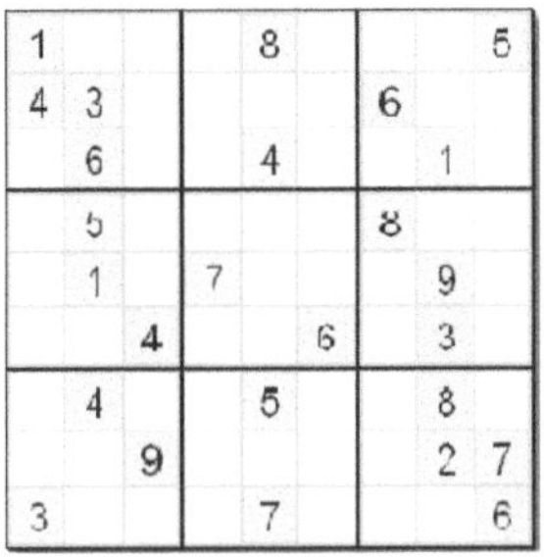

Fig. 1. First puzzle

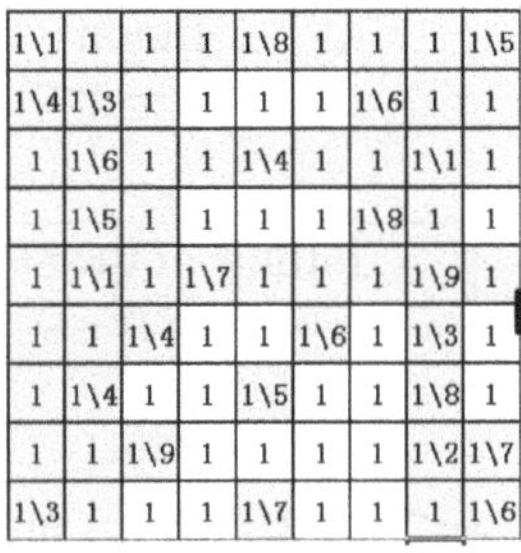

1\1	1	1	1	1\8	1	1	1	1\5
1\4	1\3	1	1	1	1	1\6	1	1
1	1\6	1	1	1\4	1	1	1\1	1
1	1\5	1	1	1	1	1\8	1	1
1	1\1	1	1\7	1	1	1	1\9	1
1	1	1\4	1	1	1\6	1	1\3	1
1	1\4	1	1	1\5	1	1	1\8	1
1	1	1\9	1	1	1	1	1\2	1\7
1\3	1	1	1	1\7	1	1	1	1\6

Fig. 2. $T_1 \cap G_1$

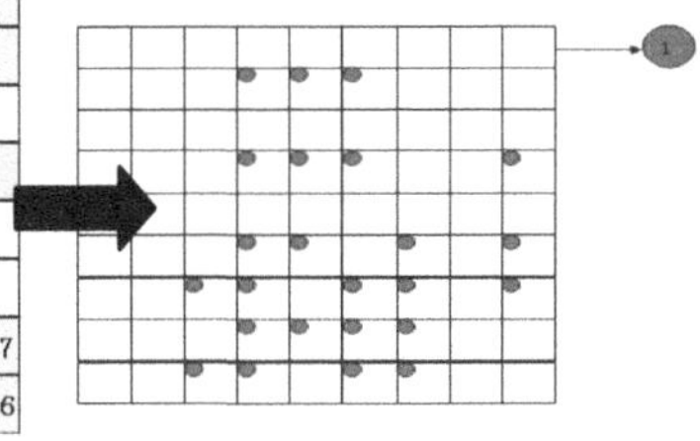

Fig. 3. TG_1

By doing so, we can obtain $TG_1,...,TG_9$ .If we can determine the digit in each grid at the moment, then there is no empty grid which ends the game. The randomly-obtained first puzzle now is the entrance level of the game and is stored in the game bank. If we cannot determine the digit in each grid, the first puzzle will not be stored into the game bank and the game continues. Combine the above nine sheets of TG_i and digits may appear in each grid, that it feasible cell $M_1 = TG_1 \cup TG_2 \cup ... \cup TG_9$ in Figure 4.

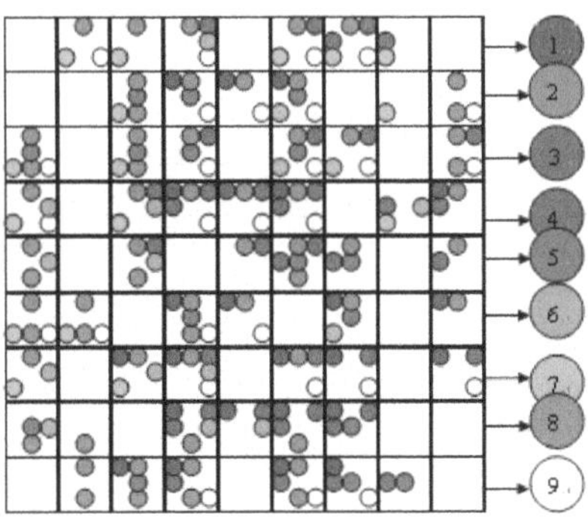

Fig. 4. $M_1 = TG_1 \cup TG_2 \cup ... \cup TG_9$

If there is only one possible digit in a grid, then the grid is called key grid. In Figure 5, only the digit 8 can be filled into the grid a_{82}, we can decide that $a_{82}=8$. By the same token, we can decide that $a_{28}=7$ .If there is no key grid, then the randomly-obtained first puzzle is without solution or without a single solution and cannot be stored into the game bank. The flow chart is as follows.

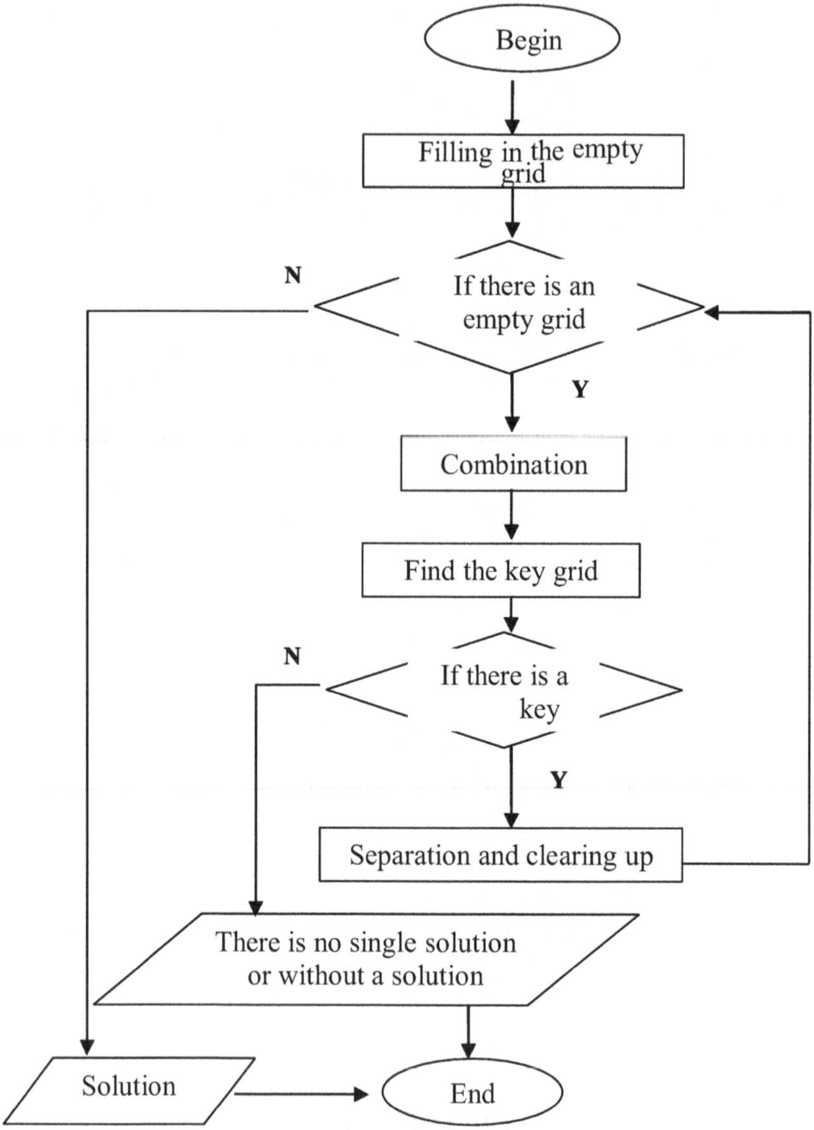

Fig. 5. Flow chart

3 Define the Difficulty Level of the Game

The difficulty levels of Sudoku Puzzle are defined in accordance with the game steps and are listed in the following:

Entrance level: if we can obtain solution by filling in the empty grid, then the first puzzle can be defined as entrance level.
Primary level: if we can obtain solution by combination and finding key grids, then the first puzzle can be defined as the primary level.
Secondary level: if we can obtain solution by separation, then the first puzzle can be defined as the secondary level.
Advance level: if we can obtain the solution after two or more times of combination and separation, then the first puzzle can be defined as the advanced level.

A unique solution is required to the question, that is to say, if there is one put, there should also be one output. If the number of empty grid is zero, then the first puzzle has one unique solution. If the number of key grid is zero, then the first puzzle has no single solution or no solution. With this algorithm, a unique solution to the question should be found out. Because of time limit, we did not consider the scenario when first puzzle has no single solution or no solution.

4 Analysis of the Algorithm Complexity

The complexity of the algorithm is mainly related to the scale of input matrix N. Through the analysis of algorithm, we find out the game rules and the first puzzle interact with the matrix of the same digit G_i once while filling in the empty grids. If the number of empty grids is s, the complexity of the game is $T = O[18\times(81-s)]$. After combination, feasible cells interact with game rules once and the complexity of the game is $T = O[(81-s)^9]$. After separation, the TG_i of each digit interact with the game rules once and the complexity of the game is $T = O[9\times(81-s)]$. Finally, repeat the process of combination and separation then the complexity of the game is related to the times of repetition.

The algorithm in this paper constructs nine TG_i in accordance with the first puzzle of Sudoku puzzle and the game rules. By observing the figure, we tell the key grids. After finding one more key grid and in accordance with the game rules, we can decrease the possible point in the TG_i and lay a foundation for finding another key grid.

References

[1] Visiting Date (February 15, 2008), `http://baike.baidu.com/view/7527.htm`
[2] Palomino, M., Martí-Oliet, N., Verdejo, A.: Playing with Maude. In: Abdennadher, S., Ringeissen, C. (eds.) Fifth International Workshop on Rule-Based Programming, RULE 2004. Electronic Notes in Theoretical Computer Science, vol. 124. Elsevier (2005)

[3] Yato, T., Seta, T.: Complexity and Completeness of Finding Another Solution and Its Application to Puzzles. IEICE Trans. Fundamentals E86-A (5), 1052–1060 (2003)
[4] Viher, B., Dobnikar, A., Zazula, D.: Cellular Automata and Follicle Recognition Problem and Possibilities of Using Cellular Automata for Image Recognition Purposes. International Journal of Medical Informatics 49, 231–241 (1998)
[5] Rules and history from the Nikoli website, http://www.nikoli.co.jp/puzzles/
[6] Clavel, M., Durán, F., Eker, S., Lincoln, P., Martí-Oliet, N., Meseguer, J., Quesada, J.F.: Maude: pecification and Programming in Rewriting Logic. Theoretical Computer Science 285(2), 187–243 (2002)
[7] SuDoku Solver by Logic v1.4 (Javascript), http://www.sudokusolver.co.uk/

Multi-commodity Logistics Network Design Based on Heuristic Algorithm

Dashen Xue[1,*], Zhaohui Li[1], and Nan Xue[2]

[1] Transportation Management College,
Dalian Maritime University, Dalian, 116026, China
xds59@dlmu.edu.cn
[2] College of Economic Management Rouen University,
Rouen, France
Cecilycat518@hotmail.com

Abstract. This paper presented an optimization model for the generic multi-commodity logistics network design, in which the objective function was to minimize the total cost, including the location cost and inventory cost and supply and demand cost. The heuristic algorithm was developed for the model. Practical application denotes the optimization method operates rapidly and the result is rational, so it can provide a scientific decision-making support method for the design of the network, it is an effective method for solving large-scale problems, and the cost of total saving is feasibility.

Keywords: Multi-Commodity logistics network model, Distribution network, Heuristic algorithm.

1 Introduction

Logistics and distribution logistics network is a process of interrelated set of organizations and facilities. Enterprises to establish a logistics network of the ultimate goal is to meet customer needs and achieve the value of goods and enhance their competitiveness [1]. With the increasingly fierce market competition, more and more enterprises begin to pay more attention to its logistics network optimization, particularly manufacturing companies and logistics services [2-6]. These companies want to have a rapid response capability, a higher dependency and greater flexibility in the logistics service system, in order to meet changing market needs.

Many companies produce goods in its logistics network, often through the deployment of one or two times after the direct delivery to their end users. End-users in the enterprise or the location of the small number of very concentrated; this logistics network structure did not show suited [7-9]. But with the expansion of business scale, the scope has expanded its marketing to national and even global, so that its product distribution channels and their end-user location more complex, showing a small volume of commodity circulation, the trend of multiple batches [10].

[*] Corresponding author.

D. Jin and S. Lin (Eds.): Advances in EECM Vol. 1, LNEE 139, pp. 75–80.
springerlink.com

If you are still using the original logistics network structure, would inevitably lead to a large number of small-scale enterprises to bear the transportation of goods, it will not only result in increased transportation costs companies will often times due to shipping, resulting in excessive use of road network and traffic congestion, thereby increasing social costs. Similar situation exists for a large number of enterprises, the problem is particularly evident [11,12]. It should be a reasonable multi-commodity logistics network structure, playing a logistics economies of scale, not only lowering their own costs, but also effectively suppressing the social costs.

2 Model Design

A scheme set up a business within the region there are m companies $A_i(i = 1,2, ..., m)$, the business volume of goods a_i; have n demand-points $B_j(j = 1,2 , ..., n)$, demand for the points b_j; with q alternative network addresses can be set $D_k(k = 1,2, ..., q)$, needs to set the network from the point of purchase, you can also purchase directly from the enterprise; the commodities needed p circulation, the introduction of that species subscript l, (l = 1,2, ..., p). Considered from a network user j the minimum purchase limit, assumed to E_j. Assume that the alternative network address setting of investment in infrastructure, storage costs and transportation costs are known to target to determine the lowest total cost of network layout program. The mathematical models are as below:

$$\min\sum_{l=1}^{e}\sum_{i=1}^{f}\sum_{k=1}^{f}D_{lik}X_{lik}+\sum_{l=1}^{e}\sum_{k=1}^{f}\sum_{j=1}^{g}D_{lkj}Y_{lkj}+$$

$$\sum_{l=1}^{e}\sum_{i=1}^{f}\sum_{j=1}^{j}D_{lij}F_{lij}+\sum_{k=1}^{g}(N_kH_k+D_k\sum_{i=1}^{f}\sum_{l=1}^{e}X_{lik}) \tag{1}$$

$$\sum_{k=1}^{g}X_{lik}+\sum_{j=1}^{n}F_{lij}\le a_{li},(i=1,2,\cdots,f) \tag{2}$$

$$\sum_{k=1}^{g}Y_{lkj}+\sum_{i=1}^{f}F_{lij}\ge b_{li},(j=1,2,\cdots,n) \tag{3}$$

$$\sum_{i=1}^{f}X_{lik}=\sum_{j=1}^{n}Y_{lkj} \tag{4}$$

$$\sum_{i=1}^{f}X_{lik}-QH_k\le 0 \tag{5}$$

$$\sum_{i=1}^{f}Y_{lij}-E_jI_{kj}\le 0 \tag{6}$$

$$\sum_{i=1}^{p} Y_{lkj} - QI_{kj} \le 0 \tag{7}$$

$$H_k = \begin{cases} 1, & \text{node k was chosen} \\ 0, & \text{node k was not chosen} \end{cases} \tag{8}$$

$$I_{kj} = \begin{cases} 1 & \text{,k has supply-demand relation with j} \\ 0 & \text{,k has no supply-demand relation with j} \end{cases} \tag{9}$$

In equation (1), X_{lij} is selected by network k inputs commodities l from enterprise i; Z_{lij} for the user j input resources directly from the factory i into the l; H_k for the selected-or-not network k is selected decision variables; D_{lik} outlets to be selected from the enterprise i, network k, commodities l purchase costs; D_{lkj} to be selected network k to the user j, commodities l costs; D_{lij} for the user j from the enterprise i, commodities l purchase fee; M_k selected for the selected network k infrastructure investment; D_k transit network for the selected storage unit goods management fees. Formula (2) shows that the goods transferred out of the business are not greater than the total production capacity of the enterprise. Formula (3) shows that the goods transferred to the customer are not less than the total demand for it. Formula (4) shows that the logistics network total amount of goods transferred is equal to the total output of goods. Formula (5) shows that the alternative outlets not been selected, by which the number of material transfer should be zero (where Q is a fairly large number). Formula (6) shows that the number of alternative outlets transit of goods should not be less than the customers need to demand their suppliers. Formula (7) shows that if the network and customers are not related, there is no goods flow among them.

3 Solution Based on Baumel-Wolf Method

Baumel-Wolf network distribution is a heuristic algorithm. This method can solve the nonlinear network layout problem of storage costs, it uses a nonlinear function to describe the network's storage costs, in the iterative process of the nonlinear function to piecewise linear approach that is, in every iteration using the marginal cost of that storage rates, which can be added up with unit transportation costs, processing after transportation planning can be directly calculated using solved.

Suppose that the cost of network storage relationship with the scale

$$R_k = \mu_k + \sqrt{d_k} \tag{10}$$

In (10), R_k is the storage costs of network k. d_k is the network scale. μ_k is a constant number.

Suppose marginal cost of network k is P_k, then

$$P_k = \frac{R_k}{2d_k} = \frac{\mu_k \sqrt{d_k}}{2d_k} \tag{11}$$

For network k, knowing its scale, the size of the storage fees will be obtained through formula (11).

Calculating steps of Baumel method are as follows:

First, find the initial solution.

Set the scale of the alternative nodes as 0, ie. $d_k = 0$. Of all resources and requirements to points, find the resources and requirements to the minimum cost rate between the points as $C_i^{(0)}$, With the Marginal cost $D_{ij}^{(0}$ to represent $C_i^{(0)}$,then

$$D_{ij}^{(0)} = C_i^{(0)} = \min(D_{ik}^{(0)} + D_{kj}^{(0)} + D_k^{(0)}) \tag{12}$$

$$(i = 1,2,\cdots,m;\ j = 1,2,\cdots,n)$$

Superscripts "0" indicates initial value. Formula (12) shows that node k is the through node by which company i sends goods to demand point j. The enterprise's goods and demand are well known. With the minimization of transportation cost $M^{(0)}$ to construct transportation model.

$$\min M^{(0)} = \sum D_{ij}^{(0)} X_{ij}^{(0)} \tag{13}$$

$$\sum D_{ij}^{(0)} = a_i,(i = 1,2,\cdots,m)\ \sum X_{ij}^{(0)} = b_i,(i = 1,2,\cdots,n)$$

$X_{ij}^{(0)}$represents the goods amount transported by company i through node k to demand points. By formula (12), the amount can be obtained by the network transfer $D_k^{(0)}(k=1,2,\ldots,q)$ is a group of network settings resolution:

$$\{d_k^{(0)}\} q_k^{(q)} = 1$$

Second, calculate the marginal cost of network.

$d_k^{(0)}$represents the scale of network. According to formula (11) calculate the scale of the marginal cost under the (storage rates) $D_k^{(1)}$.

$$D_k^{(1)} = \frac{\mu_{k\sqrt{d_k^{(0)}}}}{2d_k^{(0)}}, k = (1,2,\cdots,q) \tag{14}$$

Thirdly, improve the solution.

Replacing $D_k^{(1)}$ with $D_k^{(0)}$, through the same process of initial resolution, we get a new set of solution $\{d_k^{(1)}\}q_k^{(q)}=1$.

Last, choose the better of the two solutions.

Comparing $\{d_k^{(1)}\}$ to$\{d_k^{(0)}\}$, if the two solutions are identical, it is the final solution. Otherwise repeat step 2 to 4 calculations, until $\{d_k^{(n)}\}$ and$\{d_k^{(n-1)}\}$are equal.

4 Case Analysis

There are two enterprises A_1 and A_2, which can provide goods in a_1= 40 units and a_2 = 60 units. There are 6 demand points B_j(j = 1,2, ..., 6), alternative locations have been selected D_k(k = 1,2, ..., 5), relationship of network storage costs and network scale is $R_k=\mu_k d_k$, in which d_k is half amount the throughput. The point of demand, cost function of the alternative network storage as well as its source and sink rates are illustrated in Table 1.

Table 1. The fee rate

	D1	D2	D3	D4	D5	B1	B2	B3	B4	B5	B6
A1	6	5	9	8	10						
A2	10	8	4	6	7						
D1	50w					10	11	4	10	7	7
D2		85w				14	7	2	9	5	8
D3			60w			7	11	3	7	2	6
D4				45w		6	9	7	8	10	9
D5					55w	8	6	6	5	9	2

Note: in table 1,

$$w=\frac{\sqrt{d_k}}{2d_k}$$

The upper left of the table indicates that the resource factory and the freight rates between the candidate points. The bottom right indicates a candidate point and demand that the freight rates between points. The lower left diagonal of a candidate point for the marginal cost of storage costs.

Through calculation, the final solution is shown in table 2.

Table 2. The final solution

Nodes chosen	D1	D2	D3	D4	D5
Scale	0	10	20	0	70

5 Conclusion

With the expansion of business and manufacturing enterprises showing a small volume of commodity circulation, the trend of multiple batches, in order to reduce logistics costs, companies need to establish a rational multi-commodity logistics network structure. In this paper, considering the premise of multi-commodity flow, a distribution to meet a wide range of business requirements and economies of scale multi-commodity logistics network structure, to minimize the logistics network of transportation costs, storage costs and construction costs for the optimization objective established to describe the general problem of multi-commodity logistics network design model and heuristic algorithm used to solve them. Numerical examples show that the above model and algorithms to solve large-scale problems related to an effective method of manufacturing enterprises for the appropriate design of the logistics network provides an ideal decision support.

References

1. Barahona, F., Chudak, F.A.: Near-optimal solutions to large-scale facility location problems. Discrete Optimization 2(1), 35–50 (2005); Maxwell, C.: A Treatise on Electricity and Magnetism, 3rd edn., pp. 68–73. Clarendon, Oxford (1892)
2. Ferguson, D., Stentz, A.: The Delayed D* Algorithm for Efficient Path Replanning. In: Proceedings of the IEEE International Conference, pp. 2045–2050 (2005)
3. Ma, Z.-J., Dai, Y.: Optimization Model for Reverse Logistics Network Design for Product Recovery. Journal of Industrial Engineering and Engineering Management 19(4), 114–117 (2005)
4. Jayaraman, V., Guide Jr., V., Srivastava, R.: A Closed- loop Logistics Model for Remanufacturing. Journal of the Operational Research Society 50, 497–508 (1999)
5. Jayaraman, V., Raymond, A., Rolland, E.: The Design of Reverse Distribution Networks Models and Solution Procedures. European Journal of Operational Research 150(1), 128–149 (2003)
6. Parsopoulos, K.E., Vrahatis, M.N.: On the computation of all global minimizers through particle swarm optimization. IEEE Trans. Evolutionary Computation 8(3), 211–224 (2004)
7. Horner, M.W., O'Kelly, M.E.: Embedding Economies of Scale Concepts for Hub Network Design. Journal of Transport Geography 9, 255–265 (2001)
8. Burkard, R.E., Dollani, H., Thach, P.T.: Linear Approximations in a Dymanic Programming Approach for the Uncapacitated Single-Source Minimum Concave Cost Network Flow Problem in Acyclic Networks. Journal of Global Optimization 19, 121–139 (2001)
9. Hahm, J., Yano, C.A.: The Economic Lot and Delivery Scheduling Problem: The Single Item Case. International Journal of Production Economics 28, 235–252 (1992)
10. Hu, T.-L., Sheu, J.-B., Huang, K.-H.: A reverse logistics cost minimization model for the treatment of hazards wastes. Transportation Research Part E 38, 457–473 (2002)
11. Fleischmann, M., Krikke, H.R., Dekker, R., Flapper, S.D.P.: A characterization of Logistics networks for Product recovery. Omega 28, 653–666 (2000)
12. Chen, Y., Cao, Z., Zhou, G.: Application of theory of constraints to reverse logistics. Logistics Technology 29, 252–254 (2006)

S-Transform and Its Application in the Spectrum Analysis of Seismic Signal

Jicheng Liu, Dapeng Ma, and Mengda Li

School of Electric Information Engineering
Northeaset Petroleum University
DaQing, China
ljcdqpi@163.com

Abstract. S transform combines the characteristic of short Fourier transform and wavelet transform, but the windows used in S transform is invariable, so and it maybe unsuited to some non-stationary procedure such as seismic signal. Thus, generalized S transform is put forward. The effect of S transform and other spectrum analysis methods are compared and generalized S transform is used to predict thin bed. Simulation and processing results of seismic signal demonstrated that generalized S transform can predict the seismic thin bed effectively.

Keywords: Spectrum analysis, S transform, seismic signal.

1 Introduction

Spectrum analysis is a very important processing method of seismic signal and it is the basis of fine geology interpretation. Fourier transform is the most commonly used spectrum analysis method, but it is a total transform and can't express the local characteristics of signal, which isn't adapted to the non-stationary signal. Therefore, the new spectrum analysis method such as short time Fourier transform, Gabor transform, wavelet transform and Cohen time-frequency distributions are developed. But these methods have shortcomings respectively, namely the width of the time-frequency window is invariable for all the observable frequency. It is necessary to use the window of adjusted width in practice. So Stockwell proposed a new spectrum method—S transform, which synthesizes the advantage both STFT and wavelet transform, and provides the combined function of time and frequency as Cohen distribution[1]. S transform descript the energy density of signal through time and frequency, and it has higher time-frequency resolution as not having the effect of cross-term.

2 S Transform

The time-frequency resolution of S transform changes with the frequency of signal, and it has direct relation with STFT. The time window is introduce in STFT to

D. Jin and S. Lin (Eds.): Advances in EECM Vol. 1, LNEE 139, pp. 81–86.
springerlink.com

overcome the total average of signal spectrum in Fourier transform. The expression of STFT is as follows:

$$STFT(\tau, f) = \int_{-\infty}^{\infty} h(t) g(t-\tau) e^{-i2\pi ft} dt \tag{1}$$

Where $g(t)$ is weighted to $h(t)$ at $t=\tau$, which can preserve the signal within τ domain and suppress the outside . The Gauss window is defined as:

$$g(t) = \frac{1}{\sqrt{2\pi}\delta} e^{-\frac{t^2}{2\delta^2}} \tag{2}$$

Its Fourier transform is $e^{-\frac{\delta^2 w^2}{2}}$, which is also a gauss function and has better congregation performance of time-frequency. Then the following expression can be gotten form (1):

$$STFT^*(\delta, \tau, f) = \int_{-\infty}^{\infty} h(t) \frac{1}{\sqrt{2\pi}\delta} e^{\frac{-(\tau-t)^2}{2\delta^2}} e^{-i2\pi ft} dt \tag{3}$$

The width of window changes with δ , which makes the distribution have the different time-frequency resolution. But the value of δ is fix, the width of window can't be adjusted adaptively with the frequency distribution of signal. Thus, let δ to be:

$$\delta(f) = \frac{1}{|f|} \tag{4}$$

δ is the function of frequency and it is a reciprocal of frequency, then S transform can be defined as[1]:

$$S(\tau, f) = \int_{-\infty}^{\infty} h(t) \left\{ \frac{|f|}{\sqrt{2\pi}} \exp\left[\frac{-f^2(\tau-t)^2}{2} \right] \exp(-i2\pi ft) \right\} dt \tag{5}$$

S transform uses the window with variable width, so the time window is wider at low frequency band to gain higher frequency resolution, and the width is narrower at high frequency band to gain higher time resolution. τ is the center of the window and it decides the location of gauss window at time axis. By using inverse Fourier transform, the nondestructive S inverse transform is as follows:

$$h(t) = \int_{-\infty}^{\infty} \left\{ \int_{-\infty}^{\infty} S(\tau, f) d\tau \right\} e^{i2\pi ft} df \tag{6}$$

3 Comparing S Transform with Other Spectrum Analysis Methods

3.1 S Transform and STFT

The difference between S transform and STFT is the time window width of STFT is fixed and its resolution is single, but S transform belongs to multi-resolution analysis expression. As in Fig.1, a signal is composed of four sine waves with different frequency, it can be seen from the figure that the spectrum width of the different sine waves is same by using STFT. But the width is different when using S transform, which is the most distinction between the two analysis methods.

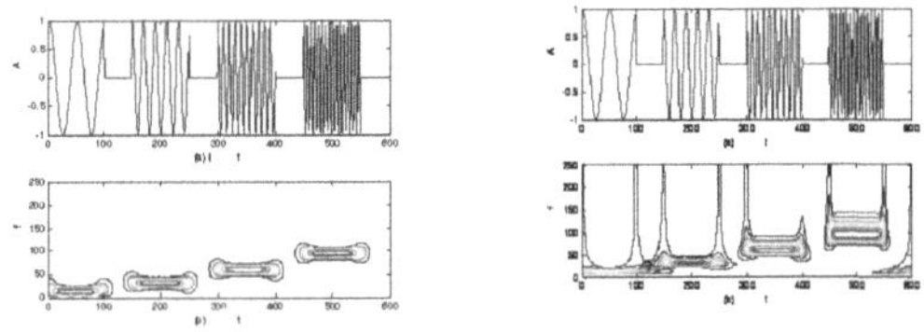

(a) Short Fourier Transform (b) S Transform

Fig. 1. S Transform and Short Fourier Transform

3.2 S Transform and Wavelet Transform

Both S transform and wavelet transform have the characteristic of multi-resolution. But frequency isn't expressed in wavelet transform, and S transform is the combination of time and frequency. Wavelet transform has multiplicity because the wavelet function has many selections. In addition, wavelet reconstruction leads to information leakage, and S transform is nondestructive and reversible. The most fundamental difference of the two transform is S transform comprising phase factor. As in Fig2, a signal is composed of two cosine wave with frequency 10Hz and 50Hz respectively, the signal length is 512ms, sampling interval is 2ms. Fig.2(a) is the scale-time distribution of wavelet transform, the wavelet bases function is Morlet wavelet, the scale range is 1~250. As the correspondence between scale and frequency, wavelet transform can also describe the frequency distribution and embodies the multi-resolution characteristic, but it isn't the direct expression of frequency. S transform in Fig.2(b) reflects the relation between time and frequency well.

3.3 S Transform and Wigner-Ville Distribution

The essential difference between S transform and Wigner-Ville distribution is the influence of cross item. As in Fig.3, signal is composed of four gauss components, its Wigner-Ville distribution is shown in Fig.3(a) and its S transform is shown in Fig.3(b). From the analysis result, it can be seen that Wigner-Ville distribution has 6 cross items(2 cross items overlap), and the cross items don't exist in S transform.

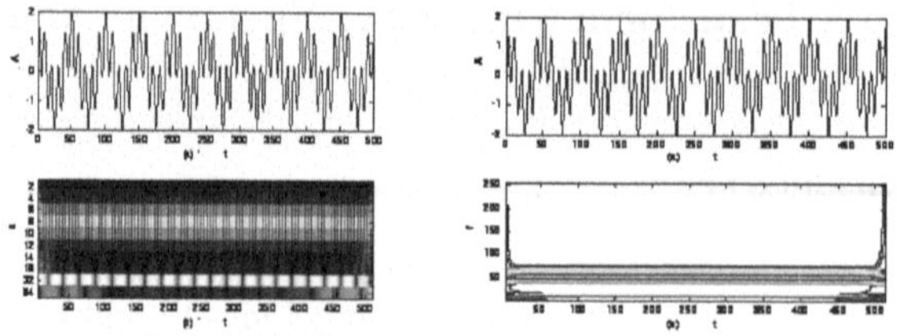

(a)Wavelet Transform (b) S Transform

Fig. 2. Wavelet Transform and S Transform

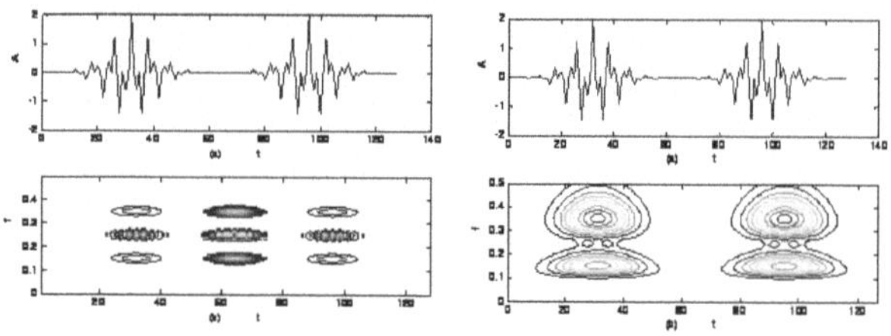

(a) Wigner-Ville Distribution (b) S Transform

Fig. 3. Wigner-Ville Distribution and S Transform

3.4 S Transform and Generalized S Transform

Generalized S transform is obtained by rebuilding the window function of S transform. The gauss window in S transform is fixed and its standard deviation is equal to the wavelength corresponding to a frequency, which limits its application. Generalized S transform overcome the default[2][3]. The generalized S transform discussed in [2] is adapted to analyze signal in the paper. As in Fig.4, signal is composed of 10Hz,50Hz,80Hz and 90Hz components, its length is 256ms, sampling interval is 1/256ms. Its S transform is shown in Fig.4(a) and Generalized S transform is shown in Fig.4(b). It can be seen from the figure that both of them can reflect the relation between time and frequency, but generalized S transform has higher frequency resolution ratio and it is more applied to non-stationary signal.

4 S Transform of Seismic Signal

The analysis of generalized S transform demonstrates that it is an effective tool applied to the high resolution of seismic signal. Fig.5 is a thin-bed model, its CDP1~50 and CDP71~128 is single reflection interface, all the reflector is 1.0, there is a thin-bed of three sampling interval between CDP51 and 70, and the reflector of top interface is 0.7[4].

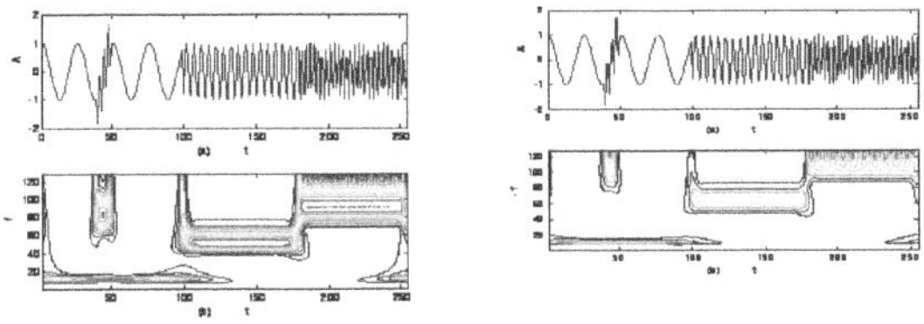

(a) (b)

Fig. 4. (a) S Transform and (b) Generalized S Transform

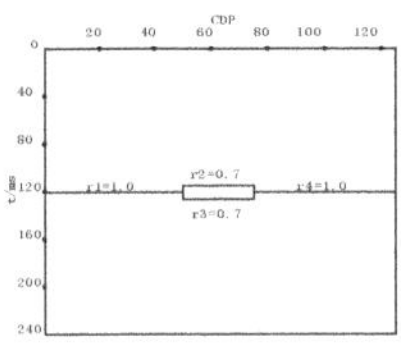

Fig. 5. Thin-bed Model

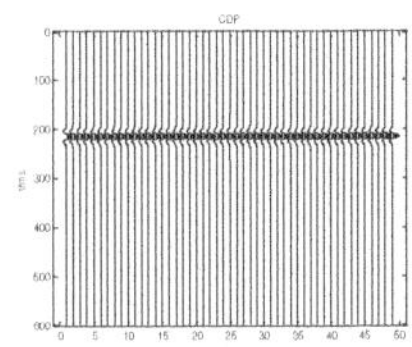

Fig. 6. Synthesized seismic profile

Fig.6 is the synthesized seismic records according to Fig.5, and the Ricker wavelet with main frequency 40Hz is selected. The top interface of thin-bed can't be distinguished directly from the records. The generalized S transform of CDP20 is shown as Fig.8, although the top-bottom interface of thin-bed can't be distinguished directly from the figure, the draw-off single frequency curve can indicate the top-bottom interface of thin-bed well, which is shown in Fig.9 that the two peak value location indicated the top-bottom interface exactly. The process based on generalized S transform to distinguish thin-bed is as follows:

(1) calculating the generalized S transform of every track;(2)extracting the single frequency curve to collect the peak value, which determines the top-bottom interface of thin-bed;(3) calculating all the tracks circularly to gain all the profile information of top-bottom interface of thin-bed;(4) demarcating the amplitude of the original profile and the calculated profile to obtain the high resolution profile based on generalized S transform.

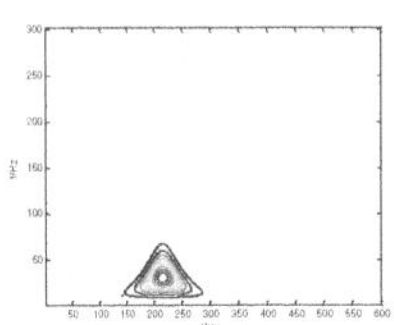

Fig. 7. The time-frequency spectrum of CDP20

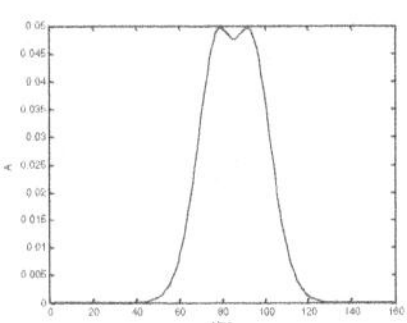

Fig. 8. The single frequency curve(100Hz)

Figure 9 is a practical seismic profile. The high resolution seismic sequence profile based on Genera S transform is shown in Figure 10, which the fine degree of lineups has been improved significantly. It is benefit to identifying and dividing the real geological horizon and to determine the thickness and the structural of geological horizon. If the logging data and the geological data are combined together, it can provide important guiding significance for the accurate interpretation of the data.

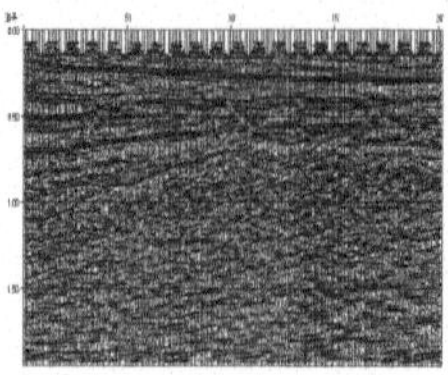

Fig. 9. The practical seismic profile

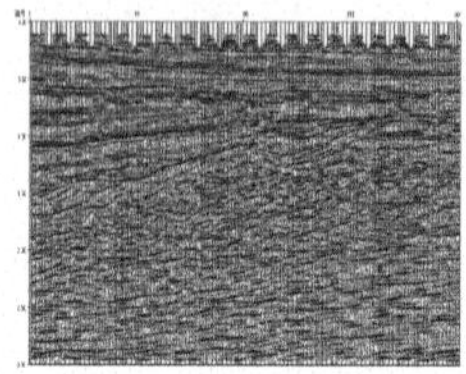

Fig. 10. The profile based on generalized S transform

5 Conclusion

Spectrum analysis is an effective method used to reflect the characteristics of seismic signal. S transform is proved having superiority to process seismic signal, and it can process the profiles of the seismic data subtly, which makes full use of the characteristic information of seismic signal. Dividing sedimentary sequences and distribution of the formation, analyzing sedimentary facies and sedimentary environment and forecasting the valuable oil and gas accumulation zones are all the important complements of seismic sequence analysis and interpretation.

Acknowledgement. The paper is Supported by the Program for New Century Excellent Talents in University of Heilongjiang province (NO.1154NCET001) and New Century Excellent Talents of Ministry of Education of The P.R.C.

References

[1] Stockwell, R.G., Mansinha, L., Lowe, R.P.: Localization of the complex spectrum: the S transform. IEEE Tran. on Signal Processing 17, 998–1001 (1996)
[2] Gao, J., Chen, W., Li, Y.: Generalized S transform and seismic response analysis of thin interbeds. Chinese Journal of Geophysics (04), 526–532 (2003)
[3] Pinnergar, C.R.: A new subclass of compex-value–transform windows. Signal Processing 86, 2051–2055 (2006)
[4] Liu, J.-C., Sun, C.-P.: Thin Bed Prediction Processing Technique Based on Generalized S Transform. In: The First International Conference on Pervasive Computing, Signal Processing and Applications, China (September 2010)

Apply Ontology and Agent to Build Web Mining Information System in E-commerce

XiangZhen Zhou*

Department of Information Management, Shengda Economics Trade & Management, College of ZhengZhou, Zhengzhou, 451191, Henan, China

Abstract. Ontologies are especially used in environments requiring sharing, reusing and interchanging specific knowledge among entities involved in different levels of manipulation of the web mining information. Recently, ontology and agent technology have been widely applied to various domains. The paper offers a methodology for building web mining and application in E-Commerce for knowledge sharing and reusing based on ontology and agent. The experimental results have shown how ontology and agent technology can be successfully applied to build web mining information systems for e-commerce.

Keywords: Agent, Ontology, Web Mining, Information Management.

1 Introduction

Type of knowledge is either explicit or tacit. Explicit knowledge can be expressed in words and numbers, scientific formula, codified procedures, and therefore, they are easily communicated. In contrast, tacit knowledge is intangible because it represents intuition, subjective insights, beliefs and expertise. The Semantic Web is an evolving extension of the World-Wide Web, in which content is encoded in a formal and explicit way, and can be read and used by software agents. Recommender systems arose in the middle 1990s to provide assistance in these searching tasks; to this aim, they automatically select those items that may be appealing to each user considering the preferences defined in his/her personal profile. When a Web service is developed, it typically has its most suitable context (e.g., platforms and devices) for the best execution performance. However, that has changed as web has users from all over the world now access the Internet, and web sites are available in virtually every different language.

The World Wide Web has become a valuable knowledge resource and the amount of information written in different languages and the number of non-English speaking users has been increasing tremendously. While some of the existing search engines support limited translation capabilities, the potential to tap into multi-lingual Web

* Author Introduce: XiangZhen Zhou ,Female,Han, Department of Information Management, Shengda Economics Trade&Management College Of ZhengZhou , Research area: Information Management System.

D. Jin and S. Lin (Eds.): Advances in EECM Vol. 1, LNEE 139, pp. 87–92.
springerlink.com

resources has not been materialized. In other words, data storage, data management, data transmission and even analysis rely on computer and network technology. While the issues for exploring environmental sustainability are well rehearsed and known, the issues that should form the social dimension are less appreciated and addressed by stakeholders involved in the development process. As new services appear with high performance requirements, mechanisms to ensure quality of service and metrics to monitor this quality become necessary.

Ontology is a conceptualization of a domain into a human understandable, machine-readable format consisting of entities, attributes, relationships, and axioms[1]. In the information security infrastructure, intrusion detection has become an indispensable defense line in face of increasing vulnerabilities exposed in today's computing systems and Internet. Effectively integrating the information and material flows within the demand and supply process is the main concern for Supply Chain Management (SCM). A knowledge base is a special database type for representing domain expertise. The repository comprises collections of facts, rules, and procedures organized into schemas.

Concepts are described both by terms, their synonyms, and also by the relations they have to other concepts. Whether a natural language definition of each concept is needed can be decided from case to case. To facilitate information sharing among different research projects, the proposed framework develop ontology which includes standardized, community-accepted descriptions of the information. WSMX tackles and addresses the requirements occurring in B2B and B2C collaborations, serving as a reference implementation for WSMO.

Finally, web mining information system in E-Commerce is proposed based on ontology and agent. Web services represent a step forward in enabling collaborations between various entities on the web and in overcoming the interoperability problems that may appear. B2B (Business to Business) partners can benefit by allowing business entities to expose their capabilities and to make use of their functionality. The paper offers a methodology for building web mining and application in E-Commerce for knowledge sharing and reusing based on ontology and agent. Recently, ontology and agent technology have been widely applied to various domains. First is to provide an environment which is easy to obtain and share information with the assistance of ontology and multi-agent systems via dynamically discover and invoke semantically enriched Web mining. The experimental results indicate that this method has great promise.

2 Using Ontology to Construct Web Mining Information System

The Semantic Web is an evolving extension of the World-Wide Web, in which content is encoded in a formal and explicit way, and can be read and used by software agents[2]. The similarity metrics employed limit the quality of the offered recommendations, because they are based on more or less sophisticated matching techniques that miss a lot of knowledge during the personalization process.

Consequently, modern business entities are challenged with identifying effective means of reducing production costs, improving product and service quality, reducing time-to-market delivery and accelerating response to customer requirements. A review of existing literature on the representation of manufacturing knowledge highlighted the need for application guidelines on the classification of knowledge for optimum reuse. Web mining essentially has many advantages which makes this technology attractive to corporations including the government agencies. This technology has enabled ecommerce to do personalized marketing, which eventually results in higher trade volumes. Web Mining is the extraction of interesting and potentially useful patterns and implicit information from artifacts or activity related to the World Wide Web. There are roughly three knowledge discovery domains that pertain to web mining: Web Content Mining, Web Structure Mining, and Web Usage Mining.

The construction method of ontology is mainly divided into two kinds at present: One is manual ontology building, in which the problem is: (1) In a complicated field it is time-consuming and strenuous;(2) We adopt different standards and modeling methods while building ontology, so that ontology is not currency;(3) It's of great subjectivity.

2.1 Information Management in E-Commerce Based Ontology and Agent

In the past few years the grid is emerging as a building infrastructure that support coordinated management and sharing of interconnected distributed hardware and software resources. Ontology provides a formal semantic representation of the objects for case representation. The information systems model can be adapted to the measurement challenges of the new e-commerce world. During the knowledge acquisition process, both expert knowledge and image samples have been provided. The role of learning is to fill the gap between symbols used during knowledge acquisition and manually segmented and annotated sample images. Equation 1 is applied for name similarity matching for two concepts The similarity calculation includes the cross-correlations among Ci^A, Cj^B, Si^A, and Si^A.

$$R(x,y)=\bigwedge_{j}\in_{J}(F_j(x)\rightarrow F_j(y))\wedge\bigwedge_{j}\in_{J}(F_j(y)\rightarrow F_j(x)) \quad (1)$$

In concept name and synonym similarity matching, concept names are first deconstructed as term sets of unit words, the similarity for term sets of two different concept names is then calculated. Berners-Lee proposed the Semantic Web as a natural evolution of the traditional Web to allow for the manipulation of content by applications with the capacity to interpret the semantics of information. Bearing in mind the results achieved in the Semantic Web, our proposal includes reasoning about the semantic descriptions of the items available in the recommender system (formalized in a domain ontology), and inferring implicit semantic relationships between them [3].

A histogram showed that the data were peaked with a high kurtosis and distinctly non-normal (kurtosis K=6.736066, Shapiro–Wilks test p_{SW}=7.077e-11, Kolomgorov–Smirnoff test p_{KS}=1.538e-06). Consequently, ontology mapping this study conducts similarity matching for concept names and considers the similarities of essential information and relationship to precisely identify the similarity between concepts. Ontologies are used for various purposes, from support for software development processes, or any other activity carried out by geographically dispersed teams, to support for information systems in runtime.

The first category treats the load pattern as a time series signal and predicts the future load using the already mentioned time series analysis techniques. Ontology provides common understanding of the domain knowledge and confirms common approbatory vocabulary in the domain, as well as gives specific definition of the relation between these vocabularies from formal model of different levels.

2.2 The Application of Ontology in Building Web Mining System

In customer relationship management (CRM), Web mining is the integration of information gathered by traditional data mining methodologies and techniques with information gathered over the World Wide Web. Nowadays, it is one of the most active and exciting areas of the database research community. When several agents are put to use in the same environment, this group of agents is usually conceived as a Multi-Agent System (MAS), as the successful completion of their tasks is subject to the decision and actions of other agents. Ontology extraction and similarity calculation are the two main components of the prototype system.

In this paper, the term ontology refers to an explicit description of concepts and relations of an application domain, including a vocabulary of terms employed in this domain and a set of axioms that express the restrictions for the interpretation of this vocabulary. Data analysis is therefore implemented to mine the useful relations in the database. The ontology, which is good at the management and reuse of relations, is used to save the results of data analysis. Algorithm 1 gives the detailed steps to build web mining information system based on ontology and agent, Algorithm 1.

Algorithms:Ontology and Agent to building Web Mining.

BEGIN

InputOntology $i \in I(c)$, i instantiates property p, ontology o.

Input the context relation set $CRS = \{CR_{IO1}, CR_{IO2}, \ldots, CR_{IOK}\}$.

Initialize the context information set $CIS = \varphi$.

Initialize $i = 1$

DO UNTIL $(i > K)$

Infer the context information CI_{IOi} between SR_{IOi} and CR_{IOi} by the fuzzy inference engine.

Add CI_{IOi} to CIS set.

increment i

END DO UNTIL

for each p instantiated by instance i do

```
  if p does not belong to concept c or any of its
superclass or equivalent concepts, OR the value taken
by p does then
  StructValid(i,o)=false
  end if
 DO UNTIL (i > K)
  If CI_IOi is a context information with Instances
belonging to "Gourmet Food" THEN
   Add CI_IOi to the CIS_GF
  END IF
 END DO UNTIL
```

3 The Building Ontology-Based Web Mining Information System

In the highly competitive environment, the suppliers of wares within B2B and B2C collaborations need frequently to fight against similar suppliers to win orders. The Web influences almost every aspect of businesses and our daily lives. From its early years of being a technical and a social phenomenon, it has now advanced to the level of a science [4]. The construction of ontologies necessarily requires an appropriate language. This language should have a well-defined semantics, be sufficiently expressive to describe complex interrelationships and restrictions among objects, and be capable of automatic manipulation and inference, all within acceptable limits of time and resources.

Once the selection has been completed, the following information namely, Minimum Price, Quantity and Supplier Identification, is provided back to WSMX, which in turn sends notifications to the winner supplier, the other suppliers and the user about the action result. The formula 2 which calculates the similarity of Ontology and Agents is as follows.

$$R_{C_j} = \{(u, v) \in U \times U | C_i(u) = C_j(v),\ i \neq j,\ i = 1, 2, \ldots, m,\ j = 1, 2, \ldots, m\} \tag{2}$$

We adopted RDF, a standard ontology web language recommended by W3C.The system utilizes the Jena package to output the results of the RDF format. The RDF is capable of describing the resources of the World Wide Web.

4 Apply Ontology and Agent to Construct Web Mining System

As the foundation of the semantic web, ontology is a formal, explicit specification of a shared conceptual model and provides a way for computers to exchange, search and identify characteristics. It becomes an important problem on ontology application to build and present ontology. In this paper we have shown how ontology and agent technology can be successfully applied to build web mining information systems. Fig.2 shows the detailed comparison results of ontology and agent. The experimental results indicate that this method has great promise.

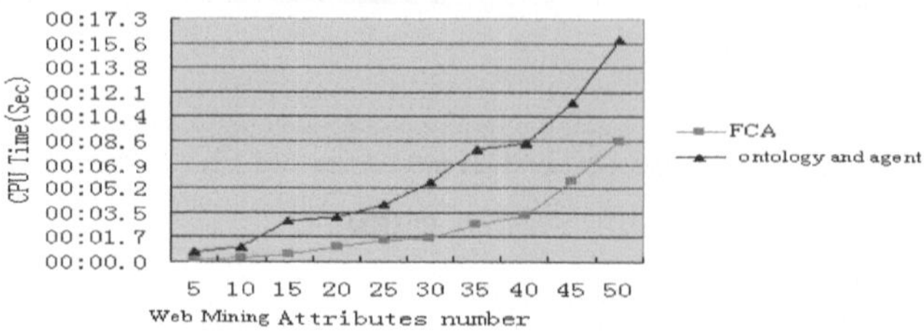

Fig. 1. Building web mining information system comparison of FCA and ontology and agent

5 Summary

As the foundation of the semantic web, ontology is a formal, explicit specification of a shared conceptual model and provides a way for computers to exchange, search and identify characteristics. The paper offers a methodology for building web mining and application in E-Commerce for knowledge sharing and reusing based on ontology and agent. We have shown how ontology and agent technology can be successfully applied to build web mining information systems.

References

1. Guarino, N., Welty, C.: Evaluating ontological decisions with OntoClean. Commun. ACM 45(2), 61–65 (2002)
2. Qin, L., Atluri, V.: SemDiff: an approach to detecting semantic changes to ontologies. International Journal of Semantic Web and Information Systems 2(4), 1–32 (2006)
3. Maedche, A., Staab, S.: Ontology Learning for the Semantic Web. IEEE Intelligent Systems 16(2), 72–79 (2001)
4. Green, P., Rosemann, M.: Integrated process modelling: an ontological analysis. Inf. Syst. 25(2), 73–87 (2000)

Seismic Response Analysis Based on Generalized S Transform with Optimized Window

Jicheng Liu, Qiuyue Zhang, and Mengda Li

School of electric information engineering
Northeaset Petroleum University
DaQing, China, 163318
ljcdqpi@163.com

Abstract. An optimized window used in generalized S transform is discussed. The optimized generalized S transform has better time-frequency concentration performance. the method is used to processed the seismic data and can provide better stratum resolving power.

Keywords: Generalized S transform, optimized windows, seismic data.

1 Introduction

Since time-frequency representation indicates variations of the spectral characteristics of the signal as a function of time, they are ideally suited for nonstationary signal analysis. The ideal time-frequency transform only provides information about the frequency occurring at a given time instant. In other words, it attempts to combine the local information of an instantaneous-frequency spectrum with the global information of the temporal behavior of the signal. The main objectives of the various types of time-frequency analysis methods are to obtain time-varying spectrum functions with high resolution and to overcome potential interferences.

The S transform can conceptually be viewed as a hybrid of short time Fourier analysis and wavelet analysis. It employs variable window length. By using the Fourier kernel, it can preserve the phase information in the decomposition. The frequency-dependent window function produces higher frequency resolution at low frequencies, while at higher frequencies, sharper time localization can be achieved. In contrast to wavelet transform, the phase information provided by the S transform is referenced to the time original, and therefore provides supplementary information about spectral which is not available from locally referenced phase information obtained by the continuous wavelet transform. For these reasons, the S transform has already been considered in many fields such as geophysics, cardiovascular time series analysis, signal processing for mechanical system, power system engineering, and pattern recognition.

Even though the S transform is becoming a valuable tool for the analysis of signals in many applications, in some cases, it suffers poor energy concentration in the time-frequency domain. Recently, attempts to improve the time-frequency representation of the S transform have been reported in the literature. A generalized S transform proposed in[], provides greater control of the window function, and the proposed

D. Jin and S. Lin (Eds.): Advances in EECM Vol. 1, LNEE 139, pp. 93–98.
springerlink.com

algorithm also allows nonsymmetric windows to be used. Several window functions are considered, including two types of exponential functions, amplitude modulation and phase modulation by cosine functions. Another form of the generalized S transform is developed in [1], where the window scale and shape are a function of frequency. The same authors introduced a bi-Gaussian window in [2], by joining two nonsymmetric half-Gaussian windows.

2 Generalized S Transform and Its Optimized Window

The S transform of signal $x(t)$ can be expressed as follows :

$$ST(\tau, f) = \int_{-\infty}^{+\infty} x(t) w(\tau - t, f) \exp(-j2\pi f t) dt \tag{1}$$

Where the Gaussian modulation function $w(t, f)$ is given by :

$$w(\tau - t, f) = \frac{1}{\sqrt{2\pi}\sigma} e^{-((\tau - t)^2 / 2\sigma^2)} \tag{2}$$

and:

$$\sigma = \frac{1}{|f|} \tag{3}$$

Then, the final expression becomes:

$$ST(\tau, f) = \int_{-\infty}^{+\infty} x(t) \frac{|f|}{\sqrt{2\pi}} e^{-((\tau - t)^2 f^2 / 2)} \exp(-j2\pi f t) dt \tag{4}$$

where f is the frequency, t and τ are both time. The generalized S transform is obtained from the original S transform equation (1),by replacing the Gaussian window with a generalized window, denoted as:

$$ST(\tau, f) = \int_{-\infty}^{+\infty} x(t) w(\tau - t, f, \mathbf{p}) \exp(-j2\pi f t) dt \tag{5}$$

In the above equation, a set of parameters that govern the shape of w are collectively denoted p. In practice, w is replaced with a specific window, and $\mathbf{p}$ is replaced with an explicit parameter list enclosed in braces. The Gaussian window was modified by Mansinha et al.(1997) and has the function form:

$$w(\tau - t, f, \{k\}) = \frac{|f|}{\sqrt{2\pi} k} e^{-((\tau - t)^2 / 2k^2)} \tag{6}$$

Where k is the only element of $\mathbf{p}$. S transform windows must satisfy the condition:

$$\int_{-\infty}^{+\infty} w(\tau - t, f, \mathbf{p}) d\tau = 1 \tag{7}$$

Condition (7) provides a link between the S transform and the Fourier transform and ensures that S transform is invertible. Then, the generalized S transform is obtained as:

$$ST(\tau, f) = \int_{-\infty}^{+\infty} x(t) \frac{|f|}{k\sqrt{2\pi}} e^{-((\tau-t)^2 f^2 / 2k^2)} \exp(-j2\pi f t) dt \tag{8}$$

where k controls the frequency resolution, If k is above 1, the Gaussian window is widened and the frequency resolution would increase. Likewise, if k is below 1, the window is narrowed and the time resolution would increase. The window functions with different values of k are shown in Figure1. The three dimension distribution of different k are shown as figure 2.

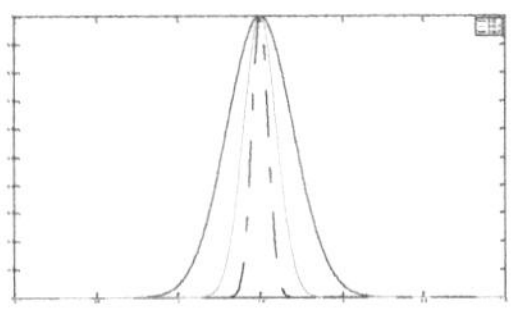

Fig. 1. Normalized Gaussian window for different values of k

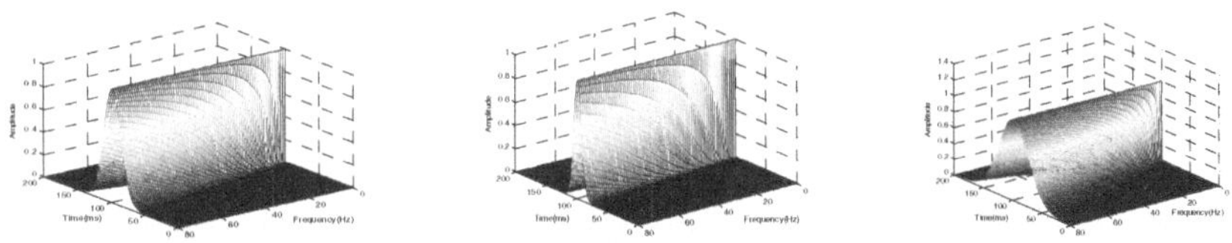

(a) k=1,the standard S transform (b)k=2 (c)k=0.5

Fig. 2. Different value of k

By finding an appropriate value of k, an improved time-frequency concentration can be obtained. In seismic high resolution processing, higher frequency resolution is contributive to distinguish thin-bed, so we discuss the optimization of $k > 1$ in the paper.

The optimal value of k will be found based on the concentration measure proposed in [3],which has some favorable performance in comparison to other concentration measures reported in [4]. The measure is designed to minimized the energy concentration for any time-frequency representation based on the automatic determination of some time-frequency distribution parameter. This measure is defined as:

$$M(k) = \frac{1}{\int_{-\infty}^{+\infty} \int_{-\infty}^{+\infty} |ST(\tau, f)| d\tau df} \tag{9}$$

Where $M(k)$ stands for a concentration measure. The algorithm for determining the optimized time-invariant value of k is defined through the following steps:

(1) For k selected from a set of $1 < k \le N$,compute S transform of the signal using (8).
(2) For each k from the given set, normalize the energy of the S transform representation, so that all of the representations have the equal energy.

$$ST_N(\tau, f) = \frac{ST(\tau, f)}{\sqrt{\int_{-\infty}^{+\infty}\int_{-\infty}^{+\infty}|ST(\tau, f)|^2 \, d\tau df}} \tag{10}$$

(3) For each k from the given set, compute the concentration measure according to (9), that is:

$$M(k) = \frac{1}{\int_{-\infty}^{+\infty}\int_{-\infty}^{+\infty}|ST_N(\tau, f)| d\tau df} \tag{11}$$

(4) Determine the optimal parameter k_{op} by

$$k_{op} = \max_k [M(k)]$$

(5) Select $ST(\tau, f)$ with k_{op} to be the final s transform:

$$ST_k(\tau, f) = ST_{k_{op}}(\tau, f)$$

As it can be seen, the proposed algorithm computes the S transform for each k and based on the computed representation, it determines the concentration measure $M(k)$. The maximum of the concentration measure corresponds to the optimal k which provides the least smear of $ST_N(\tau, f)$.

3 Testing and Conclusion

The testing signal is selected as figure 3, and its window parameter is optimized then the generalized S transform of the signal is analyzed. The signal is expressed as follows:

$$x(t) = \cos(132\pi t^2 + 14\pi t^2) + \cos(10\pi t - 2\pi t^2)\cos(30\pi t + 6\pi t\ )$$

The value of k during generalized S transform is optimized, and the results is shown as figure 4(a),where $k_{op} = 6$. The standard S transform, generalized S transform with different k and the optimized generalized S transform are demonstrated in figure 4(b)(standard S transform), (c)(the generalized S transform with k =4),(d) (the generalized S transform with k =8), and (d) (the generalized S transform with $k_{op} = 6$). It can be seen that the optimized result has better time and frequency concentration performance.

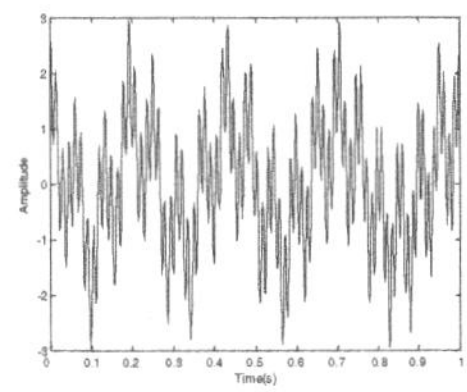

Fig. 3. Test signal

(a) optimized k value (b) standard S transform

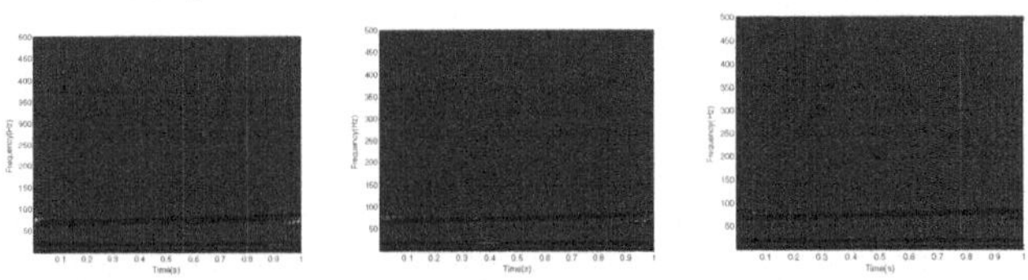

(c) generalized S transform with k =4 (d) generalized S transform with k =8 (e) generalized S transform with $k_{op} = 6$

Fig. 4. Generalized S transform with different k

A section of seismic data is selected, it is processed using the generalized S transform without optimized k and with optimized k_{op} respectively. The results are shown as figure5.From the results, it can be seen that the reconstruction profile based on generalized S transform with optimized window width have better resolution and it can provide the more legible stratum structure.

Generalized S transform is proved having superiority to process seismic signal, and it can process the profiles of the seismic data subtly, which makes full use of the characteristic information of seismic signal. The paper used the optimized window width method discussed in [3] to make the generalized S transform having better concentration performance. The processed results of seismic data demonstrate the validity and it can provide better stratum resolving power.

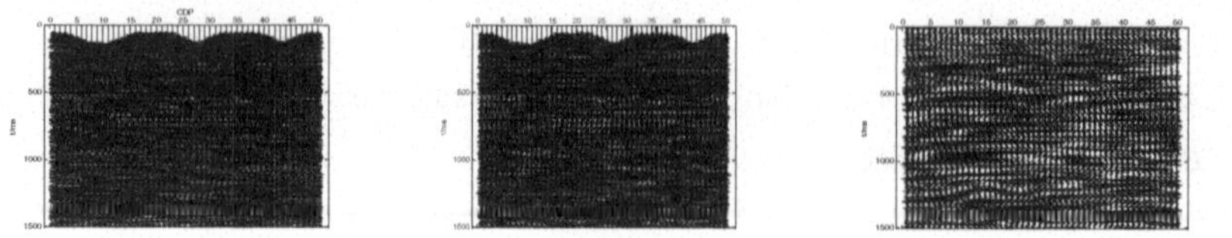

(a) seismic profile (b)generalized S transform reconstruction with k=2 (c)generalized S transform reconstruction with k_{op} =4

Fig. 5. Reconstruction of seismic profile based on the generalized S transform

Acknowledgement. The paper is Supported by the Program for New Century Excellent Talents in University of Heilongjiang province (NO.1154NCET001) and New Century Excellent Talents of Ministry of Education of The P.R.C.

References

[1] Robert Pinnegar, C., Mansinha, L.: The S-transform with windows of arbitrary and varying shape. Geophysics 68(1), 381–385 (2003)
[2] Pinnegar, C.R.: A new subclass of complex-valued S-transform windows. Signal Processing 86, 2051–2055 (2006)
[3] Sejdić, E., Djurović, I., Jiang, J.: A Window Width Optimized S-Transform. EURASIP Journal on Advances in Signal Processing, 1–13 (2008)
[4] Stankovic, L.: A measure of some time-frequency distributions concentration. Signal Processing 81, 621–631 (2001)

A Hybrid Algorithm to Solve Traveling Salesman Problem

Xiaofeng Chen, Zhenhua Tan*, Guangming Yang, and Wei Cheng

Software College, Northeastern University, Shenyang City, Liaoning, China
neucxf@163.com, tanzhenhua192@126.com, yanggm@mail.neu.edu.cn

Abstract. The TSP problem is a typical one in the field of combinatorial optimization. After study other researchers' related works, this paper presents a hybrid algorithm based on simulated annealing, ant colony and genetic in reference to previous research, in order to improve computing performance. Algorithms of this paper are used for solving traveling salesman problem, and the simulation contrast test results show that the algorithm has better convergence speed and optimal results; it also shows that the algorithm is feasible and effective.

Keywords: Simulated Annealing Algorithm, Ant Colony Algorithm, Genetic Algorithm, TSP Problem.

1 Introduction

Traveling Salesman Problem (TSP), which is referred to as the TSP, is the most basic line problem, this problem is looking for a single traveler who starts out by, after through all the needs of a given point, then back to the origin path with the least cost. At present the main methods of solving TSP are simulated annealing algorithm [1], genetic algorithm [2], heuristic search method, Hopfield neural network algorithm [3], the ant colony algorithm [4], etc. This paper has presented a hybrid simulated annealing, ant colony and genetic algorithm after comprehensive analyzing their characteristics and the algorithm is used for traveling salesman problem.

2 Hybrid Algorithm

The simulated annealing algorithm is a global optimization algorithm based on the physics background of solid annealing process. The advantages of the simulated annealing algorithm are simple computing process, universal, strong robust, suitable for parallel processing, high efficiency. The disadvantages of the simulated annealing algorithm are slow convergence speed, the long execution time, and if cooling process is slow enough, it can get better performance of the solutions, but the relative one is too slow convergence speed; (2) if the cooling process is too fast,it may not get the global optimal solution.

* Corresponding author.

D. Jin and S. Lin (Eds.): Advances in EECM Vol. 1, LNEE 139, pp. 99–105.
springerlink.com

Ant colony algorithm was put forward by Dorigo [5] in 1991, it came from the natural ants who looked for source food from their nests with the shortest path of inspiration.: (1) the positive feedback information mechanism; (2) distributed computing features; (3) general stochastic optimization method; (4)the heuristic search features. The disadvantages of ant colony algorithm are as following: shortage of initial pheromone, the slow solving speed [6].

Genetic algorithm is a kind of reference for the evolution of the biological world law (the survival of the fittest, superior bad discard genetic mechanism) evolved random search method. The advantages of genetic algorithm are as following: versatility and strong robustness,strong search global, hidden parallelism, which is very suitable for parallel computing. The disadvantages of genetic algorithm are as following: can not solve large-scale calculation problems, and it's very easy to get into "premature".

Considering the shortages of the ant colony algorithm in the early stages of the pheromone, search for a long time, easy appear stagnation, based on previous research, a hybrid simulated annealing, ant colony and genetic algorithm has put forward. That is adding to initialization pheromone reference path of fusion method in the ant colony algorithm, updating the simulated annealing algorithm and genetic algorithm, and using local optimization way to accelerate the speed of the solution.

Suppose there are n cities, first of all, generate 3n paths, choose n better paths as reference paths, then use the ant colony algorithm route to reference pheromone.Secondly, use the simulated annealing algorithm cooling process to control iteration in the most outer layers of each iteration; search the whole solution space with the ant colony algorithm in the middle layer, then generate the iteration father optimal path. Thirdly, use genetic algorithm to specify the first respectively reference path and the father of each iteration optimal path for two father generation for crossover and mutation operators in the most inner layer and generate each iteration son optimal path, then use the simulated annealing algorithm to evaluate the father and the son optimal path to assess, and take a good path as the iterative optimization path, and the iteration of the optimal path of accumulation of pheromone to update and then update the reference path to the iterative optimization path, and keep the iterative optimization of the best path and best optimal path length. Finally, when achieving annealing temperature, we can calculate the best optimal path, path length and output of the best optimal.

The simulated annealing algorithm accepts a solution which is less than the current one with a certain probability; therefore it will probably get out of this local optimal solution to achieve the global optimal solution.

(1)The Selection of Initial Temperature

T0 should ensure stable starting temperature distribution in the probability of each state that are equal. In practical calculation, we can select it according to experience; after a lot of experimental analysis's, we calculate T0 equals to 100 in this paper.

(2) The Selection of Temperature Update Function

Temperature update function, which is the method used to drop the temperature, is used to change temperature in the outer cycle. This article takes the temperature update function for:

$$T(t) = T0/(1 + \delta t) \tag{1}$$

The characteristic of temperature drop is that it is fast in the beginning, then the cooling rate is lesser, therefore optimal focus is in the low temperature area. δ can improve annealing curve shape. δ equals to 0.9 in this paper.

(3) The Algorithm is Terminated Criteria

Adopt one degree method, i.e. annealing for the final temperature of minT equals to 1, when the temperature equals to 1 or more than 1 , the algorithm will stop.

The probability of the ant k chooses a node from the current node i to the next node j in time t is as follows:

$$P_{ij}^{k} = \begin{cases} \dfrac{\tau_{ij}^{\alpha}(t) \bullet \eta_{ij}^{\beta}(t)}{\sum\limits_{s \in allowed_k} \tau_{is}^{\alpha}(t) \bullet \eta_{is}^{\beta}(t)} & j \in \text{allowed} \\ 0 & \text{otherwise} \end{cases} \tag{2}$$

Among of them: q is a unified random from 0 to 1 , q0 is a threshold parameter from 0 to 1, $\tau_{ij}(t)$ means link pheromone which is at moment (i, j) at t, α is an important degree of a pheromone, β is an important degree for inspiring information, allowedk = {0, 1,... , n-1}—tabuk means there are k ants that can select node set at moment t, tabuk is used to save the table which is the kth ant's walk path; $\eta_{ij}(t)$ is an inspiration function, $\eta_{ij}(t) = 1/d_{ij}$, and d_{ij} is a city distance between the city i and j.

When the time is from t to (t + 1) , if the node I and j are the two adjacent nodes of chosen path, pheromone— $\tau_{ij}(t+1)$ will update as follows:

$$\tau_{ij}(t+1) = (1-\rho)\tau_{ij}(t) + \Delta\tau_{ij} \tag{3}$$

$$\Delta\tau_{ij} = \sum_{k=1}^{m} \Delta\tau_{ij}^{k} \qquad \Delta\tau_{ij}^{k} = \begin{cases} \dfrac{Q(t)}{L_k} & \text{if ant k use acr(i,j) in its tour} \\ 0 & \text{otherwise} \end{cases} \tag{4}$$

Among of them: ρ is a pheromone volatile coefficient from 0 to 1, $\Delta\tau_{ij}$ is a pheromone increment at this cycle path—i and j, $\Delta\tau_{ij}^{k}$ is a pheromone which is the kth ant leaves at this cycle path—i and j, Lk is the kth ant's walking length at this cycle, Q(t) is a pheromone degree.

Real number coding uses the problem variables to code directly. The form of the chromosome X is as following: X = (x1 x2, and.. xn), xi ∈ R, i = 1, 2,... , n.Real number coding genetic algorithm has advantages of high precision, it is easy to search for large space.

Assuming that there are 10 cities, two father generation individuals: C1 = (1,2,3,4,5,6,7,8,9,10), C2 = (1,3,8,2,9,4,5,10,7,6). Firstly, crossover operation generates two random numbers such as 4 and 6 within the scope from 1 to 10, C2 is in corresponding of interval [4, 6] values for (2,9,4), the cross process is as following: insert (2,9,4) at the fourth before C1 to generate C1 '= (1,2,3,2,9,4,4,5,6,7,8,9,10), then remove the original C1 (2,9,4) of the chromosome at C1 ' to produce the offspring of C3 = (1,3,2,9,4,5,6,7,8,10). Crossover operation can be defined as follows:

$$c3 = cross(c1, c2) \tag{5}$$

There are 10 cities, the path C0 variates to produce another path C1. Select the random j1 from 1 to n at the visited cities, exchange j1 and j1 + 1 visited cities in the path C0, and the rest of the city is unchanged, and at the time the path is C1, for example, C0 = (2,3,10,4,1,5,7,9,8,6), j1 = 3, then C1 = (2,3,4,10,1,5,7,9,8,6). Variation operation can be defined as follows:

$$c1 = mutation(c0) \tag{6}$$

Initialize system, set the initial temperature, initial solution state, the basic parameters, the number of ants m, the algebraic count NC and so on; randomly generate 3 * m paths, record their length, sorted by length;

Choose m good path as the reference path Road (m, n), calculate the reference path length L_ Road (m), record the best reference path R_best, calculate the best reference path length L_best;

By (2-3) update the cumulative pheromone of the reference path Road (m, n);

Initialize taboo table;

While T>=T_min

Put the m ants into the n cities Tabu(1,n);

for j=2:n exclude the selected n cities

For each ant , according to formula (2-2) ,calculate the probability distribution of the city to be selected , using roulette wheel selection probability principles to select next city, and adding the node number of the selected variables into their own taboo tables Tabu (i, j) .

Calculate the sum of path length L_ Tabu (i,);

Record the best path T_best for this iteration, calculate the current optimal path length L_min;

Take each iteration path Tabu (i,:) cross operate with the best reference path R_best and the corresponding reference path Road(i,:) respectively.

Path(i,:)=cross(Tabu(i,:),R_best0,n);

Path(i,;)=cross(Tabu(i,:),Road(i,:),n);

Take each path Path (i,:) after cross operation to do mutation operation:

Path(i,:)=mutation(Path(i,:),n) ;

Calculate the sum of path length L_Path (i) after cross and mutation operation;

Compare simulated annealing acceptance criteria with the path length which before and after cross operation, retain the optimal selection;

If L_Path(i)< L_ Tabu(i)|exp((L_ Tabu(i)- L_Path(i))/(T))>rand

L_ Tabu(i)= L_Path(i); Tabu(i,:)= Path(i,:) ;

Update the reference path: Road(m,n)和L_ Road (m)

if L_ Road (i)< L_ Tabu(i)

L_ Road (i)= L_ Tabu(i) ;

Road(i,:)=Tabu(i,:) ;

By (2-2) to update the cumulative pheromone of this iteration path Tabu (i,:);

Retain the best route GBP (NC) in each generation, calculate the best route length GBL (NC) of each generation;

Clear taboo table; calculate the temperature T by annealing formula (2-1); and iteration counter NC plus 1.

3 The Simulation Experiment and Analysis Contrast

In order to test the effectiveness of the algorithm, choose international general TSP database —TSPLIB city31 (the best solution for 15404) and att48 (the best solution for 33522) as the experimental examples to study.

The hybrid algorithm parameters are as following: $\alpha = 1$, $\beta = 5$, $\rho = 0.7$, Q = 100, T0 = 100, minT = 1, $\delta = 0.9$, comparing with the basic ant colony algorithm,

Table 1. Several Algorithm Test Results

Algorithm	City31			Att48		
	Average	Best solution	Worst solution	Average	Best solution	Worst solution
Simulated annealing algorithm	26824	21865	34329	78047	57585	132497
Genetic algorithm	19059	17310	21033	52045	45303	58387
Basic ant colony algorithm	15621	15609	15841	35119	34579	35454
Mixed methods	15604	15503	15609	34760	34084	35537

the basic genetic algorithm and simulated annealing algorithm respectively. In the scale of each algorithm on traveling salesman problem to do 30 times of experiments, then obtain the results as Table 1 shown. Fig 1 is a solution city31 hybrid algorithm of the best solutions, the total distance is 15503. Fig 2 is a solution att48 hybrid algorithm of the best solutions, the total distance is 34084. From Table 1 we can see that the hybrid algorithm proposed in this paper has the best effect.

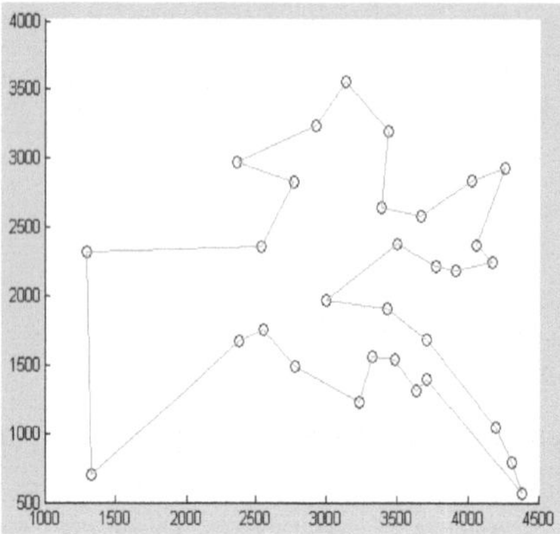

Fig. 1. The best results of city31 solution

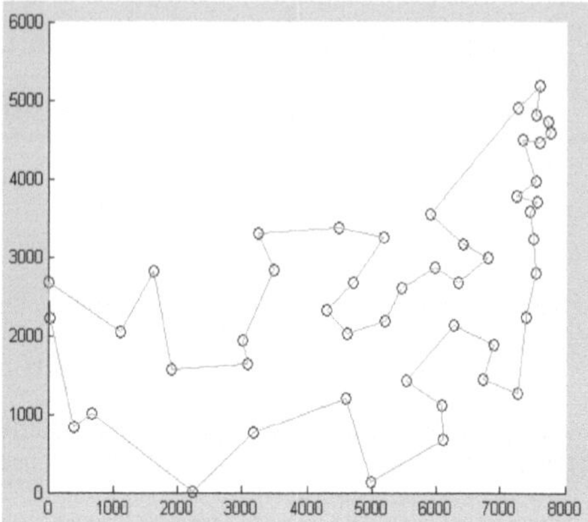

Fig. 2. The best results of att48 solution

4 Conclusions

This paper has proposed hybrid algorithm which is combined with the advantages of simulated annealing, ant colony and genetic algorithm, and overcome the long search time, search stagnation and excellent local search shortcoming of ant colony algorithm. And the method has been used for solving traveling salesman problem successfully. Through experiment analysis and comparison with hybrid algorithm, the precision of the results did raises a lot. The hybrid algorithm retains and combines with simulated annealing, ant colony and genetic algorithm, so the basic structure is not only for TSP which can provide efficient solutions, and also provide a solution of the optimization problem of application framework. In the later work, we will study the actual application of the algorithm further.

Acknowledgement. This work is supported by the National Natural Science Foundation of China under Grant No. 61070162, No. 71071028, No. 60802023 and No. 70931001; the Specialized Research Fund for the Doctoral Program of Higher Education under Grant No. 20070145017; the Fundamental Research Funds for the Central Universities under Grant No. N090504003 and No. N090504006.

References

1. Yang, W.-B., Zhao, Y.-W.: Improved simulated annealing algorithm for TSP. Computer Engineering and Applications 46(15), 34–36 (2010)
2. Lin, W., Delgado-Frias, J.G., Gause, D.C., et al.: Hybrid newton-raphson genetic algorithm for the traveling salesman problem. Cybernetics and Systems 26(4), 387–412 (1995)

3. Tan, K.C., Tang, H., Ge, S.S.: On parameter settings of Hopfield networks applied to traveling salesman problems. IEEE Transactions on Circuits and Systems 52(5), 994–1002 (2005)
4. Stutzle, T., Hhoos, H.: The MAX-MIN ant system and local search for the traveling salesman problem. In: Proceedings of the IEEE International Conference on Evolutionary Computation (ICEC 1997), pp. 309–314. Indianapolis, USA (1997)
5. Colorni, A., Dor Igo, M., Maniezzo, V.: Distributed optimization by ant colonies. In: Proc. Europ. Conf. Artif. Life, p. 1342142. Elsevier Publishing, Paris (1991)
6. Wu, Q.-H., Zhang, J.-H., Xu, X.-H.: An Ant Colony Algorithm with Mutation Features. Journal of Computer Research & Development 36(10), 1240–1245 (1999)
7. Xu, Z.-H., Song, B., Guo, Y.-Y.: Using Ant Colony Algorithm and Simulated Annealing Parallel Algorithm to Solve Traveling Salesman Problem. Journal of Hebei University of Technology 39(39), 48–51 (2010)
8. Zhang, X.-R., Gao, S.: Solving Traveling Salesman Problem by Ant Colony Optimization Genetic Hybrid Algorithm. Microelectronics & Computer 26(4), 81–84 (2009)
9. Du, Z.-Z., Liu, G.-D.: Solution of TSP problem based on hybrid genetic simulated annealing algorithm. Computer Engineering and Applications 46(29), 40–42 (2010)
10. Lee, S.G., Jung, T.U., Chvng, T.C.: An effective dynamic weighted rule for ant colony system optimization. In: Proceedings of the 2001 Congress on Evolutionary Computation, vol. (2), pp. 1393–1397 (2001)
11. Dorigo, M., Gambardella, L.M., Middendorf, M.: Guest editorial special section on ant colony optimization. In: IEEE Transactions on Evolutionary Computation, vol. (6), pp. 317–319 (2002)

A Study on Small Town Core Competency Evaluation Based on Rough Set Theory

XiangHui Li[1] and KeNing Da[2]

[1] Economic College of Shenyang University, Shenyang, China, 110044
[2] Management College of Shenyang Architecture University, Shenyang, China, 110168

Abstract. In order to overcome the subjective characteristic of the existing own core competence evaluation method, we present a rough set theory-based fuzzy comprehensive evaluation method, which has the advantage of dealing with uncertain information. We determine the weight coefficient of each evaluation index according to attribute significance. Then, we determine the weight coefficient based on attribute significant calculating of information systems, and further obtain the objective weight coefficients. This method can solve the multi-objective conflict and has good feasibility in practice.

Keywords: small town, core competitiveness, rough set.

1 Introduction

Small town is the boundary between the city and countryside in an intermediate state [1]. Small town is in the combination of urban and rural, which is also the bridge and link contacting urban and rural. It is the organization center of agriculture industrialization and the basic point of industry cluster. In addition, it becomes an important carrier of regional economy development and plays an important part in transferring rural surplus labor force nearby [2,3].

The core competency of small town is a typical complex and chaotic system. There are lots of elements and environment subsystems in different ways which constitute the small town core competence. Taking industry development force evaluation index for example, we use rough sets theory to ascertain the weight coefficient of fuzzy comprehensive evaluation. The method can effectively analyze inaccurate, inconsistent and incomplete information. In further, it can find the implicit knowledge and reveal potential regulation. Thus, it provides scientific theory for small town core competitiveness research.

Industrial development force is the important index for evaluating small town core competitiveness. It contains two secondary indexes which are industrial structure force and industrial agglomeration, and it also contains eighteen third grade indexes shown in Table 1.

D. Jin and S. Lin (Eds.): Advances in EECM Vol. 1, LNEE 139, pp. 107–113.
springerlink.com

Table 1. Industrial development force evaluation index

Industrial development force	Industrial Structure force	Annual total output value of primary industry
		The contribution rate of primary industry to GDP
		Annual average growth rate of primary industry
		Annual total output value of secondary industry
		The contribution rate of secondary industry to GDP
		Secondary industry annual average rate of growth
		The tertiary industry annual total output value
		The contribution of the tertiary industry to GDP
		Tertiary industry average rate of growth
	Industry agglomeration force	The number of Leading Industry
		Leading Industry saleroom (ten thousand yuan)
		The number of Leading Enterprises (group) contacts with Leading Industry
		The number of specialized markets contacts with Leading Industry
		Leading industry over taxes (ten thousand yuan)
		The number of agricultural production base contacts with Leading Industry
		household contacts with the Leading industry
		Leading industry contacts with farmer of net income (yuan)
		The number of intermediary service organizations contacts with Leading Industry

2 The Rough Set Theory

Rough Set Theory (RST) was put forward by Poland Z.P AWLAK in 1982, which is a kind of data analysis tool to deal with the fuzzy and uncertain knowledge. Although it was concerned by scholars in the world until 1990, now it has become the most active field of information science. Through attributes reduction, Fuzzy Comprehensive Evaluation based on RST can determine the weights of every evaluation index by attributes importance, which can avoid the subjectivity of weights determination in present evaluation methods, therefore, it is more feasible in practice.

3 Attribute Reductions of Industrial Development Force Bases on RST

3.1 Reduction of Industrial Structure Force

In Table 2, among the evaluation indexes of industrial force, from C11 to C19, they represent the primary industry annual output value and the tertiary industry annual average growth rate of industry structure force respectively (as shown in Table 1).

Table 2. Information system of small town industry structure force evaluation

Town	NO.	C11	C12	C13	C14	C15	C16	C17	C18	C19
A	1	1	1	1	3	3	1	2	1	1
B	2	3	2	3	2	2	2	2	3	1
C	3	2	3	3	1	2	2	1	2	3
D	4	3	3	2	1	1	3	1	1	2
E	5	2	1	2	3	3	2	3	1	1

Decision System $S = <U, C>$, $U = \{1,2,3,4,5\}$

$C = \{C11, C12, C13, C14, C15, C16, C17, C18, C19\}$

$U / ind(C) = \{\{1\},\{2\},\{3\},\{4\},\{5\}\}$. By undistinguishable relationship, we can get $U / ind(C13, C19) = \{\{1\},\{2\},\{3\},\{4\},\{5\}\}$

Thus, $\{C13, C19\}$ is one of the reductions, but it is not only one.

Calculate importance of each attribute:

$U / ind(C_{13}) = \{\{1\},\{2,3\},\{4,5\}\}$, $U / ind(C_{19}) = \{\{1,2,5\},\{3\},\{4\}\}$

$U / ind(C - \{C_{19}\}) = U / ind(C_{13}) = \{\{1\},\{2,3\},\{4,5\}\}$

$U / ind(C - \{C_{13}\}) = U / ind(C_{19}) = \{\{1,2,5\},\{3\},\{4\}\}$

Let $P = U / ind(C)$, $Q = U / ind(C - \{C_{13}\})$, $S = U / ind(C - \{C_{19}\})$

$POS_Q(P) = \{3,4\}$, $POS_S(P) = \{1\}$

Importance of each attribute are as follows:

$$\mu_{C_{13}} = \frac{|pos_P(P)| - |pos_Q(P)|}{|pos_P(P)|} = \frac{3}{5}, \quad \mu_{C_{19}} = \frac{|pos_P(P)| - |pos_S(P)|}{|pos_P(P)|} = \frac{4}{5}$$

Among nine factors affecting industry structure, annual average growth rate of the primary industry C13 and annual average grow rate of the tertiary industry C19 are important factor, and their weight coefficients are as follows:

$$\omega_{C_{13}} = \frac{\mu_{C_{13}}}{\mu_{C_{13}} + \mu_{C_{19}}} = 3/7, \omega_{C_{19}} = \frac{\mu_{C_{19}}}{\mu_{C_{13}} + \mu_{C_{19}}} = 4/7.$$

Then, we evaluate above five small towns by industry structure in fuzzy, the evaluation process is as follows,

$$F_1 = A * \omega = \begin{bmatrix} 103.4 & 106.6 & 106.5 & 104.0 & 104.6 \\ 107.2 & 106.6 & 112.2 & 109.7 & 107.2 \end{bmatrix}' * \begin{bmatrix} 3/7 \\ 4/7 \end{bmatrix}$$

$$= [105.5714,\ 106.6000,\ 109.7571,\ 107.2571,\ 106.0857]'$$

Then, the evaluation coefficient of five towns A, B, C, D, E are respectively

105.5714, 106.6000, 109.7571, 107.2571, 106.0857.

In order to avoid the impact of dimension, we adopt index coefficient with discretization, the result is

$$F_1' = A*\omega = \begin{bmatrix} 1 & 3 & 3 & 2 & 2 \\ 1 & 1 & 3 & 2 & 1 \end{bmatrix}' * \begin{bmatrix} 3/7 \\ 4/7 \end{bmatrix} = [1 \quad 1.86 \quad 3 \quad 2 \quad 1.43]'$$

3.2 Reduction of Industrial Agglomeration Force

In Table 3, from C21 to C29, they denote the number of leading industry and the number of service organization agent in industry agglomeration force (as shown in Table 1).

Table 3. Evaluation information system of industrial agglomeration force for small town

Town	NO.	C21	C22	C23	C24	C25	C26	C27	C28	C29
A	1	1	1	1	1	1	1	2	1	1
B	2	2	3	2	2	1	2	2	2	1
C	3	3	3	3	1	2	3	3	1	2
D	4	2	3	1	3	3	2	1	2	2
E	5	2	3	2	2	2	2	1	1	1
F	6	2	2	2	1	1	2	1	3	1

Let $C = \{C_{21}, C_{22}, C_{23}, C_{24}, C_{25}, C_{26}, C_{27}, C_{28}, C_{29}\}$

$U / ind(C) = \{\{1\},\{2\},\{3\},\{4\},\{5\},\{6\}\}$, by knowledge reduction, we can obtain one reduction $\{C_{21}, C_{24}, C_{28}\}$, which is not the only one. Take $\{C_{21}, C_{24}, C_{28}\}$ as an example, and we will calculate the importance of each attribute separately. Calculate weight as follows

$$U / ind(C - \{C_{21}\}) = \{\{1,3\},\{2\},\{4\},\{5\},\{6\}\},$$

$$U / ind(C - \{C_{24}\}) = \{\{1\},\{2,4\},\{3\},\{5\},\{6\}\}$$

$$U / ind(C - \{C_{29}\}) = \{\{1\},\{2,5\},\{3\},\{4\},\{6\}\}$$

Let

$P = U / ind(C)$, $Q = U / ind(C - \{C_{21}\})$, $S = U / ind(C - \{C_{24}\})$, $T = U / ind(C - \{C_{29}\})$

$POS_Q(P) = \{2,4,5,6\}$, $POS_S(P) = \{1,3,5,6\}$, $POS_T(P) = \{1,3,4,6\}$

The importance of attribute C21, C24, C29 are

$$\mu_{C_{21}} = \frac{|pos_P(P)| - |pos_Q(P)|}{|pos_P(P)|} = \frac{2}{6} \qquad \mu_{C_{24}} = \frac{|pos_P(P)| - |pos_S(P)|}{|pos_P(P)|} = \frac{2}{6}$$

$$\mu_{C_{29}} = \frac{|pos_P(P)| - |pos_T(P)|}{|pos_P(P)|} = \frac{2}{6}$$

Among nine factors affecting industry agglomeration force, the number of leading industry C21, the average household net income of leading industry C28 are important factors, and their weight coefficient are

$$\omega_{C_{21}} = \frac{\mu_{C21}}{\mu_{C_{21}} + \mu_{C_{24}} + \mu_{C_{29}}} = \frac{1}{3},$$

$$\omega_{C_{24}} = \frac{\mu_{C24}}{\mu_{C_{21}} + \mu_{C_{24}} + \mu_{C_{29}}} = \frac{1}{3},$$

$$\omega_{C_{29}} = \frac{\mu_{C29}}{\mu_{C_{21}} + \mu_{C_{24}} + \mu_{C_{29}}} = \frac{1}{3}$$

Then, we evaluate above six towns by industry agglomeration force in fuzzy, the evaluation process is

$$F_2 = A * \omega = \begin{bmatrix} 1 & 2 & 3 & 2 & 2 & 2 \\ 0 & 1 & 0 & 2 & 1 & 0 \\ 0 & 0 & 1 & 1 & 0 & 0 \end{bmatrix} * \begin{bmatrix} 1/3 \\ 1/3 \\ 1/3 \end{bmatrix} = [0.33,\ 1,\ 1.33,\ 1.67,\ 1,\ 0.67]$$

Thus we get evaluation coefficients of above six towns are 0.33, 1, 1.33, 1.67, 1, 0.67.

$$F_2^{'} = A * \omega = \begin{bmatrix} 1 & 2 & 3 & 2 & 2 & 2 \\ 1 & 2 & 1 & 3 & 2 & 1 \\ 1 & 1 & 2 & 2 & 1 & 1 \end{bmatrix}^{'} * \begin{bmatrix} 1/3 \\ 1/3 \\ 1/3 \end{bmatrix} = [\ 0.99 \quad 1.66 \quad 1.99 \quad 2.33 \quad 1.66 \quad 1.33]$$

3.3 Determination of Secondary Index Weight on Industry Force

By RST, we will analysis the importance of industry structure force C1 and industrial agglomeration force C2, which are two factors affecting industry development force.

Table 4. Small town industry aggregation evaluation information list

Town	NO.	C_{13}	C_{19}	C_{21}	C_{24}	C_{29}
A	1	3	1	1	1	1
B	2	3	3	3	1	2
C	3	2	2	2	2	1
D	4	2	2	2	1	1

Similar with above method, we can get one reduction $\{C_{19}, C_{24}\}$, their importance during the process of classification are $\mu_{C_{19}} = \frac{3}{4}, \mu_{C_{24}} = \frac{2}{4}$

Thus we get their weight coefficients,

$$\omega_{C_{19}} = \frac{\mu_{C19}}{\mu_{C_{19}} + \mu_{C_{24}}} = 0.6, \omega_{C_{24}} = \frac{\mu_{C24}}{\mu_{C_{19}} + \mu_{C_{24}}} = 0.4.$$

which also means the weight coefficients of industrial structure force C1 and industrial agglomeration force exponential C2 are 0.6 and 0.4.

$$F = \begin{bmatrix} F_1^{'} & F_2^{'} \end{bmatrix} * \omega = \begin{bmatrix} 1.86 & 3 & 2 & 1.43 \\ 0.99 & 1.99 & 1.66 & 1.33 \end{bmatrix}^{'} * \begin{bmatrix} 0.6 \\ 0.4 \end{bmatrix} = [1.5 \quad 2.6 \quad 1.8 \quad 1.4]$$

The evaluation index coefficient on the second level, the first level and the ranking of small towns A, B, C, D are shown in Table 5.

Table 5. Evaluation coefficient table

Town	Evaluation coefficient of industrial agglomeration Force	Evaluation coefficient of industrial structure fore	General evaluation coefficient of first level index	Ranking
A	1.86	0.99	1.5	3
B	3	1.99	2.6	1
C	2	1.66	1.8	2
D	1.43	1.33	1.4	4

3.4 Industry Force Evaluation Index Weight Table

Table 6. Evaluation index weight of core competitiveness for small town (industry development force)

Industry development force	Industrial structure fore	0.6	annual average growth rate of primary industry	3/7
			annual average growth rate of tertiary industry	4/7
	Industry Agglomerate-on force	0.4	Number of leading industry	1/3
			number of specialized markets contacts with leading industry	1/3
			household net income of farmer contacts with leading industry (yuan)	1/3

4 Summaries

Since evaluation index are at different levels in multi-index evaluation system, they have different effecting on evaluation, even though they are at the same level, every attribute may have difference in significance. Therefore, it is important to determine the weight coefficient in fuzzy comprehensive evaluation, which will affect the evaluation result directly. By the analysis in this paper, we use fuzzy comprehensive evaluation based on RST to reduce attributes of core competency for small town, the method can resolve the problem of fuzzy and multi-object conflict, it also overcomes the subjective in weight determination during evaluation.

Acknowledgements. The authors would like to thank Humanities and Social Sciences Planning Fund of Ministry of Education (no. 10YJA79009), Social Sciences Planning Fund of Liaoning Province (no. L10BJL033) for the financial support of this research. We thank the anonymous reviewers for their careful review and valuable suggestions on the manuscript.

References

1. Almor, T., Hashai, N.: The competitive advantage and strategic configuration of knowledge-intensive, small- and medium-sized multinationals: a modified resource-based view. Journal of International Management 10, 479–500 (2004)
2. Zhao, H.B.: The competitiveness of the industry of-a theory and reviewed in this paper. The Contemporary Financial 12 (2004)
3. Zhou, H.S., Wu, Y.M., Lu, W.C.: City competitiveness evaluation index and methods. The Economy 12 (2003)

Research of Distributed Index Based on Lucene

Zhuang Chen, Chonglai Zhu, Wei Cheng, Qiulin Song, and Shibang Cai

Institute of Computer Science and Engineering, Chongqing University of Technology, ChongQing, China
zhuang.ch@gmail.com, j025zcl@126.com

Abstract. Inverted index is the mainstay technology for full-text retrieval, however, there exist some problems such as low efficiency of index construction, updating and high Maintenance cost. In order to improve the retrieval performance of large scale indexing, the strategy of distributed indexing is proposed in this paper. This paper gives a detailed description of building index structures using Map Reduce, and compares the performance of building index and query between distributed search system and central search system. Experimental results show that the proposed scheme greatly reduces the time of index construction and query time.

Keywords: Inverted index, Search engine, Distributed index.

1 Introduction

In recent years, with the rapid increasing data of web, search engines are facing enormous challenges, for instance, storage capacity, and query processing time. Since the content on the Web changes extremely rapidly, there is a need to periodically crawl the Web and update the inverted index. Indexes can either be updated incrementally or periodically rebuilt, after every crawl. With both approaches, the key challenge is to handle the large whole-scale changes commonly observed between successive crawls of the Web. For efficiency and simplicity, most traditional search engine employ the rebuilding approach [1]. In this case, it is critical to build the index rapidly to quickly provide access to the new data. This procedure is easy to realize, but at the same time, causes a series of bad consequences, for example, a time lag of data updating and query, low expandability and visit bottleneck. It cannot meet the real time request of system. For large scale information searching, a traditional search engine is still far from meeting the request toward data retrieval.

In this paper, we focus on the efficiency of building index structures using Hadoop. Hadoop is an open source implementation of MapReduce. The general idea of using MapReduce for distributed index generation is described in modern textbooks such as [7].In fact, indexing has been one of the examples given in the original description of Google's version of MapReduce by Dean and Ghemawat [4]. It has been used in a relatively straightforward manner by the open-source Katta project for generating text indices for Lucene.

D. Jin and S. Lin (Eds.): Advances in EECM Vol. 1, LNEE 139, pp. 115–121.
springerlink.com

2 Lucene and Index Technology Introduction

Lucene is a Java language development with the full text search engine toolkit, provides a series of API, can filter and analysis of document pretreatment sorting indexing and retrieval. It keeps in efficient and simple features, but also can according to the actual demand, free developers customization and combination various core functions. The basic principle is index Lucene for documents and retrieving, through establishing good index files, data of improving customer searches response speed. Lucene provided is search kernel, any document just convert text format can be indexed retrieval. Processing document content includes the Internet Web page, all kinds of local document (text files, Word document, HTML or PDF files) or any can extract text message format. Lucene not a complete application, but for application provides index and search function from the overall. Lucene is a used to implement the full text retrieval of libraries.

2.1 Indexing Technology

Index is a data structure, allowing you to quickly find a certain word or phrase from a large number of documents. Indexer is a software to establish, maintain and manage index. It extracts index entries from original documents, and generates. Indexed table of document collections. When the amount of stored documents is enormous and pediocratic. Indexing can help tremendously to improve the efficiency of search. There are various methods to implement text retrieval, such as inverted file, suffix array and signature file. Though suffix array just takes a short time, it is difficult and to maintenance, and has high costs. However, suffix array is unsuited for the indexing of search engines. The signature file is a good index mode, but inverted file has higher speed and yields better performance than that. Inverted index is a high-effeciency organization index, and is an effective way to support multiple retrieval models, hence achieves high performance of information retrieval as a whole. Inverted index uses a specific word or phrase as index entry, which makes it very suitable for keyword searches. So it has been widely adopted in most current Information retrieval systems.

Inverted index also known as reverse indexing, uses keywords as the means of accessing and stores them in storage position of a document or a set of documents. It is most commonly employed as index mechanism for document or document collections. It is the most commonly used document data in Document Retrieval System.An inverted index is mainly composed of index entries and file lists. An index entry is used to store keywords, while a file list is mainly for storing the occurrences of keywords in the document and location. The data organizational structure of inverted index as shown in Figure 1.The figure at the left is a keyword named Ti in dictionary. The figure at the right is inverted list which the keyword is corresponding to, and TFi counts the number of occurrences of Ti in the document. nij is file number, position serial of Ti in the document is < hit1 ,hit2 ,…. >.

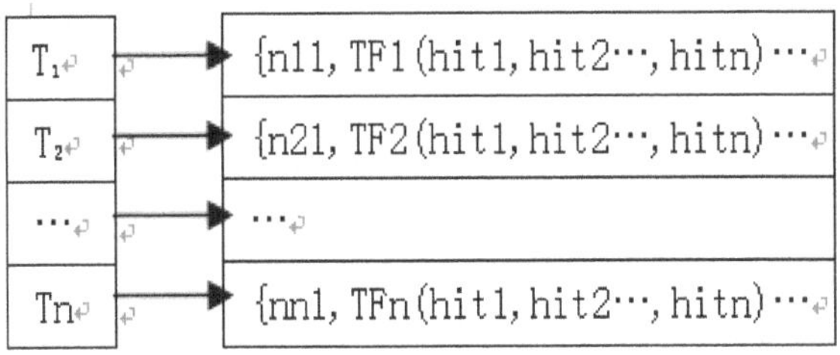

Fig. 1. The data organizational structure of inverted index

2.2 Index Construction

The fundamental concepts in Lucene are index, document, field and term. An index contains a sequence of documents. A document is a sequence of fields. A field is a named sequence of terms. A term is a string. The same string in two different fields is considered a different term. Thus terms are represented as a pair of strings, the first naming the field, and the second naming text within the field.

1)Index: In computer science, an inverted index is an index data structure storing a mapping from content, such as words or numbers, to its locations in a database file, or in a document or a set of documents.

2) Segment: The Lucene index is splitted in smaller chunks called segments. Each segment is an index in itself. Lucene search in all of them in sequence.

3) Document: Documents are the unit of indexing and search. A Document is a set of fields. Each field has a name and a textual value. A field may be stored with the document, in which case it is returned with search hits on the document. Thus each document should typically contain one or more stored fields which uniquely identify it.

4) Field: A field is a section of a Document. Each field has two parts, a name and a value. Values may be free text, provided as a String or as a Reader, or they may be atomic keywords, which are not further processed. Such keywords may be used to represent dates, urls, etc. Fields are optionally stored in the index, so that they may be returned with hits on the document.

5) Term: A Term represents a word from text. This is the unit of search. It is composed of two elements, the text of the word, as a string, and the name of the field that the text occurred in, an interned string.

2.3 Creating Index

The Lucene search engine is an open source, Jakarta project used to build and search indexes .

For instance, there are three texts:

T0= I was a student, and I came from chongqing.
T1=My father is a teacher,and I am a student.
T2= Chongqing is my hometown, but my father's hometown is Wuhan.

Firstly, uses tokenizer to obtain key words form the three texts. The processing procedure is as follows.

(1) In the preprocessing stage: Lucene segment the content into words by tokenizer, which are used as candidate keys. In Chinese, now the most widely used method is Chinese word segmentation.

(2) In the analysis stage, Lucene needs to filter out stop words, and each word is converted into capital letters. Then the indexer calls the function of addDocument(Document), passing parameters to Lucene in order to operate the index.

The processed data are stored in the index file,which are stored in the disk as data structures of inverted index.

After treated through above process, the three texts' keywords are as follows:

key0= I am student I come chongqing.
key1= my father Is teacher I am student.
key2= Chongqing is my home town my father home town is wuhan.

Now, we can establish inverted index. Inverted index structure is shown as Table 1.

After establishing the index, Lucene uses Index Reader to update or delete index.

Table 1. Inverted index structure

Term	DocID[Frequency]	Loc
I	1[2],2[1]	[1,4,5]
Am	1[1],2[1]	[2,6]
student	1[1],2[1]	[3,7]
come	1[1]	[3,7]
Chongqing	1[1],3[1]	[6,1]
My	2[1],3[2]	[1,3,5]
father	2[1],3[1]	[2,6]
Is	2[1],3[2]	[3,2,8]
teacher	2[1]	[4]
hometown	3[2]	[4,7]
wuhan	3[1]	[9]

3 Distributed Indexing Using Map Reduce

Distributed computing is a field of computer science that studies distributed systems. A distributed system consists of multiple autonomous computers that communicate through a computer network. The computers interact with each other in order to achieve a common goal. The central search system creates and queries index on one computer, which is applicable to small and medium-sized system. However, the volume of documents achieves the level of TB(when a large scope of image retrieval appears), the central search system cannot satisfy the need of index. The data of distributed search system is identical logically, and dispersed physically. It needs a series of database indexes to work together, and therefore expands the retrieved scope, and is easy to extend. Searcher uses server cluster to combine all nodes' indexes in the network .It increases concurrent efficiency of search engine, and ensures real-time response. Disturbed indexes' system structure chart is shown in Figure 2.

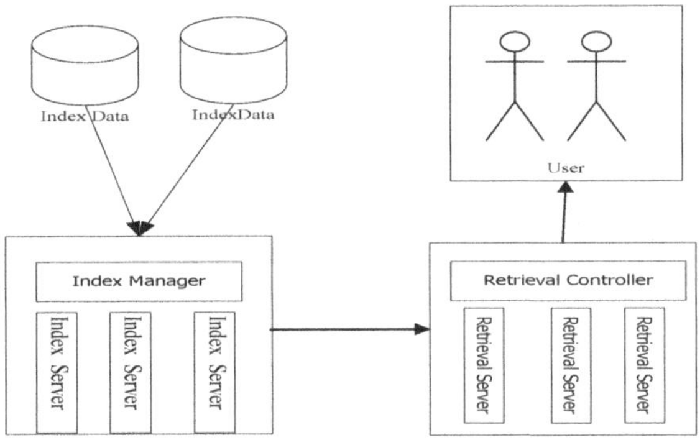

Fig. 2. Disturbed indexes' system structure

MapReduce is a framework for processing huge datasets on certain kinds of distributable problems using a large number of computers, collectively referred to as a cluster. Computational processing can occur on data stored either in a file system or within a database. The map phase can be used to load documents from the distributed file system and parse them into (key, value) pairs corresponding to terms and documents respectively. The first step in the reduce part of the framework performs precisely what is required to "invert", to collect all values belonging to the same key, corresponding to the documents that contain the same term. This is the input to the user-provided reduce function, which can write the index to the distributed file system, by using the indexing component of any IR package designed for single machine indexing. Google uses the technology of MapReduce to create index while Lucene uses it for indexing management. The index server are mainly responsible for reading, analysis, mapping and management of index.

4 Experiments

We have implemented distributed indexing using Hadoop as described above, and using Lucene as the underlying search engine. To evaluate the efficiency of the proposed scheme, all experiments were run on an Intel P4 3.0GHZ PC with 1GB DDR II memory system. The 7200-rpm sata hard disk is 160GB, and the network is 100M LAN. One of the nodes as the master, and the remaining three are used as child nodes to build a distributed environment.

4.1 Indexing Performance

During the experiment, using a machine implementation of the centralized crawling strategy, the implementation of the remaining 3 machines distributed crawling strategy. Figure 3 shows, with the increase of the size of your crawled pages, distributed memory parallel operation makes the index of increase, so crawl the increase is not linear with time, the average crawl time with the expansion of records but smaller. The experimental results show that the distributed environment to some extent on the efficiency improvement, but more important is the sharing of the index of the storage pressure, but also laid the query execution module distributed computing basis.

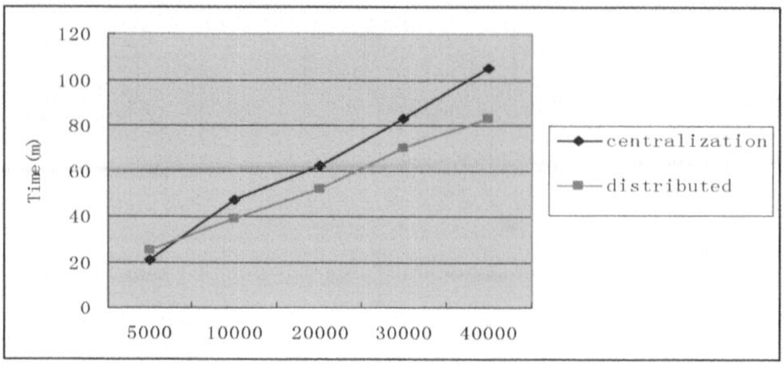

Fig. 3. Time needed for constructing new indexes

4.2 Retrieval Performance

To measure the performance of our indixes by executing the same query set. This query set contains 50000 queries sampled randomly from the queries. We focus on the relative performance of the various schemes, the absolute performance of the search engine may depend on many factors, in particular the hardware environment. We assessed performance for two kinds of queries, keyword queries and unigram queries with a field restriction.

Figure 4 shows the query execution times for keyword queries. The results showed that when the index scale is smaller than 1000, the efficiency of centralized search is higher than the distributed, but with the index scale increases, the efficiency of distributed query gradually manifested because under the in this scale are stored in sub-server memory. When the index achieves the certain scale, needs to read the index structure from the external memory, therefore needs the extra time expenses.

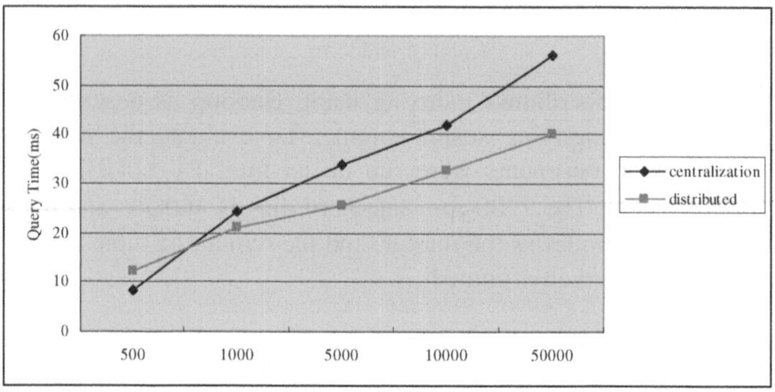

Fig. 4. Query time for varying system

5 Conclusion

This paper have elaborated on the index scheme of inverted index in detail, and analyze the implementation of building index structures for web data using Hadoop's

MapReduce. The presented work provides a preliminary analysis of the efficiency of distributed indexing using MapReduce and retrieval performance using various index structures.

This paper propose strategies of disturbed indexing which breaks through the traditional indexing strategy, ensures the high-performance of lookup, and provides personalized service for enterprises .In addition, a great number of data retrieval is repetition work,20% of keywords covers about 80% of retrieval. How to optimized strategy of cache to acquire the better search performance is an important research direction in the future.

References

1. Brin, S., Page, L.: The anatomy of a large scale hypertextual web search engine. In: 7th International WWW Conference (1998)
2. Justin, Z., Moffat, A.: Inverted Files for Text Search Engines. ACM Computing Surveys 38(2), 1–56 (2006)
3. Dean, J., Ghemawat, S.: MapReduce: Simplied Data Processingon Large Clusters. In: Proc. of OSDI 2004, pp. 137–150 (2004)
4. Rowstron, A., Druschel, P.: Pastry: Scalable, distributed object location and routing for large scale peer-to-peer systems. In: Proc. IFIP/ACM Middleware, Heidelberg, Germany (November 2001)
5. Clarke, I., Sandberg, O., Wiley, B., Hong, T.: Freenet: A distributed anonymous information storage and retrieval system. In: Proc. of the ICSI Workshop on Design Issues in Anonymity and Unobservability, Berkeley, CA (June 2000)
6. The Apache Jakarta Project: Lucene, `http://lucene.apache.org/`
7. Apache Hadoop. Hadoop [EB/OL]. [2009203206], `http://hadoop.apache.Org/`
8. Zobel, J., Moffat, A.: Inverted files for text search engines. ACM Computing Surveys 38(2), Article 6 (2006)

The Sliding Mode Variable Structure Control for Double Inverted Pendulum System Based on Fuzzy Reaching Law

SuYing Zhang[1], ShuMan Shao[1], Ran An[2], Sun Feng[1], and Yun Du[1]

[1] Hebei University of Science & Technology, Electrical Information College, Shijiazhuang, China
zhsy8985@sina.com, shaoshuman@163.com
[2] Electronic Information Department, Yanjing Vocational and Technical College, Langfang, China

Abstract. A new method about the sliding mode variable structure based on fuzzy reaching law is proposed and it can be applied to double inverted pendulum system. Not only can this method adopt fuzzy reaching law control the double inverted pendulum effectively, also it can significantly weak chattering and improve the speedingness and robustness of the system. Simulation results show the way is possible and effective. The method of using fuzzy reaching law system is better than traditional variable structures.

Keywords: Sliding mode variable structure, Double inverted pendulum system, Fuzzy control, Reaching law, Weaking chattering.

1 Introduction

In the 1950s, Emelyanov, a soviet research, first proposed variable structure control (VSC) based on sliding mode (SM).Then Utkin and Itkis ect further developed the theory. VSC with SM, which is the primary method of VSC has strong robustness and ability of anti-jamming .It has good dynamic performance for the control system that is uncertain, and without online identification. So it has drawn more and more people's attention. However, this method has discontinuous switching characteristic. When sampling time is too long, there will be superimposing on zigzag paths on the Smooth mode, and formed chattering. Owing to the chattering born of the variable structure itself, this also hinders sliding model control method widely used. Therefore, scholars from various countries use all sorts of methods ,such as fuzzy control, adaptive control, neural network control method combined with the sliding mode variable structure, in order to improve its shortcomings and performance. This paper also focuses here.

This paper mainly discusses the fuzzy control and variable structure method for combining, and designs a sliding mode controller that bases on fuzzy reaching law, then the simulation of double inverted pendulum control. The results of simulation explain that this method not only ensures fast and retains system strong robustness, but also inhibits chattering effectively.

D. Jin and S. Lin (Eds.): Advances in EECM Vol. 1, LNEE 139, pp. 123–129.
springerlink.com

2 Double Inverted Pendulum System Model

The masse of the cart: M=1.096Kg ; the masse of pendulum1 : m_1=0.05Kg ; the masse of pendulum2: m_2=0.13Kg ; the masse block m_3=0.236 Kg ; the length from the gravity center of the pendulum1 to its pivot: l_1=0.0775m ; the length from the gravity center of the pendulum2 to its pivot: l_2=0.25m ; The angles of the pendulum1and pendulum2 from their upright positions are denoted separately as θ_1 and θ_2;

Suppose no friction exists in the pendulum system. Then the dynamic equation of such a double inverted pendulum system can be obtained by Lagrange's equation. The state equation and output equation of the double inverted pendulum:

$$\begin{bmatrix} \dot{x}_1 \\ \dot{x}_2 \\ \dot{x}_3 \\ \dot{x}_4 \\ \dot{x}_5 \\ \dot{x}_6 \end{bmatrix} = \begin{bmatrix} 0 & 0 & 0 & 1 & 0 & 0 \\ 0 & 0 & 0 & 0 & 1 & 0 \\ 0 & 0 & 0 & 0 & 0 & 1 \\ 0 & 0 & 0 & 0 & 0 & 0 \\ 0 & 86.69 & -21.62 & 0 & 0 & 0 \\ 0 & -40.31 & 39.45 & 0 & 0 & 0 \end{bmatrix} \begin{bmatrix} x_1 \\ x_2 \\ x_3 \\ x_4 \\ x_5 \\ x_6 \end{bmatrix} + \begin{bmatrix} 0 \\ 0 \\ 0 \\ 1 \\ 6.64 \\ -0.088 \end{bmatrix} u \tag{1}$$

Here u is the control input, state variables: $x_1 = x, x_2 = \theta_1, x_3 = \theta_2, x_4 = \dot{x}, x_5 = \dot{\theta}_1, x_6 = \dot{\theta}_2$

By calculated, the system is observable and controllable. But because the system has right half plane poles, it is natural unstable system.

3 Sliding Mode Variable Structure Controller

The design of controller has two main parts. one is the design of sliding surface, which determine the switching vector function; the other is the design of sliding mode control law.

3.1 Design of Sliding Surface

As shown in Eq. (1), stated equation of double inverted pendulum is:

$$\dot{x} = Ax + Bu \tag{2}$$

x is state vector, $x \in R^n$, u is Control vector, $u \in R^m$.

Here, s is switching vector function. For the double inverted pendulum system, we can derive $s = Cx = 0$, $s \in R^m$. We can set $c = [c_1 \ c_2 \ c_3 \ c_4 \ c_5 \ c_6]$.

Then
$$s = Cx = c_1x_1 + c_2x_2 + c_3x_3 + c_4x_4 + c_5x_5 + c_6x_6 \tag{3}$$

We select as:

$$T=\begin{bmatrix} I_{n-m} & -B_1B_2^{-1} \\ 0 & I_m \end{bmatrix}=\begin{bmatrix} 1 & 0 & 0 & 0 & 0 & 0 \\ 0 & 1 & 0 & 0 & 0 & 0 \\ 0 & 0 & 1 & 0 & 0 & 0 \\ 0 & 0 & 0 & 1 & 0 & 1/0.088 \\ 0 & 0 & 0 & 0 & 1 & 6.64/0.088 \\ 0 & 0 & 0 & 0 & 0 & 1 \end{bmatrix} \tag{4}$$

The system is taken non-singular linear transformation as $\overline{x}=Tx$.

Equation (2) can be written as:

$$\begin{aligned} \dot{\overline{x}} &= \overline{A}x+\overline{B}u \\ \overline{s} &= \overline{C}x \end{aligned} \tag{5}$$

By Eq. (4) and Eq.(5) ,we can get

$$\overline{A}=TAT^{-1}=\begin{bmatrix} 0 & 0 & 0 & 1 & 0 & \vdots & -11.4 \\ 0 & 0 & 0 & 0 & 1 & \vdots & -75.5 \\ 0 & 0 & 0 & 0 & 0 & \vdots & 1 \\ 0 & -458.1 & 448.3 & 0 & 0 & \vdots & 0 \\ 0 & -2954.9 & 2955.1 & 0 & 0 & \vdots & 0 \\ \cdots & \cdots & \cdots & \cdots & \cdots & \vdots & \cdots \\ 0 & -40.3 & 39.5 & 0 & 0 & \vdots & 0 \end{bmatrix}=\begin{bmatrix} \overline{A}_{11} & \overline{A}_{12} \\ \overline{A}_{21} & \overline{A}_{22} \end{bmatrix} \tag{6}$$

$$\overline{B}=TB=\begin{bmatrix} 0 \\ B_2 \end{bmatrix}=\begin{bmatrix} 0 \\ 0 \\ 0 \\ 0 \\ 0 \\ \cdots \\ -0.088 \end{bmatrix}=\begin{bmatrix} \overline{B}_1 \\ \overline{B}_2 \end{bmatrix} \tag{7}$$

$$\overline{C}=CT^{-1}=\begin{bmatrix} \overline{C}_1 & \overline{C}_2 \end{bmatrix}, \text{ we set } \overline{C}_2=1 \tag{8}$$

From Eq. (5) and Eq.(8) , we have

$$\overline{s}=\overline{C}_1\overline{x}_1+\overline{C}_2\overline{x}_2=0 \tag{9}$$

From Eq.(5) and Eq.(6) ,we have

$$\dot{\overline{x}}_1=\overline{A}_{11}\overline{x}_1-\overline{A}_{12}\overline{x}_2 \tag{10}$$

From Eq.(11) and Eq.(12), sliding mode equation can be written as

$$\dot{\bar{x}}_1 = (\bar{A}_{11} - \bar{A}_{12}\bar{C}_1)\bar{x}_1 \tag{11}$$

Contrasting state equation of LQR control in the closed-loop system, it is obvious that $\bar{C}_1$ equivalent feedback matrix K. So switching surface design will be considered as the LQR problem of state feedback matrix. We select a set of the weighted matrix, which can make the double inverted pendulum system basic stably.

$$Q = \begin{bmatrix} 350 & 0 & 0 & 0 & 0 \\ 0 & 500 & 0 & 0 & 0 \\ 0 & 0 & 500 & 0 & 0 \\ 0 & 0 & 0 & 1 & 0 \\ 0 & 0 & 0 & 0 & 1 \end{bmatrix}, \quad R = 1 \tag{12}$$

State feedback matrix can be obtained:

$K = [-18.7083 \quad -18.2318 \quad 242.9403 \quad -29.2614 \quad 5.5587]$ then

$\bar{C} = [\bar{C}_1 \quad \bar{C}_2] = [K \quad 1] = [-18.7083 \quad -18.2318 \quad 242.9403 \quad -29.2614 \quad 5.558 \quad 1]$

3.2 Design of Control Law

In this paper, we choose exponential reaching law:

$$\dot{s} = -\varepsilon \operatorname{sign} s - ks, \varepsilon > 0, k > 0 \tag{13}$$

Differential is taken in both sides of $s = Cx$, and considering system mode at the same time. We can have

$$\dot{s} = C\dot{x} = CAx + CBu \tag{14}$$

Comparing (13) and (14) can be obtained:

$$u = -(CB)^{-1}[CAx + \varepsilon \operatorname{sign} s + ks] \tag{15}$$

However, due to the nature of the sliding mode variable structure with switch characteristics, there is high-frequency oscillation in sliding movement. So, in fact, chattering is bound to exist. Although the control method that uses reaching law has certain chattering inhibition. When meeting strong interference, the control effect is still not ideal.

According to the above problems, this paper introduces fuzzy control. The parameters of reaching law are adjusted by fuzzy reasoning. We design the two-dimensional fuzzy controller, which is $s, \dot{s}$ for input, ε as output. Control strategy: when S or $\dot{s}$ is bigger, make ε increase, so that get a faster approach speed; when S or $\dot{s}$ is smaller, System is close to the sliding surface, make ε decrease. So that slow approach speed makes the system into sliding mode, and reduced the chattering strength.

3.3 Fuzzy Adjustment

Fuzzy sets as follows:

$$\underline{s} = \{NB, NM, NS, ZO, PS, PM, PB\}$$

$$\underline{\dot{s}} = \{NB, NM, NS, ZO, PS, PM, PB\}$$

$$\underline{\varepsilon} = \{NB, NM, NS, ZO, PS, PM, PB\}$$

Here $NB, NM, NS, ZO, PS, PM, PB$ are separately negative big, negative, negative small, zero, positive small, middle, positive big, respectively. the fuzzy field as :

$$\underline{s} = \{-6,-5,-4,-3,-2,-1,0,1,2,3,4,5,6\}$$

$$\underline{\dot{s}} = \{-6,-5,-4,-3,-2,-1,0,1,2,3,4,5,6\}$$

$$\underline{\varepsilon} = \{-6,-5,-4,-3,-2,-1,0,1,2,3,4,5,6\}$$

Table 1. Fuzzy control rule table

ε \ $\dot{s}$ / s	NB	NM	NS	ZO	PS	PM	PB
NB	PB	PB	PB	PB	PM	0	0
NM	PB	PB	PB	PM	PS	0	0
NS	PM	PM	PM	PS	0	NS	NS
ZO	PM	PM	PS	0	0	NM	NM
PS	PS	PS	0	NS	NS	NM	NM
PM	0	0	NM	NB	NB	NB	NB
PB	0	0	NM	NB	NB	NB	NB

4 The Simulation Results and Analysis

The initial state: X=[0.1 0.1 0.01 0 0 0] ; the desired state : X=[0 0 0 0 0 0] ; the sampling time: T=0.02s ;Simulation time is 20s

1) The simulation result is exponential reaching law variable structure control without adding interference to the system as shown in figure 1.
2) The simulation result is exponential reaching law with adding interference to the system as shown in figure 2.
3) The simulation result is variable structure control that reaching law is adjusted by fuzzy reasoning with adding interference to the system as shown in figure 3.

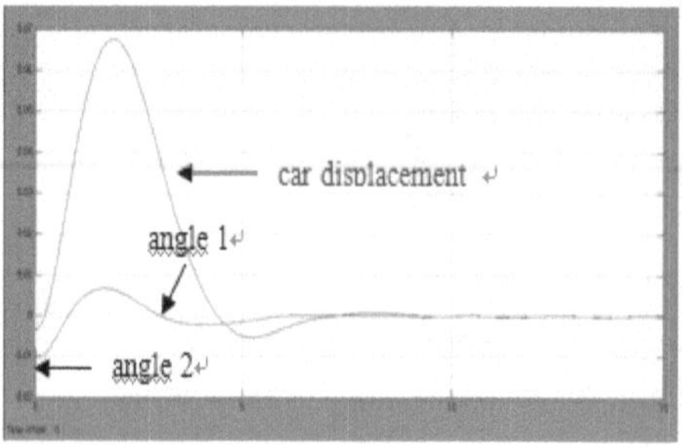

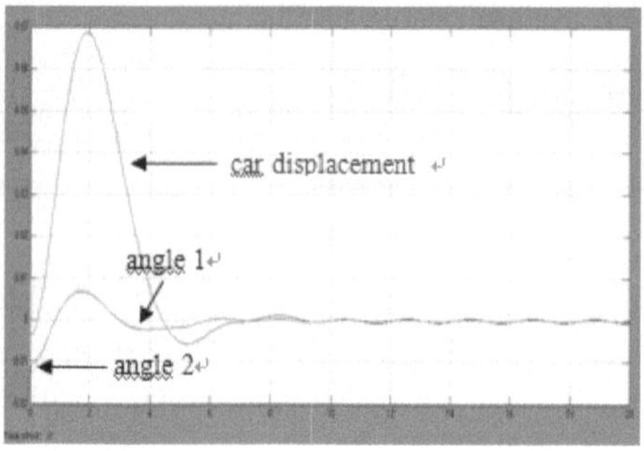

Fig. 1. Control result for exponential reaching law variable structure control without interference

Fig. 2. Control result for exponential reaching law variable structure control with interference

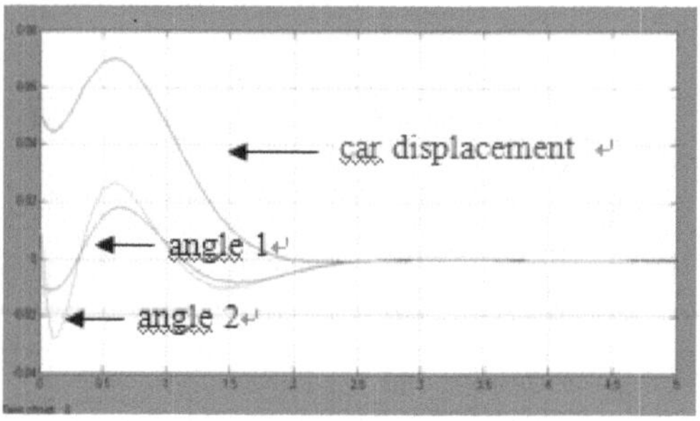

Fig. 3. Control result for fuzzy reaching law variable structure control with interference

Comparing three figures can be seen that only the reaching law of variable structure system is able to suppress chattering. But when the system is added interference, the control result is not satisfactory. And there are different levels of chattering. To solve the above problems, joining the fuzzy control adjust reaching law. So that, the ability of suppressing chattering and anti-interference is improved markedly. This shows that the method of variable structure based on fuzzy reaching law can significantly weak chattering and improves dynamic performance. when robustness of the system. When there is interference system has good robustness.

5 Conclusion

This paper design variable structure controller based on fuzzy reaching law .this control method is not only simple, but also have good control effect. The double inverted pendulum can be control effectively by this method. Through the simulation experiment of the double inverted pendulum, we can see that the method can effectively suppress chattering, and have good robustness and speedingness.

References

1. Gao, B.: Variable Structure Control Theory. China Science and Technology Press, Beijing (1990)
2. Sturegeon, W.R., Loscutoff, M.V.: Application of Modal Control and Dynamic Observers to Control of a Double Inverted Pendulum. In: Proc. JACC, Stanford, pp. 857–865 (1972)
3. Cui, P., Wen, Z.: Based on State Space Pole Placement Control of Inverted Pendulum. Laboratory Research and Exploration 22(2), 70–72 (2003)
4. Yuan, X., Jiang, X.: Based on Sliding Mode Variable Structure of The Inverted Pendulum Stability Control. Control Theory and Application 21(5), 720–723 (2004)
5. Muchammad, R., Odaka, T., Ogura, H., Takahama, T.: A Study of Knowledge Representation of Fuzzy Control for Inverted Double Pendulum System by Applying Adaptive Control with Range Scaling Method. Japan Soc. Fuzzy Theory and Systems 8(3), 576–585 (1996)
6. Saez, D., Cipriano, A.: Design of Fuzzy Model Based Predictive Controllers and Its Application to An Inverted Pendulum. In: Proceedings of the Sixth IEEE International Conference on Fuzzy Systems, July 1-5, vol. 2, pp. 915–919 (1997)
7. Yamakawa, T.: Stabilization of an Inverted Pendulum by a High-speed Fuzzy Logic Controller Hardware System. Fuzzy Sets and Systems 32, 161–180 (1989)

Study on E-satisfaction in the Consumer E-commerce Environment Based on TAM and TTF Extended Model

Yaqin Li[1,2], Jianjun Sun[1], and Yuequan Yang[2]

[1] Department of Information Management, Nanjing University, Nanjing, China
yqli@yzu.edu.cn
[2] College of Information Engineering, Yangzhou University, Yangzhou, China

Abstract. The purpose of this study is to propose an extended model of e-satisfaction in the consumer e-commerce environment. The model is based on technology acceptance model (TAM) and task-technology fit (TTF) model. The study proposes a conceptual model that the e-satisfaction is direct or indirect influenced by several constructs: perceived value, fulfilment, e-trust, social presence, perceived enjoyment besides TAM and TTF. The interrelationship between these constructs is explained. Empirical study of this extended model is expected in future further research.

Keywords: TAM, TTF, e-satisfaction, consumer e-commerce, extended model.

1 Introduction

With the rapid development of e-commerce, online shopping has been one of the most favorite shopping means because of its compelling advantages such as lower cost structure, greater flexibility, more convenience and faster transaction. However, in order to attract users repurchasing in the consumer e-commerce environment, how to keep and raise the customers' e-satisfaction is more difficult than that of the brick-and–mortar context. Because of this, the importance of e-satisfaction has been a critical and hot issue of research in the stream of information system and business, especially marketing [1,2,3,4,5,6].

To explain the factors of impacting e-satisfaction in the consumer e-commerce environment, online shopping as an example of technology adoption is viewed in this paper. Many e-commerce researches have used technology adoption as a theoretical foundation [7,8]. Two models of workplace technology adoption, which are technology acceptance model (TAM) and task-technology fit (TTF) model, are selected as our research's basis and further explain them. In order to further explore the antecedents of the e-satisfaction, some important constructs are integrated into a new theoretical extended model including perceived enjoyment, social presence, perceived value, fulfilment and e-trust. The reason why use these constructs will be given in the following study. The key purpose of this study is to build an extended model about e-satisfaction to examine users' consuming behaviours in consumer e-commerce

D. Jin and S. Lin (Eds.): Advances in EECM Vol. 1, LNEE 139, pp. 131–136.
springerlink.com

environment, further to help managers understand online customers' e-satisfaction formation and repurchase behaviour, and retain long-terms customers. Yet, to our knowledge, few studies have explored e-satisfaction with extended TAM/TTF up to now. The rest of the paper is organized as follows. We first review the TAM and TTF model, and then develop a conceptual extended e-satisfaction model with hypotheses. The final section, we draw some conclusions and point out future research.

2 TAM and TTF Model

2.1 TAM

The TAM has become one of the most extensively used theories in information science research [8,11,12]. The TAM, which is used to interpret and predict users' information technology adoption behaviors, is based on the theory reasoned action (TRA) argued that beliefs influence attitudes, which lead to intentions, and finally to behaviours. In [10], two important determinants, perceived usefulness (PU) and perceived ease of use (PEOU) are proposed as illustrated in Fig.1. In [10], Davis defined PU as "the degree to which a person believes that using a particular system would enhance his or her job performance". In our study, we argue that PU may be defined as a compositive system, which can make a customer smoothly fulfil his or her online series tasks, i.e., to meet his or her needs, such as good information quality, on-time and accurate delivery, etc. Another important determinant is PEOU. In [10], it was defined as "the degree to which a person believes that using a particular system would be free of effort". In this paper, we see it as the electronic commerce website has good information system design, facilitates customers interaction with the website and provides good quality e-service. Some researchers also suggest that the two constructs, PU and PEOU, have some external variables including information quality, website system quality, service quality [3,4,11] and perceived enjoyment [12].

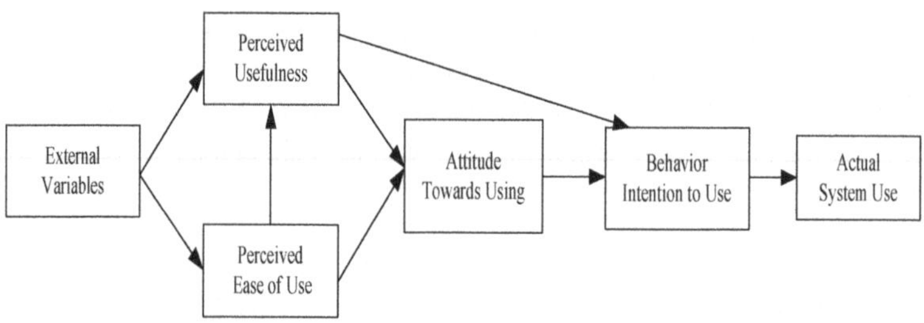

Fig. 1. Technology Acceptance Model (TAM) [9]

2.2 TTF Model

In [13], Goodhue&Thompson has proposed the TTF model, which highlights that the importance of the fit between technology and users' tasks in achieving individual performance impacts from information technology, as shown in Fig.2. In [7], fit or compatibility affects PU and attitude toward use was demonstrated. At the same time, tasks and technology affect TTF, and TTF affects PEOU and actual use [14].Furthermore, TTF model is a valuable addition to TAM for online shopping tasks [8]. Shopping task was defined as "a combination of both the purchase and the product information search activities" [8]. However, in this paper, e-commerce task is a series of courses of completing online shopping from searching for products or services information to receiving products or services which customers wanted. This concept of e-commerce tasks includes aspects of both online transaction and offline fulfilment.

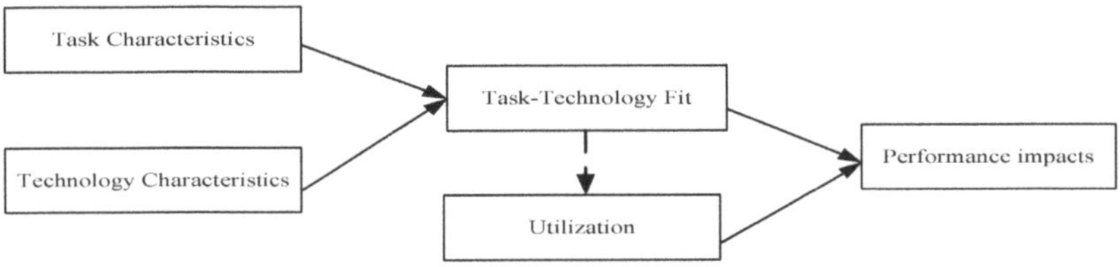

Fig. 2. Task-Technology Fit Model (TTF) [13]

3 Extended Model: Conceptual Framework and Hypotheses

Our research model is derived from TRA. The extended model views e-satisfaction as influenced by direct and indirect relationships between fulfilment, TAM, perceived value, e-trust, perceived enjoyment, TTF and social presence. As depicted in Fig.3, the model postulates that fulfilment affects e-satisfaction directly. TAM, perceived value and e-trust are believed to affect fulfilment. Furthermore, the TAM influences perceived enjoyment, TTF and social presence. The constructs and motivate the paths in the extended model will be discussed in the following section.

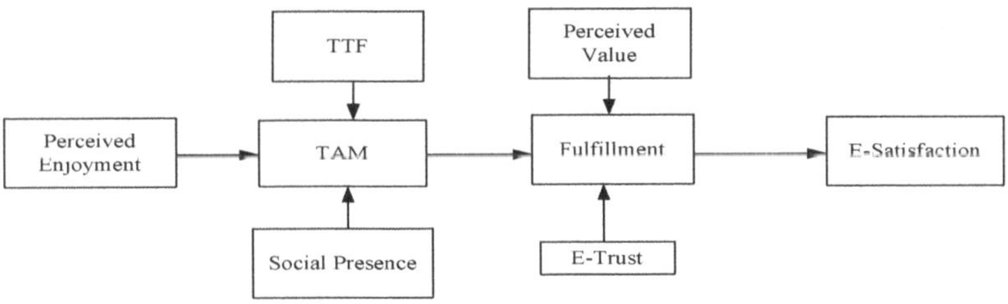

Fig. 3. Extended TAM/TTF E-Satisfaction Model

3.1 Fulfillment as Antecedent of E-satisfaction

E-satisfaction leads to repurchase intention, induces customers' e-loyalty [3]. Many e-commerce websites want to raise customers' e-loyalty through meeting e-satisfaction. Because of this, the research about e-satisfaction has been a hot spot, which induces different research fields' interest such as information system and marketing disciplines [8,11,12,15]. E-satisfaction was defined as "the contentment of the customer with respect to his or her prior purchasing experience with a given electronic commerce firm" [16].In this paper, the same definition about e-satisfaction was employed.

In the paper, fulfillment is defined as the whole purchasing course, including from the beginning searching for products/services information to the last transaction of receiving products/services which the customer wanted and after-sales services, attains customers' objection perfectly. Good fulfilment is deemed to raise customers' e-satisfaction, leads to purchasing constantly. Some researchers also found that fulfilment influences e-satisfaction deeply [3,4]. So the following hypothesis is formulated:

H1. Fulfilment has a positive effect on e-satisfaction.

3.2 TAM, Perceived Value and E-trust as Antecedent of Fulfillment

PU and PEOU are two important constructs in TAM. Three dimensions of website quality, i.e., information quality, system design quality, service quality, do influence PU and PEOU [3,11]. Furthermore, PEOU benefits customers from searching information, finishing transaction smoothly. i.e., PEOU can propel fulfilment in the consumer e-commerce. Namely, PEOU has a positive effect on fulfilment. At the same time, PU can strengthen customers to utilize e-commerce system to fulfil their needs [8]. i.e., PU has a positive effect on fulfilment. Therefore, the second hypothesis is given:

H2. TAM has a positive effect on fulfilment.

Nowadays, more and more e-commerce businesses have known that establishing and maintaining a long-term relationship with customers are very important. Trust in the e-commerce (e-trust) has been conceptualized as "customers confidence in the quality and reliability of the products/services which e-commerce websites offered" [4]. Because of e-commerce's virtuality, customers perceived a higher level risk with online shoppers than traditional shoppers. Thus, online customers would prefer to transact with e-commerce companies which they can trust [4]. E-trust has a positive effect on users' adoption intentions of online shopping [12]. Furthermore, we can suppose that e-trust can induce good fulfilment. Then the third hypothesis is proposed:

H3. E-trust has a positive effect on fulfilment.

Perceived value is defined as customers perceive products' or services' value, which is on the basis of customers' evaluation about the whole transaction course. The effect of perceived value on repurchase intentions is completely mediated through satisfaction [17]. Customers perceived value may propel fulfilment in the consumer e-commerce surroundings. In line with the argument, the fourth hypothesis is given:

H4. Perceived value has a positive effect on fulfillment.

3.3 Perceived Enjoyment, TTF and Social Presence as Antecedent of TAM

Perceived enjoyment refers to the extent to which the activity of interacting with the consumer e-commerce website is perceived to be enjoyable or playful [12]. Namely, customers have a good online shopping experience. When customers purchase in the consumer e-commerce website with a perfect purchasing experience—perceived enjoyment, they will think the e-commerce website as PU and PEOU [12]. Therefore, the fifth hypothesis is shown:

H5. Perceived enjoyment has a positive effect on TAM.

TTF model's key is the customers' task and information system fitted well. If customers' online shopping tasks fit the e-commerce systems well which websites provided, customers will produce PU and PEOU. TTF has a positive effect on PU and PEOU [8]. In line with this, the sixth hypothesis is given:

H6. TTF has a positive effect on TAM.

Social presence is defined as social communication with others, let persons not alone, i.e., being together with others. Users' feelings of social presence are formed through communication processes. In other words, social presence is a subjective measure developed over the entire course of the interactions including websites system, communication contents and their possible courses of action [12]. Social presence is particularly relevant in the consumer e-commerce environment. In order to enhance customers' social presence feelings, consumer e-commerce websites should offer customers services through reviews, discussion forums, live help and instant messaging technologies etc. to promote more social interaction. Social presence makes customers feel the e-commerce system usefulness and ease of use [12]. Therefore, the seventh hypothesis is proposed:

H7. Social presence has a positive effect on TAM.

4 Conclusions and Future Research

An extended TAM/TTF e-satisfaction model is developed to explain customers consuming satisfactory behaviour in the consumer e-commerce surroundings. In addition to the traditional information system adoption TAM/TTF variables, other constructs are proposed to develop our extended model. These constructs include fulfilment, e-trust, perceived value, perceived enjoyment and social presence. The value of this extended model may help e-commerce websites and their operation businesses further understand customers' e-satisfaction to improve their marketing strategies and websites services. However, the extended model should be examined by empirical investigations in future research. At the same time, we may further explore the e-satisfaction constructs to improve the extended model in order to better explain customers' online shopping behaviors.

Acknowledgement. This work is partially supported by the National Natural Science Foundation of China under Grants No. 60874045, and the Natural Science Foundation of the Jiangsu Higher Education Institutions of China under Grant No.10KJB510027.

References

[1] Bai, B., Law, R., Wen, I.: The Impact of Website Quality on Customer Satisfaction and Purchase Intentions: Evidence from Chinese Online Visitors. International Journal of Hospitality Management 27, 391–402 (2008)

[2] Chen, Q., Rodgers, S.: A Critical Review of the E-Satisfaction Literature. American Behavioral Scientist 52, 38–59 (2008)
[3] Lee, H., Choi, S.Y., Kang, Y.S.: Formation of E-satisfaction and Repurchase Intention: Moderating Roles of Computer Self-efficacy and Computer Anxiety. Expert Systems with Applications 36, 7848–7859 (2009)
[4] Kim, J., Jin, B., Swinney, J.L.: The Role of Etail Quality, E-satisfaction and E-trust in Online Loyalty Development Process. Journal of Retailing and Consumer Services 16, 239–247 (2009)
[5] Thirumalai, S., Sinha, K.K.: Customization of the Online Purchase Process in Electronic Retailing and Customer Satisfaction: An Online Field Study. Journal of Operations Management 29, 477–487 (2011)
[6] Wang, Y.J., Hernandez, M.D., Minor, M.S.: Web Aesthetics Effects on Perceived Online Service Quality and Satisfaction in an E-tail Environment: the Moderating Role of Purchase Task. Journal of Business Research 63, 935–942 (2010)
[7] Chen, L., Gillenson, M., Sherrell, D.: Enticing Online Consumers: an Extended Technology Acceptance Perspective. Information and Management 39, 705–719 (2002)
[8] Klopping, I.M., McKinney, E.: Extending the Technology Acceptance Model and the Task-technology Fit Model to Consumer E-commerce. Information Technology, Learning, and Performance Journal 22, 35–48 (2004)
[9] Davis, F.D., Bagozzi, R.P., Warsaw, P.R.: User Acceptance of Computer Technology: A Comparison of Two Theoretical Models. Management Science 35(8), 983–1003 (1989)
[10] Davis, F.D.: Perceived Usefulness, Perceived Ease of Use, and User Acceptance of Information Technology. MIS Quarterly 13(3), 319–340 (1989)
[11] Li, J.J., Sun, J.J.: Empirical Study on Website Quality, User Perception and Technology Adoption Behavior. Journal of the China Society for Scientific and Technical Information 30(3), 227–236 (2011) (in Chinese)
[12] Qiu, L.Y., Li, D.: Applying TAM in B2C E-Commerce Research: An Extended Model. Tsinghua Science and Technology 13(3), 265–272 (2008)
[13] Goodhue, D.L., Thompson, R.L.: Task-Technology Fit and Individual Performance. MIS Quarterly 19(2), 213–236 (1995)
[14] Dishaw, M.T., Strong, D.M.: Extending the Technology Acceptance Model with Task-Technology Fit Constructs. Information & Management 36(1), 9–21 (1999)
[15] Li, B.Q., Sun, J.J., Chen, Y.: An Approach to Users' Utilization of Networked Academic Information Resources Based on the Extended TAM. Journal of the China Society for Scientific and Technical Information 27(4), 596–606 (2008) (in Chinese)
[16] Anderson, R.E., Srinivasan, S.S.: E-satisfaction and E-loyalty: A Contingency Frame Work. Psychology & Marketing 20(2), 123–138 (2003)
[17] Patterson, P.G., Spreng, R.A.: Modelling the Relationship between Perceived Value, Satisfaction and Repurchase Intentions in a Business-to-Business, Services Context: an Empirical Examination. International Journal of Service Industry Management 8(5), 414–434 (1997)

The Application Research about Data Warehouse Based on ERP

XuWen Guo, Min Chang, YaHui Dong, and LianFeng Zhang

Economics and Management
Henan Institute of Science and Technology,
Xinxiang, China
guoxuwen308@yahoo.com.cn

Abstract. This paper describes the limitations of enterprise ERP system, on this basis, proposes the application pattern of ERP-based data warehouse, discusses the build process of ERP-based data warehouse, and summaries the implementation of the system elements.

Keywords: Data Warehouse, ERP, Decision support, OLAP.

1 Introduction

AS an important tool of enterprise information management, ERP system is a typical online transaction processing system, its database and application software is partial to transaction, another word, the business operations and business basic transactions are emphasized to ERP system. With the social development and increasing business competition, more and more companies recognize that: the correct and timely decision-making is the key to the survival and development. However, the existing ERP system can not use their accumulated large amounts of data to effectively support enterprise senior making decision. ERP-based data warehouse system is the best solution of improving corporate decision-making capacity.

2 Paper Preparation

2.1 Enterprise ERP System Limitations

With the application of enterprise ERP system, time of difficulty to enterprise information integration and lack of enterprise information has been outdated. We have entered into the decision support information explosion and increasingly urgent demand of the times. ERP systems can not meet the needs of enterprise management decision-making, its limitations mainly as follows:

The Low Use of Enterprise Data

ERP system, although the data within the enterprise to do a very good finishing and automation, has limited use of enterprise data, a lot of useful information on business resources are idle and wasted.

D. Jin and S. Lin (Eds.): Advances in EECM Vol. 1, LNEE 139, pp. 137–140.
springerlink.com

The Information Needs of Difficulty to the Senior Decision-Making

ERP system mainly supports business operations and daily transaction processing. It only provides the current business operations of the job-level data and surface information. But enterprise senior pays more attention to the comprehensive information, historical information, and information reflected the whole process of business playing in order to finding help decision-making trends.

The Unachieved Decision Support Functions of Enterprise Expects

At present, the existing ERP system is far from achieving the desired business decision support, ERP system has turned the decision-making needs of the enterprise into a variety of statistical queries, and these features into its various modules achieved. This is not conducive to ERP system upgrades and maintenance, and difficult to the use of OLAP technology.

2.2 Data Warehouse System Based on ERP

The Architecture of Data Warehouse System Based on ERP

The building of data warehouse systems based on ERP may improve decision support capabilities of ERP system by accumulating data to ERP systems as the main data source of data warehouse, and be effectively composed of the background of decision support system by using of Internet technology, OLAP technology, data mining technology and so on. The application pattern of data warehouse system based on ERP is shown in Figure 1: The application model consists of three main parts:

(1)Part of the data sources, including ERP system, the data generated in the various subsystems and external data sources and other documentation, etc

(2)Data conversion part, it may achieve the conversion processing by collecting data. During data conversion data is checked integrity and validity to ensure the quality of data stored in the data warehouse, and is loaded and updated according to the rules of plan and conduct;

(3) Decision-support section, including data warehouse, knowledge base, inference base, OLAP analysis services, DM services, which is the core of the whole system.

The application about data warehouse of based on ERP as follows: lots of data in the ERP of each subsystem are turned into analysis data, and loaded in data warehouse by extracting, cleansing, converting, loading and other processes. According to the actual needs of different sectors, the data in data warehouse is loaded into different data marts by extraction, synthesis, and the data in data warehouse or data marts is analyzed and processed by OLAP and data mining technology, analysis and processing results can be available to users, but also can be used as decision support systems knowledge into the knowledge base, knowledge into the knowledge base may generate inference results and provide for decision makers through the inference rules.

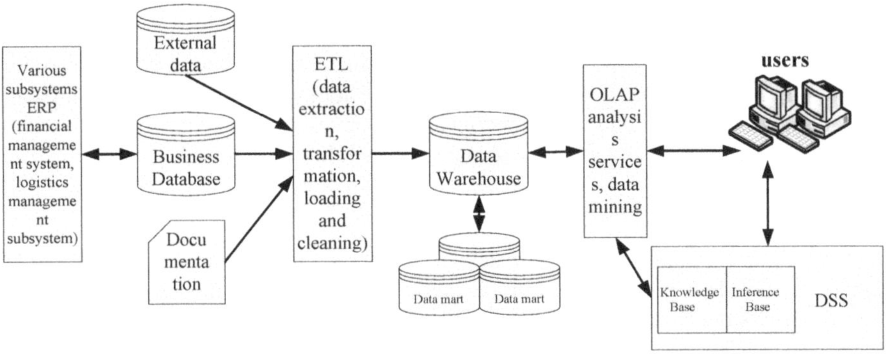

Fig. 1. The architecture of data warehouse system based on ERP

The Build Process of Data Warehouse Based on ERP

Data from the various subsystems in ERP is converted to analytical data as data base of data warehouse through data process during the Construction of ERP-based data warehouse. Data Warehouse is a continuous loop, leaving the system to further improve the feedback process o building a data warehouse. It can be divided into several steps as shown in Figure 2:

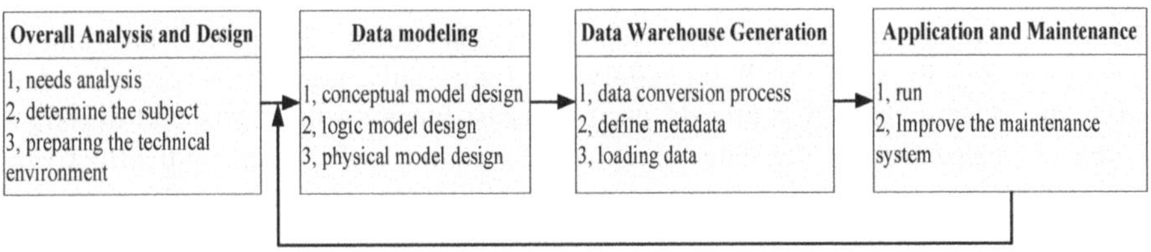

Fig. 2. The build process of data warehouse based on ERP

2.3 Problems Concerned about the Comstruction of Data Warehouse Systems Based on ERP

The data warehouse systems based on ERP are established for the purpose of turning a better transaction data in the ERP and external data into information needed for management decisions and the ability to multi-dimensional depth of these data mining. Establishment of a data warehouse based ERP system needs to pay attention to the following questions:

The Validity of Data Sources

Part of the data is poor availability in the accumulated details data through the running ERP system, Therefore, the data warehouse based on ERP system can load data using a variety of ways, may not be entirely dependent on the existing ERP system.

The Data Conversion Problems

Data conversion should follow the simple, easy principle, the general data conversion must spend a lot of time, so you can set up some temporary database used for data extraction, cleaning and loading. Data conversion means can be established in the data warehouse to do before and after the change in flexibility.

Data Warehouse Design

Data warehouse design includes the design of data warehouse model, OLAP model design and the design of data mining models. Data warehouse design is first started from the user's decision problem, the storage structure of data warehouse is designed according to decision-making theme ; OLAP model design must determine the required dimensions and variables for the decision metrics, define the relationship between decision-making model for the decision theme , and form the star structure and the snow structure, on this basis, multi-dimensional data table may be resulted, and multi-dimensional database can be established; Data mining model design need to identify variables associated with the target data, create a new variable on the basis of the original variables, train and verify for data collection, and prepare for the realization of prediction.

3 Conclusion

ERP data warehouse is an important direction about enterprise information technology development, ERP is mainly used for daily transaction process, data warehouse management is primarily used to assist business managers and decision-making. Construction of the data warehouse based on ERP system will effectively improve the management level and core competitiveness and bring great economic benefits to the enterprise.

References

1. Zhao, H., Yang, J.P.: ERP and business intelligence integration. Integrated Management (09) (2008)
2. Lin, H.: ERP data warehouse environment research. Science and Technology Information (02) (2008)
3. Li, S.-J.: Based on the ERP data warehouse system design. Accounting Monthly (04) (2009)
4. Zhang, M., Chen, W.: Business intelligence system design based on ERP. Mudanjiang Institute of Education (04) (2008)
5. Chen, Z.: Development of business intelligence data warehouse support for ERP applications. Mechanical Management and Development (04) (2009)
6. Chen, Z.: Business process optimization based on enterprise business intelligence technology. Master Thesis, vol. (11). Wuhan University of Technology (2007)

The Development of Intelligent Knowledge and Information Management System in E-commerce Based on Fuzzy Cognitive Map and Ontology

Qing Duan*

Office of Academic Affairs, The Central Institute For Correctional Police, Baoding 071000, Hebei, China

Abstract. Ontologies are formal, explicit specifications of shared conceptualizations of a given domain of discourse. Fuzzy cognitive map(FCM) is a modeling and simulation methodology describing on an abstract conceptual representation any system. Finally, the intelligent knowledge and information management system in E- Commerce is proposed based on ontology and FCM. Our experiments showed that the proposed methodology allows for the derivation of more significant ontology terms and more useful ontology rules from the information system. The experimental results indicate that this method has great effective promise.

Keywords: intelligent knowledge, ontology, fuzzy cognitive map, information management.

1 Introduction

Ontology, which is a theory about the nature of existence in philosophy, plays an important role in the integrating heterogeneous information. Ontologies also play an important role in biomedical informatics and in knowledge management. The conceptual architecture is composed of layers with different functionality representing the abstraction levels employed and the connections among them. Ontologies are frequently formalized with knowledge representation structures, e.g. frame-based systems where concepts are represented in frames and relationships between concepts as frame slots. The same web content needs to be rendered differently on different mobile devices. The same web content needs to be rendered differently on different mobile devices. The HMDT platform serves as a transcoding service broker, which uses ontology-based metadata and web services technologies.

TFIDF and its variations have been widely used in the applications such as Text Summarization. This process is hard to general users, however; so is it to the experts. Two major factors are behind this difficulty. In this work, our approach is implemented essentially by developing such an ontology which starts with a conceptualization of general systems (i.e. the first step)[1]. Cognitive maps can often

* Author Introduce: DUAN Qing(1981.12-), Male, Han, Master of education technology of Northeast Normal University, Research area: contemporary and long-range education.

D. Jin and S. Lin (Eds.): Advances in EECM Vol. 1, LNEE 139, pp. 141–146.
springerlink.com

be imprecise. We tend to classify and cluster the massively detailed cognitive spatial information we encounter using simplifications, such as the gathering of objects and landmarks into hierarchies and regions. A cognitive map (CM) is composed of concept nodes that represent the factors describing a target problem; arrows that indicate causal relationships between two concept nodes; and causality coefficients on each arrow indicating the positive (or negative) strength with which a node affects other nodes. Since fuzzy set theory and fuzzy logic have proved to be successful in handling imprecise and vague knowledge, they have been combined with the BSC leads to fuzzy balanced scorecard. Ontologies are formal, explicit specifications of shared conceptualizations of a given domain of discourse.

Domain ontology can help users locate and learn related information more effectively. This paper presents a novel method, which uses fuzzy cognitive networks (FCNs). Applied to the study of artificial intelligence (AI) ontology is a logical theory that gives an explicit, albeit, partial account of a conceptualized real-life system. Current domain-specific search engines do help users to narrow down the search scope by the techniques of query expansion, automatic classification and focused crawling; their weakness, however, is almost completely ignoring the user interests. The repository comprises collections of facts, rules, and procedures organized into schemas. E-learning is an alternative concept to the traditional tutoring system. The course tutor in a software tutoring system controls learners relatively weaker than (s)he does in the traditional one, where the (human) tutor is in charge of the contents and sequence of instructions.

Finally, this paper puts forward the methodology for intelligent knowledge and information management system in E- Commerce based on fuzzy cognitive map and ontology. FCM is a knowledge-based methodology suitable to model complex systems and handle information from an abstract point of view. The paper offers a methodology for building intelligent knowledge and information management system for knowledge sharing and reusing based on fuzzy cognitive map and ontology. FCM allow a set of identified causality coefficients to be organized in an adjacency matrix. Concepts and relationships are basic components in an ontology. Web documents are the most important source for deriving concepts and relationships.

Ontologies provide an unambiguous terminology that can be shared by all involved in a software development process. In this paper, an intelligent knowledge and information management system in E- Commerce based on fuzzy cognitive map and ontology is presented. The experimental results indicate that this method has great promise. Ontologies are used as references to annotate resources with concepts in standardized ways, e.g. in RDF. Our experiments showed that the proposed methodology allows for the derivation of more significant terms and more useful rules from the information database.

2 Ontology-Based Intelligent Knowledge and Information Management System

A class is used to represent a concept defined in the ontology. Classes are typically arranged in a taxonomy, including classes and subclasses. Both intelligent agent and semantic web service technologies are able to reach remarkable achievements and in some cases have overlapping functionalities.

Ontologies provide an unambiguous terminology that can be shared by all involved in a software development process. Historically, information retrieval was derived from well-organized search engine collections of textual based data. Ontology is one of the branches of philosophy, which deals with the nature and organization of reality. Applied to the study of artificial intelligence (AI) ontology is a logical theory that gives an explicit, albeit, partial account of a conceptualized real-life system.

The web services execution environment supports common B2B and B2C (Business to Consumer) scenarios, acting as an information system representing the central point of a hub-and-spoke architecture. Both intelligent agent and semantic web service technologies are able to reach remarkable achievements and in some cases have overlapping functionalities.

2.1 Intelligent Knowledge and Information System Based on Ontology

The research issue for intelligent information integration (III) has become ubiquitous and critically important with the increasing dependence on Internet/Intranet and information and communication technology. Since it was proposed in the early 1990s, it has demonstrated its suitability to assist decision making in management. However there could be more refined matching solutions based upon light semantic matching or heavy semantic matching for more rich ontologies, where the 'Reasoner' will come to process the logic constraints provided by the customer[2]. Therefore, there is a need for efficient KM systems that can organize and access knowledge contained in such corpora. Ontology building is a task that pertains to ontology engineers, an emerging expert profile that requires the expertise of knowledge engineers and domain experts. In this information-exploding era, the user expects to spend short time retrieving really useful information rather than spending plenty of time and ending up with lots of garbage information. First, the system calculates the TF-IDF and Entropy value of keywords, then will define the TF-IDF threshold θ_t and the Entropy threshold θ_e. Consider two concepts (E_1, I_1) and (E_2, I_2) of one or more contexts. Let n, m be the cardinalities of the sets I_1, I_2, respectively, i.e. $n = |I_1|$, $m = |I_2|$, and suppose that $n \leqslant m$. The set $\mathscr{P}(I_1, I_2)$ of the *candidate sets of pairs* is defined by all possible sets of n pairs of attributes defined as follows: Eqs1.

$$\begin{aligned}\mathscr{P}(I_1,I_2) = \{\{\langle a_1,b_1\rangle \cdots \langle a_n,b_n\rangle\} | a_h \in I_1, b_h \in I_2, \forall h = 1,\ldots,n, \text{ and} \\ a_h \neq a_k, b_h \neq b_l, \forall k,l \neq h\}.\end{aligned} \quad (1)$$

As one example of the importance of ontology quality consider the area of business process modelling. Ontology construction is a difficult and tedious work because the developing process needs vast of works, costs, and efforts of a lot of experts.

Each webpage profile reflecting a webpage describes how the webpage is interpreted by the domain ontology, while a website profile describes how a website is interpreted by the semantics of the contained webpages. Ontology based systems is more powerful if it is required to accommodate the growth of domain knowledge with time. It is for this reason that many are looking for more efficient methods for ontology reuse, extraction and extension, what we call tailoring or ontology tailoring, to be more specific.

These models are freely available as a modeling tool within the respective community and allow scientists to focus on their needs rather than building a model from scratch. In order to model the extracted domain knowledge an ontology was defined. This contains all the entities or concepts identified in the domain, together with the information about incidents and control actions mentioned above.

A customer can currently obtain services from a specific company through a variety of channels, including facsimile, email, company websites, and phone calls to the call center. The ontology is a computational model of some portions of the world. It is a collection of key concepts and their inter-relationships collectively providing an abstract view of an application domain.

2.2 Apply Fuzzy Cognitive Map to Build Knowledge Information System

In addition, a Fuzzy Cognitive Map (FCM) is proposed for modelling Critical Success Factors (CSFs). FCM is a knowledge-based methodology suitable to model complex systems and handle information from an abstract point of view. FCMs is a modeling and simulation methodology describing on an abstract conceptual representation any system[3]. FCM has been especially effective in resolving problems in which many relevant factors are causally interrelated with one another and that require decision-makers to analyze the causal relationships before solving the problems.

Our methodology requires concepts on at least three hierarchy levels in order to compare retrieval times for S1-type sentences against retrieval times for S2-type sentences. After evaluating the developed ontologies the conclusion was that they complemented each other well and that merging the ontologies would yield the best result. An FCM looks like a cognitive map, it consists of nodes (concepts) that illustrate the different aspects of the system's behavior. FCM development method is based on Fuzzy rules that can be either proposed by human experts and/or derived by knowledge extraction methods. Although formal procedures themselves are still essential to protect and safeguard the system, it is equally important for security administrators to make the presence of formal controls felt through attachment, commitment, involvement, and norms.

From an Artificial Intelligence perspective, FCMs are supervised learning neural systems, whereas more and more data is available to model the problem, the system becomes better at adapting itself and reaching a solution. Various types of studies apply CM. For instance, CM was robustly used to solve distributed decision process modelling in networks, decision analysis, stock investment analysis problems, and business process redesign.

3 Using Ontology and Fuzzy Cognitive Map Model to Build Information System

The popularity and dynamics of the Internet/Intranet is another source of widespread distribution, with information represented and stored in heterogeneous formats such as structured data, semi-structured data, and unstructured data. One of the great advantages of having situations represented in a formal language is that facts that are not explicitly stated can be derived using an inference engine[4].

Information systems executives consider the capability of IS infrastructure as one of the most important issues. Finally, the Augmented FCM approach has been adopted, because it doesn't need that experts change slightly their judgement for consensus as Delphi methodology. The equation 2 which calculates the similarity of data acquisition is as follows.

$$f = \frac{1}{1 + e^{-\lambda x}} \tag{2}$$

Where $\lambda > 0$ is a parameter determining its steepness. In this approach, the value $\lambda = 1$ has been used. This function is selected since the values A_i of the concepts, lie within [0, 1].Ontology provides common understanding of the domain knowledge and confirms common approbatory vocabulary in the domain, as well as gives specific definition of the relation between these vocabularies from formal model of different levels. Accessing the intelligent information sources separately without integration may lead to the chaos of information requested. It is also not cost-effective in EB settings. This data is re-directed by WSMX to the auction/negotiation web service that deals with the selection of the best offer. If there are common nodes, then the element w_{ij}^{Aug} in the augmented matrix is the equation 3 as follows.

$$w_{ij}^{\mathrm{Aug}} = \frac{\sum_{k=1}^{n} w_{ij}^{k}}{n}, \tag{3}$$

An FCM consists of nodes that represent the factors most relevant to the decision environment and arrows that represent different causal relationships among factors. One factor provides a direct positive or negative effect on another.

4 The Development of Intelligent Knowledge and Information Management System Based on Ontology FCM

Ontologies are formal, explicit specifications of shared conceptualizations of a given domain of discourse. The aim of ontology is to obtain, describe and express the knowledge of related domain. In this paper, we apply the technology of ontology and fuzzy cognitive map to automatically develop the concept hierarchy of knowledge and information management system in e-business and to match up the binary relation matrix of documents and terms to express the independence, intersection and inheritance between different concepts to form the concept relationship of ontology. Our contribution here is the message flow automation, matching and calculation of the winner by using a semantic web service platform. Fig.1 shows the detailed application of ontology and FCM in intelligent knowledge and information management system. The experimental results indicate that this method has great promise.

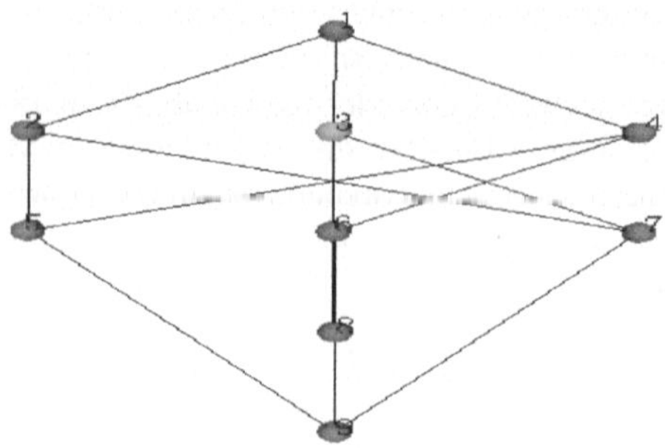

Fig. 1. The application results of ontology and FCM in knowledge information system

5 Summary

Fuzzy cognitive maps were introduced and allow us to provide fuzzy causation measures for the cognitive maps proposed. In this paper, we adopt ontology and fuzzy cognitive map to improve the performance of intelligent knowledge information from documents. The paper probes into building knowledge and information management system in e-business based on ontology and fuzzy cognitive map.

References

1. Berners-Lee, T., Hendler, J., Lassila, O.: The Semantic Web. Scientific American, 5 (2001)
2. Lee, S., Kim, B.G., Lee, K.: Fuzzy cognitive map-based approach to evaluate EDI performance: a test of causal model. Expert Systems with Applications 27(2), 287–299 (2004)
3. Chen, T.-Y., Chen, Y.-M., Su, C.-Y.T.-Y.: Designing a multiple-layer ontology-based knowledge representation model in virtual enterprises. Journal of Information Management 15(1), 239–262 (2008)
4. Tsadiras, A.K., Margaritis, K.G.: A study of the two unit certainty neuron fuzzy cognitive map. Neural, Parallel & Scientific Computations 9(1), 67–90 (2001)

Design and Fast Verification of RF Front-End Based on MAX2769

Shuaihe Gao, Lin Zhao, and Lishu Guo

College of Automation, Harbin Engineering University, Harbin, China
heu_gnss@hotmail.com

Abstract. The GPS radio frequency (RF) front-end based on Max2769 is designed and the corresponding test method is developed in this paper. The research of hardware design includes the circuit analysis and PCB realization. Then FPGA program is utilized to configure the serial peripheral interface (SPI), and to collect the intermediate frequency (IF) data immediately. Data format conversion and signal capture are implemented for the collected raw data, to verify the performance of designed RF front-end.

Keywords: RF Front-end, Data Collection, Signal Acquisition.

1 Introduction

The weak GPS signal obtained by the antenna is amplified, frequency-converted, and analog-to-digital converted in the RF front-end of the GPS receiver[1, 2], where the raw analog signal is transformed to IF digital signal. The performance of the RF front-end could directly impact the results of the following baseband processing and navigation calculation. With the development of software receiver, the RF front-end and the antenna would be the only parts realized by hardware[2, 3]. At present, there are two basic forms of RF front-end, which are the front-end based on discrete components and front-end based on integrated chip[4]. Here the realization by discrete components could bring more flexible, especially for the future multi-frequency GNSS signal processing. However, the interference is relatively high comparing to the front-end based on integrated chip, and the cost is much higher. So the integrated chip method is popular in the design of present receivers. There are many outstanding RF chip for the designer, such as the GP2015 made by Zarlink, SE4110 made by SIGE, MAX2741 and MAX2769 made by Maxim, and so on. In this paper the RF front-end scheme based on MAX2769 would be researched, and the related design of circuit and PCB would be analyzed. Meanwhile, to verify the function of the design immediately, the test program based on data collection by JTAG and signal acquisition is developed.

2 The Design of Circuit and PCB

The core chip of the designed RF front-end is MAX2769, which offers highest performance and integration at a low cost. The complete receiver chain is

D. Jin and S. Lin (Eds.): Advances in EECM Vol. 1, LNEE 139, pp. 147–152.
springerlink.com

incorporated on this chip, including a dual-input low-noise amplifier (LNA) and mixer, followed by the image-rejected filter, programmable gain amplifier (PGA), voltage-controlled oscillator (VCO), fractional-N frequency synthesizer, crystal oscillator, and a multi-bit analog-to-digital converter (ADC). The ADC could provide 1 or 2 quantized bits for both I and Q channels, or up to 3 quantized bits for I channel. Output data is available either at the complementary metal oxide semiconductor (CMOS) logic or at the limited differential logic levels. The connection to the circuit diagram could be shown in figure 1, and there are some caveats in the circuit design.

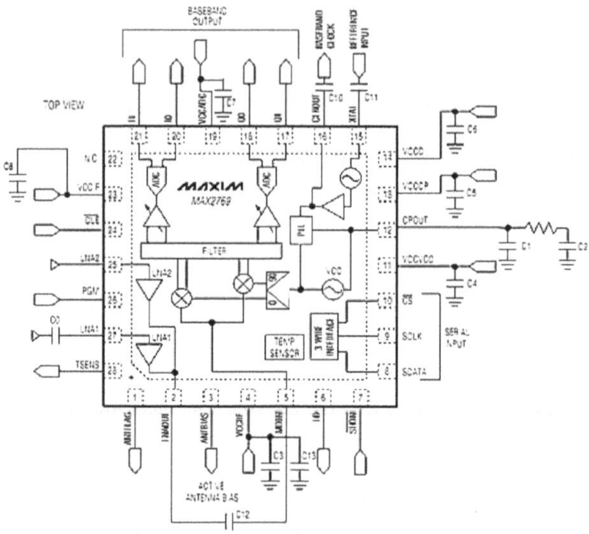

Fig. 1. Connection of the circuit diagram for MAX2769

The active antenna and passive antenna could be both selected in this designed, and be automatically detected by the current consumption to switch LNA1 and LNA2. When the high sensitivity active antenna is inserted, it could replace the applications of passive antenna. External antenna is generally connected to LNA2, and internal antenna connected to LNA1, to achieve automatically switch in the RF front-end.

To improve the performance of filter, the surface acoustic wave (SAW) filter NSVS658 is selected additionally to alternative the filter capacitor. The center frequency is 1575.42MHz and the bandwidth is 2MHz.

To promise the quality of the clock signal, the 16.368MHz temperature compensate x'tal (crystal) oscillator (TCXO) is applied in this design.

In addition, to further expand and validate the hardware platform, the clock signal pin, the IF signal output pins, hardware status pin, and the system configuration pins all leads reserved.

Based on the above design issues in the RF front-end, the following points should also be considered in the PCB design process.

The impedance matching problems between input and output existing in the antenna part and filter part. In the RF front-end hardware platform, the quality of impedance matching could directly the signal quality. Here the characteristic impedance of the microstrip transmission line is 50Ω. To satisfy the requirements, the width of the microstrip line should calculated by the following formula.

$$w / h = 5.98 / 0.8 \times e^{(\frac{Z_0 \times \sqrt{E_r + 1.41}}{87})} - t \tag{1}$$

Where, h is the dielectric layer thickness; E_r is the dielectric constant; t is the conductor thickness; and $3w$ is the width of the lower wire.

The layout of the circuit board should be considered comprehensively. For example, considering the stability of the reference clock signal would directly determine the performance of GPS receivers, the signal wiring should be laid away from other clock, the digital signal and radio frequency signal. Filter capacitor should be placed as close to the corresponding power supply pins. To ensure the stability of the power level, copper area should be increased to play a better isolation.

Wiring and drilling in the circuit board should be designed carefully. Ground and power lines should be widened, generally the more coarse the better. RF signal traces should be as short, straight, and the right angle should not taken in the same level to prevent crosstalk between the signals and the coupling between the transmission lines. Avoid taking holes in the pads, which will reduce the signal interference.

3 Data Collection Based on FPGA Signaltap

3.1 The SPI Configure for the Front-End Chip

The flexibility of MAX2769 chip is that internal SPI interface could be controlled by the related parameter configuration, which include the serial clock line Sclk_out, master output / slave input line Sdata_out, and the slave select line CS_out. When CS_out is set to low level, it will allow activation from the side. Sclk_out is the synchronous clock for communications. The activation state once started, there should be data transmission occurred, and even the data can be empty. Whether the received data is meaningful depends on the agreed application protocol identification. In this design, the MAX2769 is configured through the FPGA, and the specific control circuit is shown in figure 2.

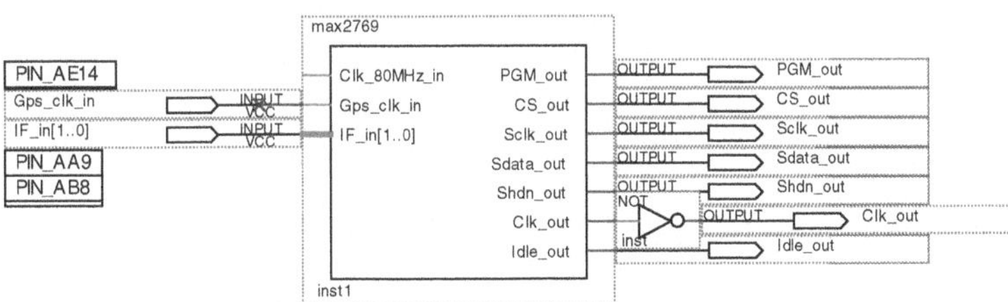

Fig. 2. Circuit diagram of SPI configure

3.2 Quick Data Collection Utilizing the Signaltap

SignalTap logic analyzer is an embedded logic analysis tool integrated in Quartus II software. Build a new "stp" file, add the necessary observations, the right trigger and sampling clock, and compile and download the program to the FPGA. Then the concept of online measurement through the graphical state of the FPGA will appear which could meet the requirements of the development of hardware debugging. The availability and controllability of SignalTap, and the reliability and simplicity of data transmission are fully utilized. In the design, the collected data is transmitted to the computer through JTAG port, where the sampling depth is 64K. The obtained output I0, I1 data flow from GPS RF front-end is shown in figure 3. Compared to the common data acquisition though serial port and USB, it avoid the complicated process of setting transport protocol, which promise the reliable of data transmission.

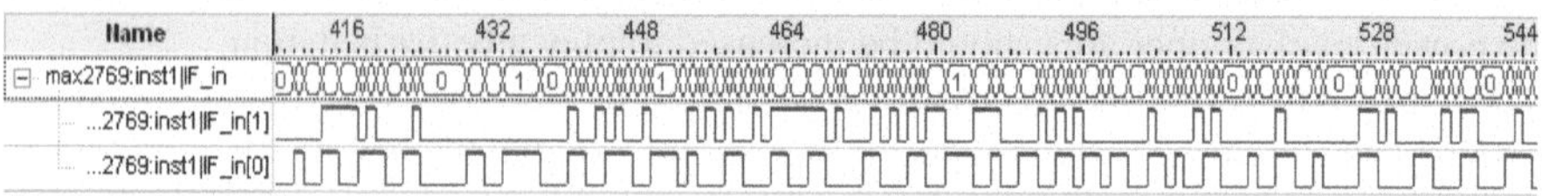

Fig. 3. Fragments of collected data though Signaltap

4 Fast Functional Verification Based on Simulation

4.1 Data Format Conversion

Before the verification of GPS IF signal, I0 and I1 data preprocess is essential. The quantitative results of the 2bit binary file should be converted to digital IF signals, the results can be shown in figure 4. Meanwhile, the power spectrum is generated.

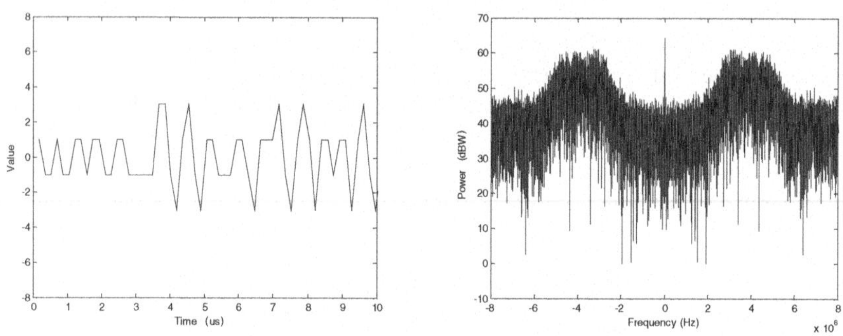

Fig. 4. GPS IF signal and its power spectrum

4.2 Simulation Verification Utilized Acquisition

The signal acquisition method based on FFT is applied in this design. Using circular correlation theory, convert the correlation of received signal and local generated signal in the time-domain to spectrum multiplying in the frequency-domain[5-7]. The data is collected at about 11:36 on 25 April 2011, the capture results are shown in table 1 and figure 5.

Table 1. The results of signal acquisition

PRN	Average correlation value	Maximum correlation peak	Scale factor	Correlation peak location
9	2503.971931	26502.37448	10.58413401	11616
12	2526.174536	14402.15390	5.701171354	4856
15	2514.608920	16059.97338	6.386668422	5485
18	2486.050408	17011.33887	6.842716788	15443
27	2581.278273	20750.60992	8.038889158	8013

According to the capture results, it is clear that the RF front-end can effectively achieve the signal frequency conversion and analog-to-digital conversion. The collected data can successfully capture five visible satellites, and the results are consistent to the high-performance commercial receiver.

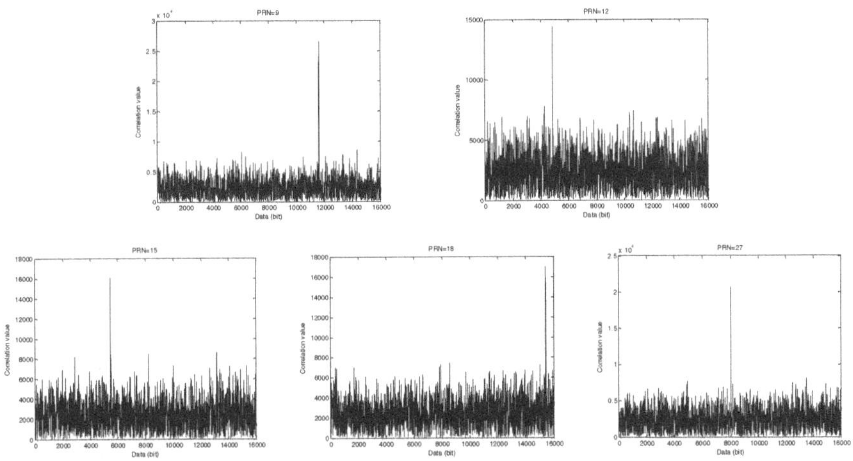

Fig. 5. The correlation peak of visible satellites

5 Conclusion

In this paper, the RF front-end in GPS receiver is designed and following verification methods are studied. The fast data acquisition utilized Signaltap can simplify the system realization, at the same time, promise high reliability. In addition, the research could provide the design reference of software GNSS receivers.

References

1. Tuttlebee, W.: Software Defined Radio: Enabling Technologies. Wiley Press (2004)
2. Dempster, A.G.: Satellite Navigation: New Signals, New Challenge. In: IEEE International Symposium on Circuits and Systems (2007)

3. Kim, J.-M., Song, H.-J., Kim, Y.-B.: Design and implementation of L1-band C/A-code GPS RF front-end chip. In: The 6th International Conference on VLSI and CAD (1999)
4. Piazza, F., Huang, Q.: A 1.57-GHz RF front-end for triple conversion GPS receiver. Solid-State Circuits 33(2), 202–209 (1998)
5. Liu, X., Li, Y.: Fast Acquisition Based on FFT for High Dynamic GPS Signals. Journal of System Simulation 19(10), 2151–2155 (2007)
6. Tang, K., Wu, M., Hu, X.: Design and validation of GPS software receiver based on RF front-end. Journal of Chinese Inertial Technology 15(1), 51–54 (2007)
7. Zhao, L., Gao, S., Ding, J., Guo, L.: Optimized FFT Algorithm and its Application to Fast GPS Signal Acquisition. Fourier Transforms - Approach to Scientific Principles. In Tech. Press (2011)

Research Progress on Satellite Navigation with Inertial Information-Aided

Shuaihe Gao, Lin Zhao, and Lishu Guo

College of Automation, Harbin Engineering University, Harbin, China
heu_gnss@hotmail.com

Abstract. Applying inertial navigation system (INS) to aid satellite navigation is considered as one of the best way to achieve high dynamic and weak signal environment. With the improvement of the GPS software receivers and the emergence of low-cost high-precision micro-electro-mechanical systems (MEMS) inertial measurement unit (IMU), IMU-aided GPS navigation system came into being. In this paper, the system framework design, accurate Doppler information assistance, multi-mode navigation data synchronization, integrated navigation filtering algorithm and other key technologies are reviewed and analyzed. Based on the summary of the existing technologies and research results, the future directions and challenges are prospected.

Keywords: integrated navigation, signal tracking, data synchronization, integrated filter.

1 Introduction

Utilizing the GPS position results to aid the initial alignment and errors correction of INS is the most popular method in the traditional coupled navigation system. However, with the development of the users' demand, to use INS information aiding GPS signal processing comes into being.[1-3] It has demonstrated plenty of advantages in improving GPS signal acquisition and tracking, which address the contradiction between errors of dynamic stress and noise in loop bandwidth. In some way, it is the evolution of ultra-tight GPS/SINS coupled system.[4-6] In early 1980s, the advantages of INS-aided GPS structure has been recognized.[7] But until the late 20th century, this structure draws more and more concern with the development of software receiver. Gustafson, and et al. define the deep integration and improve the tracking loop performance, which enhance the anti-jamming capability of receivers.[8] Gautier, and et al. furtherly deepen the concept of GPS/INS ultra-tight navigation and analyze the basic state observer equations in the coupled Kalman filter. In addition, the researchers in University of New South Wales, University of Calgary, and et al. have proposed some effective methods in system design, aided tracking loop control, integrate filter and so on.[9-13]

In this paper, the system framework design, accurate Doppler information assistance, multi-mode navigation data synchronization, integrated navigation filtering algorithm and other key technologies will be reviewed and analyzed. Based on the

D. Jin and S. Lin (Eds.): Advances in EECM Vol. 1, LNEE 139, pp. 153–158.
springerlink.com

summary of the existing technologies and research results, the future directions and challenges will be prospected.

2 System Framework Design

The typical feature of INS-aided GPS receiver is the information fusion is brought into the process of signal tracking loops. There are two basic structures for information aiding, which are shown in Fig. 1.

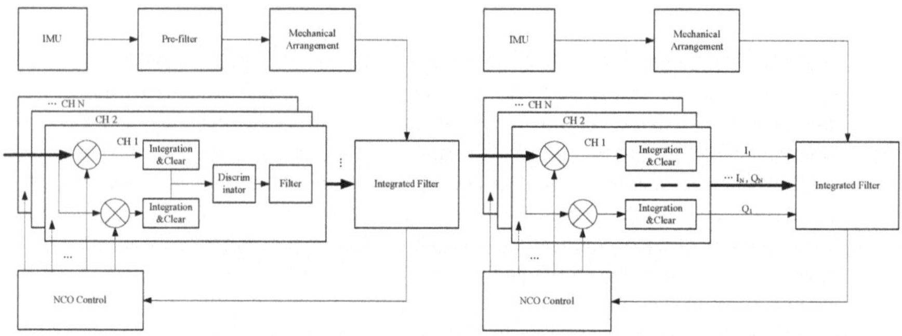

Fig. 1. Information aiding based on tight filter and direct data fusion

As shown in Fig.1, according the pseudorange and pseudorange rate information, the position and velocity will be calculated. Meanwhile, the output information and ephemeris are used to deduce the shift of Doppler frequency and control the numerical controlled oscillator (NCO), which will relieve the stress of dynamic. In this structure, the demand of hardware platform is not high.[14-15]

The structure of information aiding based on direct data fusion is shown in Fig. 1. The position results are estimated directly by the GPS I-branch and Q-branch signal and IMU data. The tracking loop in the GPS receiver and information integrated are achieved in the same filter, which is considered as the concentrated form. Usually, the frequency of I-branch and Q-branch signal processing is 1000Hz, which is relatively difficult for normal platform, so the demand of the hardware is high.[16-17]

Besides, there has been some new structure derivated. In literature [18], the combination of differential GPS and SINS is proposed, and the error could be controlled at less than 38cm in more than 80% moving area, which improve the navigation performance significantly. The structure applying vector tracking is proposed in literature [19], which use a combination of Kalman filter feedback loop instead of the traditional stand-alone and parallel tracking loop. The simulation results indicate the advantages in the low signal-to-noise ratio (SNR) environment. The combination application of GPS, GLONASS and SINS is put forward in literature [20], however, the experimental results show that the GLONASS system does not improve the system performance significantly. It is worthy to be mentioned that the integration of GPS and simplified IMU (one gyro and two or three accelerometers), where the high-dimensional positioning is not required. So the system is more simplified, of course, the main application of this structure is in the area of vehicle positioning.[21-22]

3 Accurate Doppler Information Assistance

To use the aiding information calculated by the IMU output and other data to enhance signal tracking in high dynamic environment is primary feature in INS-aided GPS system. In literature [23], tracking performances with traditional loops and aiding loops are compared, which indicate that the bandwidth could be reduced with the information aiding. In the GPS receiver, the Doppler frequency shift is mainly caused by the relative motion between the satellite and the receiver, and time errors existing in the satellite clocks and receiver clocks. Among them the relative motion is the main factor. However, the accuracy of Doppler shift estimation is directly influence the tracking performance, therefore, using model analysis to decrease the random errors of Doppler estimation is essential.[24]

In traditional situation, the random errors are always considered as Gaussian white noise, but lots of data illustrate that the errors not satisfy the white noise model. In literature[10] and [12], the modelings with Gauss-Markov (GM) and Auto Regressive (AR) are proposed respectively. They both play a role to improve the accuracy in some way. The AR(2) modeling method is reckoned as a typical method. Firstly, to determine the stationary of stochastic Doppler error sequence, and apply time series analysis method to modeling. Then, the model type and order number are selected by studying the autocorrelation function and the partial correlation function of stochastic error sequence. Use Yule-Walker method to estimate the model parameters and obtain appropriate model. However, there are two obvious defects in the modeling process:

(1) The random errors are reckoned as linear models, and sometimes they can not reflect the characteristics of the complex systems.
(2) Use the existing data to achieve modeling, which will ignore the uncertainty factors caused by system itself and external environments. The actual models of the system and the established theoretical models often not fully match.

4 Multi-mode Navigation Data Synchronization

The data fusion between different sub-systems could break out the limit in the working environment and errors accumulation over time. The primary of integration of these two systems is to promise the synchronization in data output. The time bias will bring serious influence on the coupled filtering. So to promise the time synchronization is important to the system design. In literature [25], a time synchronization solution using 1PPS (Pulse Per Second) signal from a GPS receiver as the common synchronization reference signal is proposed. In literature [26], a software method to estimate and compensate time synchronization error was proposed based on the investigation of physical concept of GPS/S1NS time synchronization error, which realized on-line estimation and compensation of time synchronization error by augmentation of filter state and modification of measurement equations. In literature [27], an accuracy time synchronization precept based on PXI-6608 Counter/Timer produced by NI Co. is presented, and its accuracy can reach 400ns. Besides, the data synchronization in INS-aided GPS system should take the rate of data into account, thence the synchronization problem is more complex in this system.

To address the impact of data synchronization, the hardware integrated form is also selected.[8] As it is known the rate of GPS tracking loop and SINS is unequal, the Interpolation is utilized to achieve data rate synchronization. According to the types of extraction with different low-pass filters, it includes poly-phase, cascaded-integration-comb (CIC) and so on. Besides, the aiding information could be interrupted for the system failure, a information predict method is proposed to compensate the missing information in literature.[28]

5 Integrated Navigation Filtering Algorithm

The design of navigation filter is one of the key technologies in INS-aided GPS system, among which the most important is the filtering algorithms. INS-aided GPS system is essentially a nonlinear system. Sometimes in order to reduce the computation and promise the real-time for system and some nonlinear factors can be ignored in the assumptions. So in the early time, the Kalman filter (KF) is mainly applied.[29-31] However, the assumptions can not satisfy actual demand, the nonlinear modeling should be established in the integrated filter. Over the years, extended Kalman filter (EKF) has been widely used in integrated navigation system to estimate the nonlinear state because of its simple, fast convergence, etc.[32] But EKF is not robust, easy-to-divergence, and only effective to approximately linear of nonlinear systems in the updated range. To overcome these shortcomings, unscented Kalman filter (UKF) is always used in integrated systems.[33-35] Simulation results validate the performance in the navigation and positioning area. It is worth to mention that particle filter (PF), which belongs the framework of Bayesian online learning Monte Carlo method, has been applied to integrated filtering, and demonstrates good performance under the dynamic systems state estimation.[36-37]

6 Conclusion and Outlook

The system framework design, accurate Doppler information assistance, multi-mode navigation data synchronization, and integrated navigation filtering algorithm are reviewed and analyzed. According to the present research results, it can be predicted that the following researches will be continued to draw the attention for the researchers.

(1) The source of aiding information. The Doppler frequency is estimated from the output inertial information, actually, ultra wide bandwidth (UWB) and other means can also be used to replace SINS. Therefore positioning based UWB and the other technologies will be utilized in information-aided GPS receivers in the future.
(2) The design of ultra-tight filter. As mentioned, the nonlinear and uncertain should be considered in the design of filtering algorithm, so further theoretical research to find adaptive nonlinear filtering algorithm to consist with the application of the integrated navigation system is the key to improve overall system performance.
(3) The receiver autonomous integrity monitoring (RAIM) in IMU-aided GPS system. The real-time, accurate receiver fault detection and isolation (FDI) through state observation is needed in the system.[38] Therefore, to explore RAIM basedon signal tracking level will greatly improve the fault detection capabilities.

References

1. Schultz, C.E.: INS and GPS integration. Technical University of Denmark, Denmark (2006)
2. Cox, D.B.: Integration of GPS with inertial navigation systems. Navigation 1(1), 144–153 (1979)
3. Jing, F., Qitai, G., Xuebin, L., et al.: Design and experiment of MINS/GPS integrated navigation system. Journal of Tsinghua University (Science and technology) 47(8), 1316–1319 (2007)
4. Schmidt, G., Phillips, R.: INS/GPS Integration Architectures. Draper Laboratory Report (2003)
5. Gustafson, D., Dowdle, J., Flueckiger, K.: A Deeply Integrated Adaptive GPS-Based Navigator with Extended Range Code Tracking. In: IEEE PLANS Conference, San Diego (2000)
6. Petovello, M.G., O'Driscoll, C., Lachapelle, G.: Carrier Phase Tracking of Weak Signals Using Different Receiver Architectures. In: ION NTM Conference, San Diego (2008)
7. Copps, E.M., Geier, G.J., Fidler, W.J.: Optimal Processing of GPS Signal. Navigation Journal of The Institute of Navigation 27(3) (1980)
8. Gustafson, D., Dowdle, J., Flueckiger, K.: A Deeply Integrated Adaptive GPS-Based Navigator with Extended Range Code Tracking [EB/OL]. Charles Stark Draper Laboratory (2000)
9. Babu, R., Wang, J.: Improving the Quality of IMU-Derived Doppler Estimates for Ultra-Tight GPS/INS Integration. In: The 2004 European Navigation Conference on Global Navigation Satellite Systems, Rotterdam, pp. 144–151 (2004)
10. Babu, R., Wang, J., Rao, G.: Algorithms for prediction of INS estimated Doppler in ultra-tight integration. In: International Symposium on GPS/GNSS, pp. 300–307. Tokyo International Exchange Center, Tokyo (2008)
11. Gao, G.: INS-Assisted High Sensitivity GPS Receivers for Degraded Signal Navigation. University of Calgary, Calgary (2007)
12. Babu, R., Wang, J.: Analysis of INS-derived Doppler effects on carrier tracking loop. The Journal of Navigation 58(3), 493–507 (2005)
13. Yang, Y.: Tightly Coupled MEMS INS/GPS Integration with INS Aided Receiver Tracking Loops. University of Calgary (2008)
14. Gao, G.: INS-Assisted High Sensitivity GPS Receivers for Degraded Signal Navigation. Department of Geomatics Engineering. University of Calgary (2007)
15. Gautier, J.D., Parkinson, B.W., Gebre-Egziabher, D.: Using the GPS/INS Generalized evaluation Tool (GIGET) for the Comparison of Loosely Coupled. In: Proceedings of ION 59th Annual Meeting Tightly Coupled and Ultra-Tightly Coupled Integrated Navigation Systems (2003)
16. Babu, R.: Ultra-tight integration of GPS/peesuolites/INS: system design and performance analysis. School of Surveying and Spatial Information Systems. University of New South Wales (2006)
17. Gustafson, D., Dowdle, J., Flueckiger, K.: A High Anti-Jam GPS based Navigator. In: Proceedings of ION National Technical Meeting (2000)
18. Godha, S., Cannon, M.E.: Integration of DGPS with a Low Cost MEMS - Based Inertial Measurement Unit (IMU) for Land Vehicle Navigation Application. In: ION GPS 2005. Long Beach (2005)
19. Xinlong, W., Jie, Y.: Deeply SINS/GPS integrated navigation method based on vector tracking. Journal of Chinese Inertial Technology 17(6), 710–717 (2009)

20. Petovello, M., O'Driscoll, C., Lachapelle, G.: Ultra-Tight Integration of an IMU with GPS/GLONASS. In: Proceedings of 13th International Association of Institutes of Navigation (2009)
21. Petovello, M.G., Sun, D., Lachapelle, G., Cannon, M.E.: Performance Analysis of an Ultra-Tightly Integrated GPS and Reduced IMU System. In: ION GNSS 2007 (2007)
22. Sun, D.: Ultra-Tight GPS/Reduced IMU for Land Vehicle Navigation. Department of Geomatics Engineering. University of Calgary (2010)
23. Babu, R., Wang, J.: Ultra-tight GPS/INS/PL Integration: A System Concept and Performance Analysis. GPS Solutions 13(1), 75–82 (2010)
24. Li., D., Wang, J., Babu, S.: Nonlinear Stochastic Modeling for INS Derived Doppler Estimates in Ultra-Tight GPS/PL/INS Integration. In: GNSS 2005, Hong Kong (2005)
25. Qian, L., Xingqun, Z., Lirui, W., et al.: Synchronizer Design in Integrated GPS/INS System. Chinese Journal of Sensors and Actuators 22(12), 1752–1756 (2009)
26. Tao, Y., Wei, W.: Software method to realize time synchronization in GPS/SINS integrated navigation system. Joumal of Chinese Inertial Technology 16(4), 436–438 (2008)
27. Kaidong, Z., Meiping, W.: A High Accuracy Time Synchronization Precept in SINS/GPS Integrated Navigation. Control Technology of Tactical Missile 55(4), 66–68 (2006)
28. Tao, Z.: Research on Key Technologies of Ultra-tight GPS.SINS Coupled System. Automation College. Harbin Engineering University (2010)
29. Li, D., Wang, J.: Kalman filter design strategies for code tracking loop in Ultra-Tight GPS/INS/PL integration. In: ION National Technical Meeting, Monterey, California (2006)
30. Li, D., Wang, J.: System Design and Performance Analysis of Extended Kalman Filter-Based Ultra-Tight GPS/INS Integration. In: 2006 IEEE/ION Symposium on Position, Location, And Navigation (2006)
31. Ding, W.: Optimal Integration of GPS with Inertial Sensors: Modelling and Implementation. University of New South Wales (2008)
32. Yu, L., Chen, X.: Application of Extended Kalman Filter in Ultra-Tight GPS/IINS Integration Based on GPS Software Receiver. In: 2010 International Conference on Green Circuits and Systems (2010)
33. Minhu, Z., Zhang, R., Chunhung, H.: Application of UKF in deeply coupled GPS / INS navigation system. Journal of Chinese Inertial Technology 17(6), 697–700 (2009)
34. Gustafson, D., Dowdle, J., Flueckiger, K.: A High Anti-Jam GPS-Based Navigator. In: ION NTM 2000, Anaheim (2000)
35. Zhou, D., Wang, Q.: Strong tracking filtering of nonlinear systems with colored noise. Journal of Beijing Institute of Technology 17(3), 321–326 (1997)
36. Fernández-Prades, C., Closas, P., Vila-Valls, J.: Nonlinear Filtering for Ultra-Tight GNSS/INS Integration. In: 2010 IEEE International Conference on Communications (2010)
37. David, B.: Particle Filtering Algorithm for Ultra-tight GNSS/INS Integration. In: ION GNSS 2008 (2008)
38. Bhatti, U.I.: Improved integrity algorithms for integrated GPS/INS systems in the presence of slowly growing errors. Department of Civil and Environmental Engineering. Imperial College London, London (2007)

Stability of a Kind of Hybrid Systems with Time-Delay

Xing-cheng Pu[1,2] and Hong-quan Zhao[3]

[1] School of Mathematics and Physics, Chongqing University of Posts and Telecommunications, Chongqing, 400065, China
[2] Key Laboratory of Network Control and Intelligent Instrument of Ministry of Education, Chongqing University of Posts and Telecommunications, 400065, China
[3] Automation College, Chongqing University of Posts and Telecommunications, Chongqing 400065, China

Abstract. This article formulates and studies the stability of a kind of Hybrid Systems with time-delay. This system is assumed to be autonomous and the state delay is time-varying. Using the Lyapunov functional approach combined with linear matrix inequality technique, neither restriction on the derivative of time-delay function nor bound restriction on nonlinear switched systems with time-varying delays is required to analyze the exponential stability of this kind of nonlinear switched systems with time-varying delays. Some delay-dependent exponential stability conditions are derived by the stability theory. A numerical example is given for illustration and interpretation of the theoretical results.

Keywords: Hybrid systems, Exponential stability, globally asymptotically stable, Time varying- delay.

1 Introduction

Linear time-invariant systems have been and continue to be the engine of control theory development. Nowadays there are many useful and interesting results on a class of linear systems that can be used to analyze and design control systems(see, e.g.[1]-[3]).

With the development of social production, many practical systems can not be modeled by linear time-invariant systems. In this case, hybrid systems are employed to model many practical systems. A hybrid system is a dynamical system with continuous dynamics, discrete dynamics, and the interaction between them(see, e.g.[4]-[6]). An area of particular interest has been the analysis of stability of hybrid dynamical systems(see, e.g.[7]-[12]).

In this paper, we investigate the stability of a kind of nonlinear hybrid systems with time-varying. Compared with existing results, techniques and methods for stability of hybrid systems with delays should be developed and explored.

2 Models and Preliminaries

To begin with, we introduce some notation and some basic definitions. Let I denote an $n \times n$ unit matrix. For $A, B \in R^{m \times n}$ or $A, B \in R^n, A \geq B$ $(A \leq B \quad , A > B \;\; A < B)$

D. Jin and S. Lin (Eds.): Advances in EECM Vol. 1, LNEE 139, pp. 159–165.
springerlink.com

means that each pair of corresponding elements of A and B satisfies the inequality“ $\geq(\leq,>,<)$ ”.Especially, A is called a nonnegative matrix if $A>0$. $PC[\hat{I},R]\triangleq \{\psi:\hat{I}\to R^n \,|\, \psi(t^+)=\psi(t)$ for $t\in\hat{I}$, $\psi(t^-)$ exists for $t\in(t_0,\infty)$, $\psi(t^-)=\psi(t)$ for all but points $t_k\in(t_0,\infty)\}$,where $\hat{I}\subset R$ is an interval, $\psi(t^+)$ and $\psi(t^-)$ denote the left- hand and right-hand limits of scalar function $\psi(t)$,respectively. Let $PC=PC([-\tau,0],R)$.

Consider the following nonlinear hybrid system with switching and time-delay .

$$\begin{cases} \dot{x}(t)=A_{\sigma(t)}x(t)+A_{\sigma(t)}x(t-\tau(t))+B_{\sigma(t)}f_{\sigma(t)}(x(t))+B_{\sigma(t)}f_{\sigma(t)}(x(t-\tau(t))),t\in[t_{k-1},t_k), \\ \Delta x(t_k)=x(t_k)-x(t_k^-)=J_k(t_k^-,x(t_k^-)),t=t_k,x(t_0+t)=\phi(t)\in PC,t\in[-\tau,0] \end{cases} \tag{1}$$

Where, $t\in R^+,x\in R^n$ is the state variable, $t_0\geq 0$ is the initial time, $\sigma(t):R^+\to I,I=\{1,2,\cdots,m\}$, R^+ is the positive real number set, the time sequence $\{t_k\}$ satisfies $0\leq t_0<t_1<\cdots<t_k<\cdots$,and $\lim_{k\to\infty} t_k=\infty$, $\sigma(t)=i$, $i=1,2,\cdots,m$ for $t\in[t_{k-1},t_k)$. Furthermore, we suppose system (1) satisfies conditions (H_1)- (H_4) at any bounded interval $\tau<t_k-t_{k-1}\leq T$;

(H_1) $J_k(t_k^-,x(t_k^-)):R^+\times R^n\to R^n\,(k\geq 1)$ are continuous functions with $J_1(t,0)\equiv 0$ for $t\in R^+$;

(H_2) There exist nonnegative constant sequence $\{L_{\sigma(t)}\}$ such that:

$$\left\|f_{\sigma(t)}(t,x(t))\right\|\leq L_{\sigma(t)}\left\|x(t)\right\|$$

Let $T_i(t_0,t)$ denote the working time of the i th subsystem during the interval $[t_0,t]$. In additional, let $\mu(T_i(t_0,t))$ denote the Lebesgue measure of the set $T_i(t_0,t)$. Then system (1) can be written as:

$$\begin{cases} \dot{x}(s)=A_ix(s)+A_ix(s-\tau(s))+B_if_i(x(s))+B_if_i(x(s-\tau(s))),s\in T_i(t_0,t) \\ \Delta x(t_k)=x(t_k)-x(t_k^-)=J_k(t_k^-,x(t_k^-)),t=t_k,x(t_0+s)=\phi(s)\in PC,s\in[-\tau(t),0] \end{cases} \tag{2}$$

Where $i\in I$ and $\bigcup_{i=1}^{m}\mu(T_i(t_0,t))=[t_0,t]$ .Suppose that System (2) satisfies assumptions H_3-H_4 besides it satisfies assumptions H_1-H_2.

(H_3)For any $i,i=1,2,...m.$, $f_i(t,0)=0$, e.g. $x(t)\equiv 0$ is a solution of System (2).

(H_4) $\left\|x(t_k^-)+J_k(t_k^-,x(t_k^-))\right\|\leq\beta_k\left\|x(t_k^-)\right\|,k=1,2,...,$ and $\|\cdot\|$ denotes the Euclidian norm of a vector or matrix.

In order to finish our results, we will use some stability definitions (see [8]) and introduce some lemmas.

Lemma 1. Suppose $x, y \in R^n$, D, E are $n \times n$ matrices, then for all $\varepsilon > 0$,

$$2x^T D^T Ey \le \varepsilon x^T D^T Dx + \varepsilon^{-1} y^T E^T Ey .$$

Lemma 2. Let ω be a nonnegative function defined on the interval $[t_0 - \tau, \infty)$ and be continuous on the interval $[t_0, \infty)$. Assume that $\dot{\omega}(t) \le -a\omega(t) + b\omega(t-\tau)$, $t \ge t_0$ where a and b are nonnegative constants satisfying $a > b$. Then

$$\omega(t) \le \overline{\omega}_0 \exp\{-\lambda(t - t_0)\}, t \ge t_0$$

And $\overline{\omega}_0 = \sup_{t_0 - \tau \le \theta \le t_0} \omega(\theta)$,and $\lambda > 0$ satisfies $\lambda - a + be^{\lambda\tau} = 0$.

Lemma 3. Let P be a positive and symmetric matrix, $x \in R^n$, then

$$\lambda_{\min}(P)x^T x \le x^T Px \le \lambda_{\max}(P)x^T x .$$

3 Stability Analysis

This section addresses the stability issues of systems (1) and (2).

Theorem 1. Suppose that assumptions H_1- H_4 hold and there exist positive and symmetric matrix P_i and positive real number a_i, b_i, and $a_i > b_i > 0$ such that

$$A_i^T P_i + P_i A_i + 3P_i P_i + L_i^2 \frac{\lambda_{\max}(B_i^T B_i)}{\lambda_{\min}(P_i)} I \le -a_i P_i , \quad \sum_{j=1}^{k} \ln d_j - \sum_{i=1}^{m} \lambda_i \mu_i (t_0, t) \le \varphi(t_0, t) \tag{3}$$

Where $\eta_k = \max\{ \sup_{t_{k-1} \le t \le t_k} e^{\lambda_i (t_k - t)} \}$, $d_k = \max\{1, \eta_i \beta_i^2 \frac{\lambda_{\max}(P_i)}{\lambda_{\min}(P_i)} \}$,

$$b_i = L_i^2 \frac{\lambda_{\max}(A_i^T A_i)}{\lambda_{\min}(P_i)} + L_i^2 \frac{\lambda_{\max}(B_i^T B_i)}{\lambda_{\min}(P_i)} , i = \sigma(t), t \in [t_{k-1}, t_k), k = 1, 2, \ldots.. \tag{4}$$

And λ_i is the unique positive root of the equation $a_i + b_i e^{\lambda_i \tau} + \lambda_i = 0$.

$\lambda_{\max}(\lambda_{\min})$ denotes the maximum (minimum) eigenvalue of the positive-definite matrix P_i, and $\omega(t_0, t)$ is a continuous function on R^+. Then $\lim_{t \to \infty} \omega(t_0, t) = -\infty$ implies that the trivial solution of system (1) and (2) is globally asymptotically stable.

Proof. Construct the following switching Lyapounov function

$$V(t) = x(t)^T P_i x(t), i = \sigma(t), t \in [t_{k-1}, t_k). \tag{5}$$

The time derivative of (5) along the solution of (1) over interval $t \in [t_{k-1}, t_k)(i = \sigma(t))$ is calculated as follows

$$\begin{aligned} \dot{V}(t) &= 2x(t)^T P_i \dot{x}(t) \\ &= 2x(t)^T P_i [\, A_i x(t) + A_i x(t - \tau(t)) + B_i f_i(x(t)) + B_i f_i(x(t - \tau(t))) \,], t \in [t_{k-1}, t_k) \end{aligned} \tag{6}$$

Applying Lemma 1($\varepsilon = 1$) to (6) yields

$$\dot{V}(t) = 2x(t)^T P_i [\, A_i x(t) + A_i x(t-\tau(t)) + B_i f_i(x(t)) + B_i f_i(x(t-\tau(t))) \,]$$
$$\le x^T(t)(A_i^T P_i + P_i A_i + 3P_i P_i)x(t) + f_i^T(x(t))B_i^T B_i f_i(x(t))$$
$$+ x^T(t-\tau)A_i^T A_i x(t-\tau) + f_i^T(x(t-\tau))B_i^T B_i f_i(x(t-\tau))$$

Applying Lemma 3 and condition (H_2), we can easily obtain

$$\dot{V}(x(t)) \le x^T(t)[A_i^T P_i + P_i A_i + 3P_i P_i + L_i^2 \frac{\lambda_{\max}(B_i^T B_i)}{\lambda_{\min}(P_i)} I]x(t)$$

$$+[L_i^2 \frac{\lambda_{\max}(A_i^T{}_i A_i)}{\lambda_{\min}(P_i)} + L_i^2 \frac{\lambda_{\max}(B_i^T B_i)}{\lambda_{\min}(P_i)}]x^T(t-\tau)P_i x(t-\tau) \le -a_i V(t) + b_i V(t-\tau) \quad (7)$$

Applying Lemma 2 to (7), we obtain $\quad V(t) \le \overline{V}(t_{k-1})\exp(-\lambda_i(t-t_{k-1}))$ (8)

Where $\overline{V}(t_{k-1}) = \sup\limits_{t_{k-1}-\tau \le \theta \le t_{k-1}} V(\theta)$, And λ_i is the unique positive root of the equation $-a_i + b_i e^{\lambda_i \tau} + \lambda_i = 0$.

Applying condition (H_4) ,note that

$$V(t_k) = [x(t_k^-) + J_k(t_k^-, x(t_k^-))]^T P_i \cdot [x(t_k^-) + J_k(t_k^-, x(t_k^-))]$$
$$\le \frac{\lambda_{\max}(P_i)}{\lambda_{\min}(P_i)} \beta_k^2 [x(t_k^-)]^T P_i [x(t_k^-)] = \beta_k^2 \frac{\lambda_{\max}(P_i)}{\lambda_{\min}(P_i)} V(t_k^-) \quad (9)$$

Since $\phi(t) \in PC, t \in [-\tau, 0]$, we can observe that there exists positive number $K > 0$ such that $\quad V(t) \le K\|\phi\| e^{-\lambda_{\sigma(t_0)}\tau}, \forall t \in [t_0 - \tau, t_0]$.

On the other hand, $\eta_k \ge \sup\limits_{t_{k-1} \le t \le t_k} e^{\lambda_{\sigma(t)}(t_k - t)} \ge e^{\lambda_{\sigma(t)}(t_k - t)}$,

Hence, $\eta_k e^{-\lambda_{\sigma(t_k)}(t_k - t_{k-1})} \ge e^{-\lambda_{\sigma(t_k)}(t_k - t_{k-1})} e^{\lambda_{\sigma(t_k)}(t_k - t)} = e^{-\lambda_{\sigma(t_k)}(t - t_{k-1})}$. (10)

By the inequality (7)、(8) 、 (9) and (10),we can derive that for $\forall t \in [t_0, t_1)$,

$$V(t) \le K\|\phi\| e^{-\lambda_{\sigma(t_0)}\tau} \cdot e^{-\lambda_{\sigma(t_1)}(t-t_0)} \le \eta_1 K\|\phi\| e^{-\lambda_{\sigma(t_0)}\tau} \cdot e^{-\lambda_{\sigma(t_1)}(t_1-t_0)} \quad (11)$$

And $V(t_1) = [x(t_1^-) + J_1(t_1^-, x(t_1^-))]^T P_{\sigma(t_1)} \cdot [x(t_1^-) + J_1(t_1^-, x(t_1^-))]$

$$\le \eta_1 \beta_1^2 \frac{\lambda_{\max}(P_{\sigma(t_1)})}{\lambda_{\min}(P_{\sigma(t_1)})} K\|\phi\| e^{-\lambda_{\sigma(t_0)}\tau} \cdot e^{-\lambda_{\sigma(t_1)}(t_1-t_0)} \le d_1 K\|\phi\| e^{-\lambda_{\sigma(t_0)}\tau} \cdot e^{-\lambda_{\sigma(t_1)}(t_1-t_0)}.$$

Therefore, for any $t \in [t_0, t_1]$, $V(t) \le d_1 K\|\phi\| e^{-\lambda_{\sigma(t_0)}\tau} \cdot e^{-\lambda_{\sigma(t_1)}(t_1-t_0)}$.

Similarly, for any $t \in [t_1, t_2]$, $V(t) \le d_1 d_2 K\|\phi\| e^{-\lambda_{\sigma(t_0)}\tau} \cdot e^{-\lambda_{\sigma(t_1)}(t_1-t_0)} \cdot e^{-\lambda_{\sigma(t_2)}(t_2-t_1)}$.

Repeating the same steps as above leads to the following results.For $t \in [t_{k-1}, t_k]$,

$$V(t) \le d_1 \cdots d_k K\|\phi\| e^{-\lambda_{\sigma(t_0)}\tau} \cdot e^{-\lambda_{\sigma(t_1)}(t_1-t_0)} \cdots e^{-\lambda_{\sigma(t_k)}(t_k-t_{k-1})}.$$

$$\leq Ke^{-\lambda_{\sigma(t_0)}\tau}\|\phi\|e^{\sum_{j=1}^{k}\ln d_j-\sum_{i=1}^{m}\lambda_i\mu_i(t_0,t)} \leq Ke^{-\lambda_{\sigma(t_0)}\tau}\|\phi\|e^{\varphi(t_0,t)}. \tag{12}$$

The remaining part of the proof is trivial and is omitted here.

Corollary 1. In Theorem 1, if there exists a positive constant α such that

$$\lambda_i > \alpha > 0 \text{ for all } i \in I \text{, and } \frac{\ln d_k}{t_k - t_{k-1}} \leq \alpha, k = 1,2,\cdots. \tag{13}$$

Then the trivial solution of (1) and (2) is globally exponentially stable.

Proof. Letting $\lambda = \min_{1\leq i\leq m}\{\lambda_i\}$, with the inequality (4), we can easily obtain

$$\sum_{j=1}^{k}\ln d_j - \sum_{i=1}^{m}\lambda_i\mu_i(t_0,t) \leq \alpha(t_k - t_0) - \lambda(t-t_0) \leq -(\lambda-\alpha)(t-t_0), t\in[t_k,t_{k+1}), k=1,2,\cdots.$$

Letting $\omega(t_0,t) = -(\lambda-\alpha)(t-t_0)$, and $\rho = \max_{1\leq i\leq m}\frac{\lambda_{\max}(P_i)}{\lambda_{\min}(P_i)}$ (14)

We derives from inequality (12)

$$V(t) \leq Ke^{-\lambda_{\sigma(t_0)}\tau}\|\phi\|e^{\varphi(t_0,t)} = Ke^{-\lambda_{\sigma(t_0)}\tau}\|\phi\|e^{-(\lambda-\alpha)(t-t_0)} \tag{15}$$

On the other hand, from the definition of $V(t) = x^T(t)P_i x(t), i = \sigma(t), t\in[t_{k-1},t_k)$

We can prove that $\|x(t)\| \leq \sqrt{\rho Ke^{-\lambda_{\sigma(t_0)}\tau}\|\phi\|}e^{-\frac{1}{2}(\lambda-\alpha)(t-t_0)}$.

The proof is thus finished.

Corollary 2. In Theorem 1, if we choose positive and symmetric matrix P_i equal unit matrix I, then inequality (3) becomes

$$A_i^T + A_i + 3I + L_i^2\lambda_{\max}(B_i^T B_i)I \leq -a_i I, \sum_{j=1}^{k}\ln d_j - \sum_{i=1}^{m}\lambda_i\mu_i(t_0,t) \leq \varphi(t_0,t) \tag{16}$$

Where $\eta_k = \max\{\sup_{t_{k-1}\leq t\leq t_k} e^{\lambda_i(t_k-t)}\}$, $d_k=\max\{1,\eta_i\beta_i^2\}$,

$b_i = L_i^2\lambda_{\max}(A_i^T A_i) + L_i^2\lambda_{\max}(B_i^T B_i)$, $i=\sigma(t), t\in[t_{k-1},t_k), k=1,2,\ldots.$.

And λ_i is the unique positive root of the equation $-a_i + b_i e^{\lambda_i\tau} + \lambda_i = 0$. $\lambda_{\max}(\lambda_{\min})$ denotes the maximum (minimum) eigenvalue of the positive-definite matrix P_i, and $\omega(t_0,t)$ is a continuous function on R^+. Then $\lim_{t\to\infty}\omega(t_0,t) = -\infty$ implies that the trivial solution of systems (1) and systems (2) is globally asymptotically stable.

The proof is same as Theorem 1.

Corollary 3. In Theorem 2, if there exists a positive constant α such that

$$\lambda_i > \alpha > 0 \text{ for all } i \in I \text{, and } \frac{\ln d_k}{t_k - t_{k-1}} \leq \alpha, k = 1,2,\cdots. \tag{17}$$

Then the trivial solution of (1) and (2) is globally exponentially stable.

4 Numeric Examples

In this section, one example will be presented to illustrate the main theoretical results proposed in above section. It is assumed that the hybrid system here has only two subsystems, and the switching sequence is $1 \to 2 \to 1 \to 2 \cdots$. In addition, we assume that $f_i(\alpha) = 0.25(|\alpha+1| - |\alpha-1|), i = 1,2$.

Through computing, we obtain
$$f_i(\alpha) = \begin{cases} -0.5, & \alpha < -1 \\ 0.5\alpha, & -1 \le \alpha \le 1 \\ -0.5, & \alpha > 1 \end{cases} \tag{18}$$

Hence, $L_i = 0.5, i = 1.2$.

Example 1. Consider a hybrid system of the form of (18) as follows

$$\begin{cases} \dot{x}(t) = A_1 x(t) + A_1 x(t-\tau(t) + B_1 f_1(x(t)) + B_1 f_1(x(t-\tau(t))), t \in [KT, KT+\delta T) \\ \Delta x(t) = b_1 x(t^-), t = KT + \delta T \\ \dot{x}(t) = A_2 x(t) + A_2 x(t-\tau(t) + B_2 f_2(x(t)) + B_2 f_2(x(t-\tau(t))), t \in [KT+\delta T, (K+1)T) \\ \Delta x(t) = b_2 x(t^-), t = (K+1)T, K = 0,1,2,\cdots. \end{cases} \tag{19}$$

Where $T = 2, \delta = 0.5, b_1 = b_2 = -4.5, \tau = 0.1$,

$$\text{And } A_1 = \begin{bmatrix} -4 & 0 \\ 0 & -4 \end{bmatrix}, B_1 = \begin{bmatrix} 1 & 0 \\ 0 & 1.2 \end{bmatrix}, A_2 = \begin{bmatrix} -4 & 0 \\ 0 & -4 \end{bmatrix}, B_2 = \begin{bmatrix} 1.2 & 0 \\ 0 & 0.8 \end{bmatrix} \tag{20}$$

Substituting A_1, A_2, B_1, B_2 into inequality (17)、(18) and (19), we obtain a feasible solution as follows $a_i = 4.64, b_i = 4.36, d_i = 1$, And $0 < \lambda_i < \frac{1}{12}, i = 1,2$.

Therefore, based on Corollary 3, the origin of (18) is globally exponentially stable, as shown in the following figure 1.

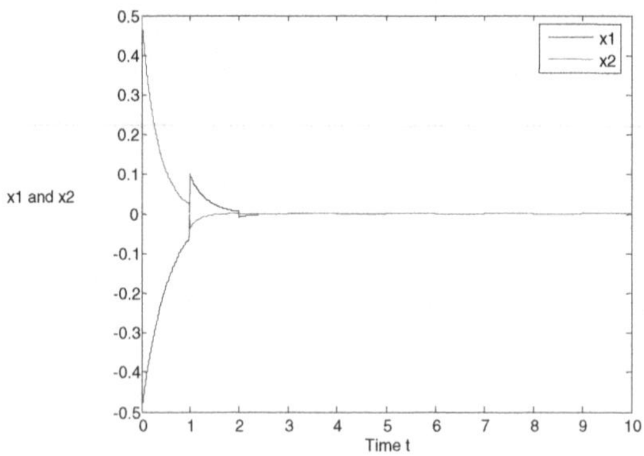

Fig. 1. Trajectories of the system (19) with (20) and the initial value $x(\theta) = [0.5, 0.5]^T$, $\theta \in [-0.1, 0]$.

5 Conclusion

The hybrid impulsive and switching systems model has been formulated based on the existing models of hybrid impulsive and switching systems. The general criteria for the asymptotic and exponential of the new model have been established by using switching Lyapunov functions and linear matrix inequality techniques, the work of this paper has improved the stability of existing results on hybrid systems.

Acknowledgements. This work was supported by the Ministry of Science and Technology Foundation of PR China under Grant No. 2010DFA12160, the Chongqing Science and Technology Commission Foundation under Grant No.2009JJ1276 and Chongqing University of Posts and Telecommunications youth Foundation under Grant No.A2009-50.

References

1. Bertsekas, D.: Note on the design of linear systems with piecewise constant feedback gains. IEEE Trans. Automat. Control 15, 262–263 (1970)
2. Molchanov, A.P., Pyatnitskiy, Y.S.: Criteria of asymptotic stability of differential and difference inclusions encountered in control theory. Syst. Contr. Lett. 13, 59–64 (1989)
3. Nair, G.N., Evans, R.J.: Exponential stabilisability of finite-dimensional linear systems with limited data rates. Automatica 39, 347–535 (2003)
4. Liberzon, D.: Switching in Systems and Control. Birkhauser, Boston (2003)
5. Liao, X., Wong, K.W.: Global exponential stability of hybrid bidirectional associative memory neural networks with discrete delay. Phys. Rev. E, Stat. Phys., Plasmas Fluids Relat Interdiscip Top 67, 042, 90 (2003)
6. Yuan, K., Cao, J., Li, H.X.: Robust stability of switched Cohen–Grossberg neural networks with mixed time-varying delays. IEEE Trans., Syst., Man Cybern. B, Cybern. 36, 1356–1363 (2003)
7. Huang, H., Qu, Y., Li, H.-X.: Robust stability analysis of switched Hopfield neural networks with time-varying delay under uncertainty. Phys. Lett. A 345, 345–354 (2005)
8. Sun, Z., Ge, S.: Switched Linear Systems Control and Design. Springer, London (2005)
9. Branicky, M.S., Borkar, V.S., Mitter, S.K.: A unified framework for hybrid control: Model and optimal control theory. IEEE Trans., Autom., Control 43, 31–45 (1998)
10. Li, X., Soh, Y., Wen, C.: Switched and Impulsive Systems, Analysis, Design and Applications. Springer, Berlin (2005)
11. Guan, Z.-H., Hill, D.J., Shen, X.: On hybrid impulsive and switching systems and application to nonlinear control. IEEE Trans., Autom. Control 50, 1058–1062 (2005)
12. Li, C., Feng, G., Huang, T.: On Hybrid Impulsive and Switching Neural Networks. IEEE Trans. on Sys., Man, and Cybernetics-part B 38, 1549–1560 (2008)

The Construction and Application of Web Services in Semantic Web Based on Ontology

ZhiPing Ding*

Department of Computer Application, Qingyuan Polytechnics,
Qingyuan, 511510, Guangdong, China

Abstract. The semantic web service ontology, OWL-S, is extended for matching user requirements with product specifications at the semantic level, with context information taken into account. Web services provide a suitable technical framework for making business processes accessible within enterprises and across enterprises. The paper puts forward the construction and application of web services in Semantic Web based on ontology. The existing problems of the web service composition in the field of searching service are discussed. Then the blueprint of a searching system based on Semantic-Aware Agent (SAA) is put forward.

Keywords: semantic web, ontology, web services, owl.

1 Introduction

An agent is an active object which possesses certain capabilities to perform tasks, and it communicates with other agents based on the organizational structure to cooperate the accomplishment of tasks. In the artificial intelligence and information integration literature, ontology is a formal, explicit specification of a shared conceptualization. One of the most important reasons is to make decisions in a certain situation most reasonable by efficiently collecting relevant knowledge from heterogeneous domains. The semantic inference engine can also use the ontology and help the blog users to build up semantic queries and set up semantic subscription.

Accessibility and knowledge are key factors for successful defibrillator data analysis. Until recently lack of accessibility to electronic information forced researchers do toilsome tasks to collect data. From the cognitive point of view, human beings recognize, learn and understand the entities and concepts in texts for a complete natural language comprehension. While intuitively quite appealing, such linguistics and proximity heuristics have so far been largely under-explored in the literature of video semantic analysis. Many recommendation mechanisms have continuously tried to develop new methods from various theories that can assist their users to seek more relevant and more effective information. Berners-Lee proposed the

* Author Introduce: ZhiPing Ding (1980.01-),Male,Han,Master of South China University of Technology, Research area: Data Mining, Application System Development, Intelligent algorithm.

D. Jin and S. Lin (Eds.): Advances in EECM Vol. 1, LNEE 139, pp. 167–172.
springerlink.com

Semantic Web as a natural evolution of the traditional Web to allow for the manipulation of content by applications with the capacity to interpret the semantics of information.

Ontology is a conceptualization of a domain into a human understandable, machine-readable format consisting of entities, attributes, relationships, and axioms[1]. To the best of our knowledge, our work is the only one at the time of this writing that produced an OWL representation of the complete FMA rather than its fragments. Existing electronic markets do without ontology-based descriptions of products, but at the expense of rigidly requiring each participant to agree a priori to conform to the standard protocol and product descriptions of the market. A knowledge base is a special database type for representing domain expertise. The repository comprises collections of facts, rules, and procedures organized into schemas. Abstract data types allowed the programmer to define a closer matching between the concept of a system and its representation in software.

Finally, this paper puts forward the construction and application of web services in Semantic Web based on ontology. To integrate the semantic Web to the Web services, new languages were developed for the specification and composition of the services. After being parsed, the web service description document will be compared with the flow process knowledge ontology to improve the accurate rate of the web service composition in searching. This system can also calculate how much are in common among different Services Composition. And at the end it is proved that this system is more effective than the traditional ones.

Ontologies are the knowledge models that provide a controlled vocabulary and semantics for describing and sharing data in a domain (data models). The experimental results indicate that this method has great promise. Therefore, such approaches require extensive standardization efforts, which is also a major problem for ebXML. The advantage of the proposed model consists in the measure is convenient to be calculated from objects and attributes classes sets, especially for the large context. An ontology that would facilitate data sharing would increase the statistical power and validity of findings thereby enhancing our understanding of psychosis and psychotic disorders.

2 Ontology-Based Web Services in Semantic Web

At the very beginning, the existing problems of the web service composition in the field of searching service are discussed[2]. A software system was developed to allow automatic case-based interpretation of images and integrates the material described in the previous sections.

Web services provide a suitable technical framework for making business processes accessible within enterprises and across enterprises. The technology of bringing together heterogeneous and distributed computer systems and establish efficient information sharing is known as information integration, especially semi-structured information integration. The ontology is increasingly seen as key to enabling semantics-driven data access and processing. Geospatial data includes maps, the numerical grid, the bathymetry, and the digital elevation model, while model data includes the state variables and all coefficients and constants, which are geo-referenced to the geospatial data sets.

Ontologies are representations of the knowledge within a domain of interest, defined via the terminology (concepts) used within the domain and the properties and relationships among domain objects. Semantic web services (SWS) technology aims to add sufficient semantics to the specifications and implementations of web services to make possible the (automatic) integration of distributed autonomous systems, with independently designed data and behaviour models. Both intelligent agent and semantic web service technologies are able to reach remarkable achievements and in some cases have overlapping functionalities.

2.1 Semantic Web Services Technology

Web services have begun to show deficiencies in the areas of service description, discovery and composition due to the lack of semantic support in Web Services Description Language (WSDL) and in the mechanism of storage and discovery services of the Universal Description, Discovery and Integration (UDDI). The eXtensible Markup Language (XML) has been the W3C standard document format for exchanging information on the Web. Ontology provides a formal semantic representation of the objects for case representation. Even though automatic ontology learning methods (such as text mining and knowledge extraction significantly support ontology engineers by speeding up their task, there is still the need of a significant manual effort, in the completion, consolidation, and validation of the automatically generated ontology. A lower number of sibling concepts is the result of the greater weight of the concept shown in equation (1). That shows the concept is more important than others. OWL-S and WSMO have the necessary characteristics for the development of semantic specifications for Web services.

$$T_{iB_1} = T'_{iB_1} + \frac{(n_b - 1) \times T_{iB}}{n_b} \tag{1}$$

T'_{iB_1} is the weight of concept B_1 and T_{iB} is the weight of its parent concept B. n_b is the number of B's child concepts, $n_b \geqq 1$. If the sibling concepts appears, the weight of concept B_1 will be enhanced. The location of concepts in web pages and the relationships between concepts are used to cluster web pages. Further, the system uses an RDF framework to describe the results of the semantic analysis(Charlotte et al,1999; klein,2001).This permits more effective querying.

Starting from this observation, we have found that component-based software engineering can be coupled with domain-specific knowledge on how to assemble and combine software components in modelling frameworks. Knowledge-based systems (KBSs) currently provide powerful and flexible means for utilizing expertise to address problems associated with a specific domain[3]. The semantic web service ontology, OWL-S, is extended for matching user requirements with product specifications at the semantic level, with context information taken into account. The framework of the workflow (Figure 1) includes five processes described as follows.

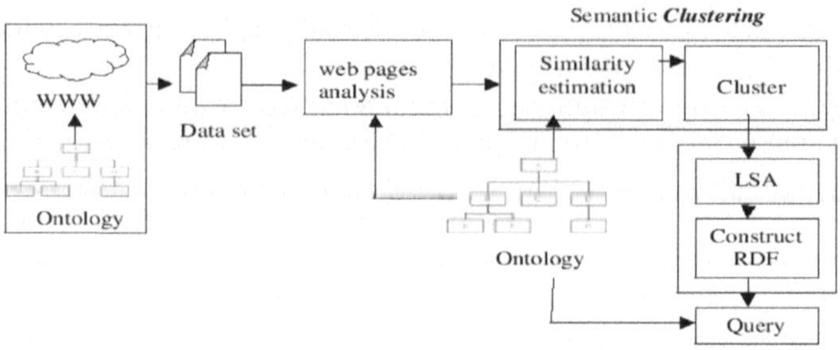

Fig. 1. The construction of web service in semantic web based on ontology

In OWL, properties are binary relations. The class at the tail of an arrow is the domain of the relation and the class at the head of the class is the range of the relation. The ontology is a computational model of some portions of the world. Since descriptions of OWL-S services are based on OWL documents, all the characteristics of the OWL domain model can be utilized to structure the service descriptions, thus facilitating the reuse of previously developed OWL ontologies.

2.2 Ontology-Based Web Service in Semantic Web

Ontologies are described using ontology languages, like The Web Ontology language (OWL), which enables explicit representation of meaning of terms in vocabularies and relationships between those terms. RDF can be used to describe the resources of a given web page, using a meaning graph of the RDF to represent a problem. The RDF accentuates the exchange and automation processing of web resources designated by a Unified Resource Identifier (URI), a string of web resources or an element of XML. The system calculates the similarity value from the relationship between web pages. The values will affect the results of clustering.

Controlled vocabulary CV = name set of concepts c with c = (name, definition, identifier, synonyms). Ontology O = G (CV, E) with $E \subseteq \mathrm{CV} \times \mathrm{CV}$ and a totally defined function t: $E \rightarrow T$, which defines the types of edges. T is the of possible edge types, i.e., the semantics of an edge in natural language.

$$S\_s(T_{12}, T_{22}) = 1 - \frac{|T_{12} - T_{22}|}{Max(T_{12}, T_{22})} \tag{2}$$

The values T_{12}, $T_{22}$are the weights of concept C_2 in the web pages d_1 and d_2 respectively. In equation (2), the level of similarity between the concepts is decided by the different weights of the same concept C_2.

3 The Application of Semantic Web Service Based on Ontology

In this context, ontologies are important artifacts for making the treatment of this heterogeneity feasible. Then the blueprint of a searching system based on Semantic-Aware Agent (SAA) is put forward. In this searching system which uses DAML as a basis, DAML-S is used to create service description document and OWL is used to create flow process knowledge ontology[4]. An OWL ontology may include relations of taxonomy among classes; data type properties and descriptions of class element attributes; object properties and descriptions of relations among class elements; class instances; and property instances.

In addition, the ontology is no longer a mere research topic. Nowadays, its relevance has been recognized in several practical fields. The similarity of the two concepts between web page d_1 and web page d_2, is calculated using the concept C_2 in d_1 and the concepts C_3, C_4, and C_5 in web page d_2 by $S\_c(T_{12}, T_{2j})$ in equation (3), respectively.

$$S\_c(T_{12}, T_{23}) = (1 - \frac{|T_{12} - T_{23}|}{Max\{T_{12}, T_{23}\}}) \times \frac{L(c_2)}{L(c_3)}. \tag{3}$$

The current proposals for the development of a standard for semantic Web services include the Ontology Web Language for Services (OWL-S), Web Service Modeling Ontology (WSMO)and Web Service Semantics (WSDL-S). Finally, we test the average cost time of distributed discovery system and centralized discovery system. Fig.2 describes the match method between DAML-S and WSDL in semantic web services.

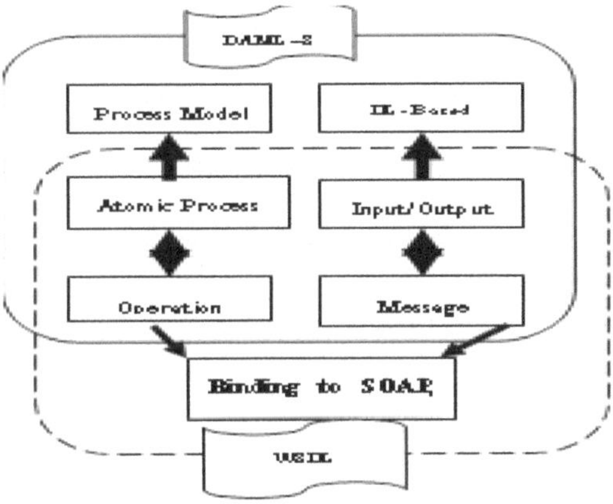

Fig. 2. The match mapping between DAML-S and WSDL

4 Construction Web Services in Semantic Web Based Ontology

This methodology combines an XML Schema to web ontology mapping, called XSD2OWL, with a transparent mapping from XML to RDF, XML2RDF. OWL-S is an ontology for services whose infrastructure, via agents, allows for the discovery, composition, invocation and monitoring of services with a high degree of automation. WSDL-S defines a mechanism for adding semantics to the Web services described in WSDL. After being parsed, the web service description document will be compared with the flow process knowledge ontology to improve the accurate rate of the web service composition in searching. This system can also calculate how much are in common among different Services Composition. And at the end it is proved that this system is more effective than the traditional ones. This paper used JENA JAVA API, provided by PHP, to automatically construct an RDF. The relationship between concepts and clusters is represented by "has" and clusters are described using the RDF "bag".

5 Summary

As the foundation of the semantic web, ontology is a formal, explicit specification of a shared conceptual model and provides a way for computers to exchange, search and identify characteristics. The use of ontologies to describe context and adaptation services constitutes a recent line of research, which is expected to gain strong momentum once the current standards that support Web Service technologies also begin to provide native support to semantic descriptions. In this paper, we discuss the construction and application of web services in semantic web based on ontology.

References

1. Haller, A., et al.: WSMX – a semantic service-oriented architecture. In: International Conference on Web Services, ICWS 2005 (2005)
2. Srini, N., McIlraith, S.: Analysis and simulation of Web services. Computer Networks 42, 675–693 (2003)
3. Fensel, D.: Ontologies: Silver Bullet for Knowledge Management and Electronic Commerce. Springer, Berlin (2000)
4. Fensel, D., Bussler, C.: The Web Services Modeling Framework. Vrije Universities Amsterdam (VU) and Oracle Corporation (2002)

Application of PID Controller Based on BPNN in Temperature Control of Electrothermal Boiler

Deying Gu[1,2] and Guoyu Wang[1,2]

[1] Department of Automation Northeastern University at Qinghuangdao, Qinghuangdao, China
[2] Graduate School of Northeastern University, Shenyang, China
{gdy0335,guoyu.wang}@163.com

Abstract. In order to achieve effective and intelligent control for Water temperature, a PID controller on BPNN-based was used in the Temperature control system of electrothermal boiler, which can adjust PID control parameters K_p, K_i, K_d automatically. While, we design a mathematical model of the electrothermal boiler with least square system identification. The simulation results prove that the Temperature control system of the electrothermal boiler not only gains the better control performance, but also acquires stronger adaptability to math model using the PID controller based on BPNN.

Keywords: electrothermal boiler, BP neural network, PID Controller, least-square system identification.

1 Introduction

The electrothermal boiler is the heat supply equipment that it transforms electrical energy to heat energy. It is the ideal energy conservation heating equipment that have high thermal efficiency, small size, pollution-free, low noise, safe and reliable operation and heat stability. It is known that heating boiler is a very complicated system with certain characteristics in the math model such as strong nonlinearity, large lag, and uncertainty of the parameters due to the work states. So, the traditional PID controller does not work well in this situation. The PID controller based on BP neural network (BPNN) is a new and effective solution. The new method perfectly combines the traditional PID controller and the BP neural network algorithm, in which the parameters of PID, K_p, K_i and K_d are adjusted according to the current model of controlled object online. The advantage of this solution can be known from the charts of the results of the simulation which is mentioned below. In this paper, we mainly use this new method to give a control value U to manage the inverter of the goal feeding roller. When the value U increases, the roller will move faster, so the temperature of the water will be risen higher. Otherwise, the temperature of the water will be deceased.

2 The Principle of PID Controller Based on BPNN

2.1 System Structure of PID Controller Based on BPNN

BP (Back Propagation) neural network, one of the typical forms of prior to network, is based on error Back Propagation algorithm of artificial neural network. The

D. Jin and S. Lin (Eds.): Advances in EECM Vol. 1, LNEE 139, pp. 173–178.
springerlink.com

structure of PID controller based on BPNN is shown in Fig.1. It makes up of traditional PID controller and BPNN. Traditional PID is used for closed-loop controlling controlled object. It takes charge of positive conducting of the control signal. BPNN is used for adjusting the PID parameters to make certain performance index optimal.

Fig.2 shows a BP neural network which has three layers. It contains the input layer, hidden layer and output layer. And the output layer neurons state represent PID controller parameters are K_p, K_i and K_d. In practical applications, when the outside environment changes, this model can make neural network output corresponding to this environment optimal PID control parameters through self-learning and weight value adjustment.

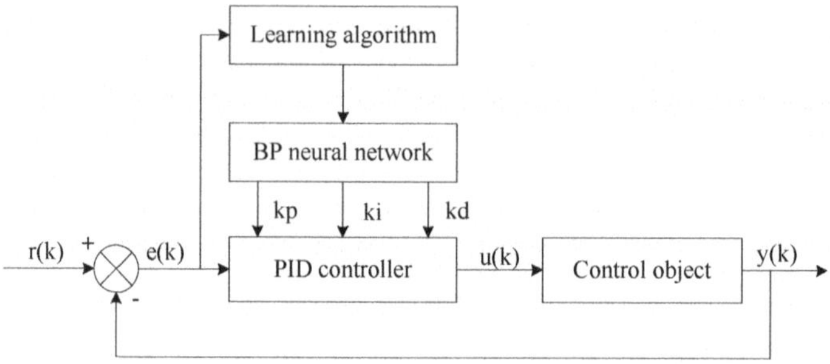

Fig. 1. Structure diagram of PID controller Based on BPNN

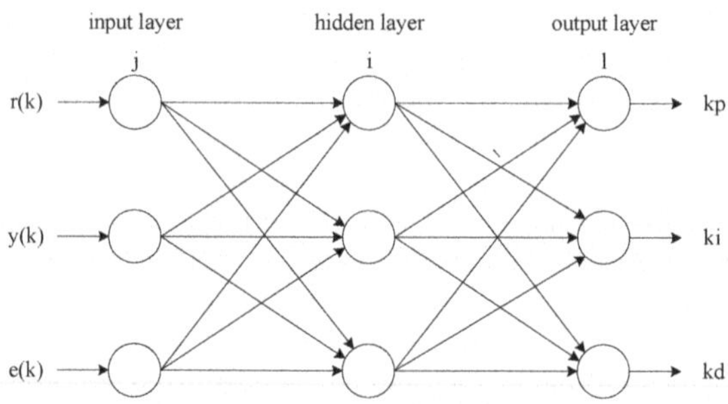

Fig. 2. Structure of BP neural network

2.2 Algorithm of PID Controller Based on BPNN

The discrete algorithm of the traditional PID is shown in (1). Where, variable e is the error between input and output. k_p is the coefficient of proportion. ki is the coefficient of integral. And kd is the coefficient of differential.

$$u(k) = u(k-1) + k_p\left[e(k) - e(k-1)\right] + k_i e(k) + k_d\left[e(k) - 2e(k-1) + e(k-2)\right] \quad (1)$$

Now introducing the BP neural network with three layers as shown in Fig. 2, the inputs of the network are:

$$O_1^{(1)} = r(k), O_2^{(1)} = y(k), O_3^{(1)} = e(k) \tag{2}$$

In this formula, the input variables are depending on the complexity of the controlled system. Each neuron of input layer is in charge of receiving messages from outside. In the application of PID control, input layer receive feedback back error messages.

The inputs and outputs of hidden layer are:

$$\begin{cases} net_i^{(2)}(k) = \sum_{j=0}^{M} w_{ij}^{(2)} O_j^{(1)} \\ O_i^{(2)}(k) = f(net_i^{(2)}(k)), i = 1, 2, \ldots\ldots Q \end{cases} \tag{3}$$

Where $w_{ij}^{(2)}$ is the weight value of hidden layer. On horn standard (1), (2) and (3) respectively represent input layer, hidden and output layer.

Hidden layer is internal information processing layer neural network, which is responsible for information transform. The activation function of hidden neurons takes positive symmetrical Sigmoid function:

The inputs and outputs of the output layer are:

$$\begin{cases} net_l^{(3)}(k) = \sum_{i=0}^{Q} w_{li}^{(3)} O_i^{(2)}(k) \\ O_l^{(3)}(k) = g(net_l^{(3)}(k)),\ l = 1,2,3 \\ O_1^{(3)}(k) = k_p \\ O_2^{(3)}(k) = k_i \\ O_3^{(3)}(k) = k_d \end{cases} \tag{4}$$

Where $O_l^{(3)}(k)$ are the three output nodes if the output layer, which give the results to external. These three output nodes represent the three adjustable parameters of PID controller, K_p, K_i and K_d. The activation function of output neurons is the non-negative Sigmoid function:

Take performance index function is:

$$e(k) = \frac{1}{2}(r(k) - y(k))^2 \tag{5}$$

Where r (k) is the input of this system and y (k) is the output. The whole neural network uses e (k) to adjust according to the negative gradient direction of the weight value and fix network weight factor. And add an inertia which makes the adjustment quickly converge to the global tiny. The inertia is (6).

$$\Delta w_{li}^{(3)}(k) = -\eta \frac{\partial E(k)}{\partial \Delta w_{li}^{(3)}} + \alpha \Delta w_{li}^{(3)}(k-1) \tag{6}$$

In this formula, η is the learning rate, while α is the inertia coefficient. That $\frac{\partial E(k)}{\partial \Delta w_{li}^{(3)}}$ can be expressed by formula (7).

$$\begin{cases} \dfrac{\partial E(k)}{\partial \Delta w_{li}^{(3)}} = \dfrac{\partial E(k)}{\partial y(k)} \cdot \dfrac{\partial y(k)}{\partial u(k)} \\ \dfrac{\partial u(k)}{\partial O_l^{(3)}(k)} \cdot \dfrac{\partial O_l^{(3)}(k)}{\partial net_l^3(k)} \cdot \dfrac{\partial net_l^3(k)}{\partial \Delta w_{li}^{(3)}(k)} \end{cases} \tag{7}$$

Where, y (k) is the transfer function of the controlled object.

And:

$$\frac{\partial net_l^3(k)}{\partial \Delta w_{li}^{(3)}(k)} = O_i^{(2)}(k) \tag{8}$$

Because $\frac{\partial y(k)}{\partial u(k)}$ is unknown, we use $\operatorname{sgn}\left(\frac{\partial y(k)}{\partial u(k)}\right)$ instead of it. This change may cause calculation imprecise, and we fix it by adjusting the learning rate η.

According to (1) and (5), we can get:

$$\frac{\partial u(k)}{\partial O_l^{(3)}(k)} = e(k) - e(k-1) \tag{9}$$

$$\frac{\partial u(k)}{\partial O_2^{(3)}(k)} = e(k) \tag{10}$$

$$\frac{\partial u(k)}{\partial O_3^{(3)}(k)} = e(k) - 2e(k-1) + e(k-2) \tag{11}$$

Through thc analysis mentioned above, we can get the learning algorithm of the output layer weighting coefficients is:

$$\Delta w_{li}^{(3)}(k) = \alpha \Delta w_{li}^{(3)}(k-1) + \eta \delta_l^{(3)} O_i^{(2)}(k) \tag{12}$$

$$\delta_l^{(3)} = e(k)\operatorname{sgn}\left(\frac{\partial y(k)}{\partial u(k)}\right)\frac{\partial u(k)}{\partial O_l^{(3)}(k)} g'(net_l^{(3)}(k)), \tag{13}$$
$$l = 1,2,3......,\ g'(\cdot) = g(x)(1 - g(x))$$

Similarly we can get the earning algorithm of the hidden layer weighting coefficients:

$$\Delta w_{ij}^{(2)}(k) = \alpha \Delta w_{ij}^{(2)}(k-1) + \eta \delta_i^{(2)} O_j^{(1)}(k) \tag{14}$$

Through the above algorithms, we can make the PID three control parameters of the corresponding network output layer adjusting according to the outside environment changes.

$$\delta_i^{(2)} = f'(net_i^{(2)}(k)) \sum_{l=1}^{3} \delta_l^{(3)} w_{li}^{(3)}(k), \tag{15}$$
$$i = 1,2,.......Q,\ f'(\cdot) = (1 - f(x)^2)/2$$

3 Simulation

In order to examine the control effect of PID controller based on BPNN, we use a simulation example to verify it.

Set the order of electrothermal boiler for 2, off-line by least squares system identification, discrete mathematics has been electrothermal boiler model.

$$G(z) = z^{-5} \cdot \frac{0.1998z + 0.0735}{z^2 - 0.5032z + 0.04979}$$

Simulation results shown in Figure 3. The simulation parameters are set as follows: The step input amplitude is 90, the number of cycles is the 150 second, the weighted coefficient is 0.5 and the first test of the learning rate is 0.3.

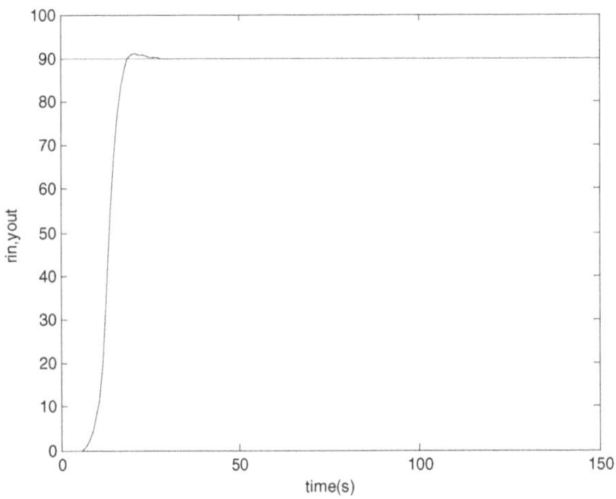

Fig. 3. Temperature curve of electrothermal boiler based on BPNN algorithm

4 Conclusion

The method which combining the traditional PID controller with BP neural network is an advanced and effective solution to operate electrothermal boiler. It not only adjusts its parameters online according to the error function, but also improves algorithm to achieve good performances. The simulation results point out that, using the BPNN PID controller, the control system could acquire better dynamic performance and stronger adaptability to math model.

References

1. Wang, Y.-P., Chen, G.-F.: Design and Application of Single Neuron PID to Burden Control System of Heating Boiler. In: ICCASM, Taiyuan, China, vol. 4, pp. 115–118
2. Nazaruddin, Y.Y., Aziz*, A.N., Sudibjo, W.: Improving the Performance of Industrial Boiler Using Artificial Neural Network Modeling and Advanced Combustion Control. In: International Conference on Control, Automation and Systems, Seoul, Korea, pp. 1921–1926 (2008)

3. Sun, Y., Lin, H.: The Research and Application of Adaptive PID Controller Based on Neural Network Predictive Model. In: Fifth International Conference on Natural Computation, Tianjin, China, pp. 454–459
4. Chen, X.-L.: Neural network adaptive controller simulation. Taiyuan University of technology, Taiyuan, China (2007)
5. Jia, Q., Guo, J.-Y., Zhao, X.-F.: Application of neural network in the boiler combustion control, Instrumentation user, Tianjin, China (2010)
6. Hou, Z.-C., Jiang, Y.-C., Guo, J.: Application of neural network for boiler combustion system identification. Harbin Industrial University Journals, Harbin (2003)

On Creative Learning of University Students

Yu Ran

Department of Art Design,
Zhengzhou Aviation Industry Management Institute,
Zhengzhou China
ranyu8888@163.com

Abstract. In order to meet the need of information age, the university students should develop creative learning approaches that include creative learning concepts, contents, methods and strategies. That is to say university students should have the ability of collaboration and communication, the sense of innovation and creativity, develop strategies and skills, and keep on reflection and evaluation. And also the universities and teachers should offer some cross-disciplinary courses and create an open, democratic learning environment.

Keywords: Information age, University students, Creative learning.

1 Conception and Characteristics of Creative Learning

The word "creativity" has been widely used in various fields. "Creativity" is a constructive action that leads to an increase in the level of organization. The same action may be a creative act when it is done for the first time and mechanical repetition of the past when it is done according to established and known rules by applying standard procedures. Nothing that is produced within the framework of an already existing system of control, whether it is work by a computer or the composition of stereotyped articles, is creativity. Creativity always goes beyond the framework of the system; it is free action. Creativity is a **metasystem** transition. [1]"

The concept of "Creative Learning" first appeared in Oxford, New York, presses in 1979. The author James W. Botkin, Mahdi E1mandjra, Mircea Malitza in the book ***No Limits to learning*** used this phrase initiatively to remind the global environmental problems and energy crisis. Creative Learning is opposed to the traditional way of learning - maintenance learning that is a fixed study to obtain insights, methods, and rules to deal with the known occurrence of the situation. It is essential for closed and fixed situation. While Creative Learning is the ability to cause changes, update, reconstruct and form a series of problem-based learning. Its main feature is integrated for an open environment and system with wide range [2].

The term 'creativity' was widely used in schools but there were variations in what was meant, ranging from an innate attribute to an approach and set of skills that could be cultivated. Creative learning was widely understood to be characterized by:

·questioning and challenging
·making connections and seeing relationships
·envisaging what might be

D. Jin and S. Lin (Eds.): Advances in EECM Vol. 1, LNEE 139, pp. 179–183.
springerlink.com

·exploring ideas, keeping options open
·reflecting critically on ideas, actions and outcomes [3].

Creative learning is an excellent study quality for university students. It should have the following characteristics:

First, collaboration and communication skills, it showed whether students were able to learn effectively in a team, whether actively participate in discussions and make recommendations, listen and respond to other people's point of view, the correct view of their own that can stick to overcome the problem, Able to communicate with others well and to put forward their own views.

Second, creativity, have imaginative and active thinking; can generate new ideas; be good at finding problems and questions; can establish contact with all aspects of the problem; have the spirit of adventurous and practice, can express their creativity artistically or scientifically.

Third, be good at using of strategies and techniques, to identify key issues and design options for a range of programs, and develop plans;

Fourth, the ability of reflection and evaluation, be able to reflect and evaluate their continue learning; be good at summing up and continue to improve and enhance, with an open mind and scattered thinking.

2 The Deficiency of University Students in Creative Learning

Currently, there are many aspects in higher education that are not conducive to cultivate the creative ability of students.

First, the courses are too numerous and narrow. From freshmen to senior students, they have to learn many public and compulsory courses, public selective courses, professional compulsory courses, professional selective courses and other courses. With the continuous updating of knowledge, the courses are increasing too. Thus, students have less and less time in independent study. They are tired of too many courses to study and examinations to take, and no time to reflect, indicate and reorganize the knowledge they learned. Therefore, what they learned are only fragment of the knowledge, without building a whole knowledge system inside their head. This curriculum is based on the name of deconstruction which simply emphasizes on the decomposition and transmission of the knowledge, but ignores the overall systematic nature of knowledge [4]. This deconstruction of the curriculum has increasingly shown its limitations, which hinders the student's divergent thinking and creativity.

Second, due to the traditional authority of educators in our country, the focus is only on the existing scientific concepts, ideas, axioms, facts, and methods accepted and understood by students on the one hand, on the other hand, their scientific interests are ignored. The students don't have enough opportunities to take the initiative to conduct scientific experiments, discussion, analysis and argumentation. Learning content is pre-determined. And the answers to teacher's questions in most cases are also determined in advance.

In such way of “accepted learning", students are not allowed to make their own discoveries [5]. Consequently, the students will lose their curiosity, pleasure and interest in exploring of knowledge. In under the influence of this traditional view of knowledge, the educational concept with imparting knowledge as its basic feature and the learning concept with accepting objective knowledge as its basic feature are built. Students are seen as immature cognitive subject that are having lack of knowledge. Therefore, they need to accumulate knowledge continually with the guidance and training of teachers to improve their cognitive ability. Students' rights of understanding uniquely, questioning and criticizing on knowledge are deprived in the learning process. They must accept the views of textbooks and teachers. The only purpose of learning is to firmly grasp or use of these so-called objective knowledge. So, it is determined that student's learning activities are in a passive position. With such long-term passive approach to learning, it is easy to make them get tired, let alone their fun of innovation.

3 Construction of Students' Creative Learning Model

To improve Students' creative learning ability, students themselves need to develop good study habits which meet the above requirements. On the other hand, universities and teachers have to work hard to offer the students to a good environment for creative learning from the following aspects:

3.1 Developing of Interdisciplinary Curriculum

Cybernetics studies have shown that, in the causal loop relationship, a complex principle is produced. The principle indicates that different causes and different results are interactive aspects of objects, which form the interwoven complexity in life. System theory has proved that the complicated system is composed by the large number of interaction factors. The interaction among them makes them more active than any single factor. These new features require the use of creative learning methods that breaks through traditional boundaries between different sciences. This learning method is developed from multi-disciplinary, multi-dimensional and complementary knowledge. In this new complex mode, a variety of disciplines form a network system with multi-link points among them [6]. In this model, the closed nature and limitation of a single discipline is overcome, which hinder the ability of students to develop creative learning ability and hinder their understanding and dissemination of truth.

Therefore, interdisciplinary curriculum can be good for students to learn creatively. Well-organized interdisciplinary curriculum will enhance the connection in all subjects, so that students can have an overall understanding of knowledge and build a whole system of it. This requires effective planning of the curriculum, and put forward higher requirements of qualifications and competence of teachers. UK research also indicated that interdisciplinary curriculum had an important positive effect on student achievement and personal development [7]. Interdisciplinary curriculum enables students to break the boundaries of traditional disciplines, to creatively apply the technology as a resource [8].

Interdisciplinary curriculum planning should be based on the basic requirements of the national curriculum. It should offer students a teaching unit or module involved in the various disciplines of knowledge through practice and skills training to have a better understanding and mastery of the basic characteristics of each subject. It constructs links between disciplines to ensure that the uniqueness of each subject isn't diminished. Interdisciplinary programmer can promote the development of a wide range of skills for the future needs of students, including the ability to work independently and ability to collaborate with others. This requires universities or discipline institutes to strengthen cooperation with the industry, so that students and teachers can participate in the research, and specific projects of industry. In addition, interdisciplinary courses also put forward higher requirements for teachers. Universities should recruit and train academic leaders and talents of interdisciplinary curriculum.

3.2 Offering an Open, Democratic and Equal Learning Environment

An open, democratic and equal learning environment can create an atmosphere of innovation and creativity, so that students feel that the likelihood of success. Teachers should make students understand that there isn't only one correct answer. They should encourage students to think creatively, to find various possible solutions, and to evaluate the effect of each solution.

However, learning environment in China has many aspects for improvement. As reported by ***"Oriental Morning Post"*** on September 25, 2010, in the article ***"A student with poor performance in China became a genius in American ",*** a student named Wang Nanzi , was in a middle school of Shanghai 8 years ago. He was considered a poor performance student by all teachers. He used to "interrupt" teachers while teaching, had many questions for them. So teachers disliked him and let him sit at the back of the classroom. In desperation, his father sent him to the United States to study. He did the same in American classroom. Once in the class he pointed out the teacher's mistakes, to his surprise, the American teacher didn't get angry and blamed his impoliteness as what Chinese teachers usually do, but appraised loudly and said to him "You're a genius!" Eight years later, he was awarded the Nation's Championship in Individual Animation Competition in the United States. Wang Nanzi is now a senior at the Philadelphia College of Art, and the best student of Animation School. Teachers and students there considered him as a genius. This case tells us something that it worthies our education authorities, teachers and educators to reflect the problems in our education. As the report pointed out, the soul and spirit of education is no longer exists nowadays. This kind of education can not even be called "education ", but only referred to as "Training for Testing"[9]. A good education has a long-term humanity influence and spirit in the classroom. Such kind of cultural humanism atmosphere is like a subtle flavor to make the students enjoy in.

Many studies at home and abroad have shown that, a good education should encourage students to ask questions, doubt, debate, experimentation, free expression, thinking critically and face the challenge bravely, which can enable students to have a personal sense of accomplishment. They will gain confidence and encourage them to be more creative and express their own views.

References

1. Patarakin, E.: Creativity and Creative Learning in the Context of Electronic Communication Network. A Framework for Analysis of Practice and Research 4, 1–13, LDI Working Paper (2003)

2. Creative Learning, `http://baike.baidu.com/view/4203870.html`
3. Ofsted. Learning: creative approaches that raise standards, `http://www.ofsted.gov.uk`
4. Peng, H.: Indian Higher engineering education reform, experience. Problems and Inspiration. Education Forum of Fudan 2, 80–84 (2008)
5. Shi, Z.: Knowledge Transition and Educational Reform, p. 120. Education Science Press, Beijing (2002)
6. Peng, H.: Engineering Education in India. Higher Education of Science 6, 57–60 (2007)
7. Creativity: find it, promote it (QCA/05/1596). Qualifications and Curriculum Authority, `http://www.qcda.gov.uk`
8. All our futures: creativity, culture and education, National Advisory Committee on Creative and Cultural Education (1 84185 034 9). DfEE (1999)
9. A student with poor performance in China became a genius in American. Oriental Morning Post (September 25, 2010)

The Principal Component Analysis in the Application of Supplier Evaluation

Qingkui Cao and Yujia Zhang

Hebei University of Engineering, 199 South Guangming Street, 056038,
Handan City, Hebei Provice, P.R. China
zhangyujia0224@126.com

Abstract. The supplier evaluation is one of core work for realizing supply strategy. This paper analysis and comprehensively evaluate the suppliers with the principal component analysis, given the sort of each product supplier equally, providing a standard to the enterprise for selecting the suppliers.

Keywords: Principal component analysis, Suppliers, Supply chain management, Evaluation.

1.1 The Standardization of Data (Data Processing)

With the sample space have m evaluation objects, index space has n index from different reflect the characteristics of the evaluation index, so as to form the n m d column with said, are as follows:

$$(x_{ij})_{mxn} = \begin{pmatrix} x_{11} & x_{12} & \cdots & x_{1n} \\ x_{21} & x_{22} & \cdots & x_{2n} \\ \cdots & \cdots & \cdots & \cdots \\ x_{m1} & x_{m2} & \cdots & x_{mn} \end{pmatrix} \tag{1}$$

$$y_{ij} = \frac{x_{ij} - \overline{x_j}}{s_j} \tag{2}$$

The principal component is by the amount of special linear combination, if each component of the different dimensions, so this kind of linear combination significance is vague, and it is difficult to give a reasonable explanation. In order to eliminate the original data index the difference between dimensional, orders of magnitude, between the index to mutual comprehensive numerical, the need to (2) type of original data standardization.

Among them, is after standardization of data, and respectively is the first j An index of the mean and standard deviation, and sample after standardization data into:

D. Jin and S. Lin (Eds.): Advances in EECM Vol. 1, LNEE 139, pp. 185–191.
springerlink.com

$$(y_{ij})_{m\times n} = \begin{pmatrix} y_{11} & y_{12} & \cdots & y_{1n} \\ y_{21} & y_{22} & \cdots & y_{2n} \\ \cdots & \cdots & \cdots & \cdots \\ y_{m1} & y_{m2} & \cdots & y_{mn} \end{pmatrix} = (\beta_1, \beta_2, \ldots \beta_n) \tag{3}$$

1.2 Establish Standardized Data Vector of the Covariance Matrix R

$$R = (r_{ij})_{n\times n} \tag{4}$$

Among them, $r_{ij} = \frac{1}{m-1}\sum y_{ki} \bullet y_{kj}$, $i, k = 1, 2, \ldots, n$

1.3 For the Covariance Matrix R the Characteristic Value of Characteristic Vector and

For R is n order real symmetric matrices, so its characteristic value are all real, it will be from big to small descending order for: $\lambda_1 \geq \lambda_2 \geq \cdots \geq \lambda_n$, The size of the characteristic value of the principal components in the description says evaluated object the size of the role. The characteristic value of the corresponding feature vector notes for

$$\lambda_i = (C_1, C_2, \ldots, C_m)^T \tag{5}$$

1.4 The Variable, Determine Contribution Calculated the Number of Main Component

The primary contribution $b_i = \lambda_i \Big/ \sum_{i=1}^{n} \lambda_i$, usually, as long as the choice before the Lord composition, make k the accumulative total contribution rate of more than 85% and can meet the requirements. Namely:

$$b_1 + b_2 + \ldots + b_k \geq 85\% \tag{6}$$

1.5 The Comprehensive Evaluation of k Principal Components

Seeking first k principal component to each of the main components of linear expression,

$$Z_i = l_{1i} \cdot \beta_1 + l_{2i} \cdot \beta_2 + \ldots + l_{ni} \cdot \beta_n \tag{7}$$

Again to k Lord ingredients of the weighted summation, weighted function for each principal component contribution rate, namely:

$$w_i = b_i \Big/ \sum_{i=1}^{k} \lambda_i \tag{8}$$

Comprehensive income:

$$Z = w_1 Z_1 + w_2 Z_2 + \ldots + w_k Z_k = \left(C_1, C_2, \ldots, C_m\right)^T \tag{9}$$

According to $C_1, C_2, \ldots, C_m$ each index of the size of the order.

2 Supplier Evaluation Applications

An enterprise has eight suppliers, according to the original score data collection (enterprise internal supplier evaluation and selection group provides) (see table 1), the application of principle component analysis compare to other suppliers comprehensive ability, quality evaluation, and then analysis to identify the core competence of suppliers and cooperation supplier.

Table 1. 8 Suppliers Comprehensive Ability Score Primitive Original Data

Indexes / Serial Numbers	The Product Quality X1	Supply Capacity X2	Supply Price X3	Technology Level X4	Rapid Response Ability X5	Geographical Position X6	Delivery Accuracy X7
Supplier 1	89	84	89	85	90	83	87
Supplier 2	84	82	79	81	97	84	78
Supplier 3	83	85	78	78	87	75	82
Supplier 4	90	89	85	88	96	90	95
Supplier 5	87	80	81	84	95	84	92
Supplier 6	91	90	75	78	91	79	91
Supplier 7	85	88	86	83	89	84	92
Supplier 8	82	87	83	84	90	85	88

2.1 Analysis Process and the Results

(1) According to the formula (2) of original data normalization, table 2

Table 2. 8 Home Supply Comprehensive Ability After Standardization Data

Index / Serial Number	The Product Quality X1	Supply Capacity X2	Supply Price X3	Technology Level X4	Rapid Response Ability X5	The Geographical Position X6	Delivery Accuracy X7
Supplier 1	0.777	-0.463	1.512	0.686	-0.514	0.000	-0.197
Supplier 2	-0.703	-1.035	-0.648	-0.469	1.407	0.226	-1.778
Supplier 3	-0.999	-0.1784	-0.864	-1.336	-1.338	-1.815	-1.075
Supplier 4	1.073	0.963	0.648	1.552	1.132	1.588	1.207
Supplier 5	0.185	-1.606	-0.216	0.397	0.857	0.226	0.680
Supplier 6	1.369	1.249	-1.512	-1.336	-0.240	-0.907	0.505
Supplier 7	-0.407	0.678	0.864	0.108	-0.789	0.226	0.680
Supplier 8	-1.295	0.392	0.216	0.397	-0.514	0.453	-0.022

(2) According to the formula (4) standardized data obtained the covariance matrix, have to table 3

Table 3. Association party gap array

1	1.0000	0.3154	0.0640	0.2214	0.2830	0.2111	0.5766
2	0.3154	1.0000	-0.0176	-0.0250	-0.3289	0.0648	0.4684
3	0.0640	-0.0176	1.0000	0.8024	-0.0593	0.5811	0.3198
4	0.2214	-0.0250	0.8024	1.0000	0.4150	0.8988	0.5247
5	0.2830	-0.3289	-0.0593	0.4150	1.0000	0.6495	0.0491
6	0.2111	0.0648	0.5811	0.8988	0.6495	1.0000	0.4725
7	0.5766	0.4684	0.3198	0.5247	0.0491	0.4725	1.0000

(3) For the characteristics of the covariance matrix root and characteristic vector and have to table 4

Table 4. Covariance matrix root and the characteristics of characteristic vector

Characteristic Value / Feature Vector	3.1808	1.6875	1.2204	0.5576	0.3143	0.0341	0.0053
1	0.2662	0.3853	0.4846	-0.5815	-0.3934	-0.0580	-0.2229
2	0.0873	0.6554	-0.0637	0.5983	-0.3820	-0.0572	0.2270
3	0.3882	-0.0971	-0.5787	-0.3088	-0.3373	0.4361	0.3251
4	0.5272	-0.1657	-0.1963	-0.0038	0.0314	-0.8053	0.0807
5	0.2734	-0.3952	0.6202	0.2094	-0.0855	0.1604	0.5546
6	0.5152	-0.1905	0.0365	0.3950	-0.0400	0.2968	-0.6717
7	0.3907	0.4389	0.0419	-0.0929	0.7587	0.2021	0.1670

(4) The variable determine contribution calculated the number of main component, getting table 5

Table 5. Main components of variance contribution

Sequence Numbers	Characteristic Value	Variance Contribution	Total Variance Contribution
1	3.1808	45.44%	45.44%
2	1.6875	24.11%	69.55%
3	1.2204	17.43%	86.98%
4	0.5576	7.97%	94.95%
5	0.3143	4.49%	99.44%
6	0.0341	0.49%	99.92%
7	0.0053	0.08%	100.00%

From table 5 visible, with three main variable accumulated explain the contribution rate of 86.98% of the total, which retain the original index 85% of the information. (5) to 3 main ingredients of the comprehensive evaluation Main components of linear expression is as follows,

$$Z_1 = 0.2662 \cdot \beta_1 + 0.0873 \cdot \beta_2 + \ldots + 0.3907 \cdot \beta_7$$
$$Z_2 = 0.3853 \cdot \beta_1 + 0.6554 \cdot \beta_2 + \ldots + 0.4389 \cdot \beta_7$$
$$Z_3 = 0.4846 \cdot \beta_1 - 0.0637 \cdot \beta_2 + \ldots + 0.0419 \cdot \beta_7$$

From the formula (8) calculation of main components of each contribution rate:

$$w_1 = 0.0746,\ w_2 = 0.0396,\ w_3 = 0.0286$$

Comprehensive income:

$$Z = w_1 Z_1 + w_2 Z_2 + \ldots + w_k Z_k$$

$$= (0.0344, -0.1304, -0.2375, 0.2636, 0.0278, 0.0458, 0.0332, -0.0369)^T$$

Table 6. 8 Suppliers new samples comprehensive ability

Principal Component / Serial Number	C1	C2	C3	Total Value	Sort
supplier 1	0.0669	-0.0059	-0.0266	0.0344	3
supplier 2	-0.0724	-0.0867	0.0286	-0.1304	7
supplier 3	-0.2270	0.0082	-0.0186	-0.2375	8
supplier 4	0.2268	0.0200	0.0169	0.2636	1
supplier 5	0.0486	-0.0439	0.0231	0.0278	5
supplier 6	-0.0861	0.0873	0.0446	0.0458	2
supplier 7	0.0381	0.0298	-0.0347	0.0332	4
supplier 8	0.0050	-0.0088	-0.0332	-0.0369	6

2.2 The Result Analysis

From table 6, which can be seen after the primary analysis of 8 suppliers after comprehensive ability from seven index dropped to three, the Lord variables seven factors transform into three factors. With this new 3 main factors on the group 8 suppliers comprehensive ability to sort through the following formula: Total value = $c_1 \times 45.44/86.98 + c_2 \times 24.11/86.98 + c_3 \times 17.43/86.98$, The sample sort, greatly reducing the workload (such as table 6).

3 The References Section

This article describes the principal component analysis method in supplier evaluation of comprehensive ability of using an enterprise application, a supplier evaluation and selection of supplier evaluation team made the original data of the comprehensive ability of 8 suppliers are evaluated and ranking, the method is proven to be practical. The method is a simple and easy of comprehensive evaluation method.

References

1. He, X., Duo, Y.: Statistical Analysis. People's University of China Publishing Press, Beijing (2004)
2. Liu, X., Li, H., Wang, C., Chu, C.: The Supplier Selection Model and Method of Review. The China Management Science (1) (2004)
3. Zhang, W.: SPSSll Statistical Analysis Tutorial. Beijing Electronic Hope Press, Beijing (2002)
4. Ma, S., Lin, Y.: Supply Chain Management. Mechanical Industry Press, Beijing (2005)
5. Wang, X., Huo, D.: The Principal Component Clustering Analysis in the Application of Coal Mine Safety Evaluation. Chinese Mining Industry 2 (2009)

AHP Method in Computing Factor Weight of the Network Learning Pattern Recognition

Xiang Zhao[1], Qi Zhao[2], and Gang Chen[3]

[1] 6th Dept. of Shijiazhuang Mechanical Engineering College Shijiazhuang Hebei, P.R. China
zxcg2001@sina.com
[2] Dept. of Information Research, Shijiazhuang Army Command Academy,
Shijiazhuang, Hebei, China
zhaoqi19781216@yahoo.com.cn
[3] HeBei Provincial Military, Shijiazhuang, Hebei, China
davidtqmm@yahoo.com.cn

Abstracts. The factors cluster analysis is needed in the Network learning pattern recognition (NLPR) process. In order to analysis factors more effectively and exactly, the paper proposed the Analytic Hierarchy Process (AHP) method to ascertain the factor weights. Firstly, the paper described AHP theory and its computing procedure. Then, combined with the factors in the Network Learning System, the paper introduced the factor weights computing approach and the result. The AHP method has higher logicality and reliability which in the analysis of the factors relative importance.

Keywords: network learning pattern recognition, weight, AHP, eigenvector.

1 Introduction

Network learning pattern recognition (NLPR) is a new evaluation method for network learning. It is used to distinguish the learners' identity and their learning mode. How to reregister and distinguish the learning factors of the network learners when they are studying is the key of NLPR system influences the accuracy of recognition directly.

The methods ago think the factors influence is same and doesn't consider different of the factor index influence degree in the clustering analysis. In fact, the influence degree is different obviously. It is needed to distinguish the primary and secondary degree. In order to reflect the factors importance degree, it's needed to adopt the factor weight of the character factors. The factor weight aggregation ai= $(a_1,a_2,a_3,\ldots,a_n)$ is called the factors weight set. The factor weight should be impersonality and avoid the subjective jamming. The w_i (i=1, 2, ……, n) is satisfied with normalization and non-negative. $\sum_{i=1}^{n} w_i = 1$, $w_i \geq 0$, (i=1, 2, ……n) . The eigenvector u_i (i=1, 2, ……n) can be regarded as to the importance degree set. It's also used to distinguish the learners' identity and their learning mode. The data which are got by the NLPR system may be used to analyze the individuation network learning mode effectively. It's also used to be

D. Jin and S. Lin (Eds.): Advances in EECM Vol. 1, LNEE 139, pp. 193–197.
springerlink.com

reliable basis to evaluate network learning mode. AHP method is a kind of mature and effective evaluation method. The paper applied AHP method to make sure the quality evaluating factors weight.

2 Analytic Hierarchy Process (AHP)

The Analytic Hierarchy Process (AHP) was proposed by the American university of Pittsburgh mathematics professor T.L.Saaty in middle of 70's. It's a qualitative analysis and quantitative analysis method. It's also a systematical and hierarchy analysis method. It divided the system into a series of seriate and hierarchy index system. According to objective judgment, the index importance degree is quantified. Finally, it solved the problem by compositor result analysis. The basic idea of the AHP method is shown in fig.1. This method compares the index relative importance degree of the evaluation indexes in pair and give out the proportion level. It constructs a judgment matrix. It gets the eigenvector by the square root method. The vector is the factor relative importance sequence. Through finding the answer of the judgment equation, the max latent root is get. Finally, it should check the judge matrix whether is satisfied consistency. The key point is the building of judgment matrix and computing the max latent root.

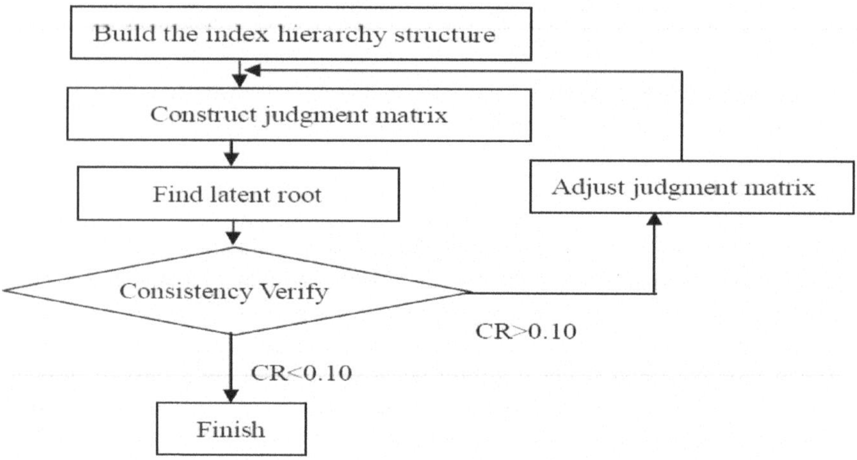

Fig. 1. Flow chart of AHP Method for Weight Computing

3 Ascertain of the Classify Learning Factors Weight Based on AHP

3.1 Build the Factor Hierarchy Structure

Firstly, we need to build the factors hierarchy structure according to the index importance degree. Five factors are extracted of the Network learning pattern recognition. They are $u_i=\{u_1, u_2,u_3,\ldots,u_n\}$, u_1(multimedia resource order program ability)、u_2(self-study control ability)、u_3(interaction control ability)、u_4(study mission finish circs)、u_5(teacher instruct circs)。 The five factors are on the same layer in the hierarchy structure.

3.2 Construct Judgment Matrix

Supposing $u_i=\{u_1, u_2,u_3,\ldots,u_n\}$ is the classify index set, $a_{ij} = f(u_i,u_j)$, if u_i is important to u_j, then $a_{ij} > 1$. If u_j *is* important to u_i, then $a_{ij} < 1, a_{ij} = \frac{1}{a_{ji}}$. At the same time $a_{ii} = 1$.

The AHP method evaluates the importance degree is used the natural number1,2,3, ……, 9, and its reciprocal number 1/2,1/3,…, 1/9. The meaning of the number is shown at the table 1.

Table 1. Table for Evaluation Measure of Evaluating Matrix

Evaluating number	Relative importance
1	u_i and u_j same important
3	u_i is a little more important to u_j
5	u_i is important to u_j
7	u_i is important to u_i
9	u_j is absolutely important to u_i
2, 4, 6, 8	important degree is between the number

Then we can get the judgment matrix. $A = \begin{bmatrix} a_{11} & a_{12} & \cdots & a_{1n} \\ a_{21} & a_{22} & \cdots & a_{2n} \\ \cdots & \cdots & \cdots & \cdots \\ a_{n1} & a_{n2} & \cdots & a_{nn} \end{bmatrix}$

In the NLPR system, we ascertain the importance degree according to the factors affiliated degree. After exacting the student learning factors, we get the importance degree that factor to the main behavior.

$r(x_0,u_1)=0.7658$, $r(x_0,u_2)=0.8567$, $r(x_0,u_3=0.7354)$, $r(x_0,u_4=0.7936)$, $r(x_0,u_5)=0.7604$

We quantified the important degree by the number rules in table 1. Finally, we get the judgment matrix of the NLPR system.

$$A = \begin{bmatrix} 1 & 0.25 & 4 & 0.33 & 2 \\ 4 & 1 & 9 & 2 & 5 \\ 0.25 & 0.11 & 1 & 0.14 & 0.33 \\ 3 & 0.5 & 7 & 1 & 4 \\ 0.5 & 0.2 & 3 & 0.25 & 1 \end{bmatrix}$$

3.3 The Computing of the Factor Weight

Stp1: according to the multiple root method $M_i = \prod_{j=1}^{n} A_{ij}$ (2)

$$W_i = \sqrt[n]{M_i} \tag{3}$$

Stp2: Normalization of the factors weight vector W

$$W_i = \frac{\overline{w_i}}{\sum_{i=1}^{n} \overline{w_i}} = (w_1, w_2, \ldots\ldots, w_n) \tag{4}$$

Here, we get the result is W= (0.15, 0.4, 0.5, 0.3, 0.1).

3.4 The Factor Weight W Consistency Verify

Stp1: Computing the max latent root $\lambda_{\max}$ of the judgment matrix

Solving the equation $AW = \lambda_{\max} W$

$$\lambda_{\max} = \sum_{i=1}^{n} \frac{(AW)_i}{n \times W_i} \tag{5}$$

Where:(AW)i is the ith component in AW vector, the computing result is $\lambda_{\max} = 5.0196$.

Stp2: Computing the consistent ration of the judge matrix

$$CI = \frac{\lambda_{\max} - n}{n-1} \tag{6}$$

n in the equation is the rank of judge matrix. Take $\lambda_{\max} = 5.0196$ and n=5 into equation (6), then we get CI=0.00497

Stp3: Computing the random consistent ration

$$CR = \frac{CI}{RI} = \frac{0.0049}{1.12} = 0.004375$$

The R.I. in the equation can be checked out from table 2, which is the average random consistency index table.

Table 2. Table of RI and Rank Value

n	1	2	3	4	5	6	7	8	9	10	11	12
RI	0	0	0.52	0.89	1.12	1.24	1.35	1.42	1.46	1.49	1.52	1.54

When C.R. is less than or equal to 0.1, we generally think that the judges matrix consistency can be accepted. If the consistency examine in the layer is dissatisfied and then it's needed to adjust the judging matrix. According to the method above, we get the five factors weight [0.15 0.4 0.05 0.3 0.1].

4 Conclusion

Comparing with the other factors weight computing method, AHP method has different application area. The importance degree analysis can be more logistical, more detailed and it can get higher Credibility after it bring into the mathematics process. This method is recommended especially when the lack of the sample data and there are many qualitative index in the evaluation.

References

1. Zhao, Q., Xue, B., Zhao, B.: Collecting System for Web Studying Factors Base on Mobile Agent. Ordnance Industry Automation (4), 35–38 (2005)
2. Satty, T.L.: The analytic hierarchy process: planning, priority setting. Megraw-Hill, New York (1998)
3. Song, B.: Research on comprehensive evaluation of equipment support quality of equipment scientific examination. The Shijiazhuang Mechanic Engineering College Academic thesis, 13–17 (March 2007)
4. Xue, B., Song, Z.-G., Zhao, B.: Research on Factor Extraction Method in Network Study Pattern Recognition Based on Mobile Agent. Journal of Ordnance Engineering College (2), 54–57 (2004)

Scale Invariant Target Recognition in Non-uniform Illuminated and Noisy Scene

Yu Zhao and Baohua Bai

Institute of Communication Engineering, Jilin University
Changchun, Jilin, China
szhaoyu@yahoo.com.cn

Abstract. A new method is presented for scale invariant target recognition under non-uniform illumination and noisy conditions. Both overlapping additive noise and nonoverlapping background noise are considered. The method includes two steps. First, the input scene is preprocessed with estimated illumination function. Then an optimum correlation filter is designed based on modified Wiener filter(WF) model, which the reference term of WF's transfer function is synthesized using the mean of training images . Simulation results verify the effectiveness of the proposed idea in terms of tolerance to scale distortion in presence of non-uniform illumination and noise.

Keywords: correlation pattern recognition, scale invariance, non uniform illumination, wiener filter.

1 Introduction

Pattern recognition using the correlation operation has been vastly applied in the last decades because it can be easily implemented in real-time optodigital processors[1][2]. Various correlation filters were provided to achieve noise robustness and distortion -invariance. However, few techniques were proposed for pattern recognition under different illumination conditions, especially non uniform illumination, which mcans the input scene is affected by multiplicative non-stationary signals[2]. Among the designs for non-uniform illumination, some perform well in the absence of additive noise[3][4]; others can accommodate additive noise but are limited to target size, which means the performance deteriorates quickly when the target size enlarges[2][5][6].

In this paper, we study scale invariant recognition under non-uniform illumination and noisy conditions. The above mentioned designs aren't appropriate for this situation when the scale changed targets embedded in additive noise, especially when the scale factor is getting bigger. In our previous work[7], inspired by Victor H. Diaz-Ramirez and Vitaly Kober's design[2], we also proposed a two step method for non-uniform illumination, which is robustness to noise and the performance isn't affected by target size. But the method provides no design for distortion problem. In this paper, to achieve scale invariant recognition, we modify the reference term of WF function in the formal article by mean of training images. So the new design shows tolerance to scale distortion under non-uniform illumination and noisy conditions.

D. Jin and S. Lin (Eds.): Advances in EECM Vol. 1, LNEE 139, pp. 199–204.
springerlink.com

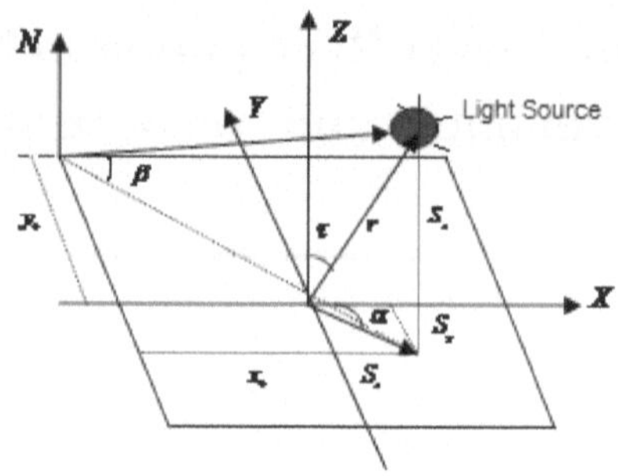

Fig. 1. Illumination model

2 Analysis

2.1 Non-uniform Illumination Models

In this case, the input scene signal depends on the light source and surface shape. In our work, we utilize the Lambertian reflectance model which reflects light equally in all directions, as shown in Fig. 1. Its reflectance map is given by [2]

$$I(x_0, y_0) = \cos(\theta) = \cos(\frac{\pi}{2} - \beta) \cdot \tag{1}$$

Where

$$\beta = \arctan\left[\frac{S_z}{[(S_x - x_0)^2 + (S_y - y_0)^2]^{\frac{1}{2}}}\right]$$
$$= \arctan\left[\frac{r}{\cos(\tau)\left[(r\times\tan\tau\times\cos\alpha - x_0)^2 + (r\times\tan\tau\times\sin\alpha - y_0)^2\right]^{\frac{1}{2}}}\right] \tag{2}$$

It is obviously that $I(x_0, y_0)$ depends on the illuminant direction parameters τ (slant angle), α (tilt angle) and r (magnitude of vector S_0). In real applications these parameters are either known or can be estimated [2][8].

2.2 Scale Invariant Target Recognition Under Non-uniform Illumination and Noise Conditions

In this section, one-dimensional notation is adopted for simplicity. Let $\{r_i(t), \quad i = 1,2,...N\}$ be a training set of scaled versions of targets of true class, and a set of the corresponding window functions is represented by $\{w_{ri}(t), \quad i = 1,2,...N\}$. $w_{ri}(x)$ is a binary function defined as zero within the target area and unity elsewhere.

Suppose that input scene $f(x)$ contains a scale distorted target in nonoverlapping background $b(x)$, and corrupted by non-illumination function $u(x)$ and additive overlapping noise $n_a(x)$:

$$f(x)=[\sum_{i=1}^{N} a_i[r_i(x-x_0)+w_{ri}(x-x_0)b_{ri}(x)]u(x)+n_a(x)^{\cdot} \tag{3}$$

Where $\{a_i, i=1,2,...N\}$ are binary random variables. $a_i=1$ denotes the existence of $r_i(t)$, which is one of the training set, and $a_i=0$ denotes its complement. Assume that the appearance of targets of training set at input scene are equal likely with the probability[5]

$$p(a_i=1)=\frac{1}{N}, \qquad p(a_i=0)=1-\frac{1}{N} \tag{4}$$

In this method, first, the input scene is preprocessed by non-illumination function $u(x)$, which equals to function given in (1). Thus it can be estimated. Then the preprocessed input scene can be written as

$$\begin{aligned} f(x)&=\sum_{i=1}^{N} a_i[r_i(x-x_0)+w_{ri}(x-x_0)b_{ri}(x)]+\frac{n_a(x)}{u(x)}. \\ &=\sum_{i=1}^{N} a_i r_i(x-x_0)+n(x) \end{aligned} \tag{5}$$

Where $n(x)$ is a new noise term defined as

$$n(x)=\sum_{i=1}^{N} a_i w_{ri}(x-x_0)b_{ri}(x)+\frac{n_a(x)}{u(x)} \tag{6}$$

Next, an optimum correlation filter is designed for $f(x)$ based on wiener filtering theory [9]. To achieve scale invariant recognition, we modify the reference term of WF function by mean of training images.

$$E[\sum_{i=1}^{N} a_i r_i(x)]=\frac{1}{N}\sum_{i=1}^{N} r_i(x) \tag{7}$$

$$R(\omega)=\frac{1}{N}\sum_{i=1}^{N} R_i(\omega) \tag{8}$$

$$H(\omega)=\frac{\frac{1}{N}\sum_{i=1}^{N} R_i(\omega)}{\left|\frac{1}{N}\sum_{i=1}^{N} R_i(\omega)\right|^2+|N(\omega)|^2} \tag{9}$$

Where $R_i(\omega)$ and $N(\omega)$ are Fourier transform of $r_i(x)$ and $n(x)$.

3 Simulation Results

Computer simulation has been carried out using MATLAB soft ware to validate the proposed recognition method.

In this paper, for simplicity, the training set includes 6 scaled images, from a range of 1 to 2 in increment 0.2.

Fig. 2 (a) and (b) show test scene of target with scale factor 1 and 1.8 in a background under nonuniform illumination. Here, nonoverlapping background is modeled by low frequency color noise. The lambertian illumination function with $\tau = 80^{\circ}$, $\alpha = 120^{\circ}$ and $r = 7$ is shown in Fig. 3. The observed scenes is obtained from the test scene degraded by additive Gaussion white noise with zero-mean and standard deviation $\sigma = 0.4$, as shown in Fig. 4 (a) and (b). Fig. 5 shows the correlation output for both targets.

It can be seen that the correlation peak is sharp and distinct and can be easily detected. To further evaluate the performance of proposed method, we calculate the values of discrimination capability (DC) for different scaled targets, as shown in Table 1. The illumination function and noise conditions are the same as Fig.3 and Fig.5.

DC is defined as

$$DC = 1 - \frac{\left|C^{B}\right|^{2}}{\left|C^{T}\right|^{2}} \tag{9}$$

Where $\left|C^{B}\right|^{2}$ is the maximum intensity in the correlation plane over the background and $\left|C^{T}\right|^{2}$ is the maximum intensity in that of the target area. Higher value of DC indicates better discrimination ability, whereas negative value of DC means that a method is failed.

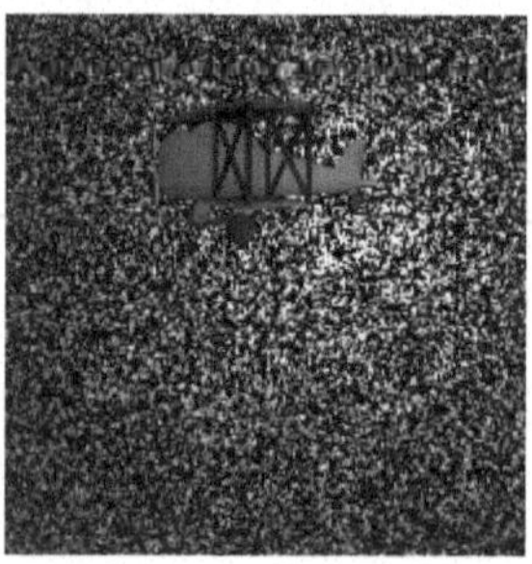

(a) original target

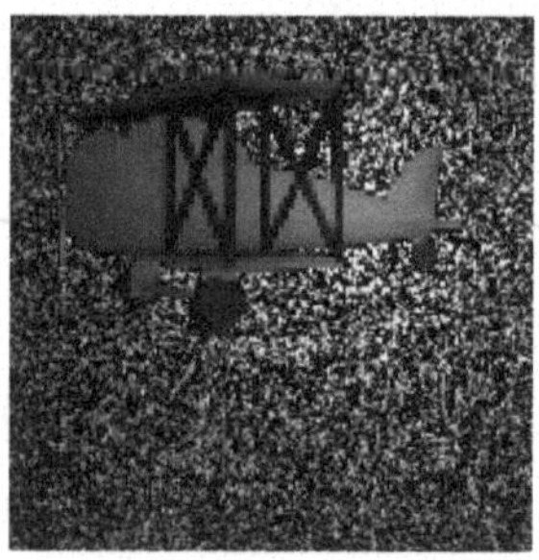

(b) scaled target with factor of 1.8

Fig. 2. Test scene

Fig. 3. The lambertian illumination model with $\tau = 80^{\circ}$, $\alpha = 120^{\circ}$ and $r = 7$

(a) original target (b) scaled target with factor of 1.8

Fig. 4. Observed scene

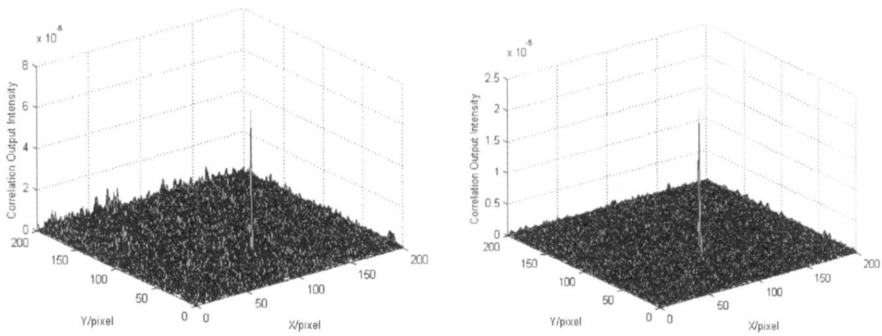

(a) original target (b) scaled target with factor of 1.8

Fig. 5. Correlation output

Table 1. DC obtained with different scaled targets

Size	1.0	1.2	1.4	1.6	1.8	2.0
DC	0.73	0.86	0.87	0.91	0.91	0.91

From Table 1 we can see that even for the worst case, the value of DC is above 0.7, which can be easily detected, as shown in Fig.5 (a). Although the DC values of scaled targets are not equal, all of them can be successfully recognized, so we could say that the proposed method has the ability of scale- invariance in a certain range of size.

Increasing noise intensity may deteriorate the performance of the proposed system, well it is not the main topic of this paper so we won't discuss it too much here.

4 Conclusion

A novel method to achieve scale invariant recognition under non-uniform illumination and noisy conditions is proposed. This method is based on estimated illumination function and modified wiener filter model. Simulation results show that the proposed system performs well for the target from the training set in a scaled range of 1.0 to 2.0.

References

1. Vijaya Kumar, B.V.K., Mahalanobis, A., Juday, R.D.: Correlation pattern recognition. Cambridge University Press, New York (2005)
2. Diaz-Ramirez, V.H., Kober, V.: Target recognition under nonuniform illumination Conditions. Appl. Opt. 48, 1408–1418 (2009)
3. Arsenault, H.H., Lefebvre, D.: Homomorphic cameo filter for pattern recognition. Opt. Lett. 25, 1567–1569 (2000)
4. Garcia-Martinez, P., Tejera, M., Ferreira, C., Lefebvre, D., Arsenault, H.H.: Optical implementation of the weighted sliced orthogonal nonlinear generalized correlation for nonuniform illumination conditions. Appl. Opt. 41, 6867–6874 (2002)
5. Javidi, B., Wang, J.: Optimum filter for detecting a target in multiplicative noise and additive noise. J. Opt. Soc. Am. A11, 2604–2612 (1997)
6. Martinez-Diaz, S., Kober, V.: Nonlinear sysnthetic discriminant function filters for illumination- invariant pattern recognition. Optical Engineering 47(6), 067201-1–067201-9 (2008)
7. Zhao, Y., Bai, B.: Target recognition in non-uniform illuminated and noisy scene. In: Proceedings - 2010 IEEE International Conference on Intelligent Computing and Intelligent Systems. ICIS 2010, vol. 2(5658310), pp. 467–470 (2010)
8. Zheng, Q., Chellappa, R.: Estimation of illuminant direction, albedo, and shape from Shading. IEEE Trans. Pattern Anal. 13, 680–702 (1991)
9. Marom, E., Inbar, H.: New interpretations of Wiener filters for image Recognition. J. Opt. Soc. Am. A13, 1325–1330 (1996)

Integration Calibration Method for Planet Rover Stereo Vision

Hongwei Gao, Guang Yang, Jianhui Song, and Xuanxuan Liu

School of Information Science & Engineering, Shenyang Ligong University, Shenyang, 110159 China
ghw1978@sohu.com

Abstract. According to the integration calibration problem for planet rover stereo vision, a transformation method form vision reference frame to rover reference frame based on theodolite is investigated in this paper which can transform the 3D measure results into rover reference frame. The integration calibration experiment results based on real planet rover show that this method possesses high localization precision and practicable value.

Keywords: Calibration, Theodolite, Coordinate transformation, 3D measure.

1 Introduction

The 21st century will be the peak for human to explore the space and the moon becomes the preferred object. By now, many of the world's scientific powers have launched a moon exploration plan. The U.S. also has proposed to restart the manned moon landing by 2020. For each country, the moon means the long-term strategic significance. The planet rover is just an effective and recent mean to achieve this goal. Realizing planet rover roaming on lunar surface is the foundation of the follow-up work. At present, the practical planet rover typically finishes all kinds of motions as planned by using human-computer interaction system which delivers parameters to the real rover. Its core technologies are stereo vision technology and virtual reality technology. The measurements of stereo vision system should be precisely done coordinate transformation so as to make the rover roam safely, accurately and reliably. In the present day, there are a few of methods of integration calibration in the multi-sensor vision measurement system. A traditional method, called "Golden Master", is to insert system as calibration basis for unifying coordinates when using a known product. Another method is "Silver Master" which periodically calibrates system by using a known but not accurate product as standard which is generated after general product sent to coordinate measurement machine. Both of the two methods are using standard as a unified calibration tool and the official measured data relative to standard has deviation. However, it is very difficult to prevent standard from damage in the process of calibration. In order to realize fast, effective and high-precision coordinate transformation, by the late eighties of last century, electronic theodolite[1] and electronic tachometer[2,3] which would be used in industrial site were invented. Electronic theodolite can establish conversion relationship between vision measurement systems easily and accurately. It has one coordinate system. Only need one calibration target and some control points which are measured in the coordinate

D. Jin and S. Lin (Eds.): Advances in EECM Vol. 1, LNEE 139, pp. 205–210.
springerlink.com

system of sensor and theodolite on it, the coordinate system of sensor can be put to the coordinate system of theodolite. Then the control points of world coordinate system will be measured by the theodolite. After these steps, every coordinate system of sensor can be unified to the world coordinate system. If the world coordinate system does not need to be established in the entity, just establish a virtual space coordinate for measurement system, the coordinate system of theodolite can be treated as the world coordinate system. This paper discusses integration calibration technology based on the coordinate of rover which transforms world coordinate system into rover coordinate and finally into manipulator coordinate. This technology can be proved feasible and precise by experiment of positioning plasterboard.

2 Method for Unified Coordinate

Usually, there are three ways of unifying coordinate: 1) With a target entity, the sensor module coordinate can be reflected. Then it would be unified through measuring the position and direction of entity. 2) Sensor as a measurement module, which is with external measurement equipment, measures the control points at the same time. The sensor module coordinate can be unified by coordinate conversion. This method is called direct coordinate unification.3) Looking camera coordinate as the agency, the sensor module coordinate can be unified by the locating camera. This is called indirect coordinate unification. The first method requires precise adjustment of the sensor module through the coordinate system and the target entities transferred to the coordinate system consistent with the definition. Thus, it is low precision and labor-intensive calibration. So the next two methods are commonly used to unify coordinate. In the following, the specific works would be discussed.

In order to achieve proper operation of navigation and positioning, the results of the three-dimensional which are measured in the world coordinate system must be transformed and shown in car coordinate system, so that the rover and manipulator can be properly droved. The specific method of camera calibration is two-step calibration of Tsai [5]. If we use electronic tachometer to make sure the location of coordinate systems, we must know the location of each coordinate system relative to the electronic tachometer, that is, rotation vector and translation vector. This allows results taking the right conversion between world coordinate system and the each coordinate system of rover.

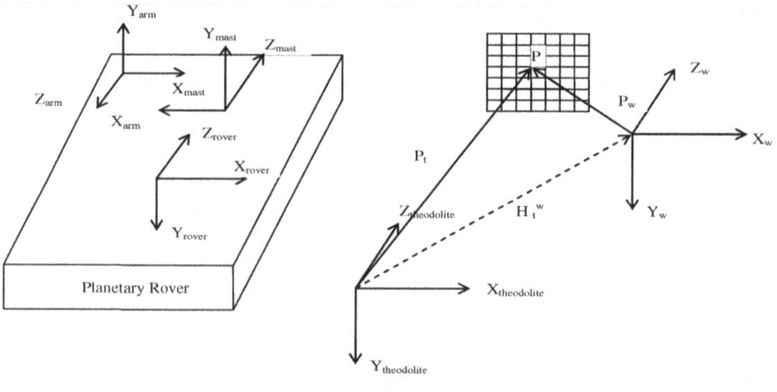

Fig. 1. Coordinate transformation scheme

As shown in figure 1, a point P in three-dimensional space can be expressed by two vectors. One is in the coordinate system of the sensor module, that is $P_w = (X_w, Y_w, Z_w)^T$; the other is given as $P_t = (X_t, Y_t, Z_t)^T$, which is measured by the theodolite system and calibrated in the world coordinate. So obeyed the unification coordinate method, the conversion usually includes translation and rotation and can be defined as:

$$P_t = R_t^w P_w + T_t^w \tag{1}$$

In Equation (1), the rotation matrix is given as and the translation vector is given as $T_t^w = (t_1, t_2, t_3)^T$. The homogeneous coordinate's form of Equation (1) is written as:

$$\begin{bmatrix} P_t \\ 1 \end{bmatrix} = \begin{bmatrix} R_t^w & T_t^w \\ 0 & 1 \end{bmatrix} \begin{bmatrix} P_w \\ 1 \end{bmatrix} \tag{2}$$

Order $H_t^w = \begin{bmatrix} R_t^w & T_t^w \\ 0 & 1 \end{bmatrix}$, called conversion matrix, includes 12 parameters. The rotation matrix R is orthogonal matrix and it satisfies the orthogonal constraint:

$$R_t^{w^T} R_t^w = \mathrm{I} \tag{3}$$

Equation (3) provides six orthogonal constraint equations. Therefore, equation (1) or (2) has only six independent parameters. Only given three control points at least Conversion matrix can be determined. With the location of the coordinate system relative to the electronic tachometer coordinate system, it would be easy to get location relationship between the world coordinate system and the car coordinate system.

3 Experimental Results

The composition of planet rover and manipulator are shown in figure 2. There are two sets of stereo vision localization system. One is on the mast, called navigation camera, and responsible for providing three-dimensional information of environment, which is useful for the planned movement of rover. The other is on the front of the rover. It can make the rover keep away from obstacles and can also offer three-dimensional surface information which guides the car manipulator for positioning and grinding work. Both of the two sets of camera use the same software. In figure 3, the vision localization system is shown throughout the whole picture, which has grinder on the cutting edge of manipulator with five degrees of freedom. It shows the car manipulator do the work for positioning and grinding to the plasterboard.

Fig. 2. Planet Rover's shape

Fig. 3. Manipulator do experiments of positioning and grinding plasterboard

In order to verify actual accuracy of vision localization system of the manipulator, only the images of the four groups of plasterboard (just the left image is given) are studied for the localization experiment, shown in figure 4. For operational safety reasons, the entire manipulator system is slowly making motions, before or after localization, as well as the process of grinding operation. It takes about 15 minutes. The obtained average location errors are shown in table 1. In the four sets of images, the overall location error in figure 4 (a) is maximum, which is due to the location of plasterboard in the image, at the edge of image. So it is easely influenced by the distortion, which leads to the increase in localization errors. In contrast, the accuracy of figure 4 (d) is higher. But overall, localization error is still less than 5 mm, the angle error between normal vector and coordinate axis is still less than 5 degrees, which can meet the requirements of engineering. This also proves that the integration calibration based on theodolite is feasible.

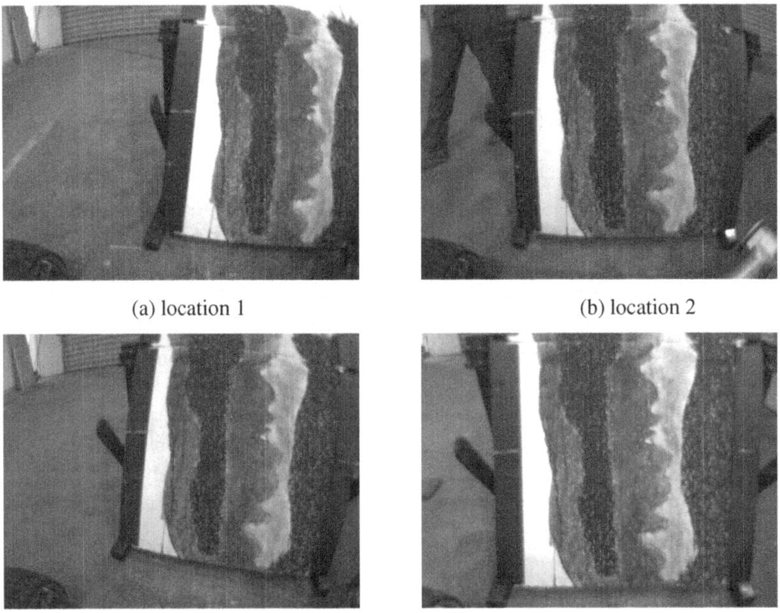

(a) location 1 (b) location 2

(c) location 3 (d) location 4

Fig. 4. Positioning experiments of plasterboard at different location

Table 1. Comparisons of Positioning accuracy between different location plasterboard

Image	positioning error (mm)			angle error between normal vector and coordinate axis (degree)		
	X	Y	Z	X	Y	Z
Fig. 5(a)	-3.02	-3.96	-4.90	-3.73	2.43	-4.47
Fig. 5(b)	2.91	-2.92	-4.66	-3.05	2.88	-4.02
Fig. 5(c)	-1.95	2.83	-4.73	3.23	2.41	-4.51
Fig. 5(d)	2.27	-1.77	-4.10	2.91	2.99	-3.94

4 Conclusion

This paper studies the integration calibration technology based on the rover bodywork coordinate system, which switch measurements in world coordinates to bodywork coordinates. Through not only the designed and combined stereo vision technology, but also the joint experiments of real and virtual manipulator based on interactive of manipulator simulation platform, the feasibility of calibration method discussed in this paper can be proved.

Acknowledgments. This work is supported by China Liaoning Provice Educational Office fund (No.20080611).

References

1. Li, Z., Li, G., Tang, T., Zhang, G.: Electronic Multi-theodolite Measuring System Applied in the Precision Installation of a Large Antenna. Hydrographic Surveying and Charting 25(1), 26–30 (2005) (in Chinese)
2. Zheng, J., Guo, Z.: A Total-station Instrument and Its Applications. Journal of Qingdao Institute of Architecture Engineering 21(4), 44–47 (2000) (in Chinese)
3. Chen, M.: Electronic Multi-theodolite's Some Applications in Industrial Measurement. Surveying and Charting on Geology and Mine 21(41), 9–11 (2005) (in Chinese)
4. Zhang, H., Duan, F., Wang, X., Ye, S.: Technique of Unifying the Coordinates of Multi-Sensor Visual Measuring System. Chinese Journal of Sensors and Actuators 19(4), 1301–1304 (2006) (in Chinese)
5. Tsai, R.Y.: A versatile camera calibration technique for high-accuracy 3D machine vision metrology using off-the-shelf TV cameras and lenses. IEEE Journal of Robotics and Automation 3(4), 323–344 (1987)

Research on Excitation Controller Based on Linear Quadratic Optimal Control

Xiaoying Zhang[1,2], Zhizhuang Cheng[1], and Cunlu Dang[1,2]

[1] College of Electrical and Information Engineering, Lanzhou University of Technology, Lanzhou, Gansu Province, China
[2] Key Laboratory of Gansu Advanced Control for Industrial Processes, Lanzhou, Gansu Province, China
zhxy525@gmail.com

Abstract. A new-style excitation adjustor based on linear quadratic optimal control is introduced in this paper. The math models of the synchronous generator excitation system are built and the control algorithm is described. A simulation model of a single-machine infinite-bus system is constructed in Matlab / Simulink. The performances of excitation system under small disturbance and short circuit are simulated. Simulation results show that the proposed control method can improve the power system stability greatly compared to the conventional PID control method.

Keywords: linear quadratic, optimal control, excitation controller, simulation.

1 Introduction

The stable operation of synchronous generator is the premise to ensure safe and reliable power supply. The excitation system is an important part of the synchronous generator. The research reveals that excitation control is the most economical and efficient method to improve and enhance power angle stability and voltage stability, and it is also the measurement of different kinds of control methods used in power system. Lots of data point out that it is the trend to use self and shunt excitation system on large and medium generator.

The traditional automatic voltage adjustor adopted PID method to regulate the deviation of feedback voltage and rated voltage, using classic control theory to confirm the parameters of the controller [1]. The PID adjustor is designed for voltage deviation signal, whose advanced phase frequency is different from low frequency surge's. The nonlinear optimal excitation controller [2,3] based on feedback could resolve stability and bad damping capacity problems under the big disturbance, but its robustness was worse when the object parameters changed greatly. The optimal control based on linear quadratic performance index is one of the design methods for control system. This method calculates simply and adjusts expediently. Because the linear quadratic obtains the control law which can realize the closed loop optimal control by state feedback, the method is used widely in industry control system [4,5]. In the single unit-infinity system, linear quadratic excitation controller corresponds to all state variables feedback. Therefore, the linear quadratic excitation controller is available in the wider frequency range and has better adaptability.

D. Jin and S. Lin (Eds.): Advances in EECM Vol. 1, LNEE 139, pp. 211–216.
springerlink.com

This paper establishes mathematical models and simulation models of the synchronous generator excitation system. Take the single-machine infinite system as an example. The performances of excitation system under small disturbance and short circuit are simulated in order to test two types of excitation controller such as traditional PID excitation controller and linear optimized excitation controller.

2 Models of Synchronous Generator Excitation System

The three order model for synchronous generator was adopted in this paper to form single-machine infinite system. Fig. 1 shows the typical configuration.

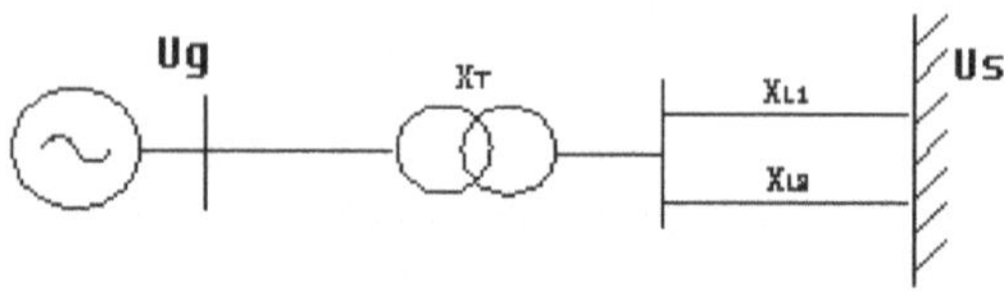

Fig. 1. The configuration of single-machine infinite system

According to the principle of the excitation system, the math models are built as

$$\dot{\delta} = \omega - \omega_0 \tag{1}$$

$$\dot{\omega} = \frac{\omega_0}{H} P_m - \frac{D}{H}(\omega - \omega_0) - \frac{\omega_0}{H} \frac{E_q' U_s}{X_{d\Sigma}'} \sin\delta \tag{2}$$

$$\dot{E_q'} = \frac{-1}{T_d'} E_q' + \frac{1}{T_{d0}} \frac{X_d - X_d'}{X_d'} U_s \cos\delta + \frac{1}{T_{d0}} U_f \tag{3}$$

$$y(t) = U_g \qquad U_g = \sqrt{U_{gd}^2 + U_{gq}^2} \tag{4}$$

$$\text{where } U_{gd} = \frac{X_q U_s \sin\delta}{X_{q\Sigma}}, \quad U_{gq} = \frac{(X_{d\Sigma}' - X_d') E_q' + X_d' U_s \sin\delta}{X_{d\Sigma}'} \tag{5}$$

$$X_{d\Sigma} = X_d + X_T + X_L, \quad X_{d\Sigma}' = X_d' + X_T + X_L, \quad X_{q\Sigma} = X_q + X_T + X_L \tag{6}$$

Equations (1) and (2) are generator's rotor movement equations; equation (3) is rotor coil's electromagnetism dynamic equation. where Us is the of the infinite system, Uf is the excitation voltage of the synchronous generator, XT is the reactance of the transformer, X_L is the reactance of the transmission line, Eq′ is the electric potential after the transient reactance xd′, δ is the angle between Eq′ and Us, ω is the mechanical angular velocity of the rotor. H is moment of inertia, D is damping coefficient of the unit, Pm is mechanical power, Td is time constant of the excitation coil. When ΔPe, Δω and ΔUg are quantities of state, after a series of mathematical transforms, state equation (7) is obtained.

$$\begin{bmatrix} \Delta \dot{P}_e \\ \Delta \dot{\omega} \\ \Delta \dot{U}_g \end{bmatrix} = \begin{bmatrix} \frac{S_E - S_u}{T_d' S_u} & S_E' & -\frac{R_u S_E}{T_d' S_u} \\ -\frac{\omega_o}{T_J} & -\frac{D}{T_J} & 0 \\ \frac{S_E - S_u}{T_d' R_u S_u} & \frac{S_E' - S_E}{R_u} & -\frac{S_E}{T_d' S_u} \end{bmatrix} \begin{bmatrix} \Delta P_e \\ \Delta \omega \\ \Delta U_g \end{bmatrix} + \begin{bmatrix} \frac{R_E'}{T_{do}'} \\ 0 \\ \frac{R_E'}{T_{do}' R_u} \end{bmatrix} \Delta E_{fd} \tag{7}$$

Where

$$S_E = \frac{E_q U_s}{X_{d\Sigma}} \cos\delta \qquad S_E' = \frac{E_q' U_s}{X_{d\Sigma}'} \cos\delta + U_s^2 \frac{X_{d\Sigma}' - X_{d\Sigma}}{X_{d\Sigma}' X_{d\Sigma}} \cos 2\delta \tag{8}$$

$$R_E = \frac{U_s}{X_{d\Sigma}} \sin\delta \qquad R_E' = \frac{V_s}{X_{d\Sigma}'} \sin\delta \qquad S_U = S_E - R_U \frac{\partial U_g}{\partial \delta} \tag{9}$$

3 Controller Design

3.1 Principle of Linear Quadratic Optimal Control

In control theory, the linear quadratic optimal control problem is probably the most fundamental optimal control problem, whose solution is unique and constitutes a linear dynamic feedback control law that is easily computed and implemented. Finally the linear quadratic optimal controller is also fundamental to the optimal perturbation control of non-linear systems. Consider the linear dynamic system as

$$\dot{X} = AX + BU \tag{10}$$

Where X represents n-dimensional state vector, U is r-dimensional control vector, A is state coefficient matrix and B is control coefficient matrix.

The system state equation eigenvalue is determined by the A matrix. The closed-loop system is formed by importing state feedback K, The feedback system's state vector is

$$U = V - KX \tag{11}$$

Where K is state feedback gain matrix. Combining equation (10) with equation (11), the equation (12) is obtained as follow

$$\dot{X} = AX + B(V - KX) = (A - BK)X + BV \tag{12}$$

In this circumstance, the closed-loop system eigenvalues will be determined by the matrix A-BK. Thus, the essence of the optimal control law is to choose the feedback matrix K and in a given control law make it achieve optimal under certain conditions. Assume that y(t) is the system's actual response, ξ (t) is expected dynamic response. The optimal control performance index should make the value of the deviation minimum. Mathematical expression of the performance index is

$$\int_0^\infty [\xi(t) - y(t)]^2 dt = J_{min} \tag{13}$$

Where J represents a functional with the function y (t) change. If X (t) represents the actual state vector, $\hat{X}$ represents the expected state vector, then the quadratic performance index that requires the state with the smallest amount of deviation is

$$J = \int_0^\infty [\hat{X}(t) - X(t)]^T [\hat{X}(t) - X(t)] dt \tag{14}$$

Although the optimal control performance index meets target, it is possible that the amount of requested control is too large and difficult to achieve. Therefore, control vector U(t) should also be limited and its expression is given by

$$J = \int_0^\infty \{[\hat{X}(t) - X(t)]^T Q[\hat{X}(t) - X(t)] + U^T(t) R U(t)\} dt \tag{15}$$

where Q is the state vector, R represents the weight matrix of the control vector.

3.2 Linear Optimal Excitation Controller Design

Firstly, we create a system of state space description, get the state coefficient matrix A and control coefficient matrix B, determine whether the matrix is full rank and whether the control system can achieve arbitrary pole configuration, and further conclude whether only true solution exists in Ricati equation. Secondly, we choose proper weight matrix R and Q, solve the equation (15) and get matrix P.

$$PA + A^T P - PBR^{-1}BP + Q = 0 \tag{16}$$

If the derived matrix is positive definite matrix P, the system is stable, otherwise, the system is unstable. Finally, calculate the optimal feedback matrix K, and get the linear optimal quadratic control law as follow

$$U(t) = -R^{-1}B^T PX(t) = -KX(t) \tag{17}$$

In the design of optimal control system, the selection of the weight matrix Q and R is an important issue. In the design of control system, in order to simplify the selection of the weight matrix, to let R = 1, Q is a diagonal matrix. So just consider the selection of the weight matrix Q, namely Q=diag(q1,q2,q3). where q1 represents state deviation Δ Pg, q2 represents state deviation $\Delta\omega$, q3 represents Δ Ug weighted.

4 System Simulation

The simulation model of the excitation controller was built by Maltlab/Simulink and encapsulated a subsystem, thus the simulation model of synchronous generator excitation system was built. The small disturbance simulation experiment and three phase short-circuit simulation experiment are carried through after setting the parameters of every element and module and adjusting the performance index. In the following simulation wave figures, the unit of the horizontal axis is time, the unit of the vertical axis is p.u.

When t=10s, 10% step disturbance signal is put on the terminal voltage reference input (Vref), the small disturbance simulations of generator excitation system are conducted under conventional PID control and linear quadratic optimal control. The simulation curves of generator electric power Pe and speed n are shown in Fig.2 and Fig.3. The comparison of the performance index is shown in table.1.

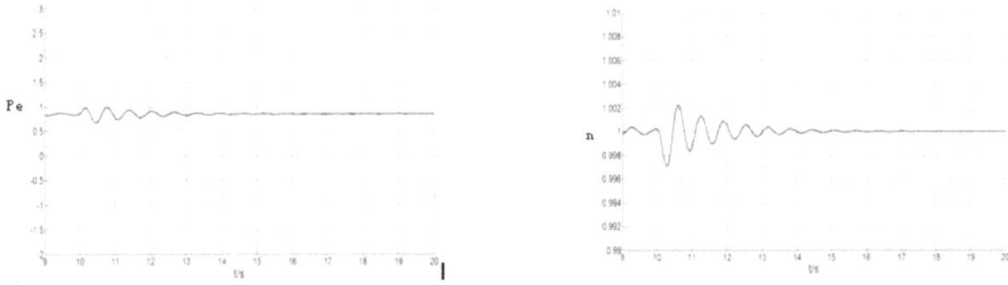

(a) The curve of generator electric power (b) The curve of generator speed

Fig. 2. The small disturbance simulation curve of conventional PID control

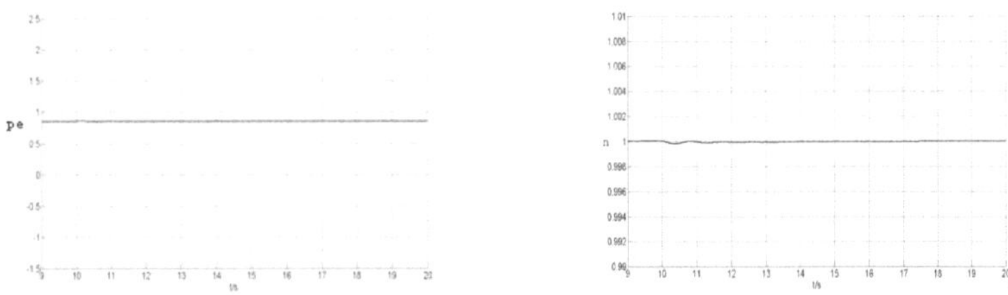

(a) The curve of generator electric power (b) The curve of generator speed

Fig. 3. The small disturbance simulation curve of Linear optimal control

Table 1. The small disturbance performance index comparison of the excitation controller

Controller type	Oscillation time	Oscillation frequency
Conventional PID	5s	8
Linear optimal	≈0s	0

When t=10s, three-phase short circuit fault is set by three-phase fault element, when t=10.2s, fault is removed. Accordingly, the three-phase short circuit fault simulation experiment is conducted. The simulation curves are shown in Fig.4 and Fig.5, The comparison of the performance index is shown in table.2.

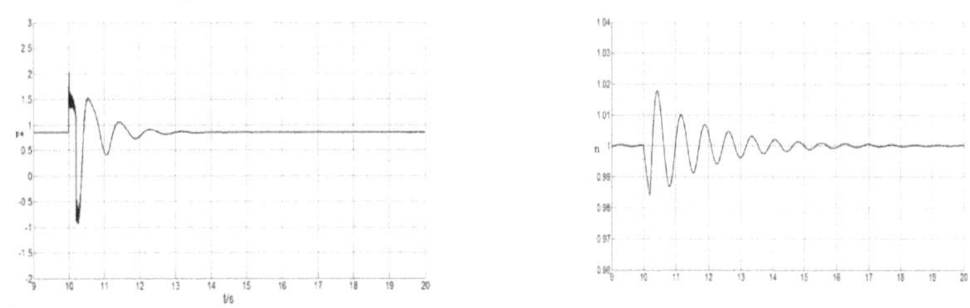

(a) The curve of generator electric power (b) The curve of generator speed

Fig. 4. The three-phase short-circuit simulation curve of conventional PID control

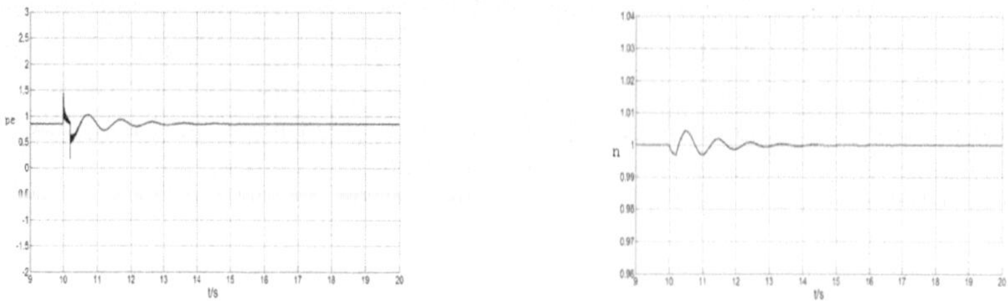

(a) The curve of generator electric power (b) The curve of generator speed

Fig. 5. The three-phase short-circuit simulation curve of Linear optimal control

Table 2. Short-circuit performance index comparison of the excitation controller

Controller type	Oscillation time	Oscillation frequency	Speed overshoot
Conventional PID	8s	10	1.9%
Linear optimal	3s	3.5	0.3%

5 Conclusions

This paper using linear quadratic optimal control theory designs a new type of synchronous generator excitation controller. Simulation results show that linear optimal excitation controller has better control performance than the conventional PID excitation controller under different operating conditions. Particularly in the static stability, PID excitation controller can not match, and in the dynamic stability, linear optimal excitation controller also shows a good control performance.

Acknowledgments. This work is supported by The National Natural Science Foundation of China (50967001) and Excellent Youth Teacher Foundation of Lanzhou University of Technology (No. Q200814).

References

1. Li, C., Guo, G., Gaolong: Nonlinear PID exciting controller for power systems. Journal of Tsinghua University (Science and Technology) 4(3), 48–50 (2000)
2. Wang, Y., Hill David, J., et al.: Transient stability enhancement and voltage regulation of power systems. IEEE Trans. on Power Systems 8(2), 620–627 (1993)
3. Zhang, J., Yu, X., Wang, B.: Improved Linear Optimal Excitation Control Tactic and Field Test. Water Resources and Power 26(3), 176–178 (2008)
4. Xie, L., Li, W.: Application of LQR in Inverted Pendulum System. Journal of Chongqing Institute of Technology (Natural Science) 22(8), 124–128 (2008)
5. Ren, W., Wen, X., Wang, W.: The Design and Simulation of the Single-neuron PID Optimal Controller Based in Linear Quadratic. Computer Applications and Software 25(5), 123–124 (2008)

Sub-matrix Summation Method for Adaptive Dimming LED Backlight

Pei Chu, Yang Li, Hua Jiang, Kanglian Zhao, and Sidan Du

Nanjing University, School of Electronic Science and Engineering,
210093 Hankou Road 22, Nanjing, China
njuchupei@gmail.com, yogo@nju.edu.cn, quantumcat@163.com,
zhaokanglian@gmail.com, coff128@nju.edu.cn

Abstract. Dynamic LED backlight technology, not only can increase the dynamic range and enhance the contrast of the LCD display, but also can largely reduce the power consumption. However, for large-scale monitor, to achieve a better display, the number of the LED units will be very large. And if we control the LED unit point by point, the drive circuit will be complicated. In this paper, a ranks-decomposition control method is proposed, which decomposes the LED brightness matrix to row and column control signals, driving the LED unit point by point. This method archives adaptive dimming effects, such as high contrast ratio and reduced power consumption, even with much less LED drivers. This paper also contains a Sub-Matrix Summation optimization algorithm implements the matrix decomposition and optimizes it.

Keywords: LED backlight, dynamic range, ranks-decomposition, Sub-Matrix Summation optimization.

1 Introduction

Liquid Crystal Display (LCD), as the most popular monitor, has many good features, such as thin, light weight and good color performance. However, since the LCD is not self-luminance display device, it requires a backlight unit. Cold cathode fluorescent lamp (CCFL) is the common backlight source. But this conventional backlight device is full luminance mode as shown in Fig. 1(a), resulting poor image contrast due to light leakage in the dark state, the best contrast ratio is 1000:1, and the unnecessary power consumption.

In recent years, lighting-emitting diode (LED) gradually replaces the CCFL, due to its low power consumption, high color gamut, and can be dynamically controlled[1,2], and is extensively used in LCD TV sets as the backlight source. Adaptive dimming LED backlight technology control the LED backlight brightness value according to the gray of the corresponding region of the image, resulting dark areas darker, bright areas brighter. This technology has two tough problems, one of which is determining a suitable backlight signal. Conventional method is "Max", which calculates the max gray level of all sub-pixel values in each backlight region. Yi-Pai Huang[3,4] etc. proposed the inverse of a mapping function (IMF) method, which achieves high contrast ratio in experiment. The other one is the design of

D. Jin and S. Lin (Eds.): Advances in EECM Vol. 1, LNEE 139, pp. 217–223.
springerlink.com

driving circuit. For large-scale LCD TVs, the number of LED units is tremendous huge. Consequently, it cost too much to control LED unites point by point. Block dimming[5] method is proposed to decrease the number of drivers as shown in Fig. 1(b). LED block dimming method can reduce the LED drivers while maintaining the high quality image display. However, due to the decrease of backlight resolution, the interface between the block and block has the non-uniform brightness. Daeyoun Cho[6] etc. proposed the X-Y Channel Driving method to decrease the drivers gradually.

In this paper, we compose a ranks-decomposition control method to get the row and column control signal. Compared with the block dimming method, it has less drivers and higher quality image display. This method abstracts row and column signals to two matrices in the form of "*0*" and "*1*", and the result of multiplying these two matrices is the image gray value. Ranks-decomposition is an NP problem. Consequently, a Sub-Matrix Summation optimization algorithm is proposed to solve it.

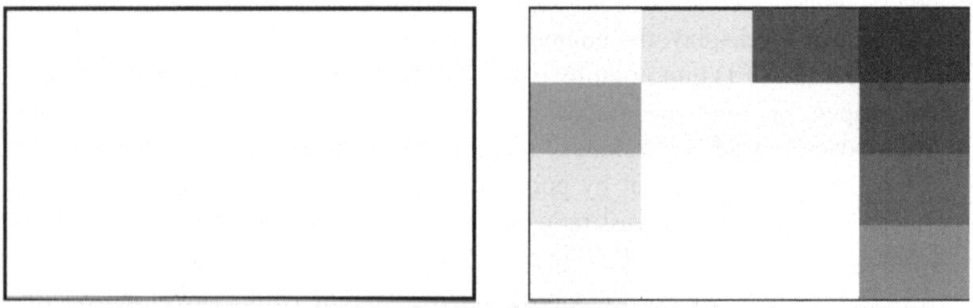

Fig. 1. Full-luminance and Block dimming. (a) Full-luminance mode, (b) Block dimming.

2 Ranks-Decomposition Control of the Backlight System

In this paper, we propose a ranks-decomposition control method. The structure of driving circuit is shown in Fig. 2(a), which consists of LED units in the crosses of rows and columns. Suppose there are M rows and N columns LED units, and the control circuit inputs M rows pulse signal A and N columns pulse signal B respectively. The total number of the drivers is $M+N$, much less than that of the conventional method with $M \times N$ drivers.

A and B is the row and column signal in the form of "*0*" and "*1*" respectively. The brightness value $I_{m,\,n}$ of any point (m, n) is decided by logical "*AND*" operation between row m signal A_m and column n signal B_n. As shown in Fig. 2(b), during a time period T, $I_{m,n}=A_m \& B_n$.

Establish the mathematical model of the structure and express as:

$$A \times B = I \quad . \tag{1}$$

Where I is the LED Backlight brightness value of $M \times N$ matrix. The elements of the matrix is from *0* to *Level*, generally *Level* is 255, covering *Level+1* gray levels. A and B is the matrix in form of "*0*" and "*1*", "*0*" is low driving signal, and "*1*" is high driving signal. A is $M \times (Level+1)$ matrix, and B is $(Level+1) \times N$ matrix. Therefore, we can use "IMF" method to determine a suitable backlight signal matrix I of the image to be display and decompose the suitable A and B according (1).

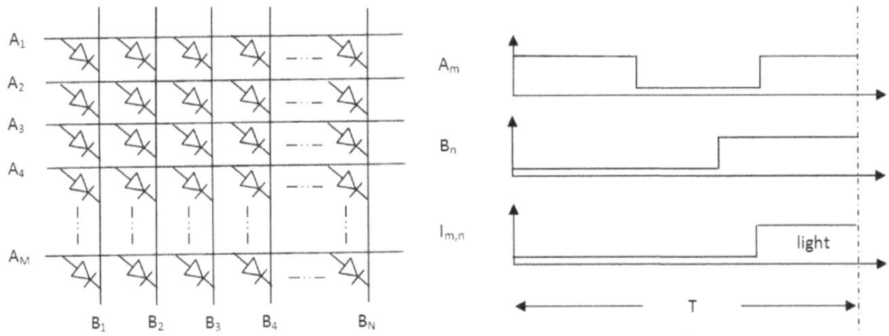

(a) (b)

Fig. 2. (a) The structure of the driving circuit, (b) Row signal "*AND*" column signal

3 Sub-matrix Summation Optimization Algorithm

According to matrix theory, not every matrix I can be decomposed to suitable A and B. So we propose the Sub-Matrix Summation optimization algorithm to solve this problem. Firstly, extend A and B to $M\times(2\times(Level+1))$ and $(2\times(Level+1))\times N$ matrix. Where, the extension of A, from $(Level+2)$ to $2\times(Level+1)$ column, is assigned "*1*". The extension of B, from 1 to $(Level+1)$ row is assigned "*1*", as expressed (2).

$$A\times B=\left[\begin{array}{c|cccc} & 1 & 1 & \cdots & 1 \\ AA & 1 & 1 & \cdots & 1 \\ & \cdots & \cdots & \cdots & \cdots \\ & 1 & 1 & \cdots & 1 \end{array}\right]\times\left[\begin{array}{cccc} 1 & 1 & \cdots & 1 \\ 1 & 1 & \cdots & 1 \\ \cdots & \cdots & \cdots & \cdots \\ 1 & 1 & \cdots & 1 \\ \hline & & & \\ & BB & & \\ & & & \end{array}\right]=I\ . \tag{2}$$

We defined $a_i=\sum_{n=1}^{(Level+1)/2} AA_i(n); b_i=\sum_{n=(Level+1)/2+1}^{Level+1} BB_i(n)$, which means a_i is the sum of row i signal of matrix AA, b_i is the sum of column i of BB. Then, extend a_i and b_i as follow:

$$A'=\begin{bmatrix} a_1 & a_1 & \cdots & a_1 \\ a_2 & a_2 & \cdots & a_2 \\ \cdots & \cdots & \cdots & \cdots \\ a_M & a_M & \cdots & a_M \end{bmatrix};B'=\begin{bmatrix} b_1 & b_2 & \cdots & b_N \\ b_1 & b_2 & \cdots & b_N \\ \cdots & \cdots & \cdots & \cdots \\ b_1 & b_2 & \cdots & b_N \end{bmatrix}. \tag{3}$$

Where, A' and B' are the $M \times N$ matrices. And (1) can be expressed as the sum of two matrices, as expressed (4).

$$A' + B' = I' \ . \tag{4}$$

Obviously, not every matrix I can be decomposed to suitable A' and B'. Therefore, we propose an optimization algorithm as following 5 steps:

1) According the "IMF" method, determine the gray value matrix of the LED backlight from the image to be displayed.
2) Fix the value of row 1 $I_{row,1}$, and calculate the difference between $I_{row,i}$, which is the value of row i, and $I_{row,1}$, and take the maximum of the difference d_i as the tolerance, as $d_i = \sum_{i=1}^{M} \max(I_{row,i} - I_{row,1})$. Then add the value of row 1 with d_i and can get the new value of row i, as $I'_{row,i} = \sum_{i=1}^{M} I_{row,1} + d_i$. Furthermore, get the new matrix I'_1, I'_1 can be decomposed to the suitable matrix A' and B', and calculate the absolute difference $\Delta I_1 = |I - I'_1|$.
3) Fix the value of row 2, 3…M, and column 1, 2…N in order, repeat step 2. Calculate the corresponding absolute difference $\Delta I_i = \sum_{i=2}^{M+N} |I - I'_i|$.
4) Select the minimum of the ΔI_i, as $\Delta I = \min \sum_{i=1}^{M+N} \Delta I_i$, and take its corresponding new matrix I'_i. Then normalize I'_i and make its maximum value is *Level*, as $I' = \begin{cases} I'_i & ;I'_i \le Level \\ Level & ;I'_i > Level \end{cases}$. The normalized matrix I' is the optimized gray value matrix of the LED backlight.
5) Let a_i equal to the minimum of the ith-row value of the matrix I', as $a_i = \sum_{i=1}^{M} \min(I'_{row,i})$, and finally we can get the matrix A'. Lastly, matrix $B' = I - A'$.

The error parameter ΔI judges the merits of our algorithm. The smaller of the ΔI, the better for our algorithm. When ΔI is equal to "0" which means that the final gray value of the LED backlight is the ideal one, as $I' = I$. Fig. 3 shows the process of the proposed algorithm. Fig. 3(a) is the source image. And Fig. 3(b) is the 8×8 matrix I, and each element is the gray value resulting from the IMF method. To demonstrate the process of proposed method conveniently, we divide the LED units into 8×8 areas. In the following experiment, we use 30×40 LED units backlight. Fig. 3(c) is the optimized gray value matrix I' with proposed algorithm. Fig. 3(d) is the row matrix A' after step 5. Fig. 3(e) is the column matrix B'.

(a)

119	114	139	133	149	160	175	177
136	147	161	164	175	183	195	199
161	170	197	197	196	201	206	208
171	209	199	195	204	210	212	217
29	194	185	192	204	213	222	241
20	167	169	186	197	206	224	254
83	129	158	151	174	185	177	247
30	30	108	153	146	165	199	249

(b)

119	147	140	133	156	167	179	229
150	178	171	164	187	198	210	260
183	211	204	197	220	231	243	293
181	209	202	195	218	229	241	291
178	206	199	192	215	226	238	288
172	200	193	186	209	220	232	282
137	165	158	151	174	185	197	247
139	167	160	153	176	187	199	249

(c)

=

119	119	119	119	119	119	119	119
150	150	150	150	150	150	150	150
183	183	183	183	183	183	183	183
181	181	181	181	181	181	181	181
178	178	178	178	178	178	178	178
172	172	172	172	172	172	172	172
137	137	137	137	137	137	137	137
139	139	139	139	139	139	139	139

(d)

+

0	28	21	14	37	48	60	110
0	28	21	14	37	48	60	110
0	28	21	14	37	48	60	110
0	28	21	14	37	48	60	110
0	28	21	14	37	48	60	110
0	28	21	14	37	48	60	110
0	28	21	14	37	48	60	110
0	28	21	14	37	48	60	110

(e)

Fig. 3. The process of the proposed algorithm.(a) source image,(b) *I*, (c) *I'* ,(d) *A'* ,(e) *B'*.

4 Simulation Result

In this section we simulate the proposed method with color image ($768\times1024\times3$) and 30×40 LED units module. The result of the simulation is shown in Fig. 4. Fig. 4(a) is the source image. Fig. 4(b) is brightness image of the 30×40 LED backlight signals. The image is divided into 30×40 areas, and the gray value of each LED is resulted with the proposed method. Obviously, the area where the image is dark, the value of corresponding LED is small, consequently, reducing the power consumption. With the determined signals, the backlight image can be convolved with a light spread function (LSF) which represents the diffusion function of the diffuser plate. To simplify the model, we use low pass Gaussian filter[7] instead of LSF. Fig. 4(c) is the backlight image which convolves the Fig. 4(b) with the Gaussian filter. Fig. 4(d) is the result image which is the result of multiplying the backlight image and input image after compensation. The result image shows that the dark area like trees and house is darker, and the light area like the sun is lighter. Consequently, the result image achieves high contrast ratio.

(a) (b) (c) (d)

Fig. 4. Simulation of the proposed method. (a) Source image, (b) LED backlight signals, (c) Backlight image,(d) Result image.

Then we choose 4 typical images, Scenery, Robot, Lily and People, to simulate with the proposed method. The experimental result images are shown in Fig. 5.

(a) (b) (c) (d)

Fig. 5. Simulation of 4 typical images. (a) Scenery, (b) Robot, (c) Lily, (d) People.

We simulate the 4 images using the X-Y Channel Driving method[6] and compare the error parameters ΔI and the power consumption of these two methods, as shown in Fig. 6. Fig. 6(a) is the comparison of the error parameters ΔI of these 4 images. For each image, ΔI of proposed method is smaller than that of the X-Y Channel[6] Driving method. Fig. 6(b) is the comparison of the power consumption of these 4 images. For each image, the proposed method reduces the power consumption gradually.

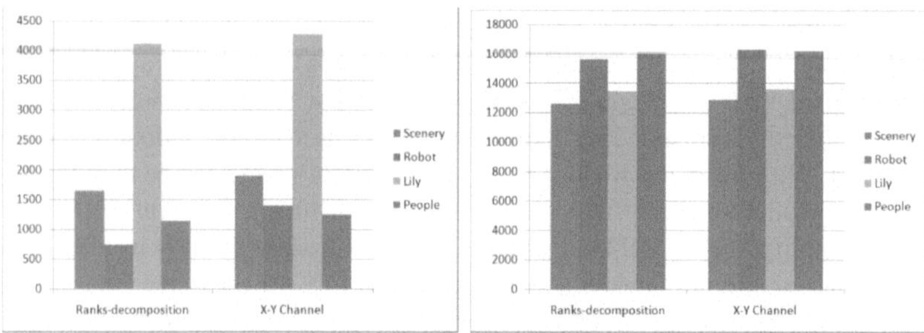

(a) (b)

Fig. 6. Comparison of different image using different method. (a) Error parameters, (b) Power consumption.

Furthermore, we simulate 50 images ($768\times1024\times3$) randomly selected using these two method and calculate the error parameters ΔI of each image and plot the curve of ΔI , as shown in Fig. 7. Vertical axis is the error parameter ΔI , and the red solid line is error distribution curve using proposed method, and the blue dotted line is the error distribution curve using the X-Y Channel[6] method. As can be seen from the Fig. 7, the red curve is mainly below the blue one, and more smooth than the later one, which means that the proposed method has lower error and is more robust than the X-Y Channel[6] method in most condition.

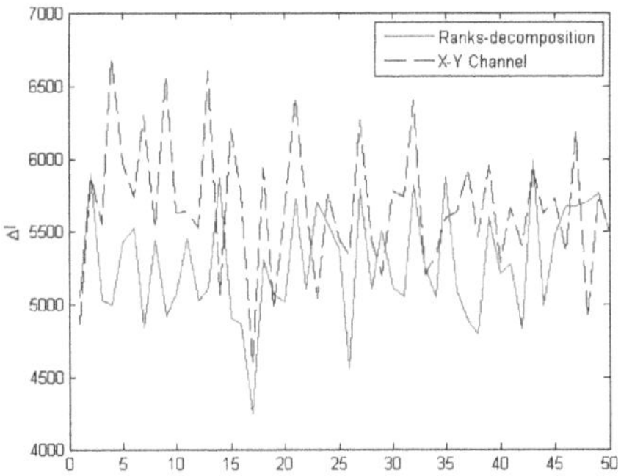

Fig. 7. Error parameter curve of two methods

5 Conclusion

The Ranks-decomposition method and Sub-matrix Summation optimization algorithm is proposed in this paper, which can reduce the LED backlight drivers, get high contrast ratio, and reduce the power consumption. Therefore, the low cost and compact design can be implemented in the large-scale LCD TVs. Compared with the X-Y Channel[6] method, the proposed method has great advantage in the error parameter control and robustness after simulating with a great number of images.

References

1. Seetzen, H., Heidrich, W., Stuerzlinger, W., Ward, G., Whitehead, L., Trentacoste, M., Ghosh, A., Vorozcovs, A.: High dynamic range display systems. ACM Trans. Graphics 23(3), 760–768 (2004)
2. Seetzen, H., Whitehead, L.A., Ward, G.: A high dynamic range display using low and high resolution modulators. SID Symp. Dig. Tech. Papers 34, 1450–1453 (2003)
3. Lin, F.C., Liao, L.Y., Liao, C.Y., Huang, Y.P., Shieh, H.P.D., Wang, T.M., Yeh, S.C.: Dynamic backlight gamma on high dynamic range LCD TVs. J. Display Technol. 4(2), 139–146 (2008)
4. Lin, F.C., Liao, C.Y., Liao, L.Y., Huang, Y.P., Shieh, H.P.D.: Inverse of mapping function method for image quality enhancement of high dynamic range LCD TVs. In: SID Symp. Dig. Tech. Papers, vol. 38, pp. 1343–1346 (2007)
5. Oh, W.S., Cho, D.Y., Cho, K.M., Moon, G.W., Yang, B.C., Jang, T.S.: A novel two-dimensional adaptive dimming technique of X-Y channel drivers for LED backlight system in LCD TVs. J. Display Technol. 5(1), 429–432 (2009)
6. Cho, D., Oh, W.-S., Moon, G.W.: A Novel Adaptive Dimming LED Backlight System With Current Compensated X-Y Channel Drivers for LCD TVs. J. Display Technol. 7(1), 29–35 (2011)
7. Oh, S.: High Dynamic Range Image Encoding for BrightSide Display. Winter PSYCH221. Applied Vision and Image Systems Engineering (2006-2007)

A BB-PSO Image Reconstruction Algorithm for Electrical Capacitance Tomography System

Xingwu Sun[1], Yu Chen[1,*], and Deyun Chen[2]

[1] Northeast Forestry University, Harbin, China
[2] Computer Institute Harbin University of Science and Technology, Harbin, China
lg_chenyu@yahoo.com.cn

Abstract. The abstract To solve the electrical capacitance tomography(ECT) technology "soft field" effect and pathological problem, a new BB-PSO algorithm for electrical capacitance tomography is presented. On the basis of analyzing ECT system measurement principle, constructing corrector formula of secant approximation algorithm in second-order information items of objective functions. The feasibility of using this algorithm for ECT problems is also discussed.It shows that it is easy to meet the convergence condition and error of image reconstruction is small. Experimental results and simulation data indicate that the algorithm can provide high quality images and high speed compared with LBP , Steepest Descent algorithm and Tikhonov algorithm and this new algorithm presents presents a feasible and effective way to research on image reconstruction algorithm for Electrical Capacitance Tomography System.

Keywords: electrical capacitance tomography, image reconstruction, PSO, Barzilai-Borwein.

1 Introduction

Flow imaging is a new technique developed rapidly in recent years, which has great developmental potential and wide industrial application prospect[1-2]. Having many distinct advantages such as low cost, wide application field, simple structure, non-invade and better safety, Electrical Capacitance Tomography(ECT) has being the most popular research direction and the main development in flow imaging technique[3]. Due to ECT's inherent nonlinear characteristic, and the quantity of available independent measured capacitance values(projection data) is limited, much smaller than the pixel quantity needed for image reconstructing, no analytical solution exist for inverse problem. Meanwhile, because of the nonlinearity and "soft-field" effect, and because of ECT system's poor stability of solution and severe ill-condition caused by the error of measurement, it brings great difficulty to image reconstruction[4-5]. Image reconstruction algorithm has always been the main difficulty for practicing and further developing ECT technique[6-7], exploring good image reconstruction algorithm is important. At present, common methods used in

* Corresponding author.

D. Jin and S. Lin (Eds.): Advances in EECM Vol. 1, LNEE 139, pp. 225–231.
springerlink.com

ECT image reconstruction include: linear back projection algorithm (LBP), regularization[8], Landweber iterative method, projected-Landweber method and conjugate gradient (CG), etc.

2 Image Reconstruction Algorithm Based on BB-PSO for ECT

At present, most of ECT imaging algorithm is the linear model which is based on the mapping from the dielectric constant to the capacitance, the model by discretization, linearization and normalization can be expressed as follows:

$$C = SG \tag{1}$$

Where $C \in R^m$ is the normalization capacitance vector, $S \in R^{m \times n}$ is the coefficient matrix (sensitivity matrix), and $G \in R^n$ is the normalization medium distribution image vector. Where the task of ECT image reconstruction is as the given electrical capacitance value C for dielectric constant distribution G.

BB(Barzilai-Borwein) method is a special kind of steepest descent method and this algorithm is super-linear convergence. The basic idea is to use the current iteration point and a bit further information to determine the step length factor, that is, α_k step length factor to meet the BB-step and the method can be used to solve the following unconstrained optimization problem.

$$\min q(x) \tag{2}$$

Where $q : R^m \to R$,q is the quadratic function and the exprss is follow:

$$\min q(x) = \frac{1}{2} x^T Bx - b^T x \tag{3}$$

Where $B \in R^{m \times m}$ is Symmetric positive definite matrix, Therefore, equation (3) is a strongly convex quadratic programming problem. To make $x_k, k = 0,1,\cdots$ column for the current iteration point, g_k as a function of q in the x_k-point gradient and g_k can be express as follow:

$$g_k = Bx_k - b \tag{4}$$

Equation (4) on the ECT problem can be written as follow:

$$g_k = S^T SG_k - S^T C \tag{5}$$

Where $B = S^T S, b = S^T C$ .Further equation (3) quasi-Newton equation is expressed as:

$$y_k = Bs_k \tag{6}$$

Where $y_k = g_{k+1} - g_k$, $s_k = G_{k+1} - G_k$, B by $\alpha^{-1} I(\alpha > 0)$ approximation, Minimum norm residual expression is as follows:

$$\min \left\| y - \alpha^{-1} I s \right\| \tag{7}$$

In order to type (7) minimum, select α_k satisfy the following formula:

$$\alpha_k = \arg\min_{\alpha} \left\| y_{k-1} - \alpha^{-1} I s_{k-1} \right\|^2 \tag{8}$$

By equation (8) can be:

$$\alpha_k = \arg\min_{\alpha} \left\| \alpha I y_{k-1} - s_{k-1} \right\|^2 \tag{9}$$

Also because the next formula:

$$s_{k-1} = -\alpha_{k-1} g_{k-1}, y_{k-1} = S s_{k-1} \tag{10}$$

By equation (10) were the following expression:

$$\alpha_k = \frac{(g_{k-1}^T g_{k-1})}{g_{k-1}^T S^T S g_{k-1}} \tag{11}$$

PSO is a swarm intelligence-based heuristic global optimization algorithm, The specific mathematical description is as follows:

In the D-dimensional search space have m-particle, position of particle $i(i = 1, 2, ..., m)$ is $X_i = (x_{i1}, x_{i2}, ..., x_{iD})$, it experienced the optimal location is recorded $P_i = (p_{i1}, p_{i2}, ..., p_{iD})$, also called individual extremum P_{best}; Groups experienced the best of all particles for the $P_g = (p_{g1}, p_{g2}, ..., p_{gD})$, also called global extremum; the speed of particle i with $V_i = (v_{i1}, v_{i2}, ..., v_{iD})$ to represent. Then for every generation, by tracking two particles are extreme to update their own, that is, particles evolve according to the following formula:

$$\begin{cases} v_{id}^{t+1} = \omega \cdot v_{id}^t + c_1 \cdot r_1 \cdot (P_{id}^t - x_{id}^t) + c_2 \cdot r_2 \cdot (P_{gd}^t - x_{id}^t) \\ x_{id}^{t+1} = x_{id}^t + v_{id}^{t+1} \qquad i = 1, 2, ..., m \qquad d = 1, 2, ..., D \end{cases} \tag{12}$$

Where t is the number of iterations, ω is Inertia weight, c_1 and c_2 is acceleration constant, set $c_1 = c_2 = 2$, r_1 and r_2 is random function within the change [0, 1].

In order to change the shortcomings of PSO algorithm to balance the global search ability of PSO algorithm and local capacity to improve the introduction of second-order particle swarm algorithm, denoted by SPSO, the particle velocity update formula was revised as follow:

$$v_{i,d}(t+1)=v_{i,d}(t)+c_1r_1(p_{i,d}-2x_{i,d}(t)+x_{i,d}(t-1))+c_2r_2(p_{g,d}-2x_{i,d}(t)+x_{i,d}(t-1)) \quad (13)$$

Solutions to consider before using ECT imaging method for imaging, and then determine the boundaries of two media area, the border region selection method is to select grayscale image similar to some units. First post here on the image gray value matrix setting filtering threshold expressed as t, select the regional parameters θ. the first the imaging unit of gray value is greater than $t+\theta$ demand convex geometry, then every point within the convex hull marking, denoted by S. Then on the gray value of the imaging unit demand greater than $t-\theta$ hull geometry, and then on each point within the convex hull marking, denoted by S', into two convex part is surrounded by the border region (Fig. 1).

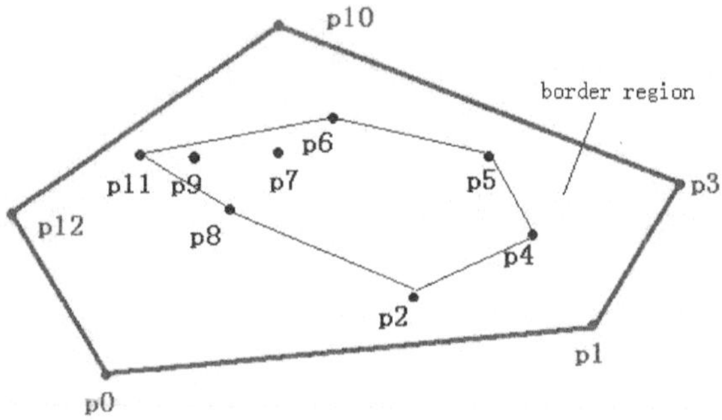

Fig. 1. convex hull to determine the boundary region

In the algorithm process, the set G_k for the ECT imaging algorithm executed by the pixel area, set $G_k^{'}$ for our calculated boundary pixels area, there are $G_k^{'} \subseteq G_k$. So each update particle velocity and displacement, update the gray in G_k its counterpart $G_k^{'}$, according to equation (1) incremental norm to determine the minimum position of particle it is optimal.

$$_{\min}\left\| C - SG_K \right\| \quad (14)$$

Further, We are going to put an amendment to image at each iteration process in accordance with the physical meaning, lead into a the estimated value $x_k^{'}$ within the scope of at 0 and 1 at each cycle. Especially when $x_k^{'} > 1$, its value becomes 0, to ensure that particles continue to update.The equation (17) has been modified into the following iterative projection:

$$\begin{aligned} v^{t+1}{}_{i,d}(t+1) &= v^{t}{}_{i,d}(t)+c_1r_1(p^{t}{}_{i,d}-2x^{t}{}_{i,d}(t)+x^{t}{}_{i,d}(t-1))+c_2r_2(p^{t}{}_{g,d}-2x^{t}{}_{i,d}(t)+x^{t}{}_{i,d}(t-1)) \\ x^{t+1}{}_{i,j} &= P_+[x^{t}{}_{i,d}+v^{t+1}{}_{i,d}] \quad i=1,2,\cdots,m \quad d=1,2,\cdots D \end{aligned} \quad (15)$$

Where P_+ is non-negative convex projection, given by the following formula:

$$P_+[x_k^{'}]=\begin{cases}0 & if \quad x_k^{'}\le 0\\ x_k & if \quad 0<x_k^{'}\le 1\\ 0 & \quad\quad x_k^{'}>1\end{cases} \tag{16}$$

3 Simulations and Analysis of Experimental Result

In order to verify the efficiency of algorithm, we perform an experiment on a 12-electrode system. The cross section of pipeline can be divided by a mesh of 32×32 when imaging, we can totally get 1024 pixel, which the effective area is 856 unit of imaging. To typical flow regime: stratified flow, core flow and bubbly flow, which carried on to prepare a constitution experiment, when imaging we adopted statistical filter threshold. The BB-PSO algorithm of this text elaborates by numerical simulation to reconstruct image, and carries on a comparison bet ween the linear back-projection algorithm(LBP), Steepest Descent algorithm(SD), Barzilai-Borwein(BB)and Tikhonov algorithm, the simulation calculation is based on MATLAB which is on one computer that has PIV3.0G CPU and 512M memory.

The experiment results shown in Table 1 (black area is water, the white area is the transformer oil).

Table 1. Results of image reconstruction

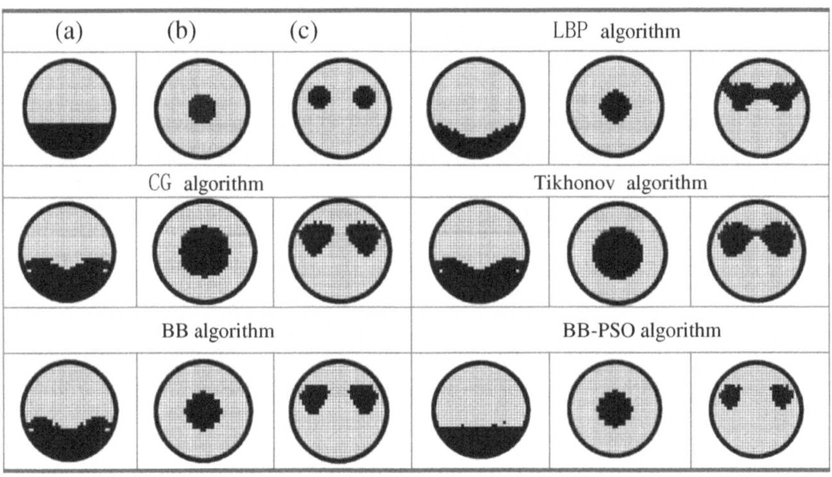

From Table 1 and Table 2 can be seen that BB-PSO algorithm is very close to the original flow pattern for the core flow and laminar flow. For complex bubbly flow pattern, we can see the results of the BB-PSO algorithm is closed original flow,but Tikhonov and the LBP algorithm for the reconstruction of flow pattern appeared ambiguous situation. Relative to the LBP,SD and BB algorithm, iterative steps of

BB-PSO algorithm is large, and iterative steps of Tikhonov algorithm is largest number.From the above analysis we can see that the use of the BB-PSO image reconstruction algorithms, for simple flow patterns and complex flow patterns of its imaging accuracy and quality than LBP, Steepest Descent algorithm (SD), Barzilai-Borwein(BB) and Tikhonov algorithm is better, but the iteration step number is large.

Table 2. Image error (%) and number of iteration

prototype	(a)	(b)	(c)	(a)	(b)	(c)
	Image error (%)			number of iteration		
LBP	40.32	74.61	98.07	1	1	1
SD	34.67	172.73	86.53	26	24	16
Tikhonov	32.25	190.91	117.31	310	300	300
BB	31.45	54.55	48.08	28	21	13
BB-PSO	20.16	54.54	63.49	100	100	100

4 Conclusions

This paper presents a BB-PSO algorithm for electrical capacitance tomography, Based on the analysis of the basic principles of electrical capacitance tomography, deduced three conjugate gradient method of iteration formulas and calculation steps and discussed the application of the algorithm is the feasibility of ECT. This algorithm programs simply, need low memory capacitance,which has the advantage of image high precision and easy satisfied with convergence condition.The numerical experiment shown that BB-PSO algorithm image reconstruction quality is better than LBP algorithm, Numerical experiments show that the algorithm of image reconstruction quality is far better than the LBP algorithm, better than the Steepest Descent algorithm (SD),Barzilai-Borwein(BB) and Tikhonov algorithm, which reconstructing image more close to raw flow regime,thus shown a new effective method of ECT image reconstruction.

Acknowledgement. This work is supported by national science foundation of China (60972127), Central Colleges basic scientific research projects special fund(DL10AB04),Northeast Forestry University Students Innovation Fund(111022541),Heilongjiang Provincial Department of Education Science and Technology Research Project(12513016),Postdoctoral Fund of Heilongjiang Province.

References

1. Liu, S., Yang, W.Q., Wang, H., Jiang, F., Su, Y.: Investigation of square fluidized beds using capacitance tomography: preliminary results. Measurement Science and Technology 12(8), 1120–1125 (2002)

2. Xi, S., Zhao, F.: Computation Methods for Optimization, vol. 7, pp. 113–117. Shanghai Scientifie & Technical Publishers, Shanghai (1983)
3. Yang, W.Q., Peng, L.H.: Image reconstruction algorithms for electrical capacitance tomography. Meas. Sci. Technol. 14, R1–R3 (2003)
4. Wang, H., Zhu, X., Zhang, L.: Conjugate gradient algorithm for electrical capacitance tomography. Journal of Tianjin University 38(1), 1–4 (2005)
5. Liu, S., Fu, L., Yang, W.Q., et al.: Prior-online iteration for image reconstruction with electrical capacitance tomography. IEE Proc. Sci. Meas. Technol. 151(3), 195–200 (2004)
6. Zhao, J., Fu, W., Li, T.: Image reconstruction new algorithm for electrical capacitance tomography. Computer Engineering 30(8), 54–57 (2004)
7. Nashed, M.Z., Walter, G.G.: General sampling theorems for functions in reproducing kernel Hilbert spaces. Math. Control Signals Systems (4), 363–390 (1991)
8. Barzilai, J., Borwein, J.: Two-point step size gradient methods. IMA Journal of Numerical Analysis (8), 141–148 (1988)

The Golden Tax Project: A Review of Problems and Solution

Fanghong Cai

The School of Public Finance & Taxation and Public Administration,
Jiangxi University of Finance and Economics
3816361@163.com

Abstract. There are some problems in the Golden Tax Project (GTP). First, managing tax by invoice cannot prevent the development of underground economy. Second, GTP cannot collect tax effectively. Third, the system can't find the cheat in deduction paper invoice. Fourth, the united networks that include public finance, tax, bank, and treasury cannot monitor taxpayer. Finally, GTP enlarges the incomes difference between local governments based on the tax-sharing system. There are some suggestions for the reform of GTP. The first suggestion is to implement integrated management that includes bank, invoice, and account in GTP. The second one is that the central government should develop tax-controlling POS that links the GTP. The third one is setting up a network invoice subsystem. The fourth one is to promote the united networks that include public finance, tax, bank, and treasury. The last one is to reform tax-sharing system so that the GTP can promote a balanced allocation of funds between local governments.

Keywords: Golden Tax Project, underground economy, retail trade, electronic commerce.

1 Introduction

GTP is the system of collecting value added tax (VAT). GTP includes a network and four subsystems and has four levels of wide area networks including county revenue, city revenue, province revenue, and state administration of taxation (SAT) in the national tax system. The four subsystems are the anti-counterfeiting tax billing system, the tax authentication system, the computer audit system, and the invoice investigation system, respectively. The process of collecting VAT has following steps. The first step is that a seller makes out a value added invoice using the tax security authentication system and declares dutiable goods using IC card. The second one is that a buyer declares dutiable goods using the VAT deduction paper invoice. The third step is that county revenue authenticates the invoice using the tax security authentication system and transfers to province revenue the result of comparing with seller's invoice information together. When the result shows it is success, the invoice can be used to reduce tax, otherwise, the paper invoice will be sent to the invoice investigation system.

D. Jin and S. Lin (Eds.): Advances in EECM Vol. 1, LNEE 139, pp. 233–237.
springerlink.com

In fact, GTP manages tax by invoice. It has a rule that a buyer can reduce tax when a seller finishes rate payments and the number of reduced tax equals to the number of rate payments. GTP has made a great contribution to the rapid growth of fiscal revenue. VAT tax average revenue is more than 35% of the national tax revenue. The average growth rate of VAT revenue in the past five years is more than 20% and is bigger than that of GDP. The growth rate of VAT was down in 2009 due to the structural tax cuts.

There are still some problems in GTP. These shortcomings not only undermine the country's financial strength and fair competition among enterprises but also result in an unreasonable financial profile appeared in the local governments. They make VAT rate remain high so that legitimate business enterprises bear a heavy tax. Therefore, it has a great practical value to study the shortcomings of GTP and propose methods to deal with these defects.

2 Analysis of Problems in the GTP

GTP is a tax collection and management information system. It has both design and management defects. In addition, there are some defects caused by tax-sharing system and united networks that include public finance, tax, bank and treasury.

2.1 Controlling Tax by Invoice Cannot Stop the Development of Underground Economy, Which Is the Design Ideal of GTP

Referring to foreign methods [1], GTP adopts the method of reducing tax by invoices. First, GTP cannot levy a tax on unlicensed operators since they have no license. They are not subject to the tax authority's administration and do not issue VAT invoices. Some operators do business on behalf of others or under the protection of the law enforcement agencies at the grassroots' level of management and so on. Second, some taxpayers make cash transactions and do not need to issue value added tax invoices. GTP cannot collect taxation on those case transactions.

2.2 GTP Is Not an Effective Management of Tax Evasion for the Retail and Small-Scale Taxpayers

First, the tax law shows that general taxpayers of retail commercial enterprises cannot make out VAT invoices for selling tobacco, alcohol, food, clothing, shoes and hats (not including labor fee), cosmetics, and other consumer goods. Retail enterprises have a large number of tax deduction invoices but they do not need to make out VAT sale invoices. GTP cannot tax to the retail business by VAT invoice [2]. Second, a small-scale VAT taxpayer pays a fixed amount of tax estimated by subjective so it is not accurate enough. GTP has no way to control tax to those taxpayers [3].

2.3 Seven Data Collected by the GTP Cannot Prevent the False Deduction Paper Invoice in the Fact

Since the limited technology of icon identification, the scan authentication system of GTP only collects seven key data currently, including the invoice number, invoice code, taxpayer identification number of both buyer and supplier, date of invoice, sale and tax amount, cannot check detail information of taxpayer business. Therefore, some taxpayer change sale items such as goods name, price and quantity etc. through applying the seven key data, and pass authentication, decrease their tax, nowadays, there are more and more the same cases. The data flow is a convert process of "electron data → paper invoice→ electron data", the paper invoice is a platform of transport data, but it do not achieve sharing of data on the real time, and it may make difference invoice information between the seller's stub invoice and the deduction paper invoice.

2.4 The United Networks That Include Public Finance, Tax, Bank and Treasury Are Not so Mature That E-commerce Revenue Is Out of Control

With the development of the GTP III, the united networks make great achievements in many aspects but there is a great distance from sharing information fully. First, Internet transactions should be the best conditions of controlling tax because the majority of transactions need the banking system. As long as the tax authorities are associated with the banking system, it will be able to clearly obtain the information of online transactions. However, banks restrict government departments' right of getting the information about customer transfer process. Banks do not agree tax authorities auditing accounts unless they have legal documents. Therefore, GTP cannot levy value added tax to many e-commerce behaviors. Second, there are not administrative subordinate relations between the national tax system and the local public financial department. The exchanges of data about public finance taxation carry out through treasury. The activities of treasury funds are later than fiscal behavior such as tax arrears, VAT refund, and financial return. The public financial department access to enterprise information is different from the revenue department, which makes GTP difficult to manage taxpayers.

2.5 GTP Expanded the Income Gap between Regions Due to Tax-Sharing System

The tax-sharing system is the budget management system of allocating the management right of taxes based on the division of responsibilities and powers between the central and local governments. In such a system, there exists a competition of tax benefits not only between the central and local governments but also among local governments. GTP enlarges the income gap among local governments. Developed areas often get more unreasonable taxes but backward areas may lose more tax revenue through GTP. For example, Changhong Company in Sichuan province sells color TV sets to electrical appliance stores in Jiangxi province; those stores in Jiangxi have to pay 170 Yuan to Sichuan SAT for selling a 1000 Yuan color TV set. Due to many reasons, Sichuan Changhong Company may require Jiangxi stores sell a color TV set for 900 Yuan but reward Jiangxi stores 200 Yuan each for the profit of stores. Therefore, the amount of output VAT of each color TV set in Jiangxi is only 153 Yuan. Clearly, the amount of input VAT is bigger than that of output VAT so Jiangxi SAT cannot receive the VAT from selling those color TV sets and other goods' output VAT is reduced by the input VAT of those color sets. In order to protect revenue, SAT in backward regions collects a fixed amount of tax to some sale enterprises.

3 Proposal of the Reform of GTP

3.1 The Collection and Administration of VAT in the GTP Should Combine Bank and Invoice with Account

To strengthen tax collection and management, many people suggest that we should learn foreign experiences and manage invoice by account. But many transactions of underground economy have not accounts and invoice. Therefore, the tax management should combine bank and invoice with account. If the tax management includes bank, invoice, and account, many underground transactions will be found because they should connect with the legal economy and bank transfers. Account and tickets are the basic contacts so the GTP will be able to enhance the collection and management of VAT.

3.2 We Should Develop Tax-Control POS That Link the GTP

Building tax-control POS (fiscal cash register) can improve VAT chain so that we can resolve the problem of collecting VAT from retail industry [4]. In addition, building the tax-control POS is also beneficial to enhance information technology of VAT, does not need to authenticate and compare the input VAT of paper invoice, and reduce the cost of management for both governments and enterprises. It can solve the problem of the inaccuracy decision that a fixed amount of VAT to a small-scale taxpayer is estimated by subjective. Because a small scale taxpayer can reduce the amount of the output VAT using the input VAT invoice like a general VAT taxpayer. It is benefit to implement equally tax to all taxpayers and reduce the rate of VAT when governments get plenty of fiscal funds.

3.3 Write Invoices Online and Collect Invoice Information Real Time

VAT general taxpayers get the electron cipher code relating the invoice system from the taxation authority when they acquire the qualification of invoicing VAT, and login in the taxation service website of state tax office by VPN or internet with cipher code, get the front office of the network version invoice subsystem. By building fire wall and other security login system, Realize safety exchange website data with tax office intranet data, and achieve buying invoice, using invoice, choosing deduction invoice and online filing etc functions on the online tax service platform. Basis on the online invoice mode, taxpayers buy electron invoice code and number, they can print tax information on any paper. When cancel paper invoice, paper voucher and physical seal in the traditional tax action will disappear, then we need the electron form voucher. 《Electron signature law》 that takes effect in 2005 provides law to favor for the electron invoice. We can carry on a safe authentication to the electron invoice by learning the CA authentication of the third party now [5].

3.4 Develop the United Networks That Include Public Finance, Tax, Bank, and Treasury to Get Effective Tax Administration to e-commerce

In order to develop Internet, e-commerce should get some preferential tax policy but it should be managed appropriately, otherwise, it will affect fair competition and cause the loss of tax source. Therefore, we should enhance the united networks to monitor public finance funds and tax of enterprises. It is benefit to ensure that enterprises tax cost justification.

3.5 Reform Tax-Sharing System so That the GTP can Promote a Balanced Allocation of Funds between Local Governments

Since GTP is a national network and has a reasonable allocation of public finance funds between local governments, we should reform the tax-sharing system. The main idea is that the VAT revenue of the consumption region is more than that of production region for encouraging consumption and promoting the reasonable distribution of industry.

References

1. Xie, Y., Man, H.: Perfection of China' VAT System through the Comparison between Domestic and Foreign VAT. Liaoning Financial College Journal 3(5), 17–19, Sum No.19 (2001)
2. Li, Q., Liu, H.: My views of the Golden Project System. Metallurgical Account (11), 22–23 (2003)
3. Niu, Y.: The choice of business enterprise tax saving. Tax collection Magazine (12), 35–36 (2002)
4. Jian, X.: The Golden Project System: start a new prospects of the industry of tax-control POS. Smart Cards World (6), 25–26 (2004)
5. Sunliping, Heguangjun, Zhangjinmei: Walk out the difficult of scanning and authenticating paper invoice. China Taxation News 11 (March 9, 2011)

Intuitionistic Fuzzy Reasoning under Quotient Space Structure

QianSheng Zhang[1] and ShiHua Luo[2]

[1] School of Informatics, Guangdong University of Foreign Studies, 510420 Guangzhou, China
[2] School of Statistics, Jiangxi University of Finance and Economics, Nanchang 330013, China
zhqiansh01@126.com

Abstract. This paper presents a theoretical basis of intuitionistic fuzzy reasoning under quotient space structure induced by an intuitionistic fuzzy equivalence relation. By constructing intuitionistic fuzzy set (IFS) representations of different grain-size spaces, the corresponding intuitionistic fuzzy reasoning method in quotient space and original universe space are studied and their relationships are then discussed.

Keywords: Intuitionistic fuzzy set, Quotient space, Intuitionistic fuzzy equivalence relation, Approximate reasoning.

1 Introduction

Fuzzy set concept, proposed by Zadeh, has been widely used in fuzzy reasoning [7] and fuzzy decision making [6]. Later, Atanassov presented intuitionistic fuzzy set and developed intuitionistic fuzzy logic [1, 2]. Also, Ciftcibasi [3] investigated two-sided (intuitionistic) fuzzy reasoning, but his method of intuitionistic fuzzy reasoning employed many fuzzy proposition logic and different implication rules, so it is very complicated and not effective in some cases. Thus, in order to further reduce the computational complexity, we need to reduce problem searching space or reduce the grain-size of universe space. Since quotient space proposed by Zhang [8] can reduce the grain-size of universe space, we try, in this paper, to construct quotient space theory by intuitionistic fuzzy equivalence relation and investigate the intuitionistic fuzzy reasoning under the quotient space structure and discover the close relationship between the ultimate intuitionistic fuzzy reasoning consequences under quotient space and under original universe space.

By introducing the intuitionistic fuzzy concept into quotient space theory and combining the hierarchical problem solving ability of quotient space theory with the ability of computing with words of intuitionistic fuzzy set theory, the proposed new intuitionistic fuzzy reasoning method under the quotient space structure greatly enhances the ability of intuitionistic fuzzy reasoning and problem solving.

D. Jin and S. Lin (Eds.): Advances in EECM Vol. 1, LNEE 139, pp. 239–244.
springerlink.com

2 Quotient Space Induced by Intuitionistic Fuzzy Equivalence Relation

Definition 2.1 [1]. An intuitionistic fuzzy relation R on a finite non-empty universe X is a function defined as $R: X \times X \to L_*$, $R(x,y) = (u_R(x,y), v_R(x,y))$, $\forall (x,y) \in X \times X$, where $L_* = \{(u,v) \in [0,1] \times [0,1] / 0 \le u+v \le 1\}$. Also, the partial order relation "$\le$" and intersection, union operators are given by

$$(u_1,v_1) \le (u_2,v_2) \quad \text{iff} \quad u_1 \le u_2,\ v_1 \ge v_2;$$

$$(u_1,v_1) \vee (u_2,v_2) = (u_1 \vee u_2,\ v_1 \wedge v_2);$$

$$(u_1,v_1) \wedge (u_2,v_2) = (u_1 \wedge u_2,\ v_1 \vee v_2).$$

Definition 2.2. An intuitionistic fuzzy relation R on a finite universe X is said to be an intuitionistic fuzzy equivalence relation, if the following conditions are satisfied.

(1) $\forall x \in X$, $R(x,x) = (1,0)$,

(2) $\forall x, y \in X$, $R(x,y) = R(y,x)$,

(3) $\forall x,y,z \in X$, $R(x,z) \ge \vee_{y \in X} \{R(x,y) \wedge R(y,z)\}$.

Let $IFE(X \times X)$ be the set of all intuitionistic fuzzy equivalence relations on X.

Definition 2.3. Let $R \in IFE(X \times X)$, we define an ordinary equivalence relation $\sim_R$ on X as $x \sim_R y \Leftrightarrow R(x,y) = (1,0)$.

Definition 2.4. A function $d: [X]^2 \to L_*$ is called an equicrural distance on quotient space $[X]$, if it fulfils the three following conditions,

$$d([x],[x]) = (0,1), \qquad d([x],[y]) = d([y],[x]),$$
$$d([x],[z]) \le d([x],[y]) \vee d([y],[z]), \qquad \forall [x],[y],[z] \in [X].$$

Proposition 2.1. Let $[X]_R = \{[x]_R \mid x \in X\}$ be the quotient space of X under $\sim_R$ induced by the intuitionistic fuzzy equivalence relation R, and we let

$$d([x],[y]) = (v_R(x,y),\ u_R(x,y)), \quad \forall x \in [x], y \in [y] \in [X]_R.$$

Then d is a normalized equicrural distance on $[X]_R$.

Proof. It is straightforward.

Definition 2.5. Let R be a given intuitionistic fuzzy equivalence relation on X, and assume that $\overline{\lambda} = (\lambda^*, \lambda_*) \in L_*$, by the equivalence relation $R_{\overline{\lambda}} = \{(x,y) |$

$|u_R(x,y) \geq \lambda^*, v_R(x,y) \leq \lambda_*\}$ the corresponding quotient space is defined as $[X]_{R_{\overline{\lambda}}} = \{[x]_{R_{\overline{\lambda}}} | x \in R\ \}$ and we can also define a function $d_{\overline{\lambda}}$ on $[X]_{R_{\overline{\lambda}}}$ as

$$d_{\overline{\lambda}}([x],[y]) = \begin{cases} (\mathbf{0,1}) & , \quad \forall \overline{\lambda}, (x,y) \in R_{\overline{\lambda}}; \\ \left(\frac{v_R(x,y)}{2+\lambda^*-\lambda_*}, \frac{u_R(x,y)}{2+\lambda^*-\lambda_*}\right) & , \quad \textit{otherwise}. \end{cases}$$

Theorem 1. The above defined function $d_{\overline{\lambda}}$ is indeed a normalized equicrural distance on $[X]_{R_{\overline{\lambda}}}$.

Proof. For any $[x],[y] \in [X]_{R_{\overline{\lambda}}}$, suppose $\overline{\lambda} = (\lambda^*, \lambda_*) \in L_*$, we can obtain that

(1) $d_{\overline{\lambda}}([x],[y]) \leq (1,0)$.

(2) For any $[x],[y],[z] \in [X]_{R_{\overline{\lambda}}}$,

If $R(x,y) \geq \overline{\lambda} = (\lambda^*, \lambda_*)$, $R(y,z) \geq \overline{\lambda} = (\lambda^*, \lambda_*)$, $R(x,z) \geq \overline{\lambda} = (\lambda^*, \lambda_*)$, then it is evident that $d_{\overline{\lambda}}([x],[z]) \leq d_{\overline{\lambda}}([x],[y]) \vee d_{\overline{\lambda}}([y],[z]) = (1,0)$.
Otherwise, we have

$$d_{\overline{\lambda}}([x],[y]) = \left(\frac{v_R(x,y)}{2+\lambda^*-\lambda_*}, \frac{u_R(x,y)}{2+\lambda^*-\lambda_*}\right),$$

$$d_{\overline{\lambda}}([y],[z]) = \left(\frac{v_R(y,z)}{2+\lambda^*-\lambda_*}, \frac{u_R(y,z)}{2+\lambda^*-\lambda_*}\right),$$

and $d_{\overline{\lambda}}([x],[z]) = \left(\frac{v_R(x,z)}{2+\lambda^*-\lambda_*}, \frac{u_R(x,z)}{2+\lambda^*-\lambda_*}\right)$.

Since R is an intuitionistic fuzzy equivalence relation, then for any $y \in X$.

$$\frac{v_R(x,z)}{2+\lambda^*-\lambda_*} \leq \frac{v_R(x,y)}{2+\lambda^*-\lambda_*} \vee \frac{v_R(y,z)}{2+\lambda^*-\lambda_*} \text{ and}$$

$$\frac{u_R(x,z)}{2+\lambda^*-\lambda_*} \geq \frac{u_R(x,y)}{2+\lambda^*-\lambda_*} \wedge \frac{u_R(y,z)}{2+\lambda^*-\lambda_*},$$

Thus, $d_{\overline{\lambda}}([x],[z]) \leq d_{\overline{\lambda}}([x],[y]) \vee d_{\overline{\lambda}}([y],[z])$.

Proposition 2.2. Let R be a given intuitionistic fuzzy equivalence relation on X, the quotient space family $\{[X]_{R_{\overline{\lambda}}} | \overline{\lambda} = (\lambda^*, \lambda_*) \in L_*\}$, or denoted by $\{X(\overline{\lambda}) | \overline{\lambda} = (\lambda^*, \lambda_*) \in L_*\}$, forms an order chain or a hierarchical structure on X.

Proof. For any $\overline{\lambda}_2, \overline{\lambda}_2 \in L_*$, and let $\overline{\lambda}_2 = (\lambda_2^*, \lambda_{2*}) \leq \overline{\lambda}_1 = (\lambda_1^*, \lambda_{1*}) \leq (1,0)$, one can easily see that $\forall x, y \in X$, $(x,y) \in R_{\overline{\lambda}_1} \Rightarrow (x,y) \in R_{\overline{\lambda}_2}$.
So, $R_{\overline{\lambda}_2}$ is coarser than $R_{\overline{\lambda}_1}$, i.e., $[X]_{R_{\overline{\lambda}_2}}$ is the quotient space of $[X]_{R_{\overline{\lambda}_1}}$.

3 Intuitionistic Fuzzy Reasoning under Quotient Space Structure

In [4,5], Lin considered the fuzzy set concept on quotient set. And in [8] Zhang introduced the theory of quotient space. But, now, we mainly investigate the intuitionistic fuzzy representation and reasoning under quotient space structure .

Definition 3.1[1]. An intuitionistic fuzzy set A on the given universe $X = \{x_1, x_2, \cdots, x_n\}$ can be defined by $A = \{x_i, < u_A(x_i), v_A(x_i) > / x_i \in X\}$, with two mappings $u_A, v_A : X \to [0,1]$, where $u_A(x_i)$, $v_A(x_i)$ represent the degree of membership and non-membership of object x_i to the set A, respectively, and the condition $0 \leq u_A(x_i) + v_A(x_i) \leq 1$ must hold for any $x_i \in X$.
We denote by $IF(X)$ the set of all intuitionistic fuzzy sets on universe X.

Below, we focus on investigating the intuitionistic fuzzy representations and logic reasoning in quotient space structure. Suppose that $[X]$ is the quotient space of X, then the induced intuitionistic fuzzy set $[A]$ on $[X]$ of A on X can be defined by

$$u_{[A]}([x]) = \bigvee_{x \in [x] \in [X]} \{u_A(x)\}, \quad \forall [x] \in [X] \qquad (1);$$

$$v_{[A]}([x]) = \bigwedge_{x \in [x] \in [X]} \{v_A(x)\}, \quad \forall [x] \in [X] \qquad (2).$$

It is obvious that $0 \leq u_{[A]}([x]) + v_{[A]}([x]) \leq 1 \quad \forall [x] \in [X]$.

Let us consider the following single-input-single-output (SISO) intuitionistic fuzzy reasoning modus ponens under quotient space stucture.

Suppose we have an intuitionistic fuzzy reasoning rule: if p is $[A]$, then q is $[B]$; and given an input "p is $[A^*]$", where p, q are linguistic variables, X and Y are the ranges of p and q, respectively; and $[A]$, $[A^*]$ are IFSs defined on quotient space $[X]$, $[B]$ is an IFS defined on quotient space $[Y]$. Then we obtain the corresponding output "q is $[B^*]$" as

$$[B^*] = [A^*] \circ R([A],[B]) \in IF([Y]),$$

where $u_{[B^*]}([y]) = \bigvee_{[x] \in [X]} \{u_{[A^*]}([x]) \wedge (u_{[A]}([x]) \wedge u_{[B]}([y]))\}$ (3);

$$v_{[B^*]}([y]) = \bigwedge_{[x] \in [X]} \{v_{[A^*]}([x]) \vee (v_{[A]}([x]) \vee v_{[B]}([y]))\} \qquad (4);$$

and the intuitionistic fuzzy relation from $[X]$ to $[Y]$ uses Mamdani min implication

$$R(A,B)=(A\to B)=A\cap B=<u_A\wedge u_B, v_A\vee v_B> .$$

Proposition 3.1. Let $A,B\in IF(X)$ and $[X]$ be a quotient space of X , if $[A]$ and $[B]$ are the induced IFSs on $[X]$ of A,B according to the above formulae (1) (2), respectively, then we have

$$[A]\cup[B]=[A\cup B], \qquad [A]\cap[B]\supseteq[A\cap B].$$

Proof. For any $[x]\in[X]$, we have

$$u_{[A]\cup[B]}([x])=\max(u_{[A]}([x]),u_{[B]}([x]))=\max(\vee_{x\in[x]}u_A(x),\vee_{x\in[x]}u_B(x))$$
$$=\vee_{x\in[x]}(u_A(x)\vee u_B(x))=\vee_{x\in[x]}\{u_{A\cup B}(x)\}=u_{[A\cup B]}([x]),$$
$$v_{[A]\cup[B]}([x])=\min(v_{[A]}([x]),v_{[B]}([x]))=\min(\wedge_{x\in[x]}v_A(x),\wedge_{x\in[x]}v_B(x))$$
$$=\wedge_{x\in[x]}(v_A(x)\wedge v_B(x))=\wedge_{x\in[x]}\{v_{A\cup B}(x)\}=v_{[A\cup B]}([x]);$$

and $u_{[A]\cap[B]}([x])=\min(u_{[A]}([x]),u_{[B]}([x]))=\min(\vee_{x\in[x]}u_A(x),\vee_{x\in[x]}u_B(x))$

$$\geq\vee_{x\in[x]}\{\min(u_A(x),u_B(x)\}=\vee_{x\in[x]}\{u_{A\cap B}(x)\}=u_{[A\cap B]}([x]),$$
$$v_{[A]\cap[B]}([x])=\max(v_{[A]}([x]),v_{[B]}([x]))=\max(\wedge_{x\in[x]}v_A(x),\wedge_{x\in[x]}v_B(x))$$
$$\leq\wedge_{x\in[x]}\{\max(v_A(x),v_B(x)\}=\wedge_{x\in[x]}\{v_{A\cap B}(x)\}=v_{[A\cap B]}([x]).$$

Theorem 2. Let $A,A^*\in IF(X)$, $B\in IF(Y)$, and let $[X],[Y]$ be the quotient spaces of X and Y, respectively. Assume $[A],[A^*]\in IF([X])$, $[B]\in IF([Y])$ are the induced IFSs of A,A^*,B according to the above formulae(1)(2), respectively. Provided that $[B^*]\in IF([Y])$ is the conclusion obtained from $[A^*]$ by the known intuitionistic fuzzy rule $[A]\to[B]$; while $B^*\in IF(Y)$ is the corresponding consequence of A^* by the given intuitionistic fuzzy reasoning rule $A\to B$, then we can get

$$[B^*]([y])\supseteq\bigcup_{y\in[y]\in[Y]}B^*(y)\quad,\qquad\forall[y]\in[Y].$$

Proof. (1) According to formulae (3) , (4), for any $[y]\in[Y]$, we have

$$u_{[B^*]}([y])=\vee_{[x]\in[X]}\{u_{[A^*]}[x]\wedge(u_{[A]}[x]\wedge u_{[B]}[y])\}$$
$$=\vee_{[x]\in[X]}\{(\vee_{x\in[x]}u_{A^*}(x))\wedge[(\vee_{x\in[x]}u_A(x))\wedge(\vee_{y\in[y]}u_B(y))]\}$$

$$= \bigvee_{[x]\in[X]} \{[(\bigvee_{x\in[x]} u_{A^*}(x)) \wedge (\bigvee_{x\in[x]} u_A(x))] \wedge (\bigvee_{y\in[y]} u_B(y))\}$$

$$\geq \bigvee_{[x]\in[X]} \{[\bigvee_{x\in[x]} (u_{A^*}(x) \wedge u_A(x))] \wedge (\bigvee_{y\in[y]} u_B(y))\}$$

$$= \bigvee_{y\in[y]} \bigvee_{[x]\in[X]} \{(\bigvee_{x\in[x]} (u_{A^*}(x) \wedge u_A(x))) \wedge u_B(y)\}$$

$$= \bigvee_{y\in[y]} \bigvee_{x\in X} \{(u_{A^*}(x) \wedge u_A(x)) \wedge u_B(y)\}$$

$$= \bigvee_{y\in[y]} \bigvee_{x\in X} \{u_{A^*}(x) \wedge (u_A(x) \wedge u_B(y))\} = \bigvee_{y\in[y]} u_{B^*}(y) ,$$

$$v_{[B^*]}([y]) = \bigwedge_{[x]\in[X]} \{v_{[A^*]}[x] \vee (v_{[A]}[x] \vee v_{[B]}[y])\}$$

$$= \bigwedge_{[x]\in[X]} \{(\bigwedge_{x\in[x]} v_{A^*}(x)) \vee [(\bigwedge_{x\in[x]} v_A(x)) \vee (\bigwedge_{y\in[y]} v_B(y))]\}$$

$$= \bigwedge_{[x]\in[X]} \{[(\bigwedge_{x\in[x]} u_{A^*}(x)) \vee (\bigwedge_{x\in[x]} v_A(x))] \vee (\bigwedge_{y\in[y]} v_B(y))\}$$

$$\leq \bigwedge_{[x]\in[X]} \{[\bigwedge_{x\in[x]} (v_{A^*}(x) \vee v_A(x))] \vee (\bigwedge_{y\in[y]} v_B(y))\}$$

$$= \bigwedge_{y\in[y]} \bigwedge_{[x]\in[X]} \{(\bigwedge_{x\in[x]} (v_{A^*}(x) \vee v_A(x))) \vee v_B(y)\}$$

$$= \bigwedge_{y\in[y]} \bigwedge_{x\in X} \{(v_{A^*}(x) \vee v_A(x)) \vee v_B(y)\}$$

$$= \bigwedge_{y\in[y]} \bigwedge_{x\in X} \{v_{A^*}(x) \vee (v_A(x) \vee v_B(y))\} = \bigwedge_{y\in[y]} v_{B^*}(y) .$$

Acknowledgments. This work is supported by the National Natural Science Foundation of China (Nos. 61070061, 60974019, 60964005), and the Guangdong Province Natural Science Foundation under Grant 9451009001002686, Guangdong Province Planning Project of Philosophy and Social Sciences (09O-19).

References

[1] Atanassov, K.T.: Intuitionistic fuzzy sets. Fuzzy Sets and Systems 20, 87–96 (1986)

[2] Atanassov, K.T.: Intuitionistic fuzzy logic. C. R. Acad. Bulgare Sc. 43, 9–12 (1990)

[3] Ciftcibasi, T.: Two-sided (intuitionistic) fuzzy reasoning. IEEE Trans. Systems Man Cybernet. A-28, 662–677 (1998)

[4] Lin, T.Y.: Context Free Fuzzy sets and Information Tables. In: European Congress on Intelligent Techniques and Soft Computing, pp. 76–80 (September 1998)

[5] Lin, T.Y.: Granular computing on binary relations 1: data mining and neighborhood systems, manuscript. Department of Mathematics and Computer Science, San Jose, California, USA (1998)

[6] Zadeh, L.A.: Outline of a new approach to the analysis of complex systems and decision process. IEEE Trans. Systems. Man Cybernet. 3, 28–44 (1973)

[7] Zadeh, L.A.: The concept of a linguistic variable and its application to approximate reasoning - II. Information Sciences 8, 301–357 (1975)

[8] Zhang, L., Zhang, B.: Fuzzy reasoning model under quotient space structure. Information Sciences 173, 353–364 (2005)

The Logistics Demand Prediction Research of Hebei Province Based on Information Technology

QingKui Cao* and XueLi Tan

College of Economics and Management, Hebei University of Engineering, 056038, Handan, China

Abstract. In recent years, with the development of economy in Hebei province, the logistics industry showed strong growth potential, but how to carry on the total demand forecasting work to improve the logistics industry of scientific planning and make the industry development.This paper first introduces the general situation of the development of economy and the logistics industry in Hebei province.Second, this paper gives a prediction on the next eight years of logistics demand in Hebei province based on information technology.Finally, it puts forward the path of the choice to promote the development of the logistics of hebei province based on information technology.

Keywords: Hebei Province, Logistics Demand, Prediction, Information Technology.

1 Introduction

In recent years, the economic development of Hebei maintained a good momentum, this powerful support for the development of the logistics industry in Hebei province. In 2008, Hebei area production (GDP) for 1.618861 trillion yuan, a year-on-year increase of 18.1%. The whole economic development in Hebei province going well for logistics industry in Hebei province to provide enough development space, Hebei province logistics demand has increased, but how to do well the total demand forecasting and then the scientific planning of the logistics industry development and a current outstanding problems.

In 2008, the core of the logistics industry in Hebei province traffic transportation industry has made a great progress for the development of the logistics industry in Hebei province, a firm foundation. In 2008, the investment, the province construction of highway, water, local railways and other transportation industry on fixed assets investment is 41.543 billion yuan, a year-on-year increase of 11.7%. The province highway reaching a total of 149500 kilometers, an increase of 2200 km; local railway construction investment of RMB 2.151 billion yuan, and local railway extend the mileage, up to 1605.35 km, compared to the 82.66 km last year; in addition, the

* The author: Qing-kui Cao (1963-), male, Ph.d., professor, school of Hebei University of Engineering, engaged in the research of logistics engineering and supply chain management.

D. Jin and S. Lin (Eds.): Advances in EECM Vol. 1, LNEE 139, pp. 245–249.
springerlink.com

province's civil aviation, waterway and pipeline transport obtained the fast development, for the development of the logistics industry plays an important role of the security. Whether from investment from a transportation construction or production ways, the development of logistics in Hebei province are showing the strong demand in Hebei province, and puts forward logistics demand forecast research available for more benign and fast logistics industry in Hebei province, but also beneficial to the development of Hebei province from on macroscopic logistics industry to make more reasonable planning, in order to promote the sustainable development of the logistics industry in Hebei province.

2 The Demand Forecast Analysis of Hebei Province Based on Information Technology

At present, the forecast of traffic of the scholars emerge in endlessly, most scholars pay attention to research the deep and complex model and algorithm, such as stochastic time series model, gray model, mark chain of the grey model was improved, and its application to logistics demand forecasting.This article is based on information technology application in hebei province freight volume forecast, it can quickly get accurate prediction results and make perfect freight traffic volume forecast figure.

Although the comprehensive quantity of shipment data can not fully reflect the comprehensive traffic, but the amount of comprehensive logistics in certain areas, certain period can reflects the comprehensive logistics to some degree. Because in the whole logistics, comprehensive freight volume decided the other amount of logistics activity work, the change also reflects the change of comprehensive logistics amount. Hebei province 2001-2008 comprehensive traffic situation is shown in table 1.

Table 1. The quantity of shipment data of Hebei province from 2002 to 2008 Million tons

Year	Quantity of shipmen	Year	Quantity of shipmen
2001	80835	2005	91330
2002	84315	2006	96784
2003	80551	2007	104188
2004	87265	2008	111383

Based on information technology, according to forecasts of the hebei province, and forecasts logistics quantity of shipment data of 2009 to 2016 is shown in table 2.

Table 2. The SPSS forecast data Million tons

Year	The original data	Forecast value	Year	The original data	Forecast value
2001	80835	76832.83	2009	——	111686.64
2002	84315	81189.56	2010	——	116043.37
2003	80551	85546.29	2011	——	120400.10
2004	87265	89903.01	2012	——	124756.82
2005	91330	94259.74	2013	——	129113.55
2006	96784	98616.46	2014	——	133470.27
2007	104188	102973.19	2015	——	137827.00
2008	111383	107329.92	2016	——	142183.73

The original data and data such as predicted in the information clipped to the original data curve and predicted curve, finally the results as shown in figure 1 shows.

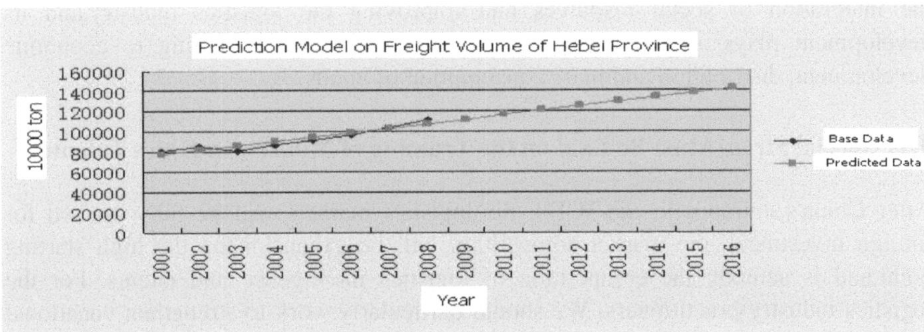

Fig. 1. The quantity of shipment predict models of Hebei province

Based on information technology,it can be concluded from above results that combines these two methods of EXCEL and SPSS to forecast the quantity of shipment , can not only make the two methods complement each other, but also obtain more scientific, objective, complete and accurate data on the regional economy reseach.First, overcome the calculation EXCEL forecast predicted trouble curve drab can't point-to-point accurate prediction reflect the faults, and second, overcome SPSS the professional degree is deep the disadvantage factors, make full use of SPSS accurate prediction for easy. The advantages of EXCEL is convenient and fast.

3 Advices to Promoting the Development of Logistics in Hebei Province Based on Information Technology

3.1 Developing the Third Party Logistics Companies

Improving the professional level of logistics enterprises and actively expanding logistics services, increasing training efforts for third-party logistics companies in the province-wide. Change the business model of traditional logistics management, improve the understanding of logistics and management, fully develop "the third profit source" initiative to promote the business of logistics outsourcing, concentrate on their core business and increase demand for third-party logistics, pace logistics a virtuous development path.

3.2 Enhancing the Coordination of the Fourth Party Logistics Planning and Management

The fourth party logistics play a crucial role in achieving a real increase the logistics industry of Hebei Province. Fourth-party logistics has an extremely important role in the integration of social resources and improving the logistics industry,and its development plays an important role in promoting or constraining to economic development, distribution and mass consumption of goods.

3.3 Special Stress Must Be Laid on the Training of Modern Logistics Talented

After China's entrance to the WTO, the logistics markets will be fully opened for foreign investment, the market competition will be expansion on the high starting point,and is actually the competition of logistics intelligence and talents. For the logistics industry practitioners, We should particularly work to strengthen vocational training and raise the level of business in various forms, created a number of professional and modern management logistics talents ; Enforce on the logistic certification, pursue the training and authentication of international assistant logistics manager, logistics manager and advanced logistics manager.

4 Conclusion

Based on information technology, this article uses EXCEL and SPSS softwares and gives a prediction on the logistics demand from 2009 to2016 of Hebei province,use the actual data of 2009 to test the forecasting data, find out the model is ideal and acceptable, the result is in accordance with the real situations. The logistics demand from 2009 to 2016 is forecasted, and will give a big impetus to logistics development in Hebei,and pus forward some relevant developing production and will be a great boost to the economy.

Acknowledgement. Fund project: Chinese National Natural Science Foundation (60075013),Hebei province National Natural ScienceFoundation(F2005000482).

References

1. Wei, S., Zhang, J.: The demand forecast analysis of material distribution. Logistics Technology (3), 19–20 (1999)
2. Li, Q., Liu, K.: The Forecasts of the volume freight traffic Based on Fractal Theory. Railway Institute General 25(3), 19–24 (2003)
3. Liu, T.-T., Deng, K.-T., Ma, C.-X.: The non-linear combination forecast of fuzzy artificial neural network theory in Demand Forecasting of the volume freight traffic. Railway Transportation Economy 30(9), 91–94 (2008)

Full-Bridge High Step-Up DC-DC Converter with Two Stage Voltage Doubler

Hyun-Lark Do

Department of Electronic & Information Engineering,
Seoul National University of Science and Technology, Seoul, South Korea
hldo@seoultech.ac.kr

Abstract. A full-bridge high step-up DC-DC converter with two stage voltage doubler is presented in this paper. The proposed converter achieves zero-voltage-switching (ZVS) all power switches and zero-current-switching (ZCS) of output rectifying diodes. The proposed converter can provide high voltage gain and the voltages across the semiconductor devices are effectively clamped. Steady-state analysis of the proposed converter is presented. A prototype of the proposed converter is developed, and its experimental results are presented for validation.

Keywords: Full-bridge, DC-DC converter, voltage-doubler, high step-up.

1 Introduction

Current-fed converters are often used in high step-up applications due to their inherent low input current ripple characteristic and high voltage gain [1]. However, voltage stresses of the switches are serious. In order to clamp the voltages across the switches, active snubbers are often employed. The snubbers require additional switches and cause additional conduction losses. As a result, the system efficiency decreases. On the other hand, the voltage-fed converters such as phase-shift full-bridge (PSFB) converters, which have been widely used, show features such as low voltage stress, fixed switching frequency, and ZVS of power switches. However, they have some drawback such as large conduction loss due to circulating current, duty cycle loss, and the voltage spikes across output rectifiers.

The asymmetrical pulse-width-modulation (APWM) technique was introduced and full-bridge and half-bridge converters with APWM control were suggested in [2]. They provide various advantages such as zero switching loss, no conduction loss penalty, and fixed switching frequency. Their major drawback is that the maximum duty cycle is limited to 0.5 and a large turn ratio of a transformer is required to obtain high voltage gain. In order to overcome these problems, a full-bridge high step-up DC-DC converter is proposed as shown in Fig. 1. The APWM technique is applied to the proposed converter to eliminate switching losses and maintain low conduction loss. The limitation of the maximum duty cycle disappears in the proposed topology. The proposed converter features high voltage gain, fixed switching frequency, soft-switching operations of all power switches and output diodes, and clamped voltages across power switches and output diodes.

D. Jin and S. Lin (Eds.): Advances in EECM Vol. 1, LNEE 139, pp. 251–254.
springerlink.com

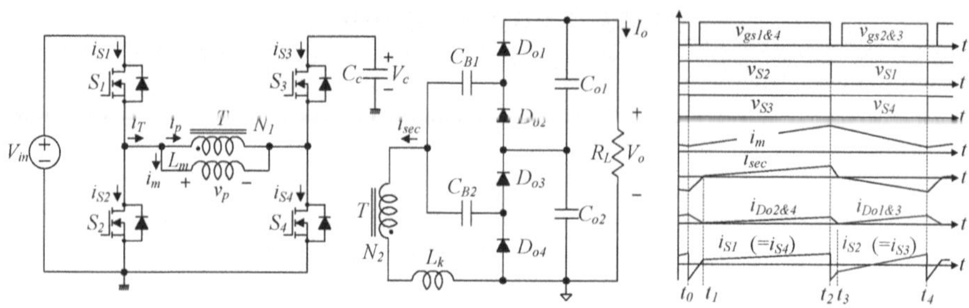

Fig. 1. Circuit diagram and key waveforms of the proposed converter

2 Analysis of the Proposed Converter

The circuit diagram and key waveforms of the proposed converter is shown in Fig. 1. The operation of the proposed converter during a switching period T_s (=$t_4 - t_0$) is divided into four modes. The switch S_1 (S_4) and the switch S_2 (S_3) are operated asymmetrically and the duty cycle D (=($t_2 - t_0$)/T_s) is based on the switch S_1 (S_4). The voltages across S_1 and S_2 are confined to the input voltage V_{in}. The voltages across S_3 and S_4 are confined to the voltage V_c across C_c. The two-stage voltage doubler consists of DC-blocking capacitors C_{B1} and C_{B2} and output diodes D_{o1} through D_{o4}. The transformer T is modeled as the magnetizing inductance L_m, the leakage inductance L_k, and the ideal transformer which has a turn ratio of 1:n (n= N_2/N_1). Before t_0, S_2, S_3, D_{o1}, and D_{o3} are conducting. At t_0, i_m and i_{sec} arrive at their minimum values, respectively.

Mode 1 [t_0, t_1]: At t_0, S_2 and S_3 are turned off. Then, the energy stored in the magnetic components starts to charge/discharge the parasitic capacitances of all switches. Since these parasitic capacitances are very small, this time interval can be ignored. When v_{S1} and v_{S4} arrive at zero, the body diodes of S_1 and S_4 are turned on. Since v_{S1} and v_{S4} are zero before S_1 and S_4 are turned on, zero-voltage turn-on of S_1 and S_4 is achieved. In this mode, i_m and i_{sec} increase linearly.

Mode 2 [t_1, t_2]: At t_1, the currents i_{Do1} and i_{Do3} arrive at zero and D_{o1} and D_{o3} are turned off. Then, i_{sec} changes its direction and D_{o2} and D_{o4} are turned on. Since the changing rates of i_{Do1} and i_{Do3} are controlled by L_k, their reverse-recovery is significantly alleviated. In this mode, i_{sec} increases linearly. The current i_m increases linearly with the same slope as in mode 1.

Mode 3 [t_2, t_3]: At t_2, S_1 and S_4 are turned off. Similar to mode 1, the parasitic capacitances are charged /discharged. Also, this time interval can be ignored. When v_{S2} and v_{S3} become zero, their body diodes are turned on. Since v_{S2} and v_{S3} are zero before S_2 and S_3 are turned on, zero-voltage turn-on of S_2 and S_3 is achieved. With the turn-on of S_2 and S_3, i_m and i_{sec} decrease linearly from their maximum values. The diodes D_{o2} and D_{o4} are still on.

Mode 4 [t_3, t_4]: A t_3, i_{Do2} and i_{Do4} arrive at zero and D_{o2} and D_{o4} are turned off. Then, the output diode D_{o1} and D_{o3} are turned on. Since the diode currents are controlled by L_k, its reverse-recovery problem is significantly alleviated. At the end of this mode, i_m and i_{sec} arrive at their minimum values, respectively.

By applying the volt-second balance law to v_p, V_c is derived by $DV_{in}/(1\text{-}D)$. The voltages v_{S1} and v_{S2} are confined to V_{in} and v_{S3} and V_{S4} are confined to Vc. Since V_c depends on D, the voltage stresses of S_1 and S_2 can be varied according to D and V_{in}. The voltage gain M of the proposed converter is given by

$$M = \frac{V_o}{V_{in}} = \frac{2n(1-2k)D}{(D+(1-2D)k)(1-D-(1-2D)k)}, \tag{1}$$

$$k = \frac{1}{2}\left(1-\sqrt{1-\frac{8L_k I_o}{nDV_{in}T_s}}\right). \tag{2}$$

Due to the two stage voltage doubler structure, the voltages across all the output diodes are confined to $V_o/2$.

3 Experimental Results

The performance of the proposed converter was verified on a 150W prototype. The prototype was designed to operate from a 48V input voltage and provide 800V output voltage. Its operating frequency was 100kHz. The turn ratio n was selected as 3. The proposed converter provides the voltage gain of 16.7 with the turn ratio of 3. The inductance L_m was selected as 100uH. The leakage inductance L_k was 67uH. The capacitor C_c was chosen as 220uF. The blocking capacitors C_{B1} and C_{B2} were selected as 6.6uF. The output capacitors C_{o1} and C_{o2} were selected as 22uF. Fig. 2 shows the measured key waveforms of the proposed converter. It agrees with the theoretical analysis. The voltage stresses of S_1 and S_2 are around 48V and the voltage stresses of S_3 and S_4 are around 110V. The ZVS operations of S_1 and S_3 are also shown in Fig. 2.. Since the voltages across the switches go to zero before the gate pulses are applied to the switches, the ZVS turn-on of the switches is achieved. Also, S_2 and S_4 operate with ZVS. The ZCS of the output diodes are also shown in Fig. 2. After the diode currents fall to zero, the voltages across the diodes rise toward $V_o/2$. The proposed converter exhibits the maximum efficiency of 94.4% at full load. Due to its soft-switching characteristic and alleviated reverse-recovery problem, it shows a higher efficiency than the conventional PSFB converter.

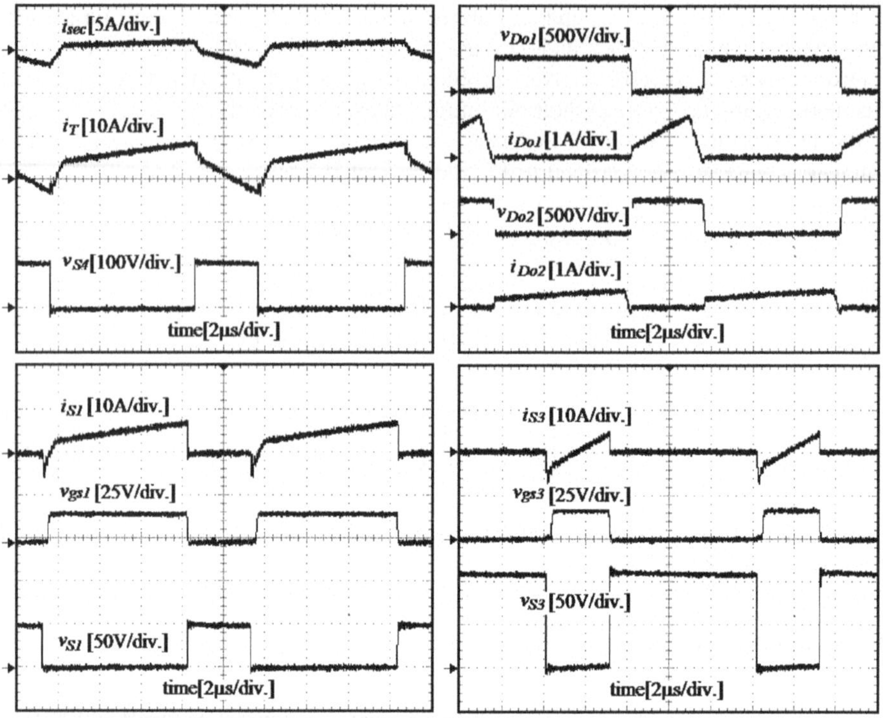

Fig. 2. Experimental waveforms

4 Conclusion

A soft-switching high step-up DC-DC converter with single magnetic component has been proposed. Both power switches operate with soft-switching and reverse-recovery problem of output diodes is dramatically alleviated by the leakage inductance. With the transformer turn ratio of 3, the voltage gain of 17 has been achieved. Due to its soft commutation of power semiconductor devices, it shows a high efficiency.

References

1. Zhu, L., Wang, K., Lee, F.C., Lai, J.S.: New start-up schemes for isolated full-bridge boost converters. IEEE Trans. Power Elec. 18, 946–951 (2003)
2. Imbertson, P., Mohan, N.: Asymmetrical duty cycle permits zero switching loss in PWM circuits with no conduction loss penalty. IEEE Trans. Industry Applications 29, 121–125 (1993)

The Research of Smart Grid Applications Based on IOT

Shengli Bao

Chengdu Institute of Computer Applications,
Chinese Academy of Science
Chengdu City, 610041, China
Happyzg3@163.com

Abstract. The applications of Internet of Things (IOT) have been more and more popular with the development of information technology. The IOT provides people with more integrated and intelligent services. Now it has been applied in some fields. But it is still not applied in the fields of grids. This paper briefly introduces the related concepts of IOT and smart grid. Meanwhile, the design of smart grid based on the IOT is proposed. The smart grid interconnection architecture is given. In the end, it shows deeply intelligent grids is the urgency of industry development wherever from personal, public service and the government needs and it can provide mankind with better services.

Keywords: Smart Grid, EPC (Electronic Product Code), WSN (Wireless Sensing Network), RFID (Radio Frequency Identification).

1 Introduction

RFID (Radio Frequency Identification) technology is a kind of wireless automatic identification technology, also known as electronic label technology, RFID technology has many advantages, is widely used in transportation, logistics, safety, security and other fields [1,2], as the bar code recognition technology upgrade replacement product, in recent years the huge market demand develops RFID quickly.

And the connotation of Internet of things is originated by a RFID of objects representing and using the network to exchange data of this concept, and constantly expand, extend, improve and gradually form Smart Grid, EPC Networking (Electronic Product Code Network) concept was put forth and was attended immediately by governments, enterprises and academic This paper briefly introduces the related concepts of IOT and smart grid. Meanwhile, the design of smart grid based on the IOT is proposed. The smart grid interconnection architecture is given. In the end, it shows deeply intelligent grids is the urgency of industry development wherever from personal, public service and the government needs and it can provide mankind with better services.

2 The Concept of Internet of Things and Smart Grid

2.1 The Concept of Internet of Things

According to the EU seventh framework RFID and networking research group issued a research report definition in September 15, 2009: the Internet of things is a part of the

D. Jin and S. Lin (Eds.): Advances in EECM Vol. 1, LNEE 139, pp. 255–259.
springerlink.com

future of the Internet, can be defined as based on standard and interoperable communications protocol. And it has ability of configuration, dynamic global network infrastructure. The Internet of things in the "matter" have mark, the physical properties and the essence of personality, using an intelligent interface to realize seamless integration with information network.

The organization's main purpose of the study is to facilitate European interior is different between RFID and networking project network; coordination including RFID networking research activities; to professional and technical balance, so that the study of maximum effect in the project cooperation mechanism.

2.2 The Concept of Smart Grid

Through the means of information, energy resources, conversion (power), transmission, distribution, power supply, electricity sales and electricity grid system, each link, intelligent communication, to achieve precise power supply, complementary power supply, raise energy utilization rate, power supply safety, save electricity cost target. The power network mentioned above called smart grid.

According to the United States Department of energy in modern power grid development report, the present paper generally seven characteristics of smart grid:

It has the ability to repair itself (self-healing); It can stimulate the user actively participate in power grid operation (incentive);It can Resist attacks (safety);It can provide high quality electric energy, reducing the economic loss (high quality);It can accommodate a variety of power generation and power storage form (new energy involved);It can provide us with prosperity of electric power market; It can optimize the equipment operation, reduce network operating costs.

To create an open system and the establishment of the model of sharing information based, integrated system of data, can optimize the network operation and management, make more intelligent power grid network, from three levels to improve power grid reliability, management efficiency and service level through the IOT. In order to realize the sustainable development of energy sources, to improve the stability of system and the sources of energy to use efficiency, smart grid in urgent need and networking technology combination.

3 The Development of Smart Grid Based on Internet of Things

3.1 Internet Development in China

At present, China has become the center of world manufacturing industry, the rapid development of information technology, networking technology research has been imminent. The Internet of things can be divided into a sensing layer, network layer and application layer and the 3 layer system, as shown in fig. 1.

Comprehensive perception, reliable transmission, intelligent information processing is the core competence of Internet of things. Comprehensive perception refers to the use of RFID, two-dimensional code, GPS, cameras, sensors, sensor networks and other perceived, capture, measurement technique whenever and wherever possible objects of information collection and acquisition. Reliable transmission is through a variety of communication network and the Internet, the object access information network, whenever and wherever possible for reliable information interaction and sharing. Intelligent processing refers to the use of cloud computing, fuzzy recognition of various intelligent computation technology, mass cross-regional and information analysis, promotion to the physical world, economic and social activities and changes of insight, realizing the intelligent decision and control [3].

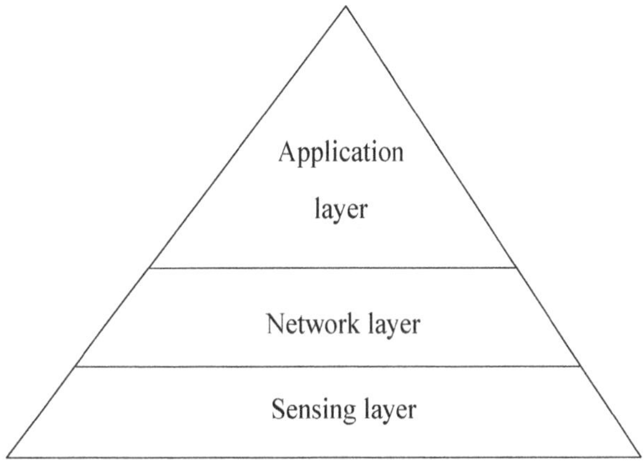

Fig. 1. The reference architecture of Internet of things

3.2 The Development of Smart Grid in China

China's smart grid study started relatively late, from big electrified wire netting and low voltage network input in 2 angles at the same time, has been fully mature and industrialization of the results are relatively few. But in many aspects of the research results have been as smart grid development laid certain foundation. China's power grid construction situation in the two outstanding contradictions:

A. Along with China's rapid economic and social development, the power consumption is growing very fast, meet the electricity needs of the task.

B. China's energy resource distribution and economic development is extremely lopsided, at the same time, the existing power grid structure is not strong enough, which restricts the economic development impact.

Consideration of our country power grid construction present situation, in the construction of Chinese characteristic of power production should refer to the experiences,formulate feasible route and scheme of construction of strong smart grid of Europe and the United States.

4 The Applications of the Smart Grid Based on Internet of Things

Smart grid WSN perception layer comprises a two-dimension code label and reader, the RFID tag and reader, cameras, sensors, sensor network (defined by a large number of various types of sensor nodes of the autonomous network, self-organization, self healing characteristics), WSN perception layer plays a major role in perception and object recognition, collection and capture information. Figure 2 shows the architecture of smart grid networking information service.

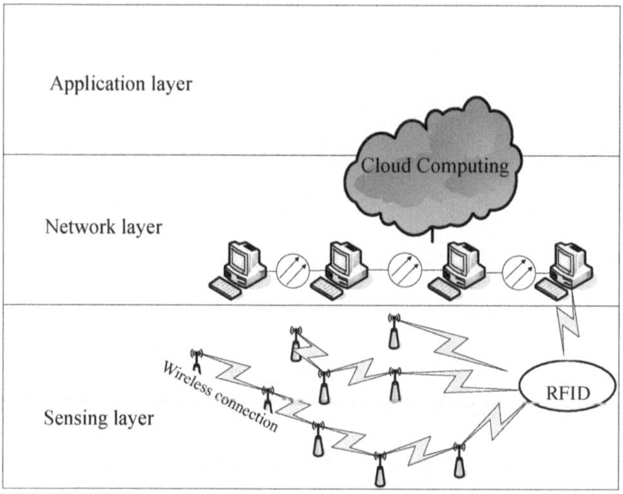

Fig. 2. The architecture of smart grid information service

Smart grid optical network layer includes a fiber optic cable, optical components optical network, network management center, information center and other use of network capacity for massive information intelligent processing part. That is to say the optical network layer should not only have the ability of the network, but also improve the ability of information processing. At the same time, the mass of the Internet to provide data and information analysis, promotion to the smart grid insight, realize real intelligent.

Intelligent network application layer is the networking technology and smart grid needs to combine, to realize intelligent network application solutions. Smart grid by application layer finally realizes the information technology and the integration of the smart grid, smart grid development has a broad impact. The application layer is the key in the process of informatization in order to meet the grid system in all aspects, intelligent communication, to achieve precise power supply, complementary power supply, raise energy utilization rate, power supply safety, save electricity costs to target each element of information demand analysis and information share.

5 Conclusions

The Internet of things technology in current digital substation there has been a big part of technology and adapt, and the current national power grid communication network is a network and telecommunications network connection, safety has very strong security. With the smart grid and the Internet of things technology further development, the Internet of things technology will penetrate further into the smart grid development and construction, has important economic value and research significance, this article for the Internet of things technology application in smart grid provides a certain reference.

References

[1] Yang, H.P., Ma, Z.H., Ning, H.S.: Double RFID label smooth upgrade of highway network toll collection system. Smart Cards and Electronic Labels (5), 36–38 (2006)

[2] Sun, X., Guo, C.: The Internet of Things networking technology and application. Science and Technology Information (26) (2010)

[3] Liu, W., Wang, H., Xiao, Q., Yang, J.: The concept of networking analysis. Telecommunications Technology (01), 5–8 (2010)

[4] Junhua: Based Intelligent the Internet of Things Digital Campus Research and Design. Wuzhou University (03) (2010)

[5] Jie, F., Zhen, X.: The Internet of Things to explore network architectures. Information and Computer (Theory) (08) (2010)

[6] Memorial. Development of The Internet of Things. Nanjing University of Posts and Telecommunications (Social Science Edition) (02) (2010)

Study of WSNs Security Route Based on Trust

HaoYu Wu

Jilin Business and Technology College,130062 Changchun, China
ccxxylw@126.com

Abstract. In order to find optimized the route communication protocol, lengthened WSNs the service life, this article introduces trust in the directed diffusion route protocol foundation to make the improvement, use trust substitution gradient, causes it to become one kind to have the security feature and the high load stabilization route protocol, through improves the DD route protocol, causes it to become a highly effective security the route protocol. The paper first introduced the security route's design principle, then proposed based on DD improvement program DD-T, finally through the code simulation realizes two kind of protocols, the result indicated that the improved route protocol, enhanced the network safety performance.

Keywords: WSNs, Trust, Directed Diffusion, DD-T.

1 Introduction

In WSNs, the node not only undertaking the data monitor duty, moreover is also undertaking the data route repeater. The WSNs main feature is take the data as the center, the data fusion technology is uses for to solve the data when retransmits after the route produces implosion and overlap question, but the route security is all WSNs security problem the most important question. How to use the existing energy resources, found optimized the route communication protocol, lengthened WSNs the service life, and was also a WSNs research important topic. The traditional route protocol design is to realize high Qos and the high band width use factor to a great extent, has neglected the energy consumption question, some traditional wireless network, for example Cellular-Network, MANET, SRWN route protocol, is also because the WSNs energy limited this characteristic, cannot be suitable.

2 WSNs Security Route Related Research Work

Defines 1: network life cycle is refers to from the network deployed that completes the normal work, to the first node because the energy exhausts the time withdrawal network experiences.

At present in the WSNs research, already proposed many route protocols, but majority had not considered that route's security problem, WSNs is facing the wormhole attack now, the Sybil attack, the sinkhole attack, hello floods and so on safe hidden danger. As shown in Table 1.

D. Jin and S. Lin (Eds.): Advances in EECM Vol. 1, LNEE 139, pp. 261–266.
springerlink.com

Table 1. All insecure on Routing

Route protocol	Security threat
TinyOS beaconing	False route information, Selective forward, sinkholes,sybil,wormhole,hellofloods
Directed diffusion and multipath variant	False route information, Selective forward, sinkholes,sybil,wormhole,hellofloods
GPSR,GEAR	False route information, Selective forward
the lowest cost forward	False route information, Selective forward, sinkholes,wormhole,hellofloods
the route	False route information, Selective forward, sinkholes,sybil,wormhole
Route protocols Based cluster	Selective forward,hellofloods

The existing several kind of security route, like SPKI/SDSI, SPIN, INSENSE and so on, has certain security, but still did not have more perfect route protocol. SPKI/SDSI (simple public key system/simple distributional system), this protocol's shortcoming lies in the authentication method to be simpler, is only a folk remedy certificate authority process, has not used the interactive authentication and the authorized behavior.

3 Security Route's Design Principle

In WSNs, the route security is assures the network security an important method measure. How design safety's route, did we think that should satisfy the following principle:

1) High energy efficiency. Because the sensor node energy is limited the characteristic and lengthens the network life cycle as far as possible the goal, had decided the route protocol must be the energy is highly effective.

2) Routing information security. Namely in the network routing information security, prevents the aggressor to intercept the routing information, analysis network topology [1].

3) Communications security. Namely the data which transmits to the network in carries on the encryption.

4 Directed Diffusion Route Protocol Based Improvement Trust

4.1 Directed Diffusion Route Protocol

Thought is to names to the network data with group of attributes, in the protocol has defined the interest and the gradient two concepts, what the interest expressed is inquires the duty, the gradient including the attribute value and direction [2]. What the protocol uses inquires actuation data transfer pattern, as shown in Figure 1.

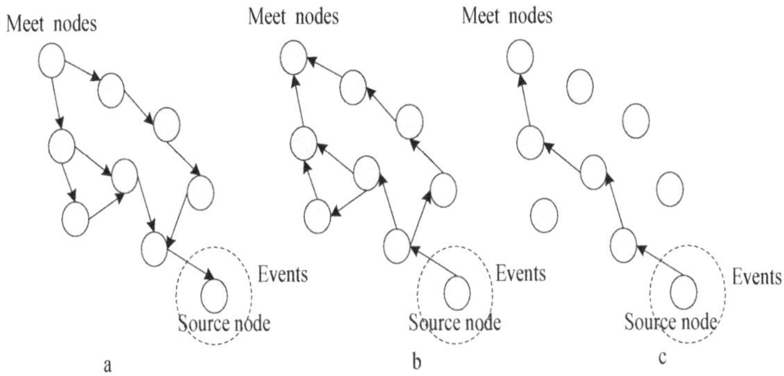

Fig. 1. The principle of directed diffusion

In Figure 1 may see that the DD route protocol may divide into 3 stages: Interest proliferation (or inquiry proliferation), gradient field establishment and data transfer. In the interest proliferation stage, the Sink node uses the pan-Hong's method, put interest information broadcasts to entire network or partial region all nodes. At the same time, the gradient field's establishment is also carrying on, after discover interested the goal, along the gradient direction transmission data, the Sink node possibly will receive the data which the different way will come, by now needed to choose a data transmission rate quick way, other achievement redundant way. Finally is the data transmission. Close with the DD protocol or the improvement protocol which has EAR and GBR. EAR[3] (Energy Aware Routing) is similar to the DD protocol, what is different is it maintains many ways, but is not looks like the DD protocol to use a gradient value biggest way, the way choice decides based on the way energy situation. GBR [4]

4.2 DD-T Plans

DD-T (Directed Diffusion based Trust route protocol) the core thought is establishes when the gradient field uses the node trust to substitute in the gradient the data transmission rate. Uses the node trust to replace the gradient have the following reason:

1.The node trust's computation. Not only considers the data transmission rate moreover also have other each kind of factor, including performance attribute and security attribute and so on, can the complete response node overall conditions. 2. The node trust's renewal is dynamic, although the gradient value is also the dynamic updating, but the gradient value's renewal is only according to the neighboring node the data transmission which rate assigns in the interest record, the geographical position and so on quite sole condition carries on, But the node data transmission rate and as well as geographical position attributes so on , the change scope within certain amount of time too will not be big, will not change, Therefore the gradient value's change cannot be too big, The DD-T plan flow chart is shown as Figure 2.

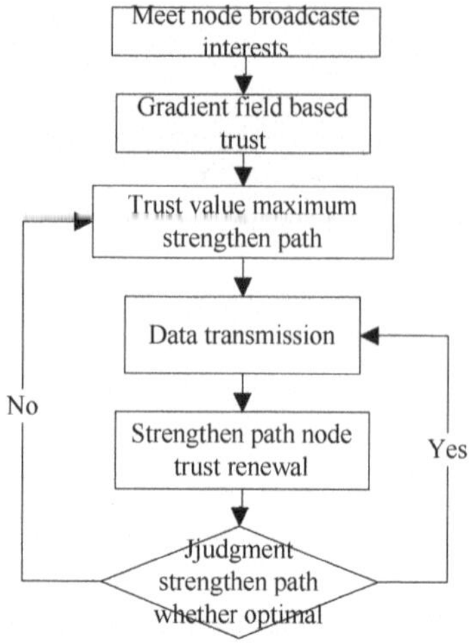

Fig. 2. The Process of DD-T

In the DD protocol, after the gradient field construction, first it is transmission data by the low gear rate way, and then the node after receiving the data will transmit one which have a greater gradient to strengthen the choice letter, finally forms a greatest gradient value to strengthen the way. In the DD-T improvement program, in the interest proliferation's pan-Hong process, the node to carry on the repeater which choice is higher than the trust valve value the neighbor node （lower than trust valve value node for not to trust node, like Figure 3), Therefore the establishment gradient field is based on the trust. In strengthens the way in the choice process, the node will strengthen the readout to the trust biggest neighbor node transmission, finally will obtain the trust in a big way to strengthen way.

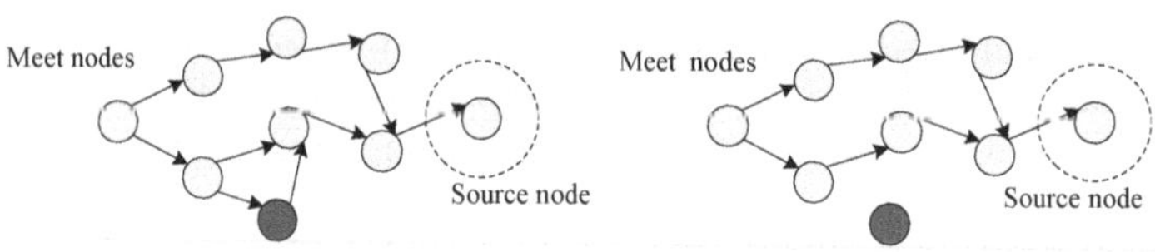

Fig. 3. Detection and isolation node which below trust threshold

4.3 Scheme Analysis

Security Analysis: Deals with the network internal and external attack the aspect, supposition enemy side node use illegal way, passed the status authentication, here name the malicious node, to the network implementation act of sabotage, chooses the repeater attack particularly, this aggressive behavior is very difficult the safety mechanism which has to survey. Therefore, we use this article to mention based on the node behavior trust appraisal mechanism, DD-T use based on the node behavior trust appraisal mechanism, its neighbor node needs to act according to its behavior to carry on the penalty, according to the trust characteristic, the malicious node's trust fast will drop, is lower than the trust valve value to kick the network. The trust high node, means the high performance, high security. The trust biggest strengthens the way, is also the degree of security highest way. Therefore trust joined, has provided the guarantee safely for the route..

Energy Analysis: In the trust computation, the node excess energy is also one of influencing factors; therefore node excess energy's how many are proportional with the trust size. Like this is advantageous to the network load stabilization.

Simulation Analysis: Simulation parameter establishment: 1,the 100*100 region space laying aside lays aside 80 nodes stochastically; 2,The node communication range is 30; 3,The nodes maximum range of communication send data consume 0.1J / bit, receiving data energy consumption negligible. It does not consider other factors on nodes' trust degree calculation of impact, residual energy of node trust effect. 4, not considering the malicious nodes.

Simulation process: In the node energy, in the position same situation, counts two algorithms in the movement process, when 10 node energies exhaust the withdrawal, the execution inquiry number of times, carries on 10 simulation, take the averaging value. Simulation arithmetic as shown in Table 2

Table 2. Simulation Arithmetic

```
Init( );
 Search_ne();
Whlie(i==10)
   {  Route();
        While {
     Trans();
}
Del_node();
}
node_einit( );
Search_ne_t();
Whlie(i==10) {
        Route_t();
  Trans_t();
  Del_node_t();
}
```

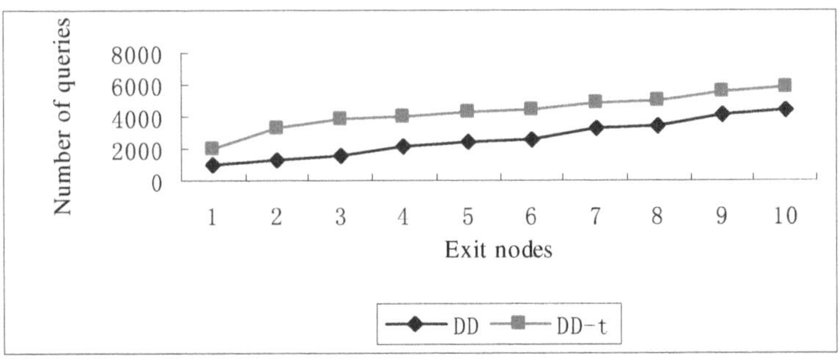

Fig. 4. Compare the inquires times of DD and DD-T

As shown in Figure 4, first the DD-T plan is surpass DD in the inquiry number of times, next in network load stabilization aspect, DD from the 1st node withdraws until the 10th node withdraws, period inquired about 3000 times, DD-T inquired about 4000 times, DD-T is surpass DD.

References

1. Stephan, O., Xu, Q.: Information Assurance In Wireless Sensor Networks. In: IPDPS 2005 (2005)
2. Intanagonwiwat, C., Govindan, R., Estrin, D.: Directed Diffusion: A Scalable and Robust Communication Paradigm for Sensor Networks. In: Proceedings of the Sixth Annual International Conference on Mobile Computing and Networks (MobiCom 2000), pp. 56–67. ACM Press, Boston (2000)
3. Shah, R.C., Rabaey, J.: Energy aware routing for lowenergy ad hoc sensor networks. In: IEEE Wireless Communications and Networking Conference (WCNC), Orlando, pp. 350–355 (2002)
4. Schurgers, C., Srivastava, M.B.: Energy efficient routing inwireless sensor networks. Proceedings of Communications for Network Centric Operations: Creating the Information Force, 357–361 (2001)
5. Xing, M., Li, L., He, Y.: WSN security route protocol research Based on prestige mechanism. Wuhan University of Science and Technology Journal 31(6), 896–900 (2009)
6. Wang, C., Jia, X., Lin, Q.: Wireless sensor network security routing algorithm based on trust. Journal of China Institute of Communications 29(11), 105–112 (2008)
7. Wu, P., Li, L.: Wireless sensor network credible route discovery algorithm (OL), `http://www.paper.edu.cn`

Study of Control Strategy Based Dual-PWM Converter under Unbalanced Input Voltage Condition*

Haijun Tao[1] and Di Hu[2,3]

[1] School of Electrical Engineering & Automation, Henan Polytechnic University, Jiaozuo City,Henan Province 454000, China
taohj99@hpu.edu.cn
[2] College of Electrical & Electronic Engineering, Huazhong University of Science and Technology, Wuhan City, Hubei Province 430074, China
[3] School of Electronic, Electrical & Computer Engineering, University of Birmingham, Edgbaston, Birmingham, B15 2TT, UK
963222472@qq.com

Abstract. The study of control strategy based dual PWM converter under unbalanced input voltage is mainly discussed in this paper. The operation of dual-PWM converter is analyzed, then the mathematical model of PWM rectifier when the power supply voltage is unbalanced is established, for the problem existing in the negative sequence grid voltage feed-forward control strategy, the closed loop control strategy to suppress negative sequence current is proposed, which can suppress various kinds of disturbances of the system, there isn't completely negative sequence current in the system. The simulation results verify the correctness of the proposed control strategy.

Keywords: Dual-PWM converter, Unbalanced voltage, Positive and negative current.

1 Introduction

With the development of the AC frequency control technology, the converter drives are widely used in the mine hoist, fans, pump etc. Especially the dual PWM inverter that the network side harmonic current are minor the energy can flow in the two-way gets more and more applications in the high-power applications. The study of three-phase VSR control system is usually assumed to be balanced with three-phase grid voltage [1,2], but when the actual grid voltage is unbalance, due to the existence of the negative sequence current and voltage in AC side, bus current bound to generate secondary non-characteristic harmonic currents to produce secondary harmonic in the bus voltage[3,4]. The harmonic voltage generates 3rd harmonic voltage in the network side by PWM modulation, then the 3rd harmonic current generated [5], so again, the grid voltage is unbalanced, 2,4 ··· times non-characteristic harmonic will appear in three-phase bus voltage, and 3,5 ··· times characteristic harmonic appear in

* This work is supported by Henan Province Key Project #082102240008 and The Natural Science Foundation of Henan Province Grant #2008A470004.

D. Jin and S. Lin (Eds.): Advances in EECM Vol. 1, LNEE 139, pp. 267–272.
springerlink.com

the network side current. More serious imbalance in supply voltage, these characteristic harmonic VSR will seriously affect the control performance of three-phase [6-8].

2 The Operation of Dual PWM Converter

Figure 1 shows a circuit diagram of dual-PWM inverter. Assuming the grid voltage is ideal voltage, while only consider the fundamental component in the network side voltage of the PWM rectifier bridge, so the network side Emf vector $\dot{U}_s$ as a reference, through controlling the voltage vector in the network side of the rectifier, it make the dual PWM converter operating in the four cases.

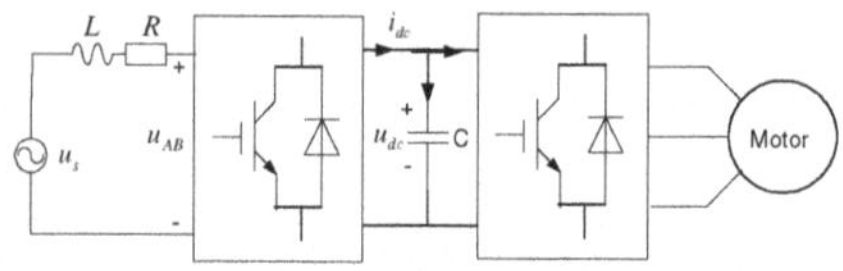

Fig. 1. Double PWM inverter circuit diagram

3 Mathematic Model of VSR under Unbalanced Voltage

Three-phase VSR topology is shown in Fig.2.

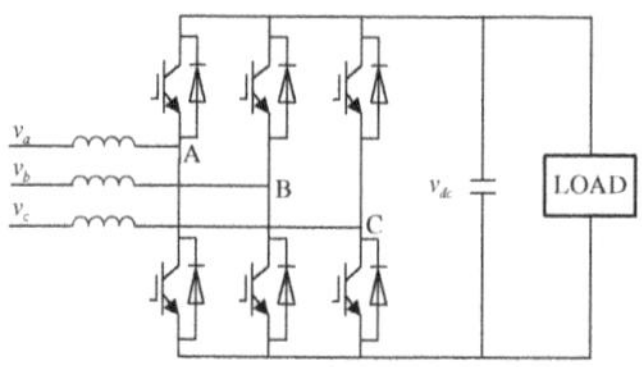

Fig. 2. The topology of AC-side converter expressed as switching function

The grid voltage contains both fundamental positive sequence component, fundamental negative sequence component and may contain zero sequence components under unbalance voltage, t, it can be expressed as

$$\begin{bmatrix} v_a \\ v_b \\ v_c \end{bmatrix} = \begin{bmatrix} V_+ \sin(\omega t + \varphi_+) \\ V_+ \sin(\omega t + \varphi_+ - 2\pi/3) \\ V_+ \sin(\omega t + \varphi_+ + 2\pi/3) \end{bmatrix} + \begin{bmatrix} V_- \sin(\omega t + \varphi_-) \\ V_- \sin(\omega t + \varphi_- + 2\pi/3) \\ V_- \sin(\omega t + \varphi_- - 2\pi/3) \end{bmatrix} + \begin{bmatrix} V_0 \sin(\omega t + \varphi_0) \\ V_0 \sin(\omega t + \varphi_0) \\ V_0 \sin(\omega t + \varphi_0) \end{bmatrix} \quad (1)$$

v_a、 v_b、 v_c --Each port voltage; V_+ -- Peak voltage of positive sequence component; φ_+ -- Initial phase of positive sequence component; V_- -- Peak voltage of negative sequence component; φ_- -- Initial phase of negative sequence component; V_0 -- Peak voltage of zero-sequence component; φ_0 -- Initial phase of zero-sequence component;

The rectifier circuit does not have zero-sequence path Shown in Fig.2, so the mathematical model for the rectifier is

$$\begin{bmatrix} \frac{di_a}{dt} \\ \frac{di_b}{dt} \\ \frac{di_c}{dt} \end{bmatrix} = \frac{1}{L}\begin{bmatrix} V_+ \sin(\omega t + \varphi_+) \\ V_+ \sin(\omega t + \varphi_+ - 2\pi/3) \\ V_+ \sin(\omega t + \varphi_+ + 2\pi/3) \end{bmatrix} + \frac{1}{L}\begin{bmatrix} V_- \sin(\omega t + \varphi_-) \\ V_- \sin(\omega t + \varphi_- + 2\pi/3) \\ V_- \sin(\omega t + \varphi_- - 2\pi/3) \end{bmatrix} - \frac{1}{L}\begin{bmatrix} v_a \\ v_b \\ v_c \end{bmatrix} \quad (2)$$

According to the principle of superposition, equation (2) can be decomposed into

$$\begin{bmatrix} \frac{di_{a+}}{dt} \\ \frac{di_{b+}}{dt} \\ \frac{di_{c+}}{dt} \end{bmatrix} = \frac{1}{L}\begin{bmatrix} V_+ \sin(\omega t + \varphi_+) \\ V_+ \sin(\omega t + \varphi_+ - 2\pi/3) \\ V_+ \sin(\omega t + \varphi_+ + 2\pi/3) \end{bmatrix} - \frac{1}{L}\begin{bmatrix} v_a \\ v_b \\ v_c \end{bmatrix} \quad \begin{bmatrix} \frac{di_{a-}}{dt} \\ \frac{di_{b-}}{dt} \\ \frac{di_{c-}}{dt} \end{bmatrix} = \frac{1}{L}\begin{bmatrix} V_- \sin(\omega t + \varphi_-) \\ V_- \sin(\omega t + \varphi_- + 2\pi/3) \\ V_- \sin(\omega t + \varphi_- - 2\pi/3) \end{bmatrix} \quad (3)$$

where,

$[i_{a+} \quad i_{b+} \quad i_{c+}]^T$ - Current of positive sequence voltage excitation;

$[i_{a-} \quad i_{b-} \quad i_{c-}]^T$ - Current of negative sequence voltage excitation.

According to (3), in general, the reactor inductance L is a very small value, then the negative sequence voltage in the grid excitation current will be very large, almost equal to short-circuit current. So the operation of the converter can not be controlled using traditional control strategy under unbalanced voltage. Therefore it needs to study the control strategy to inhibit negative sequence current.

4 Control Strategy of Inhibiting Negative Sequence Current

4.1 The Negative Sequence Grid Voltage Feed-Forward Control Strategy

Summary, when using the traditional control strategy, the AC side converter in the power grid, inspired by the negative sequence voltage produces a very large negative sequence current, which causes the converter AC-side over-flow, there May damage the device of the converter in the worst case. Therefore, we must discuss the control strategy inhibiting the negative sequence current in this case.

According to equation (3), the reason that power will not produce a very large positive sequence current is because the positive sequence voltage in the converter AC side under unbalanced voltage, and current closed-loop control the difference between the positive sequence voltage in AC side and the positive sequence voltage of the power to make the positive sequence in the AC side does not exceed the system power allows. Therefore, considering the AC side converter and the grid to generate a negative sequence voltage close to the negative sequence voltage to offset the negative sequence voltage excitation power, without the negative sequence voltage excitation naturally there will be no negative sequence current. Negative sequence grid voltage feed-forward control strategy for the basic idea is to detect the negative sequence grid voltage converter AC side in the modulation of the negative sequence voltage with the same power grid to offset the grid voltage negative sequence voltage.

In this control strategy, according to the principle of superposition equation (2) can be decomposed

$$\begin{bmatrix} \frac{di_{a+}}{dt} \\ \frac{di_{b+}}{dt} \\ \frac{di_{c+}}{dt} \end{bmatrix} = \frac{1}{L}\begin{bmatrix} V_+ \sin(\omega t+\varphi_+) \\ V_+ \sin(\omega t+\varphi_+ - 2\pi/3) \\ V_+ \sin(\omega t+\varphi_+ + 2\pi/3) \end{bmatrix} - \frac{1}{L}\begin{bmatrix} v_{a+} \\ v_{b+} \\ v_{c+} \end{bmatrix} \quad \begin{bmatrix} \frac{di_{a-}}{dt} \\ \frac{di_{b-}}{dt} \\ \frac{di_{c-}}{dt} \end{bmatrix} = \frac{1}{L}\begin{bmatrix} V_- \sin(\omega t+\varphi_-) \\ V_- \sin(\omega t+\varphi_- + 2\pi/3) \\ V_- \sin(\omega t+\varphi_- - 2\pi/3) \end{bmatrix} - \frac{1}{L}\begin{bmatrix} v_{a-} \\ v_{b-} \\ v_{c-} \end{bmatrix} \tag{4}$$

The strategy control block diagram shown in Fig.3:

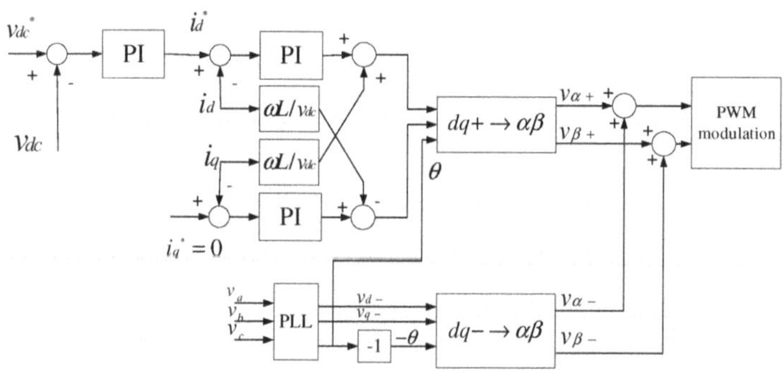

Fig. 3. The negative sequence voltage feed-forward control strategy diagram

4.2 The Closed Loop Control Strategy for Zero-Negative Sequence Current

Summary, the negative sequence voltage feed-forward control strategy is simple, but it is an open-loop method, the various disturbances that exist in the system did not inhibit, the compensation may not be able to achieve the desired effect. The best way to restrain disturbance is to use closed-loop control. Similar with the positive sequence system, negative sequence current under closed-loop control system is also studied in the rotating coordinate, just rotating frame under the negative sequence is synchronous with the negative sequence grid voltage vector of negative sequence, the negative sequence rotating coordinates are come from the two-phase-locked loop.

To transform equation (4) into positive and negative sequence rotating coordinates, the average model are

$$\begin{bmatrix} \frac{d\bar{i}_{d+}}{dt} \\ \frac{d\bar{i}_{q+}}{dt} \end{bmatrix} = \frac{1}{L}\begin{bmatrix} \bar{v}_{d+} \\ \bar{v}_{q+} \end{bmatrix} - \begin{bmatrix} 0 & -\omega \\ \omega & 0 \end{bmatrix}\begin{bmatrix} \bar{i}_{d+} \\ \bar{i}_{q+} \end{bmatrix} - \frac{\bar{v}_{dc}}{L}\begin{bmatrix} d_{d+} \\ d_{q+} \end{bmatrix} \quad \begin{bmatrix} \frac{d\bar{i}_{d-}}{dt} \\ \frac{d\bar{i}_{q-}}{dt} \end{bmatrix} = \frac{1}{L}\begin{bmatrix} \bar{v}_{d-} \\ \bar{v}_{q-} \end{bmatrix} - \begin{bmatrix} 0 & -\omega \\ \omega & 0 \end{bmatrix}\begin{bmatrix} \bar{i}_{d-} \\ \bar{i}_{q-} \end{bmatrix} - \frac{\bar{v}_{dc}}{L}\begin{bmatrix} d_{d-} \\ d_{q-} \end{bmatrix} \tag{5}$$

From equation (5) the inverter mathematical model under the conditions of positive and negative sequence is exactly the same, so the negative sequence current dq closed-loop control is the same with positive sequence conditions, PI controller parameters are the same with positive sequence Current loop PI parameters .The control strategy block diagram shown in Fig.4:

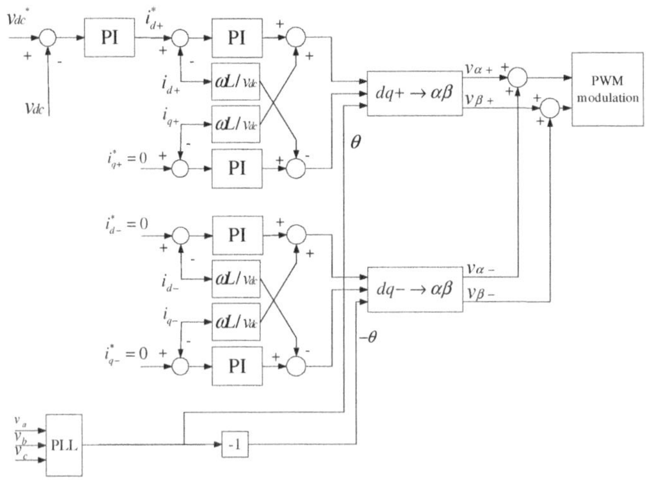

Fig. 4. Zero negative sequence current closed loop control strategy

5 Simulation and Results

Single-phase grid voltage drop are the most common faults, the simulation is in single-phase grid voltage drops. According to earlier, these results can be generalized to other unbalanced fault mode.

Simulation conditions: Power line voltage 380V, DC voltage : 600V, DC side load current 200A, 50% of C-phase voltage sag at 0.5 seconds.

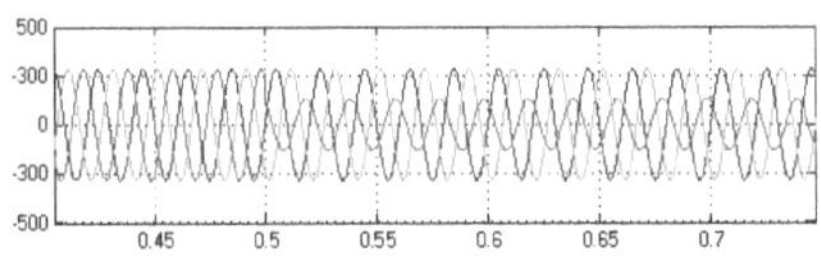

Fig. 5. The waveform of grid voltage(V)

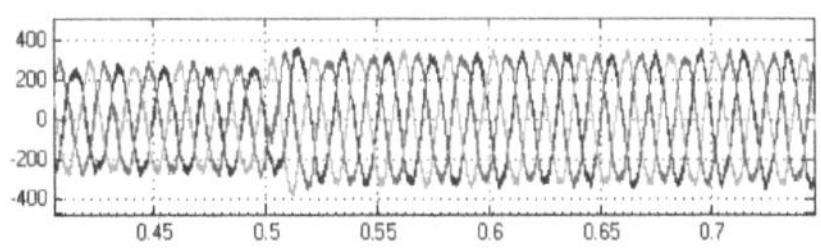

Fig. 6. The waveform of AC current(A)

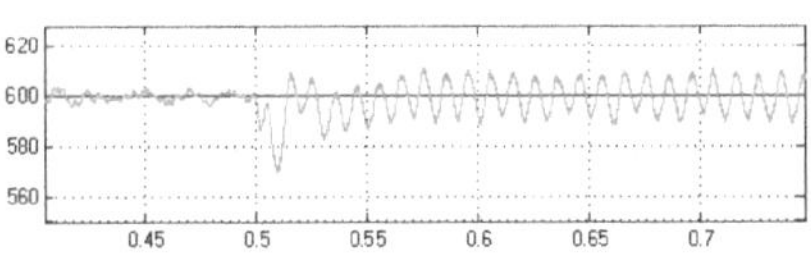

Fig. 7. The waveform of DC voltage(V)

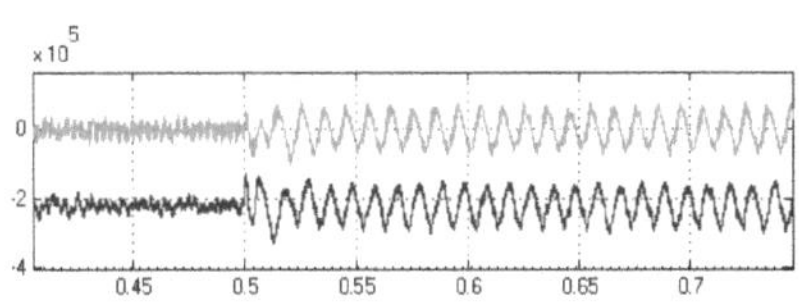

Fig. 8. The waveform of active and reactive (W,Var)

In the control strategy inhibiting negative sequence current, due to the presence of negative sequence voltage, three-phase active and reactive power contains second harmonics. DC bus voltage at the same time also contains second harmonics.

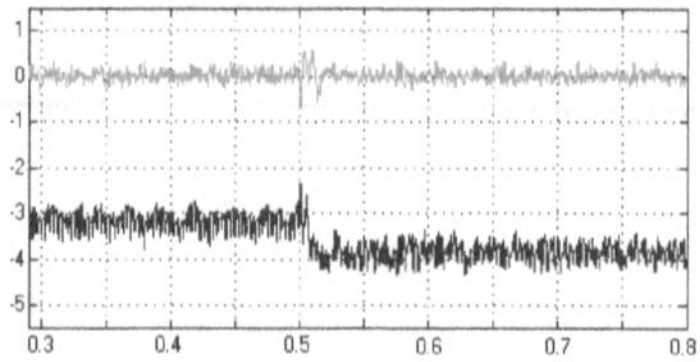

Fig. 9. The waveform of Positive current (A)

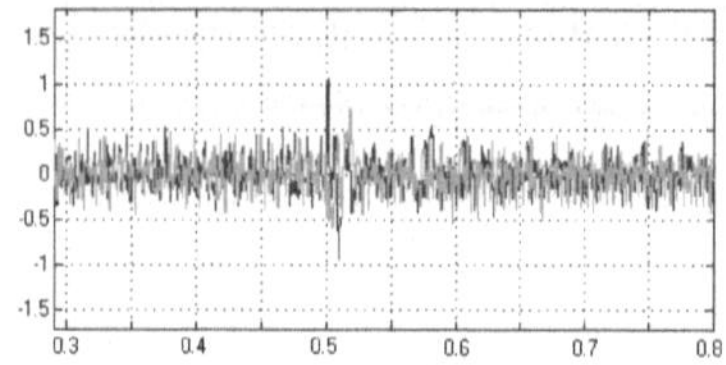

Fig. 10. The waveform of negative current (A)

Because of the negative sequence current closed-loop control, the control performance is well, the system is completely non-negative sequence current.

6 Summary

Firstly, the operation of dual-PWM converter is analyzed, then the mathematical model of PWM rectifier when the power supply voltage is unbalanced is established, for the problem existing in the negative sequence grid voltage feed-forward control strategy, a zero or negative sequence current loop control strategy is proposed, which can suppress various kinds of disturbances of the system, there isn't completely negative sequence current in the system. The simulation results verify the correctness of the proposed control strategy.

References

1. Zhang, X., Ji, J., Yu, Y., et al.: Study of low Voltage stress space vcctor PWM control for current source PWM rectifier. Proceedings of the CSEE 24(2), 144–149 (2004)
2. Zhang, X., Zhang, C.: Study on a new space Voltage Vector control method about reversible PWM converter. Proceedings of the CSEE 21(10), 102–105 (2001)
3. Zhang, X., Zhang, C.: PWM rectifier and it's control. Machinery Industry Press, Beijing (2003)
4. You, X., Li, Y., Valouch, V., et al.: SAPF control strategy under the condition of non-ideal source voltages. Proceedings of the CSEE 24(2), 55–60 (2004)
5. Stankvic, A.V., Lipo, T.A.: A novel control method for input output harmonic elimination of the PWM boost type rectifier under unbalanced operating conditions. IEEE Trans. Power Electronics 16(5), 603–611 (2001)
6. Park, K.-S., Ahn, S.-C., Hyun, D.-S.: New control scheme for 3-phase PWM AC/DC converter without phase angle detection under unbalanced input voltage conditions. In: APEC 2000, Fifteenth Annual IEEE, New Orleans, LA, USA (2000)
7. Jeon, S.-J., Lee, F.C.: Three control strategies for a thee-leg AC-DC converter under unbalanced AC voltage condition. In: Proceedings of the 25th Annual Conference of the IEEE, IECON 2003, Roanoke, Virginia, USA (2003)
8. Zheng, Z., Wang, C.: A new calculation algorithm of positive and negative sequence component of PWM converter under voltage unbalance condition. Engineering Journal of Wuhan University 43(4) (2010)

A Novel Learning Classification on Pattern Recognition

Yu-jie Sang

School of Science, Shenyang Aerospace University, Shenyang, 110136 China
syj1964sau@126.com

Abstract. In this paper we propose a novel approach to model pattern classification based on tangent distance within a statistical framework for classification. Statistical characteristics of classifier data are analyzed by the parametric estimates in the regression equation. We research on feature extraction for the training data in statistical pattern recognition by transformation and optimal Bayesian decision rule. Finally, we give our experiments on automatic speech recognition. The experimental results show the effectiveness of our approach.

Keywords: Optimal Bayesian decision, parametric estimates, pattern classification, Pattern Recognition.

1 Introduction

Automatic recognition and classification of patterns are of great importance in a variety of engineering and scientific disciplines such as computer vision and artificial intelligence [1]. The design of recognition system requires pattern representation, feature extraction, classifier design and performance evaluation. These relationships are often described in the form of several parameters; these parameters are adjusted until a reasonable measure of fit is attained. Statistical pattern recognition has been used successfully to design a number of commercial recognition systems [2]; at the same time, automatic pattern recognition systems are rising enormously due to the availability of large databases and stringent performance requirements. Consequently, combining several sensing modalities and classifiers is now a commonly used practice in pattern recognition.

2 Tangent Distance and Corresponding Approximations

Let $t \in \Re^D$ be a pattern and $y(t,\alpha)$ denote a transformation of t that depends on a parameter $L-tuple\ \alpha \in \Re^L$, where we assume that t does not affect class membership. The set of all transformed patterns now comprises a manifold $M_t = \{ y(t,\alpha) : \alpha \in \Re^L \} \subset \Re^D$ in pattern space [4]. The distance between two patterns can then be defined as the minimum distance between the manifold M_t of the pattern x and the manifold M_δ of a class specific prototype

D. Jin and S. Lin (Eds.): Advances in EECM Vol. 1, LNEE 139, pp. 273–279.
springerlink.com

pattern δ, which is truly invariant with respect to the regarded transformations (cf.Fig.1):

$$d(t,\delta) = \min_{\alpha,\beta\in\Re^L} \{\|y(t,\alpha) - y(t,\beta)\|^2\} \tag{1}$$

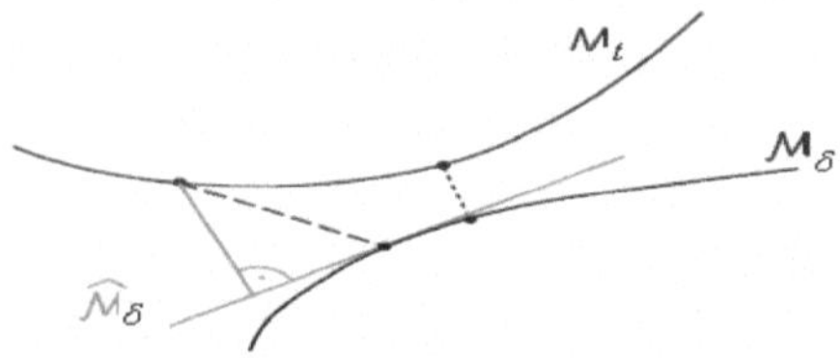

Fig. 1. Illustration of the Euclidean distance between an observation t and a reference (dashed line) in comparison to the distance between the corresponding manifolds (dotted line). The tangent approximation of manifold of the reference and the corresponding (one-sided) tangent distance is depicted by light gray lines.

The tangent vectors t_l that span the subspace are the partial derivatives of the transformation t with respect to the parameters $\alpha_l = (l = 1,\cdots,L)$, i.e. $t_l = \partial y(t,\alpha)/\partial\alpha$.

Thus, the transformation $y(t,\alpha)$ can be approximated using a Taylor expansion around $\alpha = 0$:

$$y(t,\alpha) = t + \sum_{l=1}^{K} \alpha_l t_l + \sum_{l=1}^{K} \mathrm{O}(\alpha_l^2) \tag{2}$$

The set of points consisting of all linear combinations of the pattern t with the tangent vectors t_l forms the tangent subspace $\hat{M}_t$, which is a first-order approximation of M_t:

$$\hat{M}_t = \{t + \sum_{l=1}^{K} \alpha_l t_l : \alpha \in \Re^L\} \subset \Re^D \tag{3}$$

Using the linear approximation $\hat{M}_t$ has the advantage that distance calculations are equivalent to the solution of linear least square problems or equivalently projections into subspaces, which are computationally inexpensive operations. Using the squared Euclidean norm, the TD is defined as:

$$d_{2S}(t,\delta) = \min_{\alpha,\beta\in\Re^L} \{\left\|(t + \sum_{l=1}^{L} \alpha_l t_l) - (\delta + \sum_{l=1}^{L} \beta_l \delta_l)\right\|^2\} \tag{4}$$

In order to reduce the effort for determining $d_{2S}(t,\delta)$ it may be convenient to restrict the calculation of the tangent subspaces to the prototype vectors.

3 Statistical Characteristics of Classifier Data

To classify a classifier data $t \in \Re^D$, we use the Bayesian decision rule [5]:

$$t \to r(t) = \arg\max_k \{p(k) \cdot p(t|k)\} \tag{5}$$

Where $p(k)$ is the a priori probability of class k, $p(t|k)\}$ is the class conditional probability for the observation t given class k and $r(t)$ is the decision of the classifier. We restrict our considerations to the case where the observations x are normally distributed with expectation and covariance matrix. Using the first-order approximation of the manifold M_δ for a mean vector δ; we obtain the probability density function for the observations:

$$p(t|\delta,\alpha,\sum) = N(x|\delta + \sum_{l=1}^{L}\alpha_l\delta_l, \sum) \tag{6}$$

The integral of the joint distribution $p(t|\delta,\alpha,\sum)$ over the unknown transformation parameters then leads to the following distribution:

$$\begin{aligned} p(t|\delta,\alpha,\sum) &= \int p(t,\alpha|\delta,\sum)d\alpha \\ &= \int p(\alpha|\delta,\sum)\cdot p(t,\alpha|\delta,\sum)d\alpha \\ &= \int p(\alpha)\cdot p(t|\delta,\alpha,\sum)d\alpha \end{aligned} \tag{7}$$

As only the spanned subspace determines the modeled variation. Hence, it is always possible to achieve the condition $\delta_l^T \sum^{-1} \delta_m = \varepsilon_{l,m}$; where $\varepsilon_{l,m}$ denotes the Kronecker delta [6]. We assume that α is independent of δ and $\sum$, i.e. $p(\alpha|\delta,\sum) \equiv p(\alpha)$. Furthermore, $\alpha \in \Re^L$ is assumed to be normally distributed with mean 0 and a covariance matrix $\gamma^2 I$, i.e. $p(\alpha) = N(\alpha|0,\gamma^2 I)$, where I denotes the identity matrix and γ is a hyper parameter describing the standard deviation of transformation parameters.

The evaluation of the integral in Eq. (7) leads to the following expression:

$$p(t|\delta,\sum) = N(t|\delta,\sum{}')$$

$$=\det(2\pi\Sigma')^{-\frac{1}{2}}\exp(-\frac{1}{2}[(t-\delta)^T\Sigma'^{-1}(t-\delta)]) \tag{8}$$

$$\Sigma' = \Sigma + \gamma^2\sum_{l=1}^{L}\delta_l\delta_l^T, \Sigma'^{-1}$$

$$=\Sigma^{-1} - \frac{1}{1+\frac{1}{\gamma^2}}\Sigma^{-1}\sum_{l=1}^{L}\delta_l\delta_l^T\Sigma^{-1} \tag{9}$$

Note that the exponent in Eq. (9) leads to the conventional Mahalanobis distance [7] for $\gamma \to 0$ and to TD for $\gamma \to \infty$ .The probabilistic interpretation of TD can also be used for a more reliable estimation of the parameters of the distribution.

Now we consider the two-class classification problem with equal prior probabilities, and a d -dimensional multivariate Gaussian distribution with the identity covariance matrix for each class [8]. Note that the features are statistically independent and the discriminating power of the successive features decreases monotonically with the first feature providing the maximum discrimination between the two classes:

(1). The mean vector m is known. In this situation, we can use the optimal Bayes decision rule to construct the decision boundary.

The probability of error as a function of d can be expressed as:

$$P_e(d) = \int_{\sqrt{\sum_{i=1}^{d}(\frac{1}{i})}}^{\infty}\frac{1}{\sqrt{2\pi}}\cdot e^{-\frac{1}{2}z^2}dz \tag{10}$$

It is easy to verify that $\lim_{d\to\infty} P_e(d)$. In other words, we can perfectly discriminate the two classes by arbitrarily increasing the number of features d .

(2). The mean vector m is unknown and n labeled training samples are available. Now the probability of error which is a function of both n and d can be written as:

$$P_e(n,d) = \int_{\theta(d)}^{\infty}\frac{1}{\sqrt{2\pi}}\cdot e^{-\frac{1}{2}z^2}dz\ ,\ \text{where}\ \theta(d) = \frac{\sum_{i=1}^{d}(\frac{1}{i})}{\sqrt{(1+\frac{1}{n})\sum_{i=1}^{d}(\frac{1}{i})+\frac{d}{n}}}$$

4 Feature Extraction in Statistical Pattern Recognition

In the training mode, the feature extraction module finds the appropriate features for representing the input patterns. The decision making process in statistical pattern recognition can be summarized as follows:

A given pattern is assigned to one of c categories $\omega_1, \omega_2, \cdots, \omega_c$ based on a vector of d feature values $t = (t_1, t_2, \cdots, x_d)$ .The pattern vector x belonging to class ω_i is viewed as an observation drawn randomly from the class-conditional probability function $p(t|\omega_i)$ .The optimal Bayesian decision rule [9] for minimizing the risk can be stated as follows: Assign input pattern x to class ω_i for which the conditional risk $R(\omega_i t) = \sum_{j=1}^{c} L(\omega_i, \omega_j) \cdot P(\omega_j |t)$ is minimum, where $L(\omega_i, \omega_j)$ is the loss incurred in deciding ω_i when the true class is ω_j and $P(\omega_j t)$ is the posterior probability. The conditional risk becomes the conditional probability of misclassification [10].

$$L(\omega_i, \omega_j) = \begin{cases} 0, i = j \\ 1, i \neq j \end{cases} \quad (11)$$

For this choice of loss function, the Bayes decision rule can be simplified as follows (also called the maximum a posteriori (MAP) rule): Assign input pattern x to class ω_i if

$$P(\omega_i |t) > P(\omega_j |t) \text{ for all } i \neq j \quad (12)$$

The optimal Bayesian strategy in this situation requires additional information in the form of a prior distribution on the unknown parameters. As a result, the performance of a classifier depends on the number of available training samples.

Let the training data be given by $t_{n,k}, n = 1, \cdots, N_k$ training patterns of $k = 1, \cdots, K$ classes. Assuming that the number L of tangent vectors is known, we consider the log-likelihood as a function of the unknown tangent vectors $\{\delta_{kl}\}$ [11]:

$$\begin{aligned} F(\{\delta_{kl}\}) &= \sum_{k=1}^{K} \sum_{n=1}^{N_k} \log N(t_{n,k} | \delta_k, \Sigma_k) \\ &= \frac{1}{1 + \frac{1}{\gamma^2}} \sum_{k=1}^{K} \sum_{n=1}^{N_k} \sum_{l=1}^{L} (t_{n,k} - \delta_k)^T \Sigma^{-1} t_{kl})^2 + const \\ &= \frac{1}{1 + \frac{1}{\gamma^2}} \sum_{k=1}^{K} \sum_{l=1}^{L} \delta_{kl}^T \Sigma^{-1} S_k \Sigma^{-1} \delta_{kl} + const \end{aligned} \quad (13)$$

with $S_k = \sum_{n=1}^{N_k} (t_{n,k} - \delta_k)(t_{n,k} - \delta_k)^T$ as the class specific scatter matrix. Σ and S_k can be regarded as covariance matrices of two competing models. The class specific tangent vectors δ_{kl} maximizing Eq. (13) have to be chosen such that the vectors $\sum^{-1/2}\mu_{kl}$ are those eigenvectors of the matrix $\sum^{-1/2}S_k(\sum^{-1/2})^T$ with the largest corresponding eigenvalues. As the above considerations show, two different models have to be determined for the covariance matrices $\sum$ and S_k. Note that features are extracted from the sensed data and some of the extracted features with low discrimination ability are discarded [12]. The choice between feature selection and feature extraction depends on the application domain and the specific training data.

5 Experiments

To show the practical value of the approach, we present results from experiments. For speech recognition, a suitable measure is the word error rate, which is defined as the ratio of the number of incorrectly recognized words to the total number of words to be recognized. The recognition system is based on Hidden Markov Models using continuous emission densities.

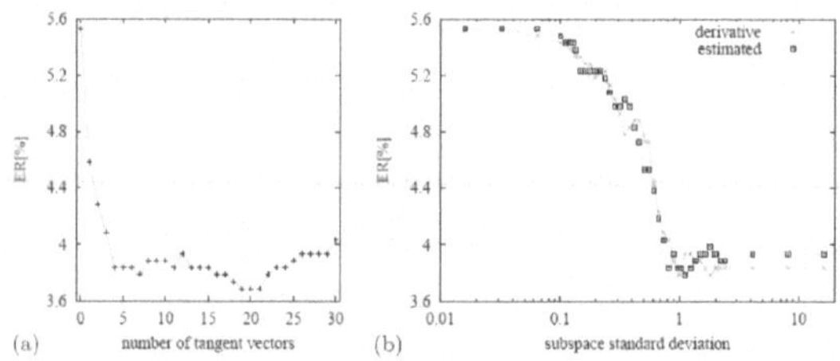

Fig. 2. (a) ER w.r.t. number of estimated tangents. (b) ER w.r.t. subspace standard deviation γ for $L = 7$ derivative and estimated tangent vectors.

Fig. 2. (a) Shows the changes of the error rate for different number of tangent vectors. Here, the tangent vectors were estimated using a local, class specific covariance matrix obtained from the set of local nearest neighbors for each training pattern. This shows that the presented method can be effectively used to learn the class specific variability on this dataset. Fig. 2. (b) Shows the error rate with respect to the subspace standard deviation γ for derivative tangents and estimated tangents using $L = 7$ each. It can be seen that, on this data, no significant improvement can be obtained by restricting the value of γ, while there may be improvements for other pattern recognition tasks. The new model proved to be very effective for pattern recognition, including the combination with globally operating feature transformations as the linear discriminant analysis. Thus, theoretical findings are supported by the experimental results.

6 Conclusions

In this paper we presented an approach for modeling and learning variability for statistical pattern recognition, embedding tangent distance into a probabilistic framework. In contrast to principal component analysis, the model disregards the specific variability of the patterns when determining the distance or the log-likelihood, respectively, which leads to an incorporation of transformation tolerance and therefore improves the classification performance. We research on feature extraction for the training data in statistical pattern recognition by transformation and optimal Bayesian decision rule. Our approaches are important in this respect, since the new model proved to be very effective for pattern recognition, including the combination with globally operating feature transformations as the linear discriminant analysis.

References

1. Dahmen, J., Keysers, D., Ney, H., Güld, M.O.: Statistical Image Object Recognition using Mixture Densities. J. of Mathematical Imaging and Vision 14(3), 285–296 (2001)
2. Yan, R., Zhang, J., Yang, J., Alexander, G.H.: A discriminative learning framework with pairwise constraints for video object classification. IEEE Trans. Pattern Analysis and Machine Learning 28(4), 578–593 (2006)
3. Keysers, D., Dahmen, J., Theiner, T., Ney, H.: Experiments with an Extended Tangent Distance. In: Proceedings 15th International Conference on Pattern Recognition, Barcelona, Spain, vol. 2, pp. 38–42 (2001)
4. Olshausen, B.A., Finch, D.J.: Emergnece of simple-cell receptive field properties by learning a sparse code for natural images. Nature 381, 607–609 (1996)
5. McDermott, E., Hazen, M.J.: Discriminative training for large-vocabulary speech recognition using minimum classification error. IEEE Trans. Audio, Speech, and Language Processing 15(1), 203–223 (2007)
6. Silverman, B.W.: Smoothed functional principal components analysis by the choice of norm. The Annals of Statistics 24(1), 1–24 (1996)
7. Bue, B.D., Stepinski, T.F.: Automated classification of landforms on mars. Computers & Geosciences 32(5), 604–614 (2006)
8. Benediktsson, J.A., Sweinsson, Arnason, K.: Classification and feature extraction of AVIRIS data. IEEE Transactions on Geoscience and Remote Sensing 33(5), 1194–1205 (1995)
9. Hammer, B., Villmann, T.: Generalized relevance learning vector quantization. Neural Networks 15, 1059–1068 (2002)
10. Kim, M., Pavlovic, V.: Discriminative learning of mixture of Bayesian network classifiers for sequence classification. In: Proc. of the 2006 IEEE Computer Society Conf. on Computer Vision and Pattern Recognition (CVPR 2006), New York, vol. 1, pp. 268–275 (2006)
11. Biem, A.: Minimum classification error training for online handwriting recognition. IEEE Trans. on Pattern Recognition and Machine Intelligence 28(7), 1041–1051 (2006)
12. Liu, X.B., Fu, H., Jia, Y.D.: Gaussian mixture modeling and learning of neighboring characters for multilingual text extraction in images. Pattern Recognition 41(2), 484–493 (2008)

Evaluation on Operational Cash Flow at Risk for China's Real Estate Listed Companies in the Event Window of Financial Crisis

Yu-jie Sang

School of Science, Shenyang Aerospace University, Shenyang, 110136 China
syj1964sau@126.com

Abstract. The operational cash flow at risk (CFaR) of China's real estate listed companies in this paper is considered in the event window of financial crisis that is the second half of 2008, this cash flow is a reality index, which discloses debt-paying ability and cash-making ability and explains the condition of no-financial enterprises in terms of cash. That means operational cash flow can increase the ability of resistance on risk and be critical to operation and development of the real estate listed companies and determines the effect of risk management particularly during the financial crisis. Then we study the cross-section data and panel data, compare the changes of the data is compared for giving evidence of risk management and simulate the operational cash flow in order to evaluate the financial control status for China's real estate listed companies and give prediction when facing financial crisis. Monte Carlo simulation method is used in order to present two types of empirical fitting analysis with the actual data. After fitting operational Net Cash Flow (NFC) and Cash Flow of Asset (CFoA) of listed real estate companies, we get operational CFaR by A-D test, χ^2 test and F test for the good fitness with curve-fitting method in practice. We analyze data for trends and seasonal variations, the empirical results show that operational cash flow has been affected in financial crisis. However, China's real estate listed companies operating cash flows at risk have been controlled effectively as a result of China's timely and effective macroeconomic policy and measures.

Keywords: Event window of Financial Crisis, Cash Flow at Risk, Real Estate Companies, Hypothesis testing, Distribution-Fitting.

1 Introduction

In China, the direct contribution of investment in real estate development to economic growth to the indirect one is about 2:3. Statistical analysis show that 25% of the steels output, 70% of glass and cement, 40% of wood and 25% of plastic products are used in real estate industry each year. China's real estate industry has been regarded as the stanchion industry of economy, and become the new increasing point. During 2004-2008, the average the real estate development, the total investment in fixed assets and annual GDP are 24.3%, 20.9% and 9.1%, respectively (see Fig.1). The great increasing investment in the real estate brought the fast growth of investment in fixed

D. Jin and S. Lin (Eds.): Advances in EECM Vol. 1, LNEE 139, pp. 281–287.
springerlink.com

assets, which drove further growth of GDP, enlarged investment multiplier effect, and then improved the contribution of the real estate industry to GDP from 6.3% to10.7%, averaged 9.28%.

However, after 29th Beijing Olympic Games in the second half of 2008, China stock market, especially for listed real estate companies, has slumped into depression since the international financial crisis in the second half of 2008. China's Real Estate Index decreased 68.03% compared with the year of 2007 when it became booming. Many big investor assets shrink nearly to 70% in 2008, more investors paid attentions to the security market. In the beginning of 2009, China stock market fluctuates drastically.

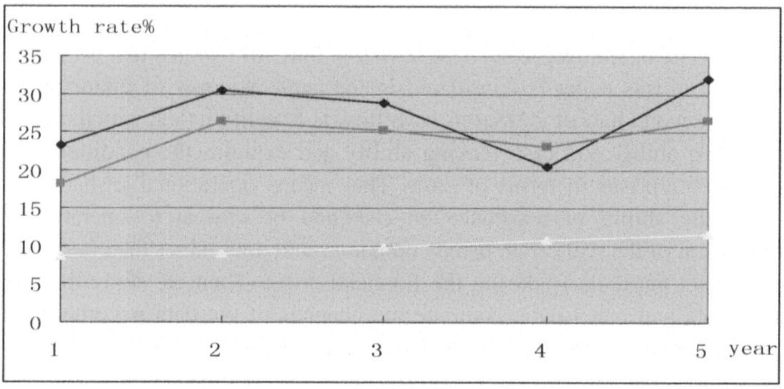

Fig. 1. Investment in real estate, total investment in fixed assets and Growth rates of GDP (Data sources: China Statistical Bureau) Blue, pink, yellow lines stand for growth rate of investments in real estate, total fixed assets and GDP respectively from 2004 to 2008

2 Related Researches on Cash Flow at Risk at Home and Abroad

The real estate is one of the most important blocks of stock market. As to the future of China stock market, the real estate industry will be the core of the A-share market. When facing RMB exchange rate fluctuation and Fluidity Surplus in China, the listed real estate companies would have more chances and higher profits. Furthermore, our country's urbanization brings about great market demand, while risk accompanies profit with the existence of market venture.

Operational cash flow is a reality index, which discloses debt-paying ability and cash-making ability and explains the financial condition of enterprises in terms of cash. If one company wants to increase production, sustain its business, and improve its superiority, it needs lots of cash flows. That means operational cash flow can increase the ability of resistance on risk and be critical to the existence and development of the company, particularly during the financial crisis.

The international research on Cash Flow at Risk of non-financial enterprises just started in recent years. Shimko(1998)[1] put forward to a measuring and calculating method, which is similar to calculate VaR. Lee & Kim (1999)[2] proposed the homoplastic method independently. Stein.& Usher[3](2000) directly investigated the cash flow's volatility of non-financial enterprises, which reflected the comprehensive effect of risk. In academia, the method based VaR is defined as exogenous model and the other named endogenous model which directly give a distribution of the cash flow.

3 Analysis of Operational Cash Flow at Risk of the Real Estate Industry

We study the enterprise financial status of operational cash flow at risk and show the new development of the real estate industry objectively, which brings investors evidence in investing the real estate industry. Enough cash flows can avoid financial crisis and show the well status of business. Combining industrial standards and peculiarities, listed real estate companies can guard against risks early with cash inflow and outflow.

Because of the great contribution of the real estate industry to GDP, many local governments keep the growth rate of investment in real estate to some degrees, then speed up the local economy development, just for the short-term benefits, they neglect the macro-policy adjustment, remise and sell the usufruct of urban land, and bring more revenue. But the price of the real estate is pushed up indirectly, so the real estate industry develops quickly, and there are more cash flows in real estate companies. Real estate industry is a long period, high risk and capital intensity industry[4]. It needs enormous cash flow to support. In recent years, some international scholars and researchers proposed the concept and measurement of Cash Flow at Risk[5], which is as an equivalent of non-financial enterprises on Value at Risk. Cash Flow at Risk means the maximum possible outflow[6] loss caused by the difference between the expected cash flow (E(C)) and the actual cash flow (C) in some probabilities:

$$pro[E(C)-C>CFaR]=p\%$$

3.1 Sample-Choosing

We will combine merits of the above two techniques and analyze the operational cash flow at risk empirical evidence from the listed companies of real estate in our country[7]. We do not hypothesize the concrete distribution of cash flow.

We choose 39 listed companies' operational cash flows with 5 years' data at least as sample from 2004 to 2008. Thus we get 300 observation- values (Data sources: WIND). After fitting operational Net Cash Flow (NCF) and operational Cash Flow of Asset (CFoA)[8] of listed real estate companies, we do A-D test, χ^2 test and F test for the good fitness to get operational CFaR. The cross-section data and panel data are studied in the article[9], and we compare the changes of the data for giving evidence of risk-analysis.

Table 1 shows the operational NCF[10] results of fitting the cross-section data and panel data. Table 2 is the operational CFoA results. All data is analyzed, using mathematical statistics in 5% and 1% quantile. Figure 2 and 4 describe the trends of operational CFaR on NCF and CFoA. Figure 3 and 5 are panel graphs of probability density from 2004 to 2008 respectively.

Table 1. Statistics of NCF

Year \ Quantile	α =0.05	α =0.01	Distribution
2008	−6.16912180	−18.40918875	Student's t
2007	−10.71967667	−54.19717737	Student's t
2006	−6.28289422	−88.3384161	Gamma
2005	−4.96385114	−13.26108667	Logistic
2004	−4.71312905	−18.92714957	Student's t

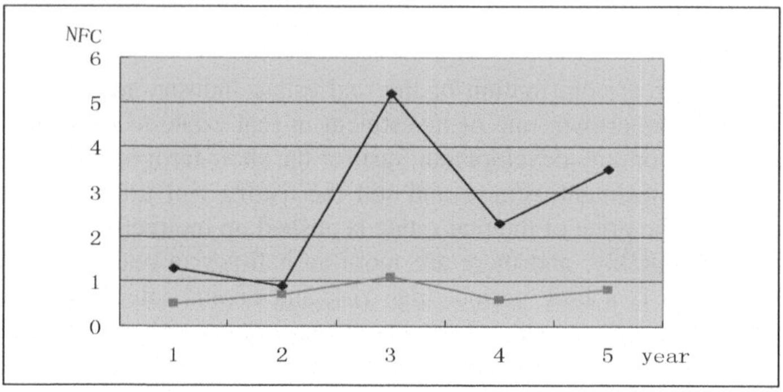

Fig. 2. Trend of CFaR of NCF from 2004 to 2008 (unit: billion Yuan) Blue lines indicates quantile a=0.05, pink line indicates quantile a=0.01

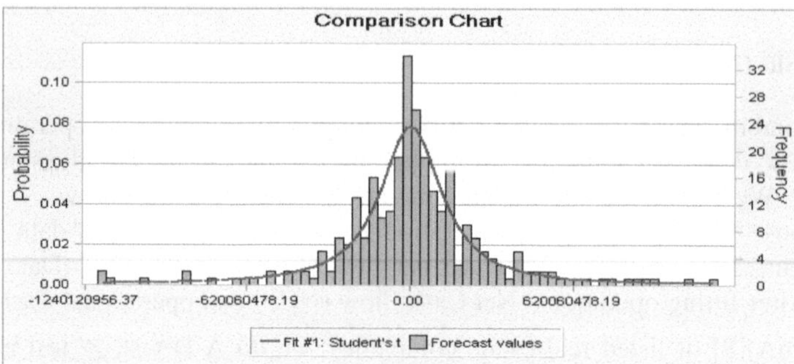

Fig. 3. Graph of probability density of NCF

3.2 Cross-Section Analysis

On operational NCF of the 5% quantile, the CFaRdecreases slowly year by year, but it increases in 2008 as to financial crisis, and then the CFaR ascends and meets inflexion point in 2006. On the whole, the CFaR of NCF still maintain in a stable level[11]. In the 1% quantile, the risk fluctuates largely and obviously. On operational

Table 2. Statistics of CFoA

Quantile / Year	α=0.05	α=0.01	Distribution
2008	−0.16	−0.25	Logistic
2007	−0.15	−0.26	Student's t
2006	−0.12	−0.24	Maximum extreme
2005	−0.46	−2.41	Minimum extreme
2004	−0.37	−1.93	Logistic

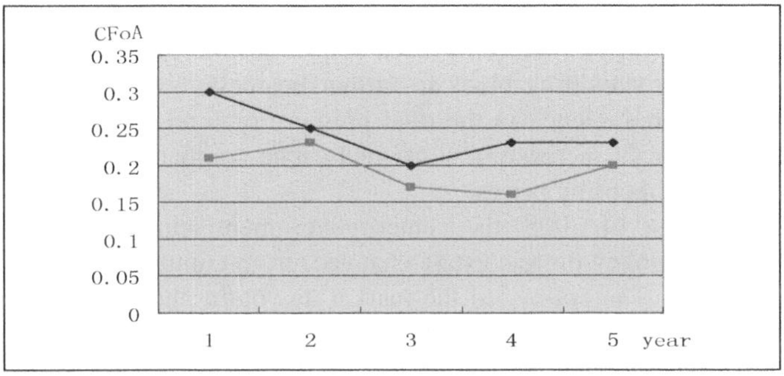

Fig. 4. Trend of CFaR of CFoA from 2004 to 2008
Blue lines indicates quantile a=0.05, pink line indicates quantile a=0.01

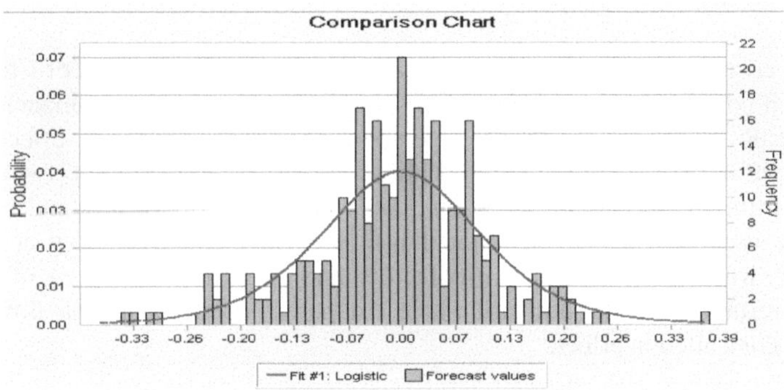

Fig. 5. Graph of probability density of CFoA

CFoA of the 5% quantile, the CFaR has no obvious abnormality excluding 2005 and 2006. In the 1% quantile, The CFaR jumps suddenly in 2007 and 2008.

There are many differences among the real estate companies[12], such as the listed capital power, investment benefit and scale, the holding proportion of cash flow and so on. So their reactions and adjustments are different when facing new policy and market risk. From the view of the data distribution, T-distribution is closed to normal distribution, so it appears a little approaching to normal distribution in the 5% and 1% quantile. The other distributions above have fat trails compared with normal distribution and obey fractal distribution, so the difference value is not obvious.

As to the real estate policy and the actual situation, the consumption structure upgrading forced the real estate to develop rapidly. The aim was to control the real estate industry consolidation in all aspects and restricted the local real estate overheating symptoms, so did the operational cash flow[13]. After 2005, real estate industries grew in super-high-speed and brought higher housing prices. In the second half of 2007 the government introduced a series of policies to guide and control the real estate market with excessive growth. These policies included coordinating the development areas, strictly controlling land management and managing mortgage and property tax. Therefore, companies benefited and housing prices rose rapidly, inducing the operating performance of listed companies increased significantly. As to 2008, land became a stumbling block to further business, and the lagged effects of macro-control appeared. Hence in the dual pressure of fund shortage and land cost increasing, the net profit increase of some real estate companies listed was a record high, so did the mean housing prices.

In the second half of 2008, the center government adjusted macro-control to housing prices. The policy on real estate changed, from controlling the investment to limiting the demand. The policy led the market downturn and decreased operational cash flow indirectly. But the most real estate companies adjusted management strategy in time and avoided risk. In the end of 2008, the companies purchased less land whereas more houses sold. The operational cash flows reduced relatively, but CFaR became safe.

4 Conclusion

In this paper, it is analyzed that operational cash flow at risk (CFaR) of China's real estate listed companies since financial crisis in order to assess their financial control status. CFaR is crucial to firm's operation and determines the effect of risk management in the event window of financial crisis. We present two types of empirical fitting analysis with the actual data, the distribution-fitting and hypothesis testing are used to analyze operational net cash flow and cash flow at risk, and meanwhile we consider the specific factors and quantitative indices of risk, the empirical results show that the real estate company's operational cash flows at risk have been controlled in effect.

References

[1] Shimko, D.: Cash before Value. Risk 11(7), 45 (1998)
[2] Lee, A.Y., Kim, J., Malz, A.M., Mina, J.: Corporate Metrics: The Benchmark for the Corporate Risk Management. Risk Metrics Group. J.P. Morgan, New York (1999)
[3] Darryll, H.: Evaluation of Value-at-Risk Models Using Historical Data. Economic Policy Review-Federal Reserve Bank of New York 2, 39–69 (1996)

[4] Stein, J., Usher, S.E., La Gattuta, D.: Cash flow at Risk and Financial Policy for Electricity Companies in the New World Order. The Electricity Journal, 15–20 (2000)
[5] Henrard, M.: Comparison of cash flow maps for value-at-risk. Journal of Risk 3(1), 57–71 (2001)
[6] Bekaert, G., Claude, B.E., Campell, R.H., Viskanta, T.E.: Distributional characteristics of emerging market returns and asset allocation. Journal of Portfolio Management 24, 102–115 (1998)
[7] Jorion, P.: Value at risk: The New Benchmark for Managing Financial Risk, 2nd edn. McGraw-Hill (2000)
[8] Stein, J.C., Usher, S.E., La Gattuta, D., Youngen, J.: A Comparables Approach to Measuring Cash flow-at-Risk for Non-financial Firms. Journal of Applied Corporate Finance 13(4), 100–109 (2000)
[9] Brender, A.: Cash-flow valuation and value at risk. North American Actuarial Journal 3(2), 26–29 (1999)
[10] Breuer, T., Krenn, G., Pistovcak, F.: Stress Tests: Maximum Loss, and Value at Risk (2002)
[11] Orshansky, M.: The Challenges of Creating a Successful Back-End Manufacturing Operations Flow for Fables Companies in the Era of Deep Sub-Micron Silicon Technologies. Silicon Corporation White Paper (2002)
[12] Hull, J., White, A.: Value at Risk When Daily Changes in Market Variables are not normally distributed. The Journal of Derivatives 5, 9–19 (1998)
[13] Longin, F.M.: Beyond the VaR. The Journal of Derivatives 8, 36–48 (2001)
[14] Leon Li, M.-Y., William Lin, H.-W.: Estimating Value at Risk via Markov Switching ARCH Models-An Empirical Study on Stock Index Returns. Applied Economics Letters 11, 671–679 (2004)
[15] Wu, L., Bao, X., Wang, Y.: How much Cash Flow be Managed——Empirical from Listed Companies in China. Finance Research 3, 162–174 (2007) (in Chinese)

Government Promotion to Implement Internationalization of TCM Industry of Gansu

ZhongHua Luo and LiXin Yun

Department of Economics and Business Management, Gansu College of Traditional Chinese Medicine, Lanzhou, 730000, China
luozhpsx@sohu.com

Abstract. Gansu province is one of the provinces famous for their abundant traditional Chinese medicinal materials. However, because of the constriction of many factors, Traditional Chinese Medicine （TCM） industry in Gansu cannot exploit its advantages, with a lower market share. It is a very important means for TCM industry of Gansu to promote its internationalization. Analyzing its connotation and the barriers in the course of implementing, the paper proposes the essential path for the internationalization of TCM industry promoted by government.

Keywords: traditional Chinese medicine, internationalization, modernization, safety production, government promotion.

1 Introduction

Recently, there exists an upsurge of return to nature all over the world. The market share of natural medicine is increasing at the rate of about 15%, enjoying the most popular, which makes internationalization of TCM face the rare chance. As one of the most outstanding of self-owned intellectual property rights in our exports, traditional Chinese medicine takes significant strategic position in the international market. With the biggest production of Chinese medicine, Gansu province will be famous for its products, which is advantageous to implementing the strategy of its internationalization. Since the high-quality Chinese medicine known by all over the world, its competitiveness is promoted, which is benefit for economic and social development of Gansu.

2 Connotation of Internationalization of TCM

Under the globalization background, internationalization of TCM industry has become a topical subject. However, what on earth does it mean, different experts give the different definitions [1,2,3,4,5]. The definition we conclude is that science and technology and production of TCM industry can enter international market abiding by the international rules and also communicate and cooperate between nations about economic and trade, science and technology and education to server people's health. So the internationalization of TCM industry covers many factors as follows.

D. Jin and S. Lin (Eds.): Advances in EECM Vol. 1, LNEE 139, pp. 289–294.
springerlink.com

(a) Supplying people all over the world with the healthiest Chinese medicine. The primary aim of internationalization is to provide the safest and the most effective medicine, especially to develop new products based on the theory of TCM to cure those diseases that western medicine failure to treat.
(b) Producing good products harmless to people and environment. It means the whole process including producing, sale and use must be safe and the course of circulation and consuming cannot bring harm to human health and destroy ecological environment.
(c) Producing abiding by international quality standards. It is inevitable for TCM industry if it wants to enter international market, which is the guarantee of superior quality.
(d) Circulation abiding by the rules of international trade. Since the industry of TCM involves a broad and gradual process, including international cooperate, developing and producing new products and sale, even the deployment for all types of resources, it is an exchange about science and technology and education as well as an international economic trade exchange abiding by the rules.

3 Trade Barrier

The foundational form embodying internationalization is to enter international market with the super products. Some developed countries usually prevent TCM from entering international market, with non-tariff barrier to protect their market share including technical barrier to trade and green trade barrier. As for TCM industry, technical barrier means that some developed countries forbid TCM from entering international market by means of establishing themselves technical standards and rules for medicine. While green barrier means some nations establish trade barrier to hinder TCM from entering international market by means of making excuses of protecting environment, human security and sustainable development. It has many ways of showing its face such as green tariff, entering rules for green market, green anti-subsidy, green mark compulsorily, certificate of the ISO14000 system, troublesome inspection process and means and the recovery for junk handling.

The developing trend of international trade shows that trade barriers have transformed to the non-tariff barrier and trade negotiations take the focus on the confirming the entrance criterion involving environmental standard and human health and security. TCM industry has to promote international competition ability by means of perfecting itself by breaking the barriers and increasing technology content and ensuring quality, which makes scientific and technological innovations as the only way to accomplish the modernization of TCM industry.

4 Practical Dilemma

Because of its favorably endowed climate of high-cold land, dank weather, arid soil, great difference in temperature, and strong UV radiation, Gansu province creates innumerable kinds of super quality TCM, some of which have been famous at home and abroad. Taking 2009 for example, the entire area for TCM and herbal medicine amounts to 2.25 million mu, the output is 400 thousand tons, and annual value RMB26.4 billion. However, TCM industry of Gansu has a lower market share because of low level internationalization influenced by multiple factors, which leads to output 1.5 thousand tons and export income lesser than ten million Yuan in 2009. Some factors as follows account for this situation.

(a) At low standardized production level and instable quality
At present, Chinese medicines in Gansu are planted dispersedly and selecting breed and planting lack for expert guide, the technology of producing pollution-free medicine cannot be popularized and there are very few of the plant base reaching the international standard. Some peasants determine what kinds of medicine they propagate and plant by themselves, which make some kinds of medicine disappear gradually. Pesticide and chemical fertilizer are abused, and soil structure is being destroyed, which makes medicine quality degrade and cannot reach the international standard.

(b)Weak ability of adding value with a low level process
Since the investment from enterprises is inadequate during the process and the process is so rough that the increase of product value decreases. Only two of the enterprises in Gansu are mainly occupied with extraction, while others are busying cleanout and packing up as well as making medicinal slices. The processing quantity of rough working medicine is only 30 percent of the total output, while many enterprises sale for raw medicines.

(c) Unsound circulation system
Most places of process medicine are in countryside where traffic is undeveloped and the infrastructure is backward. Uneven distribution of infrastructure for storage, sort and cutting slices with unsound logistics affect the circulation system of TCM.

(d) Failing to build quality control system and market supervision disorders
Since it is difficult to implement the entrance system in circulation without general quality standard systems and specialized agencies, most TCM products have no marks of medical active ingredient content, pesticide residue and secondary pollution brought by so2. The fake and inferior commodities strongly influence the internationalization of TCM industry without effective market supervisions.

(e)Poor brand consciousness and intellectual property rights protection
Many famous brands of Gansu are copied because of poor brand consciousness and without protecting intellectual property, which makes the prestige of the famous TCM brands lower in international markets and influences the competitiveness and market shares.

5 Analyzing the Means of Internationalization of TCM Industry

It is the objective requirement of internationalization to make and to process the TCM materials according to international standards, and overcoming and removing trade barriers as well as protecting intellectual property are all of importance, some factors as follows are coming to the concrete measures.

(a) Clean production based to realize modernization of TCM of Gansu

The production modernization of TCM is the necessity to overcome the international technology trade barriers as well as the basis and prerequisite. Clean production is the key point under guidance of modern technology during the process of producing good TCM[6].

It is essentially a plan and management to lower energy consumption, to use fewer materials, and to lower or even to clear out the rubbish from production. The entire implementing process of clean production is as shown in Fig.1 [7].

Firstly, the standard TCM materials source base should be built with the help of modern science and technology according to Good Agriculture Practice (GAP). Scientific management should be improved, ensuring the water, soil and air not polluted and the fertilizer, feed and medicines to pests and diseases should be safety to TCM products. Breeding and selecting high quality seeds and rescuing the famous kinds of TCM in danger of extinction should be enhanced to improve all the process of production.

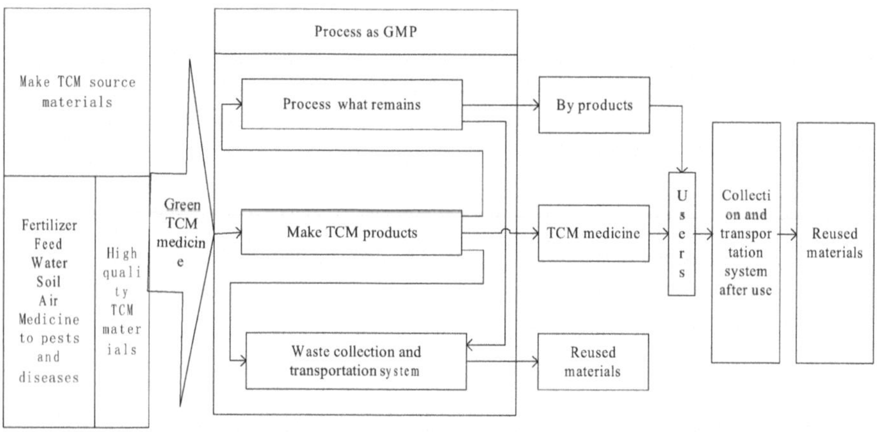

Fig. 1. Entire process of green production

Secondly, high quality TCM materials namely trueborn materials selected and medicine should be made under the condition of modern science and technology. Infrastructural facilities and production and process technique should be modernization according to GMP ensuring TCM products from raw materials to finished products safety to the markets and the consumers.

Finally, the utilization rate of resources should be raised during clean production, using resources as sustainable as possible. Resources and energy should be economized without producing harmful substance of environmental pollution. The rubbish after use should be removed avoiding making pollution.

(b)Strengthen cooperation and exchanges to build new general quality standards for TCM industry.

There are no general quality standards for TCM in the international markets, so TCM industry in China has to abide by the standards built by ourselves. Many other countries and regions often prohibit TCM products from importing their markets on the pretext of failing to come up to their own required standards, taking TCM prohibited by the European Union in 2011 for example, it made the market shares of TCM products lower than ever. Government should strengthen cooperation and build and improve quality standards suited to conditions in China as soon as possible, and then make some responding measures to make the standards approved by the markets overseas as early as possible, which will provide a basis for TCM production, process and circulation and guarantee fair competition.

(c) Improving propaganda and promotion TCM products and strengthening the influence in international markets.

Although Gansu has much high quality TCM, it has not raised its international fame and influence. Of all the high quality products of Gansu, surveys show that only two or three kinds of TCM have their fames and others drowned in so many kinds of medicine. Government and enterprises should make joint efforts to improve propaganda and promotion in various forms such as by media and advertising and organizing communicating meeting and TCM knowledge training.

(d)Protect famous high quality TCM brands and promote their values in name of the intellectual property rights.

Focus is lack of enough attention to the high quality TCM of Gansu, which make some famous brands be bad-faith registered overseas and be copied, and the course of internationalization is seriously affected. Also, since no efforts are vigorously made to create and breed new TCM products, the brands from Gansu are few and have weak competition. The intellectual property rights should be used adequately, especially in the name of protecting appellations of origin. Enterprise's creating new and famous brands should be encouraged, and policies should be made by government to protect the famous brands from being copied.

6 Conclusion

The internationalization of TCM industry is of great significance to the social economy development of Gansu, it makes TCM products follow some general international rules during the course of production, process and sale. Implement modernization should be based on green production. It also needs government promotion and government should propagate and supervise the building of GAP base, green production and establish quality standards as well as protect the intellectual property rights.

Acknowledgments. Our studies is supported by the Gansu College of TCM funds for Youths No.BH2010-075 and the Gansu Planning Program of Philosophy and Social Science,2010.

References

1. Lipsey, R.E., Weiss, M.Y.: Foreign production and exports in manufacturing industries. Reviewof Economics and Statistics 63, 488–494 (1981)
2. Gruber, W., Mehta, D., Vernon, R.: The R&D factor in international trade and investment of United States. The Journal of Political Economy 75, 20–37 (1967)

3. Young, S.: International Market Entry and Development. University of Strathclyde Publishers (2003)
4. Arndt, S., Kierzkow, H.: New Production Patterns in the World Economy. Oxford University Press, Oxford (2001)
5. Gary, G.: Beyond the Producer-Driven /Buyer-Driven Dichotomy: An Expanded Typology of Global Value Chains, Special Reference to the Internet 09 (2000)
6. Information on, `http://www.un.org/zh/development/progareas/global/agenda.shtml`
7. Jao, Z., Hongzhong, Z.: A Study on the Internationalization and modernization of TCM. Medicine and Society 4 (2004)

The Application of Ranking Algorithm of Optimization of Web Site

Sun Jianhua, Lindeqiang, Zhangying, and Zhushijie

Beijing Union University, Beijing, 100191, China
{helen,linqiang}@ygi.edu.cn

Abstract. About web site optimized, also SEO, people must know, in order to make money, there some companies developed some business to increase the ranks of website. At the beginning of the web site ranking algorithm to see what content for SEO contains: basic ranking knowledge, PR value, search for a page of the basic method, they also provide important procedure source code to read in interfaces and feedback, and web pages inside data flows by the java code, using some tools of SEO, Web address analysis, Web search Html analysis etc. At the same time, to build a website in future, it will provide constructive comments. And all of this should be under the framework of the law.

Keywords: ranking algorithm, hits, optimization, Page ranking.

1 System Overview

The 21st century has entered into the web marketing era. In china with the growing popularity of internet use, more and more people are shopping on the internet, and the internet shopping as a main marketing. The web site as before has an image to display only; it is also a marketing function now. Search engines just like a gate of the numerous business website, the network was divided into two parts, client and commerce web site. To make more online web surfer browse their website and Ranking algorithm of optimization play a key important role. This subject is to analysis, compare the various factors of ranking on the web site, and optimize model established.

About web site optimized, also SEO, people must know, in order to make money, there some companies developed some business to increase the ranks of website. At the beginning of the web site ranking algorithm to see what content for SEO contains: basic ranking knowledge, PR value, search for a page of the basic method, they also provide important procedure source code to read in interfaces and feedback, and web pages inside data flows by the java code, using some tools of SEO, Web address analysis, web search HTML analysis, optimized web structure etc. At the same time, to build a website in future, it will provide constructive comments.

By a large number of experiments to observe the simulation of site-specific PR value, to a large number of major search engine algorithm, the experimental and the calculated correction given in the ranking algorithms, and design of system implementation. Modify the layout of the program, including fonts, web link changes, a built-in links to the practical application of the black hat, and the IP address and

D. Jin and S. Lin (Eds.): Advances in EECM Vol. 1, LNEE 139, pp. 295–299.
springerlink.com

other links redirect the application, and website design to promote a video upload system.

To further enhance the visibility of the site, also designed a forum BBS optimization, link optimization, and website content optimized 3U collection of software, website optimization model of the Trinity.

Baidu Summit 2008, in Wuhan, Chairman and CEO Lihongyan said, for the entire industry, the business community, search engine advertising is a very big revolutionary change. I think that each company's marketing department should have an SEO department, or SEM sector after several years, it wants to promote their products through such channels is bound to, and such a channel must be the most effective and can be measure, the most know how to spend their money out of the channel. So it means that you should take a way to promote website ranking while you build your web site.

2 Status of Search Engine Optimization

SEO, Search Engine Optimization, also means search Optimization. Associated SEM and SEP(Search Engine Marketing and Search Engine Profit),its perfect combination, but SEM and SEP in different, SEM is passive, and SEP is active to face to users, so that SEP can get profit even if few traffic. SEO technology becomes more universal.

It's principle of work: After the search engine spiders grab the page, the indexing process inverted index calculated, the search engine is ready to handle the user searched at any time.

Users enter keywords in the search box, the top library program calls the index data to calculate the rankings to the users, and ranking process is a direct interaction with the user. Top search terms of work consists mainly of processing, file matching, the choice of the initial subset, correlation calculations, filter and adjust the ranking, rankings, search the cache, the query and click on the log and so on.

In the country, PR value has almost become a site to exchange links and judge the only indicator of the weight. Methods used are: building site outside the link of quality, create original content to enhance the site, control the export number of links, adding important classification catalog. If the site can be added to this manual review of important categories, then that is more a quality site, to improve the PR value of the site plays a significant role. SEO in China is the initial and ongoing exploration stage. Want to improve your site's ranking, SEO is the most convenient and cheap way. Although SEO is not free, but the cost is relatively low, especially when the owners to master technology. SEO is scalable, as long as the master keyword research and content of extension methods, the site could go on and increase the target keywords and traffic. SEO is scalable, as long as the master keyword research and content of extension methods, the site could go on and increase the target keywords and traffic.

Abroad, Search engine optimization has become an industry. Development in this field abroad early, as early as 97 years or so, some people engaged in related work. They have professional staff, professional company (and very large) in the search engine research, optimization and so on.

From the current situation at home and abroad, the competition is fierce, and Survival of the fittest.

3 A Model of Algorithm Optimization

The basic idea of Website Optimization is: Through the web site functionality, structure, layout, content and other key elements of rational design, makes the site features and expressions to achieve the best results; you can fully demonstrate the site online marketing capabilities. Website Optimization consists of three levels of meaning: the optimization of user access to information, optimize the network environment, as well as optimization of operation and maintenance of Web site.

Search engine optimization is not the focus of ranking rules, even more important for users to facilitate access to information and services. The basic method of search engine optimization: find the specific factors; Increase web site weight; Web site re-structure optimization of internal and external links.

3.1 Ideas

1) Determining factors: the same time repeated the same words in the different search engines (clear history), and record the position and display the information observed a few weeks inference. One by one comparing different keywords and found the idea of optimizing the whole point. Post News with high click rate in the BBS, and contains External links with advertising information. Discussed and analysis above, summed up effectiveness and the role hierarchy.
2) Weight increase: According to the above factors to re-modify the site, the introduction of new activities, changes in site layout. Records and revise the modified site in a search engine to enhance the level of assessment tests to reach for the standards and results. TO achieve automatically modified webpage through the program, to carry out fast authentication, by effective time interval post, and recognize the authentication code to rise weights. Posted by BBS, an increase leave words in micro-blog portal with external link, and using the QQ mass to change weight.
3) Finally, the re organize and classify; of the enterprise's internal Web site information, then increase business links within the site tightness. Such as increasing the classification of the new label content ... to save the changes to modify and optimize the site every day. And Show site in a search engine rankings.

3.2 Improved Algorithm

Damping factories is an element to affect PR value. It is the link to another site, the actual PR values obtained, is typically 0.85. First we look at formula PR(A)=(1-d)+d(PR(t1)/C(t1)+ ...+ PR(tn)/C(tn)), where PR (A) that t1 on the site from external links, according to Pagerank system to increase your website PR value. PR (t1) indicates that the external linked Web sites score their own PR. C (t1), said the sites have external links number of external links. The value of the voting rights of a website is the website PR value * damping. That is the number of external links the site more, PR but will score lower, they are inversely related. Therefore, more important is timely corrected algorithm according to the number of external connections.

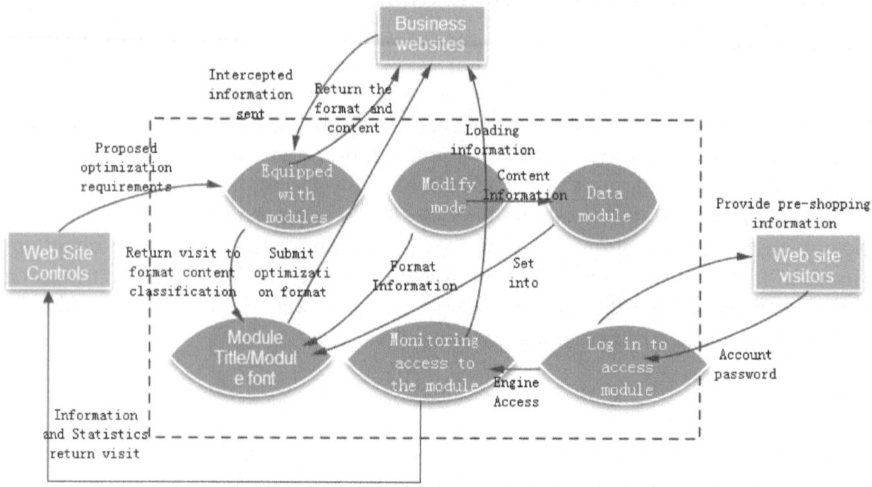

Fig. 1. This figure shows the structure of the site access patterns

4 Website Structure Optimization

4.1 Clear Navigation Design

Text navigation: make full use of the most common HTML code text navigation; do not use images, JavaScript, flash as navigation links. Click away: A good navigation system to make all pages and page hits closer the better. A website is not very high weight, content pages and click on the home page is best not more than 5 times. Anchor text contains keywords: Navigation System category page for links are often the most important source of internal links, a huge amount of anchor text of its relevance to the target page has a large impact, so the directory name should contain keywords.

4.2 Webpage Optimization

Page optimization includes, Optimization of internal links, JS optimization, Web page optimization, Image optimization method, Iframe Optimization, META, Keyword, FAQ pages, Establishment of site map, Establishment of error page 404 and so on.

There are also good ways to increase site weight, such as, important applications within the link - black hat SEO. Application of external links -IP collected and redirected, Video Upload System.

5 Conclusion

Today is the Internet age, Network brings to people's lives great changes, the network information dissemination has become one important part of life, and hard to imagine what you can do without a network. No doubt, Internet is today the most convenient and efficient communication of information means. Almost all companies have created their own websites. A business, organization or institution, whether profit or nonprofit, as long as public service and hope that their concerns by the public. Therefore, the website will undoubtedly become to enhance the public image of the Government to improve the corporate propaganda, and expand sales volume and improve the company's Web site an important way to influence.

References

1. Ou, C.: SEO wisdom - search engine optimization and website marketing revolution. Senior Search Marketing Strategist
2. Liuzhenpeng, lvning, Zhangguoxu, Zhangyajing, Wangpei: DHT VSM based Web services. elected policy Journal of Guangxi Normal University (natural) 2 (2007)
3. Huangdecai, Qihuachun: Page Rank Algorithm research. Computer Engineering 4 (2006)
4. Lishuchun: Combined with content analysis of Page Rank algorithm. Information 1 (2005)
5. Qihuachun, Huangdecai, Zhengyuefeng: PAGERANK algorithm improved with time feedback. Journal of Zhejiang University of Science 3 (2005)
6. Zhaojiahe: Network based on semantic analysis and application of information acquisition algorithm. Journal of Dalian University of Technology (2006)
7. Liuweiwei: Research and Implementation of Focused crawler search engines. Journal of Nanjing Technology University (2006)
8. Chenjiehui: Search engine ranking algorithm. Haihe University (2007)
9. Yangchunwei: Web mining and its application of Web search engine. China University of Petroleum (2007); [12] About ranking web URLS, http://www.jianzhanzhe.com/seo/wangzhanyouhua/Index.html, http://cn.alexa.com/, http://wenku.baidu.com/view/fa5e1be2524de518964b7d81.html, http://www.chinarank.org.cn/, http://baike.baidu.com/view/1518.htm, http://pagrank.info/

A Collaborative Project Management System for Heavy Gas Turbine

Ying Sun

School of Economic and Management, Shenyang Ligong University
110159 Shenyang, China
379073114@qq.com

Abstract. The design and development of heavy gas turbine is a complex, no similar precedent, it must be managed and organized by projects. In this paper, we study the collaborative project management method and the information support technologies of the heavy gas turbine development. Firstly, the organization model of the collaborative project management is given according to the characteristics of gas turbine projects. Secondly, the main business processes is analyzed. Finally, the system architecture of collaborative project management is presented.

Keywords: project management, project management information system, collaboration.

1 Introduction

Heavy gas turbine project is a complex, no similar precedent in product development and prototype trial projects. The project must be endeavor very long make-span and involve more R&D organizations and multiple vendors. Meanwhile, many of the key products, processes and manufacturing technology should be gradually resolved in the process of the project. Based on the above characteristics, It can be expected that the project implementation process is not smooth and it requires a lot of communication and continuous control. To ensure project delivery time primary objective to deliver quality and cost-constrained project conditions, the dynamic control of the entire project is necessary. The overall management of the project should be managed by project management information system with digital, visual collaboration and alliances management support platform.

The project management systems currently employed in the Manufacturing industry can be divided into two types. The first one is off-the-shelf commercial software, where projects are managed using Gantt Charts, the Program Evaluation and Review Technique (PERT) [1] and the Critical Path Method (CPM) [2]. These management techniques have quickly spread into many private enterprises. Thus, a lot of the related commercial software packages cater for the aforementioned techniques; examples include Microsoft Project, Primavera Project Planner and SAP. The second type of project management system is custom in-house software, when commercial software does not meet the particular requirements of an engineering project or firm; some firms will develop custom in-house project management software to meet their needs. Examples of this include Bechtel [3, 4], Kajima [5, 6], and CTCI [7, 8].

D. Jin and S. Lin (Eds.): Advances in EECM Vol. 1, LNEE 139, pp. 301–305.
springerlink.com

As a result of the issues mentioned above, the research presented in this paper designed a solution of project management information system according to the characteristics of Heavy Gas Turbine. This solution can assist firms and teams to cope with issues of multi-dimensional information integration, management, and visualization of engineering project information, as well as achieving general project objectives of scope, quality, time, and cost.

2 The Function Model

For completing the gas turbine project, XXX company considers the following ideas to adjust the project management structure:

(1) Adapt to the technological transformation of enterprises, the flat organizational structure should be achieved.
(2) Adjust organizational downsizing, reducing the number of manager.
(3) Regroup and optimize existing operations, reduce management layers, removing repetitive work and reduce turnover link.
(4) Strengthen the application of information technology, both the division of labor so that all departments, co-ordination, information sharing, to reduce costs.

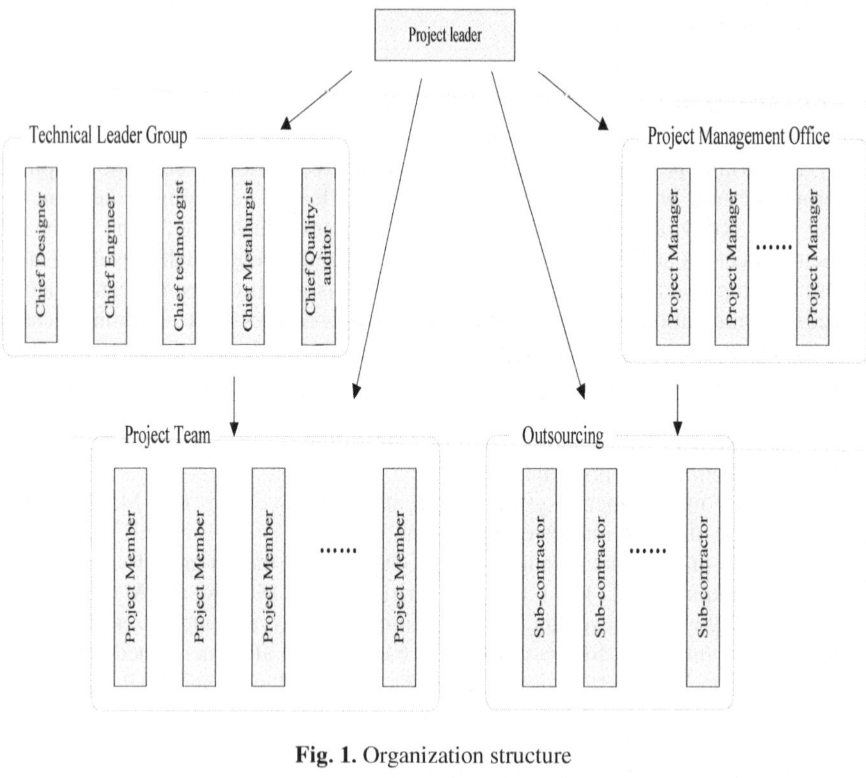

Fig. 1. Organization structure

According to the current management mode as well as the characteristics of heavy gas turbine project, the projects plan to break the traditional organizational structure, the implementation of the project manager responsibility system. Project manager is the project's main responsibility, control project expenses. All the participating collaborative organizations are regarded as collaborative organizations to complete specific tasks. The organization structure is shown as Fig.1.

Project manager is responsible for project planning, organizing, directing and coordination of the overall project plan. The chief engineer is responsible for the unit design, manufacturing coordination and command of the gas turbine. The chief designer is responsible for the coordination and command of the gas turbine. The chief technologist is responsible for the overall process technologies, including dealing with major technical issues, high-tech applied research organizations, scientific research, process reengineering, etc. The chief metallurgist is responsible for the overall work of metallurgical technology, including process research and processing technical problems. The chief quality-auditor is responsible for the overall quality of project work, the establishment of quality assurance system, the hardware and software implementation of quality control. Project management office devotes to deal with all the kinds of project management work, and there are several project managers in it.

3 Main Business Processes

For heavy gas turbines development project, a clear project management processes, including the operation of the overall project control processes and specific project planning / execution control flow.

Project overall control process should take into account the critical needs of the progress of the project development process and quality. In large-scale project development, the various technical and business progress in advance according to their possible, lack of standardized assessment and recognition, often in the project area as part of the business or key product component problems, the entire project delays and waiting, affect the project's progress and quality. In addition, the lack of uniform evaluation criteria, there are often all aspects and areas of technical standards and deliver standardized knowledge inconsistent, affecting the efficiency and quality of coordination. For the overall control flow of heavy gas turbine project, we used the international advanced management concepts development - stage door (Stage-Gate) management.

4 The System Architecture Design of Collaborative Project Management

The architecture of the Collaborative Project Management information system(CPM) consists in the following six parts: collaborative portal, decision support, project process management, project lifecycle management, design knowledge management and collaborative service bus. Among them, the collaborative service bus is the core of the CPM, where the agree service, interface service, deliver service, computing service, management service, supplier service et al are realized. The system architecture of project management of Heavy Gas Turbine is shown as Fig.2.

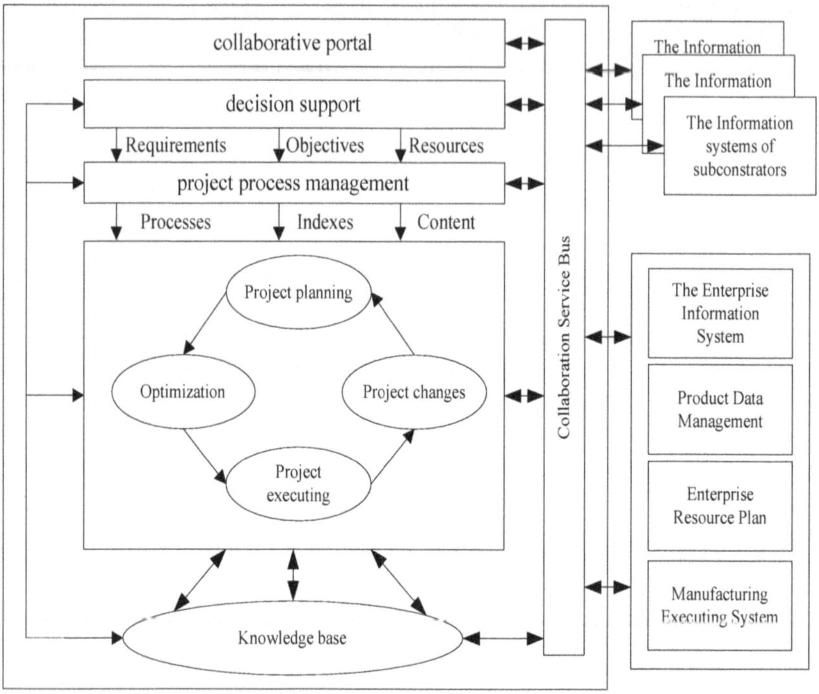

Fig. 2. System Architecture

5 Summary

In this paper, we studied the collaborative project management method and the information support technologies of the heavy gas turbine development. The organization structure model of the collaborative project management was presented according to the characteristics of gas turbine projects. The main business processes was analyzed and reengineered. At last, the system architecture of collaborative project management was built.

Acknowledgment. This research was funded by National Natural Science Foundation of China under Grant No.71071100.

References

1. Kerzner, H.: Project Management: A Systems Approach to Planning, Scheduling, and Controlling, 9/e. Wiley, USA (2005)
2. Woolf, M.B.: Faster Construction Projects with CPM Scheduling. McGraw-Hill, USA (2007)

3. Schmitz, J.: Boston Central Artery/Tunnel (CA/T) project. Interface management. In: Proceedings of the Project Management Institute Annual Seminars & Symposium, Dallas, TX, USA, pp. 312–316 (1991)
4. Parsons, B.K., James, J.E., Reuss, M.C.: Automation Integration Over the Project Lifecycle for the Pueblo Chemical Agent-destruction Pilot Plant. In: Proceedings of the ASME Design Engineering Technical Conference and Computers and Information in Engineering Conference, Salt Lake City, UT, USA, pp. 925–934 (2004)
5. Nagasaki, T., Banno, K., Wakasugi, K., Yasaka, A., Tanimura, T.: Kajima Linkage of Information for New Construction System. In: Proceedings of the Eight International Conference on: Computing in Civil and Building Engineering, Stanford, CA, USA, pp. 518–525 (2000)
6. Pena-Mora, F., Tanaka, S.: Information Technology Planning Framework for Japanese General Contractors. Journal of Management in Engineering 18, 138–149 (2002)
7. Hsieh, S.H., Chen, C.S., Liao, Y.F., Yang, C.T., Wu, I.C.: Experiences on development of a 4D plant construction simulation system. In: Proceedings of the 11 International Conference on Computing in Civil and Building Engineering, Montreal, Canada, pp. 256–267 (2006)
8. Hsieh, S.H., Chen, C.S., Liao, Y.F., Yang, C.T., Wu, I.C.: Construction Director: 4D simulation system for plant construction. In: Proceedings of the Tenth East Asia-Pacific Conference on Structural Engineering and Construction (EASEC 2010), pp. 135–140 (2006)

Design of a PWM/PFM Buck DC-DC Converter with High Efficiency

Renji Wang, Zhigang Han, and Jian Wu

School of Electronic and Information
Tongji University
Shanghai, China
rankiw@hotmail.com

Abstract. This paper presents a buck DC-DC converter with PWM and PFM mode which has high efficiency. The converter operates in PWM at nominal load current (>80mA) with the fixed switching frequency of 1 MHz, and enters PFM at light load current (<80mA). When entering PFM, the converter will shut down several modules to reduce quiescent current and make the switching frequency greatly declined. The chip was implemented using a TSMC 0.18um CMOS 1P3M mixed-signal process. As the simulation indicates, the converter can rapidly switch between PWM and PFM according to the load current, which is suitable for battery-support portable system.

Keywords: DC-DC converter, buck converter, dual-mode, PFM, PWM.

1 Introduction

DC-DC converter has been widely applied in portable electronic devices, such as mobile phone , laptop , PDA and so on. With the improvement of various performances of portable devices, the consumption corresponding to it rises rapidly but the battery technology doesn't. Therefore, how to increase the output current and the high conversion efficiency becomes the crucial problem to design the DC-DC converter [1~4]. In addition, there is a wide load current range for the portable devices with standby mode, e.g. mobile phone, mp4, so we need a kind of DC-DC converter which is able to keep high efficiency through the entire range to extend the standby time and the battery using time [2]. Chip works in fixed frequency under conventional PWM mode, so conversion efficiency will decrease significantly when the load current declined to the quiescent current of the chip. PFM mode was put forward to deal with the conversion efficiency at light load current, but it doesn't keep high efficiency at heavy load. Therefore in order to keep high efficiency through the entire load current range, we choose the PWM/PFM dual mode converter.

Conventional design is that designs the PWM/PFM circuit separately, or add the external control port onto the chip and switch the work mode according to the system status. This paper presents a PWM/PFM mode DC-DC buck converter; its maximum output current is up to 2A. The chip combined two modes with one Mode-switch module and shift between PWM/PFM automatically according to the load, which doesn't need external control port. The chip was implemented using a TSMC 0.18um

D. Jin and S. Lin (Eds.): Advances in EECM Vol. 1, LNEE 139, pp. 307–313.
springerlink.com

CMOS 1P3M mixed- signal process. The results of simulation show that the maximum conversion efficiency is up to 96% when the load current is 20uA ~ 2A and is more than 55% at 20uA light load current.

2 Circuit Design

2.1 System Architecture

The block diagram of dual-mode DC-DC converter presented by this paper with external components is shown in Fig 1. The inductance L and capacity C of the dashed part constitute the law pass filter (LPF), and make up the feedback network with R_{F1} and R_{F2}. Oscillator(OSC), Reference (Bandgap & Ibias), Over-current Protection (Izero), Under-voltage and Temperature Protection (UVLO&TSD) constitute the peripheral circuit of the chip; Error Amplifier (EA), RC Loop Compensation (RC), Slope Compensation (SLOPE), Current Sense (Current Sense), PWM Comparator (PWM Comparator), Control Logic (Control Logic) and Driver (Driver & Buffer), Power Tube SP and Rectifier Tube SN together make up the central circuit under Peak Current PWM mode; only PFM Comparator (PFM Comparator) and mode-switch module (mode switch) realize the shift between PFM and PWM, through which converter rapidly switched to PFM at light load. The chip doesn't use PFM or PWM modules separately, but shared the most part of circuit under two modes, which saved the layout area and consumption respectively. There are five package Pins on the chip. Each of them is: Supply voltage (Vin), Enable (EN), Ground Pin (GND), Feedback (FB) and Switch Pin (SW).

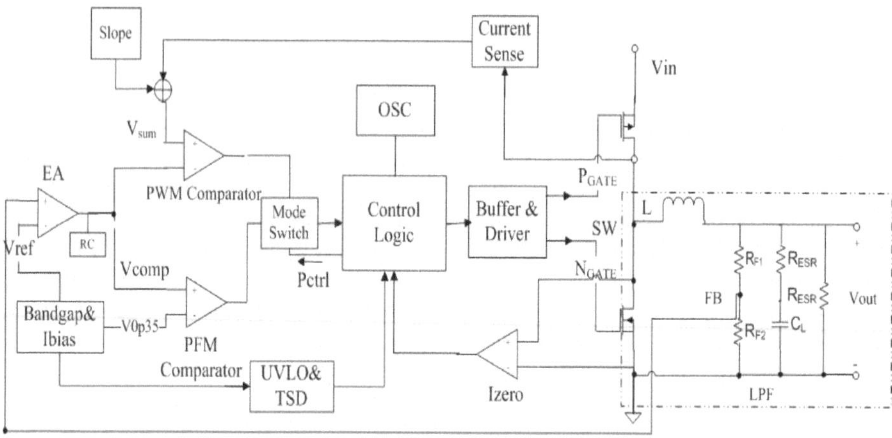

Fig. 1. Block diagram of dual-mode buck converter IC with external components

2.2 Design of the PWM/PFM Shift Module

The positive feedback mechanism presented by this paper is illustrated in Fig.2: the switching process between PWM&PFM is accelerated through set up a minimum conduction time T_{on} of transistor. At the beginning of one clock cycle, the power transistor turns on, and the signal P_{ctrl} rises up and then V_2 is outputted to turn on MN2 through passing the XOR gate and Inverse gate.

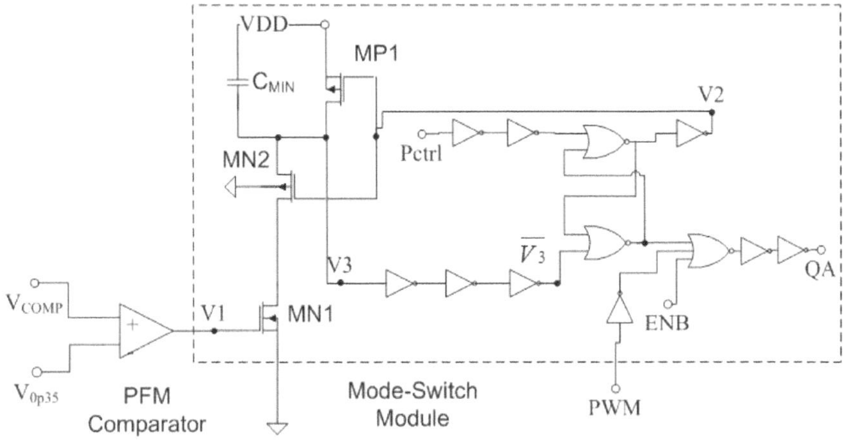

Fig. 2. Block diagram of mode-switch Module

When the converter operates in heavy load, the load current is comparatively high and the output voltage comparatively low. The V_{COMP} which comes from comparing feedback voltage and reference is much higher than the switching voltage of mode-switch module. So MN1 is fully on when V_1 is high, which make it possible to pull V3, the drain of MN2 to the ground. Then signal *QA* is determined by signal *PWM* and *QA* pulse turn off the power transistor normally. The circuit operates under PWM.

We chose V_{COMP} not FB as the detection signal for mode switching because V_{COMP} is more flatten and got a broader range which make it easier to detect. The converter is able to switch mode smoothly at load change [7].

The output voltage of PFM comparator V_1 is connected to the gate of MN1. When the MN2 is on, we could approximately consider the discharging current as:

$$I_1 = \frac{1}{2} K'(\frac{W}{L})_{MN1} (V1 - V_{TH})^2 \tag{1}$$

After time T_{on}, the voltage of capacity C_{MIN} decreased blow the cut-in voltage of upward Pmos of inversion gate.

$$T_{on} = \frac{C_{MIN} V_{THP}}{I_1} \tag{2}$$

T_{on} : the forced on time of power transistor.

During this period, V_3 is high level, so the input *PWM* signal is shielded, the power transistor is continuous on, power supply charges load, and then output voltage rises. System rapidly switches to PFM through this positive feedback mechanism.

Fig.3 shows waveforms of the process of PWM switches to PFM. $\overline{V_3}$ is the signal resulting from V_3 across the inverter chain. When $\overline{V_3}$ is high, *QA* is determined by

P_{Ctrl}; when $\overline{V_3}$ is low, *QA* is determined by *PWM*. At first, V_{comp} dropped and came across V_{0p35}, so the pulse width of $\overline{V_3}$ began to decrease. The *QA* stayed low due to signal $P_{Ctrl.}$ Power transistor is on for *Ton* time and charging the load. Some part of current flows into the capacity for current that flows through load is very small. So the output voltage continues to rise, which makes V_{COMP} further decrease and keep it below V_{SUM} and PWM signal low. The chip entered the PFM and the circuit exhibits a positive feedback.

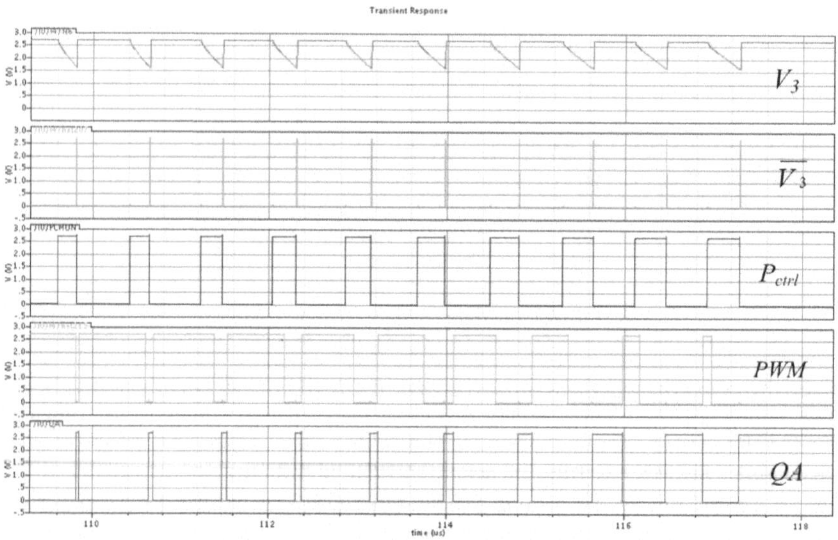

Fig. 3. Waveforms of the Mode-switch Process

2.3 Analysis of Logic Module under PFM

The internal logic module is shown as Fig.4. When the chip operates in light load and QA is high, the falling-edge-triggering D flip-flop transit the signal to output Q at one raising edge and the signal reach the OSCON port through three chained inverter. And the OSCON signal shut down the Oscillator(OSC), Current-Sense (CURRENT SENSE), Under-voltage and Temperature Protection(UVLO&TSD), and reduce the bias current of PWM Comparator, disable the slope voltage of Slope Compensation (SLOPE) to make V_{sum} stay at a relatively low value. Then the circuit enters PFM.

There are two states under PFM: standby state (t_1) and charging state (t_2). Firstly, under standby state, OSCON outputs low level one clock after *QA* stays high. External MOSFET is shut down, capacity supplies the load, output voltage decreases gradually and V_{comp} rises up; When V_{comp} has surpassed V_{sum} ,the chip entered charging state.*QA* signal falls, output P_{on} rises up and turns on power transistor directly. The entire process is shown as Fig 5.

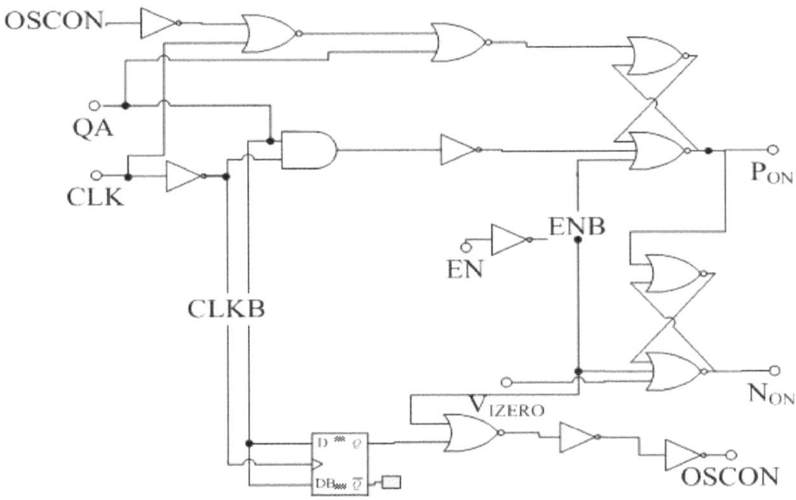

Fig. 4. Block diagram of PFM-mode controlling loop

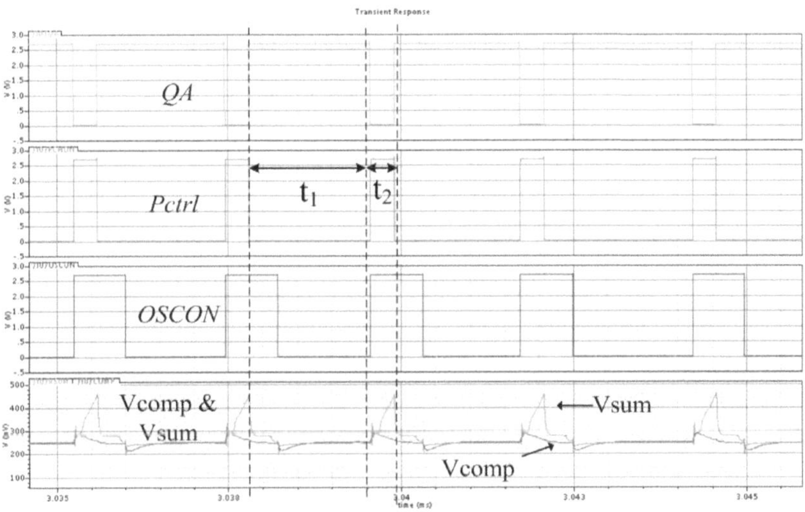

Fig. 5. Waveforms of the Nodes in PFM-mode operation

When the state of load has changed, from light load to heavy load, OSCON remains high due to the affect of QA and the circuit will stay in charging state which makes V_{comp} surpass V_{0p35} at last. The chip returns PWM.

Under PFM, the chip reduces the switching frequency and shut down part modules to save the quiescent consumption.

3 Test Results and Discussion

During the test, the typical value of load current under PWM and PFM was 600mA and 0.05mA with input voltage 2.7V. We set the output current change as 600mA-0.05mA-600mA. Fig 8 has given out the waveform of measured load transient response which corresponds to the change of current. From Fig 6 we can see the converter shift between PWM and PFM automatically according to the load current. The ripple voltage was 8mV under PWM and 18mV under PFM. The rush on voltage was 160mV and the overshoot voltage was 300mV. The settle time was 20us (PWM to PFM) and 40us (PFM to PWM).

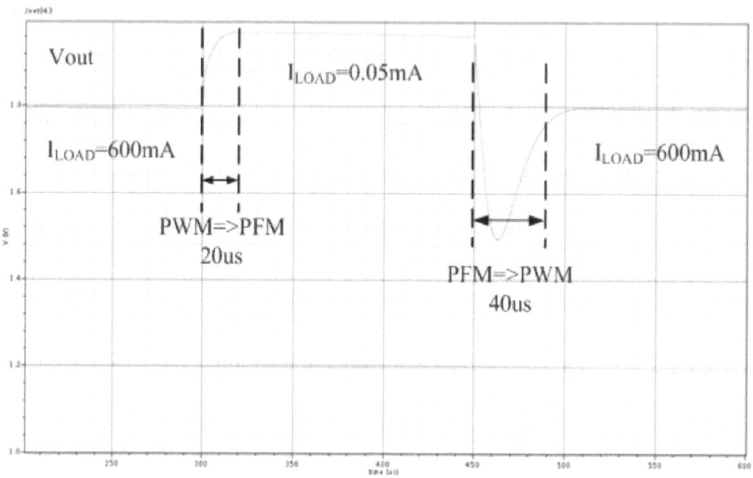

Fig. 6. Measured load transient response

4 Conclusion

Huge output current, high stability, high efficiency, buck converter with pulse-width modulation and pulse-frequency modulation has been presented. The converter operates under PWM with a fixed 1MHz frequency when load current is above 80mA and operates under PFM with variable frequency when load current is blow 80mA, which keeps high efficiency through the entire load current range (20uA ~ 2A). The chip was implemented using a TSMC 0.18um CMOS 1P3M mixed-signal process. The results shows that the circuit could switch between PFM and PWM automatically, the conversion efficiency could be up to 96%, and higher than 55% at light load. The chip with high efficiency and low quiescent consumption in broad load range is especially suitable for portable device powered by single battery such as cell phone etc.

References

1. Zhang, X.: Digital PWM/PFM controller with input voltage feed-forward for synchronous buck converters. In: Twenty-Third Annual IEEE Applied Power Electronics Conference and Exposition, APEC 2008 (2008)
2. Safu, B., Rincon-Mora, G.A.: High-efficiency dual-mode, dynam-ie, buck-boost power supply IC for portable applications. In: Proc. IEEE VLSID (2005)
3. Xiao, J., Peterehev, A., Zhang, J.: An ultra-low-power digitally-controlled buck converter IC for cellular phone applications. In: Nineteenth Annual EE APEC (2004)
4. Sahu, B., Rincon-Mora, G.A.: A low voltage, noninverting, dynamic, synchronous buck-boost converter for portable applications. IEEE Trans. Power Electron (2004)
5. Wan-Rone Liou, A.: High efficiency dual-modebuck converter IC for portable applications. IEEE Transactions on Power Electronics 23(2), 667–677 (2008)
6. Vorperian, V.: Simplified analysis of PWM converters using the PWM switch Part I: Continuous conduction mode, Part II: Dis-continuous conduction mode. IEEE Trans. Aerosp. Electron Syst., AES 26, 490 (1990)

References

Distributed Grid-Based Localization Algorithm for Mobile Wireless Sensor Networks

Can Sun, Jianping Xing*, Yuxin Ren, Yang Liu, Junchen Sha, and Juan Sun

School of Information Science and Engineering
Shandong University Jinan, China
suncan0203@gmail.com, xingjp@sdu.edu.cn

Abstract. Localization is an essential supporting technology of wireless sensor networks (WSNs). Most of traditional localization algorithms are proposed for static WSNs. There are only a few algorithms proposed for mobile wireless sensor networks (MWSNs). In this paper, we propose a range-free localization algorithm for MWSNs, named distributed grid-based localization algorithm (DGL). In this algorithm, anchor nodes can increase their transmitting power, that is to say, they can change their communication range. Each sensor node can establish rectangular coordinate system itself, and then divide coordinate system into square grids. Simulation results demonstrate this algorithm outperforms other well-known localization algorithms.

Keywords: mobile wireless sensor networks (MWSNs), distributed, range-free, localization, grid, DGL.

1 Introduction

Wireless sensor networks (WSNs) are becoming more and more popular in many fields, such as environmental monitoring, military surveillance, target tracking, health monitoring, etc. In all these applications above, location is very critical information. However, equipping GPS receivers or manually configuring locations is not cost effective for most applications in WSNs. So localization technique in WSNs is a significant supporting technique [1, 2].

According to the mechanisms used for estimating location, localization algorithms can be divided into two categories: range-based and range-free [2, 3]. Because of hardware limitation and cost of nodes, range-free localization algorithms have received more attention in research.

Centroid algorithm was proposed by Bulusu N [4]. In Centroid algorithm, the center of the locations of all anchors which is heard is seen as the estimated location. In Amorphous algorithm, sensor nodes compute their location based on the received anchors locations and corresponding hop counts [5]. Many other algorithms, such as APIT, DV-hop, ROCRSSI, and CLS are proposed [3, 6, 7, 8]. But, all the algorithms mentioned above are proposed for st0atic wireless sensor networks. And their performances for mobile wireless sensor networks are not very ideal.

---

* Corresponding author.

D. Jin and S. Lin (Eds.): Advances in EECM Vol. 1, LNEE 139, pp. 315–321.
springerlink.com

In [9], Hu and Evans proposed MCL algorithm for MWSNs. Time is divided into many discrete intervals. MCL includes three phases, initialization, prediction, and filtering. MCL is a filter combined with probabilistic models. The key idea of MCL is to represent the posterior distribution of possible locations using a set of weighted samples. MCL needs many samples, and need to filter the observed data, in addition, if a sensor node isn't in anchor's communication range, the nodes in its communication range are used as relay nodes. So the localization range is set to two hops. Node's energy is limited, and communication among nodes will cost too much energy. MCL is not energy efficient for sensors. MCL's localization accuracy is not perfect, too.

This paper describes a range-free, distributed grid-based localization algorithm (DGL) for MWSNs. This paper is organized as follows. Section 1 is the introduction of WSNs localization. We describe our algorithm (DGL) in section 2. Simulation of our algorithm and comparison with other algorithms are described in section 3. Finally, our work is been concluded.

2 Description of DGL

Algorithm DGL uses two types of nodes: anchor nodes and sensor nodes. Anchor nodes own GPS device to get accurate position information. They can increase their transmitting power in a fixed step at each time, that is to say, their communication ranges are divided into several levels. Anchor nodes broadcast information with ID, location, power level and time. Energy of anchor nodes is assumed to be unlimited. Sensor node has simple data processing and storage capabilities. It can establish rectangular coordinate system, and divide the rectangular coordinate system into square grids. It also can transfer power level into communication range. The coordinate system established should ensure that if the node can receive information of the anchor nodes with the highest power, anchor nodes should be guaranteed in the coordinate system. Both anchor nodes and sensor nodes can move themselves. The signal was transmitted in an ideal mode.

The algorithm consists of the following steps:

Step1. Initialization. Each sensor node receives and saves information transmitted from anchor nodes, and establishes a rectangular coordinate system, according to the centroid of the received location information of anchor nodes. Then the rectangular coordinate system is divided into many square grids by each sensor node, along the horizontal axis and vertical axis. If the number of anchor nodes the sensor node received is less than 2, the sensor node cannot establish the rectangular coordinate system. Then it wait the next time, until the number is larger than 2. This step is shown in Fig.1, dots represent anchor nodes, and circle represents sensor nodes.

Step2. Voting. According to the power level built-in before, anchor nodes vary communication range from small to large in a fixed step at each time point. Sensor node saves information it first received from each anchor node, and changes power level into communication range according to the program built-in. Then, if the distance between the centroid of a grid and the anchor node is in the communication range, i.e., the distance conforms the following inequality:

$$r_i < \| x - a_i \| < R_i \quad (1)$$

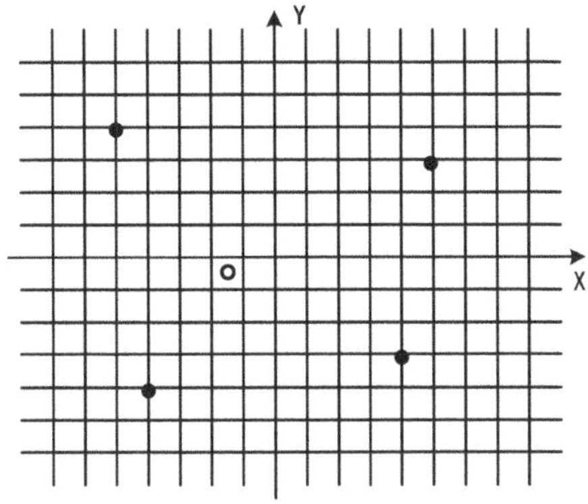

Fig. 1. Initialization

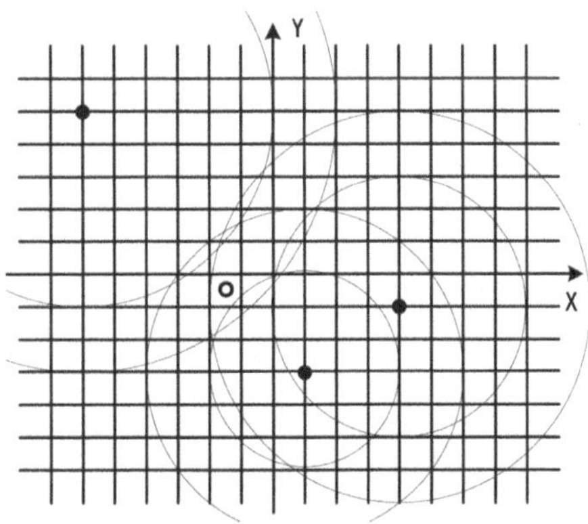

Fig. 2. Voting

the grid is regarded as a possible location of the sensor node. So the sensor node adds a vote to the grid, and saves it. Note that x is the centroid of a grid, a_i is the location of the anchor node which the sensor node received, r_i is the lower bound of communication range, and R_i is the upper bound. As shown in Fig.2, rings represent communication range of anchor nodes. If the centroid of a grid is in this ring, sensor node votes to this grid.

Step3. Localization. The sensor node selects all the grids in the rectangular coordinate system which owns the highest votes, and then, computes the centroid of all these grids. So the centroid of all these grids is regarded as the estimated local location of the sensor node. For example, as shown in Fig.3, the grids in the bold line own the highest votes, the point represents the centroid of these bold grids, and the circle represents the real location of the node. We regard the location of point as the estimated location of the sensor node.

Step4. Coordinate transformation. According to the anchor nodes' locations received, sensor node transforms its local location into global location.

Step5. The system goes to the next time point, and return to Step1.

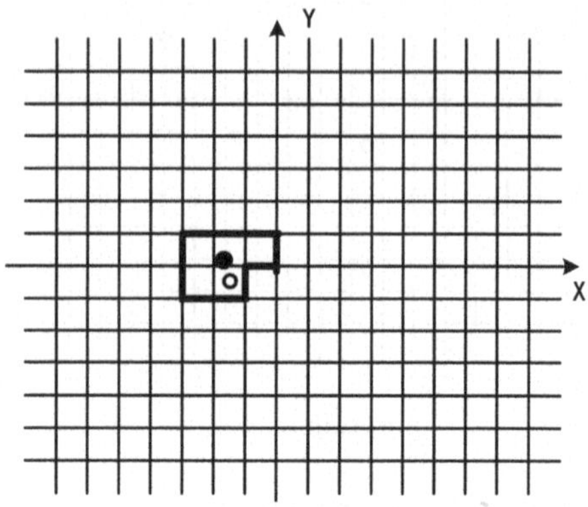

Fig. 3. Localization

3 Simulations and Results

In this section, simulation results are presented and analyzed. We compare DGL to other three localization algorithms, MCL, Centroid and Amorphous.

We consider a 2-dimensional region with a size of 500m * 500 m. We adopt a modified random waypoint mobility model [9, 10] for both sensor nodes and anchor nodes. All the nodes are deployed in this region randomly. The time is divided into discrete intervals. In the modified random waypoint model, a node randomly chooses its direction and its speed of movement and its pause time after arriving at the destination. Then, sensors estimate their location. We assume all nodes are unaware of their velocity and direction, but have a known maximum velocity. We define V_{max}, V_{min}, S_{max} and S_{min}, the maximum and minimum velocity of sensor node and anchor node, respectively. A sensor node's velocity is randomly chosen from $[V_{min}, V_{max}]$; and a anchor node's velocity is randomly chosen from $[S_{min}, S_{max}]$. The pause time is set to 0. The average localization error is defined in the following formula:

$$error = \frac{1}{N}\sum_{i=1}^{n} \| x_i - \widehat{x_i} \| \tag{2}$$

Note that x_i is the actual position of sensor node i , and $\hat{x}_i$ is the estimated position of this node, N is the number of sensor nodes, $\|\bullet\|$ denotes the Euclidean distance between real location and estimated location of sensor node.

In MCL, if a sensor node is not in the communication range of an anchor node, the sensor nodes in its communication range are used as relay nodes to estimate this sensor node's location. So the localization range is set to two hops. In many cases, sensor node's energy is limited, but anchor node's energy is unlimited. Communication among a node and other nodes will cost too much energy. So, MCL is not energy effective for sensor nodes. To save sensor node's energy, in DGL, we assume that communication range of sensor nodes is r, communication range of anchor nodes is $R = 2r$.

In the simulation, we define node density N_d, the average number of node in one hop distance r. To comparison with MCL, we also define anchor nodes density S_d, the average number of anchor nodes in one hop distant r, too, so the number of anchor nodes in all these algorithms is the same.

In DGL, we assume that if the anchor node increases transmitting power by one level, the corresponding communication range of anchor node is increased L. In the following simulation, we define L=10m. In addition, edge length of grid is set to 1m. If the number of anchor nodes' information that the sensor node received is too small, localization error will be too larger, so we assume that if the number is no less than 2, this sensor node can be localized. To further improve the localization accuracy, we also assume that if a grid's votes are less than half the number of anchor nodes information the sensor node received, it will not be localized, too.

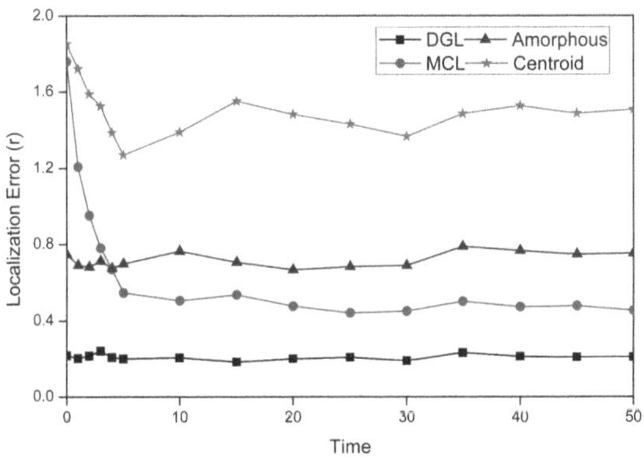

Fig. 4. Accuracy comparison. $N_d = 10, S_d = 1, V_{\max} = S_{\max} = r$

Fig.4 shows localization accuracy comparison of these four algorithms. All the algorithms are simulated 50 time points. MCL makes use of past information, so at the first five time points, it is not converged, but its localization accuracy is improved quickly. After that, its localization accuracy is nearly constant. However, DGL, Amorphous and Centroid do not exploit past information, their performances have no relationship with the time, so localization accuracy of these three algorithms change small. Simulation result shows that localization error of DGL is about 20% of the communication range, but localization error of MCL in condition of convergence is about 45%, Amorphous is about 70%, and Centroid is over 120%. So, in this condition, DGL performs much better than other three algorithms.

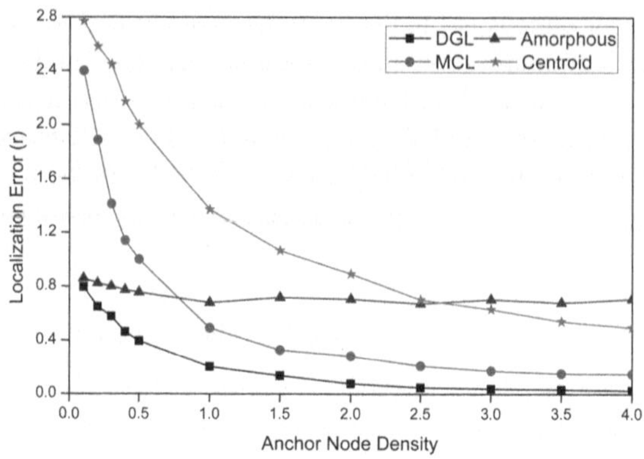

Fig. 5. Impact of anchor node density. $N_d = 10, V_{\max} = S_{\max} = 0.2r$

Increasing anchor node density makes localization more accurate, but increases hardware costs at the same time. Fig.5 shows the impact of anchor node density for these four algorithms. Anchor node density is varied from 0.1 to 4. Simulation result shows that the accuracy of DGL, MCL, and Centroid all improve as anchor node density increases, but for Amorphous technique, the localization error does not improve much, fluctuation is generated due to randomness. DGL outperforms MCL, Amorphous, and Centroid all the time when varying anchor node density. If anchor node density is small, sensor nodes cannot receive enough number of anchor nodes' information, so localization accuracy is poor. As anchor node density increases, the ratio will increase.

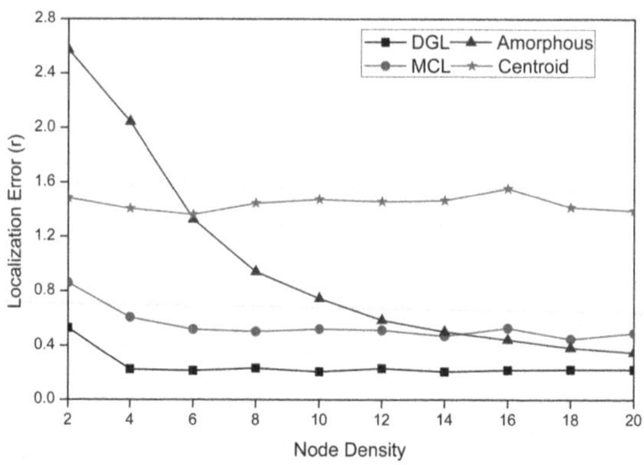

Fig. 6. Impact of node density. $S_d = 1, V_{\max} = S_{\max} = 0.2r$

Fig. 6 shows the impact of node density of these four algorithms. Node density is varied from 2 to 20. Simulation result shows that Amorphous is seriously affected by node density and it performs well when node density is high, but performs poor if the node density is low. Localization accuracy improves quickly when node density increased. Centroid is hardly affected by node density and fluctuation is generated due to randomness, but localization error is very large. However, DGL and MCL perform relative steady and accurate in the condition of not only high node density but also low node density. In all condition we simulate, DGL outperforms other three algorithms.

As mentioned earlier, we can conclude that DGL performs better than MCL, Amorphous and Centroid. It is an effective and accurate localization technique.

4 Conclusion

In this paper, a range-free distributed grid-based localization algorithm (DGL) for MWSNs is proposed. Simulation results demonstrate that this algorithm outperforms other well-known localization algorithms, and it is accurate and effective.

Acknowledgment. This work is supported by Program for New Century Excellent Talents in University (Grant No.NCET-08-0333), Shandong Province Natural Science Foundation (Grant No. ZR2011FM039), Independent Innovation Foundation of Shandong University(Grant No.2010JC011) and Independent Innovation Foundation of Shandong University (Grant No. 2010JC013).

References

1. Afyildiz, I.F., Su, W., Sankarasubramaniam, Y., Cayirci, E.: A Survey on Sensor Networks. IEEE Communications Magazine 40, 102–114 (2002)
2. Mao, G., Fidan, B., Anderson, B.D.O.: Wireless sensor network localization techniques. Computer Networks 51, 2529–2553 (2007)
3. He, T., Huang, C., Brian, B.M., Stankovic, J.A., Abdelzaher, T.: Range-Free Localization Schemes for Large Scale Sensor Networks. In: Proceedings of the Annual International Conference on Mobile Computing and Networking, pp. 81–95 (2003)
4. Bulusu, N., Heidemann, J., Estrin, D.: GPS-less low-cost outdoor localization for very small devices. IEEE Personal Communications 7, 28–34 (2000)
5. Nagpal, R., Shrobe, H., Bachrach, J.: Organizing a Global Coordinate System from Local Information on an Ad Hoc Sensor Network. In: Prodeeding of Workshop on Inforamtion Processing in Sensor Networks, pp. 333–348 (2003)
6. Niculescu, D., Nath, B.: DV based Positioning in Ad hoc Networks. Journal of Telecommunication Systems 22, 267–280 (2003)
7. Liu, C., Wu, K., He, T.: Sensor localization with Ring Overlapping based on Comparison of Received Signal Strength Indicator. In: IEEE International Conference on Mobile Ad-hoc and Sensor Systems, pp. 516–518 (2004)
8. Fretzagias, C., Papadopouli, M.: Cooperative location-sensing for wireless networks. In: Proceedings of the Second IEEE Annual Conference on Pervasive Computing and Communications, pp. 121–131 (2004)
9. Hu, L., Evans, D.: Localization for Mobile Sensor Networks. In: Proceedings of the Annual International Conference on Mobile Computing and Networking, pp. 45–57 (2004)
10. Camp, T., Boleng, J., Davies, V.: A survey of mobility models for ad hoc network research. Wireless Communications and Mobile Computing 2, 483–502 (2002)

Modified Electromagnetism-Like Method for Constrained Optimization Problems

Lixia Han and Shaojiang Lan

School of Computer Science and Technology, Chian University of Mining and Technology, Xuzhou, China
lxhan@cumt.edu.cn

Abstract. This paper proposed a novel electromagnetism-like method for solving the constrained optimization problems. A new technique was adopted to transform the constrained optimization problems to an unconstrained optimization model with bound constrained. The modified method adopted the attraction-mechanism of the electromagnetism theory to encourage the particles to converge to its vicinal attractive valleys. Compared with the existing algorithms, numerical results on five benchmark function demonstrated that the proposed algorithm was competitive for the constrained optimization problems.

Keywords: electromagnetism-like method (EM), constrained optimization problem, charge.

1 Introduction

Many real life constrained optimization problems that arise in area such as engineering design, molecular biology, physics involve nonlinear functions of many variables. In this paper, we concentrated on the constrained optimization problem in the form of:

$$\begin{aligned} \min_{x\in S} \quad & f(x) \\ s.t. \quad & g_i(x) \le 0 \quad i = 1,2,\cdots,l \\ & h_j(x) = 0 \quad j = 1,2,\cdots,m \end{aligned} \tag{1}$$

Where $f: R^n \to R$ is objective function, $g_i(x)$ and $h_j(x)$ are the inequality and equality constraints, and $S = \{x \in R^n : l \le x \le u\}$ is the decision space for constraints optimization problem (1).

Clearly, the aim of the constrained optimization is to find the global optimal solution x^* from the decision space S which satisfied all the constrained conditions. In many case, the problems (1) involve multi-modal and non-differentiable nonlinear functions, hence, the gradient based method cannot be used for such problems. In recent years, many scholars have devoted themselves to stochastic-type methods, such as genetic algorithm [1~4], particle swarm optimization and so on.

In 2003, Birbil and Fang [5] proposed a population-based stochastic optimization method known as Electromagnetism-like (EM) algorithm for unconstrained global optimization. It treats each solution as a charged particle. The method simulates the attraction-repulsion mechanism to move the particles to the vicinity of the global

D. Jin and S. Lin (Eds.): Advances in EECM Vol. 1, LNEE 139, pp. 323–329.
springerlink.com

optimum. Due to its advantages of intelligence, simplicity, global search ability and parallelism, the EM algorithm [6~8] has become a hot research area and has been successfully used for traveling salesman problem, project scheduling, learning fuzzy If-Then rules, and so on.

It known to all, the EM algorithm was designed for global optimization problems with bound constraints. Here, we are interested to extend the EM algorithm for solving the constrained optimization problems like (1).

The organization of the paper is as follows. In section 2, an unconstrained model was presented with the min-max objective function. The full description of our proposed modification was discussed in Section 3. In section 4, the numerical results and discussion are presented. Section 5 contains the conclusion and future research.

2 New Unconstrained Optimization Model

For constrained optimization problem (1), we defined the level of constraints violation as follows.

$$IF(x) = \max\{0, g_i(x), |h_j(x)| \, | i = 1 \sim l, j = 1 \sim m\} \tag{2}$$

Let $p(x) = \frac{1}{1+IF}$

According to the definition (2), a particle $x \in S$ with $p(x) = 1$ is feasible for constrained optimization problem (1), whereas $0 < p(x) < 1$ then the particle is infeasible. Therefore, the constrained optimization problem (1) can be transformed into the following unconstrained optimization problem.

$$\max_{x \in S} \max\{(1 + \frac{f(x^w) - f(x)}{f(x^w) - f(x^b)}) \cdot sign(p(x)), p(x)\} \tag{3}$$

Where $sign(x) = \begin{cases} 1 & x = 0 \\ 0 & x > 0 \end{cases}$, x^w and x^b are the worst and the best solution, respectively, in the population.

According to the scheme (1), we have the following conclusion:

1) For feasible particle $x \in S$, the objective function $Z(x) = 1 + \frac{f(x^w) - f(x)}{f(x^w) - f(x^b)}) \geq 1$;
2) Otherwise, the objective function $Z(x) = p(x) < 1$

In this way, the feasible particles dominate infeasible particles in the decision space S. There, the optimal solution x^* of the unconstrained global optimization problem (2) is the global optimum for the constrained optimization problem (1).

3 Modified EM

3.1 Charge and Force Vector Calculation

In the EM, the charge of the particle relatives the objective function and determines the power of attraction and repulsion for other particles. For the unconstrained optimization problem (2), the charge equals to the objective function value by

$$q^i = Z(x^i) \tag{4}$$

Thus, particles that have better objective function value possess higher charges and attract particles with worst objective function value. Conversely, particles that have worst objective function value have lower charges and repulse particle with better objective function. For any two particles, the force exerted on particle x^i by particle x^j is computed according to the following equation:

$$F_j^i = (x^j - x^i)\frac{q^j - q^i}{\left\|x^j - x^i\right\|^2} \tag{5}$$

For scheme (5), we can see:

(1) If $Z(x^j) > Z(x^i)$, then particle x^j attracts x^i, and the direction of the force should be $\overrightarrow{x^i x^j}$;

(2) whereas if $Z(x^j) < Z(x^i)$ then x^j repulses x^i, and the direction of the force should be $\overrightarrow{x^j x^i}$.

(3) Especially, the force equals to 0 when $Z(x^j) = Z(x^i)$.

In fact, the force computed by scheme (5) is proportional to the different of their charges and inversely proportional to the Euclidean distance between the particles. In this way, the new algorithm encourages the particles to approach to its vicinal attractive valleys. Subsequently, the total force exerted on each particle x^i by all other point in the population is determined as a vector summation of individual component vector forces,

$$F^i = \sum_{j=1}^{N} F_j^i \tag{6}$$

3.2 Movement along the Total Force

In order to maintain the feasibility, the normalized total force vector was used to update the position of the particle as follows.

$$y^i = \begin{cases} x^i + \lambda \dfrac{F_k^i}{\left\|F^i\right\|}(u_k - x_k^i) & F_k^i > 0 \\ x^i + \lambda \dfrac{F_k^i}{\left\|F^i\right\|}(x_k^i - l_k) & otherwise \end{cases} \tag{7}$$

Where the parameter λ is assumed to be uniformly distributed between 0 and 1.

3.3 Improved EM for Constrained Optimization Problem (IEM)

1) Randomly generate the initial population $P(0)$ of N particles from the decision region S. Let the generation number $t = 0$;
2) For each particle in population $P(t)$, the random line search is used to generate a improved particle. These new particle constitute a set denote by $L(t)$;

3) After evaluating the total force vector, move the particle $x^i \in L(t)(i=1,\cdots,N)$ along the direction of the force to a new position with a random stepwise. All these new particle constitute a set $M(t)$;
4) Select the best N particle form $P(t) \cup L(t) \cup M(t)$ as the next generation population $P(t+1)$. Let $t=t+1$;
5) If the termination condition is satisfied, then stop and output the best solution; otherwise, go to step 2).

4 Numerical Results and Discussion

For the proposed algorithm (IEM), computational tests were implemented on five benchmark functions. These functions are known as g01, g03, g04, g06 and g07 ([9] for completeness). For each of the benchmark function, 30 dependents runs were performed by IEM in MATLAB.

To demonstrate the effectiveness of the novel algorithm for constrained optimization problems, we compared the numerical results with 4 well-known algorithms: the VEGA [1], NPGA [2], MOGA [3] and COMOGA [4]. Fairly, the total number of function evaluation was in all cases, of 350000. Table 1~5 report the best function value (denote by Best), the worst function value (denoted by Worst), mean results (denoted by Mean) and the standard deviation (denoted by SD) obtained by EGA, NPGA, MOGA, COMOGA and IEM.

Table 1. Numerical results on g01 by VEGA, NPGA, MOGA, COMOGA and IEM

Algorithm	Best	Worst	Mean	SD
VEGA	11.136417	-9.534448	-10.249717	0.349041
NPGA	-11.007717	-4.719421	-8.033259	1.718908
MOGA	-14.504487	:-13.306435	-13.981660	0.320086
COMOGA	-4.806906	0.0000	-1.203723	1.638475
IEM	-14.997223	-14.987565	-14.992267	0.002998

For g01, the best solution found by COEM was x^b =[1,1,1,1,1,1,1,1,1, 2.99854705389,2.999415810767137,2.999260189462] with $f(x)$=-14.997223. Thus, the distance between the x^b and the global minimum $x^* = (1,1,1,1,1,1,1,1,1,3,3,3,1)$ was less than 1.7×10^{-3}. The worst function value found by IEM was -14.987565, while the best was -11.136417, -11.007717, -14.504487, -4.806906 obtained by VEGA, NPGA, COMA and COMOGA, respectively. It is also interesting to note that the mean of our results was -14.992267, and the standard deviation is the least one among the five algorithms. For g03, the best solution found by our algorithm was x^b =(0.316584189815382,0.314187792317949,0.315338346083726,0.327230546050401,0.314621351150462,0.316093563786808,0.316569552521654,0.311804474100949,0.319977268780484,0.309547152455811). As we can see, the best result found by IEM was better than the best result obtained by NPGA, MOGA and COMOGA. The VEGA was unable to reach the feasible region (denoted by N.F.). As for the benchmark function g04, the results of VEGA, NPGA, MOGA, COMOGA and IEM are appealing from table 3. To get a direct idea of the results of the five methods, we defined the error as follows.

$$error = \frac{|f(x) - f(x^*)|}{|f(x^*)|} \times 100\% \tag{8}$$

Where $f(x^*)$ is the known global optimal. For Table 1~5, we can computed the error between best objective value and the known global is less than 0.6% for COMOGA, 0.04% for VEGA, NPGA, COMA and IEM. Encouragingly, the error between mean objective value found by IEM and the known global the is less than 10^{-6} for g04. For g06, the best solution obtained by IEM was x^b =(14.095014333012333, 0.842991712906820) with $f(x^b)$=-6961.77910. In this case, each method approached the global optimum. In fact, IEM was the best among these methods, and the MOGA and NPGA followed. For g07, the best result found by COMOGA disappointed us with the best objective value 468.216675 while the global optimum was 24.3062091. As always, COMOGA occupied the last place in term of performance and IEM the top place for this benchmark function. Note our algorithm performed significantly better than other algorithms in term of the mean as well as best results. For the five

Table 2. Numerical results on g03 by VEGA, NPGA, MOGA, COMOGA and COEM

Algorithm	Best	Worst	Mean	SD
VEGA	N.F.	N.F.	N.F.	N.F.
NPGA	-0.981203	-0.884223	-0.928032	0.021715
MOGA	-0.868598	-0.263762	-0.561975	0.157852
COMOGA	-0.022967	-0.0000	-0.001121	0.004195
IEM	-0.997953	-0.969300	-0.984581	0.010171

(N. F. represents the method was unable to reach to the feasible region.)

Table 3. Numerical results on g04 by VEGA, NPGA, MOGA, COMOGA and COEM

Algorithm	Best	Worst	Mean	SD
VEGA	-30652.3300	-30625.0429	-30638.7759	5.3372
NPGA	-30659.6562	-30639.6035	-30653.0599	5.0220
MOGA	-30659.8457	-30552.6582	-30615.2475	28.2624
COMOGA	-30483.4746	-30389.0917	-30397.1336	37.6405
IEM	-30665.535	-30665.4867	-30665.5201	0.01549

Table 4. Numerical results on g06 by VEGA, NPGA, MOGA, COMOGA and COEM

Algorithm	Best	Worst	Mean	SD
VEGA	-6941.9321	-6743.4951	-6873.1396	46.6093
NPGA	-6956.9716	-6310.1254	-6776.4187	176.1811
MOGA	-6957.9506	-6845.4321	-6903.7746	29.7420
COMOGA	-6622.2802	-4859.3310	-6058.8651	436.7865
IEM	-6961.779	-6961.4557	-6961.7276	0.0931

Table 5. Numerical results on g07 by VEGA, NPGA, MOGA, COMOGA and COEM

Algorithm	Best	Worst	Mean	SD
VEGA	28.631790	35.525009	32.014552	1.736919
NPGA	26.232813	30.784266	28.296385	1.106179
MOGA	27.512201	80.891251	36.427887	10.502327
COMOGA	468.216675	1933.539917	1173.0339	610.67987
IEM	24.317977	24.430563	24.344678	0.040361

benchmark functions, the error between the mean objective value and the known global optimum is less than 1.6% for IEM. In general, our algorithm outperformed all five methods in term of best, worst, mean and standard deviation for these benchmark functions. Therefore, IEM is a competitive algorithm for constrained optimization problems.

5 Conclusion

This paper presented a novel stochastic algorithm for the constrained optimization problems. The new unconstrained optimization model was presented for the constrained optimization problem and a computation of the total force between particles is presented to create new particle. Our numerical results suggest that the IME is effective. The future work of this study includes the application of EM for combinatorial optimization problems.

Acknowledgement. This work was supported by the fundamental Research Funds for the Central Universities (No.2011QNA29).

References

1. Coello Coello, C.A.: Treating Constraints as Objectives for Single-objective Evolutionary Optimization. Engineering Optimization 32(3), 275–308 (2000)
2. Coello Coello, C.A., Mezura-Montes, E.: Handling Constraints in Genetic Algorithms Using Dominance-based Tournaments. In: 5th International Conference on Adaptive Computing Design and Manufacture, vol. 5, pp. 273–284 (2002)
3. Coello Coello, C.A.: Constraint Handling Using an Evolutionary Multiobjective Optimization Technique. Civil Engineering System 17, 319–346 (2000)
4. Patrik, D.S., Nicholas, J.R.: The COMOGA Method: Constrained Optimization by Multiobjective Genetic Algorithms. Control and Cybernetics 26, 391–412 (1997)
5. Birbil, S.I., Fang, S.C.: An Electromagnetism-like Mechanism for Global Optimization. Journal of Global Optimization 25(3), 263–282 (2003)
6. Rocha, A.M.A.C., Fernandes, E.M.G.P.: A Modified Electromagnetism-like Algorithm Based on Exploratory Moves. In: 2th Conference on Optimization Methods & Software, Prague, Czech Republic, July 4-7 (2007)
7. Alikhani, M.G., Javadian, N., Tavakkoli-Moghaddam, R.: A Novel Hybrid Approach Combining Electromagnetism-like Method with Solis and Wets Local Search for Continuous Optimization Problems. Journal of Global Optimization (July 2008)
8. Rocha, A.M.A.C., Fernandes, E.M.G.P.: Feasibility and Dominance Rules in the Electromagnetism-Like Algorithm for Constrained Global Optimization. In: Gervasi, O., Murgante, B., Laganà, A., Taniar, D., Mun, Y., Gavrilova, M.L. (eds.) ICCSA 2008, Part II. LNCS, vol. 5073, pp. 768–783. Springer, Heidelberg (2008)
9. Thomas, P.R., Xin, Y.: Stochastic Ranking For Constrained Evolutionary Optimization. IEEE Transaction on Evolutionary Computation 311, 1–17 (2000)

Comparison of Two Kinds of Distance in Research on the Method of the Extraction of Load Pattern

LiQing Liu, QiaoLin Ding, TieFeng Zhang, and Jian Chen

School of Electric and Electronic Engineering, North China Electric Power University, Hebei Baoding 071003, China

Abstract. The paper clusters power users' load curves based on fuzzy C means (FCM) clustering algorithm, and to overcome the drawback due to Euclidean distance, the paper also defines a new similarity of curves to extract the power load models which eliminates the limitations of the Euclidean distance that considers only the geometric distance as the similarity measurement of curves.

Keywords: load mode, FCM, Curve similarity measure.

1 Introduction

With the development of smart distribution network [1], more and more sensors, smart meters into the electricity network was used to obtain real-time data of network which will be used to support the operation, the transformation and maintain information of power network after they were processed by screening [2]. These applications can not do without power users' load patterns [3], through the study of the load model we can get the decision-making information to improve the power grid. Clustering the power customers to extract the user's load pattern provides important theoretical and practical values for secure, reliable and economical power grid operation.

2 The Curve Similarity Measurement

Supposing d_{ij} is the distance between the sample i and sample j, for any i and j, d_{ij} should meet three conditions [4]:

(a). $d_{ij} = d_{ji}$ (symmetry). (b). $d_{ij} > 0$. (c). When $i = j, d_{ij} = 0$.

The similarity measurement method as follows: (1) Euclidean distance. For $X_i = \{x_{i1}, x_{i2},\ldots, x_{in}\}$ and $X_j = \{x_{j1}, x_{j2},\ldots, x_{jn}\}$ are two samples, that Euclidean distance between X_i and X_j, $d_{ij} = \left\| x_{ik} - x_{jk} \right\|$ (1)

(2) Correlation Coefficient. Let n-dimensional vector X and Y, the covariance is a scale to measure how X and Y changing together. If the larger (or smaller) values of X tend to be associated with larger (smaller) values of Y, then the covariance is a larger positive. If the larger values of X tend to be associated with the smaller values of Y, then the covariance is negative [5]. Covariance value depending on the range of X and Y can

D. Jin and S. Lin (Eds.): Advances in EECM Vol. 1, LNEE 139, pp. 331–334.
springerlink.com

be eliminated by being standardized. The correlation coefficient between X and Y is as follows:

$$\rho(X,Y)=\frac{\sum_{i=1}^{n}[x(i)-\overline{x}][y(i)-\overline{y}]}{(\sum_{i=1}^{n}[x(i)-\overline{x}]^{2}\sum_{i=1}^{n}[y(i)-\overline{y}]^{2})^{\frac{1}{2}}} \tag{2}$$

In Equation (2), $\overline{x}=\frac{1}{n}\sum_{i}^{n}x(i)$, $\overline{y}=\frac{1}{n}\sum_{i}^{n}y(i)$ .Supposing Def-Distance(X, Y) is the similar distance between time series X and Y. According to the demand of the characteristics of Equation (a) to (c), the similarity distance is given as follows,

$$\text{Def-Distance}(X,Y)=1-\rho(X,Y) \tag{3}$$

For $\alpha=x(i)-\overline{x}$, $\beta=y(i)-\overline{y}$, the similarity distance formula can be simplified to

$$\text{Def-Distance}(X,Y)=1-\frac{(\alpha,\beta)}{\|\alpha\|_2\|\beta\|_2} \tag{4}$$

Following we will prove the distance satisfying the three formulas (a) to (c).

Condition 1: According to the definition, there has, Def-Distance(X, Y) =1- $\frac{(\alpha,\beta)}{\|\alpha\|_2\|\beta\|_2}=1-\frac{\alpha^T\cdot\beta}{\|\alpha\|_2\|\beta\|_2}=1-\frac{\beta^T\cdot\alpha}{\|\beta\|_2\|\alpha\|_2}$ = Def-Distance(X,Y)

Condition (b) ~ (c) can be obtained according to the Cauchy - Schwarz inequality. $|(\alpha,\beta)|\le\|\alpha\|_2\|\beta\|_2$, if and only if α is linear correlation to β, equation was established. Since the load time series is non-zero positive, Def-Distance(X, Y) ≥ 0 and Def-Distance(X, X) = 0.

Def-Distance(X, Y) = 0 $\Rightarrow X=Y$, the conclusion was proved. As can be seen from the above proof, Def-Distance similarity measurement meets the basic requirements. Time series X and Y, $\overline{x}=\frac{1}{n}\sum_{i}^{n}x(i)$, $\overline{y}=\frac{1}{n}\sum_{i}^{n}y(i)$, C_{XS} is amplitude scaling parameter, C_{SH} is translation parameters. We want to prove that Def-Distance ($C_{XS}X$+ C_{SH},Y) = Def-Distance(X, Y).

According to condition 1: Def-Distance(X, Y)=1- $\frac{\alpha^T\cdot\beta}{\|\alpha\|_2\|\beta\|_2}$,

$$\text{Def-Distance}(C_{XS}X+C_{SH},Y)=1-\frac{((C_{XS}X+C_{SH})-\overline{(C_{XS}X+C_{SH})})^T\cdot\beta}{\|(C_{XS}X+C_{SH})-\overline{(C_{XS}X+C_{SH})}\|_2\|\beta\|_2}$$

$$=1-\frac{(X-\overline{X})^T\cdot\beta}{\|(X-\overline{X})^T\|_2\|\beta\|_2}=1-\frac{(x(i)-\overline{x})^T\cdot\beta}{\|x(i)-\overline{x}\|_2\cdot\|\beta\|_2}=1-\frac{\alpha^T\cdot\beta}{\|\alpha\|_2\|\beta\|_2}=\text{Def-Distance}(X,Y).$$

3 Case Study

Choose 147 power users' (belonging to five industries) actual consumption data collected by automatic meter reading system. The collecting intervals are 30 minutes.

The original data were normalized with the following formula. x'_{ij} is normalized data and x_{ij} is original load data, $x'_{ij} = x_{ij} / \max_i \{x_{ij}\}$. Knowing from the cases, the samples wrong clustered are the 15th and the 23rd users. The membership degrees between the user load curves and the typical load profile calculated by the Def-Distance are larger. So the load modes extracted by the Def-Distance are more representative. We make the average of the two as the typical load patterns and study only on the wrong clustered samples.

Following we calculate the time series variance σ and the deviations the user 15 and user 23 between class 1 and class 2 which are shown in Table 1. Time-series variance and the deviation are calculated as follows:

(1) Calculate the difference sequence (assumed to be D) between sequence points corresponding to sample X_1 and X_2, that is $D= \{d_1, d_2,\ldots, d_n\}=\{x_{11}- x_{21}, x_{12}- x_{22},\ldots, x_{1n}- x_{2n}\}$.

(2) Convert the difference sequences into standard deviation, namely:

$$\sigma(D) = \left[\frac{1}{n-1} \sum_{i=1}^{n} (d_i - \overline{D})^2 \right]^{\frac{1}{2}} \tag{8}$$

(3) Deviation is the Euclidean distance between X_1 and X_2.

Table 1. Deviations Analysis of Incorrectly Clustered Samples

Incorrectly Clustered Samples	σ_1	σ_2	Deviation1	Deviation
User 15	1.003	0.8	1.1327	0.8748
User 23	1.068	0.8	1.4897	0.9664

From Table 1, σ_1 and Deviation 1 are larger than σ_2 and Deviation2. User15 and user23 are more similar to the second typical curve. The two users which were clustered into different classes by two algorithms are actually belonging to the electronic industry. The two user's loads between time 20 to 25 and between time 30 to 35 are contrary to the clustering center they belong to, so they are the unusual users.

4 Summary

This article has given a new distance to calculate the similarity of the curves, and changed the limitations and shortcomings that Euclidean distance only defined the similarity based on the geometric mean distance. Because the load curves have time nature which are easily noised by the time series. Similarity measurement presented in this paper overcomes this limitation, even when the translation or scaling of the load sequence occurs, it still has the standard of similarity measure.

Acknowledgements. The paper supported by "the Fundamental Research Funds for the Central Universities" (11 MG39).

References

1. Vojdani, A.: Smart Integration. IEEE Power and Energy Magazine 6(6), 71–79 (2008)
2. Chicco, G., Ilie, I.S.: Support vector clustering of electrical load pattern data. IEEE Transactions on Power Systems 24(3), 1619–1628 (2009)
3. Willis, H.L., Schauer, A.E., Northcote-Green, J.E.D., Vismor, T.D.: Forecasting Distribution System Loads Using Curve Shape Clustering. IEEE Transactions on Power Apparatus and Systems, PAS 102(4), 893–901 (1983)
4. Keogh, E.J., Pazzani, M.J.: An enhanced representation of time series which allows fast and accurate classification. In: Clustering and Relevance Feedback, Proceedings of the 4th International Conference of Knowledge Discovery and Data Mining. AAAI Press (1998)
5. Hand, D., Heikki, M., Padhraic, S.: Principles of data mining, pp. 21–23, 186–199. Massachusetts Institute of Technology (2001)

Research of Insider Threat Based on Process Profiling

Hui Wang[1,2], ChaoQin Zhang[3], DongMei Han[1], and Yang Xu[2]

[1] College of Computer Science and Technology, Henan Polytechnic University, Jiaozuo, China
[2] College of Electronic Science and Engineering, Jilin University, Changchun, China
[3] School of Computer and Communication Engineering,
Zhengzhou University of Light Industry, Zhengzhou, China
wanghui_jsj@hpu.edu.cn

Abstract. Traditional security tools are suitable to prevent external threat, but these are useless to address insider threat. Different from targeting command line processes for masquerade detection in some systems, this paper will create profiles of user process activity to identify insiders. In classification, Naive Bayesian computes the probability of each class occurring and selects the class with the largest probability.

Keywords: Insider Threat, Process Profiling, Naive Bayesian.

1 Introduction

Due to the authorized access of insiders, detecting threats posed by them is much different than for those external to the organization. Common security applications, such as Intrusion Detection Systems, Access Controls, firewalls, and so on, are suitable to prevent external threat, but in most these mechanisms are useless for insiders. Traditional external defenses are primarily concerned with network data, and while insiders can utilize organizational networks. And the focus of these defenses is on prevention of incoming traffic, not internal traffic. According to the US Secret Service and CERT jointly sponsoring the 2007 E-Crime Watch survey, 34% of e-crime damage was from insider threats[1]. It is very necessary that the research fields continue to make progress toward the development of insider threat detection systems.

2 Research of Classification of Insider Threat Systems

To mitigate insider threat, the research conducted can be divided into several categories below.

Access Control

Yu et al.[2] proposed the Display-Only File Server to fight against insider information theft. Pramanik et al. presented a security policy and framework for insider threat similar to Digital Rights Management (DRM). Park et al. improved the Role-Based Access Control (RBAC) model to better adapt the insider threat scenario by using Composite Role-Based Monitoring (CRBM).

D. Jin and S. Lin (Eds.): Advances in EECM Vol. 1, LNEE 139, pp. 335–338.
springerlink.com

Semantic Analysis
Symonenko et al.[3] combined social context, user roles, and semantics to detect insider threat, but only discussed the semantic portion of the project. Cathey et al. tried to detect misuse by clustering information from document and user queries, but only clustered documents that the user accesses.

System Calls
Liu et al.[4] attempted to analyze the features of system calls such as n-grams of system call names from traces, frequency counts of system call names, and system call parameters like the return code, but the method of using k-nearest-neighbor outlier detection produced large numbers of false alarms for insider threat. Battistoni et al. established a kernel driver for the Windows operating system which restricted calls to "dangerous" system calls, but the approach seems infeasible considering the performance problem.

Resource Usage
Shavlik et al.[5] made use of data from Windows 2000 workstations to generate user profiles, and were able to produce less than one false alarm per user per day. But this research conducted masquerade detection, not user identification.

Hybrid Systems
George Mason University is conducting the research of the Detection of Threat Behavior (DTB) project[6]. With the Multi-Entity Bayesian Networks (MEBN), the system associates intention with each user and use it in the comparisons to human behavior models.

3 User Profiling and Identification

3.1 Data Collection

Several past research about masquerade detection are also based on process data, but these systems basically target command line processes which are generally executed in succession. Different from their ways, we will create profiles of user-process activity to identify which user generated it. Assume that all data collected is from the user logged in the workstation and each user is executing their normal duties. Given the volume of collectable data is infeasible to cover everything, some subset must be chosen. As far as know, there is no existing repository of insider threat data so that the related data must be collected. So, we will mainly focus on process resource usage and monitor usage for all running processes base our system, and this is different form other researchers targeting command line processes. We can develop the related program to collect data on all running processes, including the name of the process, the process's handle count, working set, total amount of user processor time, total amount of privileged processor time, and the timestamp of each record.

3.2 Related Naïve Bayesian Algorithm

Considering the amount of computation time they required, the Naïve Bayesian algorithm and updateable Naïve Bayesian will be advisable to be selected. The algorithm should meet the below criteria.

One important reason for choosing Naïve Bayesian is that it generally requires little training data to generate a useful model. To ensure the effectiveness of training data collected, managers should supervise employees more closely whether users are using their own computer. Naïve Bayesian is also extremely fast in comparison to many classification schemes as learning time is linear in the size of the training set. This is exactly the demand of insider threat systems that administrators need to get know of attacks as soon as possible.

Table 1. Criteria Satisfied in the Naïve Bayesian Algorithm

Criteria 1	Create models of user behavior quickly and efficiently;
Criteria 2	The model of user behavior should occupy little storage space;
Criteria 3	Classify gathering data quickly and efficiently;

Maxion found that Naïve Bayesian and its updateable variant are effective in detecting masquerade attacks contrast to other methods over the same data set. So, we think it is effective with the insider data we collected. The Naïve Bayesian probability can be obtained by the following:

$$p(C = c_k | A_1, \dots, A_n) = \frac{p(C = c_k)\prod_i p(A_i | C = c_k)}{\sum_j p(C = c_j)\prod_i p(A_i | C = c_j)} \quad (1)$$

This states that the probability of class c_k occurring given the set of attributes A_1 through A_n, is equivalent to the prior probability of class c_k occurring times the product of the probabilities of each of the attributes occurring given class c_k, divided by the probability of the attributes occurring. When used as a classifier, the denominator on the right hand side of the equation can be omitted as it is not dependent on the class, c_k, and can be considered a constant. In classification, Naive Bayesian computes the probability of each class occurring and selects the class with the largest probability:

$$classify(a_1, \dots, a_n) = \arg\max_{c_k} \; p(C = c_k)\prod_i p(A_i | C = c_k) \quad (2)$$

4 Naïve Bayesian Example

Assume that there are four processes represented: Excel, Word, Acrobat and Internet Explore. If a process has a value of 1, then the process was running when the record was collected, and if it has the value 0, it was not running. As Table 2 follows, it contains twelve records for two users. Given this data set, we would like to use Naive Bayesian to find out who was using the computer when all four processes were running simultaneously.

According to the formula (2) for using Naive Bayesian in classifications, We can compute these probabilities:

P(A) = 1/2; P(IE|A) = 1/2; P(Excel|A) = 2/3; P(Word|A) = 5/6; P(Acrobat|A) = 1/2;
P(B) = 1/2; P(IE|B) = 5/6; P(Excel|B) = 2/3; P(Word|B) = 1/6; P(Acrobat|B) = 1/2;

Now, we can estimate which user is more likely to be using the computer:

P(A|IE,Excel,Word,Acrobat) = P(A)P(IE|A)P(Excel|A)P(Word|A)P(Acrobat|A)= 0.046

P(B|IE,Excel,Word,Acrobat) = P(B)P(IE|B)P(Excel|B)P(Word|B)P(Acrobat|B)= 0.023

Table 2. Naive Bayesian Example Training Set

Record #	Excel	Word	Acrobat	IE	User
1	1	0	1	1	A
2	1	0	0	0	A
3	1	1	0	1	A
4	0	1	1	1	A
5	1	0	0	0	A
6	1	0	1	1	A
7	0	1	0	1	B
8	0	0	1	1	B
9	1	0	0	1	B
10	1	0	1	0	B
11	1	0	1	1	B
12	1	0	0	1	B

Since the probability for user A is greater than that of user B, we predict that user A is using the computer when all four processes are running.

Acknowledgment. This research are supported by the Doctor Grant of Henan Polytechnic University (B2010-62), the Key Laboratory of National Defense Science and Technology, Ministry of Educational of China (Grant No. 421060711421), and the Natural Science Grant of the Education Department of Henan Province(Grant No. 2011B520015) .

References

1. CSO Magazine, United States Secret Service, CERT, and Microsoft. 2007 E-crime watch survey (September 2007)
2. Yu, Y., Chiueh, T.: Display-only file server: a solution against information theft due to insider attack. In: DRM 2004: Proceedings of the 4th ACM Workshop on Digital Rights Management, pp. 31–39. ACM, New York (2004)
3. Yilmazel, O., Symonenko, S., et al.: Leveraging one-class SVM and semantic analysis to detect anomalous content. In: ISI, pp. 381–388 (2005)
4. Liu, A., et al.: A comparison of system call feature representations for insider threat detection. In: Proceedings from the Sixth Annual IEEE SMC, pp. 340–347 (2005)
5. Shavlik, J., et al.: Selection, combination, and evaluation of effective software sensors for detecting abnormal computer usage. In: KDD, pp. 276–285 (2004)
6. Costa, P., et al.: Bayesian ontologies in AI systems. In: Proceedings of the 22nd Conference on Uncertainty in Artificial Intelligence (2006)

Simulation Comparisons between Two Real-Time Computation Methods for Harmonic and Reactive Currents

Zicheng Li

School of Electrical and Information Engineering,
Jiangsu University, Zhenjiang, 212013, China

Abstract. The direct computation method (DCM) and discrete Fourier series method (DFSM) are two very similar real-time detection methods for harmonic and reactive currents. In order to determine whose detection performance is better, using MATLAB 7.0, simulation comparisons are conducted. The results indicate that the detection precision of DCM is higher than that of DFSM when the power supply frequency fluctuates, and the power supply voltage has no distortion. The detection precision of DCM is also higher than that of DFSM when the power supply frequency has no fluctuation, and the power supply voltage distorts. Therefore, the detection performance of DCM is better than that of DFSM.

Keywords: active power filter, harmonic and reactive currents, direct computation method, discrete Fourier series, detection precision.

1 Introduction

The widespread use of power electronic appliances in power systems has caused an alarming increase in harmonic pollution. To address this, the use of active power filters (APFs) has been considered as an effective solution for compensating for the harmonics generated. Furthermore, the performance of APFs is substantially influenced by the detection methods for harmonic current which they use [1]–[3].

Akagi et al. proposed the instantaneous reactive power theory (IRPT) (i.e., *p–q* theory) in 1983. Since then, IRPT has been continuously developed and refined. At present, IRPT-based methods for three-phase APF are regarded as relatively mature methods [1], [3]. However, they require a complex circuitry to realize the transformation. For example, four to six high-precision analog multipliers are used in each phase. This makes the circuit sensitive to component parameter variations [1], [4]. For the single-phase APF, there is no mature method available. Currently, many detection methods for the harmonic current [1]–[3], [4]–[7] have already been proposed. Each method has its own strengths, but they still involve certain limitations that are difficult to overcome [2], [3].

Among these methods, the direct computation method (DCM) [3] and discrete Fourier series method (DFSM) [7] are two very similar real-time detection methods for harmonic and reactive currents of the single-phase APF. They both have the traits

D. Jin and S. Lin (Eds.): Advances in EECM Vol. 1, LNEE 139, pp. 339–344.
springerlink.com

of using a formula to compute directly, a small computation quantity, a high detection precision, etc, and their computation formulae are so similar. Therefore, it has a practical meaning to determine whose detection performance is better through contrastive studies.

2 The Proposed Question

Assuming that the N sampling (periodic sampling, the sampling cycle is T/N) values of a load current $i_L(t)$ in a cycle T are $i_L(1)$, $i_L(2)$, … , $i_L(N)$, the corresponding N sinusoidal values of $e_s(t)$ [$=\sin(\omega t)$], which have the same frequency and phase as the power supply voltage $u_s(t)$ and whose amplitude is 1 V, are $\sin\omega t(1)$, $\sin\omega t(2)$, … , $\sin\omega t(N)$.

Using the detection principle "The integral in a cycle for the absolute value of a load current subtracting its fundamental active current is the least when the load current is a periodic current," DCM is proposed in [3]. According to DCM, the amplitude of the fundamental active current at the moment of sampling $i_L(N)$ is

$$I_p = \sum_{i=1}^{N}[i_L(i)\sin\omega t(i)] \Big/ \sum_{i=1}^{N}[\sin\omega t(i)]^2 \quad . \tag{1}$$

The I_p can be computed using (1). At the moment of sampling $i_L(N)$, the fundamental active current is $I_p\sin\omega t(N)$, and the sum of the harmonic and reactive currents is $i_L(N)-I_p\sin\omega t(N)$. This is DCM.

On the basis of the Fourier series expression of the nonlinear load current, DFSM is educed in [7]. Based on DFSM, the amplitude of the fundamental active current at the moment of sampling $i_L(N)$ is

$$I_p = \sum_{i=1}^{N}[i_L(i)\sin\omega t(i)] \Big/ \left(\frac{N}{2}\right) \quad . \tag{2}$$

The I_p can be computed using (2). At the moment of sampling $i_L(N)$, the fundamental active current is $I_p\sin\omega t(N)$, and the sum of the harmonic and reactive currents is $i_L(N)-I_p\sin\omega t(N)$. This is DFSM.

Equations (1) and (2) are so similar. Both their computation quantities are very small, though the computation quantity of (1) is more than that of (2). In theory, the sameness of (1) and (2) is proved when $N\geq 3$ in [8], so DCM is essentially consistent with DFSM.

In actual applications, the power supply frequency (fixed at 50 Hz) always fluctuates, and the power supply voltage often distorts. Therefore, which method has higher detection precision when the power supply frequency fluctuates, or the power supply voltage distorts? Out of question, the higher the detection precision, the better is the method.

In theory, the contrastive studies on DCM and DFSM have definite difficulties, though they are both on the basis of the Fourier series expression of the nonlinear load current. Although it is feasible to compare the two methods through experiment, it is difficult to find their hair-like differences, and it is discommodious. Without question, digital simulation is a simple and doable valid method to compare DCM with DFSM based on (1) and (2).

3 Simulation Studies

Using MATLAB 7.0, digital simulations between DCM and DFSM were conducted. During the simulation, N was 500. The simulation results are shown in Figs. 1 to 8 and Tab. 1, where, e_s is the power supply voltage whose amplitude is 1 V, and i_L is the nonlinear load current. Furthermore, i_{c1}, and i_{e1} (i_{c1}–i_c) detected by DCM; and i_{c2}, and i_{e2} (i_{c2}–i_c) detected by DFSM are the sum of the harmonic and reactive currents of i_L, and the error, respectively. Here, i_c is the theoretical sum of the harmonic and reactive currents of i_L.

In Figs. 1, 3, 5, and 7, the zero points from negative to positive value of i_L are the same as e_s. In Figs. 2, 4, 6, and 8, the zero points from negative to positive value of i_L are in 30^0 of e_s.

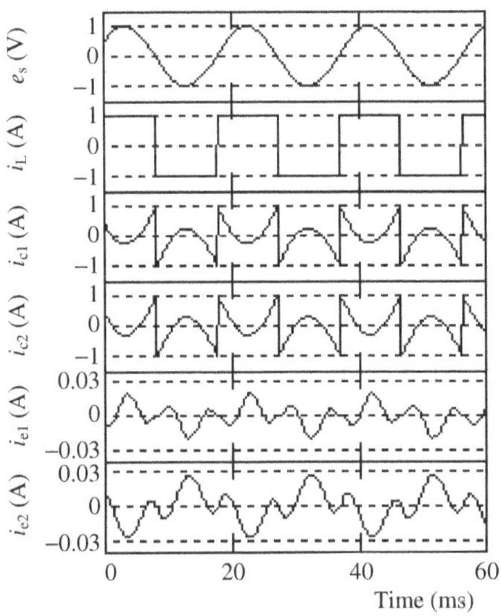

Fig. 1. Simulation results when the power supply frequency is 52 Hz, and the power supply voltage has no distortion.

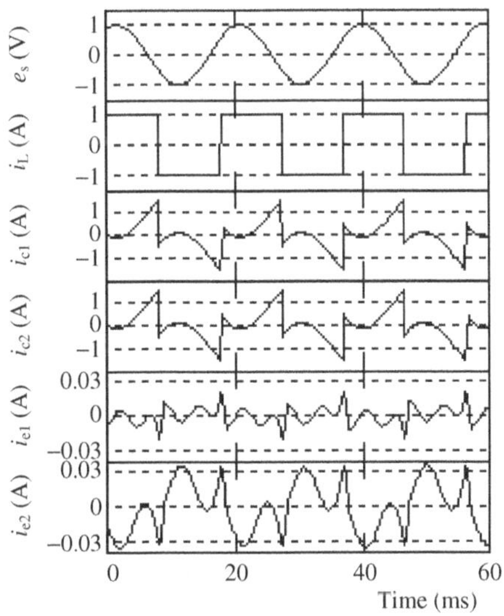

Fig. 2. Simulation results when the power supply frequency is 52 Hz, and the power supply voltage has no distortion.

The simulation results when the power supply frequency fluctuates, and the power supply voltage has no distortion are shown in Figs. 1 to 4. The simulation results when the power supply frequency is 52 HZ are shown in Figs. 1 and 2, and 48 Hz in Figs. 3 and 4. From Figs. 1 to 4, the maximums of $|i_{e1}|$ are clearly less than those of $|i_{e2}|$, respectively. Therefore, when the power supply frequency fluctuates, and the power supply voltage has no distortion, the detection precision of DCM is higher than that of DFSM.

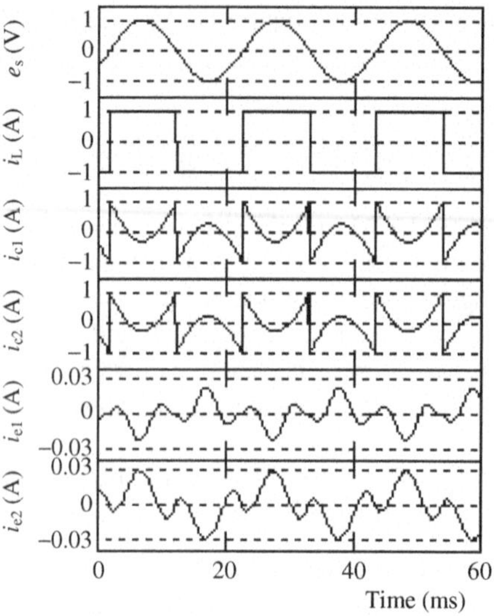

Fig. 3. Simulation results when the power supply frequency is 48 Hz, and the power supply voltage has no distortion.

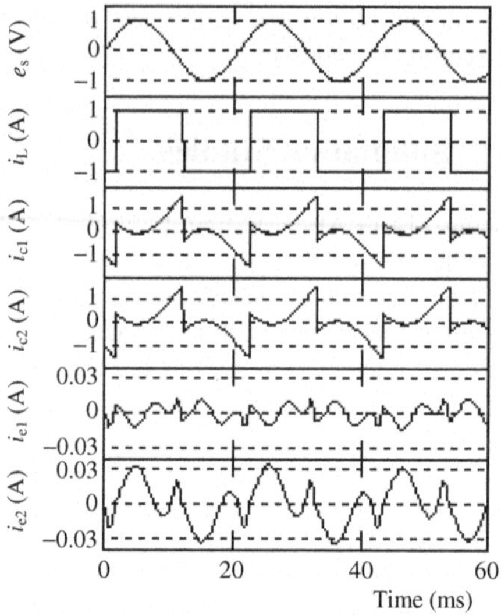

Fig. 4. Simulation results when the power supply frequency is 48 Hz, and the power supply voltage has no distortion.

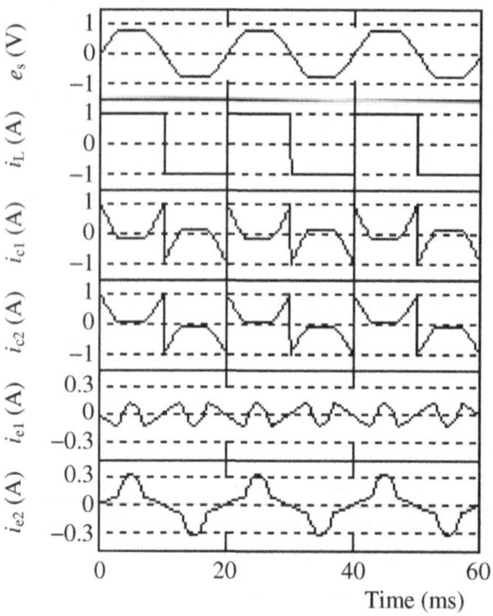

Fig. 5. Simulation results when the power supply frequency has no fluctuation, and the power supply voltage distorts.

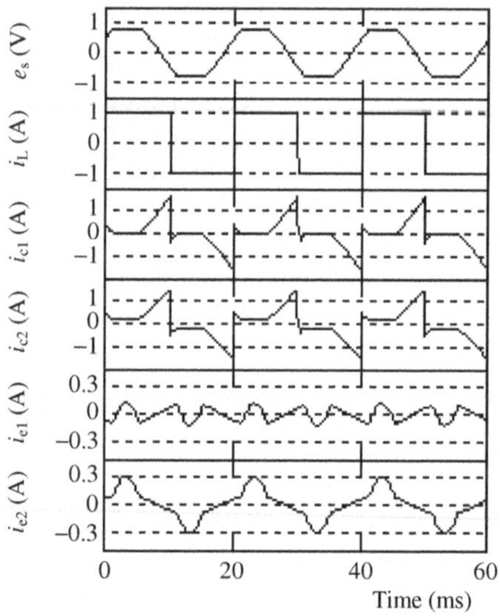

Fig. 6. Simulation results when the power supply frequency has no fluctuation, and the power supply voltage distorts.

The simulation results when the power supply frequency has no fluctuation, and the power supply voltage distorts are shown in Figs. 5 to 8. From Figs. 5 to 8, the maximums of $|i_{e1}|$ are clearly less than those of $|i_{e2}|$, respectively. Therefore, when the power supply frequency has no fluctuation, and the power supply voltage distorts, the detection precision of DCM is higher than that of DFSM.

The maximums of $|i_{e1}|$ and $|i_{e2}|$ for Figs. 1 to 8 are shown in Tab. 1. From Tab. 1, the maximums of $|i_{e1}|$ are clearly less than those of $|i_{e2}|$, respectively.

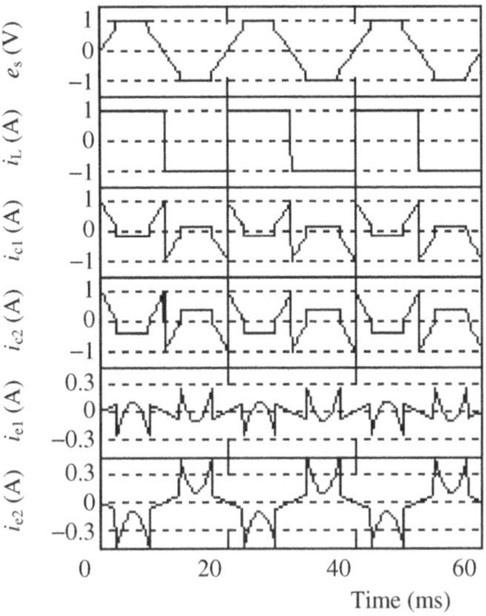

Fig. 7. Simulation results when the power supply frequency has no fluctuation, and the power supply voltage distorts.

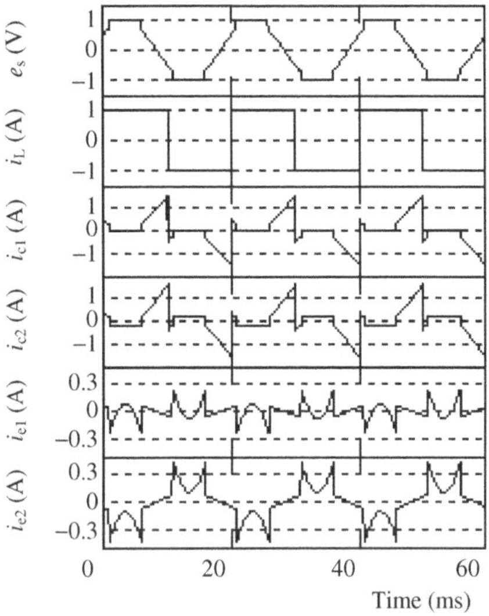

Fig. 8. Simulation results when the power supply frequency has no fluctuation, and the power supply voltage distorts.

Table 1. The maximums of $|i_{e1}|$ and $|i_{e2}|$ for Figs. 1–8

Fig. No.	The max. of $|i_{e1}|$ (mA)	The max. of $|i_{e2}|$ (mA)	Fig. No.	The max. of $|i_{e1}|$ (mA)	The max. of $|i_{e2}|$ (mA)
Fig. 1	20	28	Fig. 5	132	341
Fig. 2	20	38	Fig. 6	120	307
Fig. 3	23	30	Fig. 7	263	483
Fig. 4	15	30	Fig. 8	238	427

4 Conclusions

When the power supply frequency fluctuates, and the power supply voltage has no distortion, the detection precision of DCM is higher than that of DFSM. When the power supply frequency has no fluctuation, and the power supply voltage distorts, the detection precision of DCM is also higher than that of DFSM. Therefore, in actual applications, DCM is better than DFSM.

References

1. Zhou, L., Li, Z.: A novel active power filter based on the least compensation current control method. IEEE Trans. Power Electron 15, 655–659 (2000)
2. Wang, Q., Wu, N., Wang, Z.: A neuron adaptive detecting approach of harmonic current for APF and its realization of analog circuit. IEEE Trans. Instrum. Meas. 50, 77–84 (2001)

3. Li, Z., Sun, Y.: A new compensation current real-time computing method for power active filter based on double linear construction algorithm. Sci. China Ser. E Technol. Sci. 49, 485–512 (2006)
4. Tepper, J.S., Dixon, J.W., Venegas, G., Morh, L.: A simple frequency-independent method for calculating the reactive and harmonic current in a nonlinear load. IEEE Trans. Ind. Electron. 43, 647–654 (1996)
5. Luo, S., Hou, Z.: An adaptive detection method for harmonic and reactive Currents. IEEE Trans. Ind. Electron. 42, 85–89 (1995)
6. Mazumdar, J., Harley, R.G.: Recurrent neural networks trained with backpropagation through time algorithm to estimate nonlinear load harmonic currents. IEEE Trans. Ind. Electron. 55, 3484–3491 (2008)
7. Li, Z., Sun, Y.: The research for the real-time detection method of single-phase circuit harmonic current based on DFT. Electrical Measurement & Instrumentation 42, 20–22 (2005)
8. Li, Z., Sun, Y., Liu, G., Yang, J., Li, F.: Essence and detection performance of direct computation method. High Voltage Engineering 34, 1720–1725 (2008)

Power Load Pattern Recognition Method Based on FCM and Decision Tree

LiQing Liu[1], QiaoLin Ding[1], TieFeng Zhang[1], and JinBao Sun[2]

[1] School of Electric and Electronic Engineering, North China Electric Power University, Hebei Baoding 071003, China
[2] Shanxi Jincheng Electricity Supply Power Company, Shanxi Jincheng 048000, China

Abstract. Different from the traditional method that classified the load curves by industry, the paper has established a load pattern recognition method through FCM clustering algorithm, ID3 decision tree etc. The correctness of the method is verified by analyzing the load shape indexes, category judgment and mining, clustering evaluation, knowledge interpretation of 239 users' daily load data owned by eight sectors.

Keywords: FCM, load pattern, load shape index, data mining.

1 Introduction

In the study of demand characteristics of the electricity market, load pattern classification and recognition based on historical data is the most basic and important task [1]. It plays an important role in power grid running, trading patterns and strategies to develop [2], load control, demand-side management [3]. The clear information of load pattern is helpful for power companies to understand deeply the power companies and their users group effect.

Load pattern recognition features includes load index calculation, clustering and feature interpretation, the new load classification and so on. In this paper, we use fuzzy C-means clustering, ID3 [4] and other algorithms to extract and identify the load model.

2 Load Pattern Recognition Model

2.1 Load Pattern Recognition Process

Load pattern recognition algorithm model and flow chart shown in Figure 1.

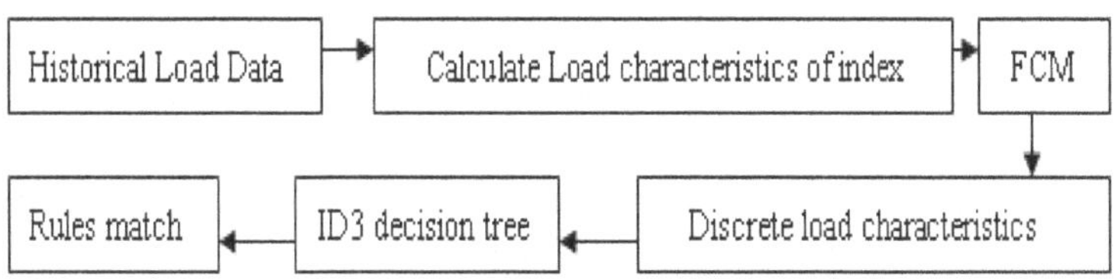

Fig. 1. Load pattern recognition algorithm flowchart

D. Jin and S. Lin (Eds.): Advances in EECM Vol. 1, LNEE 139, pp. 345–348.
springerlink.com

2.2 The Algorithm Involved

(1) Data normalization. To eliminate the impact of the load magnitude to pattern analysis, we use the extreme value normalized the load shape. x'_{ij} is normalized data and x_{ij} is original load data, $x'_{ij} = x_{ij} / \max_{i} \{ x_{ij} \}$.

(2) FCM clustering algorithm. FCM is a clustering algorithm which determines the degree that the elements belong to one class by the membership degree of each data point [8]. Dataset $X=\{x_1, x_2, \ldots, x_n\}$, its fuzzy C partition can be described by $U = [u_{ij}]$, u_{ij} element of matrix U denotes the membership degree between the jth ($j = 1,2, \ldots, n$) data points and the ith ($i = 1,2, \ldots, c$) class , u_{ij} satisfies the following conditions:

for any j, i, $\sum_{i=1}^{c} u_{ij} = 1$, $u_{ij} \in [0,1]$, $\sum_{j=1}^{n} u_{ij} > 0$.

FCM clustering algorithm is to get the partition matrix U and cluster centers V which minimize the objective function. The objective function is given in Equation (1):

$$(\min) J_m(U,V) = \sum_{j=1}^{n} \sum_{i=1}^{c} u_{ij}{}^{m} d_{ij}{}^{2}(x_j, v_i) \tag{1}$$

Where n is the number of the sample, c is the number of cluster centers, m is the weighted index of u, d_{ij} is the Euclidean distance between the sample points and the cluster centers, d_{ij} is given :

$$d_{ij}(x_i, v_i) = \left\| v_i - x_j \right\| \tag{2}$$

(3) Cluster validity evaluation L(c). Supposing c is class number, v_i is clustering center of the ith class center. The molecular of L(c) is the distance between classes. The denominator of L(c) is the distance between class center and its data. The greater the L(c) is, the more reasonable the classification is. L(c) is as follows,

$$L(c) = \frac{\sum_{i=1}^{c} (\sum_{j=1}^{n} u_{ij}^{m}) \left\| v_i - \bar{x} \right\|^2 / (c-1)}{\sum_{i=1}^{c} \sum_{j=1}^{n} u_{ij}^{m} \left\| x_j - v_i \right\|^2 / (n-c)} \tag{3}$$

(4) Interpretation of categories and application. Using entropy-based ID3 [5] algorithm to construct decision tree and recognize the load categories and explain load feature. Given a collection S, containing examples with each of the C outcomes, the entropy of S is

$$Entropy(S) = \sum_{I \in C} [-p(I) \log_2 p(I)] \tag{4}$$

Where $p(I)$ is the proportion of S belonging to class I. Note that S is not a feature but an entire set of examples. Entropy is 0 if all members of S belong to the same class. The range of entropy is 0 ("purity") to 1 ("impurity.") The next measure is an information gain that used to measure the expected reduction in entropy. For a particular feature A, $Gain(S, A)$ means the information gain of sample set S on the feature A and is defined by the following equation:

$$Gain(S,A) = Entropy(S) - \sum_{V \in A} [(|S_v| / |S|), Entropy(S_v)] \tag{5}$$

Where: $\sum$ is a summation on each possible value v of feature A, S_v = subset of S for which feature A has value v, $|S_v|$ =the number of elements in S_v, $|S|$ = the number of elements in S.

3 Case Study

Choose 239 power users' (belonging to eight industries) actual consumption data .The collecting intervals are 30 minutes. The paper use the load curve shape index [6]to identify the load model. The load curve shape indexes are in the Table.1. (av、max、peak、val、sh represents respectively average、maximum、peak、valley、flat)

Table 1. Load Shape Indexes (LSI for short)

LSI	Definition	Time	LSI	Definition	Time
a_1	$a_1=P_{av}/P_{max}$	allday	a_4	$a_4=P_{av.peak}/P_{av}$	08:00-11:00,18:00-21:00
a_2	$a_2=Q/(24P_{max})$	allday	a_5	$a_5=P_{av.sh}/P_{av}$	06:00-08:00,11:00-18:00,21:00-22:00
a_3	$a_3=(P_{max}-P_{min})/P_{max}$	allday	a_6	$a_6=P_{av.val}/P_{av}$	22:00-Next day06:00

The clustering results are shown in fig.2. $L(1)=0$, $L(2)= 143.7881$, $L(3)= 152.4827$, $L(4)= 167.0154$, $L(5)= 163.7592$, according to L(c), we cluster the load into four class.

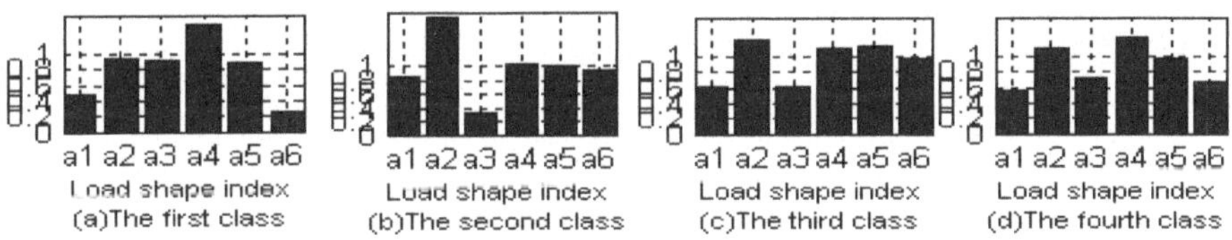

Fig. 2. Clustering results of load shape indexes

The load shape index is discreted as low (L), middle (M), high (H).

Table 2. Load Shape Indexes Disperse Interval

LSI	L	M	H	LSI	L	M	H
a_1	<0.5342	[0.5342,0.7525]	>0.7525	a_4	<1.0874	[1.0874,1.2884]	>1.2884
a_2	<1.0683	[1.0683,1.5051]	>1.5051	a_5	<0.9666	[0.9666,1.0878]	>1.0878
a_3	<0.3659	[0.3659,0.6641]	>0.6641	a_6	<0.4360	[0.4360,0.7743]	>0.7743

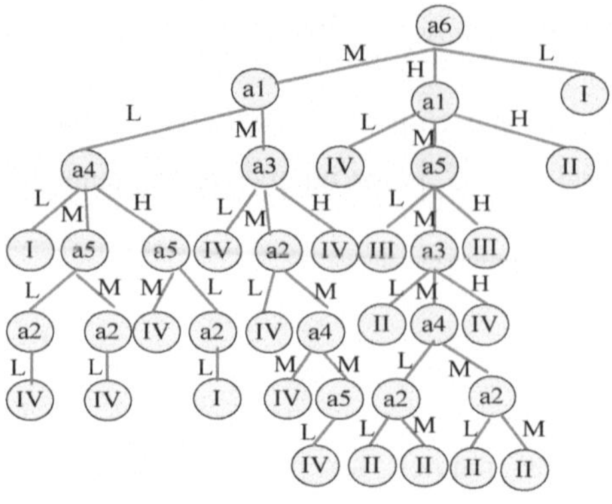

Fig. 3. ID3 Decision Tree Based on Load Shape Index
(I, II, III, IV represents respectively class number)

ID3 decision tree is shown in Fig.3. According to Fig.3 , we can classify the load fast. The decision tree is consistent with the actual load data. The method is effective.

4 Summary

Based on the idea of data mining, the paper has established the load pattern recognition model based on FCM and ID3 decision tree. The model consists of pre-treatment, extraction of the typical load pattern, load pattern recognition module. Tested by actual data, the method is effective and feasible.

Acknowledgements. The paper supported by "the Fundamental Research Funds for the Central Universities" (11 MG39).

References

1. Pitt, B., Kirchen, D.: Application of data mining techniques to load profiling. In: IEEE PICA, Santa Clare, CA (1999)
2. Chicco, G., Napoli, R., Postulache, P., et al.: Customer characterization options for improving the tariff offer. IEEE Trans. on Power Systems 18(1), 381–387 (2003)
3. Chicco, G., Napoli, R., Postulache, P., et al.: Customer characterization options for improving the tariff offer. IEEE Trans. on Power Systems 18(1), 381–387 (2003)
4. Wei, Z.-G., Zhou, X.: Improved Algorithm Based on ID3. Computer Science 37(7A), 11–12 (2010)
5. Jearanaitanakij, K.: Classifying Continuous Data Set by 1D3 Algorithm. In: 2005 Fifth International Conference on Information Communications and Signal Processing, pp. 1048–1051 (2006)
6. Zhao, Y., Li, L., Liu, J., et al.: Combinational Recognition Model for Demand Side Load Profile in Shanghai Power Grid. Power System Technology 34(1), 145–151 (2010) (in Chinese)

The Design and Implementation of Remote Controlling Based on Embedded WinCE

Bo Li and Ke Liao

Institute of Astronautics and Aeronautics of UESTC
University of Electronic Science and Technology of China, Chengdu, China
libo@uestc.edu.cn, leocole@qq.com

Abstract. The design of remote controlling is considered as an important numerical design in network environment. In this paper, aiming at the actual demand of present industrial fields, some key technologies based on embedded WinCE such as remote desktop share and remote file transfer were analyzed, studied and realized. It makes the existing embedded system migrate to the host PC smoothly, and all the data and applications of embedded system can be displayed and controlled on the remote PC in order to meet the diversified industrial controlling standard application needs. Finally, a case on WinCE platform is presented to verity the feasibility and efficiency of this scheme.

Keywords: WinCE, embedded operation system, remote desktop share, remote file transfer, operation system configuration.

1 Introduction

With the wave of information technology continues to advance, the embedded technology is now used in modern industrial production widely at an unprecedented rate. The communication mechanism based on TCP / IP provides a new way for industrial remote controlling, and the best choice to achieve the aim is using the embedded system as the bridge [1], scholars have relative researches. The proposals put forward by Jing Li, Weidong Hao [2] and Jianyang Zhao, Weihong Ding [3] are based on embedded platform, using MPEG-4 encoding standard, and the transmission are for 3G networks; Xiaoxue Yang, Lihu Wang and so on [4] and Guanbao Wang [5] propose the schemes that using DirectFB or LabVIEW to achieve remote control operation of embedded systems with windows remote desktop technology.

It can be seen that the existing embedded remote controlling schemes consume a large number of resources, its browser interface often do not have the intuition and compatibility, and the cost of flow charge is prohibitive. The remote desktop control technology which is widely used in the PC is still in its infancy in the applying on embedded systems, it needs related software to achieve the controlling of embedded devices, and there is a certain lack of reliability and stability in this scheme.

This study analysis the remote control technology based on embedded systems that customize the remote control function into their own kernel to enhance systems' reliability and stability, and it mainly for the remote synchronized collaborative design. To meet the requirements, the following two aspects are considered:

D. Jin and S. Lin (Eds.): Advances in EECM Vol. 1, LNEE 139, pp. 349–353.
springerlink.com

1. To provide a none-distance site visits environment with the remote desktop sharing technology [6]. It allows remote users to control the embedded devices real-time and synchronously. Client user operates the server's application by local mouse and keyboard, after the server's desktop is compressed and encoded [7], it is sent to the client being displayed with the protocol RFB (Remote Frame Buffers).

2. To achieve remote file transferring with FTP (File Transfer Protocol). FTP is based on client / server architecture, it integrates distributed and P2P technology [8], and achieves the synchronization updates of on-site data real-time and conveniently.

2 The Design of System Software Platform

2.1 The Development Process of Windows CE Platform

In this paper, Windows CE 6.0 is chosen as the operating system platform. It is a 32-bit, open source, scalable embedded real-time operating system, and the design focus is the custom configuration of remote desktop share and remote file transfer components on Windows CE operating system.

The Windows CE operating system is customized through the Platform Builder for CE 6.0 [9]. It can complete the customization, compilation and debugging of embedded operating system. The main development steps are described as follows: enter the new platform wizard; select BSP (the Board Support Package); select the system components; select the client's internet service; select multimedia technology; select the application and so on; and eventually generate kernel image file [10]. After the hardware and operating systems were available, what remain are some necessary applications for the platform (Fig.1).

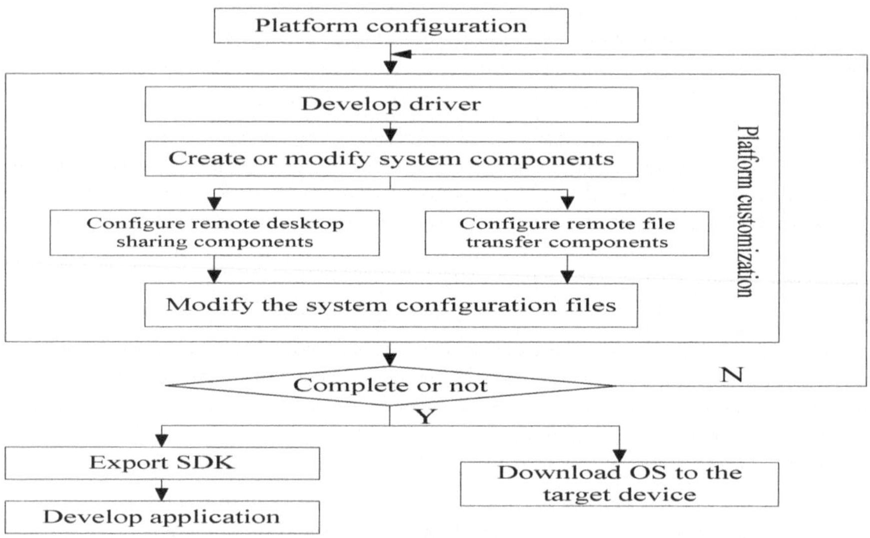

Fig. 1. The Development Process of Customizing Windows CE Platform

2.2 The Customization of Windows CE Remote Desktop

In the process of customizing BSP, the core of whole software platform design is the customization of Windows CE remote desktop components. While customizing, it should first select the components to support Windows CE remote desktop, add "Core OS"->"CEBASE"->"Core OS Services"->"Debugging Tools"->"Remote Display Application" in Windows CE project. At the same time, the telnet server is compulsory in BSP, just add "Core OS" -> "CEBASE" -> "Communication Services and Networking" -> "Servers" -> "Telnet Server". Then we can login telnet server in PC to execute the remote command after the target machine starts.

Then, we configure the host IP address of remote desktop. Open the registry file "project.reg" in Windows CE project, and then add these followings:

```
[HKEY_LOCAL_MACHINE\SOFTWARE\CERDISP]
"Hostname"="192.168.1.10" //the remote PC's IP address
[HKEY_LOCAL_MACHINE\COMM\TELNETD]
"IsEnabled "=dword: 1       //open telnet service
"UseAuthentication"=dword: 0
//login without username and password
```

2.3 The Customization of Windows CE Remote File Transfer

The customization of Windows CE remote file transfer is also an important part of system software platform development. Monitor officers need a PC program which supports file transfer protocol, to connect the remote file transfer protocol server program on the target machine, and upload or download data to do the professional analysis. In Windows CE system, we can also complete this function with customizing FTP server components of BSP.

In the Windows CE project, select and add "Core OS" -> "CEBASE" -> "Communication Services and Networking" -> "Servers" -> "FTP Server".

Then add these followings in registry file "project.reg":

```
[HKEY_LOCAL_MACHINE\Comm\FTPD]
"IsEnabled"=dword: 1
"UseAuthentication"=dword: 0
"AllowAnonymous"=dword: 1
"AllowAnonymousUpload"=dword: 1
"AllowAnonymousVroots"=dword: 1
// allow anonymous login and transfer files
"DefaultDir"="\\"    // set the default directory
```

3 The Implementation and Test of System

3.1 The Compilation and Startup of Windows CE Remote Service

After the configuration of remote service components, just compile the platform. When the compilation is finished, it generates binary image files "NK.bin" and "NK.nb0", download them to the target machine and startup.

First, we start the remote desktop program in PC. While the Windows CE system is power on, we find "cerhost.exe" in the X: \ WINCE600 \ Public \ Common \ Oak \

Bin \ I386 directory, then run this executable file. Next we start the target machine. Login the target machine with Telnet, and type the following command in the command line: cerdisp –c, then the remote desktop client starts up. So far, we can see the Windows CE desktop of target machine on the PC with cerhost.exe.

Next, open the FTP tool, enter the IP address of target machine, and wait for the connection, and then we can successfully login the FTP Server of Windows CE system to upload and download data, it is shown in Fig.2.

3.2 Image Acquisition

After compiling these programs above, we can achieve remote controlling between PC and embedded operating system in LAN. When the remote connection is successful, users can completely open and operate the complicated files according to the habits in PC, and continue to use the familiar software operating environment of server-side instead of to learn and adapt new operating methods. We have an example for remote image acquisition, install camera on the target machine, and run the camera program, we can see the image acquisition interface on the remote PC, and the experiment result is shown in Fig.3. The FTP tool is used to transfer the capture images into the remote PC client.

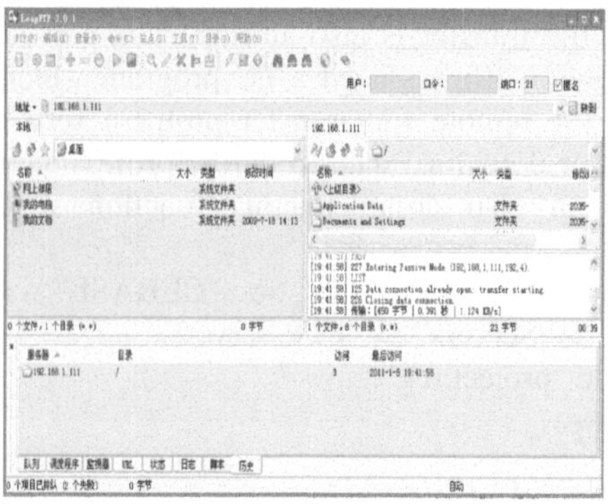

Fig. 2. The Successful Connection of Remote File Transfer

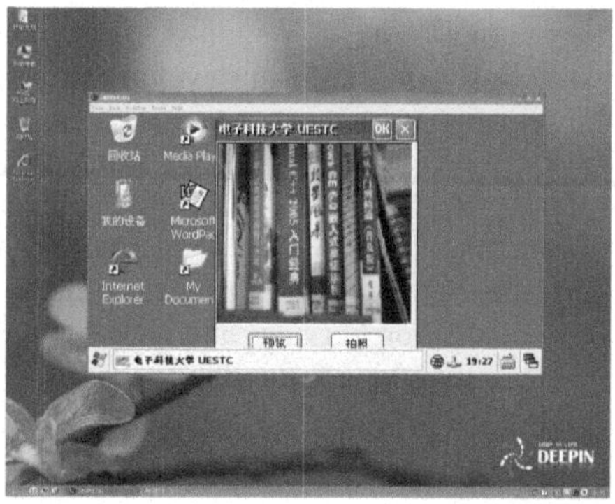

Fig. 3. The Camera Capture Interface of Remote Control

4 Conclusion

The innovation of technological and the changes of method inevitably bring up the overall efficiency and reduce the cost. Previously, it may take ten minutes or an hour, even one day to reach the industrial site. In many cases, the time cost because of the distance factors is much higher than the cost to solve the problem itself. Based on this situation, this design of the system expands the concept of the existing network of embedded controlling system to the remote server-side development, and provides a more perfect remote controlling design and implementation of embedded system. Through the remote controlling module, the existing embedded systems can migrate to the host PC smoothly, and all of the data and applications can be displayed and controlled on the remote PC. It breaks the limited of applying locations of embedded devices, and can save mass time of each problem. Experimental tests prove that this design is convenient, real-time and efficient, it is very suitable to apply in hazard areas, security areas and other industrial field, and it has wide market prospects.

Acknowledgments. This work is sponsored by the National Natural Science Foundation of China under Grant No. 51075060, supported by the Scientific and Technical Personnel Service to Enterprise Project of Science and Technology Ministry under Grant No. 2009GJE0000205, and also supported by Youth Foundation of University of Electronic Science and Technology of China (UESTC). The authors also gratefully acknowledge the helpful comments and suggestions of the reviewers, which have improved the presentation.

References

1. Liu, K., Yu, L.J.: Remote Data Acquisition Based on Embedded Linux System. Computer Applications 26(6), 264–265 (2006)
2. Li, J., Hao, W.D.: Research and design of embedded network video monitoring system based on Linux. In: International Conference on Computer Science and Software Engineering, pp. 1310–1313. IEEE Press, New York (2008)
3. Zhao, J.Y., Ding, W.H.: Research on Remote Monitor the Control System Based on 3G Streaming Media. Computer Measurement and Control 17(12), 2400–2406 (2009)
4. Yang, X.X., Wang, L.H., Ye, J.N., et al.: DirectFB Applied in Embedded Remote Desk Control System. Computer Engineering and Design 31(9), 2127–2130 (2010)
5. Wang, G.B.: Remote Monitoring and Supervision Based on LabVIEW. Industrial Control Technique 05, 91–93 (2006)
6. Simoens, P., Praet, P., Vankeirsbilck, B., et al.: Design and implementation of a hybrid remote display protocol to optimize multimedia experience on thin client devices. In: 2008 Australasian Telecommunication Networks and Applications Conference, pp. 391–396. IEEE Press, New York (2008)
7. Taubin, G., Rossignac, J.: Geometry compression through topological surgery. ACM Transaction on Graphics 17(2), 84–115 (1998)
8. Balakrishnan, H., Kaashoek, M.F., Karger, D., et al.: Looking up data in P2P systems. Communications of the ACM 46, 43–48 (2003)
9. Phung, S.: Professional Windows Embedded CE 6.0 Programming. Tsinghua University Press, Beijing (2009)
10. Jiang, B.: Windows CE.Net Programming. China Machine Press, Beijing (2006)

Study of Privacy Protection Based on SWTA Algorithm

Yuqin Wang

Jilin Business and Technology College, Department of Information Engineering,
130062, Changchun, China
41215421@qq.com

Abstract. In order to solve privacy protection excavation algorithm data protection degree is not high at present, the business database's regular loss factor is big and so on questions, this paper uses the PPARM algorithm and based on moves term the privacy protection algorithm (IMBA) which unifies strategy, proposed one kind based on connection rule privacy protection algorithm- SWTA algorithm. And through the experimental verification based on the SWTA algorithm's privacy protection algorithm had a better confidentiality and the validity compared to the original algorithm in the data protection degree.

Keywords: Privacy protection, Connection rule, PPARM, IMBA, SWTA.

1 Introduction

with the Internet technology and information technology high speed development, the researchers may obtain the massive data fast from each website, and carries on the excavation using the data mining technology to the data, but like this possibly causes the primary data important information to have the revelation, therefore conducts the research to the privacy protection's data mining, prevents the primary data and the related important sensitive information was discovered that becomes unusual important and urgent [1]. The privacy protection data mining's concept was put forward unceasingly in the recent several years, The connection rule's data mining most early is raised by R. Agrawal, nowadays already became a most important data mining technology, what it mainly solves is makes the extraction useful project relations or rule in the database to the massive item sets[2]. Privacy protects based on the connection rule excavation's to solve in the privacy protection connection rule sensitive rule and the non-sensitive regular question research has the very vital practical significance.

2 Connection Rule

The connection rule excavation generally speaking is refers to from a great quantity data discovered that the interesting incidence relation, namely distinguishes the attribute value collection which from the data set appears frequently, is also called the frequent item set connection rule basic concept as follows:

D. Jin and S. Lin (Eds.): Advances in EECM Vol. 1, LNEE 139, pp. 355–360.
springerlink.com

(1) Data item and data item set. Supposes I={i_1,i_2,…,i_m} is a m different project set, then ik(k=1,2,…,m) is called the data item, the data item set I is called the data item set, its element integer is called the data item set the length, the length is the k data item set is called the k Uygur data item set, called the k- item set.

(2) Business. Business T is a subset of data item set, each business has only identifier Tid with it correspondence, all business's overall constituted all business collection D.

(3) suppose the set {i1, 12… 1m.}Is m composition set, supposes assigns business database D, each business T is a set, causes T I, each business has only identifier Tid. Suppose A is an item set, business T contains A works as when only $A \subseteq T$. The connection rule is the shape like $A \Rightarrow B$ implication type, $A \subset I$, $B \subset I$, and $A \cap B = \varnothing$, rule $A \cap B$ establishes in business collection D, has support s, which s is in D business contains $A \cup B$ percentage. It is a probability $P(A \cup B)$. Rule $A \Rightarrow B$ has confidence c in business collection D, c is in D contains A business simultaneously also contain B the percentage. This is conditional probability P (B/A). Namely

$$\sup port(A \Rightarrow B) = P(A \cup B) \tag{1}$$

$$Confidence\,(A \Rightarrow B) = P(B/A) \tag{2}$$

A set is called the item set, contains k item set to be called the k item set. If the item set I support satisfies the pre-definition the smallest support threshold value, then I is the frequent item set. The confidence general formula is:

$$confidence(A \Rightarrow B) = \frac{\sup(A \cup B)}{\sup(A)} \tag{3}$$

The connection rule is a shape like $A \rightarrow B$ mathematical expression, A and B do not intersect item, namely $A \cap B = \varnothing$, inside the connection rule has two very important measures which is the support and the confidence. The support description assigns the item set the frequent degree. The confidence determined B in contains the number of times which A in the business appears, is also the item relates[3]. Their formula respectively is:

$$\sup port(A \Rightarrow B) = \frac{|A \cup B|}{|N|} \tag{4}$$

$$confidence(A \Rightarrow B) = \frac{|\sup(A \cup B)|}{|\sup(A)|} = \frac{|A \cup B|}{|A|} \tag{5}$$

3 Privacy Protection Technology

The privacy protection's research direction is actually the privacy demand which applies by the society needs decided. The general privacy preservation technology devotes to protects the data owner's privacy in the quite low level, this technology is mainly realizes through the introduction statistical model and the probabilistic model. But data mining's privacy preservation technology mainly wants to solve in the top level data protection, how to act according to the different data mining application characteristic, realizes to the privacy the protection.

3.1 Privacy Protection Concept

Any thing has its dual character, the data mining is not exceptional, while the data mining applies massively in the social domain produces wealth, produces along with it is the privacy revelation question. In the 1960s, a US's professor has published publicly about the privacy article, proposed "information right of privacy" the concept, deals with the impact which the information age right of privacy receives. About the privacy concept, Oliveira proposed that his viewpoint, the privacy may divide into two kinds, one kind is individual privacy, and another kind is the collective privacy.

3.2 Privacy Protection Excavation Algorithm Classification

Generally speaking, the privacy preservation technology may divide into the central data the privacy protection and the distributional data privacy protection. The vertical distribution is mainly refers to the data to distribute according to the attribute in many organizations [4]. The privacy maintains the technology is mainly divided three kinds:

① Privacy maintains technology based on the heuristic. ② Privacy maintains technology based on the information security theory. ③Privacy maintains technology based on the restructuring privacy.

4 SWTA Algorithm Realizations

4.1 SWTA Algorithm Description

In Affairs library collection {T1: ABCD, T2: CD, T3: ABCDEF, T4: BDF, T5: ABCD, T6: BCF, T7: ABD, T8: BCDEF} the stipulation smallest support is 40%, the smallest confidence is 60%. Calculation support{A $\rightarrow$ B}=4/8=50%, confidence{ A $\rightarrow$ B }=4/4=100% May know by the PPARM algorithm a=1,b=2,k=min(a,b)=1,Only needs to contain B sensitive business to delete 1 time, can cause its confidence to reduce 37.5% If using the classic Apriority algorithm obtain nine association rules, if rules A$\rightarrow$B, B $\rightarrow$F, E $\rightarrow$F is sensitive rules, then the corresponding set of sensitive items SI = {{A, B}, {B, F}, {E, F}}, Algorithms are dealing with sensitive item sets Si={A,B},At this point pending the collection of sensitive items SI,={{B,F},{E,F}}, Num(B,{A,B},SI,T1)=1,Num=(A, {A,B},SI'T1)=0. It can be seen, if removed A from the transaction T1, not SI 'sensitive items in the impact, if out of the project B, then it will affect SI' in the sensitive item set {B, F}.

4.2 SWTA Algorithm Description

Input: sensitive set Si, minimum support: minsup, the minimum confidence: minconf, library services DB

```
   Output: Business storehouse DB'
   Begin
   M [ ];W [ ];size1; size2;
   If Si. support>=minsup and Si.confidence>=minconfidence then
Compute totalDelNum();
   a= mod( |D|×(sup(x∪y)-minsup))+1;
   b= mod( |D|×(sup(x∪y-minconf×sup(x)))+1;
   k= min (a, b);
   Return k;
   For (j=1, j<=k,j++)
   I: = ItemMove ( DB );
   I: = ItemMove ( );
   Maxnum=-1;
   For each itemj in Si
   Calculate Num (itemj, Si, RI', Tj)
   If (Num (itemj,Si,,RI',Tj)>maxNum)
   Maxnum=Num (itemj, Si, RI', Tj)
   ItemMove= itemj;
   Else if (itemj, Si, RI', Tj) =maxNum;
   If ItemMove.supp > itemj.supp;
    ItemMove = itemj;
   T: =findshortesttransaction ()
   For(n=1;n<= |j|;n++)
   If (Si in Tj)
   M[n] =Tj;
   Else W[n] =Tj;
   Size1=sizeof (M [0]);
   For (m=0; m<n; m++)
   Size2=sizeof (M[m]);if (Size1>=Size2)
   Size1=Size2; mintransaction=M[m];
   Else {mintransaction= M[0];
   Return mintransaction;
   For (i=0; i<sizeof (W[n]); i++)
   If (W[i] in minstring)
   U[n]=W[i];
   Min=sizeof(minstring);
   For (k=0;k<sizeof (U[n]);k++)
   If(|sizeof(minstring)-sizeof(U[k])|< min)
   Min = sizeof(minstring)- sizeof(U[k]);string=U[k];
   Return string;
   Move(ItemMove,mintransaction,string)
   Output the updated database DB as DB'
```

5 Experiment Results and Analysis

5.1 The SWTA Algorithm Verify

In order to evaluate the usefulness of this algorithm , the test pass to many data set to carry on a great deal of experiment verification, recorded PPARM algorithm and SWTA algorithm under the parameter the same circumstance, several experiment datas set of movement result, see table 1 and table 2.

Table 1. Time-consuming compared

Item number	Support	confidence	Conceal rate	PPARM	SWTA
10	0.2	0.3	0.1	0.625	0.710
			0.2	0.714	0.826
			0.3	0.813	0.877
			0.4	0.813	0.963
50	0.3	0.4	0.1	5.179	5.362
			0.2	5.289	5.904
			0.3	5.671	6.382
			0.4	5.179	6.616
100	0.4	0.5	0.1	17.434	19.493
			0.2	18.439	22.769
			0.3	20.127	25.785
			0.4	23.576	27.974

Table 2. Loss rate compared

Item number	Support	confidence	Conceal rate	PPARM	SWTA
10	0.2	0.3	0.1	22.058	19.045
			0.2	23.064	20.947
			0.3	24.157	21.069
			0.4	25.869	22.368
50	0.3	0.4	0.1	11.265	8.953
			0.2	12.573	10.101
			0.3	14.468	11.479
			0.4	16.847	14.421
100	0.4	0.5	0.1	17.434	12.984
			0.2	18.439	14.170
			0.3	20.127	16.411
			0.4	23.576	19.188

From the table 1 and the table 2 can analyze, if give to settle same function parameter, to two algorithms, the PPARM algorithm solved the influence to the non-sensitive rule, so consume to take a lot of time more. But more important, Be compared to PPARM algorithms, But more important, compared to PPARM algorithms, SWTA mining algorithm creation the rule throw rate much lower. Compare to PPARM algorithm, under usually condition SWTA algorithm be always lower than the rule that the PPARM algorithm generates to throw lose rate. According to record several data of experiments, result such show as figure 1.

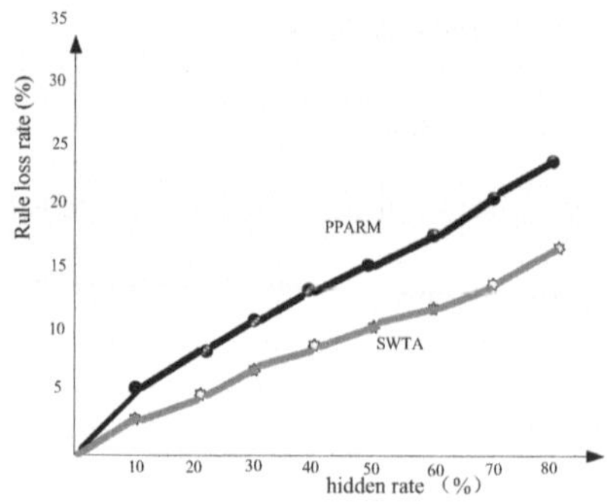

Fig. 1. The relation graph of hidden rates and rule loss rate

5.2 The Result Analysis

From the experiment get conclusion, if give two algorithms settle with the same parameter, for the different algorithm, Firstly: The SWTA algorithm is obviously expended time longer than time that the PPARM algorithm expends, this is because of the SWTA algorithm, not only consider how select delete the sensitive rule item, also need to consider how to select the hidden sensitive item that will wanted move to the best candidate business. Secondly: to the same data set, generally, under the same condition of settle parameters, and also under the same condition at conceal rate, SWTA rule's throwing to lose rate is more clearly small than the rule of PPARM algorithm to rule lose rate.

References

1. Zhou, S., Li, F., Tao, Y.: study of privacy protect faces to a data application. Journal of Computer 32(5), 847–861 (2009)
2. Agrawal, R., Imielinski, T., Swami, A.: Mining association rules between sets of items in large databases, pp. 50–65 (2005)
3. Agrawal, R., Srikant, R.: Privacy-preserving data mining. In: Proceedings of the ACM SIGMOD Conference on Management of Data, pp. 439–450. ACM Press, New York (2003)
4. Vassilios, V.S., Gkoulalas-Divani, A.: A Survey of Association Rule Hiding Methods for Privacy (2007)
5. Verykios, V.S., Gkoulalas-Divanis, A.: A Survey of Association Rule Hiding Methods for Privacy. In: Workshops Proceedings of the 6th IEEE International Conference on Data Mining (ICDM 2006), ICDM 2006, pp. 502–506 (2006)

Study on Design of Drive Circuit for Piezoelectric Actuator

Zelong Zhang[1], Zhenming Liu[2], Guangyao Ouyang[2], and Xiaofeng Li[2]

[1] Department of Information Countermeasures, AFAR, Wuhan 430019, China
[2] College of Marine and Power Engineering, Naval University of Engineering, Wuhan 430033, China
zzlong518@126.com

Abstract. According to the dynamic application require and material performance parameter of piezoelectric actuator, a type of drive circuit for piezoelectric actuator is advanced, which is based on PWM control and adopt IGBT and singlechip. The particular principle fig of the drive circuite is given. The dynamic drive experiments to the piezoelectric actuator indicate that the drive circuite has favourable dynamic response performance.

Keywords: piezoelectric actuator, drive circuit, IGBT.

1 Introduction

Piezoelectric actuator has the virtues of little volume, quick response and big force etc, and is used widely in tiny displacement output devices[1]. Recently, the application of piezoelectric actuator in injector of diesel is the research hotspot of the domain.

The performance of the piezoelectric actuator is determined extently by the performance of the drive circuit. Currently most drive circuits of the piezoelectric actuator are applied staticly, and current magnifier is used in the drive circuit, which emphasize the stability and precision of the voltage. However, in the actual application, specially in the application of the injector of the diesel engine, the piezoelectric actuator is required seasonally elongating and abbreviating dynamicly, which demand the drive circuit have excellent dynamic performance. Because the above reasons, the dynamic drive circuit for the piezolelectric is developed in this paper, which is based on PWM control type and adopt IGBT and singlechip.

2 Drive Circuit Design

2.1 Design Project

The design project of the drive circuit of the piezoelectric actuator is shown in Fig.1. The drive circuit is composed of two classes [2] .The first class is the circuit for rising the voltage. In the first circuit, the Boost voltage ascending circuit is applied to ascend the 48V/24V voltage provided by accumulator, and the capacitance is used as filter

D. Jin and S. Lin (Eds.): Advances in EECM Vol. 1, LNEE 139, pp. 361–366.
springerlink.com

and power container, and then, the regulable DC voltage of 0～200V is output. The second class is the circuit for charge and discharge. In the second circuit, the DC voltage, which is output by the first circuit is cut and then the voltage pulse is output. The voltage pulse is used to charge quickly for piezoelectric actuator.

In the Fig.1, VT_1～VT_3 are IGBT and diode VD_1～VD_3 is shunt-wound diode of the IGBT VT_1～VT_3 respectively. VD_0 is discharge diode. L_2 is the inductance which restrict the current in the charge and discharge process. The function of L_2 is to protect the switch instrument and piezoelectric actuator, which will be destroied by the big current otherwise.

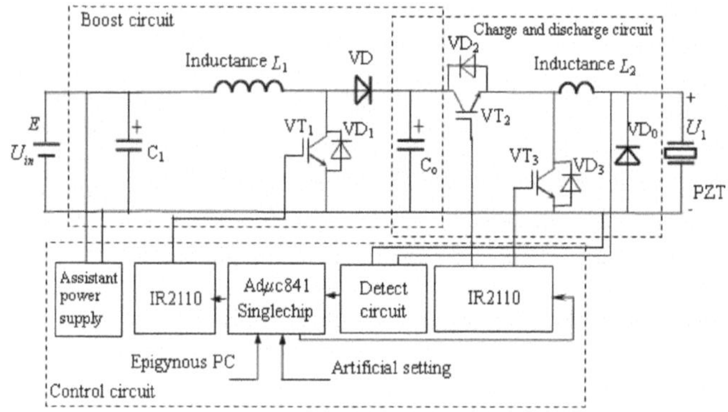

Fig. 1. The drive circuit sketch of piezoelectric injector

2.2 Charge and Discharge Circuit Design

According to the setting, which is set by man or epigynous PC, the Adμc841 singlechip give appropriate width and periodic charge PWM signal to drive ship IR2110 to control the openning of charge IGBT VT_2. The capacitance C_0 charge the piezoelectric actuator through charge IGBT VT_2 and discharge diode L_2. During the period of charge IGBT VT_2 open, the Adμc841 singlechip detect the output voltage of piezoelectric actuator, then the PI adjustor keep the output DC voltage steady.

At the edge of charge PWM signal, the Adμc841 singlechip output high level signal to drive ship IR2110 to open the discharge IGBT VT_3, and then, the piezoelectric actuator dischage through inductance L_2 and discharge IGBT VT_3.

The power storaged in piezoelectric actuator transfer to inductance L_2. When the output voltage detect circuit detect the output voltage reduce to some value(near zero), which indicate that the power storaged in piezoelectric actuator will be release over, the Adμc841 singlechip output low level signal to drive ship IR2110 to close the discharge IGBT VT_3.

The power storaged in piezoelectric actuator return to electrical source, which can reduce the wastage of circuit and enhance the efficiency of energy use. The equivalent charge and discharge circuit is shown as Fig.2.

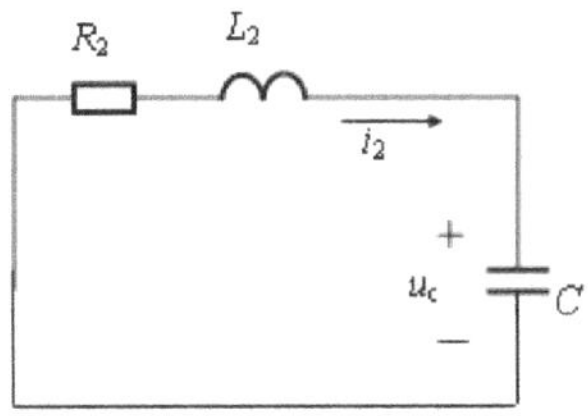

Fig. 2. The equivalent charge and discharge circuit

According to Fig.2, the equivalent equation of charge and discharge circuit is[4]:

$$U_1 = L_2 \frac{di_2}{dt} + R_2 i_2 + u_c \tag{1}$$

$$L_2 \frac{di_2}{dt} + R_2 i_2 + u_c = 0 \tag{2}$$

$$i_2 = C \frac{du_c}{dt} \tag{3}$$

The inductance L2 have great influence on the time and current magnitude in charge and discharge process, except that, decide the select of IGBT. So, the calculation is conducted for inductance L2 using MATLAB/SIMULINK. Through calculation, the equivalent resistance R2=3 Ω , the inductance L2=20μH. Using the values of R2 and L2, the time of charge and discharge process is less than 0.1ms, the max value of charge and discharge current $i_2(t)$ is 48A. Based on the calculation, the IGBT of 60A, 400V is selected.

2.3 Singlechip Control Circuit Design

The Adμc841 singlechip is selected to design in this paper[3]. The singlechip control circuit is the control kernel of the whole drive circuit, which is shown as Fig.3. The singlechip control circuit is composed of three deparments: analog input, A/D conversion and PWM signal output.

2.3.1 Analog Input Gate:P1

The P1 gate of the Adμc841 singlechip is analog input dypass when using ADC, namely ADIN0.

Through sampling, filter and insulation, the voltage U_1 is send to AD port of Adμc841 singlechip to conduct data collection. High frequency, multisampling collection and filter is selected to collect data. It is assured that the sampling signal is not anamorphic by dual manage of software and hardware. In the Fig.3, the 2.7V diode and 5817 diode are protect circuit of the A/D input port.

2.3.2 A/D Conversion

The inner ADC of the Adμc841 singlechip is composed of converters, which is based on routine approach step by step. The speed and digit of the inner ADC is 5μs and 12 respectively. The range of input analog to the converter is 0～VREF. The inner ADC provide the 2.5V fiducial voltage, which is high precision and low excursion. When the input analog is in 0～VREF, the output is binary dada, 1LSB=FS/4096 or 2.5V/4096=0.61mV。

2.3.3 PWM Output Passage

The output wave of PWM is produced by software, and three groups of the PWM signal is outputed from I/O port:P2.7, P2.6, P2.5. After cushion by 74HC540, the PWM signal is given to drive circuit to work IGBT. In the Fig.3, the MAX707 is restoration circuit, and have watchdog function.

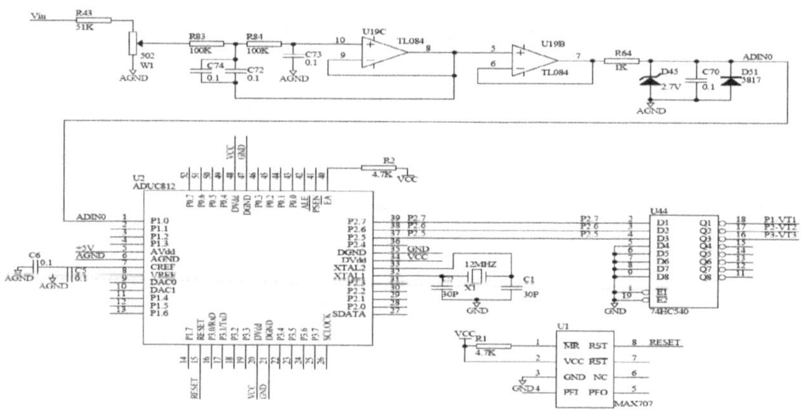

Fig. 3. The control circuit of Adμc841 singlechip

2.4 Control Signal Amplify Circuit Design

The main function of the control signal amplify circuit is to amplify the PWM signal coming from singlechip to drive VT_1, VT_2 and VT_3.

It can be learn from Fig.1 that VT_1 and VT_3 are common ground, VT_2 is suspendible. However, VT_2 and VT_3 can be regard as a leg of bridge, and VT_2 is high end and VT_3 is low end.

In order to simplify the circuit design, two pieces of high voltage suspend drive chips are selected, one of which drive VT_2 and VT_3, the other drive VT_1. The particular circuit is shown in Fig.4 and Fig.5. As shown in the Fig.4 and Fig.5, P1-VT_1, P2-VT_2 and P3-VT_3 is PWM signal from singlechip. Gi-VT_i and Si-VT_i (i=1,2,3) are given respectively to G pole and S pole of IGBT. R_{Gi}(i=1,2,3) is the corresponding resistance of the IGBT, whose value is 30 Ω. C_{11}, C_{12}, C_{13} and C_{14} is the filt capacitance for power supply, whose parameter is 10μF and 25V.

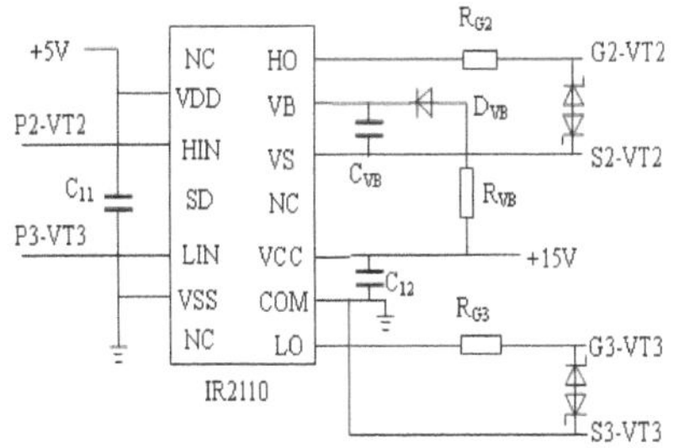

Fig. 4. Use IR2110 drive VT_2 and VT_3

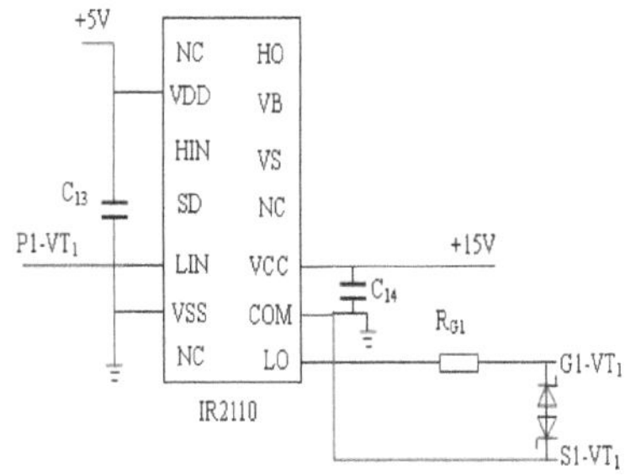

Fig. 5. Use IR2110 drive VT_1

3 Experimentation

The testing drive circuit of piezoelectric actuator is established according to segmental circuite design. The voltage and current of the piezoelectric actuator is observed by Agilent oscillograph. The time and the magnitude of the charge and discharge current are registered. In the test, the piezoelectric actuator is actived by the 180 V drive voltage and the 2.6ms pulse-width.The measured ascend graph of the drive voltage is shown in Fig.6 and the descend graph of the drive voltage is shown in Fig.7. The current graph in charge and discharge process is shown in Fig.8 and Fig.9. From the experimental results we can learn that the time of drive voltage ascend is 0.17ms, and the time of drive voltage descend is 0.14ms, the peak value of the current is 12A.

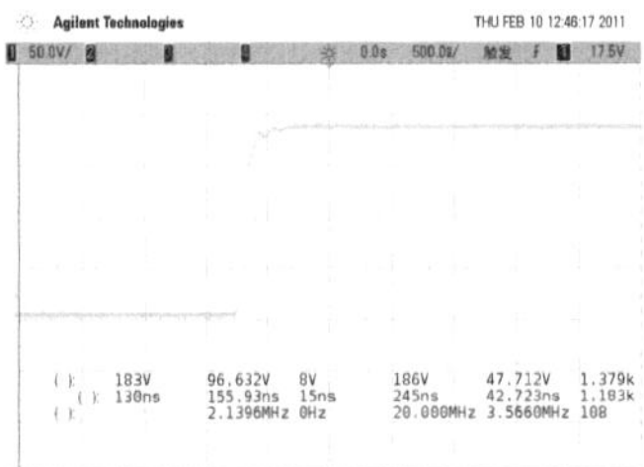

Fig. 6. The ascend graph of the drive voltage

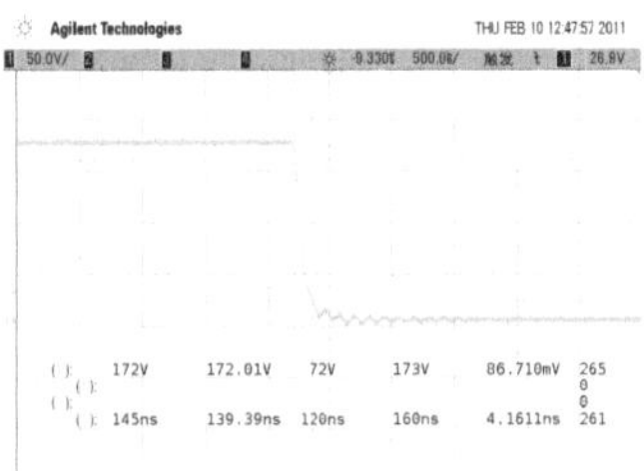

Fig. 7. The descend graph of the drive voltage

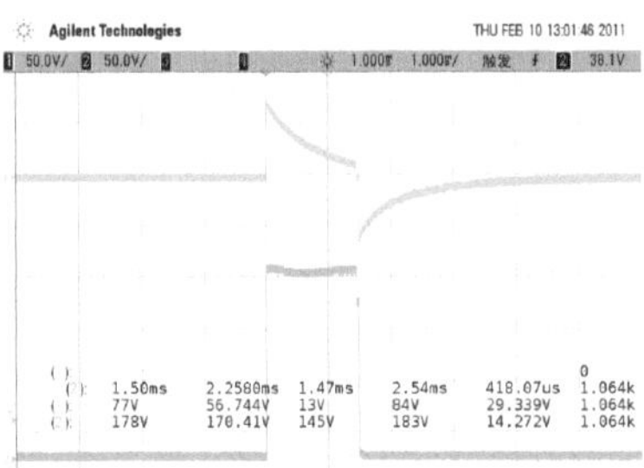

Fig. 8. The current graph of the drive voltage

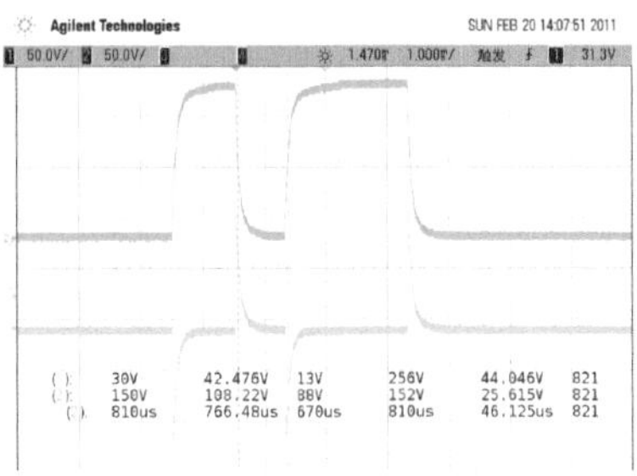

Fig. 9. The current graph of the drive voltage

4 Conlusion

Based on PWM control, a type of drive circuit, which is used for dynamic application of piezoelectric actuator is developed, using IGBT and singlechip. The experimental results indicate that the response time of drive circuit is less than 0.17ms, the peak value of chage current is 12A and the frequency is 100Hz. The drive circuit can meet the dynamic requests of piezoelectric actuator.

References

1. Li, F.: A New Power Supply for Piezoelectric Ceramic Based on PA85. Piezoelectrics & Acoustooptics 27(4), 392–394 (2005)
2. Michael Jacob, J., Jiang, X.: Power Electronics: Principles & Applications (2005)
3. Li, G., Lin, L., Su, T.: Easy Learning & Developing ADuC841 of Hign Performance SoC (2006)
4. Wang, G.: The Basis of Modern Electronic Circuit Applications (2006)

Interleaved Buck-Boost Converter with a Wide Conversion Ratio

Hyun-Lark Do

Department of Electronic & Information Engineering,
Seoul National University of Science and Technology, Seoul, South Korea
hldo@seoultech.ac.kr

Abstract. An interleaved buck-boost converter with a wide conversion ratio is proposed in this paper. Two converter cells are used in the porposed converter. One is a single-ended primary inductor converter (SEPIC) and the other is a Cuk converter. Since both converter cells have an inductor and a switch, the input current of the proposed converter has low current ripple by utilizing an interleaving technique. The SEPIC converter cell provides a positive step up/down output and the Cuk converter cell provides a negative step up/down output. Therefore, the proposed converter has also step up/down capability and a wide conversion ratio. The proposed converter has a low input current ripple, step up/down capability, and a wide conversion ratio. Theoretical analysis and performance of the proposed converter were verified on an experimental prototype operating at 100 kHz switching frequency.

Keywords: Buck-boost converter, SEPIC converter, Cuk converter, interleaving technique.

1 Introduction

Recently, an interleaved technique has been vastly used due to its advantages well introduced in [1]. One of major advantages is low current ripple. The current ripple is an important factor in many applications. It is because large current ripple may shorten the lifetime of the sources and filter capacitors and induce the use of additional filters. Especially in the fuel cell systems, reducing the input current ripple is very important because the large current ripple shortens fuel cell lifetime as well as decreasing performances [2].

In this paper, an interleaved buck-boost converter with a wide conversion ratio is proposed in this paper. The proposed converter has features such as step up/down capability, low input current ripple, and a wide conversion ratio. Input current is equally shared by two converter cells of a SEPIC converter cell and a Cuk converter cell and has small ripple component due to interleaving operation of two converter cells. Since each converter cell has step up/down capability, the proposed converter also has step up/down capability. The SEPIC converter cell provides positive output and the Cuk converter cell provides negative output. Therefore, the conversion ratio of the proposed converter is two times wider than the conventional buck-boost converter.

D. Jin and S. Lin (Eds.): Advances in EECM Vol. 1, LNEE 139, pp. 367–370.
springerlink.com

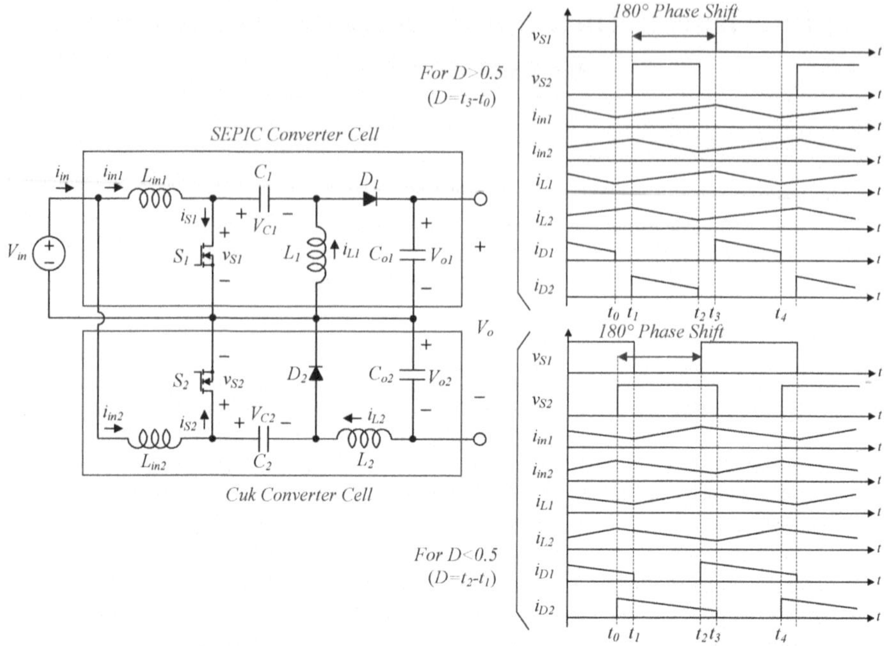

Fig. 1. Circuit diagram and key waveforms of the proposed converter

2 Analysis of the Proposed Converter

Fig. 1 shows the circuit diagram and key waveforms of the proposed converter. To simplify the steady-state analysis, it is assumed that the voltages across the capacitors C_1, C_2, C_{o1}, and C_{o2} are constant and all the semiconductor devices are ideal. Two converter cells are operated at the same frequency but the phases are displaced to each other by 180°. The operation principle of each converter cell does not change according to the duty cycle D. Only to show that the switch voltages and diode currents are overlapped for $D<0.5$, key waveforms in both cases of $D<0.5$ and $D>0.5$ are shown in Fig. 1. Since the operation for $D<0.5$ is basically identical to the operation for $D>0.5$, the operation principle of the proposed converter is provided only for $D>0.5$.

2.1 Operation Principle

The operation in one switching period T_s can be divided into four modes.

Mode 1 [t_0, t_1]: At t_0, the switch S_1 is turned on. Then, D_1 is turned off. Since V_{in} is applied to L_{in1}, the current i_{in1} increases linearly with the slope of V_{in}/L_{in1}. With the turn on of S_1, V_{C1} is applied to L_1. Then, i_{L1} increases linearly with the slope of V_{C1}/L_1. In this mode, S_2 is on D_2 is off. Therefore, i_{in2} increases linearly with the slope of V_{in}/L_{in2} and i_{L2} increases linearly with the slope of $(V_{C2}-V_{o2})/L_2$.

Mode 2 [t_1, t_2]: At t_1, the switch S_2 is turned off. Then, D_2 is turned on. Since $-(V_{C2}-V_{in})$ is applied to L_{in2}, the current i_{in2} decreases with the slope of $-(V_{C2}-V_{in})/L_{in2}$. Since $-V_{o2}$ is applied to L_2, the current i_{L2} decreases linearly with the slope of $-V_{o2}/L_2$. The operation of the SEPIC converter cell does not change in this mode.

Mode 3 [t_2, t_3]: At t_2, the switch S_2 is turned on. Then, D_2 is turned off. Therefore, i_{in2} increases linearly with the slope of V_{in}/L_{in2} and i_{L2} increases linearly with the slope of $(V_{C2}-V_{o2})/L_2$. Similar to mode 2, the operation of the SEPIC converter cell does not change in this mode.

Mode 4 [t_3, t_4]: At t_3, S_1 is turned off and D_1 is turned on. Since $-(V_{C1}+V_{o1}-V_{in})$ is applied to L_{in1}, the current i_{in1} decreases linearly with the slope of $-(V_{C1}+V_{o1}-V_{in})/L_{in1}$. Also, i_{L1} decreases with the slope of $-V_{o1}/L_1$. The operation of the Cuk converter cell does not change in this mode.

2.2 Design Parameters

By applying volt-second balance law to the voltage waveforms across the inductors, the capacitor voltages can be easily obtained by $V_{C1}=V_{in}$, $V_{C2}=V_{in}/(1-D)$, $V_{o1}=V_{o2}=DV_{in}/(1-D)$. Since the output voltage V_o is the sum of V_{o1} and V_{o2}, the voltage conversion ratio M of the proposed converter is given by

$$M=\frac{V_{o1}+V_{o2}}{V_{in}}=\frac{2D}{1-D} \quad . \tag{1}$$

From (1), it can be seen that the conversion ratio of the proposed converter is two times wider than the conventional buck-boost converter. Each converter cell can be designed according to the design criteria of the conventional SEPIC and Cuk converters. The inductor current ripples Δi_{in1} and Δi_{in2} are given by

$$\Delta i_{in1}(=\Delta i_{in2})=\frac{V_{in}DT_s}{L_{in}} \quad , \tag{2}$$

where $L_{in}=L_{in1}=L_{in2}$.
The input current ripple Δi_{in} is given by

$$\Delta i_{in}=\frac{V_{in}|2D-1|T_s}{L_{in}} \quad . \tag{3}$$

3 Experimental Results

A prototype is implemented with specifications of V_{in}=48V, V_o=100V ($V_{o1}=V_{o2}$=50V), P_o=120W. The circuit parameters are $L_{in1}=L_{in2}$=240uH, $C_1=C_2$=13.2uF, $C_{o1}=C_{o2}$=220uF, $L_1=L_2$=120uH, f_s=100kHz. The experimental waveforms of the prototype are shown in Fig. 2. The measured duty cycle is 0.51. The voltage stresses of S_1 and S_2 is around 100V. It can be seen that the experimental waveforms agree with the theoretical waveforms and the analysis. Due to the interleaved operation, the ripple component of the input current is significantly reduced. The measured efficiency curve is also shown in Fig. 2. The proposed converter exhibits the maximum efficiency of 92.5% at 100W. The power consumed in the control circuit is ignored in the efficiency curve.

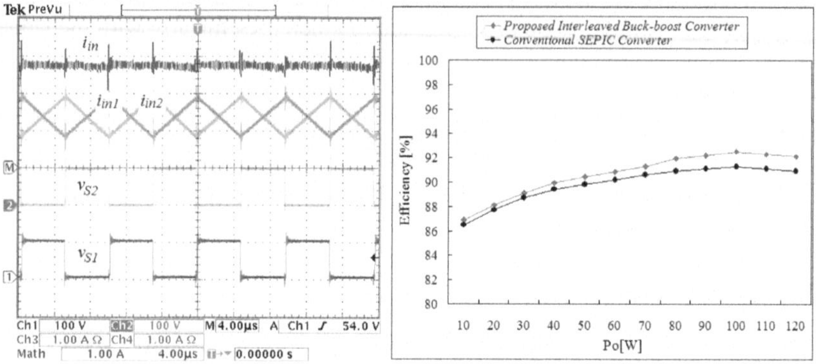

Fig. 2. Measured input current waveforms and efficiency curve

4 Conclusion

An interleaved buck-boost converter with a wide conversion ratio has been proposed. It can provide a low ripple input current by utilizing the interleaving technique. Also it has step up/down capability and its conversion ratio is wider than the conventional buck-boost converter. Due to these features, the proposed converter can be a good candidate for non-isolated step up/down DC-DC converter.

References

1. Giral, R., et al.: Interleaved Converters Operation Based on CMC. IEEE Trans. Power Elec. 14, 643–652 (1999)
2. Kong, X., Khambadkone, A.M.: Analysis and Implementation of a High Efficiency, Interleaved Current-Fed Full Bridge Converter for Fuel Cell System. IEEE Trans. Power Elec. 22, 543–550 (2007)

Research on the Incompletely Insuring Condition for Risk-Seeking Insureds

Bao-Long Li[1,2]

[1] The Engineering Management Department at Luoyang Institute of Science and Technology, Luoyang City, 471023, China
[2] Management School of Tianjin University, Tianjin City, 300072, China
libaolong@yahoo.com.cn.

Abstract. Because the incompletely insuring condition deduced from risk-averse insureds cannot stop the insurance crisis happening, there is a need to restudy the incompletely insuring condition under the influence of adverse selection in insurance market. First of all, this paper demonstrated the bilateral information asymmetry on the insurance market. Following that, through the application of Karush-Kuhn-Tucker Theorem, this paper deduced the incompletely insuring condition newly. Finally, this paper draws the conclusion that insurance company should control the compensation amount through the design of insurance contract.

Keywords: Information asymmetry, KKT Theorem, Risk-seeking, The incompletely insuring condition.

1 Introduction

Insurance is a tool to transfer risk, but the risk level is still high enough to create insurance crisis. In terms of minimizing the risk and stabilizing the insurance market, the insureds's behaviors have to be restricted by some rules, e.g. the incompletely (partially) insuring condition. In the insurance literature, as to incompletely insuring condition, most researchers hypothesize that the insureds are risk-averse. According to [1-3], the insurer's expected profit is higher when insureds are risk-averse clients. However, the reality is not like this, and the incompletely insuring condition for risk-averse insureds cannot stop the insurance crisis happening. So, there must be that some insureds are risk-seeking. Therefore, this is a need to study the incompletely insuring condition for risk-seeking insureds.

2 Application of Model

2.1 The Hypothesis of the Model

On the assumption that insurance companies compete in contracts, one risk-seeking individual has a strictly convex and increasing Von Neumann-Morgenstern utility function $u(.)$ ($u'(.)>0$, $u''(.)>0$) depending on wealth level alone and has the state independent initial wealth endowment W. This individual faces a possible loss Y

D. Jin and S. Lin (Eds.): Advances in EECM Vol. 1, LNEE 139, pp. 371–375.
springerlink.com

($0<Y<W$), with probability of loss P ($0<P<1$). The insurance company would recompense the loss of this person by the price π ($0<\pi<1$). Let X stand for the insurance amount, then the premium on insurance is πX. When compensation takes place, the insurance company would pay X for the loss of this policyholder, and expected compensation of insurance company is PX, due to $(1-P)0+PX=PX$.

In the case of perfect market, this person would seek to be fully insured, which leads to $X=Y$. In addition, the income of the premium of insurance company should be equal to the expected compensation of insurance company, which leads to $\pi X=PX$, namely $\pi=P$. Of cause, perfect market does not exist in economic life, and partial insurance is the spontaneous result of adverse selection. So, the result should be $X<Y$, which means the sum of compensation of the insurance company should be less than the total loss of this person or the amount insured by insurance company should be less than the amount needed by this person.

2.2 The Utility Maximization Framework for 'Risk-Seeking'

According to the above demonstration, it is easy to see that adverse selection leads to $X<Y$. After knowing this fact, the next action adopted by the policyholder would be the interest of this paper. Due to $X<Y$, a utility maximization framework is implemented to model the choice faced by this policyholder, which is as following:

$$\begin{gathered} \max \mathrm{E}\,[Pu(W-Y-\pi X+X)+(1-P)u(W-\pi X)] \\ \text{s.t. } X<Y \end{gathered} \tag{1}$$

Then according to the KKT Theorem, the optimality equation can be written as:

$$L=Pu(W-Y-\pi X+X)+(1-P)u(W-\pi X)-\lambda(X-Y) \tag{2}$$

Let

$$\partial L/\partial X=(1-\pi)Pu'(W-Y-\pi X+X)-\pi(1-P)u'(W-\pi X)-\lambda=0 \tag{3}$$

And

$$\lambda(X-Y)=0 \tag{4}$$

In fact, due to the constrained condition $X<Y$, equation (4) could be replaced by $\lambda=0$. In this case, equation (3) is rewritten as:

$$(1-\pi)Pu'(W-Y-\pi X+X)=\pi(1-P)u'(W-\pi X) \tag{5}$$

Obviously, $W-Y-\pi X+X<W-\pi X$, also considering $u'(.)>0$ and $u''(.)>0$, those imply $0<u'(W-Y-\pi X+X)<u'(W-\pi X)$, which is equivalent to $(1-\pi)P>\pi(1-P)$. Solving the inequation $(1-\pi)P>\pi(1-P)$, the result can be got as following: $P>\pi$. $P>\pi$ means that the insurance price is lower than the possibility of this person's loss in the case of risk-seeking. So, the income of the premium of insurance company is lower than the expected compensation ($\pi X<PX$), which implies that insurance company would make a loss on business with 'risk-seeking'. However, as mentioned earlier, the income of insurance premium should be equal to the expected compensation, i.e. $\pi X_1=PX_2$. X_1

represents the insurance amount insured for the policyholder; X_2 represents the compensation amount paid by insurance company. On account of $P>\pi$, X_1 must be bigger than X_2 (i.e. $X_1>X_2$), which means that the insurance amount insured for policyholder could be bigger than the compensation amount paid by insurance company.

Generally speaking, in the case of 'risk-seeking', $X<Y$ means the sum of compensation of the insurance company should be less than the total loss of this person, and $P>\pi$ means the compensation amount paid by insurance company should be less than the insurance amount insured for policyholder. Therefore, the incompletely insuring condition should be symbolized as $X_2<X_1<Y$. X_2 represents the compensation amount paid by insurance company; X_1 represents the insurance amount insured for policyholder; Y represents the total loss of this policyholder (or insurance amount needed by policyholder).

In reality, the situation is far more complex. People maybe 'risk-averse' and 'risk-seeking', and the transition between 'risk-averse' and 'risk-seeking' always happen. In addition, insurance company may find people with different level of 'risk-averse' or with different level of 'risk-seeking'. The above statements provide great complexity and difficulty for analysis. Because, it is extremely difficult for insurance company to treat 'risk-averse' and 'risk-seeking' respectively, in order to resist vast risk, insurance company should adopt the more stringent incompletely insuring condition (i.e. $X_2<X_1<Y$) deduced from 'risk-seeking', which means that insurance company need not only restrict the quantity of insurance bought by this person (i.e. $X<Y$), but also restrict the quantity of insurance compensated by the insurance company (i.e. $X_1>X_2$).

3 The Analysis of Decision Making

According to section 2, it is easy to see that the policyholder can only decide X_1 and the insurance company can only decide X_2. It could be viewed as that both the policyholder and the insurance company have more information in different way. Therefore, the insurance market shows us a structure of bilateral information asymmetry, and the policyholder would react to the incompletely insuring condition by his information advantage in order to deal with the information advantage of the insurance company. Based on this understanding, we rewrite the model in section 2, in order to see the influence of the incompletely insuring condition on the policyholder or (and) the insurance company.

3.1 The Analysis of Decision Making for Policyholder

Under the incompletely insuring condition, a utility maximization framework is implemented to model the decision making faced by this policyholder, which is as following:

$$\begin{gathered}\max \mathrm{E}\left[Pu(W-Y-\pi X+X)+(1-P)u(W-\pi X)\right] \\ \text{s.t.}\quad X_2<X_1,\quad X_1<Y\end{gathered} \tag{6}$$

The result could be got as following:

$$-Pu'(W-Y-\pi X_1+X_2)=(1-P)u'(W-\pi X_1) \tag{7}$$

Due to $u'(.)>0$, $0<P<1$, $-Pu'(W-Y-\pi X_1+X_2)<0$, but $(1-P)u'(W-\pi X_1)>0$. Obviously, equation (7) cannot be founded, which means that this policyholder cannot realize utility maximization.

3.2 The Analysis of Decision Making for Insurance Company

Under the incompletely insuring condition, a utility maximization framework is implemented to model the decision making faced by insurance company, which is as following:

$$\begin{gathered}\max\ E[Pv(W_1+\pi X_1-X_2)+(1-P)v(W_1+\pi X_1)]\\ \text{s.t.}\quad X_2<X_1\end{gathered} \tag{8}$$

v represents the utility function for insurance companies. When insurance company is risk-neutral, $v''(.)=0$; When insurance company is risk-averse, $v''(.)<0$. W_1 represents the state independent initial wealth endowment for insurance company. X_2 represents the compensation amount paid by insurance company and X_1 represents the insurance amount insured for policyholder. The result could be got as following:

$$Pv'(W_1+\pi X_1-X_2)=0 \tag{9}$$

Due to $v'(.)>0$, $0<P<1$, equation (9) cannot be founded, which means that insurance company cannot realize utility maximization.

3.3 The Analysis of Decision Making for Both Policyholder and Insurance Company

According to A and B in section 3, conclusion could be drawn: on insurance market, under the incompletely insuring condition, the policyholder and insurance company cannot realize utility maximization respectively, which would minimize the damage of information asymmetry to both the policyholder and insurance company. In this case, a utility maximization framework could be implemented to model the decision making faced by both the policyholder and insurance company, which is as following:

$$\begin{gathered}\max E\ [Pu(W-Y-\pi X+X)+(1-P)u(W-\pi X)\\ +Pv(W_1+\pi X_1-X_2)+(1-P)v(W_1+\pi X_1)]\\ \text{s.t.}\quad X_2<X_1,\ X_1<Y\end{gathered} \tag{10}$$

The result could be got as following:

$$\frac{v'(W_1+\pi X_1)}{u'(W-\pi X_1)}=1 \tag{11}$$

Equation (11) implies that, when compensation of insurance company does not happen or loss of policyholder does not happen, the marginal utility of the policyholder and insurance company is equal. Also, as long as equation (11) is founded, the policyholder and insurance company can realize utility maximization together. If the insurance amount insured for the policyholder is X_{1j}, equation (11) could be rewritten as:

$$\frac{v'(W_1 + \pi X_{1j})}{u'(W - \pi X_{1j})} = 1 \tag{12}$$

When j=1 or j=2, equation (12) could be rewritten as:

$$\frac{v'(W_1 + \pi X_{11})}{u'(W - \pi X_{11})} = \frac{v'(W_1 + \pi X_{12})}{u'(W - \pi X_{12})} = 1 \Rightarrow \frac{v'(W_1 + \pi X_{11})}{v'(W_1 + \pi X_{12})} = \frac{u'(W - \pi X_{11})}{u'(W - \pi X_{12})} \tag{13}$$

Obviously, equation (13) is the Pareto optimum condition. Therefore, under information asymmetry, due to the incompletely insuring condition founded by this paper, the Pareto optimum could be realized.

According to above demonstration, on insurance market, under the incompletely insuring condition founded by this paper, because the policyholder can only decide X_1 and the insurance company can only decide X_2, the policyholder and insurance company cannot realize utility maximization respectively. However, the policyholder and insurance company can realize utility maximization together, and also the Pareto optimum could be realized.

4 Conclusion

In insurance literature, the incompletely insuring condition receives little attention. The reality is that incompletely insuring condition for risk-averse insureds cannot stop the insurance crisis happen. Therefore, this is a need to study the incompletely insuring condition for risk-seeking insureds. This paper provides a brand new thought of the incompletely insuring condition. Based on the Karush-Kuhn-Tucker Theorem, the analysis reveals that the incompletely insuring condition should be expressed as $X_2<X_1<Y$, in which, X_2 represents the compensation amount paid by insurance company, X_1 represents the insurance amount insured for policyholder, and Y represents the total loss of this policyholder or insurance amount needed by policyholder. Under this incompletely insuring condition founded by this paper, because the policyholder can only decide X_1 and the insurance company can only decide X_2, the policyholder and insurance company cannot realize utility maximization respectively. However, the policyholder and insurance company can realize utility maximization together, and also the Pareto optimum could be realized.

References

1. Iwaki, H., Kijima, M., Morimoto, Y.: An economic premium principle in a multiperiod economy. Insurance: Mathematics and Economics 28, 325–339 (2001)
2. Viaene, S., Veugelers, R., Dedene, G.: Insurance bargaining under risk aversion. Economic Modelling 19, 245–259 (2002)
3. Li, B.-L., Wang, X.-Q., Fan, Z.-Q.: The bilateral information asymmetry on insurance market. In: Proceedings 2009 IEEE 16th International Conference on Industrial Engineering and Engineering Management, pp. 750–752 (2009)

Panoramic Virtual Field Geological Information System Based on RIA

Chunyan Deng[1], Linfu Xue[2,*], Jinxin He[2], Wenqing Li[2], and Zhiguo Zhou[3]

[1] College of Computer Science and Technology, Jilin University, Changchun, China, 130012
dengcy@jlu.edu.cn
[2] College of Earth Sciences, Jilin University, Changchun, China, 130061
xuelinfu2001@yahoo.com.cn, he_jinxin@126.com,
liwenqingboy@163.com
[3] School of Computer Science and Information Technology, Northeast Normal University, Changchun, China, 130024
zhouzg281@nenu.edu.cn

Abstract. The technology of Virtual Reality (VR) could improve the visualization and the interactivity of the existing Geological Information System effectively, but the real-time performance becomes the bottle-neck of such integration system because of both the large and complex geoscience data and the huge computation of constructing virtual environment. Rich Internet Applications (RIA) provides a new solution to this problem. The author analyzes the opportunities brought by these two computer technologies and implements a three-dimension panoramic virtual field Geological Information System as a demonstration system for the XingCheng field geological teaching and studying base of Jilin University, which could manage, analyze and manipulate the panoramic geological images and related geoscience data. Practice shows that it not only improves the visualization and the interactivity of the existing Geological Information System but also solves the bottle-neck of real-time performance efficiently.

Keywords: IBR, RIA, Geoscience Data, Geological Information System.

1 Introduction

Geological Information System is an interdiscipline of computer science, geology, geography and cartography, etc. Geological Information System has developed rapidly in recent years, but with the rising requirements of the geological workers, now more and more users make demands for high visualization and interactivity of the system. The technology of Virtual Reality (VR) could improve those performances of the existing system, but the real-time performance becomes the new bottle-neck because of the large and complex geoscience data and the huge computation when constructing virtual environment. A new technology in the computer field, Rich Internet Applications (RIA), offers a new solution to the bottle-neck and makes it possible to

* Corresponding author.

integrate VR to the existing system. In this thesis, the author studies a panoramic virtual field Geological Information System based on RIA in order to explore an efficient solution to above bottle-neck. The contribution could provide technology support for the integration of VR and Geological Information System. It also has great significances on implementing the share of geoscience data and promoting the infomationization progress of natural resource [1].

2 Research Status and the Related Technologies

In recent years, Geological Information System has developed greatly. It has implemented the share of geoscience data and offered decision support for the geological workers to some degree. But there are still some disadvantages such as visualization, interactivity and the real-time performance in the existing system. The development of two new computer technologies, which are VR and RIA, brings great opportunities to the further development of the Geological Information System.

VR is a new integrated information technology of recent years, which enable users to observe and manipulate the virtual world interactively through the interactive User Interface (UI) implemented with computer [2]. Because of the features of immersion, interactivity and imagination, VR has been applied in many fields such as aeronautics and astronautics, machinery design, scientific computing, film amusement, chemical medicine and military training, etc. With the presentation of the idea of Digital Earth [3] and the implementation of digital China [4], VR has been applied in the filed of geology to some extent. For example, three-dimension geological modeling [5] and the visualization of geological information [6]-[7] have both improved the visualization of the existing system greatly. In theory, the technology of VR could also improve the visualization of the existing Geological Information System. However, because of its owner data features of the Geological Information System, the real-time performance becomes the bottle-neck when constructing VR, which has prevented the developing progress of the integration of VR and the Geological Information System. There are no related studies at home and abroad until now.

There are three methods to implement VR, which are Model-Based Method [8], Image-Based Rendering (IBR) [9]-[10] and Image-Based Modeling and Rendering (IBMR). The relationship of these three methods [11] is displayed in Fig. 1. Considering about the features of geoscience data and the function requirements, the author finds that using panoramic images of IBR to build the system is a feasible and efficient solution to integrate VR to the existing Geological Information System.

The technology of VR could improve the visualization and the interactivity of the existing Geological Information System greatly, but the problem such as real-time performance and the image loading speed caused by constructing VR should be solved firstly. And RIA provides a perfect solution to above question.

RIA is the core technology of Web2.0, and it focuses on combining the responding speed of desktop application with the extension of Web application [12]. It offers high interactivity, flexible user experiences and powerful client application. Because of those huge advantages, RIA has been applied quickly recently. Flex and Asynchronous JavaScript and XML (Ajax) are both the main technology of RIA. Flex has great advantages on interactivity and interface displaying, and Ajax could shorten responding time through its asynchronous technology so as to improve user experiences. If the advantages of both could be combined to solve the problem of slow image loading speed caused by constructing VR, the visualization and interactivity of the existing Geological Information System could be improved perfectly.

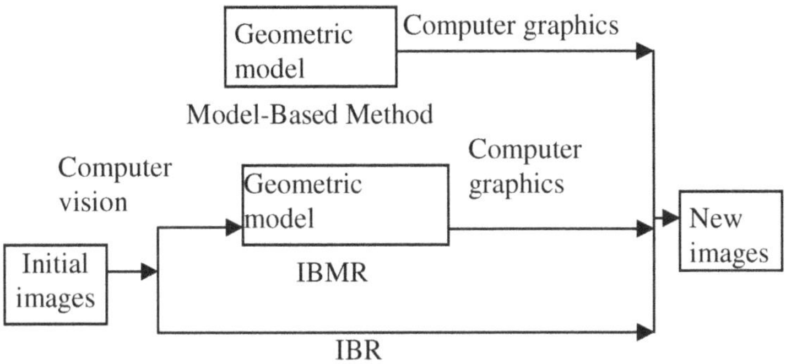

Fig. 1. The relationship of three methods of implementing VR

3 System Design

To illustrate the integration solution, the author implements a three-dimension panoramic virtual field Geological Information System as a demonstration system for the XingCheng field geological teaching and studying base of Jilin University. The system design is described as following.

3.1 The Basis Construction Procedure of the Integration System

Firstly, the initial images acquired from field are uploaded to the system and preprocessed to get panoramic images. Then the programmers use these panoramic images to construct three-dimension virtual field environment. Finally, appropriate organization methods are adopted to manage the data of scenes and the panoramic virtual field Geological Information System is completed.

3.2 Key Technologies of the System

(1) Data acquisition and image preprocessing
The XingCheng field geological teaching and studying base of Jilin University lies in the Diaoyutai beach of the east in the XingCheng city of Liaoning Province, and it locates at 40º37´N and 120º42´E, the geographical and geological conditions along which are complex. So the Car-Carrying method and the manual work method are both adopted.

In the area having convenient communications, PIAS-1 [13], a Car-Carrying acquisition device developed and operated by the project team, is adopted to capture initial images, geographical and geological information at the speed of 30km/h. In the area where the geographical and geological conditions are very complex, geological workers collect data with a handheld device along the linear teaching route, and the space is set to 20 meters in the teaching area and 100 meters out of the teaching area. The space could be adjusted dynamically.

Because the data of initial images is very large in the system, more than 10 photos are needed to get one panoramic image if ordinary camera lens are adopted. It would overburden the job of capture and the latter image splicing. So the author adopts the fish-eye lens to decrease the amount of initial photos. PTGUI is used to splice the spherical panoramic images and Pano2VR is used to get the panoramic image for cube model, which is the model of three-dimension virtual field environment.

(2) Constructing three-dimension virtual environment

When constructing the virtual environment, the real-time performance and the image loading speed are as important as the visualization and the interactivity. There are three modes to construct panoramic virtual environment, which are the cylinder mode, the spherical mode and the cube mode. The cylinder mode has some limitations on the vision range in vertical direction, and there is distortion in the spherical panorama mode. However, the cube mode overcomes the shortages of both them. Furthermore, the cube mode has good regularity and it is easy to program. Considering about that the project team not only need to implement a three-dimension virtual environment but also need to improved the real-time performance through RIA, which need lots of programming, so the cube mode is selected as the implementation mode of the system in this paper.

The interface of the system is developed with FLEX [14]-[15], and the three-dimension panoramic virtual field environment is developed on the basis of PV3D [16], which is an open source software. Its mapping arithmetic is improved to implement partial texture mapping in order to diminish the huge computation. And the technology of Ajax is integrated to the system to implement the asynchronous image loading so as to reduce the waiting time of users and enhance the user experiences. Now the website of Google provides the maps application programming interface (Google Maps API) to use the tool [17], which lightens the programming jobs greatly.

(3) Managing the data of scenes

In the system, the scenes are organized according to the filed teaching routes. Each route includes several nodes, and each node is designed to a scene in the virtual environment. The scene is implemented with a panoramic image, and users could roam horizontally or vertically in different virtual scene and even could zoom rapidly and nimbly. To implement it, XML is adopted to organize the scene data. Each XML file manages a certain route data, and the related data such as the geographical coordinate, geological information and the location of the panoramic image are defined in it, then the application accesses the corresponding XML file to implement different virtual scene.

4 The Implementation of System

The system has been applied to the field practice teaching system of College of Earth Sciences of Jilin University and run well.

5 Conclusions

The technology of VR could improve the visualization and interactivity of the existing Geological Information System greatly, but the real-time performance caused by constructing virtual environment becomes the bottle-neck of the new system. The author implements a three-dimension panoramic virtual field Geological Information System based on RIA, which provides a new solution to the integration of VR and Geological Information System. Theory analysis and the practice application prove that the solution of constructing the three-dimension panoramic virtual field Geological Information System based on RIA is feasible, efficient and innovative. In the respect of the technology, this contribution could offer the technology support for the integration of VR and Geological Information System and promote the interdisciplinary studies. In the respect of the application, it could set up a panoramic field geological database and lay the foundations for the further resource investigation and cooperation. Furthermore, because panoramic images have higher resolution than remote sensing images, the combination of panorama and remote sensing has great application potential in many filed such as natural resource investigation, universal geology education, the evaluation of geological environment and geological hazard forecast, etc.

Acknowledgment. This work was supported by the Foundation for Science Front and Crossing Discipline of Jilin University (Jinxin He, 2009) and the Scientific and Technical Development Plan of Jilin Province of China (Zhiguo Zhou, Grant No.201101003).

References

1. Jiang, Z.: Present Status and Features of ormationization of the Full Process of Regional Geological Survey at Home and Abroad. Geological Bulletin of China 27(7), 956–964 (2008)
2. Zhao, M.: Study of Algorithm in Generating Image-based Cylinder Panoramic Image, p. 4. Guizhou University, Guiyang (2006)
3. Wu, L., Liu, Y., Zhang, J., Ma, X., Wei, Z., Tian, Y.: Geographical Information System—theory, method and application, p. 7, pp. 289–299. Science Press, Beijing (2003)
4. Digital China, `http://baike.baidu.com/view/30603.htm?fr=ala0_1`
5. Yang, Y.: Application of 3D geological modeling technology in buried hill reservoir research—Taking Caotai Buried Hill for example. Complex Hydrocarbon Reservoirs 2(3), 40–43 (2009)
6. Song, H., Wang, S., Liu, J., Zhong, D.: Three-dimensional visualized analysis and application of geological information of large underground cavern groups. Advances in Science and Technology of Water Resources (6), 80–84 (2007)

7. Chen, Z., Wang, L., Bi, L.: Application and Development of 3D Visual Model For Deposit. Modern Mining (9), 59–62 (2009)
8. Rogers, D.F.: Procedural elements for computer graphics, 2nd edn., pp. 20–30. MacGraw-Hill Companies, New York (1998)
9. Yang, H.: The Research of Image-based Rendering for Rendering of Virtual Scene, p. 20. National University of Defense Technology, Changsha (2005)
10. Yang, W.: The research of virtual reality system based on panoramas, p. 31. Guizhou University, Guiyang (2007)
11. Tang, J.: Omnidirectional Image Based on IBR, pp. 5–30. Central South University, Changsha (2002)
12. Hu, J.: Flex Tutorial, pp. 1–5. China Machine Press, Beijing (2009)
13. PIAS, http://zhuanli.baidu.com/pages/sipo/20082007/24/7b67427d340d201cde913daa6c56e531_0.html
14. Qiu, Y.: FLEX First step: FLEX 2 Application Based on ActionScript 3.0, pp. 311–319. Tsinghua University Press, Beijing (2007)
15. Noble, J., Anderson, T.: Flex 3 Cookbook, pp. 20–34. Publishing House of Electronics Industry, Beijing (2009)
16. Lively, M.: Professional Papervision3D, p. 1. John Wiley & Sons (2010)
17. Google Maps API Web Services, http://code.google.com/intl/zhCN/apis/maps/documentation/directions/

mGlove: Enhancing User Experience through Hand Gesture Recognition

Yong Mu Jeong[1], Ki-Taek Lim[1], and Seung Eun Lee[2,*]

[1] SoC Research Center, Korea Electronics Technology Institute
Sungnam-si, Gyeonggi-do, Korea
{ymjeong,limkt}@keti.re.kr
[2] Dept. of Electronic & Information Technology
Seoul National University of Science and Technology, Seoul, Korea
seung.lee@seoultech.ac.kr

Abstract. In this paper, we propose hand gesture recognition equipment (mGlove) for a user interface which provides more effective experience compare to conventional devices such as a keyboard and a mouse. Communication module establishes the communication channel between mGlove and a host PC. Experimental results demonstrate the feasibility of our proposal for enhancing user experience by gaining full control of an avatar through hand gesture recognition.

Keywords: Gesture Recognition, User Interface, zigbee.

1 Introduction

Over the last decade, smart interface becomes a major issue that is how to support more natural and immersive user interface. The user interface has evolved from simple controllers such as a pointing device or a button, to a tangible input device. The tangible interface [1] could be able to interact with the virtual space through a physical behavior. Emerging smart interface applications such as gesture recognition, motion tracking, and speech recognition are quickly entering the mobile domain and realistic game systems.

Gesture recognition technology is applied to the game controller such as Nintendo Wii [2] or Microsoft Kinect [3]. Nintendo Wii controller, wiimote, could perceive the user's gesture with infrared emitter in monitor and infrared camera in it. And wii remote plus controller is possible to track the location of controller with acceleration sensor. Microsoft's Kinect consists of the infrared projector and infrared camera. Kinect is able to track user's body based on vision processing.

In order to applicable to realist user interface, a gesture recognition system requires fast response time with adequate accuracy. Previous studies on hand gesture recognition are classified into two categories: 1) optical method based on image processing with a camera, and 2) mechanical method based on measuring joint motion

* Corresponding author. Tel. no: +82-2-970-9021

D. Jin and S. Lin (Eds.): Advances in EECM Vol. 1, LNEE 139, pp. 383–386.
springerlink.com

with a data glove. Due to the accurate classification capability and fast response time of data glove based gesture recognition compared to the vision based recognition, it has been adopted to many systems.

In this paper, we propose hand gesture recognition equipment (mGlove) to enhance user experience, which provides more effective experience compare to conventional devices such as a keyboard and a mouse. Communication module establishes the communication channel between mGlove and a host PC. Experimental results demonstrate the feasibility of our proposal for enhancing user experience by gaining full control of an avatar with hand gesture.

The rest of paper is organized as following: section 2 describes our mGlove and gesture recognition framework in detail. Section 3 explains the experimental result when applying the proposed system to a real system and we conclude in section 4 by outlining the direction for the future work on this topic.

2 Hand Gesture Recognition

mGlove captures the user's hand gesture by using flex sensor and forwards data to the host PC through Zigbee communication channel. The gesture recognition and management of interaction between recognized gesture and host PC are realized in software.

2.1 mGlove

An Embedded processor (Em357) controls mGlove and establishes communication channel with a host PC. A flex sensor measures angle displacement by detecting changes in resistance and the measurement data is forwarded to gesture recognition unit in order to complete hand gesture recognition. When flex sensor is bended, resistance of flex sensor increases. We implemented two flex sensors in mGlove to control the avatar in game through recognizing two-finger gesture. The battery cell is a 160mAh Li-polymer. Fig.1 shows mockup case of mGlove.

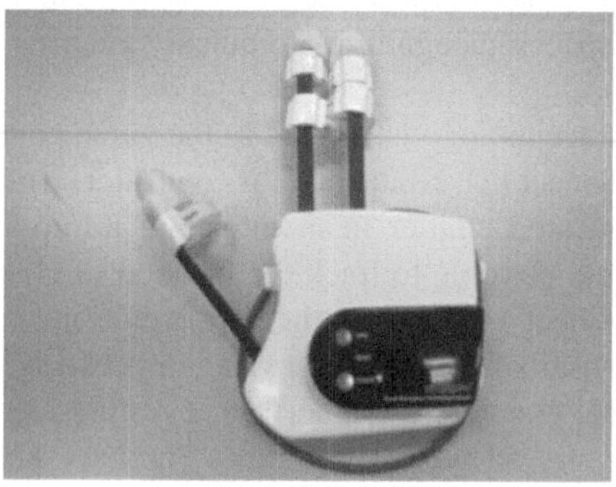

Fig. 1. Mockup case of the mGlove

2.2 Gesture Recognition Module

Gesture Recognition module classifies user's gesture based on collected sensor data and transmits the result to the interaction manager. The gesture recognition module consists of aggregator, signal classifier, sequence tracker and gesture recognizer. The interaction manager controls the game server' I/O. The classification is completed by comparing feature vector with predefined DB. In order to match input feature vectors, we use a brute force match algorithm that exhaustively compares a pair of feature vectors from each gesture based on the Manhattan (a.k.a. L1) distance. It should be noted that several other match algorithms are available, but the brute force match is simple to implement and has sufficient accuracy for the usage model of interest. The gesture recognition module requires the following blocks as shown in Fig. 2:

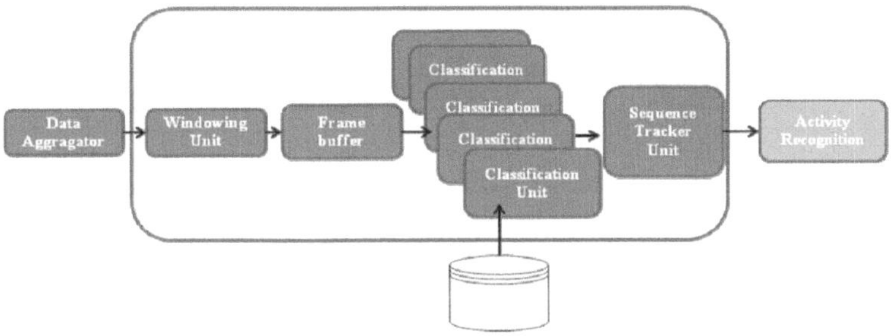

Fig. 2. Diagram of gesture recognition module

Windowing Unit: The continuous signals from the sensors should be divided into smaller granularity for the recognition purpose. This module extracts feature vectors from the acquired sensor signals in the predefined time window which is dynamically adjusted for the efficient hand gesture recognition.

Classification Unit: Classification unit requires computation of the Manhattan distance between each pair of feature vectors (one vector from query data and one form the data base) for the brute force matching. For each feature vector of the query data, performing the distance function on all feature vectors of a data based image give us the rank of candidate gestures for the query data within the time-window. We have multiple classification units corresponding to each sensor.

Sequence Tracker Unit: The sequence and combination of motions from the sensors form the gesture. The sequence tracker unit keeps track the ranked results of the classification units and completes the gesture recognition.

3 Experimental Results

We set up the metaverse server by using opensim [4] and used hippo viewer [5] as a client (See Fig. 3). Interaction manager provide the way to control the avatar in the server based on the result from the gesture recognition module. In order to demonstrate the performance and enhanced user experience, we built an environment adopting the proposed mGlove. The recognized gesture information is translated to the command for the client, hippo viewer, through the interaction manger, realizing the control of the avatar in the host PC only using the gestures. The accuracy and response time of the gesture recognition is suitable to experience the realistic control of the avatar.

Fig. 3. A demonstration of hand gesture recognition using our mGlove

4 Conclusions

In this paper, we proposed hand gesture recognition equipment for enhanced user experience. We accomplished to gain full control of the avatar in the host PC by using the hand gestures, enhancing the user experience. The future work in this area is as follows. We plan to improve the gesture recognition time for the real-time synchronization between a user and an avatar. We also plan to establish bi-direction communication channel to feedback the experience of the avatar to the user. We expect that enhancing the user experience on virtual world such as realist game will bring forth a new spectrum of novel usage models for smart devices, bio-medical equipments, entertainment system for automobile and robotics.

Acknowledgement. This research was supported by R&D Program of MKE (Ministry of Knowledge Economy) in Korea, [10032108, Development of Multiverse Game Platform Technology with Multimodal Interaction].

References

1. Ullmer, B., Ishii, H.: Emerging frameworks for tangible user interfaces. IBM Systems Journal 39(3-4), 915–931 (2000)
2. Nintendo Wii, http://www.nintendo.com/wii
3. Microsoft Xbox Kinect, http://www.xbox.com/en-US/Kinect
4. Opensim, http://opensimulator.org/wiki/Main_Page
5. Hippo viewer, http://mjm-labs.com/viewer/

Dynamic Departure and Landing Scheduling of Military Planes Based on Receding Horizon Optimization

XunQiang Hu[1,2], XiaoFang Xie[1], DongXin Liu[1], and Jie Liang[1]

[1] Naval Aeronautical and Astronautical University, Yantai, 264001
[2] Naval Arms Command Academy, Guangzhou, 510430, China
navalboy@163.com

Abstract. The actual and planned departure and landing time of military planes are never identical exactly. To reduce unfavorable inference on air combat brought by the time difference, this paper designs a dynamic departure and landing scheduling policy based on receding horizon optimization. For plane queue falling into each predictive window, minimal sum of weighted earliness and tardiness.is defined as the objective function. An algorithm based on discrete particle swarm optimization is designed to sequence the plane queque. Results of simulation show the advantage of the method.

Keywords: dynamic scheduling, receding horizon optimization, discrete particle optimization.

1 Introduction

Before an air combat planned departure and landing time are always assigend to each military plane in order to guarantee the best battle result. Unluckily in fact military planes rarely take off or land according to the planned time. It's important to reduce the gap between planed and actual departure/landing time for air force with some scheduling method. Considering at the same airport the departure or landing planes queue maybe gain new member or eliminate old ones at any time, the depature and landing scheduling is a dynamic course [1]. In section 2, the mathmatical model of departure and landing scheduling is built. In section 3, a method integrating receding horizon optimization with discrete PSO is introduced to solve aforesaid model. In section 4, simulation is executed to analyze and validate the model and method. Finally, the conclusion is drawn in Section 5.

2 Departure and Landing Scheduling Model

2.1 Time Window Contraint

Departure and landing time should have a lower bound and an upper bound which define a time window. If a military plane takes off too early, resource (e.g. fuel) may be wasted ineffectively. Contrarily too late departure time usualy leads to failed mission.

The earliest landing time means the plane always flies at its highest speed when it executes the task. The last landing time means the plane can land in safe mode with left fuel [2].

D. Jin and S. Lin (Eds.): Advances in EECM Vol. 1, LNEE 139, pp. 387–394.
springerlink.com

2.2 Wake Vortex Seperation Constrain

According to aerodynamics a plane will produce wake vortex in its flight path. The vortex will bring disturbance to another following plane flying too closely in the same flight path. International Civil Aeronautics Organization (ICAO) has classified all planes into heavy-duty plane, medium-sized plane and light-duty plane according to maximum take-off weight. The standard vortex seperation between two planes is prescribed by ICAO as shown in Table 1.

Table 1. Standard Seperation between Two Planes (Dimension: Second)

Type of Plane		Posterior Plane					
		1	2	3	4	5	6
Anterior Plane	1	60	90	120	90	90	90
	2	60	90	90	90	90	90
	3	60	60	60	90	90	90
	4	90	90	90	60	90	120
	5	90	90	90	60	90	90
	6	90	90	90	60	60	60

Note: 1, 2 and 3 denote landing heavy-duty plane, medium-sized plane and light-duty plane respectively; 4, 5 and 6 denote takeoff heavy-duty plane, medium-sized plane and light-duty plane respectively.

2.3 Mathematical Model

For a plane queque Q to be scheduled, define variables as below:

n : amount of planes in Q ; t_j : planed departure/landing time of plane j ; e_j :earliest departure/landing time of plane j ; l_j : last departure/landing time of plane j ; S_{ij} : the standard vortex seperation between plane i and j ; g_j : influence coefficient if plane j departure/landing earlier than t_j ; h_j : influence coefficient if plane j departure/landing later than t_j ; x_j : the actual departure/landing time of plane j ; $E_j : \max\{0, t_j - x_j\}$; $T_j : \max\{0, x_j - t_j\}$.

Considering the constraints S_{ij} and e_j , at the same time the operation rate of airport have to be increased. So x_j can be achieved as queation (1):

$$x_j = \begin{cases} e_j (j=1) \\ \max\{x_j + S_{ij}, e_j\} (j=1,2,\cdots i-1, i>1) \end{cases} . \quad (1)$$

Define $penalty_j$ as breach measure when plane j can't meet l_j constraint. Finallly the mathematical model of departure and landing scheduling can be defined as equation (2):

$$f(Q) = \min \sum_{j=1}^{n} \left(g_j E_j + h_j T_j + penalty_j \right).$$
$$penalty_j = \begin{cases} 1, & x_j > l_j \\ 0, & else \end{cases}. \tag{2}$$

3 Dynamic Scheduling Method

Because size of departure/landing queue maybe changes at each sample time, departure/landing scheduling is a dynamic course. If amount of planes within the departure/landing queue is too large, the calculating overhead will increase badly. In this section, we introduce a dynamic scheduling method based on receding horizon optimization and discrete particle swarm optimization.

3.1 Schediling Policy Based on Receding Horizon Optimization

In view of time, receding horizon optimization isn't accomplished in an action, by contraries it's an on-line optimizing course time after time. If time is sampled, for every sampling time two time windows are defined. The two time windows are respectively named predictive time window (PTW) and receding time window (RTW). When the kth sampling time arrives, optimal decision is made according to the information within current PTW and is executed in current RTW [4]. Usually one PTW involvves some RTWs.

Assume T_r is defined as the length of RTW. One PTW involves N RTWs, i.e. the length of PTW is NT_r. At the kth sampling time, a plane queue Q arrives to be scheduled with a kind of optimizing algorithm. In queue Q, the planed departure or landing time of each plane falls into PTW $\left[kT_r, (k+N)T_r\right]$. When the scheduling is finished, those planes which actual departue or landing time falls into RTW $\left[kT_r, (k+1)T_r\right]$ are elimitated from Q. At the $k+1$th sampling time, the rest of Q and the new arrived planes make up of a new plane queue Q'. For queue Q', scheduling course is repeated. Fig. 1 shows the dynamic scheduling policy.

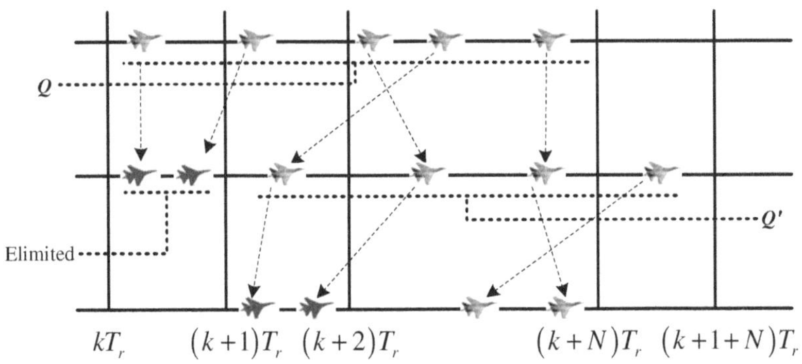

Fig. 1. Dynamic Scheduling Policy

3.2 Schediling Algorithm Based on Discrete PSO

PSO is an efficient optimizing algorithm with few parameters and computation overhead. In a PSO system, a few particles move in the search space and each particle represents a possible solution of a special problem. A particle has two properties: velocity and position. In standard PSO, velocity and position are updated with equation (3) and (4) [5]:

$$v^{k+1} = \omega \cdot v^k + c_1 \cdot r_1 \cdot \left(p_{best}^k - x^k\right) + c_2 \cdot r_2 \cdot \left(g_{best}^k - x^k\right). \tag{3}$$

$$x^{k+1} = x^k + v^{k+1}. \tag{4}$$

Where v^k and x^k is the velocity and position of a particle in k th interation; $\omega \in [0,1]$ is an inertia factor and decreases with the iteration going; c_1 and c_2 are constants called learning factor, in general, $c_1 + c_2 = 4$; r_1 and r_2 are random numbers wich submit to average distribution within $[0,1]$. Equation (3) and (4) are applicable with solving optimal problems in real number field. More details about standard PSO can be achieved in reference [5].

A plane queue Q can be regarded as a particle and the value in the jth dimension of particle Q is the order number of the jth plane in queue Q after scheduling is finished. Obviously the particle is coded with natural number. So a new algorithm named DPSO is designed to change updating equations to adapt to number field. DPSO use equation (3) to update velocity, but position equation is changed.

Firstly, a mapping named "value to address" (VTDM) is defined as $f : S \to N$, where $S = \{s_1, s_2 \cdots s_n\} \subseteq R$, $N = \{1, 2, \cdots n\}$. If m is the position of s_j in newly ascending reordered S, $f\left(s_j\right) = m$. During the $k+1$th iteration s_j^{k+1} is get with equation (5):

$$s_j^{k+1} = \frac{\left(\max\{g_j, 1/h_j\}\right)^{\alpha}}{\left(1 + \exp\left(-v_j^{k+1}\right)\right)^{\beta}} + \gamma \cdot rand_1. \tag{5}$$

In equation (5), $\max\{g_i, 1/h_j\}$ is heuristic information. The larger g_i is, the greater earliness penalty is and plane j should be put off. The larger h_i is, the greater tardiness penalty is and plane j should be moved up. To avoid local optimizatiom, DPSO introduces a random number $rand_1$ wich submit to average distribution within $[0,1]$ as disturbance. α 、 β 和 γ are three non-negative real number the adjust weight of heuristic information, main function and disturbance.

Equation (6) is achieved with appling VTDM to set $S^{k+1} = \{s_1^{k+1}, s_2^{k+1}, \cdots s_n^{k+1}\}$.

$$y_j^{k+1} = f\left(s_j^{k+1}\right) = k_1. \tag{6}$$

Ulteriorly equation (7) is achieved as below. In equation (7) $P \in (0,1)$ is a given threshold; $rand_2$ and $rand_3$ are random numbers wich submit to average distribution within $[0,1]$.

$$z_j^{k+1} = \begin{cases} x_j^k + y_j^{k+1} + rand_2, & rand_3 < P \\ x_j^k + rand_2, & otherwise \end{cases} . \tag{7}$$

New position of particle j, i.e. the new order number of plane j can be achieved with appling VTDM to set $Z^{k+1} = \{z_1^{k+1}, z_2^{k+1}, \cdots z_n^{k+1}\}$ as Equation (8) below::

$$x_j^{k+1} = f(z_j) = k_2 . \tag{8}$$

3.3 Flow Process of Dynamic Scheduling Method

For convenience the dynamic departure/landing scheduling method is name RH-DPSO aalgorithm. Flow process of RH-DPSO is shown in Fig. 2.

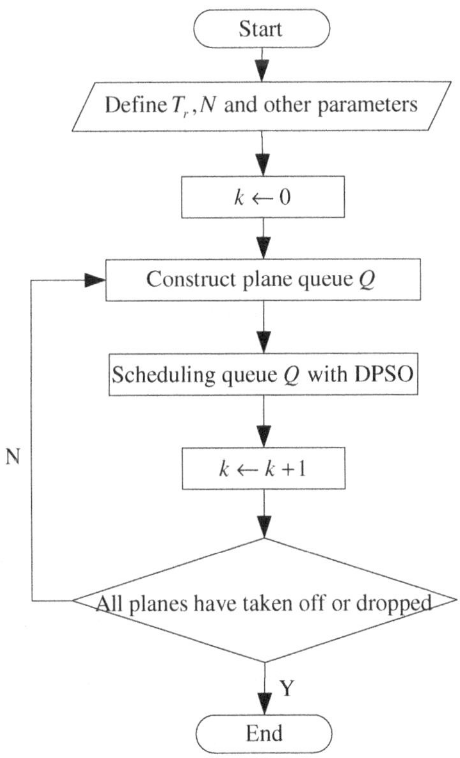

Fig. 2. FlowProcess of RH-DPSO

4 Simulation and Analysis

Considering an airport with single runway, amount of plane to be scheduled is 20. Data of simulation is shown in table 2.

Table 2. Data of Simulation

Plane No.	Type	t_i	e_i	l_i	g_i	h_i
1	4	17	0	517	0.5	0.5
2	2	100	0	600	0.2	0.3
3	5	217	0	617	0.2	0.7
4	5	279	0	979	0.5	0.6
5	2	343	143	743	0.3	0.6
6	4	424	324	1424	0.6	0.2
7	3	517	217	717	0.9	0.8
8	5	610	210	1010	0.2	0.5
9	2	688	488	1588	0.3	0.1
10	3	770	470	2270	0.5	0.4
11	2	850	50	1850	0.7	0.7
12	4	905	205	2005	0.7	0.3
13	2	1126	1026	1926	0.4	0.2
14	6	1541	741	2241	0.7	0.2
15	2	1554	1254	1854	0.3	0.8
16	1	1790	990	2590	0.1	0.7
17	1	1848	1048	2748	0.5	0.8
18	3	2051	1451	2651	0.5	0.8
19	4	2069	1669	2469	0.6	0.5
20	1	2172	1872	3000	0.9	0.6

Parameters of DPSO are shown in table 3. Length of RTW is 300 seconds and 3 RTWs involved in one PTW.

Table 3. Parameters of DPSO

ω	c_1	c_2	α	β	γ
1.0	2.0	2.0	0.0125	0.01	0.1

Departure/landing scheduling based on RH-DPSO and DPSO is programmed and simulated with Matlab. Simulation results are shown in table 4.

As table 4 shows, compared with DPSO, RH-DPSO achieves similar optimizing degree with shorter calculating time. So RH-DPSO has advantage of rapidity of convergence. As the scale of a special optimal problem increases, this advantage becomes more evident [4].

Table 4. Simulation Results

Simulation Data		RH-DPSO		DPSO	
Plane No.	t_i	Scheduled No.	x_i	Scheduled No.	x_i
1	17	1	0	1	17
2	100	2	80	2	100
3	217	5	190	3	217
4	279	4	260	5	445
5	343	3	332	4	515
6	424	6	428	7	585
7	517	8	524	6	657
8	610	7	705	8	753
9	688	10	835	11	981
10	770	11	925	10	1071
11	850	12	1005	12	1151
12	905	9	1075	9	1221
13	1126	13	1175	13	1321
14	1541	15	1490	15	1554
15	1554	14	1562	14	1626
16	1790	16	1743	16	1807
17	1848	17	1873	17	1937
18	2051	19	1988	19	2069
19	2069	18	5060	18	2141
20	2172	20	2241	20	2322
Objective function		1965		2055	
Calculating time (sec.)		9.07		11.97	

5 Conclusion

Frequency of departure/landing scheduling of military planes will increase with the increasing scale of air combat in the future. It beacomes more important for departure/landing to emphasize rapid dynamic scheduling. In this paper, RH-DPSO is introduced as an attempt to deal with this problem. Future research has two aspects: One is intergating RH-DPSO with other optimizing algorithms to decrease the possibility of local optimizatiom. The other aspect is analyzing and optimizing parameters of RH-DPSO to shorten the calculating time forwardly.

References

1. Wang, L.L., Chen, Y.X.: Demand Model for Operational Aircraft Allocation/Support Based on Airdrome Capacity. Electronics Optics & Control 19, 54–56 (2010)
2. Beasley, J.E., Krishnamoorthy, M., Sharaiha, Y.M., et al.: Scheduling Aircraft Landings The Static Case. Transportation Science 34, 180–197 (2000)
3. Cao, L., Deng, Y.X., Wang, X.H., et al.: Optimization Method for Airport Flow Based on Aircraft Types. Journal of Nanjing University of Aeronautics &. Astronautics 40, 646–650 (2008)

4. Wang, B., Xi, Y.G., Gu, H.Y.: An Improved Rolling Horizon Procedure for Single-machine Scheduling with Release Times. Control & Decision 20, 257–265 (2005)
5. Eberhart, R.C., Kennedy, J.: New Optimizer Using Particle Swarm Theory. In: 6th IEEE International Symposium on Micro Machine and Human Science, pp. 1942–1948. IEEE Press, Nagoya (1995)

Prediction and Estimation to the Target in Opto-electronic Tracking System Based on Set Membership Estimation

Junwei Lv, Ning Guo, and Jihong Yu

Department of Control Engineering
Naval Aeronautical and Astronautical University Yantai, China
13356995902@163.com

Abstract. During the course of prediction and estimation to the target in opto-electronic tracking system, stochastic noises statistical character affects the precision badly. To solve this problem, set-membership estimation is introduced to the target prediction in this paper. Considering the non-linear feature of opto-electronic tracking system, the paper analyzes extend optimal bounding ellipsoid (EOBE). Result of simulation shows that, compared with extend kalman filter (EKF), the prediction precision of EOBE is almost as good as that of EKF. Especially when the noise is bounded but not white and Gaussian, EOBE has better robustness.

Keywords: opto-electronic tracking, state estimation, set membership estimation, ellipsoidal bounding.

1 Introduction

Design of prediction and estimation algorithm to the target is always based on the assumption that the noise is white and Gaussian. The assumption is reasonable when detect frequency is lower than bandwidth of noise obviously. Whereas, with the development of opto-electronic detector, detect frequency of opto-electronic enhanced accordingly, and time relativity of observe noise can not be ignored any more[1]. To solve this problem, reference [2] and [3] eliminate the noise in the signal, then it view the worked signal as the input of kalman filter. Actually, during the course of noise elimination, new noise is introduced and the effect isn't as ideal as anticipation.

Based on the assumption that the noise is unknown but bounded (UBB), set membership estimation shows us a new way to deal with the non-white noise. The soul of set membership estimation is[4]: with UBB assumption, a set membership which always contains the true parameter or transferor function was given based on measurement data. The range of set membership dwindles with the number of data increases. When the number is infinite, the set member converges at the true parameter. When the UBB assumption is true, the true parameter of transferor function belongs to set membership surly. Compared with traditional statistical description, set

D. Jin and S. Lin (Eds.): Advances in EECM Vol. 1, LNEE 139, pp. 395–400.
springerlink.com

membership estimation need less restriction and be more similar with real system. So, it has great engineering significance for us to introduce the set membership estimation to the prediction and estimation of the target in opto-electronic system.

2 Optimal Bounding Ellipsoid Algorithm

The shape of set membership can be mainly described as ellipsoid, box, polyhedron, parallelotope, zonotope, etc. People pay more attention to the optimal bounding ellipsoid (OBE) algorithm because it can be calculated recursively and the process is easier than the others. Idea of OBE is introduced as below.

Consider the general discrete, linear state-space system:

$$x(k) = F(k-1)x(k-1) + G(k-1)w(k-1) \ . \tag{1}$$

$$z(k) = H(k)x(k) + v(k) \ . \tag{2}$$

where $x(k) \in R^n$ is the state vector, $z(k) \in R^m$ is the measurement. $F(k-1) \in R^{n\times n}$ is the non-singular state transfer matrix, $G(k-1) \in R^{n\times l}$ is the input matrix of disturbance, $H(k) \in R^{m\times n}$ is measure matrix whose row is full rank. $w(k-1) \in R^l$ is the disturbance, $v(k) \in R^m$ is the sensor noise, assuming that the noise is unknown but bounded, and they falls into the following ellipsoid respectively:

$$W_k = \{w(k) : w^{\mathrm{T}}(k)Q^{-1}(k)w(k) \le 1\} \ . \tag{3}$$

$$V_k = \{v(k) : v^{\mathrm{T}}(k)R^{-1}(k)v(k) \le 1\} \ . \tag{4}$$

where $Q(k) \in R^{l\times l}$ and $R(k) \in R^{m\times m}$ are both known positive definite matrix. $x(0)$ is the known initial state which is bounded by an ellipsoid given as

$$E_0 = \{x_0 : (x-\hat{x}_0)^T P_0^{-1}(x-\hat{x}_0) \le 1\} \ . \tag{5}$$

where $\hat{x}(0) \in R^n$ is the center of ellipsoid, $P(0) \in R^{n\times n}$ is positive definite matrix that describe shape of ellipsoid. State feasible set X_N is the set which is consistent with measurement vector sequence $\{z(k)\}_{k=1}^N$, ellipsoid E_0 state equation (1).measurement equation (2), ellipsoid sequence $\{W(k-1)\}_{k=1}^N$ and $\{V(k)\}_{k=1}^N$. Idea of ellipsoid set member filter is to calculate boundary of the ellipsoid and count center of the set as estimation state. Similar with kalman filter, OBE algorithm has two steps: time update and measurement update. Mission of time update is to acquire the feasible set of one step state estimation by calculating the vector sum of state noise ellipsoid set and state transfer matrix ellipsoid set; Mission of measurement update is to acquire the feasible set of estimated state by calculating the intersection between measurement set and state transfer feasible set. The details can be found in reference [5], this paper won't describe any more.

3 Extend Optimal Bounding Ellipsoid Algorithm

Transformation of reference frame is unavoidable during the course of opto-electronic tracking and measurement, which turns state equation or measurement equation into a non-linear one. Extend optimal bounding ellipsoid (EOBE) algorithm is used to the analysis of non-linear system. Step of EOBE is [6]: firstly, bound the high-rank items in the Taylor extended equation with ellipsoid by interzone analysis technical, then plus the "high-rank item" ellipsoid with noise ellipsoid, and finally predict state of target with linear OBE algorithm.

Considering the general discrete, non-linear state-space system as

$$x(k+1) = f[x(k)] + w(k). \tag{6}$$

$$z(k) = h[x(k)] + v(k). \tag{7}$$

where $x(k) \in R^n$ is the state vector, $z(k) \in R^m$ is the measurement. $f(\cdot)$ and $h(\cdot)$ are non-linear state equation and measurement equation respectively, both of them are two-rank diffentiable function. The definition of $w(k)$, $v(k)$ and $x(0)$ are as same as equation (3)to(5). Label the following ellipsoid as $\mathrm{E}(a,P)$:

$$\mathrm{E}(a,P) = \{x \in R^n \left| (x-a)^T P^{-1} (x-a) \le 1 \right.\}. \tag{8}$$

where a is the center of $\mathrm{E}(a,P)$, P is the boundary matrix of $\mathrm{E}(a,P)$ and it is positive definite. Define the state ellipsoid set at k as $x(k) \in \mathrm{E}(\hat{x}(k), P(k))$, then recursive step from k to $k+1(k=0,1,\cdots)$ is

1) Calculating the distributed range of each state vector at k

$$X^i(k) = [\hat{x}^i(k) - \sqrt{P^{i,i}(k)}, \hat{x}^i(k) + \sqrt{P^{i,i}(k)}]. \tag{9}$$

where $i = 1,2,\cdots,n$, $\hat{x}^i(k)$ is i th item of state vector at k, $P^{i,j}(k)$ is (i, j) item of $P(k)$.

2) Bounding the high-rank items with interzone analysis technical. To a system with signal state, expand the non-linear state equation at point $\hat{x}(k)$

$$x(k+1) = f(\hat{x}(k)) + \left(\frac{\partial f(\hat{x}(k))}{\partial x}\right)^T (x - \hat{x}(k)) + \frac{1}{2}(x - \hat{x}(k))^T \frac{\partial^2 f(X(k))}{\partial x^2(k)}(x - \hat{x}(k))) \tag{10}$$

then distributed range of Lagrange high-rank item is

$$X_{R_2}(x-\hat{x}(k),X(k))=\frac{1}{2}(x-\hat{x}(k))^T\frac{\partial^2 f(X(k))}{\partial x^2(k)}(x-\hat{x}(k)) \tag{11}$$

3) Calculating boundary of the error caused by linearization, and covers the error with ellipsoid $E(0,\overline{Q}(k))$, where $[\overline{Q}(k)]_{R(k)}^{i,i}=2(X_{R_2})^2, [\overline{Q}(k)]_{R(k)}^{i,j}=0, (i\neq j)$
4) Calculating the fictitious state error ellipsoid,

$$\hat{w}(k)\in \mathrm{E}(0,\hat{Q}(k))\supset \mathrm{E}(0,Q(k))\oplus \mathrm{E}(0,\overline{Q}(k))$$

and we can have the fictitious measure error ellipsoid $\hat{v}(k+1)\in E\left(0,\hat{R}(k+1)\right)$ with the similar step.
5) Forecast the boundary of state ellipsoid with linear OBE.

$$\hat{x}(k+1|k)=f\left(\hat{x}(k|k)\right) \tag{12}$$

$$P(k+1|k)=\Phi(k)\frac{P(k|k-1)}{1-\beta(k)}\Phi^T(k)+\frac{\hat{Q}(k)}{\beta(k)}, \beta(k)\in(0,1) \tag{13}$$

where $\Phi(k)=\frac{\partial f(x(k))}{\partial x(k)}\Big|_{x(k)=\hat{x}(k)}$ is the Jacobian matrix of state model.

6) Update the boundary of state ellipsoid $E(k+1)=E\left(\hat{x}(k),P(k+1)\right)$ with linear OBE, and new state is the center of ellipsoid.

4 Simulation and Analysis

To test performance of EOBE, a simulation of tracking experiment was given as below with EOBE and EKF respectively. Assuming that the opto-electronic tracking apparatus is on origin, original state of target is $X_0=[0,5000,5000,300,0,0]$, sampling cycle is $T=0.02s$, the target fly with constant velocity. The simulation was carried out with two different kind of assumption. Firstly, we assume that $w(k)$ and $v(k)$ are both Gaussian and white. $\sigma_w^2=[50\ 50\ 30\ 10\ 0.1\ 0.1]$, $\sigma_v^2=[20\ 20\ 10\ 5\ 0.1\ 0.1]$. Boundary of original ellipsoidal was defined by 3σ rule. The estimation error of position and velocity of x axes were shown as figure 1 and figure 2. Secondly, we assume that the disturbance is bounded non-white noise and the distributed region is $[-1,1]$. The result was shown as figure 3 and figure 4. Figure 5 shows position estimation of x axes and its upper and lower boundary when the disturbance is bounded non-white noise. The smaller graph in figure 5 is the detail of boundary.

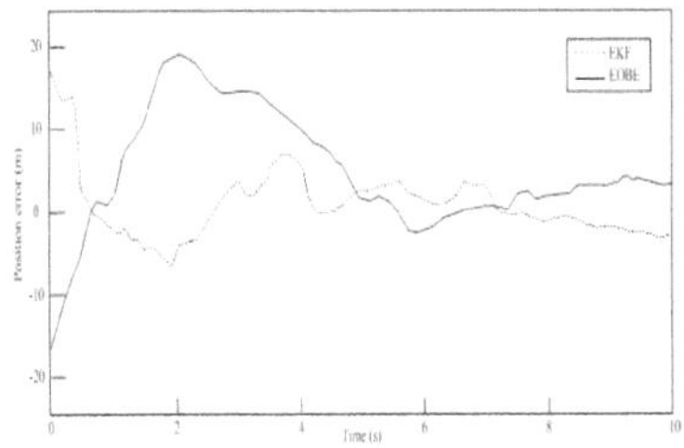

Fig. 1. Position error of x axes (noise 1)

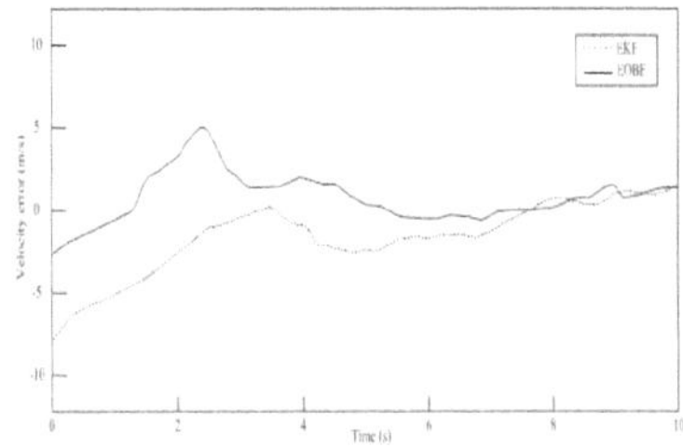

Fig. 2. Velocity error of x axes(noise1)

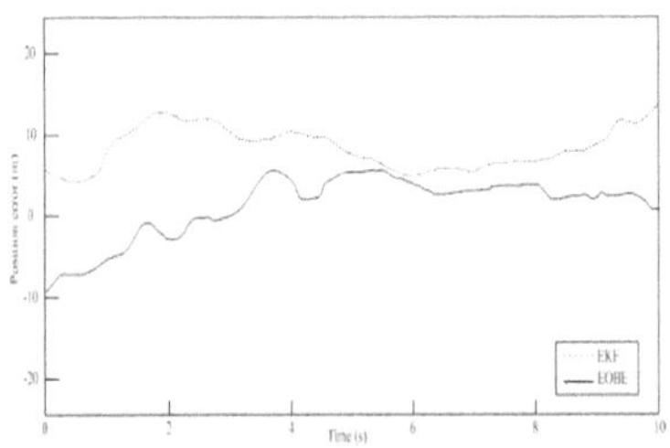

Fig. 3. Position error of x axes (noise 2)

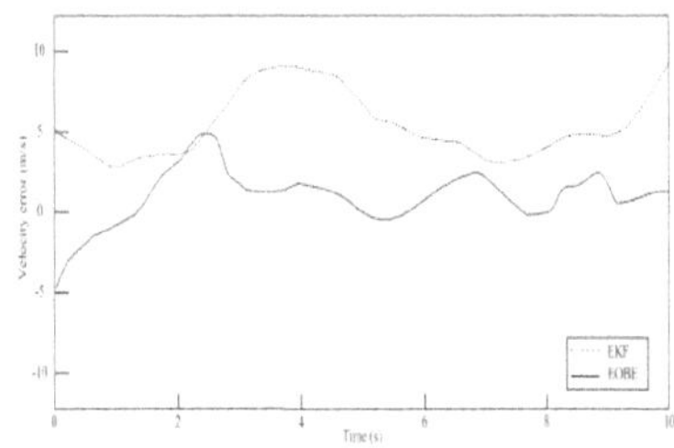

Fig. 4. Velocity error of x axes (noise 2)

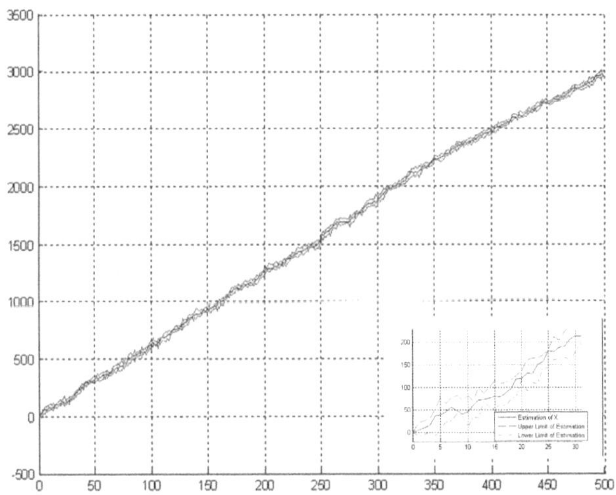

Fig. 5. Position estimation in x axes and its boundary

Figure 1 and figure 2 show that with the first assumption, the precision of EKF and EOBE is similar with each other. But with the second assumption, the estimation error of EKF is larger than EOBE's obviously. And with time past, error of EKF showed in figure 3 and figure 4 has ascending trend.

5 Conclusions

With the development of opto-electronic detect frequency, time relativity of noise can not be ignored any more and performance of traditional filter was affected in large degree. Based on the analysis to set membership estimation, the paper introduced EOBE to the prediction to the target in Opto-electronic tracking system. Result of simulation shows that when the disturbance is bounded non-white noise, stability of EOBE is better than EKF's, so EOBE has stronger practicability. In this paper, to simplify the analysis, target is assumed to fly with constant velocity. To the maneuver target, study on the performance of EOBE is necessary.

References

1. Li, C.: Research on estimation performance of optic-electric tracking systems with intermittent observations and filters development. Nanjing University of Science and Technology 58 (2009)
2. Chenglin, W.: Measurement multi-resolutional preprocessing method in signal denoising. Journal of Electronics and Information Technology, 296–301 (2002)
3. Chenglin, W., Chenmin, S., Quan, P.: Measurement multiscale preprocessing method in signal processing. Journal of Henan University (Natural Science), 5–9 (2000)
4. Zhigang, N., Renhuang, W., Yi, T.: Summary on set membership estimation. Journal of Guangdong non-ferrous Metals, 63–66 (2005)
5. Qing, H.: Research on dynamic GPS filtering and application of embedded system on GPS terminal. Postdoctoral report of Zhejiang University (2005)
6. Qing, H., Zhang, J.: A square root extended set membership algorithm with applications to nonlinear system estimation. In: IEEE Procceding on International Conference. Intelligence Compute Technic Automatic, Changsha, pp. 559–562 (2008)

ZVS Buck Converter with a Self-driven Synchronous Rectifier

Hyun-Lark Do

Department of Electronic & Information Engineering,
Seoul National University of Science and Technology, Seoul, South Korea
hldo@seoultech.ac.kr

Abstract. A zero-voltage-switching (ZVS) buck converter with a self-driven synchronous rectifier is proposed in this paper. An additional circuit consisting of LC and an auxiliary switch is added to a conventional buck converter and a self-driven synchronous rectifier is used instead of a buck diode. Due to the ZVS feature, the switching loss is significantly reduced. Also, the conduction loss is reduced by applying a synchronous rectifier. The operation principle and steady-state analysis of the proposed converter are provided. A prototype of the proposed converter is developed, and its experimental results are presented for validation.

Keywords: Zero-voltage-switching, buck converter, synchronous rectifier.

1 Introduction

Synchronous buck converters are adopted in many applications. However, due to their hard switching operation, the switching loss is large and the efficiency is relatively low especially at heavy load [1]. In addition, the reverse recovery problem of the anti-parallel body diode of the synchronous switch exists. In order to remedy this problem, ZVS control scheme for a pulse-width modulation buck converter under boundary conduction mode was suggested in [2]. ZVS control scheme can reduce the switching loss. However, it significantly increases the ripple component of the inductor current. Consequently, the output voltage ripple is increased and an additional filter stage may be required. Also, the additional filter stage may cause additional conduction loss and the efficiency will go down.

A ZVS buck converter with a self-driven synchronous rectifier is proposed. It utilizes an additional circuit consisting of an inductor, a capacitor, and an auxiliary switch to provide ZVS of the power switches. The ZVS feature solves the reverse recovery problem of the anti-parallel body diode of the synchronous switch and reduces the switching loss. To reduce the conduction loss, a self-driven synchronous rectifier is utilized instead of a buck diode. The theoretical analysis is provided in the following section. The theoretical analysis is verified by an experimental prototype with 100V-to-48V conversion.

D. Jin and S. Lin (Eds.): Advances in EECM Vol. 1, LNEE 139, pp. 401–404.
springerlink.com

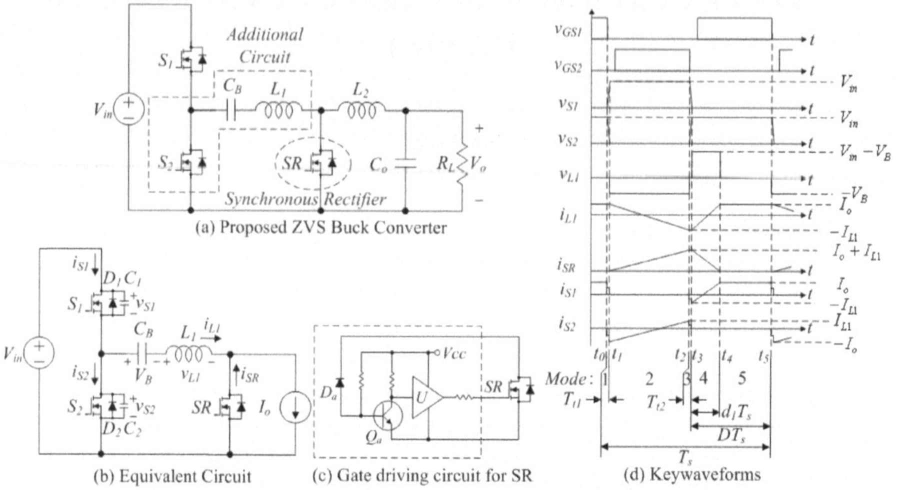

Fig. 1. Circuit diagrams and key waveforms of the proposed converter

2 Analysis of the Proposed Converter

Fig. 1 shows the circuit diagrams and key waveforms. Compared to a conventional buck converter, the switch *SR* is used instead of a buck diode. Before t_0, the switch S_1 is conducting. Since the switch *SR* is turned off, $i_{S1}=i_{L1}=I_o$.

Mode 1 [t_0, t_1]: This mode begins with turn-off of the switch S_1. At t_0, the current i_{L1} is I_o. The energy stored in L_1 starts to charge/ discharge C_1 and C_2. Therefore, the voltages v_{S1} and v_{S2} rise/ fall. With an assumption that the capacitors C_1 and C_2 have very small values, all the currents can be considered as constant.

Mode 2 [t_1, t_2]: At t_1, the voltage v_{S2} across S_2 arrives at zero and its body diode D_2 starts to conduct. Then, the gate pulse for S_2 is applied. Since the voltage v_{S2} is maintained as zero at the moment of the turn-on of S_2, zero-voltage turn-on of S_2 is achieved. Since the voltage v_{L1} across the inductor L_1 is $-V_B$, the current i_{L1} decreases linearly from I_o. Since the current i_{L1} decreases linearly from I_o, the difference between Io and i_{L1} flows through the switch *SR*. The current i_{SR} increases linearly.

Mode 3 [t_2, t_3]: This mode begins with the turn-off of S_2. At t_2, the energy stored in L_1 starts to charge C_2 and discharge C_1. Similar to Mode 1, all the currents can be considered as constant.

Mode 4 [t_3, t_4]: At t_3, the voltage v_{S1} across S_1 arrives at zero and its body diode D_1 starts to conduct. After that, the gate pulse for the switch S_1 is applied. Since the voltage v_{S1} is maintained as zero at the moment of the turn-on of S_1, zero-voltage turn-on of S_1 is achieved. In this mode, the voltage v_{L1} across L_1 is $V_{in}-V_B$. Therefore, the current i_{L1} increases linearly from $-I_{L1}$. Since $i_{SR}=I_o-i_{L1}$, i_{SR} decreases linearly.

Mode 5 [t_4, t_5]: At t_4, the current i_{L1} arrives at I_o. Consequently, the current i_{SR} arrives at zero and *SR* is turned off. In this mode, i_{L1} is maintained as I_o.

The proposed self-driven synchronous rectifier is a voltage sensing type. By detecting the voltage across *SR*, the driving signal is generated. At t_1, the body diode of *SR* is turned on. The voltage of drain terminal of *SR* is lower than the source terminal by a diode voltage drop. Since the base charge of Q_a is removed by turn-on of D_a, Q_a is turned off. Then, "High" signal is applied to a gate driver *U*, which is a conventional single-channel MOSFET driver, and the gate signal is applied to *SR*. Now, *SR* is fully turned on and all the current is flowing through the channel of *SR*. At t_4, i_{SR} decreases to zero. When i_{SR} changes its direction, the voltage of drain terminal of *SR* is higher than the source terminal. Then, D_a is turned off and Q_a is turned on. Then, "Low" signal is applied to the gate driver and *SR* is turned off rapidly.

By referring to v_{L1}, the capacitor voltage V_B is obtained as follows:

$$V_B = \frac{d_1}{1-D+d_1} V_{in} \ . \tag{1}$$

Since the average voltage across L_1 and L_2 should be zero under a steady state, the average value of v_{S2} is equal to V_B+V_o. Therefore, the voltage gain of the proposed converter is derived by

$$\frac{V_o}{V_{in}} = \frac{(1-D)(D-d_1)}{1-D+d_1} \ . \tag{2}$$

When d_1 is zero, the voltage gain is equal to that of the conventional buck converter under a continuous conduction mode.

Since the average capacitor current should be zero under a steady state, the average value of i_{SR} is I_o. Therefore, I_{L1} is derived by

$$I_{L1} = \frac{1+D-d_1}{1-D+d_1} \cdot I_o \ . \tag{3}$$

The *SR* current reset timing ratio d_1 is derived by

$$d_1 = \frac{2L_1 I_o}{(1-D)V_{in}} \cdot \frac{1}{T_s} \ . \tag{4}$$

3 Experimental Results

The prototype of the proposed converter is implemented with specifications and parameters of V_{in}=100V, V_o=48V, L_1=16uH, L_2=370uH, C_B=6.6uF, C_o=470uF f_s=100kHz, P_o=35*W*. Fig. 2 shows the experimental waveforms. The conventional synchronous buck converter is also implemented with the same circuit parameters except for the additional circuit. It can be seen that the experimental waveforms agree with the theoretical analysis. The switches S_1 and S_2 operate with ZVS. The efficiency of the proposed converter is compared with the conventional synchronous buck converter. Due to its soft-switching characteristic, the efficiency of the proposed converter is improved by 1.5% compared with the conventional one. Since the conventional synchronous buck converters can operate with ZVS at light load, its efficiency is higher than the proposed converter. At light load, the conduction loss due to the additional components in the proposed converter is relatively higher.

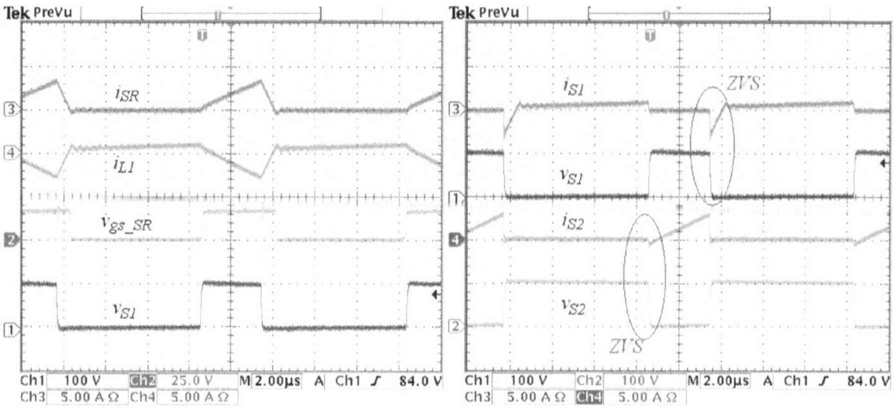

Fig. 2. Experimental waveforms

4 Conclusion

A ZVS buck converter with a self-driven synchronous rectifier has been proposed. By utilizing the auxiliary circuit, both switches can operate with ZVS regardless of load condition. Soft-switching operation improves the efficiency at medium and heavy loads. Due to additional components, the light load efficiency is lower. However, ZVS characteristic is still maintained at light load.

References

1. Li, W., et al.: A family of interleaved boost and buck converters with winding-cross-coupled inductors. IEEE Trans. Power Elec. 23, 3164–3173 (2008)
2. Chiang, C.Y., et al.: Zero-voltage-switching control for a PWM buck converter Under DCM/CCM boundary. IEEE Trans. Power Elec. 24, 2120–2126 (2009)

Rural Logistics Finance Distributed Interactive Simulation System Based on High Level Architecture

Dou Zhi-wu[1] and Lu Lin[2]

[1] School of Business, Yunnan University of Finance and Economics
650221 Kunming, Yunnan
[2] Human Resources Department, Yunnan Tobacco Leaf Company
650218 Kunming, Yunnan
zhiwudou@163.com

Abstract. At present, it is hard to borrow from the financial institutions for the rural logistics subject. In order to reveal the relationship between the rural logistics financial system elements and its operation rules, to provide support for solving the problem of the funds flow fracture in rural logistics, a distributed multi-targets rural logistics finance system was put forward, and constructed the distributed interactive simulation of the rural logistics finance system based on high level architecture, through the example of simulation data, revealing the relationship between system elements and its operation rules, the results proved the method is effective.

Keywords: Logistics finance, Rural logistics finance, High level architecture, Distributed interactive simulation.

1 Introduce

According to investigation, a farmer had to tearfully kill the sheep bought with agriculture loan due to lack of money to buy feed in our country western rural; the number shows, due to the fruits, vegetables and dairy products rotting our country has loss 750 billion yuan every year only during the transportation. The products loss rate of Fruit and vegetable is at an average of 5% in developed country, 1%-2% in the United States, but 25%-30% in our country. The low rate of rural logistics efficiency and the high rate of product loss have been a serious impact on China's new rural construction. Investigating its reason, it is to difficult to evaluate the credit rating of rural logistics subject with the distributed, small scale and poor condition for financial institutions, which lead financial institutions are not like to lend money to the rural logistics subjects, causing village logistics capital chain rupture. The solution to these problems may not rely on the supporting agriculture loans or subsidies from government only. How to ensure the smooth capital flow of rural logistics is the key to solving the problem. In order to fundamentally solve this problem, we have to first solve the problem to combine the rural logistics with the funds flow -- namely rural logistics finance. This paper presents a distributed multi-targets rural logistics finance system, and a distributed interactive simulation based on high level architecture (DIS/HLA) to study the system, aims to reveal the relationship between the elements of the system and its operation rules, which proposed targeted solutions to provide support to solve various problems in rural logistics finance.

D. Jin and S. Lin (Eds.): Advances in EECM Vol. 1, LNEE 139, pp. 405–410.
springerlink.com

2 Rural Logistics Finance and DIS/HLA Research

Logistics finance concept was suggested by domestic scholars, in the foreign logistics finance research and has not formed a complete and independent system [1]. At present, the domestic study on operation mode of logistics finance is mainly along the bank of goods, warehouse receipt pledge and confirmed warehouse business [2]. Financing warehouse concept and mode of operation is presented by Zhu Dao-li [3]. Logistics finance concept was developed by Zou Xiaopeng andTang Yuanqi in 2004 May [4] for the first time. Experts and scholars that put on their main focus on the concept and theory of financial logistics operation mode, the research from the relationships of system elements and operational rules is scarce.

The High Level Architecture (HLA) [5] was developed by the U.S. Defense Modeling and Simulation Office (DMSO) to provide simulation interoperability and reusability across all types of simulations and was adopted as an IEEE standard. The Runtime Infrastructure (RTI) is an implementation of the HLA standard that provides the platform for distributed simulations [6].

In HLA, a simulation is called a federation and a simulation component is called a federate [7,8]. Federates communicate with their peers by sending interactions or updating object attributes. Federates do not communicate directly with each other and all communication is administrated by the RTI based on the interest of each individual federate. HLA/RTI aims to provide a software platform for simulation interoperability.

3 Rural Logistics Finance System and Its Distributed Interactive Simulation System

3.1 Rural Logistics Finance System

Rural logistics finance refers the financing and settlement service activities to add the value of rural logistics in the rural logistics operation process, which goods is media, logistics enterprises and individual is subjects, government is supervision and guarantee, finance is tools. Rural logistics finance system is the organic whole composed with the rural logistics finance main body, it is a distributed multi-targets system, involving many interests body, such as the government, financial institutions, logistics enterprises, production enterprises and farmers. The simplified system structure as shown in fig. 1:

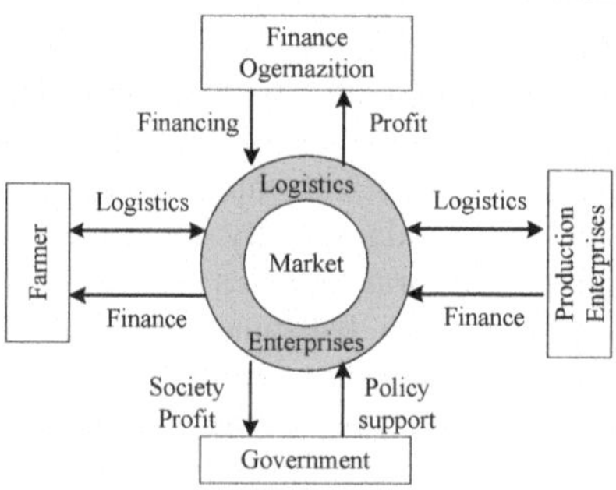

Fig. 1. The simplified system structure

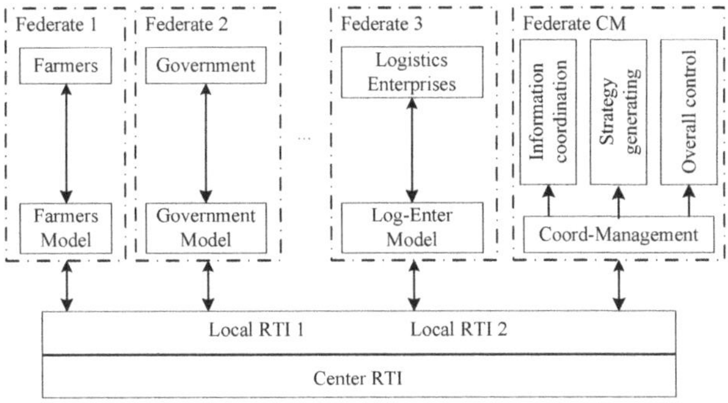

Fig. 2. The rural logistics finance system simulation structure

3.2 Its Distributed Interactive Simulation System

Fig.2 is the rural logistics finance system simulation structure diagram based on DIS/HLA. In the simulation system each rural logistics finance body is modeled as a simulation federate. The simulation system implements a complex distribution rural logistics finance system experiment, with which the different subjects, various functions, and the system properties can be research. The system mainly includes RTI service module, simulation coordinated management module, simulation federate module and the network infrastructure, in which each rural logistics financial subjects is developed to a federation, each federation establishes its data model and simulation federation model, and then transform into simulation federate to realize their simulation function. The simulation system can be used as a separate functional modules operating independently, and can also be used as a simulation federation to join larger simulation system. Each module specific functions are described below:

3.3 Simulation Modules Function

The system mainly completes 4 modules function:

(1) RTI service

In addition to execute 6 management functions (federation management, declaration management, object management, data distributed management, time management, ownership management), RTI service module need to manage computing resources from different organizations and the availability of these computing resources that may change during the execution of the simulation.

(2) Network information

Network information takes charge of the simulation resource monitoring, simulation estate evaluation, load balancing, federation transferring and code transferring. From the beginning of the first federation establishing, the network information module begins to monitor the whole simulation processes and the estate of every federate. The information is the base of the simulation decision.

(3) Federation module

The function of the module is to establish federation, registry federate instance, start-up federate process and delete federate exited federation.

(4) Network basis establishment

The network basis establishment is the physics basement of simulation running.

4 Mathematics Models and Experiment Result Analysis

4.1 Mathematics Model

Here taking the pricing of financial institutions as an example, this article only gives the income model of logistics financial institutions and farmers'. Whether rural logistics finance system normally run, or that the parties can participate in the rural logistics finance system, is decided by the income of the participating body composed the system. The experimental simulate financial institutions and each participant's behavior to reveal the relationship between parties and the operation law of the system.

In the paper, we suppose the rate of loan of financial institution for other subjects (except outside the government) i in rural logistics financial systems is r_i:

$$r_i = r_0 + c_i + P_i\lambda + k_i(\delta - r_0) \tag{1}$$

In formula (1), r_0 : Loan cost rate of capital (this can be regarded as the savings rate of the national provisions); c_i : Financial institutions un-fund management cost; P_i : The default probability of loan body; λ : Loss given default; k_i : The credit index of loan body; δ : Risk premium index;

Then, the income of financial institution j is R_i:

$$R_j = \sum_i^n A_i r_i - r_0 \sum_i^n A_i \tag{2}$$

In formula (2), A_i: the loan number of the i loan body.

Then , the income of the i loan body R_i is:

$$R_i = R_0 - C_i + R_s \tag{3}$$

In formula (2), R_0: the income without loan; C_i: the cost of loan; R_s: the income with loan.

Using the mathematical model of main body to establish simulation model, and then to build a distributed interactive simulation federate model, to establish simulation federation, undertake rural logistics finance system simulation, the paper study the restrict relationship between system elements, and reveals the system operation rules.

4.2 Experiment Results Analysis

Tab. 1 shows part of simulation results. Simulation used in two ways, one is the main body of the rural logistics system using the rural logistics finance financing and operation, the other is not using. The experiment is carried out 1000 times with the original data from a financial institution nearly 5 years. Most likely reflect the system's authenticity.

Table 1. Parts index of results

	Using rural logistics finance	Un-using rural logistics finance
Pricing of finance institutions	Reducing 20%	Rising 30%
The default probability of loan body	Reducing 50%	Rising 35%
The credit index of loan body	Rising 37%	Reducing 30%
Income of system	Rising 32%	Reducing 25%
Successful loan rating	90%	43%

Tab. 1 data shows, using the rural logistics finance mode, financial institutions pricing and the main loan risk tended to decrease, the profit of the loan and system subjects and the successful loan rates tend to increase, rather than the results in exactly the opposite. a in-depth analysis of the causes of the results, due to the use of rural logistics finance mode, the guarantee mechanisms and the government credit is increasing to the entire system, and if using the supply chain mode in the same time, the sharing level of enterprise credit will be more improved, which greatly improves the credit level of loan body and decreases the default probability, thus makes the default risk of financial organization reduce and tend to reduce price. All of these are the reason of the successful loan rate and system income increasing.

5 Conclusion

According to the analysis and application of DIS/HLA, we know the method is a very applied, available and convenient method. The conclusion of simulation experiment point that:

- The distributed quality convenient system design, development.
- The reuse quality reducing the investment and time cost.
- The RTI smoothing the simulation process and optimizing the results of experiment.

In summary, it is valuable to study the application of DIS/HLA in rural logistics finance, some inner algorithm and management mechanism need to deep research, and how to integrate with grid computer is to a hot point of study.

Acknowledgement. Natural Science Foundation of Yunnan Province (2009ZC087M).

References

1. Luo, Q., Zhu, D.-L., Chen, B.-M.: Third party logistics service innovation: Financing warehouse and its operation model. Chinese Circulation Economy (2), 11–15 (2002)
2. Zhu, D.-L.: Research on the key technologies of logistics financial innovation. China logistics innovation congress lecture (2004)

3. Shi, D.-L.: Financing warehouse and Logistics financial services. China logistics innovation congress lecture (2004)
4. Zou, X.-F., Tang, Y.-Q.: Logistics finance: New research in the field of logistics. China Logistics and Purchasing (17), 42–44 (2004)
5. Zhou, Y., Dai, J.-W.H.: Simulation programming design, pp. 4–13. Publishing house of electronics industry, Beijing (2002)
6. Dou, Z.-W., Deng, G.-S., Mao, H.-J.: Dynamic grid-based method research in data distributed management. Systems Engineering-Theory & Practice 27(1), 137–142 (2007)
7. Dou, Z.-W., Li, H.-W.: Research on self-adaptived dynamic grid-based method in data distributed management. Systems Engineering-Theory & Practice 28(12), 128–132 (2008)
8. Dou, Z., Li, H.: The production system optimizing simulation in Petrochemical industry based on HLA. In: 2009 Second International Symposium on Electronic Commerce and Security (May 2009)

A Study of the Deployment Solution of Education Resource Sharing System Based on ZooKeeper*

JianShe Huang**

Zhejiang Business Technology Institute, 315012, Ningbo, China

Abstract. This paper proposes a way to deploy the education resource sharing system based on Zookeeper distributed file system technology. By using the technologies of real-time document pushing and weighted resource election algorithm, the system also provides the latest and the most advanced education resources to users. It can avoid some problems that occur during the development of education resources when schools or educational institutions conduct the development in isolation without sharing best practices, updating on a timely fashion, being systematic, applying consistent standard, and even guaranteeing the quality of key course offerings. Therefore, the system can improve the utilization of education resources, and reduce wastes of human, material and financial resources introduced by the duplication of education resource development. From the technical perspective, this paper provides new ideas and methods for the development and sharing of education resources.

Keywords: Curriculum developing; education resource sharing; Zookeeper.

1 Introduction

The education resources are the most important and valuable asset for everyone in the education industry, especially institutions such as schools. Regarding to the curriculum reform and development, although every teacher is actively participating, he or she might have different understanding, so the curriculum materials produced also vary. This leads to large varieties and quantities of education resources which are extremely dispersed with little consistency in quality. The traditional education management method stores education resources through addition of huge volumes of storage devices, which introduces enormous cost for schools. Furthermore, it also wastes the teachers' time unnecessarily because they develop their own education resources in isolation without collaboration and communication with each other.

In many foreign countries, the high quality resource sharing system is relatively systematic and mature with high levels of utilization. Literature [1-2] pointed out: The

---

* This study was one of the results of science of education planning research subject "Education resource Sharing System Deployment Researching based on Zookeeper"(item number: YGH127) in 2011, Ningbo.

** Author: (1968 -), female (the Han nationality), from Hubei, master's degree, teacher, lecturer, engineer, main research direction: networking, distributed system, data mining.

D. Jin and S. Lin (Eds.): Advances in EECM Vol. 1, LNEE 139, pp. 411–416.
springerlink.com

core of a high-quality educational resource sharing organization is a governmental institution or a professional association, with the funds mainly from the government and the technical support by a company primarily. Besides the above organization, universities, as another main entity for knowledge distribution, also contribute significantly to high-quality education resource sharing. For example, the United States is one of the countries with the richest education resources, and its government, universities and other organizations invest a lot of human and financial resources to the construction of the education network. In particular, ERIC, the US Educational Resources Information Center, supported by the U.S. Department of Education, Academy of Education and National Library of Education, can efficiently help users retrieve the information needed due to its thorough resource construction plan, complete functionality, rich content, user-friendly search engine, Email response system and expert consultation and so on; Another example is Canada. In 2002, the Canadian government initiated the eduSource project targeting at sharing network resources across the whole nation and making the educational resources developed available to all Canadians.

In China, because of regulations, attitude and other factors, education resources are developed by individual schools or educational institutions without unified management, resulting in duplication of effort, inconsistent quality, and low utilization. With the push from the Ministry of Education, quality courses are partially shared. But the sharing is still not systematic, nor did it engage the team strength of all the good teachers. As a result, the so-called quality courses are not of highest quality. For example, the major components of the curriculum, such as teaching methods and plan, might be of high quality, but the teaching courseware might not be the best. Because of this, there are some expert studies on China's education resource sharing, literature [3-4] explored ideas and practices of building a good education resource sharing platform, but realization of such a platform still has a long way to go.

2 Education Resource Sharing System Architecture Based on Zookeeper

Zookeeper is a highly reliable, consistent, and high-performance file system with built-in notification mechanism making it an ideal choice for the education resource sharing system. Its architecture mainly consists of: the education user, the expert groups, Zookeeper service platform, Zookeeper cluster groups, resources directory server and the Agent and so on, as following:

A. Education users: schools and educational institutions connected to the system.

B. The expert groups: expert groups of different expertise and geographies, or comprehensive education experts.

C. Zookeeper service platform: First, it provides an interactive interface for users to request services. The user can register, login, customize services and configure and manage user through a browser. Different users can customize the services, for example, the educational institution user A only cares about courses related to Economics, he can configure his service to get only the Economics course materials;

Second, it also provides an interactive interface for experts to upload scores record, experts can register, login and upload resource score record sheet through a browser.

D. Zookeeper cluster group: the core module of Zookeeper service platform. There are five servers per cluster group, ensuring the reliability, consistency and stability of the entire Zookeeper distributed file system. These cluster groups also manage the communication with Agent, store information that each Agent needs to monitor, and are used for the education resource and script update.

E. Resource directory server: it is the list of services that the user can select or customize after obtaining the appropriate permissions.

F. Agent: Standard configuration of all servers using the service, which initiates along with system start. It is used to monitor and handle changes associated with its own node and trigger corresponding applications to realize Zookeeper service functionalities.

3 Education Resource Automated Deployment Services

3.1 Zookeeper Service Node Architecture

Zookeeper service node of each educational user's computer connected to the system has the following architecture:

A. Every computer has one node directory on server named by its own IP address.

B. Under ZY root directory of Zookeeper server, there is a directory called EduInformationList. Documents under this directory supply the education resource information this machine must configure, such as Course, Syllabus, and so on. As shown in Figure 1.

C. Agent can monitor and control node status of education resources according to the corresponding names in EduInformationList. For example, if "Syllabus" resource exists in the computer, Agent will generate "Syllabus" directory under the directory corresponding to its own IP, and set the initial "status" as "listen", which means listening. Status can be "listen", "install", "uninstall" and so on. As shown in Figure 2.

D. "install", "uninstall" under education resource directory(such as Syllabus) store the path to the scripts of operations like installation, uninstall. When status changes, it will execute the corresponding script according to status value.

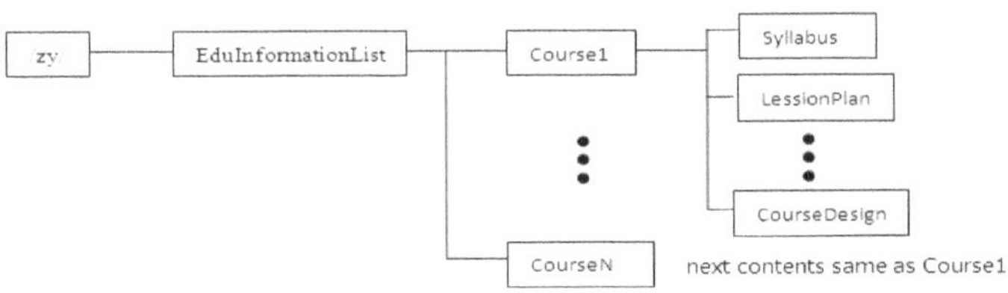

Fig. 1. Resource Directory Architecture in Zookeeper

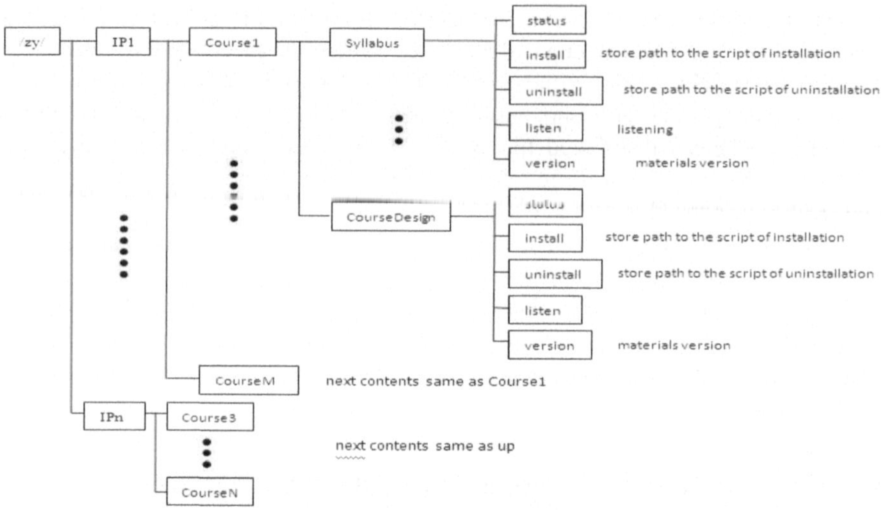

Fig. 2. Each Educational Constitution Customize Resource Directory Architecture

3.2 Automate the Deployment Process of Education Resources

The automated deployment process of education resources is shown in Figure 3.

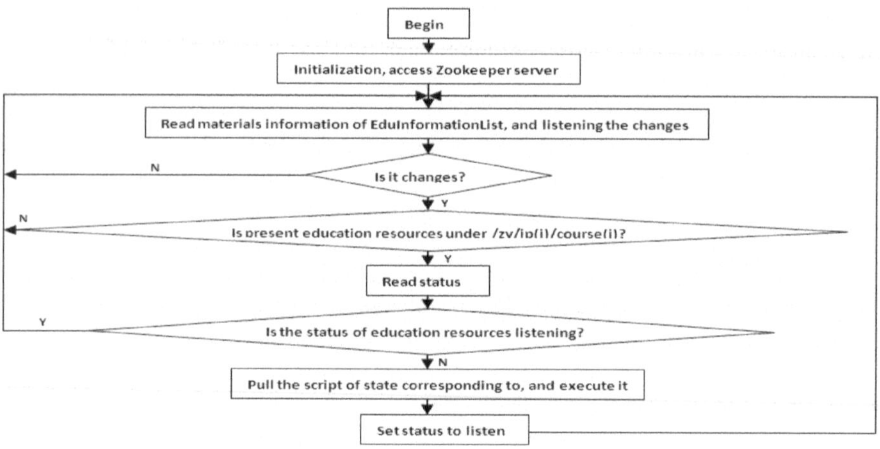

Fig. 3. The Automated Deployment Process of Education Resources

4 Core Technologies of Education Resource Sharing Systems

4.1 Zookeeper Introduction

Yahoo! open-source ZooKeeper system is a distributed small file system with high consistency guarantee. According to literature [5], Zookeeper has the following advantages:

A. Interface: ZooKeeper provides an interface similar to one that handles distributed file system operations, and easy to package an external application. Languages supported include c、java、python、and so on.

B. Rich applications: ZooKeeper supports many kinds of application scenarios, including mass deployment, configuration updates, name service, queue mechanism, and so on. No matter what kind of application, however, an external package is required.

C. High- Reliability, High-Consistency and high-Availability: ZooKeeper is deployed in the form of server clusters, and the information will be immediately synced up and backed up as soon as a server gets started, thus high reliability is guaranteed; In addition, the "read" operation from the user who connects to the system depends on the local database of each server. Once the node data changes, the user will be notified to ensure the obtained data is consistent. ZooKeeper's high availability is shown through its wide range of applications and services.

D. Stability: ZooKeeper has been deployed in large scale in Yahoo!, which fully demonstrates the stability of its performance.

4.2 Weighted Resource Election Algorithm

When the educational constitution's terminal in the access system requests for resource optimization, the education materials supplied must be uploaded in accordance with the rules. For instance, the rule may specify that not only the education materials be uploaded, but also the type of materials---- In "$UiA_1A_2A_3A_4A_5A_6A_7A_8A_9A_{10}A_{11}A_{12}BiCi$" format in which, Uk represents user k(k=0,1,----x), A_1A_2represents the type of the educational constitutions, such as higher vocational schools or secondary vocational schools; A_3A_4 represents geographic areas, such as Ningbo or Hangzhou; A_5A_6 represents the major that courses belongs, such as Computer major; $A_7A_8A_9A_{10}$ represents Curriculum name; $A_{11}A_{12}$ represents material name, such as lesson plans or Syllabus; BiCi is reserved for the experts' judgment, Bi shows the similarity between this material and the best previous material. If the similarity is high, it does not get selected; Ci is expert assessment score of the material. The final score of the education material will be decided based on a weighted algorithm, and the material with the highest score will be selected. The weight differs for different experts. For course materials belong to the same major, the experts in adjacent areas will have higher weight than those in areas further away. Comprehensive expert scoring weight is also different in different situations. In addition, the time when the expert gave the score is also considered. If the time is same as the last evaluation, it means that the expert did not give a core this time, and the result is invalid. If invalid result is more than 5%, this evaluation is invalid, and corresponding application needs to be triggered again to remind the experts to give their assessment in time.

Supposing there are n experts, represented by P = {P0, P2, ... Pn-1}, and W (Pi) is the weight of expert Pi's assessment of the material, ranging from 0 to 1. C counts the number of times the material evaluation time is the same as the last evaluation time. The final score of user Uk's curriculum material UkSj (j=0,1, ... m) is calculated as:

IF $(\sum_{i=0}^{n-1}[W(Pi) * Bi] < 0.5)$ // When similarity between this material and best previous material is low, proceed with evaluation.//

{ IF $(\frac{C}{N} <= 0.5\%)$ // When more than 95% of the experts participate in the evaluation, the score is valid.//

$$UkSj = \sum_{i=0}^{n-1}[W(Pi) * Ci];$$

} //The material with the highest UkSj is selected.//

5 Summary

This paper proposes to deploy the education resource sharing system based on Zookeeper distributed file system technology, and provides a deployment solution. The innovation in this article lies in:

A. Proposed the weighted algorithm for education resource election.

B. Easy sharing of education materials.

C. Highly effective resource-sharing environment can make educational researchers more focused and effective.

D. This file system has high reliability, consistency and stability.

Problems: free sharing and intellectual property of education materials may need more consideration.

References

1. Liu, Y., Wei, N.: Research and Inspiration of Foreign High Quality Distance Education Shared Resource. Modern Distance Education 06 (2004)
2. Wang, F.-Z., Zhu, J.-J.: Resource Sharing in P2P Networks in the Present. Computer Knowledge and Technology 06(36) (2010)
3. Ren, W.-M.: Few Observations on the Development of Modern Distance Education and Construction of the Education resources. China Distance Education, 05 (2000)
4. Li, F.-Y., Li, C.-L., Li, H.: Grid Multi-Agent Authentication Proxy and Its Application in Education resources Share. Computer Technology and Development, 08 (2009)
5. White, T.: In: Zeng, D.-R.(trans.) Hadoop:the Definitive Guide. Tsinghua University publishing house, 05 (2010)

A Comparison Method of Function Transformations to Reduce the Class Ration Dispersion

Neng-yan Wei* and Yong Wei**

College of Mathematics and Information, China West Normal University, 637002, Nanchong, Sichuan, China
3306866@163.com

Abstract. According to monotone transformation which reduces the class ration dispersion, this paper mainly researches the comparison principle between the transformation of two monotone transformations and each of them on reducing the class ration dispersion. Theoretically, we prove that the transformation of two monotone transformations is better than any of them in raising the smooth degree of the original data series under certain conditions, we also prove the feasibility and validity of the method by some examples.

Keywords: grey model, class ration dispersion, smooth degree, function transformation.

1 Introduction

Since the gray system theory was found, nearly 30 years, it has been used in industry, agriculture, economic and other fields. The grey model GM (1,1) is the most widely used as the basic model. So people has been studying new modeling technique from the practice in order to improve the GM (1,1) model of forecasting accuracy. There are many function transformations, introduced in the references[1], such as

$$f(x) = \ln x (x > e),\, x^{\frac{1}{T}} (x > 1, T \geq 1),\, x + C (C > 0),$$
$$a^{-x} (0 < x < \frac{1}{\ln a}, a > 1).$$

The deficiency of the existing smoothness condition is pointed out in the reference[2] and it defines a new smooth degree respectively according to the classification. In the reference [5], it gives the necessary and sufficient conditions to reduce the class ratio dispersion, and proves that reducing the class ratio dispersion and raising smooth degree are equivalent. This paper mainly uses the criterion for comparing the function transformations which gives in the reference [4] to construct a new method. For two increasing functions, two decreasing functions or one of the functions is increasing and the other is decreasing they all

* Neng-yan Wei is master graduate student of department of mathematics and information, china west normal university, major study is grey system analysis.

** Corresponding author: Yong Wei is doctor ,master tutor, professor of department of mathematics and information, china west normal university, major study is grey system analysis.

D. Jin and S. Lin (Eds.): Advances in EECM Vol. 1, LNEE 139, pp. 417–423.
springerlink.com

can reduce the class ration dispersion, but reducing the class ratio dispersion ,the transformation of two monotone transformations is more effictive than any of them.

2 The Basic Concept

Definition 1[6]. If $\frac{b_k}{\sum_{i=1}^{k-1} b_i} < \frac{a_k}{\sum_{i=1}^{k-1} a_i} (k = 2,3,\cdots n)$ for the monotone decreasing series $\{a_k\}_{k=1}^n, \{b_k\}_{k=1}^n$,then we consider $\{a_k\}_{k=1}^n$ to be smoother than $\{b_k\}_{k=1}^n$.

Definition 2[7]. Let $\{b_k\}$ be monotone decreasing series ,and $\{a_k\}$ be monotone increasing series,if $1 \le \frac{1}{k-1} / \frac{b_k}{\sum_{i=1}^{k-1} b_i} \le \frac{a_k}{\sum_{i=1}^{k-1} a_i} / \frac{1}{k-1}$ $(k = 1,2,3,\cdots n)$ then we deem the monotone decreasing series $\{b_k\}$ to be smoother than the monotone increasing one $\{a_k\}$.

Theorem 1[4]. The monotone increasing function transformation $F(x)$ can strengthen the smooth degree of the increasing series if and only if $\frac{F(x)}{x}$ is monotone decreasing for $\forall x \in [\alpha, \beta]$.

Theorem 2[4]. The monotone decreasing function transformation $F(x)$ can strengthen the smooth degree of the increasing series if and only if $F(x)$ is monotone increasing for $\forall x \in [\alpha, \beta]$.

Theorem 3[4]. If $F(x), G(x)$ are monotone increasing function transformation, then $F(x)$ is better than $G(x)$ on reducing the class ratio dispersion if and only if $\frac{F(x)}{G(x)}$ is monotone decreasing for $\forall x \in [\alpha, \beta]$.

Theorem 4[4]. If $F(x), G(x)$ are monotone decreasing function transformation, then $F(x)$ is better than $G(x)$ on reducing the class ratio dispersion if and only if $\frac{F(x)}{G(x)}$ is monotone increasing for $\forall x \in [\alpha, \beta]$.

3 Construction of the New Function Transformations

Theorem 5. If $F(x), G(x)$ are monotone increasing function transformations and $F(x)$、 $G(x)$ can reduce the class ration dispersion for $\forall x \in [\alpha, \beta]$ For arbitrary nonnegative a, b

a. $aF(x)+bG(x)$ is better than $F(x)$ on reducing the class ratio dispersion if and only if $\frac{F(x)}{G(x)} \le \frac{F'(x)}{G'(x)}$.

b. $aF(x)+bG(x)$ is better than $G(x)$ on reducing the class ratio dispersion if and only if $\frac{F(x)}{G(x)} \ge \frac{F'(x)}{G'(x)}$.

c. $aF(x)+bG(x)$ is better than $F(x)$ and $G(x)$ on reducing the class ratio dispersion if and only if $\frac{F(x)}{G(x)} \equiv \frac{F'(x)}{G'(x)}$.

Proof: a. Necessity: $F(x), G(x)$ are monotone increasing function transformations and $F(x)$、$G(x)$ can reduce the class ration dispersion for $\forall x \in [\alpha, \beta]$,so $\left[\frac{F(x)}{x}\right]' < 0, \left[\frac{G(x)}{x}\right]' < 0$ and

$$\therefore \left[\frac{aF(x)}{x} + \frac{bG(x)}{x}\right]' = \left[\frac{aF(x)+bG(x)}{x}\right]' < 0,, \left[aF(x)+bG(x)\right]' > 0$$

From theorem 1,we can get the $aF(x)+bG(x)$ can reduce the ratio dispersion. Because of $\frac{F(x)}{G(x)} \le \frac{F'(x)}{G'(x)}$, we can get the

$$\therefore \left[\frac{aF(x)+bG(x)}{F(x)}\right]' = \frac{bF(x)G'(x)-bF'(x)G(x)}{F^2(x)} \le 0$$

From theorem 3 we know that $aF(x)+bG(x)$is better than $F(x)$ on reducing the class ratio dispersion.

Sufficiency: The $aF(x)+bG(x)$is better than $F(x)$ on reducing the class ratio dispersion for $\forall x \in [\alpha, \beta]$,and $aF(x)+bG(x)$is monotone increasing function transformation. Then from theorem 3 ,we can get

$$\left[\frac{aF(x)+bG(x)}{F(x)}\right]' = \frac{bF(x)G'(x)-bF'(x)G(x)}{F^2(x)} \le 0$$

and $F(x)G'(x) - F'(x)G(x) \le 0$ .so we can deem $\frac{F(x)}{G(x)} \le \frac{F'(x)}{G'(x)}$.

b. Similarly, we can get that the $aF(x)+bG(x)$ is better than $G(x)$ on reducing the class ratio dispersion if and only if $\frac{F(x)}{G(x)} \geq \frac{F'(x)}{G'(x)}$.

c.Integrating a with b we can get the result.

Theorem 6. If $F(x), G(x)$ are monotone decreasing function transformations and $F(x)$、$G(x)$ can reduce the class ration dispersion for $\forall x \in [\alpha, \beta]$, For arbitrary nonnegative a, b

a. $aF(x)+bG(x)$ is better than $F(x)$ on reducing the class ratio dispersion if and only if $\frac{F(x)}{G(x)} \geq \frac{F'(x)}{G'(x)}$.

b. $aF(x)+bG(x)$ is better than $G(x)$ on reducing the class ratio dispersion if and only if $\frac{F(x)}{G(x)} \leq \frac{F'(x)}{G'(x)}$.

c. $aF(x)+bG(x)$ is better than $F(x)$ and $G(x)$ on reducing the class ratio dispersion if and only if $\frac{F(x)}{G(x)} \equiv \frac{F'(x)}{G'(x)}$.

Proofing method is consistent with theorem 5.Here mainly used theorem 4 to proof two decreasing function transformations on reducing the class ratio dispersion.

Theorem 7. If $F(x)$ is monotone increasing function transformation and $G(x)$ is monotone decreasing function transformation,which all can reduce the class ration dispersion for $\forall x \in [\alpha, \beta]$. For arbitrary nonnegative a, b, the $aF(x)+bG(x)$ is better than $F(x)$ on reducing the class ratio dispersion if $aF(x)+bG(x)$ is monotone increasing function transformation, then it can reduce the class ratio dispersion.

Proof: $F(x)$ is monotone increasing function transformation,so $F'(x) > 0$. $G(x)$ is monotone decreasing function transformation,so $G'(x) < 0$ .Above all we can get $F(x)G'(x) - F'(x)G(x) \leq 0$ .However , $F(x)$、$aF(x)+bG(x)$ are monotone increasing function transformations and they can reduce the class ratio dispersion for $\forall x \in [\alpha, \beta]$,so there is ,

$$\left[\frac{aF(x)+bG(x)}{F(x)}\right]' = \frac{bF(x)G'(x)-bF'(x)G(x)}{F^2(x)} \le 0 \text{ and } \frac{aF(x)+bG(x)}{F(x)}$$

is monotone decreasing function transformation.Then from theorem 3,we can get $aF(x)+bG(x)$ is better than $F(x)$ on reducing the class ratio dispersion.

Theorem 8. If $F(x)$ is monotone increasing function transformation and $G(x)$ is monotone decreasing function transformation,which all can reduce the class ration dispersion for $\forall x \in [\alpha, \beta]$. For arbitrary nonnegative a, b, the $aF(x)+bG(x)$ is better than $G(x)$ on reducing the class ratio dispersion if $aF(x)+bG(x)$ is monotone decreasing function transformation too, then it can reduce the class ratio dispersion.

Proofing method is consistent with theorem 7.Only different is that those two functions are monotone decreasing function transformation.

4 Example

(1). If $a>1, b>0$ and $x>e$, $F(x)=ax+b$, $G(x)=e^{-x}$ can reduce the class ratio dispersion.The $F(x)+G(x)$ is better than $F(x)=ax+b$ on reducing the class ratio dispersion.

Proof: If $a>1, b>0$,we can know the $F'(x)=a>0$ and $\left[\frac{F(x)}{x}\right]' = \frac{-b}{x^2} < 0$ .From theorem 1 we deem that $F(x)$ can reduce the class ratio dispersion.If $x>e$,we know

$$[G(x)x]' = \left[\frac{x}{e^x}\right]' = \frac{e^x(x-1)}{e^{2x}} > 0$$. From theorem 2 we deem that $G(x)$ can reduce the class ratio dispersion.Beceause of $[F(x)+G(x)]' = (ax+b+e^{-x})' = a-e^{-x} > 0$,the $F(x)+G(x)$ is monotone increasing function transformation,and $\left[\frac{F(x)+G(x)}{x}\right]' = \frac{-xe^{-x}-e^{-x}-b}{x^2} < 0$,so from theorem 3 we know that $F(x)+G(x)$ can reduce the class ratio dispersion. From the

$\left[\frac{F(x)+G(x)}{F(X)}\right]' = \frac{-axe^{-x}-be^{-x}-ae^{-x}}{(ax+b)^2} < 0$ we can know $\frac{F(x)+G(x)}{F(X)}$ is monotone decreasing function, so $F(x)+G(x)$ is better than $F(x)$ on reducing the class ratio dispersion.

(2). If $0<a<1, b>0, 0<x<\ln a^{-1}$,

$F(x)=ax+b$ and $G(x)=e^{-x}$ can reduce the class ratio dispersion. The $F(x)+G(x)$ is better than $G(x)$ on reducing the class ratio dispersion.

Proofing method is consistent with example 1 that mainly used theorem 8

(3). There are X={2.4， 2.5， 2.6， 2.7,2.8}， $F(x)=2x+1$， $G(x)=e^{-x}$,through calculation $F(x)+G(x)$ is really better than $F(x)$ on reducing the class ratio dispersion.

(4). There are X={0.6， 0.8， 1.0， 1.2 }, $F(x)=0.3x+0.1$ and $G(x)=e^{-x}$,comparing $F(x)+G(x)$ and $G(x)$ which is better on reducing the class ratio dispersion.

5 Conclusion

This paper constructs a new method of function transformation mainly using the theorem which use the function transformation to reduce the class ratio dispersion in literature[5]. From theorems 5,6,7 and 8,we can get the criterion of comparing the two function transformation to reduce the class ratio dispersion and we also prove the feasibility and validity of the method by some examples.

References

1. Liu, S.F., Deng, J.L.: The range suitable for GM(1,1). Systems Engineering-Theory and Practic e(5), 121–124 (2002) (in Chinese)
2. Wei, Y., Hu, D.H.: Deficiency of the Smoothness Condition and Its Remedy. Systems Engineering-Theory and Practice (2002)
3. He, B., Meng, Q.: Study on Generalization for Grey Forecasting Model. Systems Engineering-Theory and Practic (9), 138–141 (2002)(in Chinese)
4. Wei, Y.: An Essential Characteristic of Function Transformation to Reduce the Class Ration Dispersion and Its Application. In: Proceedings of the 7th World Congress on Intelligent Control and Automation, Chongqing, China, June 25-27 (2008)

5. Wei, Y., Zhang, Y.: The Necessary and Sufficient Condition of the Monotone Decreasing Function Transformation to Raise The Smooth Degree of Monotone Series. Journ. of Grey System(UK) (4) (2008)
6. Wei, Y., Zhang, Y.: A Criterion of Comparing the Function Transformations To Raise the Smooth Degree of Grey Modeling Date. The Journal of GreySystem(UK) (1), 91–98 (2007)
7. Wei, Y., Zhang, Y.: An Essential Characterictic of the Discrete Function Transformation to Increase the Smooth Degree of Date. The Journal of Grey System(UK) (3), 293–300 (2007)

A 3D Simulation of Gas Emission in Working Face Based on Lattice Boltzmann Method

Qiuqin Lu, Shaomin Yang, and Guangqiu Huang

School of Management, Xi'an University of Architecture and Technology,
Xi'an 710055, Shaanxi, China
luqq911@sohu.com, huangnan93@sohu.com, luqiuqin88@yahoo.cn

Abstract. A new dynamic model for simulation of gas emission in working face was established based on double distributed lattice Boltzmann method(LBM). This model was used to simulate different initial gas press, and many simulation results such as gas emission law and relationship of gas source press and emission velocity in working face were obtained. The conclusion is that as the gas source press increase, earth stress increase, the gas burst velocity increase. The simulation results demonstrate that we can see clearly the gas distribution in working face after gas emitting and can get the rule of working face gas emission. Also LBM provides a new method in further studying of coupling theory between coal and gas, outburst mechanism of gas and gas drainage.

Keywords: gas migration, gas emission, block coupling method, lattice Boltzmann Method.

1 Introduction

Gas emission in working face is the main factor that menace mine safety production. The gas emission block mining, gas burst has great effect to mine safety production. Therefore it has enormous realistic meaning to synthesis manage gas emission. This paper considering the action of rock stress and gas press simulate gas emission and burst based on lattice Boltzmann Method, it have reference and instruction meaning for searching and synthesis managing gas.

2 LBM Simulation Model of Gas Migration in Coal Seam

2.1 Modeling Condition

There are many factors influence the flowing of coal seam, for simplifying a question, some hypothesizes are made in gas flow model[1-5]. Fig.1 is gas emission model.

(1)The gas penetrate from the coal seam is easier than from the motherboard and ceiling, and we can consider the motherboard and ceiling haven't gas;

(2) The porous ratio is not influence by the change of gas pressure;

(3) The gas presses the isothermal process transaction when flow in the coal seam, laminar flow saturation, obedient to Darcy law.

D. Jin and S. Lin (Eds.): Advances in EECM Vol. 1, LNEE 139, pp. 425–430.
springerlink.com

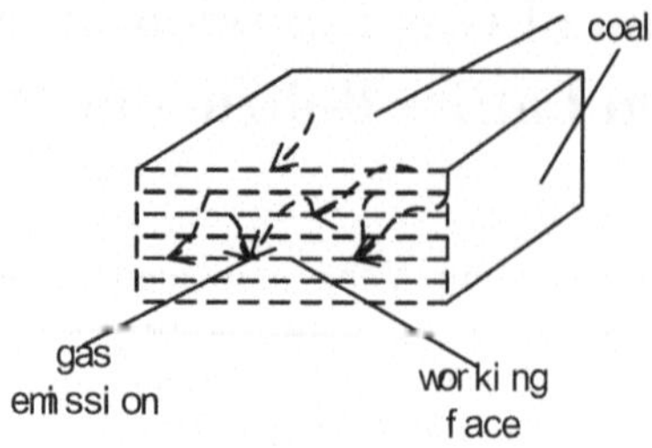

Fig. 1. Gas emission model

2.2 LBM Model of Coal Seam Migration

Velocity model. Evolving equation of velocity model[6-9] is

$$g_i(x+c_i\Delta t, t+\Delta t) - g_i(x,t) = -\frac{1}{\tau_g}\left[g_i(x,t) - g_i^{eq}(x,t)\right] + F_i\Delta t,\quad i = 0 \sim 14$$

Where g_i is The density distributes function, c_i is particle velocity, τ_g is laxation time of density distributes function come to the equilibrium state distributes, $g_i^{(eq)}$ is the equilibrium state density distributes function. Equilibrium state density distributes function is as follow.

$$g_i^{eq} = w_i\rho\left[1+\frac{c_i\cdot u}{c_s^2}+\frac{uu:(c_ic_i-c_s^2I)}{2\phi c_s^2}\right] \qquad i = 0 \sim 14$$

Where w_i is coefficient of D3Q15 model, ϕ is pore ratio of coal seam, ρ is density of lattice, c_s is sound velocity of D3Q15 model.

$$F_i = w_i\rho(1-\frac{1}{2\tau_g})[\frac{c_i\cdot F}{c_s^2}+\frac{uF:(c_i\cdot c_i-c_s^2I)}{2\phi c_s^4}] \qquad i=0\sim 14$$

Where $F=-\frac{\phi\upsilon}{K}u-\frac{\phi F_\phi}{\sqrt{K}}|u|u+\phi G$, where $G=-g\beta(C-C_0)$, g is acceleration of gravity, β is The coefficient of volume expansion of concentration; the relation between shape factor of coal porous seam F_ϕ and hole degree adopt Ergun[9] empirical formula: $F_\phi = 1.75/\sqrt{150\phi^3}$, $K=\phi^3 d_p{}^2/\left[150(1-\phi)^2\right]$

The stress seepage is:

$$K = K_0 e^{a_1\sigma_v + a_2\sigma_H + a_3\sigma_h + a_4\tau}$$

Where $\sigma_v = \sum_{i=1}^{n} r_i h_i = \bar{r}H$, $\sigma_H = \sigma_{hv} + \sigma_{Hmax} + \sigma_{hh1}$, $\sigma_h = \sigma_{hv} + \sigma_{Hmin} + \sigma_{hh2}$

$$\sigma_{hv} = \lambda(\sigma_v - \alpha p) \approx \frac{v}{1-v}(\sigma_v - \alpha p),\ \sigma_{hh1}=\sigma_{hmin}v,\ \sigma_{hh2}=\sigma_{Hmax}v$$

where σ_{hh1} is the sub stress at the maximum horizontal stress direction generated by the minimum horizontal construct stress(σ_{hmin}), σ_{hh2} is the sub stress at the minimum horizontal stress direction generated by the maximum horizontal construct stress(σ_{Hmax}), the shearing force is: $\tau=(\sigma_H-\sigma_h)/2$. In the velocity's field, volume average density and velocity can obtain from distribute function g_i.

$$\rho=\sum_i g_i,\quad u=\upsilon/(c_0+\sqrt{c_0^2+c_1|\upsilon|})$$

Where $c_0=\frac{1}{2}(1+\phi\frac{\upsilon\Delta t}{2K}),\ c_1=\phi\frac{F_\phi\Delta t}{2\sqrt{K}}$, assistant velocity is:

$$\upsilon=\sum_i g_i\frac{c_i}{\rho}+\frac{\Delta t}{2}\phi G.$$

Making use of many size spread methods can reduce the macroscopic Brinkman-Forchheimer-Darcy equation proposed by Nithiarasu from the equation:

$$\begin{cases}\nabla\cdot u=0\\ \dfrac{\partial u}{\partial t}+u\cdot\nabla\dfrac{u}{\phi}=-\dfrac{1}{\rho}\nabla(\phi p)+\upsilon_e\nabla^2u+F\end{cases}$$

Where υ_e viscidity coefficient which is determined by τ_g: $\upsilon_e=(\tau_g-0.5)c_s^2\Delta t$, p gas pressure, and $p=c_s^2\rho/\phi$, where ϕ is coal seam hole degree.

Pressure field model. Evolving equation of pressure model is

$$g_i(x+c_i\delta_t,t+\delta_t)-g_i(x,t)=-\frac{1}{\tau}[g_i(x,t)-g_i^{(eq)}(x,t)]+\delta_tF_i$$

The equilibrium state distribution function is:

$$g_i^{(eq)}=\omega_i[\frac{\phi p}{c_s^2\rho}+\frac{c_i\cdot u}{c_s^2}+\frac{uu:(c_ic_i-c_s^2I)}{2\phi c_s^4}]$$

$$p=\frac{c_s^2\rho}{\phi}\sum_i g_i,\quad u=\frac{v}{c_0+\sqrt{c_0^2+c_1|v|}},\quad v=\sum_i c_ig_i+\frac{\delta_t}{2}\varepsilon G$$

A macroscopic control equation can be evolved by the spread technique of Chapman-Enskog.

$$\begin{cases}\dfrac{1}{c_s^2}\partial_t(\varepsilon p)+\nabla\cdot u=0\\ \partial_tu+u\bullet\nabla(\dfrac{u}{\varepsilon})=-\dfrac{1}{\rho}\nabla(\varepsilon p)+v_e\nabla^2u+F\end{cases}$$

2.3 D3Q15 Calculation Model of Coal Seam Migration

Fig.2 is D3Q15 calculation model of coal seam migration.

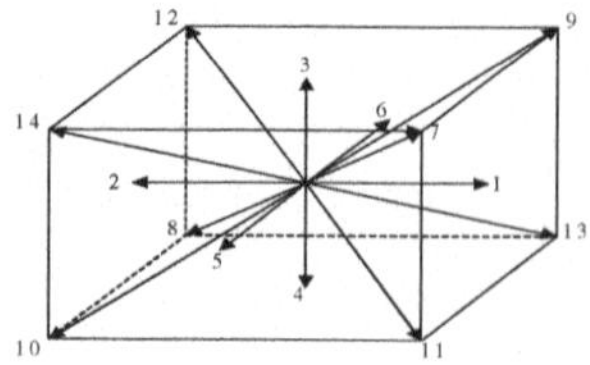

Fig. 2. D3Q15 model

The disperse velocity set (E)of the model and right coefficient W is:

$$E = \left(\begin{pmatrix} 0 \\ 0 \\ 0 \end{pmatrix}, \pm \begin{pmatrix} 1 \\ 0 \\ 0 \end{pmatrix}, \pm \begin{pmatrix} 0 \\ 1 \\ 0 \end{pmatrix}, \pm \begin{pmatrix} 0 \\ 0 \\ 1 \end{pmatrix}, \pm \begin{pmatrix} 1 \\ 1 \\ 1 \end{pmatrix}, \pm \begin{pmatrix} 1 \\ 1 \\ -1 \end{pmatrix}, \pm \begin{pmatrix} 1 \\ -1 \\ 1 \end{pmatrix}, \pm \begin{pmatrix} 1 \\ -1 \\ -1 \end{pmatrix} \right) c$$

$$\omega_i = \begin{cases} 2/9 & i=0 \\ 1/9 & i=1\sim6 \\ 1/72 & i=7\sim14 \end{cases}$$

3 Case Study

The simulation parameters are: the 3D grid is 65×35×35, gas density is 0.5, coefficient of migration viscidity is 0.08, Rey=10. In the following, the influence of initial press on gas emission are researched, the simulation result of initial press p=1.0, 2.5, 4.0 are as Fig.3 to Fig.5 respectively.

Fig.3 is the simulation result of gas emission at initial press p=1.0 at time step t=5000, 10000.

Fig.4 is the simulation result of gas emission at initial press equal to 2.5 at time step t=5000, 10000.

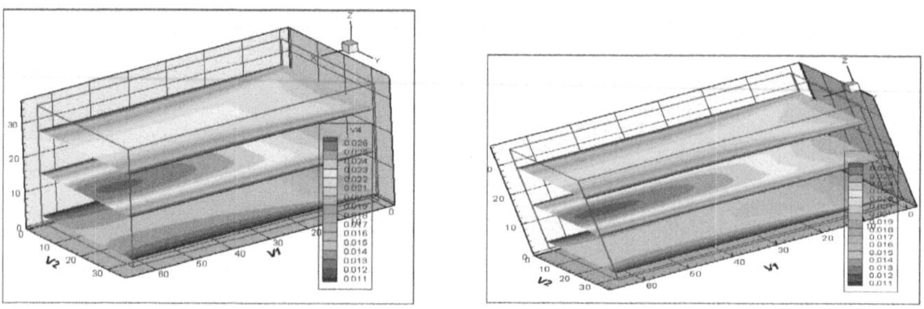

(a) t=5000 (b) t=10000

Fig. 3. Simulation result of gas emission at initial press p=1.0

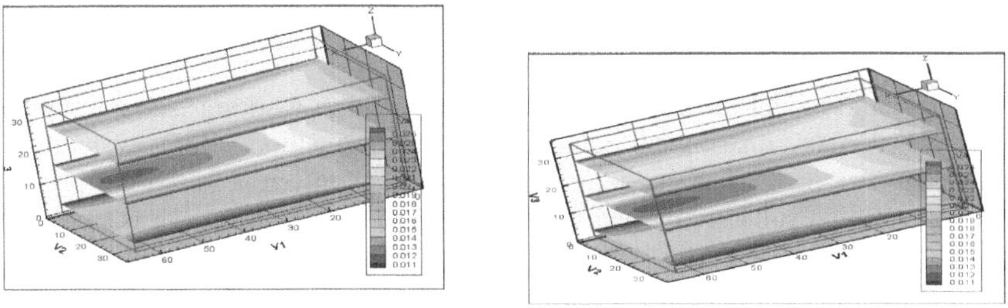

(a) t=5000 (b) t=10000

Fig. 4. Simulation result of gas emission at initial press p=2.5

Fig.5 is the simulation result of gas emission at initial press equal to 4.0 at time step t=5000, 10000.

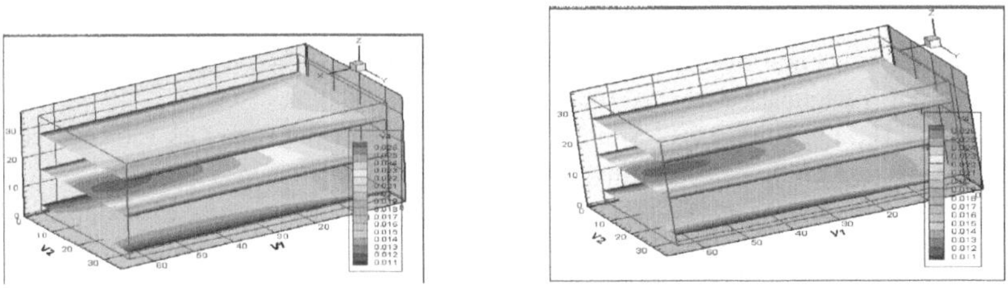

(a) t=5000 (b) t=10000

Fig. 5. Simulation result of gas emission at initial press p=4

Different presses from gas source have different gas emission effect. According to the different initial press of gas source, the relationship of gas source press and gas emission in working face is shown in Fig.6 by simulation.

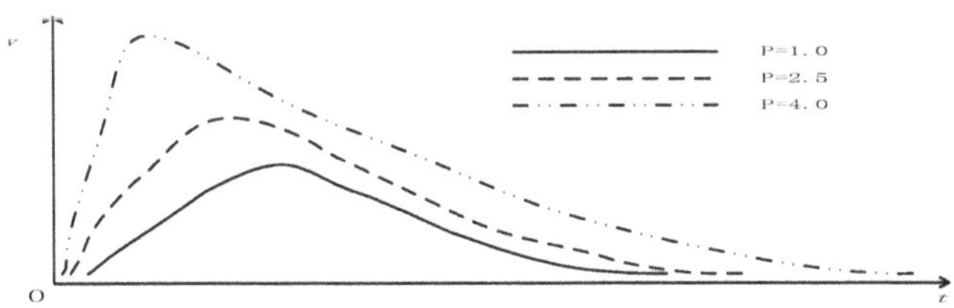

Fig. 6. The relationship of gas source press and gas emission in working face

As is shown in Fig.6, the emission velocity and time length is closely related to initial press. Initial press is more great, the time take to the biggest gas emission is more short, but the total time of the gas emission is the longest.

4 Conclusion

A new dynamic model for simulation of gas emission in working face is established based on double distributed lattice Boltzmann method(LBM)—velocity and concentration gradient. This model is used to simulate different initial gas press, and many simulation results such as gas emission law and relationship of gas source press and emission velocity in working face are obtained. This simulation method can observe the gas emission directly, it provides new way of thinking for explore further the couple mechanism of gas and coal, the burst mechanism of coal and gas, and the design of gas drawing method.

Acknowledgments. This work is supported by Research and Development Project of Science and Technology of Shaanxi Province under Grant 2011K06-08, and by Nature Science Foundation of Shaanxi Provincial Department of Education under Grant 11JK0772 and 09JK524.

References

1. Bai, F.: Experimental Research on the Problem of Methane Ascension in the Gob. Journal of Liaoning Technical University 16(4), 412–416 (1997)
2. Dong, G., Hu, Q.: COSFLOW Simulation Prediction of Coal Face Gas Emission and Gob Gas Drainage. Mining Safety & Environmental Protection 34(2), 4–7 (2007)
3. Li, Z.-X., Wu, Q., Xiao, Y.-N.: Numerical Simulation of Coupling Mechanism of Coal Spontaneous Combustion and Gas Effusion in Goaf. Journal of China University of Mining & Technology 37(1), 38–42 (2008)
4. Ye, Q., Lin, B., Jiang, W.: The Study of Methane Outflow Law in Coal Mining Face. China Mining Magazine 15(5), 38–41 (2006)
5. Wang, L.-Z., Du, C.-Z., Bu, W.-K.: Numerical Simulation of Gas Emission from Borehole Wall in Coal Seam. Mining Safety & Environmental Protection 35(6), 4–7 (2008)
6. Yan, W.-W., Liu, Y., Xu, Y.-S.: Study on the natural convection heat transfer in porous media using LBM. Journal of Xi an Shiyou University(Natural Science Edition) 22(2), 149–152 (2007)
7. Guo, Z.L., Shi, B.C., Zheng, C.G.: A coupled lattice BGK model for the Bouessinesq equation. Int. J. Num. Fluids 39(4), 325–342 (2002)
8. Guo, Z.L., Zheng, C.G., Li, Q., Wang, N.C.: The Lattice Boltzmann method on Fluid Fluid Dynamics. HuBei Science and Technology Publishing House (2002)

Simulation of Working Face Gas Emission Based on LBM

Qiuqin Lu, Shaomin Yang, and Guangqiu Huang

School of Management, Xi'an University of Architecture and Technology, Xi'an 710055, Shaanxi, China
luqq911@sohu.com, huangnan93@sohu.com, luqiuqin88@yahoo.cn

Abstract. A lattice Boltzmann method(LBM) model for simulation of gas emission in working face was established which was used with the ventilation system of model U to simulate gas transport in fissured coal and gas diffusion in tunnels after emitting. And the simulation results demonstrate that we can see clearly the gas distribution in working face after gas emitting and can get the rule of working face gas emission. Also LBM provides a new method in further studying of coupling theory between coal and gas, outburst mechanism of gas and gas drainage.

Keywords: working face, gas emission, simulation, Lattice Boltzmann Method.

1 Introduction

Gas accident is one of the severe disasters of coal mine, while in working face with high gas, there are many hidden danger for the large quantity of gas emission. The law of working face gas emission can influence the prediction result of gas emission in working face, thus determine the design of ventilation and gas taking out. The quantity of gas emission influence the rationality of the working face ventilation, the validity of gas dispensary measure and effectiveness production with safety, therefore it has great meaning to research the gas emission, control and manage. Many researches on the migration and emission of gas are based on numeric, simulation and road test[1-3]. The advantages of these methods are solve and model by traditional method easily, while these method can not reflect the relationship of the complexity of total system and the simple movement of flow particle and they are difficult in processing complex boundary. This paper use a kind of direct simulation method to simulate the gas migration in coal seam and diffuse in U-type working face ventilation system after emission from coal seam by Lattice Boltzmann Method(LBM).

2 LBM Model of Gas Emission

2.1 Modeling Conditions

Coal is a kind of porous medium. There are many factors influence the flowing of coal seam, modeling conditions are: The gas penetrate from the coal seam is easier

D. Jin and S. Lin (Eds.): Advances in EECM Vol. 1, LNEE 139, pp. 431–436.
springerlink.com

than from the motherboard and ceiling, and we can consider the motherboard and ceiling are not penetrate gas and without gas. The porous ratio is not influence by the change of gas pressure. The gas content in the coal seam is composed by dissociated gas and adsorbed gas. The gas is an ideal gas. The gas presses the isothermal process transaction when flow in the coal seam, laminar flow saturation, obedient to Darcy law. Because of working face flow is belong to a turbulent flow, gas directly drive by flow, neglect the probability of gas entering the adopted zone. A gas emission model of U type ventilation system in coal working face is constructed based on the above hypothesizes.

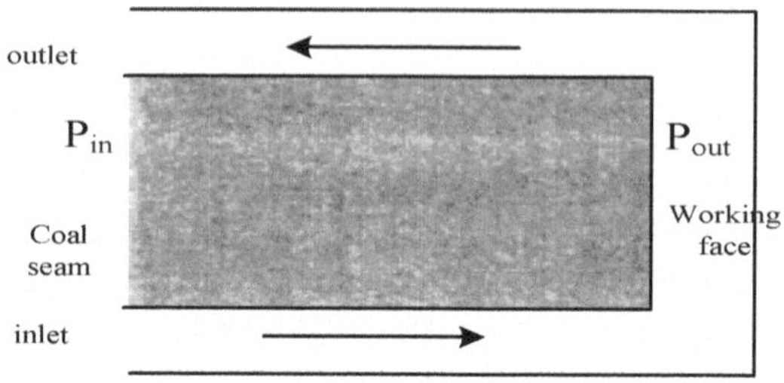

Fig. 1. The kinetics model of gas emission

2.2 LBM Equation of Gas Migration in Working Face

Gas flows in the synthesize effect of concentration gradiant and air flow after emission. A double model of velocity field and concentration field which is coupled by Boussinesq[6] are adopted. Velocity field model adopt model of D2Q9, concentration field adopts model of D2Q4.

(1) Velocity model. Evolving equation of velocity model is [6]:

$$g_i(x+c_i\Delta t, t+\Delta t) - g_i(x,t) = -\frac{1}{\tau_g}[g_i(x,t) - g_i^{(eq)}(x,t)] \quad i = 0\cdots 8 \tag{1}$$

Where g_i is The density distributes function, c_i is particle velocity, τ_g is laxation time of density distributes function come to the equilibrium state distributes, $g_i^{(eq)}$ is the equilibrium state density distributes function. Equilibrium state density distributes function is as follow.

$$g_i^{(eq)} = \begin{cases} -\dfrac{4\sigma p}{c^2} - \dfrac{2u^2}{3c^2} & i{=}0 \\ \dfrac{\beta p}{c^2} + \dfrac{c_i \cdot u}{3c^2} + \dfrac{(c_i \cdot u)^2}{2c^4} - \dfrac{u^2}{6c^2} & i{=}1,2,3,4 \\ \dfrac{\gamma p}{c^2} + \dfrac{c_i \cdot u}{12c^2} + \dfrac{(c_i \cdot u)^2}{8c^4} - \dfrac{u^2}{24c^2} & i = 5,6,7,8 \end{cases}$$

Where α, β, *model* parameter, u,p are respectively macroscopic velocity and press of fluid, it is:

$$u = \frac{1}{\rho}\sum_{i=0}^{8} c_i g_i \ , \ p = \frac{c^2}{4\delta}[\sum_{i=1}^{8} g_i + \frac{4}{9}(-1.5\frac{u^2}{c^2})] \ , \rho = \sum_{i=0}^{8} g_i \tag{2}$$

The eq.1 can derivative Navier-Stokes equation by making use of a Chapman-Enskog method.

$$\nabla \cdot u = 0, \frac{\partial u}{\partial t} + u \cdot \nabla u = -\nabla p + \upsilon \nabla^2 u$$

Where $\upsilon = (2\tau - 1)(\Delta x)^2 / (6\Delta t)$ is coefficient of movement viscosity, $\Delta x, \Delta t$ are respectively lattice width and time to tread lengthways.

(2) Concentration field. Model of concentration field evolution equation is [6]:

$$C_i(x + c_i\Delta t, t + \Delta t) - C_i(x,t) = -\frac{1}{\tau_c}[C_i(x,t) - C_i^{(eq)}(x,t)] \quad i=1\ldots4 \tag{3}$$

Where C_i is the concentration distributes function, τ_c is laxation time of concentration distributes function come to the equilibrium state distributes, its definition is: $C_i^{(eq)} = \frac{C}{4}[1 + 2\frac{c_i \cdot u}{c}]$.

Where C is gas macro-concentration, $C = \sum_{i=1}^{4} C_i$ .eq.3 can derivative control equation by making use of a Chapman-Enskog spread method. $\frac{\partial C}{\partial t} + \nabla \cdot (uC) = D\nabla^2 C$, Where $D = (2\tau_c - 1)(\Delta x)^2 / (4\Delta t)$ is coefficient of mass diffusion.

2.3 LBM Equation of Gas Migration in Coal Seam

The gas flow in coal seam is in a laminar flow status, and is generally oozed streaming running in the hole and the crack by laminar flow status to the low-pressure zone by high pressure zone, obey Darcy law.

(1)velocity field. According to the gas flow in the coal seam, the following equation is adopted to simulate its velocity field, an evolvment equation [6] is:

$$g_i(x + c_i\Delta t, t + \Delta t) - g_i(x,t) = -\frac{1}{\tau_g}\left[g_i(x,t) - g_i^{eq}(x,t)\right] + F_i\Delta t, \quad i = 0,\cdots,8 \tag{4}$$

Where $c=\Delta x/\Delta t$, Δx and Δt are respectively lattice width and time to tread lengthways. τ_g is relax time; g_i is distribute function; $g_i^{(eq)}$ is the equilibrium state distributes function, equilibrium state distributes function is :

$$g_i^{eq} = w_i \rho \left[1 + \frac{c_i \cdot u}{c_s^2} + \frac{uu : (c_i c_i - c_s^2 I)}{2\varphi c_s^2}\right] \qquad i = 0,\cdots,8$$

Where $w_0 = 4/9, w_i = 1/9, i = 1,\cdots,4; w_i = 36, i = 5,\cdots,8$. Mechanics item in eq.5 is

$$F_i = w_i \rho (1 - \frac{1}{2\tau_g})[\frac{c_i \cdot F}{c_s^2} + \frac{uF:(c_i \cdot c_i - c_s^2 I)}{2\phi c_s^4}] \quad i = 0, \cdots, 8$$

Where $F = -\frac{\phi \upsilon}{K} u - \frac{\phi F_\phi}{\sqrt{K}} |u| u + \phi G$, where $G = -g\beta(C - C_0)$, g is acceleration of gravity, β is The coefficient of volume expansion of concentration; the relation between shape factor of coal porous seam F_ϕ and hole degree adopt Ergun empirical formula:

$$F_\phi = 1.75 / \sqrt{150\phi^3}, K = \phi^3 d_p{}^2 / \left[150(1-\phi)^2\right]$$

Volume average density and velocity can get from distribute function g_i

$$\rho = \sum_i g_i, \; u = \upsilon / c_0 + \sqrt{c_0^2 + c_1 |\upsilon|}$$

Where $c_0 = \frac{1}{2}(1 + \phi \frac{\upsilon \Delta t}{2K})$, $c_1 = \phi \frac{F_\phi \Delta t}{2\sqrt{K}}$, $\upsilon = \sum_i g_i \frac{c_i}{\rho} + \frac{\Delta t}{2} \phi G$.

(2) Concentration field. Model of concentration field evolution equation is

$$C_i(x + c_i \Delta t, t + \Delta t) - C_i(x,t) = -\frac{1}{\tau_c}[C_i(x,t) - C_i^{(eq)}(x,t)] + \Delta t \frac{R}{b} \quad i = 1 \cdots 4 \qquad (5)$$

Where C_i is the concentration distributes function, τ_c is laxation time of concentration distributes function come to the equilibrium state distributes, R source item of diffuse, $R = a\sqrt{p}$, $C_i^{(eq)}$ is equilibrium state distribute function, its definition is: $C_i^{(eq)} = \frac{C}{4}[1 + 2\frac{c_i \cdot u}{c}]$.

Where C is micro concentration of flow, $C = \sum_{i=1}^{4} C_i$.

3 Case Study

Take Fig.1 as example, initial condition: t=0, u_x=u_y=0, ρ_0=1.0, p_0=*1/3*; u_0=0.05, *Renold*=2500. co *highConcentration*=*500*; *lowConcentration*=10; *Renold*=10, ϕ=0.3,0.6,0.9,p_0=1.1, a=10. The velocity and concentration of every block are calculated respectively, and very block are coupled in the public boundary as above mention. Boundary are processed by wall rebound.

Fig.2(a) is distribution of streamtrace and velocity in direction x. Laws of velocity in direction x is: the velocity in the inlet air lane is positive while in the outlet air lane is negative.

Fig. 2(b) is the contour map of velocity in direction y (u_y). The maximum and minimum velocities in direction y are in the working face. In the working face the velocity range is 8.5 to -0.2. In the inlet and outlet air lane the velocity in direction y is zero.

Fig.3(a)~Fig.3(b) are the tunnel concentration distribution at t=6000,10000 time step respectively. In the initial stage in of gas emission, the quantity of gas contents in the coal seam is larger, the quantity of burst gas is smaller. As time move on, the contents of coal seam gas become index decrease, a great deal of the gas gush, diffuse at work surface. After gas gushes, gas concentration gradual increase from inlet to outlet, while the gas concentration from inlet to central range change gradually, while from central range to outlet change quickly, especially at 30m of the side of outlet gas concentrations is more great. At the same time, we can also discover that from the coal wall to goaf gas changes from high, higher and low distribution.

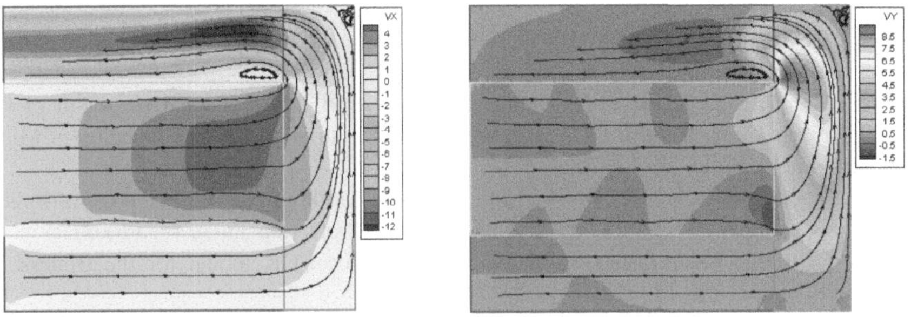

(a) Velocity *of x* direction(u_x) (b) Velocity *of y* direction(u_y)

Fig. 2. Velocity at t=10000

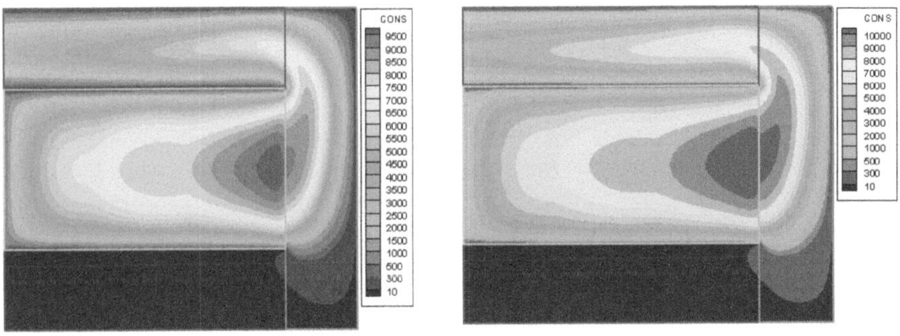

(a) Concentration at t=6000 (b) Concentration at t=10000

Fig. 3. Tunnel concentration distribution at t=6000, 10000 time step

4 Conclusion

According to the double velocity-concentration lattice Boltzmann model, the gas migration in the coal seam and diffusion in working face after emission are created respectively. Block coupling algorithm are used to simulate gas migration in coal seam and defuse in working face in U-type ventilation system. This simulation method can observe the gas emission directly, it provides new way of thinking for explore further the couple mechanism of gas and coal, the burst mechanism of coal and gas, and the design of gas drawing method.

Acknowledgments. This work is supported by Research and Development Project of Science and Technology of Shaanxi Province under Grant 2011K06-08, and by Nature Science Foundation of Shaanxi Provincial Department of Education under Grant 11JK0772 and 09JK524.

References

1. Dong, G., Hu, Q.: COSFLOW Simulation Prediction of Coal Face Gas Emission and Gob Gas Drainage. Mining Safety & Environmental Protection 34(2), 4–7 (2007)
2. Li, Z.-X., Wu, Q., Xiao, Y.-N.: Numerical Simulation of Coupling Mechanism of Coal Spontaneous Combustion and Gas Effusion in Goaf. Journal of China University of Mining & Technology 37(1), 38–42 (2008)
3. Wang, L.-Z., Du, C.-Z., Bu, W.-K.: Numerical Simulation of Gas Emission from Borehole Wall in Coal Seam. Mining Safety & Environmental Protection 35(6), 4–7 (2008)
4. Guo, Z.L., Shi, B.C., Zheng, C.G.: A coupled lattice BGK model for the Bouessinesq equation. Int. J. Num. Fluids 39(4), 325–342 (2002)
5. Guo, Z.L., Zheng, C.G., Li, Q., Wang, N.C.: The Lattice Boltzmann method on Fluid Fluid Dynamics. HuBei Science and Technology Publishing House (2002)

Data Integration and Modeling Based on the Implementation of Informatization in the Power Engineering Enterprises

Xiaohua Song, Lixiao Wang, and Pie Zu

School of Economics and Management, North China Electric Power University, Beijing, 102206, China

Abstract. Power engineering enterprise has all various kinds of data, large amount of data, and the relationship between the data is complicated. Therefore, the key point of power engineering enterprise information system implementation is building a complete set of data model and summing up a set of data preparation methods and way which can be applied to the enterprises. Based on the present situation and development of a power engineering enterprise, this article proposed a suggestion on data modeling steps of information system implementation, discussed the methods of data collection, analysis and preparation when implementing information system, established unified data and information classification and coding system, constructed a unified data model which was verified finally.

Keywords: power engineering enterprise, informatization, data collection, data modeling.

1 Introduction

Informatization is the key to achieve modernization and digital management of power engineering. At present, the informatization of project management is rare, the level of informatization needs to be raised and engineering data scatter in different departments or subsidiaries. Therefore, it is significant for the enterprises to build an information system of project management based on power engineering company. It will improve capability of project management and enhance response capacity of market. But the construction of information system is not just the use and planning of itself, but more is the improvement of foundation preparation.

2 Data Collection

The work contents of this part include data-related collection, analysis and sorting of the key projects and product.

D. Jin and S. Lin (Eds.): Advances in EECM Vol. 1, LNEE 139, pp. 437–441.
springerlink.com

2.1 Process Data Collection

All the enterprise's management and business activities perform for a variety of processes. So, processes form a variety of products data. Each process is not only input and output, but there are some important intermediate data. These data can affect implementation of PLM systems directly. So, data must be collected in detail.

2.2 Research Data Collection

This process needs to undergo three stages of development: digital design, modular design, and collaborative design. The three stages are evolving. Each stage has its own needs of the data.

First, the digital product design. The purpose of the digital of product model (as shown in Table 1) is to improve designs to enhance efficiency of product development and products' reliability.

Table 1. Digital design

Digital design	Introduction
Digital product	The digital description of product in the product's full life cycle.
Digital assembly	Using computer technology to simulate product assembly process to test their assembly performance based on the digital product definition..
Digital simulation analysis	Using computer and network technology to achieve product functions of every stage in the product life cycle.
Digital processing	Using computer to model the actual processing trajectory to consider manufacturability of design stage of the product.

Second, the Modular Design. Modular technology is designed for deploying product rapidly. It makes enterprises by combining existing modules or new module to produce products quickly and cost-effectively which can adapt market and technological changes.

Third, the Collaborative Design. Collaborative design involves design processes, designers, computer collaboration and other entities. The relationship between design entities is shown in table 2.

Table 2. The relationship between design entities

Design entities	Introduction
Collaborative of product process	Coordinating the product process of all stages to achieve optimal objective.
Synergy between the designers	Solving cooperation between the personnel of different areas and the same field design personnel of different levels.
Synergy between designers and computer	To make computer systems more closely with the designers, we need to strengthen the study of man-machine cooperation.
Collaborative of design environment	Coordinating transformation of understanding and expression about knowledge between designers under heterogeneous environment.

2.3 Data Collection of Products' Functional Structure

The electric power enterprise is a complex system which contains a lot of secrets people do not know. The technical data is the key point of achieving product functions. It relates to product structure, simulation, evaluation and security.

3 Data Analysis

The main target of information systems is data processing. After data collection phase, the data collected is messy and therefore need to collate and analyze. According to the general relationship between data in the product life cycle given in figure 1, data can be easily classified and collected to form new processes data meeting new product life cycle in the future.

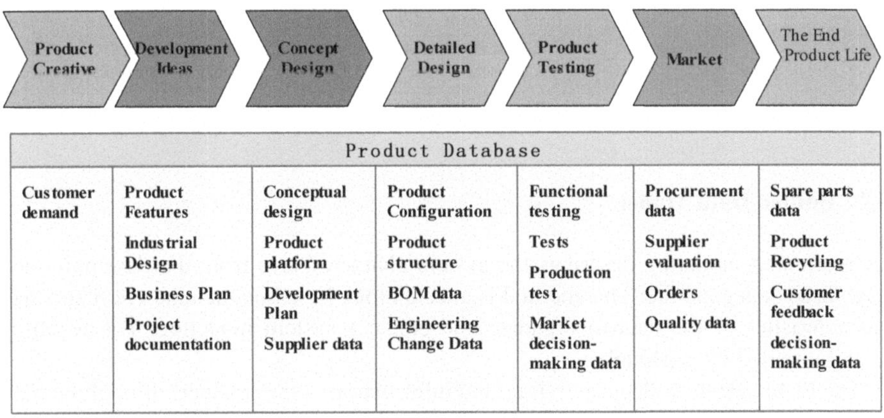

Fig. 1. Data relationships of various stages in product life cycle

4 Data Modeling

In planning the new information system, "the overall data plan" should be as the core of data modeling. Then it establishes a unified classification and coding system of data and information to integrate data resources spread in various information systems of single functions. We can build data model of the product life cycle and the association relationship of data model between the various stages according to data classification. (Data Map, shown in Figure 2).

4.1 Unified Code

Unified code is very important to data exchange and share between the various aspects of product development. It can reduce duplication of information collection and determine the corresponding relationship between Information and concepts to ensure the reliability, comparability, usability and traceability of the information. Therefore, we need to have a clear coding system and design encoding rules to give the product and document an effective, unique and reasonable identification number.

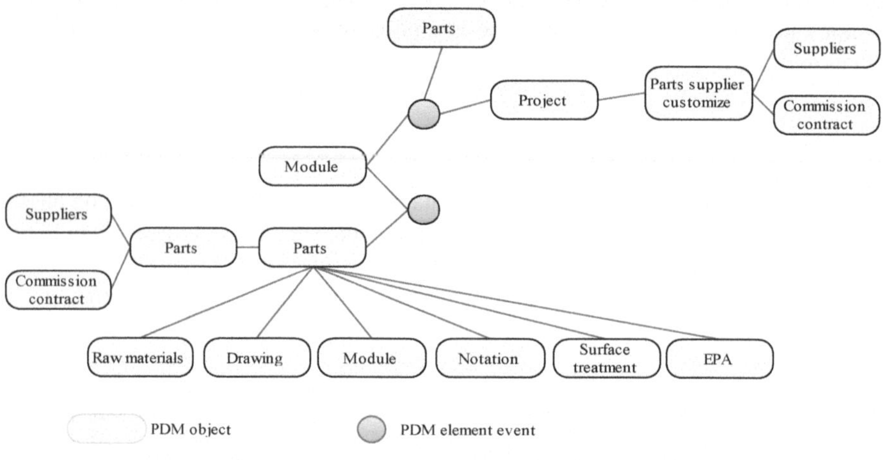

Fig. 2. Data Map

4.2 Unified Data Model

Data-oriented modeling describes the system characteristics from the data point and how these data generate. This method is a technology based on the inherent nature and characteristics of the data to organize and create a system modeling. The resulting model is called Data Model.

The PLM system collects activities and information of the products throughout their life cycle. There are many independent systems in this system .They have their own format data. How to provide data services for various departments quickly and accurately and ensure consistency of the data is the key issues of product data management.

5 Model Verification

According to the order of data collection, analysis and modeling, we select two or three typical product components to determine the feasibility and correctness of the model. The main tasks include:

(1) Data collection: according to the selected two or three parts, we collect processes data, research data and technical data, and the existing basic data.

(2) Data analysis: analysis, classification, and integration the collected product data.

(3) Data modeling: preparing uniform codes, collecting the table of data and establishing a unified data model.

(4) Analyzing and sorting: according to the implementation of the above three steps, amending the proposals.

6 Conclusions

This study sums up a set of methods of data preparation applied to power engineering companies based on a preliminary understanding of informatization of power engineering enterprise from the informatization perspective. Draw the following conclusions:

(1) The methods of data preparation relying on information technology platform lay the foundation for the implementation of information system, improves the enterprise's performance of product development and enhances the ability of the accumulation of knowledge about product development and product innovation.

(2) This study gives proposals on the data modeling when implementing information system. It describes how to collect, analyze and sort data, how to establish a unified data and information coding system and how to set up a unified data model. Finally, we validate the model.

References

1. Schwinn, A., Schelp, J.: Design Patterns for Data Integration. Journal of Enterprise Information Management 4(18), 14–15 (2005)
2. Wang, J.: Informatization Construction and Management of Power Enterprise. The Power of Information Technology 7(2), 17–19 (2009)
3. Ma, Z., Wang, X.: Digital Combination Fixture Management System. In: The Proceedings of the China Association for Science and Technology, vol. 4(2), pp. 302–305 (2008)
4. Wainwright, D., Sambrook, S.: The ethics of data collection: unintended consequences? Journal of Health Organization and Management 3(24), 7–10 (2010)

Efficient 3D-Visualization Methods for Electrical Exploration Data

Jie Hua[1], Tingyan Xing[1,*], Xiaoping Rui[2], Yanyun Sun[3], and Haizhi Zhang[1]

[1] Institute of Information Engineering, China University of Geosciences, 100083 Beijing, China
[2] College of Resources and Environment, Graduate University of Chinese Academy of Sciences, 100049 Beijing, China
[3] Institute of Geophysics and Information Technology, China University of Geosciences, 100083 Beijing, China

Abstract. Electrical exploration is one of the most important methods for geophysical exploration. In this paper, the authors explore efficient 3D-visualization methods for electrical exploration data by using computer graphics technology. This paper proposes a high-performance method of data organization according to the data storage and distribution, and then builds the exploration line model by using a method of generating TIN based on the data points in exploration direction. In order to explain the spatial circumstance of exploration area comprehensively, this paper proposes a partial interpolation algorithm to realize the modeling of data field rapidly. These models show the electrical exploration data vividly and comprehensively, providing reliable bases for related geological interpretation.

Keywords: Electrical exploration, 3D visualization, Data simulation, Virtual reality, Spatial information.

1 Introduction

As an effective method for obtaining underground 3D information, geophysical exploration is based on the various kinds of different physical properties of rocks and minerals [1]. Electrical exploration is one of the geophysical exploration methods which based on the differences of rocks and minerals' electromagnetic properties. Traditionally, to start analyze of geophysical exploration data, we need to analysis from point, line, profile and etc., and then we can interpret, deduce and forecast. Owing to the complexity of underground structures [2], the traditional ways of displaying 2D image have revealed some deficiencies in the process of geological interpretation. With the rapid developments of virtual reality and scientific visualization, the boring data can be converted into 3D graphics, and help people insight into the inherent relationships and rules of data [3].

* Corresponding author.

D. Jin and S. Lin (Eds.): Advances in EECM Vol. 1, LNEE 139, pp. 443–448.
springerlink.com

This paper takes as one of exploration method called CASMT (Controlled source audio magneto-telluric method) as an example, explores and implements a group of algorithms for the data visualization, and builds related models.

2 High-Performance Organizations for Electrical Exploration Data

During the store procedure of electrical exploration data, the information of different exploration line is separated, as well as the information of the surface and the underground. Each exploration line's information is composed of two files. The file as the left part of Table 1 showing stores the surface information of electrodes, and another file as the right of Table 1 showing stores the underground exploring information, which is explored form the midpoint of the two adjacent electrodes. Obviously, electrical exploration data is a kind of real 3D spatial information which need to process data organization and model by the ways of real 3D spatial information.

Table 1. Surface and Underground Information file of electrical exploration data

No.	X	Y	Height	Lng	Lat		No.	PosZ	Value
0	4523388	375024	1663	40. 50061	97. 31059		20	1660	2. 839
40	4523423	375044	1658	40. 50072	97. 31068		20	1657	2. 839
80	4523458	375064	1677	40. 50084	97. 31076		20	1657	2. 749
...							...		

The information explored by electrical exploration is so much and separated that the data organization requires both improving the algorithms and reducing storage redundancy. This paper proposes a high-performance method for organization of electrical exploration data. The process of the data organization for each exploration line is as follows: Firstly, get the information of every electrode in Surface Information file, then average the features in every two adjacent points to create a new data structure which stores the surface information of exploration point. Secondly, get the information of every exploration data point in Underground Information file, and compose a exploration point data structure by combining the a series of underground exploration data points with a surface exploration point which have the same Point No.. Thirdly, compose the more exploration points into the information of one exploration line. Finally, organize the information of several exploration lines.

3 Build Exploration Lines Model Based on Exploration Direction Data Points That Can Generate TIN

After making the data organization, we can create the 3D visualization model by exploration data points themselves. Due to the different number of exploration data points explored by each two electrodes, model electrical exploration data by using the method for modeling regular data is not possible. Therefore, this paper proposes a method based on exploration direction data points that can generate TIN. As Fig.1 shows, the general ideas for building this kind of TIN are as follows: Build triangulation by the order of surface exploration points, regarding a number of data points explored by two adjacent exploration points as the modeling unit, and down to generate TIN gradually.

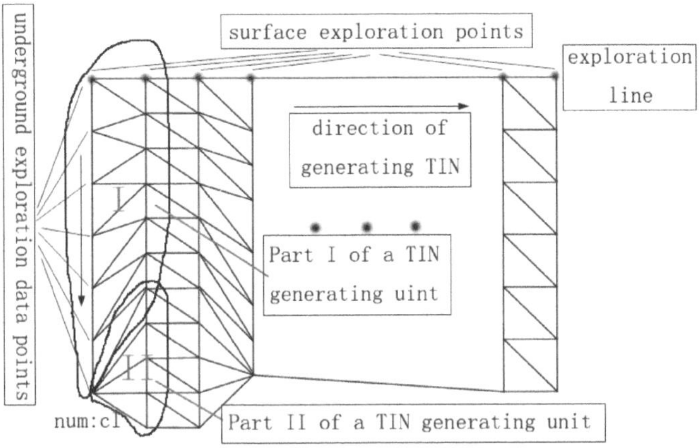

Fig. 1. Schematic diagram for the algorithm of generating TIN based on exploration direction data points

What Fig.2 shows is the 3D visualization model under this method. The warmer colors in this model indicate that these areas have greater apparent resistivity values. On the contrary, cooler colors indicate smaller apparent resistivity values.

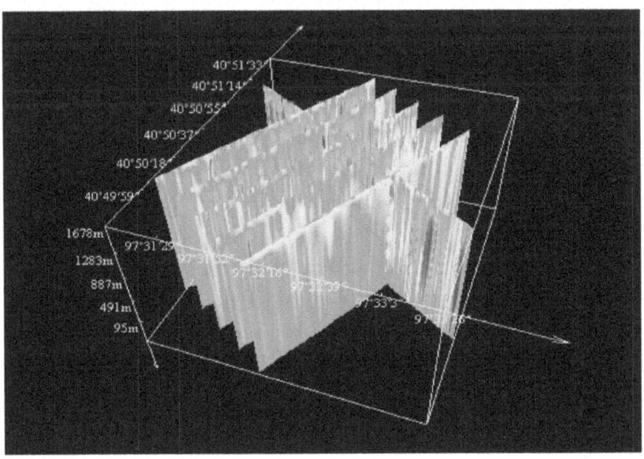

Fig. 2. The 3D visualization model for data of electrical exploration line

4 Build Field Model Based on the Fast Partial Interpolation Algorithm

It can be clearly seen that the exploration density of electrical exploration is quite high, and forms the data field in a certain underground range. In order to model easily, the original data need to be spatially interpolated into regular field data [4].

The closer two objects are, the more similar properties they have. This study interpolates the original exploration data by this principle. The general formula of Inverse Distance Weighted interpolation is [5]:

$$\hat{Z}(S_0) = \sum_{i=1}^{N} \lambda_i Z(S_i) \tag{1}$$

In this formula, $\hat{Z}(S_0)$ stands for forecast value of S_0; N stands for the number of samples used in the forecast calculation; λ_i stands for weight of each sample in the calculation, which will reduce with the distance of the sample and forecast point increasing; $Z(S_i)$ stands for the measured value of the sample. The formula of determined weights is:

$$\lambda_i = d_{i0}^{-p} / \sum_{i=1}^{N} d_{i0}^{-p} \sum_{i=1}^{N} \lambda_i = 1 \tag{2}$$

Among the formula, d_{i0} stands for the distance between each sample and forecast point; p is a parameter.

In order to minimize the complexity of the algorithm, we take p a value of 2in this study. Therefore, the formula for interpolating electrical exploration data field from irregular to regular by IDW is:

$$\hat{V}(S_0) = \sum_{i=1}^{N} d_{i0}^{-2} V(S_i) / \sum_{i=1}^{N} d_{i0}^{-2} \tag{3}$$

According to the spatial distribution character of electrical exploration data, we propose an improved algorithm for IDW, which achieves the balance of interpolation accuracy and algorithm complexity. Ideas for the arithmetic are as follows: Each forecast point in the regular data field need to calculate its forecast value by IDW; the forecast points in the regular data field is close to exploration line, and the distribution is $S \times D$; the samples' distributions of the original exploration line data at the line direction and the depth direction are S_s and D_s. In this way, to every element of $S \times D$, when sampling the surrounding samples, it is just necessary to pick partial samples explored by corresponding exploration line instead of the whole samples.

Specific algorithms are as follows: give the range ratios for those two interpolation directions as Sr and Dr; to most of the forecast points, the ranges of samples involved in the interpolation will be with the results which calculated by the given ratios; as for peripheral forecast points, if the ranges calculated by the given ratios are beyond the scope of the original data field, the excess parts of the ranges will be limited by the scope of the original data field. So calculate begin index $S_0$and theend index S_e for interpolations of each forecast point in exploration line direction. Similarly, we can get the begin indexD_0 and the end index $D_e$in the depth direction. So resistivity $Value_{j,k}$of each forecast point is:

$$\mathrm{Value}_{j,k} = \frac{\sum_{l=S_0}^{S_e} \sum_{m=D_0}^{D_e} d_{i0}^{-2} V(S_i)}{\sum_{l=S_0}^{S_e} \sum_{m=D_0}^{D_e} d_{i0}^{-2}} \quad d_{i0}^{-2} = 1/((x_i - x_0)^2 + (y_i - y_0)^2 + (z_i - z_0)^2) \quad (\text{while } 0 \le j \le S-1,\ 0 \le k \le D-1,\ j \in N,\ k \in N) \qquad (4)$$

This paper takes both *S* and *D* values of 30, and *Sr* and *Dr* values of 0.05.The result shows that these values can not only guarantee the interpolation accuracy, but also minimize algorithm complexity. Figure3 shows the visualization result of field model (slices and surfaces) with this method.

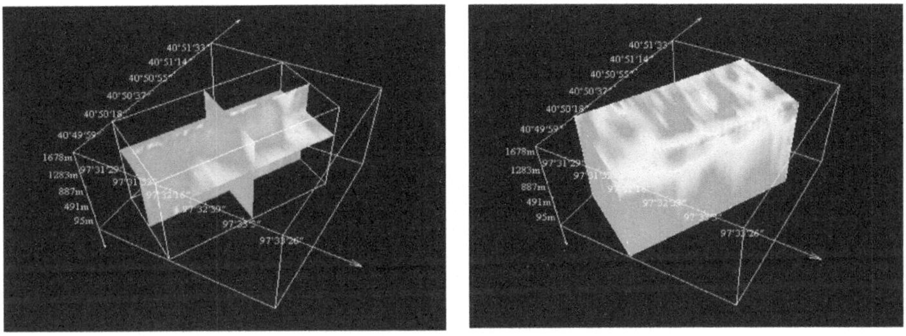

Fig. 3. The 3D visualization model of electrical exploration data field

5 Conclusions

In recent years, the visualization of geophysical data has become an important study content in the field of geophysics. This paper takes the CSAMT data as an example, implements the efficient visualization methods for electrical exploration data. Firstly, this paper creates the 3D visualization model of the exploration line by using the method based on exploration direction data points generating TIN. Then this paper studies a high-performance interpolation algorithm for the data, and interpolates the original data to regular data field by using an improved IDW which can interpolate the data fast and partially. After that, this paper uses a modeling method which compromises modeling face elements and volume elements, and builds the slices and surfaces 3D visualization models. These models show the apparent resistivity of underground rocks and minerals by different shades in 3D space, so that we can delineate the geological body more easily via resistivity anomalies in 3D space. This study demonstrates that creating electrical exploration data as 3D models by visualization technology can interpret data better and service for planning and decision-making, and has good prospects for popularization.

Acknowledgments. This work was supported by a grant from Special Fund of Central Collegiate Basic Scientific Research Bursary.

References

1. Marrett, R., Laubach, S.E., Olson, J.E.: Anisotropy and beyond: Geologic perspectives on geophysical prospecting for natural fractures. The Leading Edge 26(9), 1106–1111 (2007)
2. Ohno, N., Kageyama, A.: Scientific visualization of geophysical simulation data by the CAVE VR system with volume rendering. Physics of the Earth and Planetary Interiors 163(1-4), 305–311 (2007)
3. Cox, D.J.: Using the Supercomputer to Visualize Higher Dimensions: An Artist's Contribution to Scientific Visualization. Leonardo 41(4), 391–400 (2008)
4. Hack, D.R., Sirakov, N.M., Feves, M. L. et al.: Use of a 3D reconstruction and visualization system to evaluate subsurface geophysical data in advance of gravel mine development. In: Bandopadhyay, S. (ed.) Application of Computers and Operations Research in the Mineral Industry, Society for Mining, Metallurgy, and Exploration, Phoenix, AZ , pp. 569–576 (2002)
5. George, Y., Lu, D.W.: Wong:An adaptive inverse-distance weighting spatial interpolation technique. Computers & Geosciences 34(9), 1044–1055 (2008)

The Research of Web2.0 Interface and Interaction

Bei Wang

Department of Eastern International Art College, Zhengzhou University of Light Industry, 450008 ZhengZhou, China
estedu@yahoo.cn

Abstract. The proliferation of the impressive web technologies provides developers with a way to produce performance efficient, usable rich and interactive web based applications. A set of new technologies is gaining the web developer's interest. In this paper we talk about the web2.0 interface and Role of Ajax in enhancing the user experience on the Web for interaction. All of our visualization components run in standard web browsers and provide rich interaction and have good interface.

Keywords: Web2.0, User Interface, Interaction, Ajax, XML.

1 Introduction

There is an increasing demand in making web user interfaces richer and more interactive. Traditional web applications also known as web1.0 work on a client-server model. The new model enabled by a set of technologies that are broadly called Web 2.0 eliminates the start-stop-start-stop nature of web applications. In an interactive Web service, the pages contain forms with information to the server, and the reply is generated dynamically. Compared to static HTML pages, interactive Web services can provide: up-to-date information (replies generated at time of request), ailor-made information (reply generated dynamically by program based on user input and current server state) and two-way communication (client can also send data to server) [1].

Research is divided into two main lines of the concept is first to Web2.0 site interface design and interactive visual design carries on the analysis, the study of goals and principles, and discussed from the level of technology to choose and network technology for Web2.0 performance specification, so that to create a good user experience.

2 Research Status

Interactive Web interface is important because it greatly affect the end user's use, the impact of the Web site application, affected users to aggregate and share information and even affects people's work and life. In the Web site in the process of dynamic interaction, Web interface appearance alone is not enough, must be user-centric user maximum efficiency of information exchange, balancing the aesthetic principles, therefore, the Web interface has become one of the more important parts of the Web site development.

D. Jin and S. Lin (Eds.): Advances in EECM Vol. 1, LNEE 139, pp. 449–454.
springerlink.com

Interface design of meaning was ignored, is General of from design of some aspects such as color, and layout, aspects to told page design, no highlight page interface design in artistry Shang of unique sexual, while ignored has user experience of important meaning, lack specific in-depth of analysis, design out of interface often only attention visual effect, and user using up but very difficulties, exists many enough humanized of local, no to user of using habits to design interface constitute. Combined with the progress of the times Web2.0 interactive Web interface design for the era of new contents have been added.

Web interface design of good web site, is considered the multimedia characteristics, color design and information arranged, just more effective must order to convey the contents of the user. Only in the format design and colour combination to the correct understanding, to design and other websites have obvious difference, impressive interface. In the human interface with web design, artistic and technical is the perfect combination[2].

3 Web Interface Design

From the day the birth of humanity between people, between man and nature and between man and tool of communication on everywhere. As long as there is communication, there will be "interface" problems. Come and go, is the two-way communication. The exchange is the nature of the interface.

Interface is refers to people and information interaction media, is the carrier of information transmission. In the design of the interface research field, the definition is: interface is a kind of the colour, text, images, symbol visual elements and multimedia elements mainly constitution, the convey information for the specific purpose of man-machine communication among the media.

3.1 Information of Web Interface Stare at Rates

Stanford University use tracking technology leads people to view the web interface browsing process. A study found that when a screen, the article text or message textswill be the first one to catch the visitor's line of sight, and then your visitors will transfer to the line of sight of Photos or Graphics , in some cases according to the user web interface tonal whole or whole format and then make a decision of the content browsing interface. Studies have looked at rate case table1 shows.

Table 1. Information of Web Interface Stare at Rates

Articles texts	92%
Briefs	84%
Photos	64%
Banners ads	45%
Graphics	22%

In the results of this study, text as the best expression to attract visitors to gaze of Visual information.Full of text, images, the color of the Web interface at its design, how to effectively communicate Visual information, is should think carefully about the issues.

3.2 Global Web Interface Design

Web interface design is not just an art, it is a science.Web interface designed to be deep and sophisticated, comprehensive look at several aspects need to start thinking.

① Consistency.

Consistency means that the web interface should look after the whole people of the same sense.Design consistency throughout the guiding principles of a main line, all the web interface design activities are the main principles to follow.Studies show that when the screen elements, such as title, menu and other elements on a computer screen changes from time to time, user think time is almost doubled.Specifically, for the operation of the same functions with an object in the image and format to the pursuit of coherence.

A consistent interface sign. For example, the company's Web site often used as the site's logo sign.If a site with different pages showing different sizes of company logo, the user may suspect that these pages belong to different sites and reduce the site's credibility.

Consistent high-level screen layout.Interface because of its different features may need to take graphics, text or graphics, and other expressions, but these interfaces high-level "package" style should be the same.For example, the interface displays all the contents of specific restrictions in the same display area, etc.

A consistent navigation bar and browsing mechanisms users in a certain part of the web site using a browser on the way should be equally applicable to other parts of the website, the main navigation bar on the type of content should always be consistent.Any inconsistencies are likely to cause customer confusion.

② Balance

Balance refers to the web interface design must be emphasized that demand has been better than the speed of processing requirements.During the design will meet the needs of users and processing speed between the conflicting requirements of the problem in dealing with this conflict, the interface designer must weigh the options, and then based on accuracy, time consumption and ease of useneeds to make a decision.For the interface designer to design a gorgeous and beautiful but ignore the convenience of the user web interface may be easy, but the user is intolerable.

③ Logic

Web interface layout and visual design for the purpose of the website and content services, the logical relationship between the content of the interface is designed for the most fundamental basis.

The most important content should be the most eye-catching way to show, for example, placed near the center of the place, use a larger font and prominent color.

Should be important to avoid the right side and bottom on the screen.Because the right side and bottom of the content is often ignored by users, but also in the browser screen is small, the need to scroll to see sense.With the font size, color, indentation, etc. to express dependencies between the content.For example, information about the pyramid structure, the natural expression is the font larger or more obvious display a higher level of content, with a smaller font that lower-level content.

4 Web Interface Design

Ajax (pronounced as one word; shorthand for Asynchronous JavaScript and XML) is a group of interrelated web development techniques used on the client-side to create interactive web applications. With Ajax, web applications can retrieve data from the server asynchronously in the background without interfering with the display and behavior of the existing page. The use of Ajax techniques has led to an increase in interactive or dynamic interfaces on web pages[citation needed]. Data is usually retrieved using the XMLHttpRequest object. Despite the name, the use of XML is not actually required, and the requests do not need to be asynchronous[3].

Like DHTML and LAMP, Ajax is not a technology in itself, but a group of technologies. Ajax uses a combination of HTML and CSS to mark up and style information. The DOM is accessed with JavaScript to dynamically display, and to allow the user to interact with, the information presented. JavaScript and the XMLHttpRequest object provide a method for exchanging data asynchronously between browser and server to avoid full page reloads.Ajax incorporates:

- standards-based presentation using XHTML and CSS;
- dynamic display and interaction using the Document Object Model;
- data interchange and manipulation using XML and XSLT;
- asynchronous data retrieval using XMLHttpRequest;
- JavaScript binding everything together.

Ajax user interfaces are highly responsive. It introduces multiple segments of interactivity on the same page. The way users use Ajax applications is very different from their traditional web experience. In these applications the concept of reloading the web pages breaks down. In some applications the response may result in changes to small parts of the current view. The URL in the address bar remains unchanged.

Ajax interfaces are a key component of many Web 2.0 applications. Google, Yahoo, Microsoft, Amazon and many others have embraced Ajax. Ajax user interfaces are highly responsive giving users the feeling that changes are instantaneous.

AutoComplete: It is an ASP.NET Ajax extender that can be attached to any TextBox control, and will associate that control with a popup panel to display words that begin with the prefix typed into the textbox. The dropdown with candidate words supplied by a web service is positioned on the bottom left of the text box. In the sample above, the textbox is associated with an AutoCompleteExtender that pulls words that start with the contents of the textbox using a web service. When you have typed more content than the specified minimum word length, a popup will show words or phrases starting with that value. Caching is turned on, so typing the same prefix multiple times results in only one call to the web service.

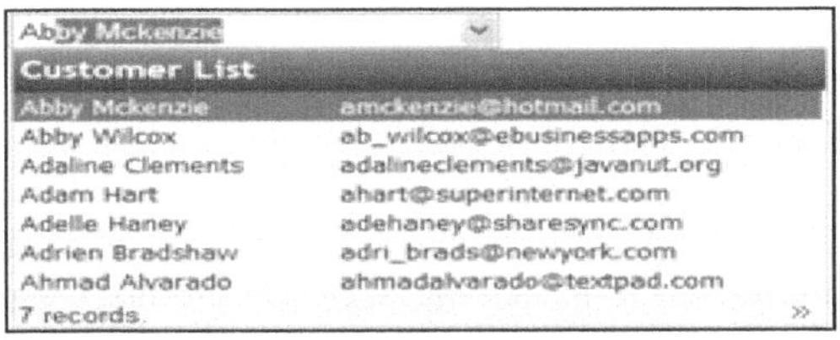

Fig. 1. The Ajax AutoComplete

Auto-refreshing: Using Ajax, it is possible to regularly auto-refresh a section of a webpage containing dynamic content without reloading the whole page.

Instant Search: it uses Ajax technology to allows you to see the search results immediately after typing in a character. Results are displayed without the need of reloading the page[4]. This saves time for users performing a search and greatly improves website performance. It reduces the need for Web clients to reconnect to a Web server every time information is downloaded. Yahoo's Flickr and instant search use Ajax.

Ajax spell checker: spell check text / form content. JavaScriptSpellCheck's Ajax feature (Asynchronous JavaScript And XML) allows our to perform live spellchecking as a web page runs. With Ajax Spell Check we can build rich, interactive spelling applications. The synchronous spellCheck function for JavaScript allows you to check spelling and get spelling suggestions with ease.

Ajax virtual desktop: Some of its features are search, news, maps, email integration, instant messenger, contact management tool etc. More features can be included through the use of third party 'Gadgets'.

Max's Ajax website preloader: Max's Ajax website preloader is a simple Ajax framework which allows you to display an animation during your website is loading. Installation and usage is very easy, only takes some line of copy and paste. An example code and detailed installation instructions are also attached.

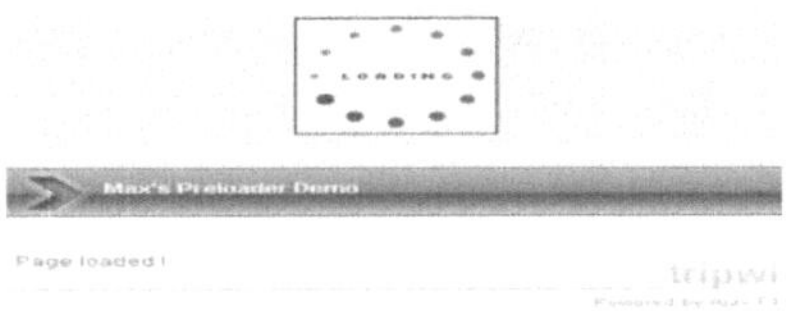

Fig. 2. The Max's Ajax website preloader

Form Validation: Forms are such a common element on the Internet we tend to blunder through them without too much thought[5]. However, if the web site has added a few nice touches that make the process easier, it tends to speed up the process and reduce any frustration in finding our preferred username. JavaScript may be used on the client-side to validate the format of simple data such as email addresses, phone numbers, dates, or credit card numbers. However, client-side processing is limited in that it cannot apply business rules or access server-side data sources to perform that validation.

Ajax IM: The Ajax-based Instant Messenger ("asynchronous javascript and xml instant messenger") is a browser-based instant messaging client. It uses Ajax to create a near real-time IM environment that can be used in conjunction with community, intranet, and social websites. No refreshing of the page is ever needed for this "web application" to work, as everything is updated in real-time via JavaScript.

Ajax is playing a crucial role in making Web 2.0 promises a reality. Some of the features of web 2.0 are:

- Use of Web as the platform
- Software delivered as a service instead of packaged software
- Cost-effective scalability
- Architecture with user participation

Ajax interfaces are a key component of many Web 2.0 applications. The way users use fully Ajaxed applications is very different from their traditional web experience.

5 Conclusion

The interactive web pages to the load is more complex than books and magazines and large amount of information, it also makes the web page of format design is the most important task of reasonable arrangement each layout form position, size, shape and color, layout form the function of the one is because the difference between classified information or effective, easy to browse and find the information[6].

Creating a rich user experience has been a goal for web developers. Ajax user interfaces are highly responsive giving users the feeling that changes are instantaneous.With increasing processor power and wireless network speeds Ajax is sure to play an important role in enhancing mobile user experience.

References

1. Crane, D., Pascarello, E., James, D.: Ajax in Action. Manning Publication Co. (2005)
2. Asleson, R., Schutta, N.T.: Foundations of Ajax, NY, USA (2005)
3. Horton, W., Taylor, L., Ignacio, A., Hoft, N.L.: The Web page Design Cookbook. John Wiley, New York (1996)
4. Palson, L.D.: Building rich Web applications with Ajax. Computer (10), 14–17 (2005)
5. Stockus, A., Bouju, A., Bertrand, F.: Web2based vehicle localization. In: Intelligent IEEE Vehicles Symposium, vol. (10), pp. 436–441 (2000)
6. Zaihuicao, Yu, D.: ETP/IITA Conference on System Science and Simulation Engineering (SSSE 2010), pp. 93–96 (2010)

SCM-Oriented Dynamic Service Architecture and Collaborative Application for Internet of Things

Tingbin Chen[1], Xi Yu[2], Qisong Zhang[3], and Jun Wang[4]

Neusoft Institute of Information, Dalian, 116024
{chentingbin,yuxi,zhangqisong,wangjun}@neusoft.edu.cn

Abstract. With the emergence and application of Internet of things, the problems of information sharing and process optimization are resolved for Supply Chain Management (SCM) Chain. Based on the analysis of basic elements of the Internet of things for supply chain, adopting dynamic services, hierarchical system structure of Internet of things is used. By the application of middleware technology and multi-layer key technology integration for Internet of things, the mass data of collection, transmission, processing and integration is realized. The service oriented architecture (SOA) and Web Service technology is applied to support distributed SCM application software which is using middleware technology to realize dynamic framework and collaborative application in heterogeneous environment for Internet of things.

Keywords: Web Services, Supply Chain, Internet of things, Collaborative application.

1 Introduction

The technology of internet of things is playing an important role in the supply chain management, and is key link to promote the development of manufacturing services. But in the supply chain there are so many different applications, information island and knowledge fragments. And the supply chain collaboration is now mainly based on distributed component technology which can't break through the platform difference and protocols limitation. Besides it can not meet the changing business knowledge sharing, dynamic and various demands in the business environment. So how to realize the visualization of supply chain has become an urgent problem to solve under such environments of supply chain management.

In order to realize the integrated supply chain collaboration platform and information sharing, and constructs business requirements of The Internet of things (IOT), key technologies for the integration system are adopted in this paper, the multi-layer structure is applied based on the dynamic services framework which is divided into four layers consists of perception layer, the transport layer, information integration layer and application service layer. By the application of middleware technology and multi-layer key technology integration for Internet of things, the mass data of collection, transmission, processing and integration is realized. Through the encapsulation and combination of dynamic services, quick service reconstruction and information transformation to the knowledge are achieved for the business values.

D. Jin and S. Lin (Eds.): Advances in EECM Vol. 1, LNEE 139, pp. 455–462.
springerlink.com

2 Supply Chain Visibility Structure Based on the Internet of Things

Vernon Francis presented the concept of supply chain visibility structure in Journal of Supply Chain Management. That is "real-time provide all kinds of the key events occurred in the supply chain which generate the information to managers, and analyze these events and give specific suggestions for managers decision-making". So the framework of internet of things is divided into show layer, data collection layer, the application layer and data layer. As shown in figure 1.

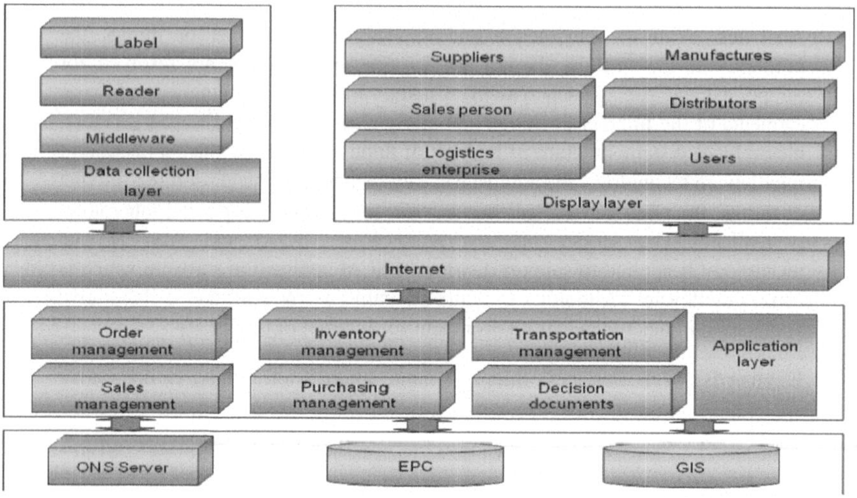

Fig. 1. Visibility structure of internet of the things based on supply chain

Show layer is system user interface, which allows users to use business functions on the Interne through the web browser. It allows the users to get the information that is approved by supply chain team through the Internet. This layer provides the input and output management for application layer. The users can accord their identity to use different kinds of service in system.

The network server includes many web pages, which provide different functions for different user groups. These pages are composed by HTML. When the network server receives a requirement of ASP file that is from the web pages, it deals with program which is contained in the document to set up HTML pages, and then sends HTML pages to the browser. In the visual internet of things architecture, business function goes through the COM which is on the application layer to execute. When the network server receives the request that is from the browser, these functions call DCOM that is initialed by ASP.

Data collection layer use RFID technology to collect EPC label information of object, and use of middleware technology to transfer information to the Internet.

The application layer mainly realizes logistics functions and transmission logistics services which have designed. It provides all kinds of business functions, such as the quotation, order arrangement, running state tracking or customer invoices etc, which are realized by business composition module and data component modules. The system can also connect with EMAIL server and inform customers' business information by the form of E-mail, such as invoice or shipping documents.

Data layer is a data storage area, which saves and searches those business data that are needed to further operation. It saves different kinds of business data, such as customer information, suppliers, business processing, and product data, etc.

3 Key Technology of Internet of Things for Dynamic Service

3.1 Cooperative Perceive Technology of Self-hierarchical of Internet of Things

Based on internet of things expanded theory system model research, through in the perception, transmission, integration, service and so on each level of the self-optimization and cross the layer scheduling, realize the optimization goal of the overall system performance[4]. There are two mainly parts as following: (1) Research on internet of things of the system optimization mechanism. Mainly by using decoupling, constraint optimization theory and technology to realize optimal utilization of the system of physical, communication and computer resources, and the optimization schedule and distribution mechanism of the radio spectrum of internet of things and reliable data of internet of things transmission mechanism under the uncertain environment. (2)Research on communication protocol stack bases on across layer collaboration that. Establish across the layer cooperation of internet of the things of communication protocol stack, and the users' terminal applications can actively select transmission path according to the business needs, and optimize transmission quality to achieve self-organizing, self-adapting to achieve internet of things every level's alliance optimization.

Internet of things gets mass data by the RFID, need to pass through the transport layer to transmission processing, and provide resources for the application layer [5].

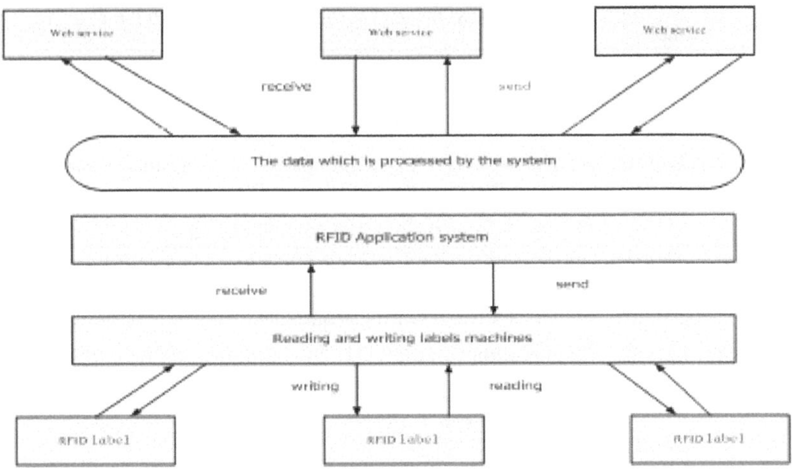

Fig. 2. Data processing flow of RFID web service

In order to enhance visibility of the items with RFID mark, need to create Web services with gateway of RFID, make internal Web services connect to the integration server, and then realize services interconnection by SOA. Combine RFID and Web services that can cross enterprise in between heterogeneous networks to realize information collection, exchange and sharing (figure 2).

Combine the RFID technology and Web Services to construct a Message-Oriented Middleware(MOM) , RFID define data interface and describe mass data information through the XML and PML technical, not only including transfer information, but also security, error recovery and interpreting data, data buffer, data broadcast, positioning network resources and other senior service. The fusion technology can make endpoints which are in the internet share different real-time data and the transaction more easily.

3.2 Based on the Content Networking Combination Technology of Dynamic Service

QoS expressed the ability which is used to satisfy the needs of users for a product or service explicitly. It had played a certain role in Web services discovery, combined, calls and integration process. With the rising number of Web service, under the condition that the service functions have meet the needs of the service requestors, the requirement to the service quality will be more and more increasingly, it is necessary to describe the QoS of Web service clearly (Tables).

Table 1. Individual service of QoS model

index	The single service index function
Execution cost	$q_{\text{price}}(s) = q_{\text{price}}(s, op)$ $q_{\text{price}}(s, op)$ ——Service s operation op of execution cost
Execution time	$q_{du}(s) = q_{du}(s, op) = T_{process}(s, op) + T_{urans}(s$
Service reputation	$q_{rep}(s) = \frac{\sum_{i=1}^{n} R_i}{n}$ R_i ——The valuation of users on the service reputation n —— The number of users
reliability	$q_{ret}(s) = N_c(s)/K$ $N_c(s)$ ——On the requirement time of the number of execution success K ——The total of call number
usability	$q_{ov}(s) = T_a(s)/\theta$ $T_a(s)$ ——Service s on the θ time of the available time

Table 2. QoS model of Combination service

index	Index function of Combination service
Execution cost	$Q_{price}(p)=\sum_{i=1}^{N} q_{price}(s_i,op_i)$
Execution time	$Q_{du}(p)=CPA(q_{du}(s_i,op_i),\ldots,q_{du}(s_N,op_N))$ CPA——The Algorithm of Critical path
Service reputation	$Q_{rep}(p)=\frac{1}{N}\sum_{i=1}^{n} q_{rep}(s_i)$
reliability	$Q_{rel}(p)=\prod_{i=1}^{N}(e^{q_{rel}(s_i)\times z_i})$ $Z_i=\begin{cases}0, & s_i \text{ not on the critical path}\\ 0, & s_i \text{ on the critical path}\end{cases}$
usability	$Q_{ov}(p)=\prod_{i=1}^{N}(e^{q_{ov}(s_i)\times z_i})$ $Z_i=\begin{cases}0, & s_i \text{ not on the critical path}\\ 0, & s_i \text{ on the critical path}\end{cases}$

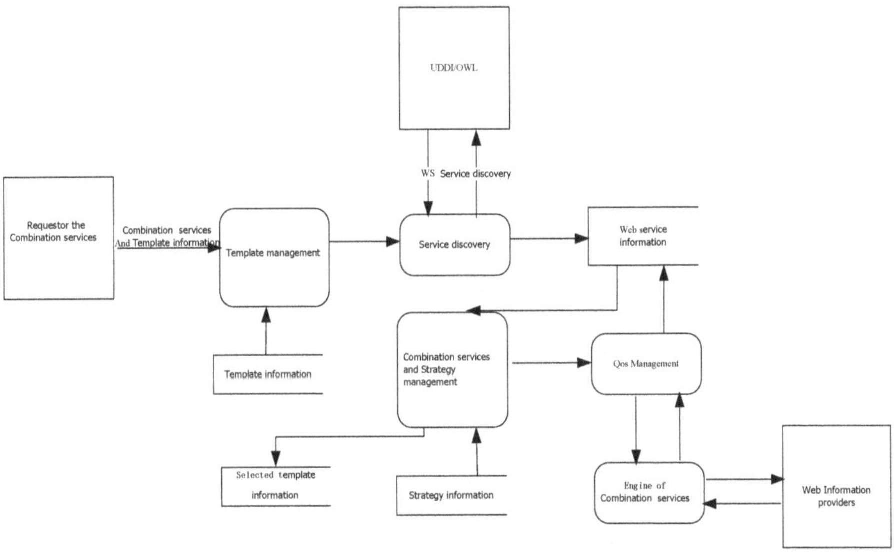

Fig. 3. Internet of things application layer service model

The technology and method of dynamic Web services combination facing IOT will research the technology of dynamic Web services combination based on fixed strategy and rules. In dynamic Web service combination process, the technology of dynamic Web services combination based on fixed strategy used a single rule strategy, the technology of dynamic Web services combination based on rules could choose many rules. The paper realized the integration of the dynamic information by Web services discovery and matching, through the expansion to the UDDI center finished the service discovery and matching to meet the needs of users by the service discovery and matching based semantic, and provide support to the dynamic Web service combination technology (figure 3).

The paper presented a method of Web service combination based on QoS index and output parameters of Web service, which setting the rules for web service combination based on QoS index and output parameters of web service. By selecting a group of web services combination from the candidate web services, got its QoS index and output parameters of Web service and judged whether they would meet the rules of Web service combination which were set before, to determine the group of web service combination which were selected to be the web service combination we needed. The process was shown in figure 4.

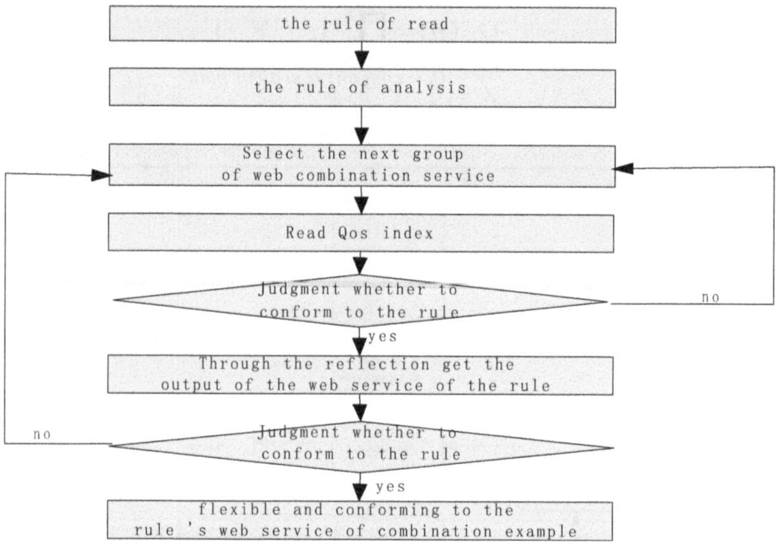

Fig. 4. The process of Web service

According to the system, the paper provided a Web service function based on QoS and output of Web service which implemented by Web services combination apparatus. The Web server combination apparatus included storage unit used to store Web service combination rules, this method is:

Read rules of Web service combination from the above storage unit, which is set in advance based on the QoS indexes which users needed and output parameters of Web service;

Read the rules of Web service combination and analyze ;

Select a group of web services combination from the candidate web services;

Get the QoS indexes of every web service from a group of web services combinations gotten by the Web server combination apparatus;

Get the output parameters of every web service from a group of web services combinations gotten by the reflection technology;

Judge the QoS indexes and output parameters of web service of every web service which we have gotten whether they matched the rules of Web server combination which have been analyzed, to determine the group of Web services selected whether is needed.

The evaluation to web services is the key for realizing the goal of dynamic Web services. Using the method of intelligence and fuzzy evaluation to evaluate the web services can improve the efficiency of Web services combination effectively to support to the business synergy of supply chain.

4 Supply Chain Enterprise Business Cooperation Application

The process of business synergy of supply chain needs many enterprises to participate in, including: the core enterprise (manufacturing enterprise), suppliers, distributors and retailers, the third party logistics enterprise. Business synergy is a complex process of internal system and external system of enterprise interacted. Here will explain the process of business synergy and detailed scene based on IOT of manufacturing business.

For example, during the process of purchasing cooperative management, the production based on supply chain must cooperative with upstream and downstream enterprise. The key of coordination of production of supply chain is the production plan and the synergy of production control. The cooperation mode between supply chain enterprises shows for the four models: synchronous mode, asynchronous mode, distributed synchronous mode, distributed asynchronous mode. So, the changing of the production of supply chain from the traditional hierarchical control organization mode to flat network organization, needs collaborative technology of business based on web services and IOT to realize collaborative production management of networking supply chain. Coordination management driven by business included the data collection, transmission and processing of raw materials, semi-finished products, products and the final products by using IOT technologies of manufacturing (figure 5).

Purchasing cooperative needs to complete the whole function of purchasing process, including: the interaction between suppliers and management collaborative services; the collaborative services of trading; the collaborative services of purchase contract; VMI receiving; the collaborative services of storage and supply management; material master data and the collaborative services of detailed information, etc. It is necessary to build extensible stratification system structure of IOT facing SCM. Therefore it is the development direction of the future logistics technology facing IOT and the trend of realizing of modern logistics that build service platform of intelligence logistics distribution of manufacturing enterprise supply chain, using technologies of IOT.

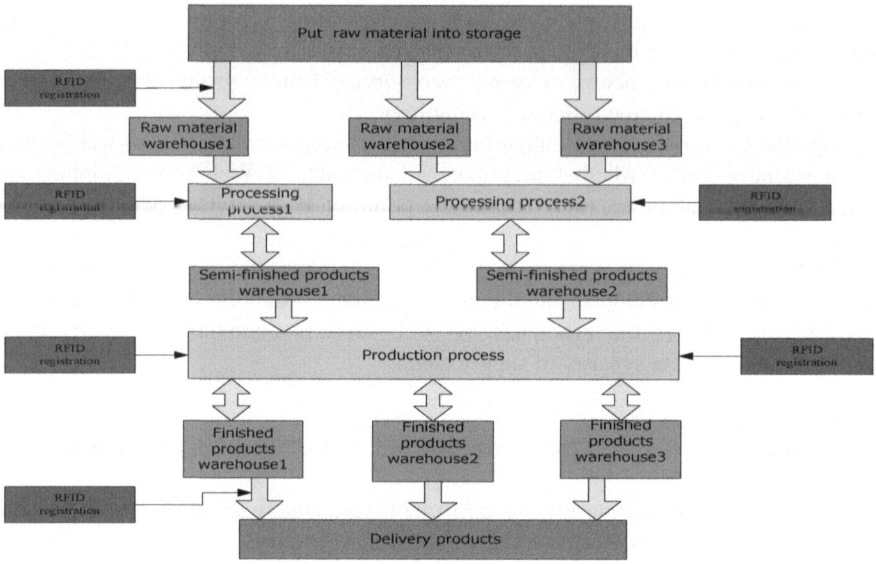

Fig. 5. Manufacturing data acquisition and processing of Internet of things

References

1. Wei, G.: The ESB-based Supply Chain Knowledge Management System Architecture. Journal of Harbin University of Science and Technology (3) (2008)
2. Chen, T.: Research on Methods of Services-oriented Integration for Supply Chain Collaboration. In: IEEE International Conference on Wireless Communications, Networking and Mobile Computing,WinCOM 2007, pp. P4701–P4707 (2007)
3. Wu, H.: Review on Internet of Things application and challenges. Journal of Chongqing University of Posts and Telecommunication, Natural Science Edition (5) (2010)
4. Gu, J., Chen, S.: Wireless Sensor Network-Based Topology Structures for the Internet of Things Localization. Chinese Journal of Computers (9) (2010)
5. Zhang, Y., Cai, W.: ASCM: SOA-based embedded adaptive safety critical middleware. Application Research of Computers (3) (2010)

Analysis of Allocation Deviation in Multi-core Shared Cache Pseudo-partition

Zhibin Huang[1,2], MingFa Zhu[1,2], and Limin Xiao[1,2]

[1] School of Computer Science and Engineering, Beihang University, Beijing 100191, China
[2] State Key Laboratory of Rail Traffic Control and Safety, Beijing Jiaotong University, Beijing 100044, China
Huangzhibin@cse.buaa.edu.cn, {zhumf,xiaolm}@buaa.edu.cn

Abstract. Allocation deviation is a commonly encountered problem in cache partition mechanism, especially pseudo partition mechanism, such as PIPP. We augment some bits to the line's status field to store the source core Id of incoming cache requests and sample the whole cache, then quantitatively analyze allocation deviation of multi-core pseudo-partition in shared last-level-cache. And we emphasize some factors that influence allocation deviation, such as the cache quota, the contention of concurrent workingset etc. Furthermore we discuss flexible handling of allocation deviation to benefit to the whole performance according to the cache utility characteristics of the benchmarks. Through our experiments and analysis, we conclude that in pseudo-partition, allocation deviation happens frequently due to the contention and improper cache quota, and it needs to be more flexibly handled.

Keywords: shared LLC, allocation deviation, cache partition, pseudo-partition, multicore.

1 Introduction

Large capacity and high-associativity shared last-level-cache is common in modern mainstream commercial multi-core processors. For example, IBM POWER7 has a 32MB on chip eDRAM highly associative shared L3 [1], Intel Core i7-900 has an 8MB shared L3 based on Intel Advanced Smart Cache [2, 3]. The shared LLC design only maintains one copy of data and has no physical region segment to the whole cache capacity so that it has more flexibility to tolerate working set of different sizes to alleviate thrashing and reduce wasting of capacity. However, unconstrained and unmanaged sharing exacerbates cache pollution because of contention between the cores for cache space, resulting in unfair usage of cache resources or impairment to overall performance. Cache partition mechanism [4-8] is considered to be an effective means to manipulate pollution and contention in shared LLC and has been comprehensively researched.

Most of cache schemes consist of three major policies: Profile, Partition Quota Selection and Partition Enforcement. Cache partitions can be classified into two categories according to the cache space occupancy: 1) Strict cache partition, which is strict constraints on the cache quota of each core, e.g. [4], [5], [6]. 2) Pseudo cache

D. Jin and S. Lin (Eds.): Advances in EECM Vol. 1, LNEE 139, pp. 463–470.
springerlink.com

partition, which is weak constraints on the cache quota, e.g. [7], [8].According to the definition in [7], a core's partitioning allocation deviation is the absolute difference between the average number of ways occupied and the target number of ways allocated. In a strict cache partition, allocation deviation occurs mainly during the process in which the allocation quota changes by the cache partition algorithm and partition quota selection mechanism whereas in a pseudo cache partition, allocation deviation occurs when the contention of cache requests between the cores is severe.

In most cache partition schemes, there is little discussion about the allocation deviation. In strict cache partition[4][9],the cache occupancy is inconsistent with the deserved cache quota when the cache quota is changed by the cache partition algorithm, therefore, the actual occupancy of a certain core is larger than its quota while that of another certain core is less than its quota. Then in the following cache victim line selection, the lines from the core larger than its quota are the candidate victim lines no matter what these blocks are useful. In Pseudo cache partition [7] [8], allocation deviation frequently occurs, therefore, we must pay more attention to it. PIPP [7] defines the allocation deviation and discusses the ratio, but has no investigation to the source and the influence on the whole performance. In this paper, we implement PIPP and augment some bits to the line's status field to track the source core id and the whole cache is periodically sampled. And then we quantitatively the factors that influence allocation deviation, such as cache quota and the contention of concurrent workingset etc. Furthermore, we discuss the relationship between the handling of allocation deviation and the whole performance.

Table 1. Baseline configuration

Processor core	2 Cores; scalar in-order;1 hardware context per core; L1-I cache and L1-D cache: 32kB, 64B line-size, 4-way, 3-cycle delay, LRU. L1 are private to each core.
Unified Shared L2 Cache	2M,64B line-size,8-way,11-cycle delay,LRU,32-entry MSHR, 128-entry store buffer.L2 Cache L2 cache is shared among all the cores, MESI protocol,

In this paper, 21 benchmarks from Spec Cpu2000/2006 are chosen and classified to 6 categories according to their MRC and miss rate. And 21 two-mixture workloads are constructed in a more diverse and representative manner. Shared LLC are managed by LRU and PIPP and sampled periodically. Furthermore, allocation deviation of each set is calculated. The occupancy of the cache managed by LRU is formed by the free contention of the cache request from concurrent workingsets [11].That is to say, the ratio of occupancy in LRU reflects the contention. On the other hand, we set the cache quota of each core by static allocation in PIPP to observe its allocation deviation. Combine these two experimental results and analyze quantitatively, we can see that: 1) Cache quota has remarkable influence on the allocation deviation. More proper cache quota has lower ratio of allocation deviation. 2) Allocation deviation is influenced by the contention of concurrent workingsets deeply. More intensive contention causes higher ratio of allocation deviation. Furthermore, allocation deviation does not always damage the whole performance, but has relationship with the cache utility characteristics of workloads. Modest contention helps to improve the cache utility and the whole performance; this is also one of the reasons that the performance of PIPP is higher than that of UCP. Consequently, allocation deviation needs more flexible handling to benefits to the whole performance rather than damage it. Our main contribution is to analyze the factors that influence the allocation deviation and more flexible handling of it.

2 Experimental Methodology

Table 1 shows the parameters of the baseline configuration used in our experiments. We simulate a 2 core processor based on the cycle-based full-system simulator, Simics [9]. The operating system of the simulator is Solaris 10. For each workload, we warm up the caches and branch predictors for 200 million instructions, and then perform detailed simulation for 400 million instructions per benchmark and for each 100 million instructions; the statistic is performed on the various attributes of L2.

Table 2. Categories of Benchmarks Selected from Spec Cpu2000/Cpu2006

	Low MissRate	High MissRate
Smooth Change	164.gzip;168.wupwise;171.swim;173.applu;177.mesa;464.h264ref;482.sphinx3;183.equake (A)	433.milc;(D)
Continuous Change	181.mcf (B)	188.Ammp;179.Art (E)
Threshold Change	172.Mgrid(T=4);197.parser(T=4); 200.sixtrack(T=2);445.gobmk(T=4) 255.vortex(T=2);256.bzip2(T=6) 300.twolf(T=4) (C)	175.Vpr(T=6) 301.aspi(T=6) (F)

Table 3. Mixture Workloads of Two benchmarks

ID	Details of Benchmarks	ID	Details of Benchmarks
AA	168.Wupwise+173.Applu	AB	164.Gzip+181.Mcf
AC	464.H264ref +300.Twolf	AD	171.Swim +433.Milc
AE	173.Applu + 179.Art	AF	464.H264ref +301.Aspi
BB	181.Mcf +181.Mcf	BC	181.Mcf +172.Mgrid
BD	181.Mcf +433.Milc	BE	181.Mcf +188.Ammp
BF	181.Mcf +175.Vpr	CC	172.Mgrid +445.Gobmk
CD	172.Mgrid +433.Milc	CE	445.Gobmk+179.Art
CF	197.Parser +301.Aspi	DD	433.Milc +433.Milc
DE	433.Milc +188.Ammp	DF	433.Milc +301.Aspi
EE	179.Art +188.Ammp	EF	179.Art+175.Vpr
FF	175.Vpr +301.Aspi		

For each benchmark, while all the other settings of L2 were kept intact, the only change is the cache set associativity which varies from 1, 2, 4, 6, 8, 10, 12, and 14 to 16 respectively as the cache size changes from 256KB to 4MB. Then its miss rate, IPC and MPKI are sampled. And categorize these benchmarks into 6 classes, which are labeled with A-F according to the order in which they appear in columns and rows, as shown in Table 2.

Two-mixture workloads are conducted in a more diverse and representative manner by choosing 2 classes of benchmarks out of 6, one workload from each of the two chosen classes. And a total of 21 workloads are constructed (refer to Table.3).

We execute these two-mixture workloads under LRU managed shared LLC and PIPP managed shared LLC respectively. We modify the gcache module of Simics and add core id to each line status field to identify the source of the line's data. For these two-mixture workloads, we set their cache quota by static allocation differently according to their contention and characteristic achieved by unconstrained LRU so as to analyze the source of the cache allocation deviation.

3 Analysis of Allocation Deviation

Fig.1 presents allocation deviation of all 21 two-mixture workloads in PIPP under different static cache quota allocation. And the plot with hollow circle is the ratio of the cache occupation managed by unconstrained LRU, which is considered as the free contention metrics between the two concurrent workingset. The plot with dotted line is the cache quota ratio by arbitrary static allocation; therefore, we can see the cache allocation deviation clear.

The average of the allocation deviation ratio for all 21 workloads is 27.8%, and the maximum is 76.5%, which occurs to BD.181.Mcf.433.Milc; and the minimum is 1.5%, which occurs to BB.181.Mcf .181.Mcf. From Fig.1, we can see that:

1) The occupation of the PIPP is different from that of the LRU due to the insertion policy and promotion/demotion policy of PIPP. For AC, AF, BB, BC, CD, CE, DD, DE, DF and FF, the ratio of occupation in PIPP is close to that in LRU. And the ratio of other two-mixture workloads is different.

2) The contention of these two concurrent workingset has remarkable influence on the allocation deviation. When the contention is intense, that is, the ratio of LRU occupancy of one benchmark is very high or very low; the ratio of allocation deviation is often high. For example, in CD.172.Mgrid.433.Milc, LRU occupancy of 172.Mgrid is very low, only 7.6%.Most of the cache space is occupied by 433.Milc, but the reused ratio of 433.Milc is very low, and its miss rate is remaining to 11.3%. Another example is BD. 181.Mcf.433.Milc.The contention from the cache request of 433.Milc interferes with other concurrent workingset's cache request heavily.

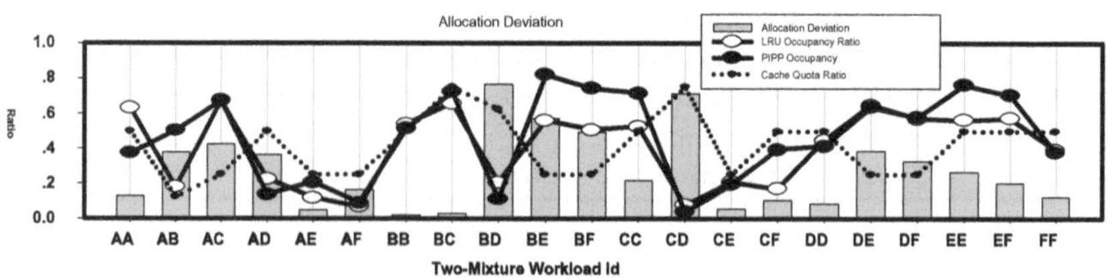

Fig. 1. Allocation Deviation for 21 two-mixture workloads

3) Improper cache quota has also influenced on the allocation deviation. For example, in BE.181.Mcf.188.Ammp and BF. 181.Mcf .175.Vpr, the occupancy of 181.Mcf is near 50% and the contention is not intense, but their allocation deviation is high due to the not proper cache quota of 181.Mcf, which is only 25%(the ratio of ways is 2:6). On the other hand, in BC.181.Mcf .172.Mgrid, its allocation deviation is very low, only 2.5% because its cache quota (75%) is close to the free contention formative occupancy (65.7%). Another example is BB.181.Mcf .181.Mcf, its cache quota is 50%, which is very close to the free contention formative occupancy, 54.1%, and therefore, its allocation deviation is only 1.5%.

Therefore, we can conclude that in PIPP the allocation deviation is influenced by two factors: 1) cache quota of each core; 2) the contention of the cache request from each concurrent workingsets. If the cache quota is unreasonable, then the allocation deviation is high and the whole performance is invaded. On the other hand, due to no explicit mechanism to protect and retain the cache space occupancy of each core in PIPP, the contention is prone to wreck the cache quota so that influences the whole performance.

On the other hand, if allocation deviation happens, how to deal with allocation deviation can benefits to the whole performance rather than damage it? In PIPP, the smallest cache quota is inherently invaded under proper cache quota because their insertion position is closest to LRU position and their newly loaded data block is prone to be driven to LRU position and becomes victim line.

In fact, the handling of allocation deviation has a direct impact on the fairness and performance.

Assume that there is an application T that has already occupied T_A (T_A>=1 and T_A<=A, A is the associativity of cache) way(s) through LRU replacement algorithm. Without loss of generality, denote these T_A cache lines by LRU_{T0}, LRU_{T1},…,$LRU_{T(TA-1)}$, among which the distance from LRUT0 to the top of stack is minimum and the distance from $LRU_{T(TA-1)}$ to the top of stack is maximum and the number of hits contributed respectively by these cache lines are $CountT_{LRUT0}$, $CountT_{LRUT1}$,…, $CountT_{LRUT(TA-1)}$. According to the properties of LRU stack replacement algorithm, if the ways of application T occupancy deceases t ways then its hit numbers will decrease by the sum from $CountT_{LRUT(TA-t-1)}$ to $CountT_{LRUT(TA-1)}$ which is closely related to the value of T_A which is the number of cache lines already allocated to application T.

We have chosen 21 benchmarks from Spec Cpu2000/Spec Cpu2006 to perform the experiments. Although all SpecCup2000 benchmarks are tested, some benchmarks were excluded from the selection list due to the fact that in L2, their MissRate are extremely low due to its locality and the filter effect of L1 Cache (e.g. 177.mesa, 183.equake, 482.sphinx3 etc.). The number of cache set is kept constant to stabilize address mapping while the associativity of set varies from 1-way to 2-way, 4-way, 6-way, 8-way, 10-way, 12-way, 14-way to 16-way. Then a statistic on MissRate and IPC is performed. The change in IPC is also calculated for T_A=4, 6,8,10,12,14,16 (refer to Fig.2). Details on the parameters of the experiments can be found in the Section 4 of this paper. From the Fig.2, we can see that, the change feature of MissRate Curve (MRC) can be categorized into 3 types:

1. Smooth Change: during the process in which associativity varies from 1 to 16, MissRate remains relatively stable. This is exhibited by 164.gzip, 168.wupwise, 171.swim, 173.applu, and 464.H264ref and 433.milc.When cache size changes, MissRate changes by less than 1%, as shown in Fig2.

2. Threshold Change: during the process in which associativity varies from 1 to 16, when associativity is below a certain threshold value, MissRate continues to decrease whereas when associativity is above a certain threshold value, MissRate remains relatively stable. For example, for 197.Parser, when associativity is equal to 1, MissRate is near 5% and when associativity is equal to 8, MissRate is approaching zero, so it is denoted by T=8; for 179.Art, when associativity is equal to 1, MissRate is as high as 29% but when associativity is equal to 4, 10, MissRate decreases to 16%, and 5% respectively; for 179.Art when associativity is more than 12, MissRate is approaching 0,so denoted by T=12, as shown in Fig3.

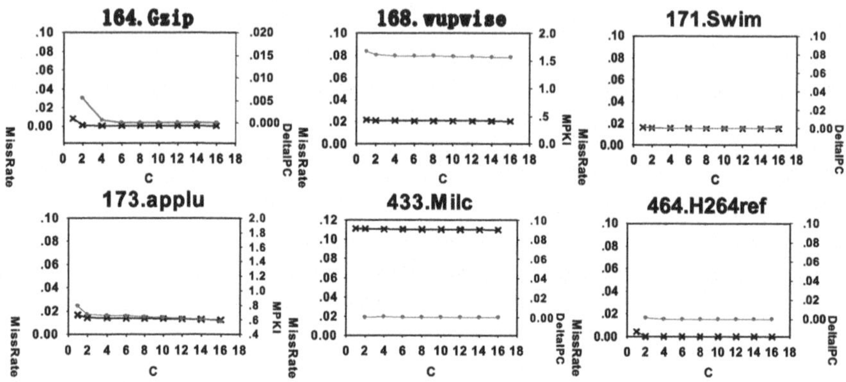

Fig. 2. Smooth Change MissRate Curve when associativity varies from 1 to 2,4,6,8,10,12,14,16 and Delta IPC Curve when T_A=4 to 16. Dots marked by filled circle form Delta IPC Curve. Dots marked by cross form MissRate Curve.

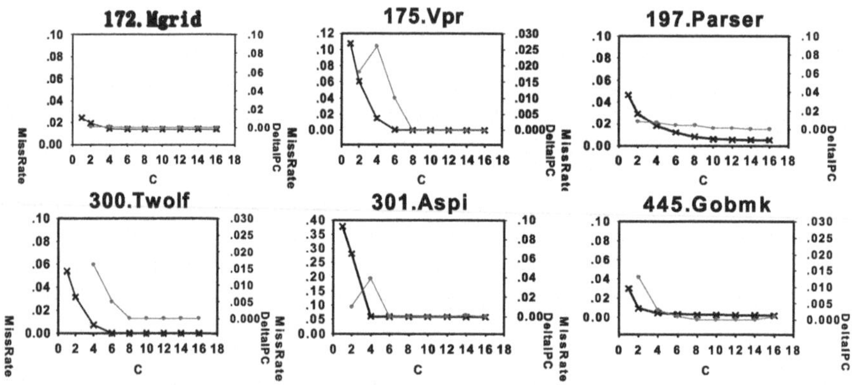

Fig. 3. Threshold Change MissRate Curve when associativity varies from 1 to 2,4,6,8,10,12,14,16 and Delta IPC Curve when T_A=4 to 16. Dots marked by filled circle form Delta IPC Curve. Dots marked by cross form MissRate Curve.

3. Continuous Change: during the process in which associativity varies from 1 to 16, MissRate continues to decrease as in 181.Mcf. During the process of changing the threshold value, T, if T exceeds a certain value then it is considered to be a continuous change. In this paper, 179.Art and 188.Ammp are categorized as continuous change due to the fact that L2 cache adopts 8-way set design in the baseline configuration, ,as shown in Fig4.

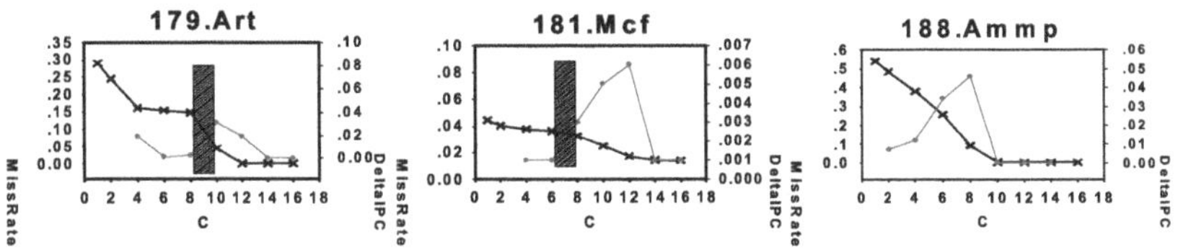

Fig. 4. Continuous Change MissRate Curve when the associativity varies from 1 to 2,4,6,8,10,12,14,16 and Delta IPC Curve when T_A=4 to 16

When handling allocation deviation, the three different types of change of the benchmarks can be treated differently by different methods. For smooth change workloads, when already been allocated with T_A-way cache, adding or reducing cache allocation has little impact on MissRate and IPC. Therefore, when handling allocation deviation for smooth change workload, it only needs to ensure no more than its quota. For threshold change workloads, they exhibit a relatively large change in IPC when the number T_A is smaller than its threshold so that it is necessary to keep its cache space occupancy from being influenced by allocation deviation and ensure at least or more than its quota. Meanwhile, when the number T_A is larger than its threshold they have a smooth change it can ensure no more than its quota. For continuous change workloads, such as 188.Ammp, that has a relatively large working set, when the allocated cache space is smaller than its working set, allocating additional cache space will result in a continuous decrease in MissRate. At the same time, the improvement in IPC accelerates as the allocated cache space is close to the threshold of its working set due to the cumulative gain of decrease in MissRate. On the other hand, thrashing has occurred to 181.Mcf and 179.Art. If the MissRate Curve is divided into subsections at the place where thrashing occurs, then it is observed that 181.Mcf exhibits similar IPC change as 188.Ammp and the allocation deviation handling method for 181.Mcf is similar to that of 188.Ammp. For 179.Art that exhibits an obvious pattern in which a relatively huge change in IPC is experienced when T_A is small, therefore, when handling allocation deviation, if T_A is small, it is necessary to keep its cache space occupancy from being influenced by allocation deviation.

From the above analysis, it is obvious that when dealing with allocation deviation, it should not be biased toward favoring cores that have obtained relatively large cache space because there is a significant impediment to the performance of cores that have obtained relatively small cache space, harming the overall fairness and performance as well. Consequently, different policies should be adopted according to the difference in workload characteristic. However, the replacement victim policy of traditional LRU stack is unable to provide an appropriate alternative approach to address the problem. Based on the live-time of the cache line this paper has proposed a novel approach that creates protections during the live time, establishes a more diverse and versatile allocation deviation handling policy. By setting up protection values for the different lengths of live-time, this new approach can fully support and implement different allocation deviation handling policies according to the difference in change types: smooth change, threshold value and continuous change.

4 Conclusion

We quantitatively analyze allocation deviation of multi-core pseudo-partition in shared last-level-cache, especially some factors that influence allocation deviation heavily, such as the cache quota, the contention of concurrent workingset etc. Furthermore, we discuss flexible handling of allocation deviation to benefit to the whole performance according to the cache utility characteristics of the benchmarks.

Acknowledgments. We thank Zhang Han for their helpful discussions, and the anonymous reviewers for their comments and suggestions on the paper. This work was supported by NSFC 60973008, and the Doctoral Fund of Ministry of Education of China under Grant No. 20101102110018 and the fund of the State Key Laboratory of Rail Traffic Control and Safety (Contract No. RCS2008K001), Beijing Jiaotong University.

References

1. Kalla, R., et al.: Power7: IBM's Next-Generation Server Processor. IEEE Micro. 30(2), 7–15 (2010)
2. Swinburne, R.: Intel Core i7 - Nehalem architecture dive (2008), `http://www.bit-tech.net`
3. Doweck, J.: Inside Intel Core Microarchitecture and Smart Memory Access. White paper, Intel Corporation (2006), `http://download.intel.com/technology/architecture/sma.pdf`
4. Qureshi, M., Patt, Y.: Utility-based cache partitioning: A low-overhead, high-performance, runtime mechanism to partition shared caches. MICRO 39 (2006)
5. Kaseridis, D., Stuecheli, J., John, L.K.: Bank-aware Dynamic Cache Partitioning for Multicore Architectures. In: ICPP (2009)
6. Kim, S., Chandra, D., Solihin, Y.: Fair Cache Sharing and Partitioning in a Chip Multiprocessor Architecture. In: PACT 2004, pp. 111–122 (2004)
7. Xie, Y., Loh, G.H.: PIPP: Promotion/Insertion Pseudo- Partitioning of Multi-core Shared Caches. In: ISCA 2009, pp. 174–183 (2009)
8. Kedzierski, K., Moreto, M., Cazorla, F.J., Valero, M.: Adpating Cache Partitioing Algorithms to Pseudo-LRU Replacement Policies. In: IPDPS (2010)
9. Suh, G.E., et al.: Dynamic partitioning of shared cache memory. Journal of Supercomputing 28(1) (2004)
10. Magnusson, P.S., Christensson, M., Eskilson, J., Forsgren, D., Hallberg, G., Hogberg, J., Larsson, F., Moestedt, A., Werner, B.: Simics: A Full System Simulator Platform. Computer 35(2), 50–58 (2002)
11. Stone, H.S., Turek, J., Wolf, J.L.: Optimal Partitioning of Cache Memory. IEEE Trans. Computers 41(9), 1054–1068 (1992)

A Study on Trusted Internet Identity Management and Its Application

Bin Han, Yueting Chai, and Yi Liu

Department of Automation, Tsinghua University, Beijing, China
mailhub@foxmail.com, {chaiyt,yiliu}@tsinghua.edu.cn

Abstract. Internet identity management (IIDM), which is a core issue of network security, has drawn close attention from governments and IT experts all over the world. This article makes a worldwide review on the development status of IIDM at first. Then a new definition of trusted IIDM is brought forward. As a realization of the concept, we represent a creative design of IIDM mode, as well as a technique solution of online information system introducing biometrics. Our study is innovative and valuable on the research and application of IIDM.

Keywords: Network Security, Internet Identity Management, Online Information System, E-Commerce.

1 Introduction

The internet, which can break through spatio-temporal restriction, has provided modern people with rapid communication approaches and abundant information channels. They can access many web applications in anonymity or pseudonym and it is network virtuality that offers the joyful freedom of information expression, communication and acquisition. However, for some other applications such as e-commerce, true identity of transaction partners should be available in case of privacy leak and economic loss, which highlight the significance of network security.

Security problems derive basically from network virtuality, while accountability deficiency and ambiquity of online behavior allow them likely to occur, because moral and legal punishment is evadable can by unreliable identity, such as anonymity, pseudonym and false information. So it's an effective means of enhancing network security to introduce trusted internet identity management (IIDM) which can build fixed connection between every realistic person and his/her internet identity (IID). Most developed countries and regions have noticed the issue and many promotion strategies on trusted IIDM are raised for near future. Different solutions turn up and work, including European eID and Korean network real name system. As to our study, we bring forward a new definition of trusted IIDM based on four key elements, which are uniqueness, truth, legality and relevance, to set a starting point in the field. Then, a creative design of IIDM mode is represented with explanation of its operation flow as further conceptual verification and application. We also expand the mode with

D. Jin and S. Lin (Eds.): Advances in EECM Vol. 1, LNEE 139, pp. 471–476.
springerlink.com

a technique solution, which is an online information system and introduces biometrics for relevance demand. Our concept and mode design are innovative and valuable work on the field of IIDM.

2 Review on IIDM Development

IIDM is so to commerce credit, national security and social stability, that many developed countries have already established relevant national strategy in order to advance it. In Jan 2011, the American government published *National Strategy for Trusted Identities in Cyberspace*, to emphasize the requirement of an identity ecosystem supporting trusted network environment, which equals a trusted IIDM system.[1] Now concrete solutions are under intense research among public and private sectors. The European Union has raised several proposals and action items since *i2010 eGovernment Action Plan* issued in 2006, to enhance the development of e-government and cross-border e-service.[2] The solution in common use for European IIDM is called electronic identity (eID), which is a series of electronic data, including identity information (such as holder's name, birth date, address etc.) and digital signature. It is usually stored in a piece of tangible card called eID card with visible identity information on the surface and inside a chip for possible biological information (such as finger print, iris and face photo) storage. The card holder can certify his/her identity online with eID card and recognition device connected with the computer. [3] [4] [5] [6] [7] [8] As a special case, there is a means for online identity certification called BankID, which allow people to log in their e-bank account with password before the e-bank system can offer identity certification service to other web applications.[9] Basically, BankID is an primary IIDM solution based on username and password. The *e-Japan Strategy* is carried out by the Japanese government to enable citizens to access all e-government service with a so-called registration card, which can also serve as passport and e-signature.[10] Since Oct 2005, Korea has practiced an internet real name system. When people access some specific web applications, real name and ID number have to be registered first and certification by some government-appointed company is necessary.[11] China has not started for long its research and development on IIDM, so it's still dependent on digital certificate.

3 New Definition of Trusted IIDM

An IID can usually be expressed in the form of a digital information set with several important attributes for necessary functions. First of all, it can distinguish an internet user exclusively within certain extent, which we call uniqueness. To be trusted, the IID have to possess three other attributes containing truth, legality and relevance. Truth means that all information items in the set are true and accurate according to the holder's current state. Legality means the IID is recognized by law and the holder behaves lawful online. Relevance means the IID is definitely bound with the holder himself/herself so that no one else could make use of it. IIDM can be considered as a series of operations over the IID. Trusted IIDM is to make the IID trusted. So we come to our own definition of trusted IIDM, which is a series of operations over IID to make it possess the attributes of uniqueness, truth, legality and relevance. There is a question about natural person and legal person in cyber space to mention here. It is the same situation as in the real world that there are online natural person and online legal person. The former represents natural individuals using internet, while the later stands for organization users such as companies, institutions and NGOs, especially e-commerce corporations which interact economically with other web users much. Consequently, IID as well as IIDM for both individual users (IU) and organization users (OU) should be established. For instance, IID for IU may include name, ID number, birth date, address and so on, while for OU organization name, registered No. and address may not be absent.

4 New Design of Trusted IIDM Mode

In order to guarantee the four key attributes for trusted IID, our IIDM mode is consist of three main procedures, which can be defined as registration, authentication and check. A technique solution of the mode will be shown later in this section, as well as an approach for relevance.

4.1 Registration

Started by users voluntarily, registration is the first step of the mode flow. It is the procedure of collecting original IID information, which can be attained from some official fundamental data base, or inputted when users apply for IID registration. As product of registration, the information set have to contain some basic identity information of the user (such as name and ID No. for IU or organization name and registered No. for OU). Scaned pictures of important ID and permission papers should also be included for further operations. For uniqueness, ID No. of IU and registered No. of OU are necessary. All the information collected allows modification and clearance by users.

4.2 Authentication

Authentication is the procedure of examining the IID information set to assure all the items are true and accurate. A true information item means it is edited truthfully. For example the ID No. is of the user's own. An accurate item means it is formally detailedly edited without careless mistake. All the information items should be examined according to some official information channel by man power or computer of public-trusted third party (PTTP), which is also the manager of national IID system and guarantor of IID truth and legality. One of the two results of authentication is authentication-denied, which could be caused by false or wrong information offered. The user denied has to re-register if he/she wants an IID. The other result is authentication-passed, which can produce a web page of identity information. When the user is asked for identity certificate by web applications or other web users, he/she can show some specific items on the page or the whole. The web page is the IID card for the online user.

Uniqueness, truth and legality are ensured during the authentication procedure. Uniqueness is generated by the ID No. or registered No., because natural persons or legal persons have already been distinguished exclusively in this way in real world. For privacy protection, the number can be transformed into a different one by some monotonic function, which can produce different results from different inputs. The result number can be included in the information set and treated as some kind of IID No. which can also identify the user. Truth and legality are guaranteed by manual or automatic examination and authority of PTTP. All the information shown on the web page and stored into IDD data base is recognized by law.

4.3 Check

Check is a repeated authentication for authentication-passed users to assure the real-time validity of their IIDs in case of real information change. Official information channels are still main judgement reference. There are two results of check, one of which is check-passed indicating the validity of the IID. If the result is check-denied, the user should be informed with reasons and re-register. The IIDM mode is illustrated in Fig.1.

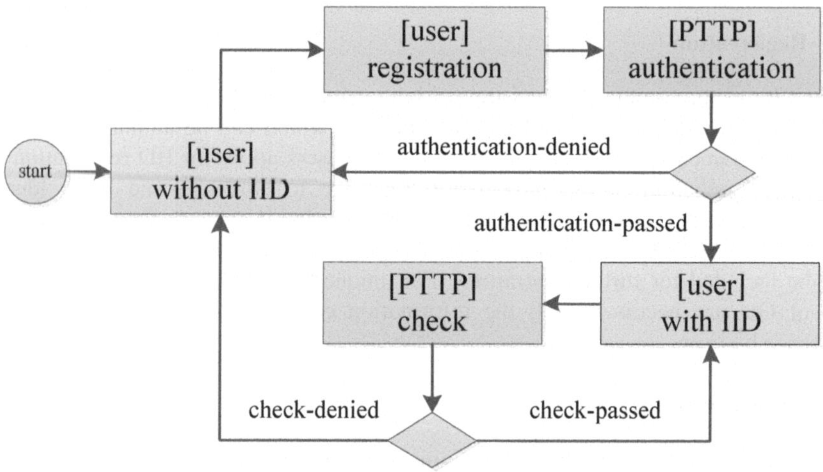

Fig. 1. The flow figure of the IIDM mode

4.4 Technique Solution

To realize the trusted IIDM mode presented above, we bring forward a technique solution of a public-trusted online information system. Main functions of the system are registration, authentication, check and query. Registration and query serve for all web users, while the rest two only for PTTP who examine and check the IID information items. The function of query enables people to search for identification of other web users. For individual privacy protection, it is proper that only organization users can be queried. The user case diagram of the system is illustrated in Fig.2.

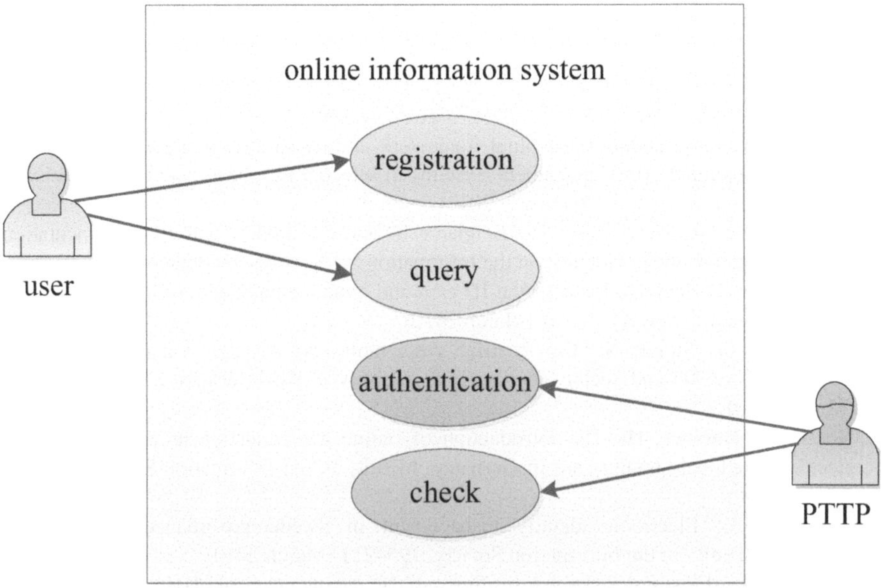

Fig. 2. The user case diagram of the information system

4.5 Approach for Relevance

To build up reliable relevance between the IID and its holder, we can add into the information set biological information (for example fingerprint), which is not difficult to obtain when recognition device is connected to computer. So users will be demand to submit their fingerprint instantaneously in the procedure of registration. During authentication, the PTTP have to determine whether the fingerprint belongs to the user described by other identity information. If it does, the relevance is set up. As trusted reference of the determination, there should be an official public data base of citizen identity information including fingerprint established by the government beforehand. For organization users, we can demand a person in the organization to submit his/her fingerprint and determine his/her adscription to the organization by offline investigation.

5 Conclusion

The trusted IIDM concept and mode design presented in this article are exploring work in the field IIDM. The four key attributes, which are uniqueness, truth, legality and relevance, guarantee the trustworthiness of IID. Compared with the eID in current use, this mode has two obvious advantages. The first is more convenient and low-cost because there is no tangible card. Second, web users can search for organization users in the online information system. It is quite helpful in e-commerce since registered website in the system can be trusted. For further study, normative IID information models should be established for different types of web users. Security of data transmission is also necessary to think about. Generally speaking, a good thread for IIDM is opened up.

References

1. Full NSTIC Strategy Document, http://www.nist.gov/nstic/
2. i2010 eGovernment Action Plan, http://europa.eu/
3. Kubicek, H.: Introduction: conceptual framework and research design for a comparative analysis of national eID Management Systems in selected European countries. Identity in the Information Society, 5–26 (March 2010)
4. Mariën, I., van Audenhove, L.: The Belgian e-ID and its complex path to implementation and innovational change. Identity in the Information Society, 27–41 (March 2010)
5. Heichlinger, A., Gallego, P.: A new e-ID card and online authentication in Spain. Identity in the Information Society, 43–64 (March 2010)
6. Aichholzer, G., Strauß, S.: The Austrian case: multi-card concept and the relationship between citizen ID and social security cards. Identity in the Information Society, 65–85 (March 2010)
7. Noack, T., Kubicek, H.: The introduction of online authentication as part of the new electronic national identity card in Germany. Identity in the Information Society, 87–110 (March 2010)
8. Grönlund, Å.: Electronic identity management in Sweden:governance of a market approach. Identity in the Information Society, 195–211 (March 2010)
9. Rissanen, T.: Electronic identity in Finland: ID cards vs. bank IDs. Identity in the Information Society, 175–194 (March 2010)
10. e-Japan Strategy, http://www.kantei.go.jp/
11. Chen, B., Zou, X., Zhou, G.: The Development Trend of the Network Identity Management. Netinfo Security, 5–8 (March 2011)

Trade Middlemen's Tax Environment Analysis: Based on General Equilibrium

Zhanxia Wu[1] and Jinhua Qin[2]

[1] Accounting School, Shanghai Institute of Foreign Trade, SIFT
Shanghai, China
wellany@163.com
[2] Management School, Shanghai Institute of Foreign Trade, SIFT
Shanghai, China

Abstract. In order to construct Shanghai Trade Center, The government issued a series of policies. Combining Coase theory and field researchs, lemen in CHINA. Futher, Through Harburg model and general equilibrium analysis, We concluded that tax environment would affect trade middlemen's transaction costs by trade convenience, which could cause their global resources to flow,till the two tax bearings are equal. Also, This article compared the tax environments in Asia's major international trading center in Shanghai, Singapore and Hong Kong, commented on the current tax reform policies in Shanghai and made suggestions on the trade middlemen's tax environment.

Keywords: General equilibrium, Trade middlemen, Tax environment.

1 Introduction

Shanghai is building international trade center. Recently, the government has issued a series of related tax policies. For trade middlemen, how is their tax environment under these new tax policies? This thesis will show trade middlemen's tax environment in Shanghai by theoretical derivation and contrast with Hong Kong and Singapore.

2 Tax Environmet Is the Importment Transaction Cost Influencing Trade Middlemen

Trade facilitation is the key link currently of the free trade zone function transformation. "Improve trade facilitation level to further build better international trade environment." As has become the consensus for Shanghai international trade center's construction. Therefore, the State Council and the Shanghai municipal government have already issued a series of tax policies in order to encourage the development of modern service industry and promote the construction of Shanghai international trade center[2].

Trade facilitation is used originally in international trade field, referring to reducing the transaction costs in the cross-border flow of goods and services, which have been caused by unnecessary administrative burden. In recent research, trade facilitation also

D. Jin and S. Lin (Eds.): Advances in EECM Vol. 1, LNEE 139, pp. 477–482.
springerlink.com

includes trading environment, domestic policies, system structure, international standards and uniform and technology infrastructure construction, etc. It can be seen from "the global trade facilitation report in 2010" issued by the World Economic Forum, among index constitutes of trade facilitation, "business environment" in a certain country has become one of the four main factors affecting facilitation index[1]. According to its definition, "business environment" contains the general efficiency of the government work, foreign investment encouragement degree, etc. Therefore, tax environment consisted of tax policies and collection efficiency through affecting "business environment" becomes one of the important trade facilitation effects. And by influencing the transaction costs, it makes a direct effect on the industrial development.

3 Trade Middlemen's Tax Effect Analysis— Based on General Equilibrium

Analysis of effects on different tax environments for enterprises' tax movement and resource allocation applies to Harburg model (1962). The basic assumptions of this model are as follows[5].

3.1 Postulate

Assume that the trade enterprise owns enterprise 1 and 2 respectively in two districts, which respectively sell commodity X and Y. Trade enterprise mainly uses two kinds of production factors of capital K and labor L. The total supply of capital K and labor L is fixed and can be made full use of. Elements can flow freely between enterprises. The market accords with completely competitive market hypothesis and is in Pareto optimality before tax.

3.2 Model Derivation

In tax-free condition, the two enterprises' cost functions respectively are $C_X(r,w,X)$ and $C_Y(r,w,Y)$, where r and w respectively stand for capital price and labor price. Due to the unchangeable scale revenue of production, we can use certain proportions of yield to describe cost functions.

$$C_X(r,w,X) = c_X(r,w)X \tag{1}$$

$$C_Y(r,w,Y) = c_Y(r,w)Y \tag{2}$$

Two commodities' prices are:

$$P_X = c_X(r,w), \tag{3}$$

$$P_Y = c_Y(r,w) \tag{4}$$

Deduce the cost functions' derivatives to labor price and capital price respectively for enterprise 1 and 2. We can know that the enterprise's demand of labor and capital respectively are

$$L_X = \frac{\partial c_X(r,w)X}{\partial w} = c_{Xw}(r,w)X \tag{5}$$

$$L_Y = \frac{\partial c_Y(r,w)Y}{\partial w} = c_{Yw}(r,w)Y \tag{6}$$

$$K_X = \frac{\partial c_X(r,w)X}{\partial r} = c_{Xr}(r,w)X \tag{7}$$

$$K_Y = \frac{\partial c_Y(r,w)Y}{\partial r} = c_{Yr}(r,w)Y \tag{8}$$

Where, c_{Xw}, c_{Xr}, c_{Yw}, c_{Yr} respectively are derivatives to labour price and capital price.

Since the total supply amount of production factors is fixed in the free trade, namely the supply of labour and capital respectively are $\overline{L}$ and $\overline{K}$. So

$$c_{Xw}(r,w)X + c_{Yw}(r,w)Y = \overline{L} \tag{9}$$

$$c_{Xr}(r,w)X + c_{Yr}(r,w)Y = \overline{K} \tag{10}$$

Let $X(P_{X,}P_{Y,}R)$ and $Y(P_{X,}P_{Y,}R)$ be Marshall demand function, then the market equilibrium points for commodity X and Y are

$$X(P_X, P_Y, R) = X \tag{11}$$

$$Y(P_X, P_Y, R) = Y \tag{12}$$

In tax conditions, if levy ad valorem duties on commodity X and Y, capital and labor, tax rates respectively are t_X、t_Y、t_{KX}、t_{KY}、t_{LX} and t_{LY}.

Then the two enterprises' price equations are

$$P_X = c_X(r(1+t_{KX}), w(1+t_{LX})) \tag{13}$$

$$P_Y = c_Y(r(1+t_{KY}), w(1+t_{LY})) \tag{14}$$

In free trade environment, both capital and labour supply keep constant in two places, so total amount is still $\overline{L}$ and $\overline{K}$. They can be described by the equations as follows.

$$c_{Xw}(r(1+t_{KX}), w(1+t_{LX}))X + c_{Yw}(r(1+t_{KY}), w(1+t_{LY}))Y = \overline{L} \tag{15}$$

$$c_{Xr}(r(1+t_{KX}), w(1+t_{LX}))X + c_{Yr}(r(1+t_{KY}), w(1+t_{LY}))Y = \overline{K} \tag{16}$$

The conditions of market equilibrium are

$$X(P_X(1+t_X),P_Y(1+t_Y),R)=X \tag{17}$$

$$Y(P_X(1+t_X),P_Y(1+t_Y),R)=Y \tag{18}$$

Clearly, the free flow of production factors will lead to the equal yield rates. If not, the elements will continue to flow until they are equal, and flow process occurs within the constant motion as tax burden flow. Resources flow caused by tax burden will result in the tax competition between governments. The degree of resources flow depends on the elasticity of tax burden of demand for resources.

4 Comparative Analysis of Trade Middlemen's Tax Environment

For ease of comparing trade middlemen's overall taxation and its influence, enterprise income tax, individual income tax and business tax in Singapore, Hong Kong and Shanghai Waigaoqiao Free Trade Zone are compared as follows[4].

4.1 Levy Objects Comparison (See Table)1

Table 1. Levy Objects Comparison

Indicators	Shanghai Waigaoqiao Free Trade Zone	Hong Kong	Singapore
Enterprise income tax	Territoriality and nationality principle.	Territoriality principle.	Territoriality and nationality principle, dividends tax-free.
Individual income tax	Territoriality and nationality principle.	Territoriality principle.	Territoriality principle.
Business tax	Income from providing labor services, transferring intangible asset, sale realty and commission.	No.	Sales of goods or supply of services.

4.2 Tax Rates Comparison (See Table 2)

Table 2. Tax rates comparison

Indicators	Shanghai Waigaoqiao Free Trade Zone	Hong Kong	Singapore
Enterprise income tax	Adjust to 25% in five years.	16.5%.	Trade middlemen are exempted from enterprise income tax.
Individual income tax	Classification rate, basically 20% except salary incomes.	15%, adjusted once annually to reduce the effect of inflation.	Six progressive tax rates for salary incomes, 3.5%~ 20%, the rest according to the scale tax rate of 20%, employment income of non-resident individuals according to the rate of 15%.
Business tax	3%~20%.	No.	5%.

4.3 Deductible Items Comparison (See Table 3)

Table 3. Deductible Items Comparison

Indicators	Shanghai Waigaoqiao Free Trade Zone	Hong Kong	Singapore
Enterprise income tax	Reasonable wages and salaries expenditure, social security fund and accumulation fund, the borrowing costs, worker welfare funds, union funds, worker education funds, business entertainment expenses, advertising fees, environmental protection special funds, property insurance premium, rent, cost contribution of the outside headquarters, public welfare donations, the intangible assets. The depreciation policy of straight-line method: 3~20 depreciation years, no initial allowance for depreciation. The longest losses cuts-even is 5 years.	The cost and expense Depreciation policy: in general using the straight-line method, initial depreciation rates can amount 20%~60%, environmental protection facilities can be accelerated depreciation. A particular accounting year losses can be carryovered to subsequent annual profits indefinitely.	50%~75%. Depreciation policy: general requirement is using the straight-line method, initial allowance for depreciation is 25%. Equipment can be completely depreciated due year, intangible assets used for intellectual property in amortize rate of 20%. Operating losses can be carryovered indefinitely backward.
Individual income tax	Unified classification excesses provision.	Considering the conditions of the taxpayer itself, the number of children and family population.	More deductible items.

4.4 Tax Collection and Administration (See Table 4)

Table 4. Tax Collection and Administrations in Singapore, Hong Kong and Shangha

Indicators	Shanghai	Hong Kong	Singapore
Enterprise income tax	Monthly prepay, annual liquidation.	In January and April prepay twice, annual liquidation.	Monthly prepay, annual liquidation.
Individual income tax	Source withhold.	Tax authorities.	Tax authorities, tax returns.
Business tax	Shortest assessable period 5 days, the longest up to a quarter.	No.	Monthly.

It can be seen from above that Hong Kong working as a duty-free paradise, tax categories are simple and the tax rate is extremely low. The mainland's tax system framework is closer to Singapore. Therefore, the tax policy reform does not depend on the variation of tax categories, but the design of tax rates and deductible items as well as the improvment of collection and administration efficiency, so as to improve the overall tax environment, improve trading convenience and attract top trade middlemen.

5 Policy Suggestions

We should draw lessons from Singapore's "global trade business plan" preferential tax ideas, basically it is around enterprise income tax preference. Therefore, preferential policies of income tax should be made earlier.Future more,We should combined with trade middlemen's advantages of large initial investment and strong brand, the most relevant income tax preferential policy should Reduce standard rate of income tax, raise the original preferential basis to keep pace with high-tech industry, diminish the income tax gap with Singapore. Try to create more convenient and efficient tax environment.

Acknowledgment. Shanghai institute of international business Project (NO. 2010-A-01).

References

1. 2010 China International Industrial Exposition Trade Facilitation Forum, Meeting documents, Shanghai (2010)
2. Management measures for business tax difference taxation, Shanghai Duty Commodity (28) (August 2010)
3. Jia, M.: The tax policies research in promoting the development of the modern services. Tax Research 2, 96–97 (2010)
4. Paying Taxes, World Bank and Pricewaterhouse Coopers. P7-112 (2010)
5. Chen, Y.: The new classical investment theory and model -relations analysis between tax and investment. Economic Problems Exploration (12) (2005)

Research of Battlefield Visualization Shadow Effects on Ogre

Fang Ye, Jingxuan Wang, and Tianshuang Fu

Information and Communication Engineering College,
Harbin Engineering University, Harbin, Heilongjiang, China

Abstract. Since shadow effects is of importance in 3D Game, this paper mainly works on real-time shadow rendering. Percentage closer filtering is one kind of technique that has proved to be an efficient solution on soft shadow; however, it suffers from aliasing problems. We come up with Gaussian Blur Filtering, which can handle soft shadow edges and solve the problem that Percentage closer filtering makes. Under the satisfaction of visual needs, we use real-time RTT technique to exchange data with GPU. After adding Gaussian Blur Filtering, we can efficiently reduce the computation of CPU and GPU computation; still it produces convincing soft shadow effects.

Keywords: Shadow Mapping, Gaussian Blur, shader, Ogre Engine.

1 Introduction

Ogre provides an excellent platform for interactive roaming in 3D games, in which real-time shadow plays a vital role in enhancing the sense of reality. Shadow volumes [Crow 1977] and shadow mapping [Williams 1978] are the two main categories of shadow algorithms. Adaptive shadow maps (ASMs) [Fernando et al. 2001], which is adopted in adaptive division, dynamically adds tree layers to enhance shadow effects. Due to the complex hardware implementation, however, it is only applicable to draw accurate shadows in simple scenes. While, Parallel-Split Shadow Maps (PSSMs) algorithm [Fan Zhang et al. 2006], which is adopted in depth division, could improve aliasing distribution and tight bounding shapes in large-scale outdoor scenes. Owning to the issue from discrete buffers in the image-space method, shadow mapping often suffers from aliasing problems. Many filtering algorithms have been proposed, such as Percentage closer filtering [Williams 1978], Variance Shadow Maps [William Donnelly 2006], Exponential Shadow Maps [Thomas Annen 2008]. Percentage closer filtering is the most simple of all these filtering techniques; Variance Shadow Maps is to "warp" the depth value when storing and reading from the shadow map; and Exponential Shadow Maps[1] is to pre-filter the shadow map, which in turn allows for high quality hardware-accelerated filtering. However, these algorithms still cannot satisfy the needs for large amounts of shadow computations, either on visual effects or speed ability. Attacking the remaining problems, this paper adds Gaussian Blur, a novel twist to William' original algorithm that improves the accuracy of bias writing a higher equality shadow map with equally less consumption.

D. Jin and S. Lin (Eds.): Advances in EECM Vol. 1, LNEE 139, pp. 483–488.
springerlink.com

2 Algorithm

This paper calculates shadow into three parts: the whole shadow, soft shadow and self-shadow. Since this paper is focus on large outdoor scenes, we choose depth division algorithm.

2.1 Scene Splitting

We use PSSMs to produce an optimized distribution of shadow map texel, which provides sufficient sampling densities for points being mapped. PSSMs scheme is to split the view frustum into several parts by expressed as [2][3] and generates shadow maps with the same resolution for each spilt parts. In view frustum part, we divide several parts:

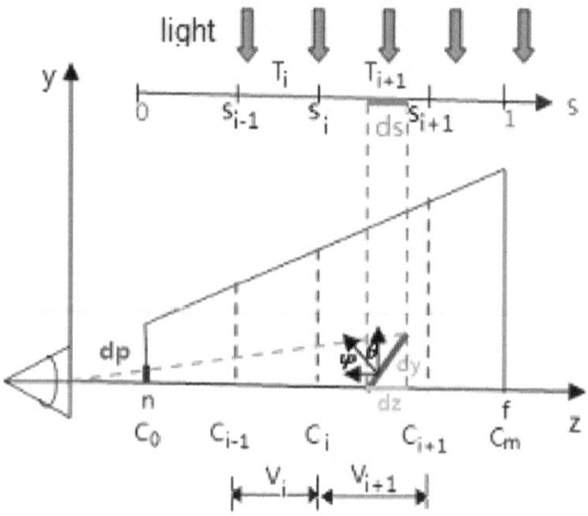

Fig. 1. Principle of PSSMs

$$\frac{dp}{ds} = n\frac{dz}{zds}\frac{\cos\varphi}{\cos\theta} \tag{1}$$

where s stands for size in normal texture space and z stands for depth in world space. φ and θ denote the angles between the surface normal and vector to the screen and shadow map plane respectively.

Using Logarithmic Split and Uniform Split, we have:

$$C_i = \lambda C_i^{uni} + (1-\lambda)C_i^{\log} \quad (0 < \lambda < 1) \tag{2}$$

where C_i^{uni} is the uniform split scheme and $C_i^{\log}$ is the logarithmic split scheme, C_i means the depth of ith split plane in the view space $(1 \le i \le m-1)$, and defines

$C_0 = n$ and $C_m = f$. The next step we use the depth partitioning [4] and the sample distribution to compute tight light-space partition frusta. By Depth Shadow Map we calculate self-shadow, which is expressed as [5]. Adding soft shadow map and self-shadow map together, we finally get the ultimate shadow effects.

2.2 Soft Shadow Problem

True soft shadow [6] is different with the changes of light intensity. We suppose the light is directional and strong, thus the shadow edge is not ambiguous. Percentage closer filtering is focus on solving aliasing problems [7] in depth shadow map and generates soft shadows. Since our approach utilizes shadow maps to improve scene-shadows qualities, we add Gaussian Blur to filter soft shadow in order to get more blurred edges. Gaussian Blur Filtering is often used to reduce image noise in pro-processing step, which using normal distribution calculates transformation of each pixel. The distance of pixel to the origin is farther, the weight is smaller. The Gaussian distribution in 1-D has the form:

$$G(x) = \frac{1}{\sqrt{2\pi}} e^{-\frac{x^2}{2\sigma^2}} \tag{3}$$

In view of Mathematics, Gaussian Blur Filtering is typically generated by convolving an image with a kernel of Gaussian values. As Gaussian distribution of every pixel in the image is non-zero, the computation of every pixel includes the entire image. In practice, the pixel which locates far away from three-sigma can be ignored, meaning sigma decides the blurred level of soft shadow in some extents.

In this paper, Gaussian Blur Filtering is used twice in two separate one-dimensional kernels, first in the horizontal direction and second in the vertical direction. We get pixels around by mapping coordinates and incoming offsets with the help of Pixel Shader. Finally, we can get the blur soft shadow effects by multiplying these pixels by an array weight of Gaussian functions.

3 Simulation Results

We run the shadow rendering tests using 1366*768 image resolutions, and generate all pictures and data on an Intel Core 2 Duo machine with a NVIDIA GEFORCE G105M using Direct3D 10.

3.1 System Platform

3D Studio Max, as a powerful model software, reads the three-dimensional model into large-scale outdoor scenes by Ogremax and operates the scenes by Ogre engine, which achieves complete cross-platform operation on Windows, Mac OSX, Linux, Windows CE with the help of the deep encapsulation of Application Program Interface: Direct3D and OpenGL. Ogre's biggest feature is to manage scenes by plugins which only calls a pointer to this scene manager when handling one large scene. The Application Program Interface communicates by Shader language, which including GLSL, C for Graphics (Cg) and DirectX High Level Shading Language (HLSL). In this paper, we use Cg to communicate with Direct3D.

3.2 Testing Results

First we use shadow map to generate real-time shadow illustrated in Fig.2 and then use the technique of PSSMs adding Percentage closer filtering shown in Fig.3. Comparing the performance of Fig.2 with that of Fig.3, we see that the algorithm of PSSMs adding Percentage closer filtering can get more authentic shadow maps. Thirdly, we add Gaussian Blur Filter shown in Fig.4, when sigma=3.

Fig. 2. Performance of shadow map algorithm (FPS:138.7)

Fig. 3. Performance of the algorithm combing Parallel-Split Shadow Maps, Percentage closer filtering(FPS:103.2)

Fig. 4. Performance of Gaussian Blur(FPS:105.6)

Comparing the performance of Fig.3 with that of Fig.4, we see the soft shadow in Fig.4 is more blurred than that in Fig.3, thus we make an conclusion that after adding Gaussian Blur Filtering, we can get more realistic soft shadows in visual effects.

Using the techniques we introduce above, we add complex models in the scene and test the ability of the system are increasing shown in Fig.5 and Table 1 offers the performance parameters in three scenes.

(i) (ii) (iii)

Fig. 5. Testing ability in different scenes

Table 1. The performance parameters in three scenes

Scene	Triangle	Average FPS	Best FPS	Batch
i	75488	59.6	189.9	64
ii	326036	33.6	188.6	125
iii	720534	20.3	188.6	232

In Computer Graphics, performance comparisons of different algorithms are based on the visual effects and the ability that hardware supports the engine. The frame rate will affect the ability. In theory, frame rate which is greater than 20 frames per second can display realistic smooth animations. Fig.6 shows the comparison the frames per second with triangles in the scene of three techniques: traditional shadow mapping(Curve1), the technique with PSSMs adding Percentage closer filtering (Curve2), the technique with PSSMs adding Gaussian Blur and Percentage closer filtering(Curve3). From Fig.6, it could see that the velocity of putting Gaussian Blur is relatively faster than using Percentage closer filtering, and we can get notable soft shadow performances.

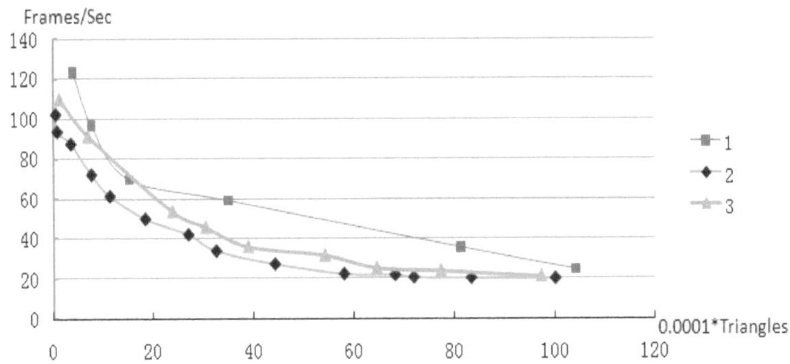

Fig. 6. Ability comparison of Percentage closer filtering with Gaussian Blur

4 Conclusion

This paper uses Cg language to simulate real-time shadow effects based on Ogre engine. As shader can enhance real-time rendering frame rates, we choose Gaussian Blur Filtering to calculate soft shadows with the help of shader, which decreases the consumption of CPU and GPU. The techniques we used in this paper are quite suitable for calculating real-time shadows in large-scale 3D Game that maintaining high performances and generating accurate shadow maps through real-time RTT technique with good expression. However, the problem which has to be considered is that in real world, whether the shadow edges are blurred has great relation with light condition. If we take light model into consideration, soft shadow changes with light changing, the performance of real-time shadow rendering can be even better and this problem will be done in the next work.

Acknowledgement. This work is supported by the Fundamental Research Funds for the Central Universities of China. (HEUCF100809).

References

1. Annen, T., Mertens, T., Seidel, H.-P., Flerackers, E., Jan Kautz, C.: Exponential Shadow Maps. In: Canadian Information Processing Society, pp. 155–161. ACM press, New York (2008)
2. Zhang, F., Sun, H., Xu, L., Kit, L., Lun, T.: Parallel-split shadow maps for large-scale virtual environments. In: 2006 ACM International Conference on Virtual Reality Continum and its Applications, pp. 311–318. ACM press, New York (2006)
3. Zhang, F., Sun, H., Oskari Nyman, C.: GPU Gems 3. In: Parallel-Split Shadow Maps on Programmable GPUs, ch. 10. Addison-Wesley Professional (2007)
4. Lauritzen, A., Salvi, M., Aaron Lefohn, T.: Sample Distribution Shadow Maps. In: Symposium on Interactive 3D Graphics and Games, New York, USA, pp. 97–102 (2011)
5. Depth Map Shadows, `http://www.okino.com/new/toolkit/1-15.htm`
6. Eisemann, E., Assarsson, U., Schwarz, M., Michael Wimmer, C.: Casting Shadows in Real Time. In: SIGGRAPH ASIA 2009, pp. 1–48. ACM press, Japan (2009)
7. Bunnell, M., Fabio Pellacini, C.: GPU Gems 3, ch. 11. In: Shadow Map Antialiasing, Addison-Wesley Professional (2004)

The Application of Information Encrypted in E-commerce Security

Zhang Li and Song Ying

Baicheng Vocational and Technical College 137000

Abstract. As the development development of e-commerce is affecting all aspects of people's lives, leading the network information security requirements to become increasingly high, so as the security of ourselves in the network activities. Application of information encryption technology will ensure the confidential requirements of the information in network activities and some relevant information. This paper begins from the introduction of encryption technology and presented to a new assume of the RSA algorithm, listing some applied examples of information encryption technology, emphasizes the importance of encryption technology in maintenaning security of network.

Keywords: Information encryption technology, E-commerce, RSA, Encryption algorithm.

1 Introduction

Information encryption is the core technology of information security. Especially with the rapid development of today's electronic commerce, electronic cash, digital currency, online banking and other Internet services. So it attracts more and more attention about how to protect information security so that is not stolen, not to be tampered with or destroyed and other issues. The key to the problem is information encryption. The so-called encryption, is to transform "the clear" human-readable information into a "ciphertext" process; and decryption is to revert the "ciphertext" back to "express". Encryption and decryption is using the cryptographic algorithms to be implemented. Cryptographic algorithm is used to hide and reveal information of the calculated process, typically the more complex algorithms; the more secure the ciphertext results. In encryption, the key is essential, the key is to follow a specific cryptographic algorithms run and produce a specific ciphertext value. The use of encryption algorithm it to protect information security so that is not stolen, not to be tampered with or destroyed.

2 Information Encryption Technologies

2.1 Encryption Model

The encryption algorithm can function as a complex transformation,

D. Jin and S. Lin (Eds.): Advances in EECM Vol. 1, LNEE 139, pp. 489–493.
springerlink.com

$$x = (y,k) . \tag{1}$$

x is behalf of ciphertext that has been encrypted sequence of characters, y represents that the sequence of characters to be encrypted, k is the key, when the encryption is completed, the ciphertext can be sent to the recipient via insecure channels, only the recipient has the decryption key can decrypt the ciphertext to be clear that the inverse transform. We will abstract out the data encryption model diagram, as the following

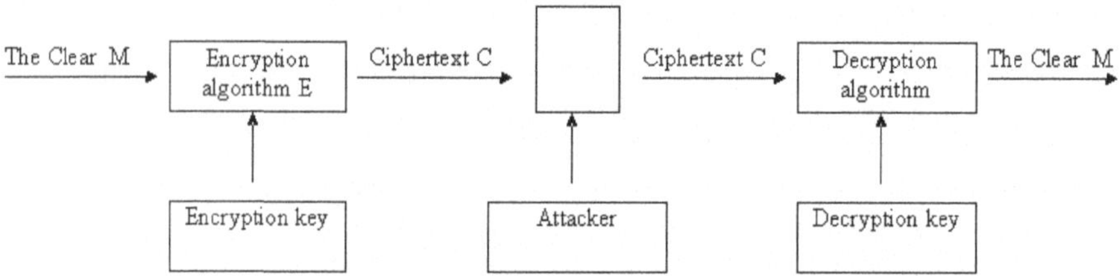

Data Encryption Model

2.2 Encryption Algorithm

Conventional encryption algorithm mainly takes advantage of symmetric algorithm, which refers to the encryption key can be derived from the decryption key and so as the opposite process. In most symmetric algorithms, the encryption / decryption key is the same. These algorithms also called secret key or single key algorithm, which requires the sender and recipient have one key before secure communications. Because the security of symmetric algorithm depends on the key, leaked key means anyone on the message encryption / decryption. As long as communications are confidential, the key must be kept confidential. Therefore, symmetric encryption and decryption algorithm refer to the process to adopt the same key, such as DES, 3DES, AES and other algorithms are symmetric algorithm. Asymmetric algorithms are also called public key algorithm, its difference from the encryption key is that used for decryption and encryption key the decryption key can not be calculated (at least within a reasonable a long time). The reason to call it as public key algorithm is that the encryption key is public and strangers can encrypt the message by encryption key, but only with the corresponding decryption key can decrypt the information. However, deriving the private key from public key is very difficult. RSA, DSA and other algorithms are asymmetric algorithms, of which RSA's most widely used, not only can be encrypted also be used to digital signature.

2.3 Improved RSA Asymmetric Encryption Algorithm

RSA encryption algorithm is currently the best known and most widely used public-key encryption algorithms; RSA's security depends on large numbers in general. It is the use of large integers modulo prime factor to solve the difficulty problem. Choose two differenf large prime numbers p and q, generally requires more than 10100; calculate

$$n = p \times q. \tag{2}$$

and

$$z = (p\text{-}1) \times (q\text{-}1). \quad (3)$$

choose a prime integer and z, denoted by d; calculated to meet the following conditions e,

$$(e \times d) \bmod z = 1 . \quad (4)$$

Here (n, e) is the public encryption key, (n, d) is a private key uj. During encryption, the plaintext block is divided into a certain size M, the encryption process is

$$C=M^e \mod n. \quad (5)$$

decryption process is

$$M=C^d \bmod n. \quad (6)$$

in RsA algorithm, the length of n is the important factor to the reliability of the algorithm, most encryption programs are used more than 1024, but the general security of RSA depends on large numbers, but whether it is equivalent to integer factorization theory has not been proven because there is no evidence to crack RSA certainly need to make large numbers decomposition. Therefore, decomposition module is the most obvious method of attack, people choose to the security of modulus greater than 10100, this would no doubt reduce the computing public key algorithm to the event and complexity. Therefore, on the basis of RSA algorithm, propose a variant of the RSA algorithm, specific ideas as follows:

User x's Public encryption transformation Ex and Secret decryption Dx's generate:

①Random select N couples prime p1、p2......pn;
②Calculate nx= p1*p2......*pn, Φ(nx)=(p1-1)*(p2-1)*......*(rj-1);
③Random select integer ex content(ex, Φ(nx)) =1;
④use euclidean algorithm to calculate dx, content ex*dx≡1 MOD Φ(nx); ;
⑤Regard nx,ex as Ex, mark Ex=< nx,ex>, keep the p1,p2,......,pn, Φ(nx) secret as Dx, as Dx=<p1,p2,......,pn ,dx,Φ(nxj) >。
Then the encryption algorithm:c = Ex(m) = mex(MOD nx); decryption algorithm:m = Dx(c) = cdx(MOD nx)。

In the RSA algorithm, contains two keys: an encryption key PK and the decryption key SK, public key encryption.

Through the certification procedures in the case of binary computing speed of 8 * 8 was significantly greater than 2 * 2 * 2 * 2 * 2 * 2, which shows the advanced nature of the RSA algorithm, as the variant of the RSA algorithm is a algorithm takes advantage of Fermat's little Theorem encryption algorithm on the basis of the original algorithm ,which can be derived by mathematical induction, there is no violation of the fundamental correctness of RSA algorithm, therefore it ensure the security of the variant RSA algorithm.

3 Information Encryption Technology in E-commerce Application

E-commerce is changing people's lives and development of the whole society, the network trade will cause a comprehensive revolution of management, work and lifestyle, as well as management's way of thinking. At the same time, robust network security environment is important on the health and sustainable development of e-commerce. The security of e-commerce platform is the main security of network and transaction information. Transaction information means to protect both sides of the transaction not be damaged, non-disclosure, and confirm the identity of parties to the transaction. Therefore, the application of data encryption in e-commerce, digital signature, digital certificate, ssl, set security protocols and other technologies to ensure the security of e-commerce transactions, electronic commerce and information with encryption technology and will certainly promote the stable development of electronic commerce. Meanwhile, the core development of e-business and key issues are the security of transactions. As the Internet itself is open, making online transactions face with dangers, and put forward the corresponding security control requirements. Information on the network has higher requirements, information security services as the following points.

①. Confidentiality of information. The information in transaction of business requires to be confidential. Such as credit card account and user name is known by the others, it may be stolen. Therefore, the information in e-commerce in general has encrypted requirements.

②. The integrity of information. Transaction file is not modified, as if it can be changed, then the transaction itself is not reliable, customers or businesses may therefore suffer danger. So the file should be able to be protectd that can not be modified in order to protect the serious and fair of the trade.

③. Information source identification. In the process of sending messages among the businessmen, there may be attacker who disguised as business and send false information, the use of cryptography, asymmetric cryptography based bookmarks system signature data can help the recipient to confirm data from a specific user.

④. Non-repudiation of information. If any party has denied the transaction, it will bring the inevitable losses to the other side. Therefore, we must ensure that communications and transactions have been carried out both operations and can not be denied.

Encrypting informationIn is the only way to ensure the security of e-commerce information, in the process of electronic transaction , the data need to be accurate, and timely exchanged, how to control the security in process of e-commerce , you can use data encryption form, as for which form, it can be selected according to the actual situation of different encryption algorithms. In any case, symmetric and asymmetric encryption. In comparison, the flaws of symmetric encryption in e-commerce applications are the management and distribution of the key is very complex. For example, payment of the shopping environment, a network with n couples of users, as every time the user uses symmetric encryption algorithms that other people do not know the unique key to ensure confidentiality of information, so the system the key has a total of

$$n (n-1) / 2. \tag{7}$$

if n equals 104, then you need to manage about 5 × 107 keys, spend a lot of storage space. Another problem of symmetric encryption is that it can not identify the trade initiator or trade the final side, because both share the same trade, the key trading parties any information through this encrypted key is transmitted to the other side, it is not used for digital signatures. Encrypted information in the process of e-commerce, multi-purpose form of asymmetric encryption. Improvements described in this article as the RSA encryption algorithm, due to its more traditional encryption algorithms faster and more efficient. It would be a better choice.

4 Conclusion

Network data encryption is an important part of network security technology. But e-commerce transaction security mechanism is not mature enough to meet the rapidly growing e-commerce. But data encryption is the basic security technologies in the network and it is more and more widely used. Password technology will penetrate into every corner of digital information; with data networks, communications systems closely link to security, to provide broader and more effective security measures.

References

1. Jing, J.: Information systems security and confidentiality. Tsinghua University Press (2005)
2. Peng, C.: The discuss about e-commerce security solutions. Hunan Radio and TV University (2009)
3. Deng, Z., Zhu, Q.: Network Security Training tutorial. People Post Press (2005)
4. Huang, Z.: Modern computer information security technology. Metallurgical Industry Press, Beijing (2005)
5. Wang, S.: On the e-commerce security data encryption technology. Communication Technology (2008)
6. Huang, Z.: Modern computer information security technology (2005)
7. Zhao, S.: On the e-commerce network security issues and countermeasures. China Scientific & Technical (2009)
8. Deng, Z., Zhu, Q.: Network Security Training tutorial. People Post Press (2005)
9. Liu, W., Xing, Y.: On the network security of e-commerce development and countermeasures. Modern shopping malls (2009)

Harmonic Suppression of High-Frequency Power Supply for ICP Light Based on Frequency Power Modulation

Bingyan Chen*, Juan Zhou, Tingwei Wu, Jinchen Wang, Yiwei Zhou, and Changping Zhu

Jiangsu Key Laboratory of Power Transmission and Distribution Equipment Technology, Hohai University
Changzhou, 213022, China
chenby@yahoo.cn

Abstract. Newly-emerging inductively coupled plasma (ICP) light has high efficiency, long life-time and other comprehensive advantages. But, it's high-frequency power supply will produce conducted electromagnetic interference (EMI) that is hardly suppressed. The paper establishes the conducted EMI model of 150W ICP light's high-frequency power supply, and analyses conducted EMI's production mechanism at the fixed working frequency. Design the modified frequency power modulation algorithm with same frequency change step δ and energy pulse duration gradual change. The conducted EMI and optical power parameters tests indicate that modified frequency modulation can solve the conducted EMI problem well. And it can improve the ICP light's luminous efficiency.

Keywords: conducted EMI, frequency modulation, testing, inductively coupled plasma (ICP), light source, algorithm.

1 Introduction

ICP light, a kind of electrode-less lamp, integrates inductively couple plasma, power electronic and electronic light source technology and it couples the electromagnetic energy to the closed glass tube through high frequency power supply. The luminous efficiency of 100W or higher ICP light can reach 75 ~ 95lm/W, having more advantage than fluorescent lamp, cold cathode fluorescent lamp, LED, etc. To distribute temperature and decrease loss, the high-frequency power supply for ICP light use fixed frequency 220kHz. With parasitic inductances and capacitances in the inverter circuit, power switches and rectifier have high di/dt and dv/dt during working. It causes strong EMI existing at the working frequency and harmonics points. Traditional methods, such as EMI filter, adding magnetic bead, using absorption circuit and so on, can hardly suppress interference. At the same time, those methods make high-frequency power supply bigger size and more heat emit.

* Corresponding author.

D. Jin and S. Lin (Eds.): Advances in EECM Vol. 1, LNEE 139, pp. 495–502.
springerlink.com

2 Analysis of EMI

2.1 The Basic Structure of ICP Light

Fig.1 indicates high-frequency power supply and ICP lamp composes the ICP system. The high-frequency power supply includes EMI filter, rectifier, inverter circuit, power-matching circuit, inverter controlling circuit, etc. The inverter circuit part shown in the fig.2 sends the high voltage VH from rectifier to the half-bridge inverter. And the half-bridge inverter is made up of power switch Q1,Q2, power matching inductance Lr, resonant capacitance Cr and DC blocking capacitance C1.

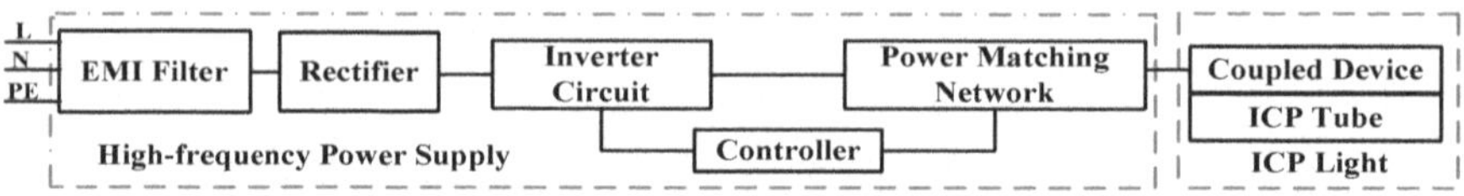

Fig. 1. Diagram of whole function of light source

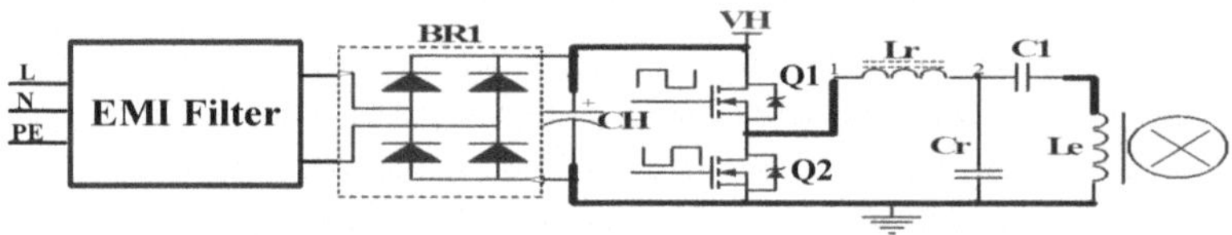

Fig. 2. Inverter circuit of high-frequency power supply

2.2 Analysis of EMI

Base on fig.1 and fig.2, the EMI source is the inverter circuit and rectifier bridge. The coupled channel of conducted EMI is a multi-port network made up by resistors, capacitances and inductances, which is a non-linear topological network. When discussing the EMI, the non-linear network can be approximated to a linear system in most cases and coupled channel's system function can be got by tests[1,3].

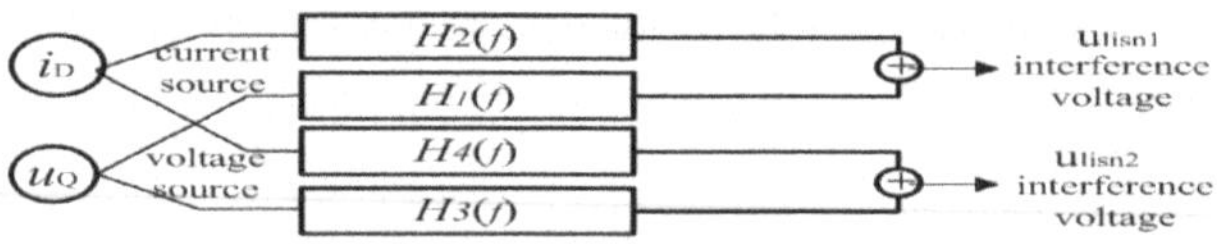

Fig. 3. Functional model of the systematic EMI propagation channel

According to the basic path of the power transfer, the voltage of power MOSFET (Q1, Q2) and the current of diode in the rectifier (BR1) are the main EMI sources. The EMI in the electric network caused by high-frequency power supply is the voltage response u_{lisn1} and u_{lisn2} interference[3-4]. The voltage responses are produced by the two interference sources above on the two 50 Ω resistors of impedance stabilization network (LISN) between the power's L and N line. The relation among the interference source and u_{lisn1}, u_{lisn2} shows in fig.3. And we can build mathematical model with the following formulas.

$$u_{lisn1}(f) = u_Q(f)H_1(f) + i_D(f)H_2(f) \tag{1}$$

$$u_{lisn2}(f) = u_Q(f)H_3(f) + i_D(f)H_4(f) \tag{2}$$

The common-mode interference voltage caused by u_{lisn1} and u_{lisn2} on the power network is $u_{cm} = 0.5(u_{lisn1} + u_{lisn2})$. With formula (1) and (2)we can get as follow:

$$u_{cm} = 0.5(u_Q[H_1(f) + H_3(f)] + i_D[H_2(f) + H_4(f)]) \tag{3}$$

Therefore, the interference voltage on the power network is caused by a combination of i_D and u_Q. In addition, the common-mode voltage component from the MOSFET is higher than that from the current jumps in the diode. In fig.1 and fig.2, the high-frequency power supply will cause higher di/dt and dv/dt when it works at fixed frequency 250kHz. It has a lot of parasitic inductance and capacitance. All those parasitic effects worse the result of formula(3). Thereby, it causes quite strong conducted EMI noises on input cable L and N. Test it with the lamp testing standard EN55015. With fixed frequency 250kHz, the system will produce quite strong and over limit-value EMI peak energy at the working frequency point and its harmonic points, but in other frequency interval, the EMI is lower than the limit-value. The EMI peak energy distributes within the bandwidth ranged from hundreds kilohertz to decades megahertz at the fixed frequency point 250kHz and its integral multiple harmonic. We take measures as follow: firstly, use magnetic core with better high-frequency characteristics and change the common-mode and differential-mod inductances and the capacitive parameter of X and Y within the IEC and FCC standards[3-5]. Secondly, add peak energy absorbing circuit on the power switches Q1, Q2 and the diode in rectifier to reduce the EMI from the parasitic effect of circuit. After repeated parameter changing and EMI experiments shows that when the ICP light system works at fixed frequency, changing EMI filter and reducing distributed parameter can hardly reduce the EMI energy. The highest EMI peak energy is 30dBuV higher than the limit-value. Meanwhile, the two methods will cause the high-frequency power supply oversized, higher cost, lower efficiency, etc.

3 The Strategy of Modified Frequency Modulation of Inverter

3.1 Frequency Modulation and Strategy of EMI Peak Suppression

From the EMI test results above, we can learn that except the fixed frequency point 250kHz and its harmonic points, the peak value and the average value of EMI energy are both much lower than the standard's limit-value. Change the inverter's switching frequency reasonably, move EMI energy to the adjacent frequency band of low EMI energy when switching Q1 and Q2, making EMI energy lower than limit. This is the strategy of modified frequency modulation[6]. In this paper the strategy set f_o as the center operating frequency and δ as the frequency change's step, increase and decrease linearly within the limit $\pm\Delta$ in the modulation period T_M, shown in fig.4.

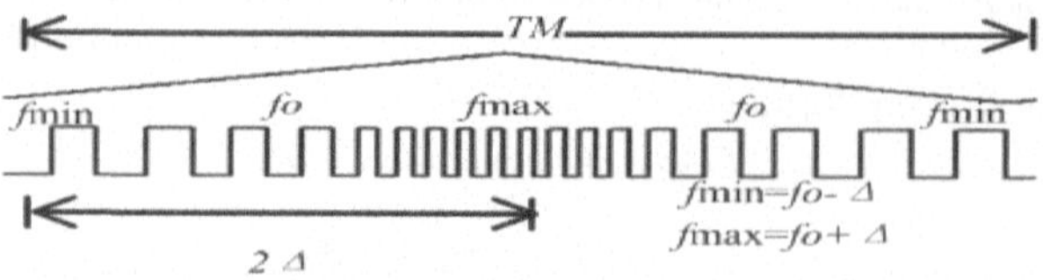

Fig. 4. Schematic diagram of modified frequency modulation

So the relationship among frequency change step δ , frequency change time n , the maximum frequency $f_{\max}$ and the minimum frequency $f_{\min}$ is $f_{\min} + n\delta = f_{\max}$. When the frequency change region is Δ, the required frequency change time $n = 2\Delta / \delta$. And when the frequency increases and decreases with the step δ in the range $\pm\Delta$, the working frequency after *i*-times change is $f_i = f_o - \Delta + i\delta$ (i=0, 1, 2, ……, $2\Delta / \delta$) .Thus, $H_e(f)$ in formula(1) and (2)can be expressed as follow:

$$H_a(f) = \frac{1}{\Delta}\sum_{i=0}^{n} H_e(f_i) = \frac{1}{\Delta}\sum_{i=0}^{n} H_e(f_o - \Delta + i\delta) \tag{4}$$

Substitute $2\Delta / \delta$ and $f_i = f_o - \Delta + i\delta$ into formula(3) can get the interference voltage on the network in modified frequency modulation mode as follow:

$$u_{cm} = \frac{u_Q}{2\Delta}\sum_{i=0}^{n}[H_1(f_o - \Delta + i\delta) + H_3(f_o - \Delta + i\delta)] + \frac{i_D}{2\Delta}\sum_{i=0}^{n}[H_2(f_o - \Delta + i\delta) + H_4(f_o - \Delta + i\delta)] \tag{5}$$

Substitute f_o into formula (3) can get the interference voltage on power network in fixed frequency modulation mode as follow:

$$u_{cm} = 0.5(u_Q[H_1(f_o) + H_3(f_o)] + i_D[H_2(f_o) + H_4(f_o)]) \tag{6}$$

If the test time T_D is long enough and each frequency pulse sends equal conducted EMI energy when ICP light works. Then formula (6) shows when ICP light adopts fixed frequency modulation, the network interference energy concentrates on the point f_o, it will cause peak EMI at this point. However formula(5) shows, adopting modified frequency modulation, the network interference energy disperse between f_o - Δ and f_o + Δ dynamically and evenly, avoiding causing peak EMI at some frequency points.

3.2 Implementation of Modified Frequency Modulation Algorithm

As fig.4 shows, in the frequency modulation period T_M , set the f_o as the center working frequency, increase and decrease within the range ±Δ, each frequency point f_i emits a (a is an integer and a≥1) switching signals with conducted EMI signal, so the expression of T_M is:

$$T_M = 2a\sum_{i=0}^{n}\frac{1}{f_o - \Delta + i\delta} \tag{7}$$

According to Sampling Theorem, the sampling frequency must be at least twice as the maximum frequency, otherwise, it can't recover from sampling signal[7]. That means the relationship between measuring time T_D and frequency modulation period T_M must be satisfied as $T_D \le 0.5T_M$ .Substitute $T_D \le 0.5T_M$ into (7) and get that:

$$a\sum_{i=0}^{n}\frac{1}{f_o-\Delta+i\delta}\ge T_D \tag{8}$$

The formula (8) defines the mathematical relationship among the parameter a , Δ , f_o and T_D when using MCU to control modified frequency modulation. Usually, from 150KHz to 30MHz, the time of detecting conducted EMI's values is 1 millisecond (ms)[8]. During the tests, the parameters' values in modified frequency modulation algorithm are: center operating frequency f_o =250KHz, range of frequency modulation Δ =25kHz, step of frequency change δ =1KHz, the minimum of frequency modulation period T_M is 2ms. According to the formula (7) and (8) can get this:

$$2a\sum_{i=0}^{50}\frac{1}{250-25+i}\ge 2 \tag{9}$$

Solving the equation (9) can get the pulse number of each frequency point: $a \ge 4.8849$.Take a=5.The inverter control circuit shown in fig.1 adopts MCU series as core control unit, and embeds the frequency modulation algorithm shown in fig.5. Thus achieve the EMI peak energy suppression of ICP light.

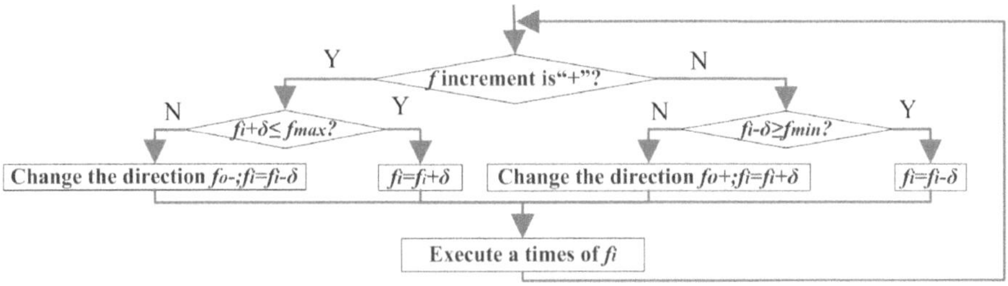

Fig. 5. Program flowchart of frequency modulation

4 The Result of Tests

4.1 The Set of Tests' Parameters

According to formula (8) and (9), set f_o =250kHz as the center operating frequency of controller. Fig.5 shows, it achieves frequency jitter from 225kHz to 275kHz when varying ±25kHz within the frequency modulation cycle of 2ms(500Hz).Each variation increment of working frequency is δ=1kHz , and each change transmits 5 square switching control signals to drive half-bridge inverter circuit.

4.2 The Tests of EMI Characteristic

Without changing the network parameters, the input power voltage and the load of the original circuit, adopt KH3939 Test-Receiver. According to lamps testing standard EN55015, test the conducted EMI of modified frequency modulation high frequency power supply from 9 kHz to 30MHz, and the frequency spectrum is shown in fig.6.

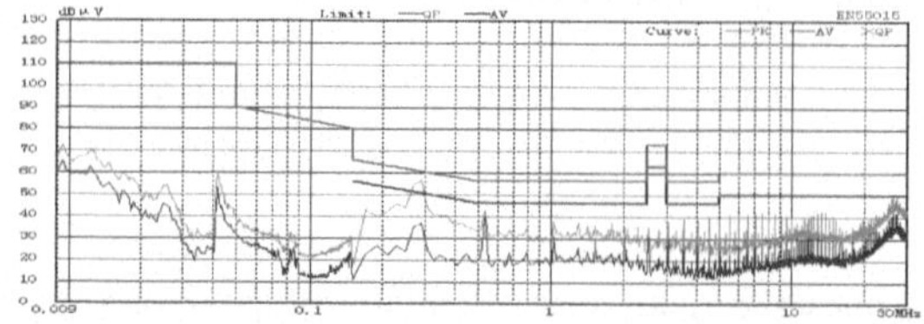

Fig. 6. The EMI testing graph of ICP light in modified frequency modulation

Throughout the frequency band of conducted EMI test, both quasi-peak(QP) value and average value(AV) don't exceed the limit, and the EMI decrease effects is striking. The harmonic amplitude that use modified frequency modulation is greatly reduced and the line is smooth without strongly peak EMI. Each harmonic approximates continuous response.

4.3 The Tests of Luminous Efficiency and Temperature Rise

Adopt high-frequency power supplies of fixed and modified frequency modulation for ICP light whose parameters are optimized, and choose ICP tube of 5700K driving color temperature and 150W powder under the same testing circumstance. Use photoelectric parameters tester, 2m diameter integrating sphere and rapid photoelectric test system HP8000LED to test the photoelectric parameters of fixed and modified frequency modulation ICP light separately. Use temperature data logging to test the enclosures temperature of fixed and modified frequency modulation high-frequency power supply separately, testing time from 0~120 minute.

The experiment showing both fixed and modified frequency modulation high-frequency power supply can start up ICP tube quickly, and reach the max luminous efficacy of light source output in one minute. But the light-electricity conversion efficiency of modified frequency modulation high-frequency power supply is 8lm/W higher than the fixed frequency modulation. The enclosure temperature rise of modified frequency modulation high-frequency power supply is 10℃ lower than the fixed frequency. It shows that compared with fixed frequency high-frequency power supply, the power dissipation of modified frequency modulation is much lower.

5 Conclusions

The true cause of the efficiency improvement in testing is that the system reduces conducted EMI energy-emission because it adopts modified frequency modulation method. Further, it simplifies the structure of filters effectively. For example, the resistances on the gate of power switches and EMI absorbing capacitance between the drain and the source can be left out. Thus, the inductance, EMI filter size and others will be reduced. All of this will make the power loss of high-frequency power supply lower. As energy saving and emission reduction, green lighting and low-carbon economy development are the hot topic nowadays.

Acknowledgements. This work was supported in part by the Fundamental Research Funds for the Central.

Universities of China under Grant 2009B31614, 2010B23914, and by the Open Projects of Jiangsu Key Laboratory of Power Transmission & Distribution Equipment Technology, Jiangsu, China, under Grant CS0906, and by the Youth Research Program of Hohai University, China, under Grant 09B001-04. And the author wishes to thank the Springer for providing this template and all colleagues who previously provided technical support.

References

1. Maniktala, S.: Switching Power Supplies A to Z. Elsevier, Singapore Pre Ltd (2008)
2. Billings, K.: Switching Power Supply Handbook, 2nd edn. McGraw-Hill Companies, Inc. (1999)
3. Part 18– Industrial, science, and Medical Equipment. FCC, July 18(1975)
4. CISPR61000-6-3:Emission standard for residential, commercial and light-industrial environment (1996)
5. IEC555: part 1: Definitions (IEC Publication 555-1, Guide); Part 2: Harmonics (IEC Publication 555-2, Standard); and Part 3: Voltage fluctuations and flicker (IEC Publication 555-3, Standard)
6. Gellérttér, S.: The effect of contaminants on the mercury consumption of fluorescent lamps. Physics D: Applied Physics 42(9) (May 7, 2009)
7. Honda, I.: Abstraction of Shannon's sampling theorem. IEICE Transactions on Fundamentals of Electronics, Communications and Computer Sciences (June 1998)
8. Krusubthaworn, A., Sivaratana, R., Ungvichian, V., Siritaratiwat, A.: Testing parameters of TMR heads affectedby dynamic-tester induced EMI. Journal of Magnetismand Magnetic Materials 316, e142–e144 (2007)
9. Zhang, X.Y., Xue, Q.: High-Selectivity Tunable Bandpass Filters With Harmonic Suppression. IEEE Transactions on Microwave Theory and Techniques 58(4), 964–969 (2010)
10. Wang, W.-X., Zhu, X.-L.: Harmonic-Suppression System Based on Fuzzy Neural Control in Power Line. In: 2010 WASE International Conference on Information Engineering (ICIE), August 14-15, pp. 93–96 (2010)
11. Guo, Y., Cheng, W., Fu, Z., Chen, L.: DCM operating characteristics and harmonic suppression study of AC-DC converter. In: Third International Conference on Electric Utility Deregulation and Restructuring and Power Technologies, DRPT 2008, April 6-9, pp. 2347–2352 (2008)
12. Takeshita, T., Goto, H., Masuda, T., Matsui, N.: Current waveform control of a high-power-factor rectifier circuit for harmonic suppression of voltage and current in a distribution system. Electrical Engineering in Japan 140(4), 62–71 (2002)
13. Xia, X.-Y., Han, X.: A Novel Active Power Filter for Harmonic Suppression and Reactive Power Compensation. In: 1st IEEE Conference on Industrial Electronics and Applications, May 24–26, pp. 1–3 (2006)
14. Chen, H., Sun, Y., Chen, W.: Harmonic suppression of network-connected distributed generation based on novel Hybrid power filter. In: 4th IEEE Conference on Industrial Electronics and Applications, ICIEA 2009, May 25-27, pp. 2914–2918 (2009)
15. Saitou, M., Matsui, N., Shimizu, T.: Modeling and Harmonic Suppression for Distribution Systems. In: The 29th Annual Conference of the IEEE Industrial Electronics Society, IECON 2003, November 2-6, vol. 2, pp. 1521–1526 (2003)

Author Biography

Bingyan Chen (M'11) received the M.Sc. degree in computer application from Nanjing University, Nanjing 210093, China, in 2010. Received the B.Sc. degree in applied physics from Huazhong Normal University, Wuhan 430079, China, in 2002.

Now he is a senior filed electronic engineer and also a PhD candidate in the Physics Experiment Center and the Jiangsu Key Laboratory of Power Transmission and Distribution Equipment Technology, Hohai Universit, China. His research interests include electric automation, power electronics and drives, electric lighting etc. He became a Member (M) of IEEE in 2011.

Mailing Address: No.200, North Jinling Road, Xinbei District, Changzhou, Jiangsu, China, Postcode : 213022, Mobile phone : 18605191221, E-mail : chenby@yahoo.cn.

Juan Zhou received the M.Sc. degree in communication and information system from Hohai University, Nanjin 210098, China, in 2007. Received the B.Sc. degree in information management and system from Hohai University, Nanjin 210098, China, in 2002.

She has been with the Information center of Hohai University, Changzhou 213022, where she is currently a communication network engineer. Her research interests include communication network and relative instruments.

Mailing Address : No.200, North Jinling Road, Xinbei District, Changzhou, Jiangsu, China, Postcode : 213022, Mobile phone : 13813546263, E-mail : zhouj@hhuc.edu.cn.

Changping Zhu (M'05) received the Ph.D. degree in communication circuit and power ultrasonic technology from China University of Mining & Technology, Xuzhou 221008, China, in 2010. Received the M.Sc. degree in computer application from Huazhong University of Science & Technology,Wuhan 430074, China, in 2002.

He is the assistant dean and professor of College of Computer and Information Engineering, Hohai University. His research interests include communication circuit and power ultrasonic technology. He became a Member (M) of IEEE in 2005.

Mailing Address : No.200, North Jinling Road, Xinbei District, Changzhou, Jiangsu, China, Postcode : 213022, Mobile phone : 13861282087, E-mail : cpzhu5126081@163.com .

Tingwei Wu , **Jinchen Wang** are the undergraduate in Electronic Science and Technology, the Department of Computer and Information Engineering Hohai University, Changzhou 213022, China.

Mailing Address: No.200, North Jinling Road, Xinbei District, Changzhou, Jiangsu, China, Postcode: 213022, Mobile phone: 15995023755, E-mail : 254322592@qq.com.

Sensor Sensitivity Analysis of Long Period Fiber Grating by New Transfer Matrix Method

Guodong Wang*, Yunjian Wang, Na Li, and Suling Wang

School of Electrical Engineering and Automation, Henan Polytechnic University, JIaozuo 454003, Henan, China
wgd@hpu.edu.cn

Abstract. The sensor sensitivity, such as axial strain sensitivity, temperature sensitivity and refractive index sensitivity, of long period fiber grating is analyzed by new transfer matrix method. The new transfer matrix method can be used to analyze the modes coupling between the core mode and multiple cladding modes. Compared with the previous method used, such as solving the coupled mode equation by the fourth order adaptive step size Runge-Kutta algorithm, the new transfer matrix method has a faster calculation speed. Theoretical results are excellent agreement with the method of solving the coupled mode equation.

Keywords: Sensor Sensitivity, New Transfer Matrix Method, Long Period Fiber Grating.

1 Introduction

Long period fiber gratings (LPFG) have found many applications in optical telecommunications such as mode converters[1], rejection filters[2], gain-flattening filters for erbium-doped fiber amplifiers[3], and optical fiber sensors for strain[4], temperature[5], and refractive index measurements[6] because of their capability of coupling power between core and cladding modes at the resonant wavelengths[7-10].

LPFG represent important alternatives to the use of FBGs in many sensor application, and show many of the same advantages in terms of an intensity-independent output to encode the measurement. In addition, the wavelength shifts of some of the key spectral features are often greater than are seen for FBGs, there by offering potentially a higher sensitivity. The sensor sensitivity of LPFG can be obtained by analyzing the variation of spectral under a certain temperature changes. The commonly methods used to analyze the spectral characteristics of LPFG and fiber Bragg gratings include transfer matrix method (TMM) and solving the coupled mode equations (SCME) [11-13]. The traditional TMM is able to analyze the uniform and non-uniform LPFG and fiber Bragg gratings when only two modes being considered [14-15]. The SCME method could obtain the spectrum of the uniform and non-uniform structure LPFG when multiple modes are considered. In this paper, a new

* Corresponding author.

D. Jin and S. Lin (Eds.): Advances in EECM Vol. 1, LNEE 139, pp. 503–509.
springerlink.com

transfer matrix method about LPFG with multiple cladding modes coupled is proposed and applied to analysis the axial strain, temperature and refractive index sensitivity of LPFG. This new TMM can analyze the modes coupled both about the uniform and the non-uniform between the core mode and the multiple cladding modes. And the new TMM is simple to implement, almost always sufficiently accurate, and generally faster than that of SCME.

2 New Transfer Matrix

The new transfer matrix of long period fiber grating can be expressed as[16]

$$T = e^{FL} \tag{1}$$

Where

$$F = j\begin{bmatrix} k_{0,1-01}^{co-co} & \frac{m}{2}k_{1,1-01}^{cl-co} & \frac{m}{2}k_{1,2-01}^{cl-co} & \cdots & \frac{m}{2}k_{1,v-1-01}^{cl-co} & \frac{m}{2}k_{1,v-01}^{cl-co} \\ \frac{m}{2}k_{1,1-01}^{cl-co} & -2\delta_{1,1-01}^{cl-co} & 0 & \cdots & 0 & 0 \\ \frac{m}{2}k_{1,2-01}^{cl-co} & 0 & -2\delta_{1,2-01}^{cl-co} & 0 & \cdots & 0 \\ \vdots & \vdots & 0 & \ddots & 0 & \vdots \\ \frac{m}{2}k_{1,v-1-01}^{cl-co} & 0 & \vdots & 0 & -2\delta_{1,v-1-01}^{cl\ co} & 0 \\ \frac{m}{2}k_{1,v-01}^{cl-co} & 0 & 0 & \cdots & 0 & -2\delta_{1,v-01}^{cl-co} \end{bmatrix} \tag{2}$$

and L is the grating length.

The new TMM is exactly enough on analyzing the transmission spectrum of LPFG and the calculating speed using the new TMM is faster than that by SCME[16].

3 Sensor Sensitivity Analysis

3.1 Axial Strain Sensitivity Analysis

When the grating is held under the tension, an axial strain $\varepsilon(z)$ will be produced along the grating length, which can be given by

$$\varepsilon(z) = \frac{F}{EA(z)} \tag{3}$$

where $A(z)$ is the cross-section area of grating at position z, E is Young's modulus of fiber. Because of the existence of E, the period of the grating and the effective index of transmission modes in fiber will be changed along the grating length, and can be expressed as:

$$\Lambda(z) = \Lambda_0 \left[1 + \varepsilon(z)\right] \quad (4)$$

$$n(z) = n_{eff} - \chi\varepsilon(z) \quad (5)$$

where Λ_0 is the grating period at position $z = 0$, $\Lambda(z)$ is the grating period at position z, χ is the elasto-optical coefficient of fiber and n_{eff} is the effective index of transmission mode in fiber.

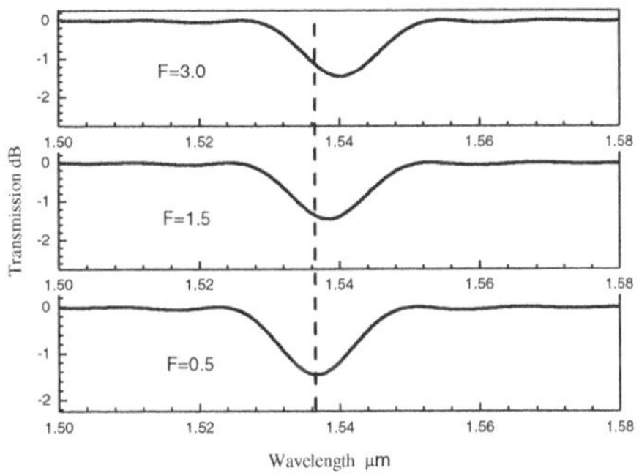

Fig. 1. The transmission spectral of LPFG when different tension F is applied to it

Fig. 1 shows the transmission spectrum of LPFG when different tension F is applied to the grating. The resonance shown in Fig.1 is associated with coupling between the core mode and the $v = 7$ cladding mode. The coupled wavelength will

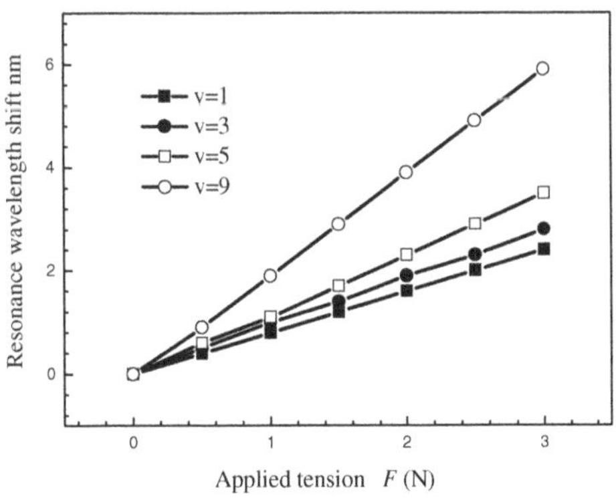

Fig. 2. The resonance wavelength of LPFG when different tension F is applied to it

shift to the longer wavelength when a certain tension is applied to the grating. The more details on the wavelength shift about more cladding mode are demonstrated in Fig. 2. From these figures we can find that the larger the tension applied to the grating, the farther wavelength shifts.

3.2 Temperature Sensitivity Analysis

When the ambient temperature varies, the grating period and the effective refractive index will be changed and can be showed as

$$\Lambda(\Delta T)=\Lambda(T_0)(1+\alpha) \tag{6}$$

$$n_{co}^{eff}(\Delta T)=n_{co}^{eff}(T_0)(1+\xi_{co}) \tag{7}$$

$$n_{cl,m}^{eff}(\Delta T)=n_{cl,m}^{eff}(T_0)(1+\xi_{cl}) \qquad m=1,2,\cdots \tag{8}$$

where T_0 is the room temperature, α is the coefficient of thermal expansion of fiber, ξ_{co} and ξ_{cl} are the thermo-optical coefficient of core and cladding material.

Figure 3 shows the transmission spectral of LPFG when ambient temperature varied. The single resonance that is visible with in the wavelength range plotted is associated with coupling to the $v=9$ cladding mode. This figure shows that the coupled wavelength will shift to the longer wavelength when the ambient temperature rised. The more details on the wavelength shift about more cladding mode is illustrated in figure 4. From these figure we can find that the larger the ambient temperature rised, the farther wavelength shift.

3.3 Refractive Index Sensitivity Analysis

When the ambient refractive index varies, the cladding modes of the fiber will be changed. Figure 5 shows the transmission spectral of LPFG when refractive index varied. The single resonance that is visible with in the wavelength range plotted is associated with coupling to the $v=9$ cladding mode. This figure shows that the coupled wavelength will shift to the shorter wavelength when the refractive index rised. The more details on the wavelength shift about more cladding mode are illustrated in figure 6. From these figure we can find that the larger the ambient index rised, the farther wavelength shift to the shorter.

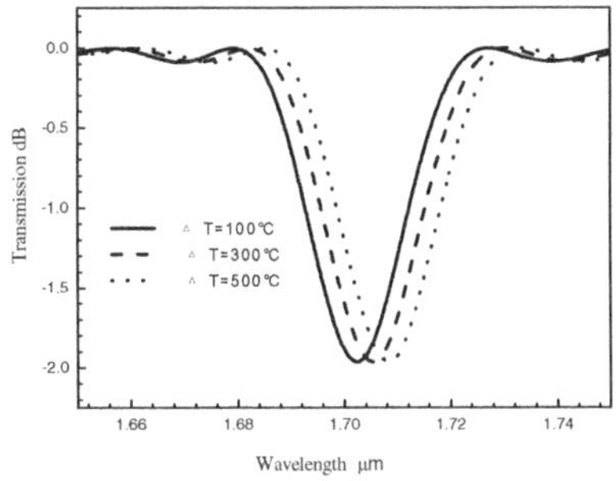

Fig. 3. The transmission spectral of LPFG when ambient temperature varied

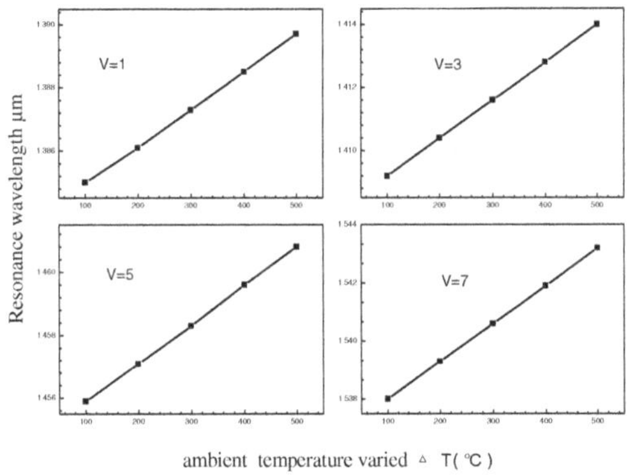

Fig. 4. The resonance wavelength of LPFG when ambient temperature varied

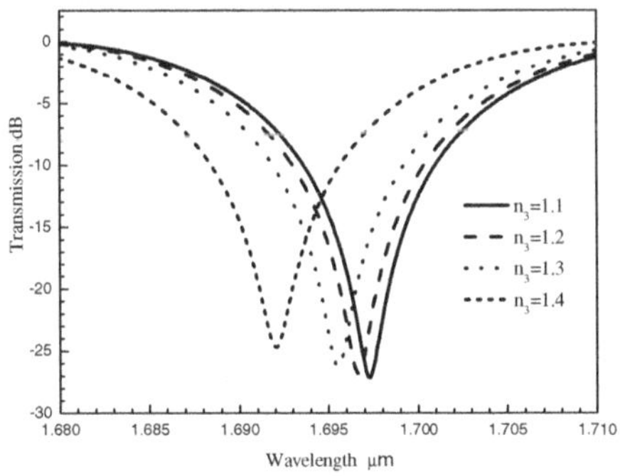

Fig. 5. The transmission spectral of LPFG when ambient refractive index varied

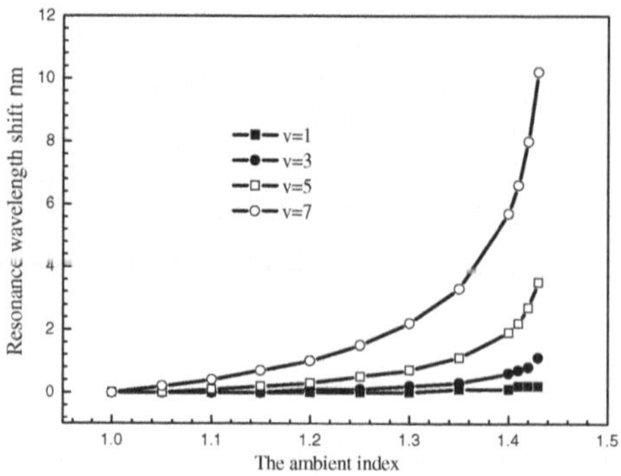

Fig. 6. The resonance wavelength of LPFG when ambient refractive index varied

4 Conclusion

In this paper a new transfer matrix method about long period fiber grating with multiple cladding modes is proposed. The new method can be used to analyze the sensor sensitivity of long period fiber grating. Compared with the usually used method, such as solving the coupled mode equation by the fourth order adaptive step size Runge-Kutta algorithm, the new transfer matrix method is of a faster calculation speed. The theoretical results are exactly agreed with the method of solving the coupled mode equation.

Acknowledgement. This work is supported by the Program of the National Natural Science Foundation of China (No. 61040016), by the Open Foundation from Henan Provincial Open laboratory of Control Engineering Key Disciplines China (No.KG2009-16) and by the Doctor Foundation from Henan Polytechnic University China (No. 648393).

References

1. Andermahr, N., Fallnich, C.: Optically induced long-period fiber gratings for guided mode conversion in few-mode fibers. Optics Express 18(5), 4411–4416 (2010)
2. Chen, K., Sheng, Q., Dong, X.: Band-rejection and band pass filters based on mechanically induced long-period fibergratings. Microwave and Optical Technology Letters 42(1), 15–17 (2004)
3. Ni, N., Chan, C.C., Tan, K.M., Tjin, S.C., Dong, X.Y.: Broad-band EDFA gain flattening by using an embedded long-period fiber grating filter. Optics Communications 271, 377–381 (2007)
4. Kang, J., Dong, X., Zhao, C., Qian, W., Li, M.: Simultaneous measurement of strain and temperature with a long period fiber grating inscribed Sagnac interferometer. Optics Communications 284, 2145–2148 (2011)

5. Venugopalan, T., Sun, T., Grattan, K.T.V.: Temperature characterization of long period gratings written in three different types of optical fiber for potential high temperature measurements. Sensors and Actuators A 160, 29–34 (2010)
6. Martinez-Rios, A., Monzon-Hernandez, D., Torres-Gomez, I.: Highly sensitive cladding-etched arc-induced long period fiber gratings for refractive index sensing. Optics Communication 283, 958–962 (2010)
7. Harhira, A., Guay, F., Daigle, M., Lapointe, J., Kashyap, R.: Long-period fiber gratings fabricated with a CO2 laser beam and phase mask. Optics Communication 283, 4633–4638 (2010)
8. Eggen, C.L., Lin, Y.S., Wei, T., Xiao, H.: Detection of lipid bilayer membranes formed on silica by double-long period fiber grating laser refractometry. Sensors and Actuators B: Chemical 150, 734–741 (2010)
9. Nam, S.H., Lee, J., Yin, S.: Control of resonant peak depths of tunable long-period fiber gratings using over coupling. Optics Communication 284, 961–964 (2011)
10. Jiang, M., Feng, D., Sui, Q.: Characteristic research on mechanically induced long-period fiber gratings. Chinese Optics Letters 7(2), 112–114 (2009)
11. Wang, G.-D., Xie, B.-B.: Improving the performance of chirped fiber grating with cladding being etched sinusoidal function. Optik 122, 557–559 (2011)
12. Dong, X., Feng, S., Xu, O., Lu, S., Pei, L.: Add/drop channel filter based on two parallel long period fiber gratings coupler. Optik 120, 855–859 (2009)
13. Shao, L.-Y., Laronche, A., Smietana, M., Mikulic, P., Bock, W.J., Albert, J.: Highly sensitive bend sensor with hybrid long period and tilted fiber Bragg grating. Optics Communication 283, 2690–2694 (2010)
14. Erdogan, T.: Cladding-mode resonances in short- and long-period fiber grating filters. J. Opt. Soc. Am. A 14, 1760–1773 (1997)
15. Erdogan, T.: Fiber grating spectra. Journal of Lightwave Technology 15, 1277–1294 (1997)
16. Wang, G., Wang, Y.: New transfer matrix method for long-period fiber gratings with coupled multiple cladding modes. Chinese Optics letters 9, 090605-1–090605-3 (2011)

Thermal Deformation Analysis of High-Speed Motorized Spindle*

Shenbo Yu and Feng yi Xiao

School of Mechanical Engineering Shenyang University of Technology, Shenyang, P.R. China
yushenbo@126.com, xiaofengyi2008@163.com

Abstract. This paper analyzes the influence of temperature rising on the spindle accuracy in the CNC high-speed spindle unit. The analysis through the heat source of the spindle, the spindle unit is modeled of by Pro/E software to study the influence of temperature on the spindle axis and spindle deformation distribution. On the basis of the understanding the relationship between generating calorific value and heat deformation, methods and technologies reducing spindle heat are taken to ensure the stability of rotational precision of spindle. Therefore, the machining accuracy of the high-speed machine tools is improved.

Keywords: High-speed, motorized, spindle, temperature, thermal distribution.

1 Introduction

With the rapid development of science and technology, modern machinery manufacturing towards high efficiency, high speed and high precision. Mechanical machining accuracy is also the important factor insuring the product quality. With the wide application of high speed machining, the requirements of high-speed and high-performance motorized spindle is increasing day by day. Generally, the machining error of the high speed machine tool is produced by the following reasons: machine thermal deformation, machine tool parts and structure of the geometry , cutting force, the tool wear and other factors. Thermal deformation error is the maximum error in all kinds of error in the machine tool, about 70% of the total errors[1]. Because motor windings in motorized spindle generate much calorific value, the temperature of the spindle raises, leading to the heat deformation increasing. Therefore, it is necessary to study heat deformation issue in depth.

2 The Analysis of the Main Heat Loss of the Motorized Spindle

2.1 Mechanical Loss

In the motorized spindle, the mechanical loss has two main parts. First, bearing produces the loss when bearing rotating; second the loss is due to the high speed

* The project has been supported by National Natural Science Foundation of China (51175350).

D. Jin and S. Lin (Eds.): Advances in EECM Vol. 1, LNEE 139, pp. 511–517.
springerlink.com

rotating rotor parts with the friction by air. In order to improve the rigidity, enhance the ability of bearing carrying complex load and reduce the volume of the motorized spindle, angular contact ball bearings in pairs of installation are widely used in motorized spindle. The loss of rolling bearing can be calculated by following formulation:

$$p = k\frac{G}{d}\omega \tag{1}$$

where G——Comprehensive load of the bearing
d——The diameter of circle of the distribution center of balls or rollers
ω——The circumferential velocity of the spindle
k——Coefficient, k=0.002~0.003 for small and medium-sized motor

Because of the spindle speed increasing, the bearing friction is also very serious. The main unknown quantity is temperature in thermal analysis. Temperature results in heat deformation, which consists of three direction components. Therefore, no matter what type of spindle unit, in thermal analysis of the finite element model only need to be considered as a node, a degree of freedom [2]. To simplify, convenient, so the internal heat of the bearing transfer, the phenomenon that the following steady heat transfer. According to Palmyra formula, bearing rolling body and rolling friction between the contact way calorific value calculation formula.

$$\Phi = \omega M \tag{2}$$

where Φ is calorific value, J; M total friction torque, NM; ω angular velocity of bearing rolling body, $r \cdot min^{-1}$.

The loss of the bearing occupies a large scale in the total loss of the motorized spindle. It also is the main factors affecting the service life of motorized spindle.

The loss of friction of the rotor parts and air in high-speed operation of the motorized spindle is also considerable heat source. Therefore, the loss of air turbulence friction in air gap between the stator and rotor can be expressed as below[3]:

$$p = \frac{\pi}{16 \times 102} C_f \rho \omega^3 D^4 l \tag{3}$$

where D, l——the diameter and length of the rotor
ω——Rotor angular velocity
C_f ——the coefficient of flow resistance, it can be calculated by Alzheimer's formula

$$C_f = 0.05(\frac{h}{\delta} + \frac{100}{\mathrm{Re}})^{0.25} \tag{4}$$

where h——height of obstacle projecting from the rotor surface
δ——air gap **Re——Reynolds number**
$\mathrm{Re} = \delta\omega / r$; r kinematic viscosity of medium

2.2 Electric Loss

In operation of the machine tool, the calorific value resulting from the power consumption of the idling is the mainly heat source. The effective input power of the motor is *P*.

$$p = \sqrt{3}UI\cos\varphi \tag{5}$$

The voltage *U* and current *I* can be measured. The relative level of effective power to actual power decided by phase φ.The calorific value of motor stator and rotor originates from motor loss. The loss of the motor is divided into four categories: mechanical loss, electric loss, magnetic loss and stray loss [4]. The first three kinds of the motor loss are primary loss. The stray loss in the total loss shares about 1% ~ 5%.

2.3 Magnetic Loss

The losses from magnetic hysteresis and eddy, P_1, P_2, in the stator and rotor core are main losses. It is caused by the effective core periodically magnetized by main magnetic flux. Reversal of magnetism is divided into the following three magnetization, static magnetization generated from slow changing of magnetized current within limits; rotation magnetization produced from rotating of iron core of the armature in magnetic field; cyclic magnetization induced by alternating current.

$$p_1 = CfB^2{}_{\max} \tag{6}$$

where C ——constant concerned with core material f ——magnetized frequency

$B_{\max}$ ——the maximum magnetic induction

Eddy current loss can be pressed by type calculation:

$$p_2 = \frac{\pi}{6\rho} \bullet \frac{\delta^2}{\gamma_c}(fB) \tag{7}$$

where δ ——The thickness of a silicon steel sheet of the laminating

γ_c, ρ ——The density and resistivity of the core

3 Simulating Research of Thermal Deformation for the Motorized Spindle Unit

The calorific value generating and temperature rising for spindle unit are a dynamic process. When the spindle runs at different load and speed, there are some differences in the heat generating and temperature rising of spindle. The greater is the load, the more high speed and the more large heat. Temperature is obvious rising. Therefore, simulation analysis of the spindle unit is completed by the Pro/E finite element software in the stable state up to 12000 r/min.

3.1 Calculated Formula of Heat for Spindle Motor

The model of spindle motor at rated power 30 kW is established . All loss of power are translated into calorific value. The influence of the calorific value on thermal deformation in four kinds of conditions is analyzed as it is shown in table 1.The calculated parameter of calorific value for motorized spindle is shown in table 2.The calculated formula is as follows[5].

$$p_o = p_i \eta \tag{8}$$

$$p_i = \sqrt{3} UI \cos\varphi \tag{9}$$

$$p_s = p_i - p_o = p_i(1-\eta) \tag{10}$$

$$Q = p_s t \tag{11}$$

where: p_i is the input power, kW; p_o the output power, kW; p_s the loss of power, kW; Q the heat, J; η the efficiency of the motor; $\cos\varphi$ is the power factor of the motor; t is running time.

Table 1. Performance parameter of electric spindle

Load rate	0	50%	60%	100%
Power factor	0.2	0.75	0.83	0.88
Efficiency	0	0.82	0.85	0.85

Table 2. Calculated parameter of heat for electric spindle

Load rate	0	50%	60%	100%
Running time /S	600	900	1800	3600
Heat /J	3650	3785	7235	15867

The calculated formula of temperature difference between the internal and external spindle unit is obtained according to the basic theory of thermal conductivity [6].

$$\Delta t = \delta\Phi / (\lambda A) \tag{12}$$

where, Δt is temperature difference of thermal conductivity, °C; δ wall thickness, m; Φ heat, J; λ thermal conductivity, W/（m·℃）; A the section area of thermal conductivity, m^2.

The temperature difference of thermal conductivity in four kinds of conditions have been calculated, respectively: Δt_1 = 29.39℃ Δt_2 =31.43℃ Δt_3 =41.26℃.

The temperature data, the highest surface temperature of spindle 80 °C, the inside 40 °C, the air 20 ℃ in the state of the load rate 60% is analyzed by finite element method.

3.2 Finite Element Analysis for Heat Deformation and the Thermal Stress of the Spindle

Heat deformation of the spindle includes axial deformation and radial deformation. The front of the spindle unit are fixed. Axial deformation of the spindle moves backward. The influence of the backward moving of the spindle on machining precision in machine tools is a little. However, the radial distortion of the spindle is the main deformation. Thermal field of the spindle model, outside diameter 160 mm , inner diameter 80 mm, is analyzed in a cross section 1/4 geometric model. Thermal stress is gained. The material performance parameters of the spindle, are shown in table 3.The results of temperature field distribution are shown in figure 1. The results of the thermal stress distribution are shown in figure 2.

Fig. 1. Spindle temperature field distribution

Fig. 2. Spindle thermal stress distribution

The kinds of materials	density/(kg·m^{-3})	specific heat capacity _/ [J(kg·℃)$^{-1}$]	coefficient of thermal conducti v-ity	thermal conducti-vity λ/[W·(m·℃)$^{-1}$]	elastic modulus E/ Gap	poisson's ratio υ	coefficient of linear expansion α/(℃)$^{-1}$
silicon steel	7800	470	15	43.2	206	0.3	1.0×10^{-6}

From above analysis, it can see that calorific value is transferred from external surface of the spindle to internal, as shown in figure 1; As spindle is applied by radial tension, the trend of deformation is outward as shown in figure 2. The distortion of the motorized spindle results from calorific value in high-speed operation, which influences machining accuracy of machine tool. Therefore, some methods and technologies should be taken to reduce the thermal deformation and guarantee machining precision of the machine tool.

4 The Measures of Reducing Fever and Enhance the Heat Dissipation Performance of Electric Spindle

(1) The method of tandem oil mist supplying is changed into the method of parallel oil mist supplying in lubrication of the motorized spindle bearing. The lubrication condition of bearings away from the oil mist source in the motorized spindle is improved.

(2) Air-gap flux density in the motorized spindle unit is changed from 0.7T to 0.5~0.6T. The flux density in tooth and yoke is properly reduced to decrease metal loss.

(3) The stator slots with half-closed width opening are chosen to reduce leakage resistance, improve the overload ability of working characteristic.

(4) High current density should be chosen in stator to enhance power. However, lower current density should be chosen in the rotor to reduce heat productivity.

5 Conclusion

Mechanical loss, electric loss and magnetic loss of the motor in the motorized spindle are primary heat sources. In mechanical source the relationship between the spindle speed and the load rate and thermal value is determinated. In electric sources the thermal value in motor stator and rotor originates from motor winding loss. In magnetic source the losses from magnetic hysteresis and eddy in the stator and rotor core are main reason resulting in thermal value. The influence of the change of temperature rising on thermal deformation of the motorized spindle is analyzed. The measures of improving motor performance have been summarized. As the power of high-speed spindle motor is improved, the application will further be expanded in the future. The research of heat exchanging will also become more thorough.

References

1. Zhang, B.L., Yang, Q.D., Chen, C.N.: High speed cutting technology and application. Mechanical industry press, Beijing (2002)
2. Kim, D., David, J.W.: An improved method for stability and damped critical speeds of rotor-bearing system. Transactions of ASME, Journal of Vibration and Acoustics 112, 112–119 (1990)
3. Li, L.Y., Cui, S.M., Song, K., Cheng, S.K.: Ha bin Institute of technology, Harbin 150001, China

4. Ha, K.-H., Hong, J.-P., Kim, G.-T., Chang, K.-C., Lee, J.: Orbital analysis of rotor due to electromagnetic force for switchedreluctance motor. IEEE Transactions On Magnetics 36(4), 1407–1411 (2000)
5. Xie, L.M., Shao, K.P., Jin, L., Zhang, H.J.: Thermal deformation research of high-speed electric Spindle, 1000-4998, 03- 0041- 03 (2011)
6. Dai, G.S.: Heat transfer. Higher education public., Beijing (1999)

The Research of a Quantitative Evaluation Model for Data Integrity QEMI in Smart Grid

ShaoMin Zhang, Peng Gao, and BaoYi Wang

School of Control and Computer Engineering, North China Electric Power University, 071003 Baoding, China
gaopeng870514@163.com, {zhangshaomin,wangbaoyi}@126.com

Abstract. With the development and increasing openness of Smart Grid, network security, especially data integrity of Smart Grid has been paid more attention. But the efficiency of existent methods is slow in Smart Grid which has a huge amount of data. Based on the definitions of integrity internationally accepted, by describing its state transition process, this paper gives a quantitative evaluation model for data integrity named QEMI which doesn't need to directly operate on the data, and provides different instances which are satisfied with different security mechanisms of Smart Grid. At last, we prove this evaluation model can provide significant results with an example.

Keywords: Quantitative evaluation, Information Integration, Smart Grid.

1 Introduction

By integrating broadband communication with control systems, using distributed intelligence and data unified management widely, Smart Grid achieves information integration and information sharing of power system. Power producers and users are no longer isolated by power companies, the electricity market is turning planned production to open competition gradually, and the power companies' businesses are becoming open, standardized and interconnected.

However, with increasing openness, calls and interactions are more frequent, the relationships of electric power system automation software are increasingly close. "Security through Obscurity" has not been a choice of safe strategies [1].

In Smart Grid, due to various systems are integrated into one platform, calls between them make security management further complex. Once the integrity of data is destructed, the loss will be unpredictable. Therefore, this paper focuses on the quantitative evaluation for data integrity in Smart Grid.

2 Present Situation and Difficulties of Research

Integrity has been studied for many years, the works of integrity research focus on definitions of integrity, integrity assurance technologies, integrity assurance model and its applications [2].Existing methods of evaluation for data integrity are mostly data check [3-5], their efficiency is low Smart Grid which has a great lot data.

D. Jin and S. Lin (Eds.): Advances in EECM Vol. 1, LNEE 139, pp. 519–525.
springerlink.com

Data integrity evaluation methods include quantitative evaluation and qualitative evaluation. Unlike the qualitative evaluation dependent on people's subjective experience and consciousness, the quantitative evaluation can reflect the essence and states of things more objectively. In contrast, quantitative evaluation can afford more objective results to the users.

Quantitative evaluation for data integrity is complex, due to the following reasons:

(1) There is no uniform view of the definitions for integrity. It makes that it's difficult to evaluate from a unified point of integrity.

(2) There are too many factors which can undermine integrity, especially the uncertainty of software correctness and the complexity of malicious attacks, which makes the evaluation for integrity more difficult [6].

3 A Quantitative Evaluation Model for Data Integrity QEMI

Literature [6] proposed a model of evaluation for data integrity, and its states transition diagram is shown in Figure 1. It brings new idea for us: evaluating the integrity of data does not mean that operate data directly. This idea would solve slow productivity problem. However, the model has problems to be resolved as follows:

(1) According to the definitions of integrity by Sandhu-Jajodia: modifications must be proper if data is modified in any states. Proper here also means be able to ensure the correct operation of power systems;

(2)Integrity definition by Ravi Sandhu is: Data can not be modified or at least any modifications should be detected. The original model does not take into account whether all the modifications will be detected;

(3)Marketization reform of power industry brings the great increasing of internal users who have the ability and knowledge of potential attacks (intentionally or unconsciously), so both internal and external security threats should be given by the results of integrity evaluation.

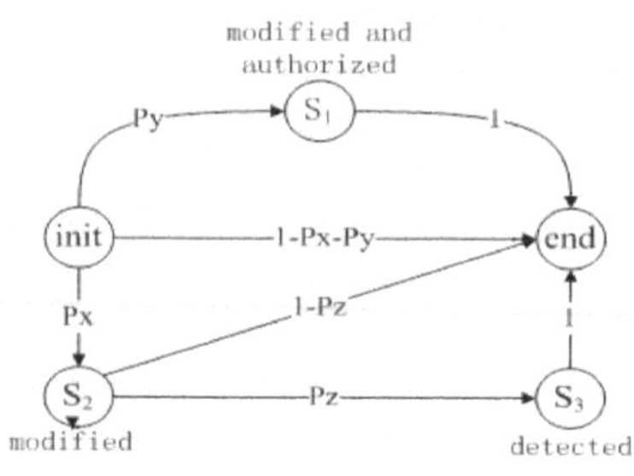

Fig. 1. State transition process of the model in literature [6]

According to the three points above, this paper presents a quantitative evaluation model for data integrity QEMI (Quantitative Evaluation Model for Data Integrity), and its states transition diagram is shown in Fig. 2.

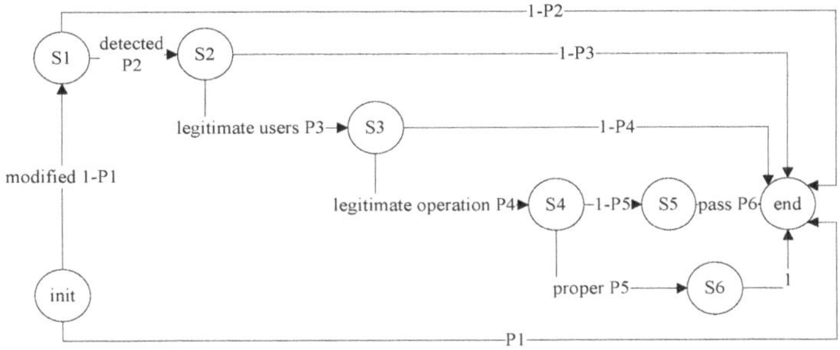

Fig. 2. State transition process of QEMI

It contains eight states: *init* is the initial state of the database; *end* is the end state of the database, $p1$ denotes the probability of transition from *init* to *end*; S_1 is the state of modified, 1-$p1$ denotes the probability of transition from *init* to S_1; S_2 is the state of detected modifications, the probability of transition from S_1 to S_2 is $p2$,and 1-$p2$ denotes the probability of transition from S_1 to *end*; S_3 is the state of modifications by legitimate users, the probability of transition from S_2 to S_3 is $p3$; 1-$p3$ denotes probability of transition from S_2 to *end*; S_4 is the state of legitimate modifications, the probability of transition from S_3 to S_4 is $p4$,and the probability of transition from S_3 to *end* is 1-$p4$; S_5 is the state of improper modifications, the probability of transition from S_4 to S_5 is 1-$p5$; S_6 is the state of proper modifications, the probability of transition from S_4 to S_6 is $p5$; From S_5 to *end* is the transition of improper modifications which passed the error test, its probability is $p6$;The value of pi(i =1, 2…6)can be calculated by certain mathematical method.

Instance1: Assumes the database is entirely reliable, and then integrity has been protected entirely.

Instance 2: Assumes the database is unreliable, there are no attacks and no error test mechanisms ($p2$=$p4$=$p6$=1). In addition, we assume that the integrity goal is data not to be modified or any modifications should be proper. This situation is shown in Fig.3.

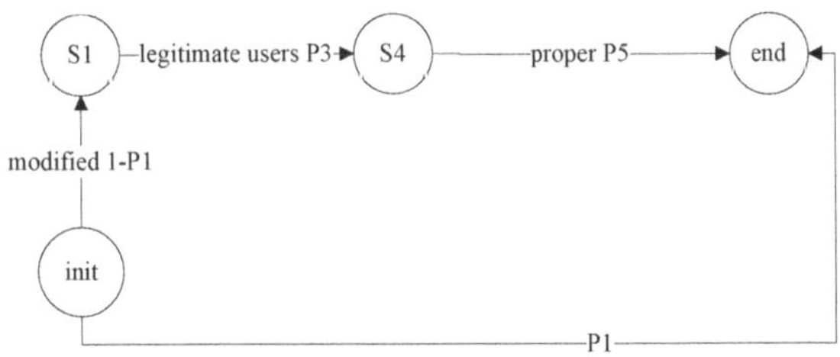

Fig. 3. State transition process of instance 2

Because there is a path through S_5 between S_4 to *end*, in other words it has improper modifications. So the integrity goal is not satisfied. We name the probability automaton *Automata*. If we adopt the probabilistic definitions, when the probability of data unmodified and the rational modification greater than or equal to p, integrity is satisfied. That is when $p1+(1-p1)p5 \geq p$, Formula (1) is established.

$$Automata \mid = \square(\neg\, modified \vee (modified \rightarrow \blacklozenge proper)) \geq p\,. \qquad (1)$$

Instance 3: Assumes the database is unreliable and there is no attacker, but it has some error detection mechanisms ($p2=p3=1$), all the modifications will be detected. We assume that the integrity goal is data not to be modified or any improper modification can not pass the error test. This situation is shown in Fig.4.

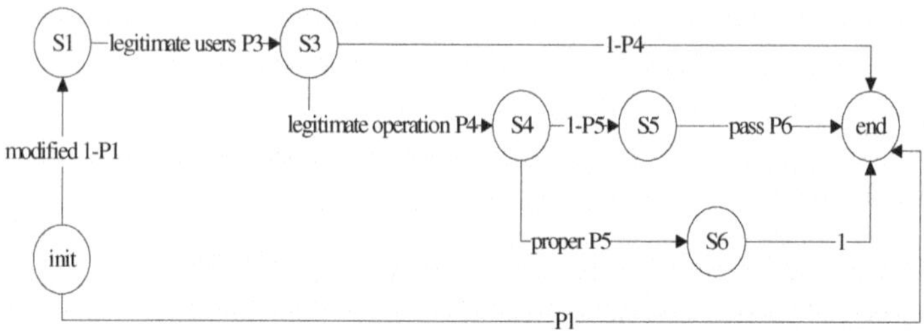

Fig. 4. State transition process of instance 3

There are a path between S_3 to *end* and a path between S_5 to *end*, so the integrity goal is not satisfied. If we adopt the probabilistic definitions, when $p1+(1-p1)p4p5+(1-p1)p4(1-p5)(1-p6) \geq p$, integrity is satisfied, Formula (2) is established.

$$Automata \mid= \square(\neg modified \vee (modified \rightarrow \blacklozenge(legitimate\ operation \rightarrow \blacklozenge proper)) \vee (improper \rightarrow \blacklozenge not\ \ pass)) \geq p \qquad (2)$$

Instance 4 : Assumes that the database is unreliable and there are external attackers and error test mechanisms, all the modifications can be detected (p2=p3=1), and the integrity goal is data not to be modified or all the modifications are proper modifications by legitimate users. Instance 3 and Instance 4 has the same security mechanism, but we assume different integrity goals. Because there are a path between S_3 to *end* and a path between S_5 to *end*, so the Automata is not satisfied the integrity goal. If we adopt the probabilistic definition, when, integrity is satisfied, that is when $p1+(1-p1)p4p5 \geq p$, Formula (3) is established.

$$Automata \mid= \square\ (\neg modified \vee (legitimate\ operation \rightarrow \blacklozenge(modified \rightarrow \blacklozenge proper))) \geq p \qquad (3)$$

4 Example Analysis

The 57th Technical Committee (IEC TC 57) of the International Electrotechnical Commission (IEC) which drafts the power system control and communication standards has drafted a series of standards. In which, 13th working groups (WG 13) is responsible for developing CIM and CIS standards related EMS, which named IEC 61970. It makes the application software of EMS modular and open, "plug-in" and interoperability, and reduces the cost of system integration and protection of user resources, and provides a basis for information integration of Smart Grid. IEC61970 standard series is divided into five parts. CIM is the foundation and core part of IEC61970 standard [7]. CIM is an abstract model, depicting all the major objects for typical EMS information model contained in the power companies, these objects contain public classes, attributes and relationships between objects, etc., used in development and integration of power engineering, planning, management, operating and financial application.

This makes the Smart Grid can transfer information in a unified format, shield heterogeneous, achieve a wide range of communication between the various parts, and provide the participants in the power system with adequate data. Smart Grid need meet the demand of interaction between providers and customers. With XML-based Smart Grid data exchange mechanism, users and producers can pass data, market and dispatch departments can get the rate of change capacity, the trend of blocking and other information. At the same time, users can get real-time electricity price information. The databases in node server of Smart Grid communicating network store data constructed in CIM structure, we call it CIM database, shown in Fig. 5. However, it makes Smart Grid not only increase openness, but also increase information security risks, data integrity is difficult to be ensured. This paper will evaluate the integrity of the database in node server of CIM network.

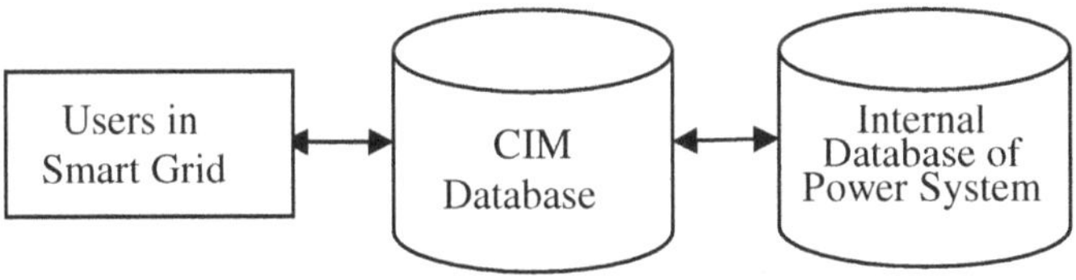

Fig. 5. Communication between users and the market and scheduling departments

We assume that CIM database is not reliable. There are attacks, but no error test. Not all the modifications will be detected. The integrity goal is protecting data not to be modified, or all the modifications are detected and proper.

According to the model proposed in this paper, we set the probability automaton of CIM database as *Automata* (*C*), the model probability of states as *pc*i (1, 2, 3 ...). According to the state transition process, we know that, when $pc1+(1\text{-}pc1)\,pc2pc3pc4pc5 \geq pc$, the integrity of CIM database is satisfied.

After a long period statistics, we get $pc1=0.936$, $pc2=0.962$, $pc3=0.882$, $pc4=0.914$, $pc5=0.971$, $pc=0.986$.

Calculated:

$$pc1+(1\text{-}pc1)\,pc2pc3pc4pc5=0.98419356538214<0.986\,. \quad (4)$$

So the integrity of CIM database has not been satisfied.

On this basis, if we want the data integrity of the CIM database is greater than 0.986, we can change the $pc1$, $pc2$, $pc3$, $pc4$, $pc5$ to achieve it.

Assume we can only change one of the indicators, when we can't change anyone but $pc1$, and we set the amount to change as $x1$, get Formula (2):

$$pc1+x1+(1-pc1-x1)\,pc2pc3pc4pc5 \geq pc\,. \tag{5}$$

Substitute the values into the Formula (2):

$$(0.936+x1)+(1-0.936-x1)\times 0.962\times 0.882\times 0.914\times 0.971 \geq 0.986\,. \tag{6}$$

We got $x1 \geq 0.002$, while $p1+x1 \leq 1$, is proper. Similarly, we set the increasing value of $pc2$, $pc3$, $pc4$ and $pc5$ as $x2$, $x3$, $x4$ and $x5$ respectively. In the same way, $x2 \geq 0.036$, $x3 \geq 0.033$, $x4 \geq 0.034$, $x5 \geq 0.036$. $p5+x5 \geq 1$, is irrational, so only increasing $pc5$ (improving the performance of error test system) can't achieve satisfied integrity.

Furthermore, we are able to target the economic purpose that is the least resources cost for enhancing the data integrity of CIM database.

We set the cost of resources for the value of $pc1$ increasing by 0.1 as 20000(including human and material resources), the cost of resources for the value of $pc2$ increasing by 0.1 as 1200, the cost of resources for the value of $pc3$ increasing by 0.1 as 1100, the cost of resources for the value of $pc4$ increasing by 0.1 as 800, the cost of resources for the value of $pc2$ increasing by 0.1 as 900.

We set the total cost of system for improving integrity as $w1$, $w2$, $w3$, $w4$, $w5$. We can get $w1 \geq 40$, $w2 \geq 43.2$, $w3 \geq 36.3$, $w4 \geq 27.2$, $w5 \geq 32.4$. We realize increasing $pc4$ costs the least resources, so do not consider $x5$ overflow problems. For acquiring maximum economic benefits, firstly we should spend the resources on increasing pc4. So, we should standardize the operations of personnel who maintain the database to reduce the improper operations of legitimate users.

5 Conclusion

This paper presents a model of quantitative evaluation model for data integrity QEMI, and describes security states and a variety of threats integrity at great length, to meet the special requirements of Smart Grid. The result of quantitative evaluation can provide network security managers with objective reference. At the same time, the model takes the differences among the various security mechanisms into account. By using an example, we prove the results of this model can provide information of reference value.

References

1. Long, L., Li, J., Liu, L.: Research on Communication Security of Substation Automation System Based on IEC62351. Journal of Changsha Telecommunications on Technology Vocational College 3(9), 1–6 (2010) (in Chinese)
2. Zhang, S., Hou, J., Wang, B.: A Model of Quantitative Evaluation for Integrity of the Server. In: 2010 International Conference on Computer Application and System Modeling (ICCASM 2010), pp. 658–661 (2010)
3. Chen, L., Wang, G.: An Integrity Check Method for Fine-grained Data. Journal of Software 20(4), 902–909 (2009)
4. Cheng, H., Feng, D.: An Integrity Checking Scheme in Out sourced Database Mode. Journal of Computer Research and Development 47, 1107–1115 (2010) (in Chinese)

5. Chen, L., Fang, X., Wang, G.: Integrity Check Method for Fine-Grained Data Based on Complex Rotary Code. Journal of Southwest Jiaotong University 44, 667–671 (2009) (in Chinese)
6. Yin, L., Guo, Y.: Research on Quantitative Evaluation for Integrity. In: Fifth International Conference on Information Assurance and Security, Shanghai, pp. 689–692 (2009)
7. Zhang, S., Liu, G.: Introduction of Standard IEC 61970. Automation of Electric Power System 26, 1–6 (2002) (in Chinese)

Research on Application of Interaction Firewall with IDS in Distribution Automation System*

BaoYi Wang, HaiPeng Yang, and ShaoMin Zhang

School of Control and Computer Engineering,
North China Electric Power University, Baoding, China
yanghaipeng8709@163.com, {wangbaoyi,zhangshaomin}@126.com

Abstract. For the clear storage of database file in distribution automation system ,information transmitted in plaintext and weak identity authentication and other information security issues, we propose the project of combining the firewall based on active defense and the intrusion detection based on passive defense.Finally, we analysis the safety property of this scheme, the analysis shows that this scheme meets the security requirement of distribution automation system, forever ensure the security of the power distribution automation system mutual communication better.

Keywords: Distribution automation, firewall, IDS, digital signature, security.

1 Introduction

In recent years, countries around the world accelerate the development of smart grid technologies, and effectively promote the intelligent of power grid.Smart Grid has become the new development trend of the future grid[1-2].Intelligent distribution network is an important part of Smart Grid, which should be assured operating safe and reliably. The power distribution automation system isolates the safety areas using firewall in horizontal direction, but cannot guarantee the communication security between security areas of distribution automation system communication. The reason is that intruders can falsify data to bypass the firewall, or find back doors in the firewall and intruders may be inside the firewall, at this time the firewall does not work, due to restrictions on the performance of firewall, usually it can not monitor intrusion real-time and it is a weakness of fi rewall that preventing viruses's intrusion.

Single intrusion detection system can not guarantee the security of the network, the reason is that intrusion detection system can not make up for the security flaws and vulnerabilities in security defense system and intrusion detection system can not prevent the penetration of aggressive behavior alone and intrusion detection system is mainly for analyzing and detecting network behavior, not fixing security problem existing in information resources [3].

* This work is supported by Hebei higher education science research plan funding subject of research of information integration based on IEC61970 and the security techniques of smart grid. (Z2010290).

D. Jin and S. Lin (Eds.): Advances in EECM Vol. 1, LNEE 139, pp. 527–532.
springerlink.com

Therefore, Interaction firewall with IDS makes protection system from static to dynamic, from two dimensional to three dimensional, enhances the mobility and response ability of the firewall, and enhances the blocking function of IDS.

2 Information Security Issues of Distribution Automation System

As the power companies in the computer security technology,security policy,and security measures,and other areas'inputs don't improve correspondingly.Therefore,the distribution automation system itself has a lot of loopholes or unreasonableness[4].

a) Distribution automation system database files stored in plain text.
b) The information in distribution automation system transmits in the clear.
c) Weak identity authentication of the distribution automation software.

According to the overall security strategy of secondary power system[5],security partition business system architecture of distribution automation system is shown in Fig.1.

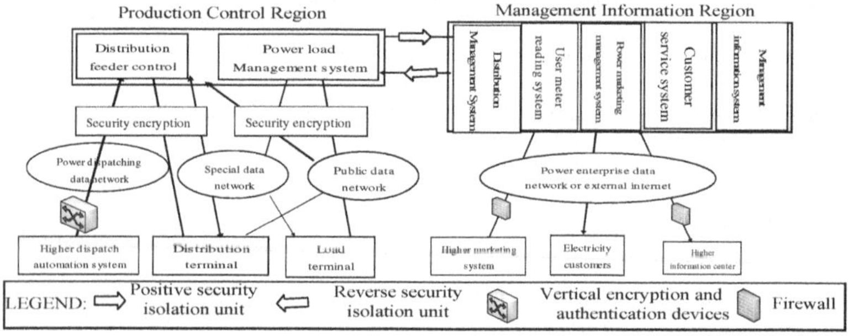

Fig. 1. Secure deployment of secondary distribution system

3 Key Technologies Analysis of Interaction Firewall with IDS

Interaction[6] is a combination of security technologies, adapts to the system better through making up for their shortcomings of each other. Interaction system model is shown in Fig.2.

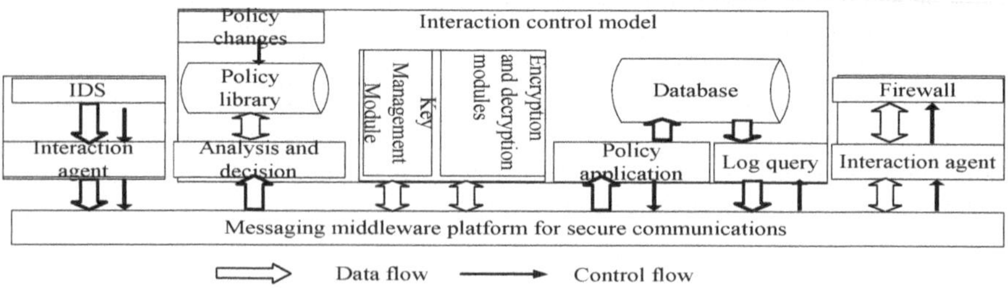

Fig. 2. Interaction model firewall with intrusion detection system

3.1 Secure Communications Platform Based on Message Oriented Middleware

According to the distribution chracteristics and business requirements of distribution automation system, this paper, data transmission through message oriented middleware platform.

Message Oriented Middleware is a efficient and reliable message delivery mechanism.which can shield characteristics of different platforms and features of communication protocols, realizing reliable, and efficient cross-platform middleware for data transmission in distributed systems. It not only can achieve systems interconnection, but also can carry out the interoperality between applications[7].

3.2 Data Encapsulation Based on XML

For example, here is the operating condition of distribution lines from power distribution terminal to the main stations of the power distribution in distribution automation system.The template desripted with XML[8] is as follow.

```
<?xml version="1.0" encoding="GB2323"?>
<env:Envelope>
<env:Header
xmlns:env=http://www.w3.org/2003/05/soapenvelope>
<m:from> CDAU </m:from>
<m:to> Distribution master station</m:to>
<m:datetime>2011-6-28</m:datetime>
</env:Header>
<env:Body>
<Tower data/>?
<Cross arm and metal data/>?
<Insulator data/>?
<Wire data/>?
<Mine facilities data/>?
<Grounding device data/>?
</env:Body>
</env:Envelope>
```

3.3 ELGamal Digital Signature

ELGamal digital signature scheme summaries as follows.

Supposing p is a large prime number, g is a non-zero elements on the finite field GF(p),signer selects a private key x random, $x \in [1,p-1]$ and satisfies $\gcd(x,p-1) = 1$.

Suppose m is the message to be signed,singer selects a random k∈ [1,p-1],then (g^x, mg^{kx}) is the digital signature of message of m.

The user receives signed message (g^x, mg^{kx}), using the private key x verifies the message, because the third party only knows g^k or g^x,so he can not achieve g^{kx}, calculating k or x from g^k or g^x is the discrete logarithm problem, and solving the discrete logarithm is very difficult.

4 The Concrete Realization of Each Module in the Interaction System

4.1 The Realization of Firewall Module

Packet filtering firewall selects packets in the network layer, according to the source address and the destination address, application or protocol, and port for each IP packet to detemine whether or not allow packets to pass through. Inside the firewall, there is a filtering rules table, which defines a varity of rules to decide whether to let the packet through. When packet comes, firewall will check every rule, if it is found matching with a rule in the rule base, then discard the packet, if it does not match any rule in the rule base, generally, the firewall will let the packet through.

4.2 The Realiztion of Intrusion Detection Module

Abnomal-based intrusion detection does not invade feature database, but tests data transmission in the power distribution automation system by identifying abnormal behavior. Anomaly detection technology establishes normal behavior model according to normal behavior statistic data, and defines the threshold of abnormal behavior in advance, then matches the model with the current user's behavior, if the current user's behavior diviates from normal behavior over the predefined threshold, the current behavior is identified as intrusion, the IDS will take relevant measures to deal with it.

4.3 The Realization of ELGamal Signature Scheme

The ELGamal signature scheme[9] needs to get the key through registing for both sides and complete the mutual identity authentication.KMC saves each registered user's information, including user ID, and signature(r,s),and maintains user registration information and distributes public key.

The process of mutual authentication of the communicating parties is the same, in the following take the USEA and USEB for example to illustrate mutual authentication process of ELGamal signature scheme.

a) In order to obtain the public key gS_b of the USEB,USEA sends message to KMC(USEA,USEB).

b) KMC checks the USEB's private key s_b, and returned gS_b to the USEA.

c) The USEA selects a random number k,k ∈ [1,p-1],calculated m=USEA ⊕ USEB, then (g^k, $mgkS_b$) sends to the USEB.

d) USEB uses the private key s_b and received message g^k for m, then obtains USEA by calculating m ⊕ USEB, namely USEB identifies that he is authenticating with USEA, then USEB sents message to KMC to obtain the USEA's public key gS_a .

e) KMC checks USEA's private key s_a and returned gkS_a to the USEB.

f) USEB sends (gkS_b , $mgkS_bS_a$) to the USEA.USEA uses private key s_a and the received news gkS_b for the m, and obtains USEB through m ⊕ USEA.Because before and after the user identity needed to authenticate is USEB and USEB's public key is gS_b ,thus USEA confirms that he is truly communicating with the USEB.Hence, USEA and USEB realise the mutual authentication.

5 Scheme Safety Analysis

The interaction of Firewall and IDS has higher security than a single security technology, comparison results are in the table 1. which, "√" means "have".

Table 1. Comparison of security between this paper scheme and other schemes

Security Technology	Firewall	IDS	Interaction of both
Enhanced security policy	√		√
Effective record online activities	√		√
Limiting exposure to user points	√		√
A security policy of the detection point	√		√
Track the attacker's attack line		√	√
Facilitate the establishment of security system		√	√
Able to seize the perpetrators		√	√
System to better grasp the situation		√	√

Table 1 shows that security has been greatly improved, but transmitting data in plaintext is unsafe, the combination of digital signature algorithm which is based on ELGamal and AES encryption algorithm allows transmitting data in ciphertext, then better ensures the safety and confidentiality of the data transmission.

The scheme has the following sccurity.

a) prevent denying. Because m in news(g^k , mg^{kS_b}) can only be worked out by the private key sb, so m in news (g^{ksb} , mg^{ksbsa})can only be worked out by the private key s_a , but solving the discrete logarithm is very complicated.

b) Prevent forgery. If USEC replaces USEA, when USEC sends news(g^u , mg^{us_b}) to USEB, of which, u∈[1, P-1],USEB can identify USEA after receiving news, so USEB sends message(g^{ksb} , mg^{ksbsa})to USEC, however USEC unable to solve m, so USEC cannot obtain private key, thus it meets failure that USEC forged into USEA.

c) Prevent fraud. If USEC replaces USEB,USEC cannot obtain consistent message,because what USEC sends to USEA is the news that USEC acquired previously,not the recent news that obtained through the KMC executes $\left(g^{k}\right)^{Sa}$ operation, so it meets failure that USEC replaces USEB.

6 Concluding

In order to defense network attacks and invasion events effectively, this paper puts forward the scheme of combining the firewall and intrusion detection system,so forms a multi-layer,safety interactive protection system,which giving full play to the advantage of both .

Although the program in this paper is proposed for the intelligent distribution network, the program based on security theoretical system can be easily applied to the other areas of smart grid information security,so as to provide a new idea and new meathod for solving the security transmission of smart grid information.

References

1. Yu, Y.-X., Luan, W.-P.: Smart grid and its implementations. Proceedings of the CSEE 29(34), 1–8 (2009)
2. Chen, S.-Y., Song, S.-F., Li, L.-X., et al.: Survey on smart grid technology. Power System Technology 33(8), 1–7 (2009)
3. Wei, M., Wang, Q., Wang, X.: Application of Intrusion Detection System to Electric Power System Information Network. The Journal of Wuhan University 39(2), 102–105 (2006)
4. Guo, Q.: Information Security Research of Distribution Automation System based on Elliptic Curve Cryptosystem. North China Electric Power University, Beijing (2009)
5. State Electricity Regulatory Commission. The Security Overall Scheme of Secondary Power System (2006)
6. Feng, Q.: Interaction Technology with Firewalls and Intrusion Detection System. Computer Application 25(12), 2763–2764 (2005)
7. Pan, G., Song, W.: Research on the Application of Message Oriented Middleware in Electric Dispatching Automatization System. Electrical Application 27(10), 61–64 (2008)
8. Wang, B.: Research on Key Techniques of Information Security in Electric Power Information System. North China Electric Power University, Hebei (2009)
9. Hu, J.-J., Wang, W., Pei, D.-L.: Double-way Authentication Scheme Based on ELGamal Digital Signature. Computer Engineering 36(6), 173–174 (2010)

Research on Secure Message Transmission of Smart Substation Based on GCM Algorithm

BaoYi Wang, MinAn Wang, and ShaoMin Zhang

School of Control and Computer Engineering,
North China Electric Power University,
Baoding Hebei, 071003, China
wangbaoyi@126.com, fork0208@126.com, zhangshaomin@126.com

Abstract. Based on IEC61850 standard smart substation system using ethernet architecture is required for higher information security. SMV, GOOSE, MMS are common messages used in smart substation , whether these messages can be transmitted secretly is critical to the security of the entire substation system. If each message uses an encryption algorithm, that is bound to increase the burden of equipment and affect its interoperability. For the security requirements, a common message authentication encryption algorithm based on GCM is proposed in this paper. By choosing different algorithm modes to achieve secure transmission of these messages, this algorithm can effectively improves the security of substation information and network performance.

Keywords: Smart substation, GCM, secure message transmission, IEC62351.

1 Introduction

With the wide application of information technology, microelectronics technology, network communication technology and microprocessors as the core of the intelligent automatic device in power grid control field, smart substation technologies develop rapidly[1]. Substation Automation System (Substation Automation System, SAS) ultimately is towards the digital, smart direction.

Smart Substation IEDs based on P2P share the LAN information. once an IED is attacked by malicious, the entire SAS may suffer from the tremendous impact. Therefore, how to effectively secure the transmission of substation information is an urgent task to solve.

2 Format Design of the Message Based on IEC62351

In order to secure the message transmission, IEC62351[2] standard expands the GOOSE message format, keep the two Reserved fields and increase of the Extension field. According to the IEC62351 suggestions, we define 20 bytes in the Extension field: 4 bytes Serial Number field (Sequence Number, SN), 16 bytes Authentication codes field (MAC).

D. Jin and S. Lin (Eds.): Advances in EECM Vol. 1, LNEE 139, pp. 533–538.
springerlink.com

On the basis of IEC62351, we redefined the Reserved Field 1. We divide the Field 1 into two small fileds: A and B, each part has one byte. A indicates the length of the Extension field, B indicates the encryption mode. The combination of the two parts can present whether a message need encryption, and which encryption mode needed, as is shown in Table.1.

Table 1. Reserved Field 1 and its related modes

Mode	A	B	F	Security Measures
NULL Mode	0	any	0	No
GCM Mode	20	255	1	Encryption and authentication
GMAC Mode	20	0	2	Just authentication

We define that when A=0, B for any value, present that the message should be sended by plaintext. A=0 means this is a non-secure messages, we need not do any encryption and authentication calculations. So we choose the NULL mode(F=0); If A=20, it presents that the message is a secure extension format, and the extension field size is 20 bytes. In this case, if B=0 means the message chooses GMAC mode (F=2), if B=255 means the message choose GCM mode (F=1). $0 < B < 255$ is kept for the future use. This definition of message format is convenient to the classification, and increases the flexibility of application.

3 Secure Messages Transmission Based on GCM Algorithm

GCM algorithm (Galois/Counter Mode, GCM) based on 128-bit block ciphers, uses the CTR(Counter Mode) mode of AES (Advanced Encryption Standard, AES) for encryption and GHASH function defined in $GF(2^{128})$ field for authentication[3]. AES_GCM algorithm with parallel processing and pipelining mechanism can achieve very high rate of throughput. The maximum throughput of GCM have now exceeded 100Gbps, reaching to 162.56Gbps[4].

Data Definition :

A: the value of MAC head field and Priority tag field as additional value A;
M: the value of APDU, such as the GOOSE data;
SN: the serial number of APDU, instead of the initial vector P;
F: the flag of GCM mode;

3.1 The Process of Encryption and Authentication by Sending IED

The ***first*** steps: the options of security types and GCM mode. As is shown in Fig.1.

Step1: Create GOOSE data set, and assemble them to the APDU, and turn them to the data link layer after ASN.1 coding;
Step2: Set the Reserved Field 1 and SN value. SeqNum is an attribute of APDU;
Step3: Judge the security types and GCM mode according to the value of A and B

```
If(value(A)==0)
  { add_Head_PriTag();//non secure message, add head and Priority Tag
```

```
    cal_CRC();//calculate the value of CRC, added to the reserved field 2
    sendGooseMessage();
    }
Else if(value(A==20 and vlaue(B)==255))
    { add_Head_PriTag();// GCM mode and secure message format
    GCM_cryption () ;//Function call, return ciphertext M' and T
      fill_M_T();//fill M' and T into field
      cal_CRC();
      sendGooseMessage();
  }
  Else if(value(A==20 and vlaue(B)==0))
    { add_Head_PriTag();// GMAC mode and secure message format
    GCM_cryption () ;//Function call, return MAC code T
      fill_M_T();//fill T into the MAC field
      cal_CRC();
      sendGooseMessage();
  }
  Else{ other operations ;}
```

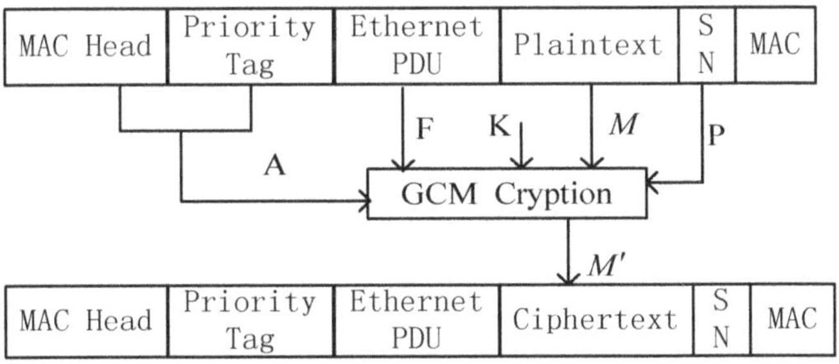

Fig. 1. Message encryption instance baseed on GCM

The ***second*** steps: GCM encryption algorithm

Step1:initialize variables: A=value(MAC Head and Priority Tag); F=value(B);

$H = E_K\left(0^{128}\right)$; $P_0 = GHASH\left(H, \{\}, P\right)$; $M = m_1 \| \cdots \| m_{i-1} \| m_n$;

Step2: ciphertext P_i XOR plaintext m_i

for (i=1, i<n,i++)

{ $Y_i = Y_{i-1}++$; $m_i' = m_i \oplus E_K\left(Y_i\right)$; … }

Step3: $m_n' = m_n \oplus MSB_{|m_n|}\left(E_k\left(Y_n\right)\right)$;//calculate the last section

Step4: $M' = m_1' \| m_2' \| \cdots \| m_n'$;//join all the ciphertext section together

Step5: $T = MSB_{16}(GHASH(H, A, M') \oplus E_K(P_0))$;//calculate the MAC

Step6: Return(M', T);

3.2 The Process of Decryption and Authentication by Receiving IED

The ***first*** steps: the options of security type and GCM mode. As is shown in Fig2.

Step1: check_CRC(), if CRC is right, go to step2;
Step2: Judge the security types and GCM mode according to the value of A and B
If(value(A)==0)
{ the message is NULL mode, propose to the receiving IED; }
Else if(value(A==20 and vlaue(B)==255))
{ If($GCM_{decryption}$ ()==true)
{ the integrity of M'' is true, propose to the receiving IED;}
Else{ the integrity of M'' is flase, drop the message;}
}
Else if(value(A==20 and vlaue(B)==0))
{ $GCM_{cryption}$ () ;//Function call, return MAC code T'
If(T==T'){the integrity of M'' is true, propose to the receiving IED; }
Else{ the integrity of M'' is flase, drop the message;}
}
Else{other operations;}

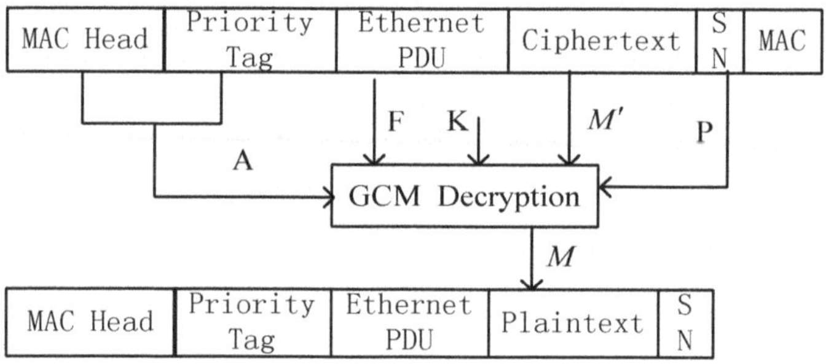

Fig. 2. Message decryption instance baseed on GCM

The ***second*** steps: GCM decryption algorithm. Firstly do hash authentication, if the hash code is right, then do decryption.

Step1: initialize variables: A=value(MAC Head and Priority Tag); F=value(B);
$H = E_K\left(0^{128}\right); P_0 = GHASH\left(H,\{\},P\right); M = m_1 \| \cdots \| m_{i-1} \| m_n$;

Step2: $T' = MSB_{16}(GHASH(H, A, M') \oplus E_K(P_0))$
If($T \neq T'$) return fail;
Else{do next;}

Step3: ciphertext P_i XOR plaintext m_i
for (i=1, i<n,i++)

{ $Y_i = Y_{i-1}$ ++ ; $m_i' = m_i \oplus E_K(Y_i)$; ... }

Step4: $m_n' = m_n \oplus MSB_{|m_n|}(E_k(Y_n))$;//calculate the last section

Step5: return $M'' = m_1 \| m_2 \| \cdots \| m_n$;

The application of the authentication encryption algorithm in data link layer, can protect the data confidentiality and integrity. Besides, the uncryption of head field and priority tag field can protect the communication parties from the influence of encryption, especially retain the convenience for the future use of VLAN technology.

4 Time Analysis of GCM Algorithm Used in Smart Substation

Assume that two IEDs are joined together and appllied authentication encryption algorithm in the data link layer. The sending IED encrypts the data and calculates message authentication code, the receiving IED decrypts the data and checks the MAC value. The sign of Δ marks the maximum time cost of GCM algorithm. Fig.3 shows the entire message delay between two IEDS.

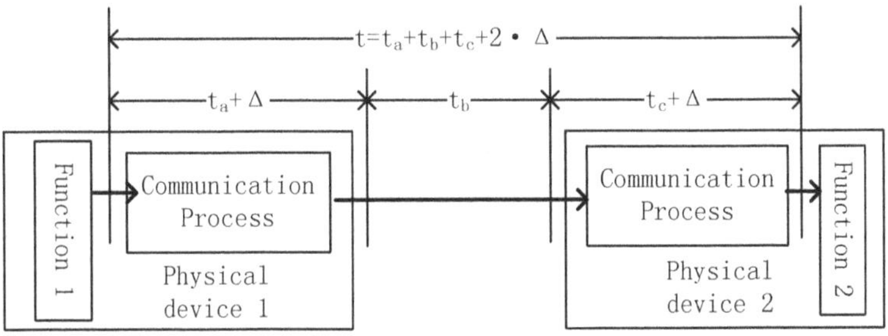

Fig. 3. Message delay based on GCM algorithm

Definition: The total delay between two IEDs is defined as t.

$$t = t_a + t_b + t_c + 2 \cdot \Delta \tag{1}$$

1) The package delay of communication protocol is defined as t_a 、 t_c ;
2) Δ present the time cost of GCM algorithm, different modes have different delay, and the GCM mode is the longest, NULL mode shortest;
3) t_b marks the transmission delay, which depends on the network bandwidth, the speed of signal transmission and the type of Ethernet. Paper[5] proves that t_b is undefinite when using shared Ethernet network, and is definite when using switched Ethernet.

The current packet size of Ethernet are 64-1518 bytes. Assuming the current rate of GCM encryption algorithm is 100Gbps, the largest time cost in a single encryption is:

$$\frac{1518B}{100Gbps} = 0.014\mu s \tag{2}$$

So $0 \le \Delta \le 0.014\mu s$, the ratio of GCM delay and maximum transmission time allowed in one comunication is:

$$\frac{\Delta_{max}}{4ms} = \frac{0.014\mu s \times 2}{4ms} = 7/10^6 \tag{3}$$

So we conclude that the time of GCM algorithm consuming is very little. With the development of technologies, GCM encryption algorithms will be faster, easier to be realized, so the application of GCM algorithm in smart substation to ensure the security of message transmission is feasible.

5 Conclusion

GCM algorithm is a high-speed parallel algorithm, and easy to achieve whatever in software or hardware[6]. Just input data one time, both authentication code and ciphertext can be achieved simultaneously. In order to achieve the secure communications within the LAN, This paper redefines the frame format, give the process of GCM algorithm used in different messages transmission, analyzes the maximum transmission delay, which theorily meets the real-time requirements of the critical messages, improves the substation information security and network performance by choosing different modes to achieve secure transmission.

References

1. Gao, X.: Digital substation application technology. China Electric Power Press, Beijing (2008) (in Chinese)
2. IEC-TS 62351-1 Power systems management and associated information exchange – Data and communications security (2007)
3. Diffie, W., Hellman, M.E.: Privacy and authentication: An introduction to cryptography. Proc. of the IEEE 67(3), 397–427 (1979)
4. Satoh, A.: High-speed parallel hardware architecture for Galois Counter Mode. In: Proc. IEEE ISCAS, New Orleans, LA, 1863–1866 (2007)
5. Sun, J.-P., Sheng, W.-X., Wang, S.-A.: Study on the new substation automation network communication system. Proceedings of the Csee 23(3), 16–145 (2003) (in Chinese)
6. Vyncke, E., Paggen, C.: LAN Switch Security, pp. 296–297. Posts & Telecom Press, Beijing (2010)

Research on Authentication Algorithm Based on Double Factor in Power Dispatching Automation System*

BaoYi Wang, SuGai Qiu, and ShaoMin Zhang

School of control and computer engineering, North China Electric Power University, Baoding, Hebei, 071003, China
wangbaoyi@126.com,qiusu135@163.com,zhangshaomin@126.com

Abstract. In view of the security requirements of power dispatching automation system and shortcomings existing in traditional identity authentication technology, a new double factor two-way identification scheme was proposed in this article, which can achieve two-way authentication between the client and the server. The scheme combines one-time password authentication with fingerprint authentication to overcome the defects of traditional identity authentication technology. It can improve the security of password and prevent replay attack, dictionary attack, intermediate attack and interception attack, etc.Moreover, it can adapt well to the existing dispatching automation system.

Keywords: Power dispatching automation system, Identity authentication, One-time password, Fingerprint authentication, Key agreement.

1 Introduction

The dispatching automation system is the foundation of secure operation of power grid and intelligent dispatching [1, 2]. Once the system becomes the attack object of inside and outside illegal workers, it will cause an oscillation of one system and large-scale blackout accident [3, 4]. Since the existing dispatching automation system has been using the traditional identity authentication way that is "user name+ password" to achieve authentication , It can confirm the user's identity effectively in certain extent, but it exists password guess, the network interception, the dictionary attack, the replay attack and the intermediate attack, etc.Therefore, it is necessary to use a stronger identity authentication technology to protect the increasing network service of dispatching automation system.

In Paper [5], the public key mechanism and Challenge/response were combined to confirm the user's identity. But, it just studied the authentication of the client by the server, though, on the contrary, it didn't mention. Paper [6] proposed an identity authentication system based on digital certificates technology, it studied two-way identity authentication between the server and the user. But, it needed the assistance

* This work is supported by Hebei higher education science research plan funding subject of research of information integration based on IEC61970 and the security techniques of smart grid. (Z2010290).

D. Jin and S. Lin (Eds.): Advances in EECM Vol. 1, LNEE 139, pp. 539–544.
springerlink.com

of a third party, which increased the server's expenditure. Paper [7] analyzed the advantage of biological recognition technology and proposed an identity authentication technology based on fingerprint. But, it just studied the authentication of the client by the server too. Paper [8] reversed the challenge direction of challenge/response mechanism, which achieved two-way identity authentication between the server and the user. But, it needed the assistance of a third party too.

Due to the characteristics of uniqueness and stability, the fingerprint recognition technology is widely applied, but it can't be revoked, once it is leaked will cause disastrous consequence. Therefore, this paper combines the challenge/response mechanism with the fingerprint recognition authentication mechanism to enhance strong points and avoid weaknesses. So a new dynamic identity authentication system based on fingerprint identification authentication was proposed. It uses agreement key generated in the process of mutual authentication between the server and the user to encrypt fingerprint information in order to prevent the information been captured or intercepted in the network transmission.

2 The Structure of the System

This system structure is described in figure1

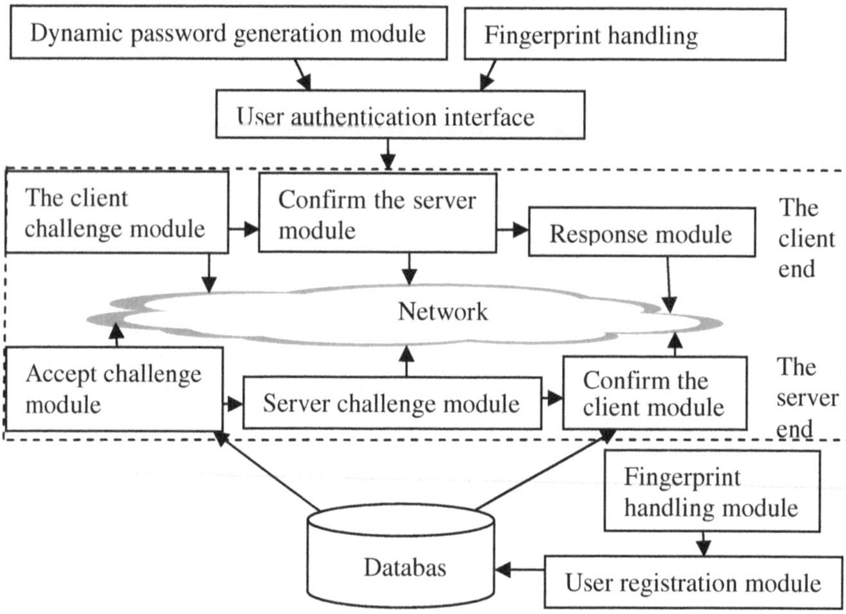

Fig. 1. The identity authentication system structure

The system is mainly composed of the fingerprint handling module, dynamic password generation module, the client challenge module, confirm the server module, response module, accept challenge module, the server challenge module and confirm the client module,etc.

3 The Working Mechanism of the Authentication System Based on One-Time Password and Fingerprint Identification

The following notation and definitions are used throughout this paper.

U:user's identity,Ru:random number generated by the client,Uid:user's name,Psd:user's password,PbKu:user's public key,PriKu:user's private key, f (Ru,Rs): key agreement function,nk:the agreement key,AS:identity of the authentication server,Rs: random number generated by the authentication server, PbKas: server's Public key,PriKas : server's Private key,+: summation operation,D_{Key} (x):uses k to decrypt x, E_{Key} (x): uses k to encrypt x.

In this article, the described identification process is divided into two periods: registration and authentication.

A. Registration period

Step1: User U logins the dispatching automation system for registering, inputs Uid .Server AS inquiries authentication database, if the Uid exists, then AS prompts U to reenter the Uid.

Step2: U inputs password psd and extracts fingerprint characteristic value T by fingerprint extraction device, then transmits them to AS, which encrypted by PbKas.

Step3: AS saves Uid, E_{PbKas} (Psd) and E_{PbKas} (T) in the database for later use.

B. Authentication period

Step1: U logins the client, inputs Uid and Psd. The client generates a random number Ru, then encrypts Uid, Psd and Ru by PbKas, the result of encryption is A, namely, A= E_{PbKas} (Uid, Psd, Ru).then the A is sent to the server.

Step2: After AS receives A, it decrypts A using its public key to get Uid, Psd and Ru. Then AS searches Uid in the database for checking, if Uid and Psd are incorrecting, the authentication fails, otherwise enters the next step;

Step3: AS generates Rs and encrypts Rs and Rs+Ru using PbKu, the result of encryption is B, namely, B= E_{PbKu} (Uid, Psd, Ru).then the B is sent to the client.

Step4: The client decrypts B using PriKas to gets Rs and Rs+Ru, then gets Ru. If the Ru is not the anticipated one, authentication fails, otherwise enters the next step;

Step5: The client calculates the symmetric key nk according to specific session key computation function f= (Ru, Rs).

Step6: According to user's fingerprint information extracted by fingerprint extraction device, the client extracts fingerprint characteristic value T1, uses nk to encrypt T1 and Rs, then sends the encrypted information M=E_{nk}(T1,Rs)to AS.

Step7:AS calculates the symmetric key nk using the same session key computation function f as that at the client, and uses nk to decrypt M to get T1and Rs,if the Rs is not the anticipated one, the authentication fails; otherwise, enters the next step;

Step8: AS compares T1 with T stored in the fingerprint template storehouse, if matches, authentication successes, otherwise, the authentication fails.

Step9: After the client received the authentication information, it can perform operations or use the symmetric session key nk to communicate with the server.

The authentication process model is described in figure 2.

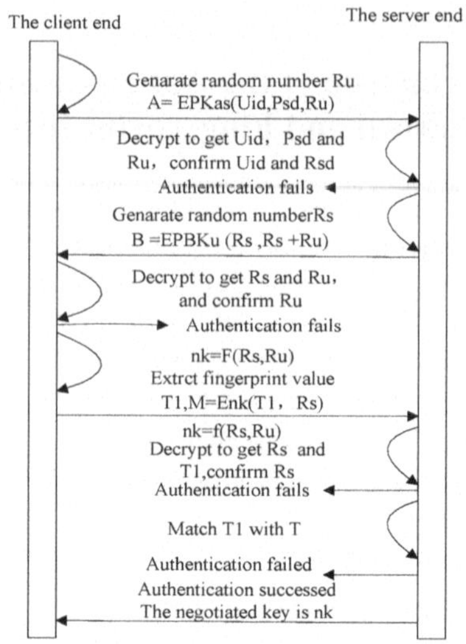

Fig. 2. The two-way authentication model

4 Security and Effectiveness Analysis

4.1 Security Analysis

(1) Resist network interception. Since user name, password and random numbers are encrypted using PbKas, even if the listener intercepts the information, because they don't have PriKas, so it is difficult to get the information in the message.

(2) Resist replay attack. The client and the server each time use random numbers as a challenge code, which ensures the uniqueness and non-repeatability of the challenge code, thus it can avoid replay attack.

(3) Resist intermediate attack. In order to destroy the security of power system, some illegal workers pretend the server or the client to steal legal worker's name and password. The intermediate pretends the server: if it intercepts the information A, because it doesn't have the server's private key, thus it is unable to make a replay. If it intercepts the user's response information M, because it doesn't have the session key, thus it can't decrypt the received ciphertext in the following communication. The intermediate pretends the client: regardless of it intercepts any information sent by the server, because it doesn't have the user's private key, so it can't response to the server, thus it can prevent intermediate attack effectively.

(4) Resist chosen ciphertext attack [9]. In this article, all encryption news is not aim at the original news but at synthetic news with random number generated by the server and the client, thus it can resist chosen ciphertext attack.

(5) Achieved two-way authentication. At present the majority identity authentication methods just emphasize the authentication of the client by the server,

so it can't prevent the server hypocrisy. This model can effectively avoid the fake between the client and server using the intersection authentication.

(6) Higher security. This article uses agreement key to encrypt the fingerprint information in order to avoid it been captured or intercepted during the transmission. Can not only remedy shortcomings existing in a single fingerprint or dynamic password, but also provide a stronger identity authentication function for the user.

The comparison between this scheme and other schemes is described in table 1, and "√"means "has".

Table 1. The comparison between this scheme and other schemes

Item	Paper [5]	Paper [6]	Paper [7]	Paper [8]	This paper
One/two way	one	two	one	two	two
Resist replay attack	√	√	√	√	√
Resist intermediate attack					√
Use fingerprint to enhance			√		√
Generate session key					√
Don't need the help of a third party	√		√		√
Resist network interception	√	√	√	√	√
Resist ciphertext attack					√

4.2 Effectiveness Analysis

(1)The entire authentication process is just the interaction authentication between the client and the server, without the help of a trusted third party involvement. So, the authentication scheme is simple and easy to implementation.

(2) High efficiency. In this scheme, random numbers are just used for authentication, they don't need other deposition, which not only greatly reduces the expenditure of the server but also rises the server's operating efficiency.

(3) Throughout the authentication process, you don't need to generate random numbers as well as large-scale operations like asymmetric encryption and decryption, unless you pass the static password authentication.

(4)Throughout the authentication process, the communication between two communication parts is carried out for only four times, which is the most effective communication number in a two-way authentication protocol [10].

(5) Login flexible. Unlike the protocol with couner, the user can't pass the authentication unless he uses a specific client. In this scheme, the user can login the system as long as there has a client end installed a recognition fingerprint device.

5 Conclusion

In recent years, along with the application of information technology and network technology in power dispatching automation system, its automated level is getting higher and higher, but it also brings a great deal of security risks at the same time. Due to the traditional identity authentication technology couldn't guarantee the security requirements of the system, therefore, a new identity authentication scheme was proposed in this article, which combines one-time password authentication with fingerprint identification authentication to overcome the defects existing in the traditional sole authentication mechanism. The scheme not only can ensure the user's security, but also can generate agreement session key, which has high reliability and doesn't require the participation of a third party. Moreover, it is simple to implement and has well reference value to identity authentication of each information system in power system.

References

1. Tan, S.-W.: Analysis and research of real-time communication current situation in power dispatch automation for China south power grid. Power System Protection and Control 38(22), 109–114 (2010) (in Chinese)
2. Zhang, R., Du, Y., Liu, Y.: New challenges to power system planning and operation lf smart development in china. In: International Conference on Power System Technology, Hongzhou, China, pp. 1–8 (2010)
3. Wang, Y., Xin, Y., Xiang, L., Lu, C., et al.: Security and Protection of Dispatching Automation Systems And Digital Networks. Automation of Electric Power Systems 29(5), 5–8 (2001) (in Chinese)
4. Hu, Y., Dong, M., Han, Y.: Consideration of Information Security For Electric Power Industry 26(7), 1–4 (2002) (in Chinese)
5. Ji, P., Zhang, Y.: Design of dynamic authentication based on digital signature. Computing Engineering and Design 29(1), 45–46 (2008) (in Chinese)
6. Lv, G.-L., Wang, D., Dai, J., Shao, Z.-R.: Advanced Authentication System Based on Digital Certificate Technology. Computer Application and Research 23(8), 144–166 (2006) (in Chinese)
7. Zhu, L., Xu, W., Liu, Y.: The Research and Design of Network Identity Certification Technology Based on Fingerprint. The Computer Engineering and Application 31(3), 171–173 (2003) (in Chinese)
8. Ding, Z.-G., Liu, D.-Z., Liu, M.-L.: Optimistic dynamic authentication based on digital signature. The Computer Engineering and Design 30(15), 3511–3513 (2009)
9. Mao, W.: Modern cryptography theory and practice. Electronic industry press, Beijing (2004)
10. Lai, X.-J.: Security Requirements on Authentication Protocols Using Challenge/Response. Graduate School of Chinese Academy of Sciences Journal 19(3), 246–253 (2002) (in Chinese)

An Applied Research of Improved BB84 Protocol in Electric Power Secondary System Communication*

ShaoMin Zhang, XiuYun Liu, and BaoYi Wang

School of Control and Computer Engineering, North China Electric Power University, 071003 Baoding, China
zhangshaomin@126.com, fliuxiuyun@126.com, wangbaoyi@126.com

Abstract. In order to ensure the security of information transmission when electric power secondary system operating, according to electric power secondary system security regulation, in the safety division we introduce the quantum cryptography in the information transfer process. Because of quantum cryptography's quantum key distribution—BB84 protocol can't identify the legal correspondents. So in our paper we proposed an improved BB84 protocol. In improved protocol we use the character of quantum to finish the identity authentication, so that we can ensure keys dynamic update dynamically and identify authentication. Based on the distinctive physical characteristics and quantum mechanics' hensenberg theory, quantum cryptography can prevent any eavesdropping and has absolutely security. In the future the quantum cryptography has great of research value for smart grid's information transmission.

Keywords: Quantum key distribution, BB84 protocol, Authenticated key.

1 Introduction

Traditional classical cryptography includes symmetric cryptography and public key cryptography, its security depends on computing security. Based on the quantum calculated theory, quantum computer has powerful parallel computing power. Because of quantum properties, quantum computer including 5000 quantum bits can solve the factorization of large numbers in 30 seconds, and traditional computer needs about 100million years, so traditional encryption technology's security greatly reduced when facing quantum computer. Quantum cryptography was first proposed by the Americans, and in 1984, G. Brassasard and CHBennett made the first quantum key distribution protocol, the BB84 protocol. In 1989 IBM Corporation completed the first quantum key distribution experiment, and realized the BB84 protocol, so quantum cryptography came into practical application from theory.

In the secondary power system, a lot of sensitive information was transported through the network, so usually we used encryption technology to protect information.

* This work is supported by Hebei higher education science research plan funding subject of research of information integration based on IEC61970 and the security techniques of smart grid (No.Z2010290).

D. Jin and S. Lin (Eds.): Advances in EECM Vol. 1, LNEE 139, pp. 545–550.
springerlink.com

Such as RSA public key algorithm's security depends on the difficulty of factoring large numbers.

However, in 1994, AT & T Labs,Peter W. Shor proposed a quantum algorithm[4],this method can decompose a large prime factors in the limited time. In 1996, BellLabs LovGrover also found a quantum search algorithm [5], it can exhaust keys of the existing algorithm quickly. Once quantum computers make a major breakthrough, its powerful computing power will make the existing encryption system 'no secrets to protect'.

Power system is complex, and the power grid becomes towards smart grid, smart grid [6-7] is an open network which is a challenge for grid's safe communication. For this, this paper presents an improved quantum cryptographic key protocol to achieve the secondary power system communications.

2 Classical Cryptography and Quantum Cryptography

2.1 Classical Cryptography

Assume A and B are communicating parties, if A wants to send a message to B, according to two sides' beforehand agreements; she encrypts the message and then sends it to B. When B receives the cipher text, he uses the key to decrypt the text and converts it to plain text. This is the classic basic communications program. Currently, asymmetric key is used widely in network and financial sector, and the shor quantum algorithm using quantum computer can easily decipher this type of encryption algorithm. Therefore, the classical encryption methods can't basically guarantee our information security, so we need to develop new technology to meet the increasing awareness of communications security requirements.

2.2 Quantum Cryptography

The biggest difference between BB84 protocol and traditional encryption techniques is that BB84 protocol can withstand any of deciphering technology and the attack of calculation tools. Because its security is based upon the laws of quantum physics, in this sense, the security of quantum cryptography has absolute advantage. About quantum cryptography's practical applications in 2007 leaded by Guo Guangcan between Beijing-Tianjin, the commercial fiber-125km achieved quantum key distribution, solved the key issues of practical fiber-optic quantum cryptography [8].

2.3 Quantum Secure Communication Protocols

Heisenberg uncertainty principle is: any of quantum system can't be measured accurately without changing its original state. In the quantum secure communication, we can use this feature to pass the keys. When eavesdropper eavesdropping we will find it quickly. In these communication programs the most famous is BB84 protocol.

BB84 protocol's main idea: the BB84 protocol under noiseless channel, in the BB84 protocol quantum communication is actually done by the two phases [9]. Firstly, they achieve key communication through quantum channel. Secondly, keys consult is finished through classic channel, and they detect the existence of eavesdropping, and then determine the final keys. So, they complete the quantum communication. Quantum communication system is shown in Figure 1.

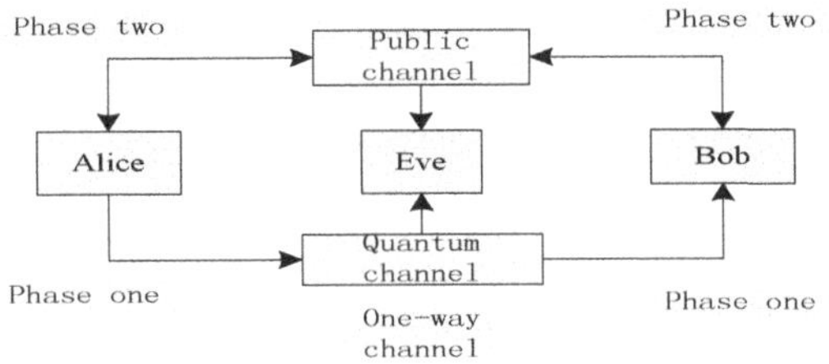

Fig. 1. BB84 quantum communication system

In order to understand easily the quantum cryptography, now we assume that the legal correspondents agree to use these four polarization states of two kinds of polarization-bases to achieve quantum key distribution, follow these steps:

1) A sends photons to B, and selects randomly one of four photon polarization states.
2) B selects randomly linear polarization or Spin polarization to measure photon's polarization, independently.
3) Polarization direction that B actually measures the polarization direction (only B knows, some states have not been detected, said a space).
4) B announces the photon's measurement base which he has detected, but he will not publish the results. Then A tells B that which measurement bases are correct and they only keep these correct photons.
5) A and B retain the same photons when detected the same measurement base, and according to the agreed rules, transforming binary sequences(Such as L-polarization and horizontal polarization are on behalf of bit'0', right-hand polarization and vertical polarization are on behalf of bit'1')

Listed in Table 1, a whole process of key distribution, the last line of the bits sequence is keys negotiated by A and B.

Table 1. The process of BB84 quantum key distribution

	1	2	3	4	5	6	7	8	9	10	11	12	13	14
1	↖	↖	↗	↓	→	↗	→	↓	↓	↖	↗	→	↖	↓
2	+	○	○	+	○	○	+	○	○	○	+	+	+	+
3		↖		↓	↗	↗	→	↖		↖	↓	→	↓	↓
4		↖		↓		↗	→			↖	↓	→	↓	↓
5		1		1		0	0			1		0		1

3 The Improved BB84 Protocol

3.1 BB84 Protocol's Problems

In the process of communication, E can neither measure nor clone information of the photons sent by A, but E can fake the identity of A and establish key—Kbe with B, while fakes B's identity and establish key—Kbe with A, so E controls the communication in the middle of two position and been not detected by legitimate traffic. For this problem, we propose an authentication based on quantum properties, through the authentication check whether the parties are legal correspondent.

3.2 An Improved BB84 Protocol

Assume that A and B have obtained a string of classical bits—Kc1 which would be as a shared authentication key. Kc1 consists of two parts: Kc1=Ka1+Kb1, Ka1 is the key that A proves itself to B, and Kb1s the key that B proves itself to A. As the authentication key, Ka1 and Kb1 only be used once in the communication, and dynamically updated every time after certification, to ensure the security of the authentication key.

This agreement consists of 11 steps, a detailed description as follows:

1)Both A and B ,respectively, converts the shared authentication key Ka1 and Kb1 to measurement base sequence When Both A and B needs to establish a new communication and identification, they firstly convert the authentication key—Kb1 and Ka1 to Measurement-based vector sequence of quantum mechanics. Conveniently, ($\otimes$ is rotation measurement base, $\oplus$ is linear measurement base), For example, if Ka1=01001110, The base sequence of the corresponding measurement are $\otimes \oplus \otimes \otimes \oplus \oplus \oplus \otimes$.

2) A and B establish the quantum channel. Quantum channel uses optical fiber as transmission medium for the transmission of quantum information.

3) A shows her identity, and B is the authenticator. B sends a string of light pulses to A and sets the direction of the polarization filter followed Ka1, if Ka1=01001110, the directions of the polarization filter are $\otimes \oplus \otimes \otimes \oplus \oplus \oplus \otimes$, so, each photon in the direction of linear polarization is polarized horizontal polarization and vertical polarization. In the direction of rotary polarization, photon is polarized left-polarization and right-polarization, B sends it to A through quantum channel.

4) A measures the sequence of quantum state that she received. A measures a total of received quantum using Measurement base sequence.

5) A tells B parts of measurement results. Parts of the results of quantum states are selected randomly. The corresponding number are encrypted by Vernam algorithm, and then transmitted to B through quantum channel.

6) B compared to A's measurements. B compared the measurements which is A's decrypt information to quantum states which sent by himself, and thus, to determine A's identity. Only both A and B own Ka1,their polarity polarization and measurement base were accordant exactly, A can get the correct quantum state measurements. Thus, for B, A's identity was proved. Once E attacked the communication, A's measurements is disaccord with B's information of quantum states, now, the certification can be terminated. A's certification results would be passed to B through quantum channel, which could guarantee the process of Information transmission safe absolutely .The Ka1 will be used only once as the authentication key. E couldn't get any available information.

7) In accordance with BB84 key distribution protocol, both A and B distributes the new authentication key Kc2, Kc2 = Ka2 + Kb2, Key will be updated after each

authentication, to ensure authentication key Kan and Kbn only be used once, so can keep the certification absolute security.

Above scenarios, t A shows her identity and B is the authenticator. Using the same way, B shows his identity, and A is the authenticator. The final step is to update the authentication keys, to ensure that keys are used only once, thus we can achieve the mutual authentication in the communication.

4 Safety Analysis of Improved BB84 Protocol

4.1 Security Analysis of Certification

Because the quantum key distribution's security is provable, E can't know the Kan and Kbn which are generated by quantum key distribution, and authentication key is updated after each authentication. So Kcn is provably secure. In steps of the process of the protocol, measurements which are sent back by A and B are encrypted with authentication key and back through quantum cryptography channel, rather than through classical channel. Only holding the authentication key, can both A and B decrypts, and ensure that the authentication information is not public, and the authentication key's absolute security [11].

4.2 Protocol Security Analysis

1) Shunt eavesdropping. The idea dose not has any problem in classical communication. But in quantum cryptography it can't be successful. According to the basic principles of quantum mechanics, they can't be separated. If E tries to intercept the photons, then B must not receive the photons. So that the photons would be discarded in the process of comparing between A and B. thus E will not get any useful information, on the contrary, the photons which B has measured must not has been intercepted, thus the keys established between A and B is certainly safe.

2) Intercept/send eavesdropping. To prevent such attacks, we introduce the authentication based on quantum properties in the initial stages of quantum key distribution to solve such problems. When E chooses the measurement basis which is different with B's, her measuring behavior will destroy the original information. At subsequent public compare stage, if there is no difference in the process of compare then we can know no eavesdropper. On the contrary, there is an eavesdropper. The more information compared the higher possibility which E detected.

3) Intercept/clone eavesdropping. The quantum no-cloning theorem tells us that any unknown quantum state can't be copied. Thus when E does not know which quantum state sent by A, she can't copy the photons.

5 Conclusions

In this paper, we studied the quantum cryptography. Because the BB84 protocol does not automatically provide for authentication, we proposed a improved BB84 protocol. In improved protocol we finish the identity authentication based on quantum properties. The protocol updated automatically the authentication key after certification every time, so it can ensure one-time pad and authentication's absolute security. The physical characteristics of quantum encryption technology also guarantee the absolute security of the encryption process. This protocol has excellent security and is easy to implement, especially for fiber-optical channel. Therefore, quantum encryption technology is significance in electric power secondary system. With the opening of the smart grid strengthen, quantum encryption technology's implementation in power systems engineering is our next research goal.

References

1. Huang, F., Liu, Y.: The Quantum Privacy Communication Searches Learns. Journal of Netinfo Security 4, 19–21 (2009)
2. Wang, R., Cao, L., Yang, D., Ma, X.: Research and Application of Network Secondary Systems in Digital Substation. Journal of Power System Protection And Control 38(12), 59–64 (2010)
3. Wang, B., Yang, L.: Security Protection of Power Secondary System Based on Security Gateway. Journal of Telcommunication for Electric Power System 29(19), 28–34 (2008)
4. Grover, L.K.: A fast Quantum Mechanica1 Algorithm for Database Search. In: Proceedings of the Twenty—eighth annual ACM symposium on Theory of computing, pp. 212–219 (1996)
5. Peter, W., Shor, P.: Proc. 35 Annu. Symp.on the Foundations of Computer Science, pp. 124–134. IEEE Computer Society Press, Los Alamitos (1994)
6. Niu, P., Kang, J., Li, A., Li, L.: New Operation form of Power Network Started by Smart Grid. Journal of Power System Protection and Control 38(19), 240–244 (2010)
7. Xing, J., Yang, H., Chen, W., Zhao, R.: Open Power Network Model Management System under Smart Grid Infrastructure. Journal of Power System Protection and Control 38(21), 227–231 (2010)
8. Guo, G.: Quantum information technology. Journal of Chongqing University of Posts and Telecommunications (Natural Science Edition) 22(5), 521–525 (2010)
9. He, X.: Quantum Cryptography and its Applications. Journal of Communication Technology 11(42), 93–95 (2009)
10. Díaz-Rodríguez, C.A., Olivares-Robles, M.A., Juárez, A.: An Overview of Quantum Cryptography:Simulation. Journal of Electrical Communications and Computers (CONIELECOMP) 2, 316–321 (2011)
11. Gong, J., Deng, Y., Chen, J.: Identity Authentication Based on the Characteristic Of Quantum. Journal of Optical Communication Technology 3, 57–59 (2010)

Non-isolated High Step-Up ZVS DC-DC Converter with Voltage Multiplier Cells

Hyun-Lark Do

Department of Electronic & Information Engineering,
Seoul National University of Science and Technology, Seoul, South Korea
hldo@seoultech.ac.kr

Abstract. A non-isolated high step-up zero-voltage-switching (ZVS) DC-DC converter with voltage multiplier cells is presented in this paper. The input stage of the proposed convert is a boost converter in a continuous conduction mode. Therefore, the input current is continuous and low ripple. Voltage multiplier cells are stacked on the output of the boost converter input stage to provide high voltage gain. Due to soft-switching characteristic of the switches, the switching loss is significantly reduced and the overall efficiency is improved. A prototype of the proposed converter is developed, and its experimental results are presented for validation.

Keywords: Soft-switching, DC-DC converter, reverse-recovery.

1 Introduction

In many applications, high step-up DC-DC converters are required. For examples, in electric vehicles, fuel cells, and photovoltaic systems, DC-DC converters act as an interface system between the low voltage sources and the load requiring higher voltage [1]. A conventional boost converter can provide a step-up function and a continuous input current. However, it requires an extreme duty cycle to obtain high voltage gain and its voltage gain is limited due to its parasitic components [2]. In order to increase voltage gain, high step-up DC-DC converter using coupled inductors have been suggested in [3]. It can provide high voltage gain but it has several disadvantages such as large voltage ringing across the semiconductor devices and low efficiency due to hard-switching operation.

In order to remedy these problems, a non-isolated high step-up ZVS DC-DC converter with voltage multiplier cells is proposed. The proposed converter features high voltage gain, fixed switching frequency, soft-switching operations of all power switches and output diodes, and clamped voltage across power switches and output diodes without any clamping circuits. Due to the soft-switching operation, the switching loss is significantly reduced. The switches operate with ZVS. The reverse-recovery loss of the output diodes is significantly reduced due to the serial inductor.

D. Jin and S. Lin (Eds.): Advances in EECM Vol. 1, LNEE 139, pp. 551–554.
springerlink.com

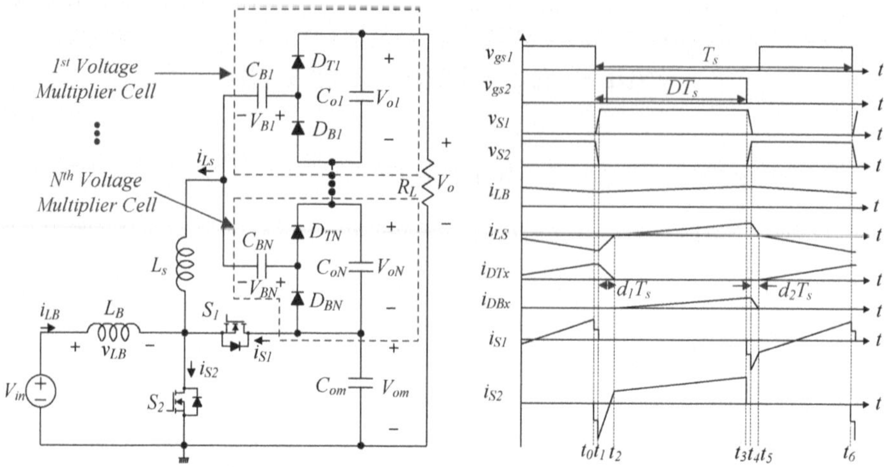

Fig. 1. Circuit diagram and key waveforms of the proposed converter

2 Analysis of the Proposed Converter

The circuit diagram of the proposed converter and its key waveforms are shown in Fig.1. The operation during a switching period T_s is divided into six modes. The switch S_1 and the switch S_2 are operated asymmetrically and the duty cycle D is based on S_2. The switch S_1 is used instead of a boost diode. Like a boost converter, the voltages across S_1 and S_2 are confined to the voltage V_{om}. The N-stage voltage multiplier cells at the secondary side raise the voltage gain without any coupled inductors and transformers. Before t_0, the switch S_1 and the upper side diode D_{Tx} ($1 \leq x \leq N$) in multiplier cells are conducting. At t_0, i_{LB} and i_{Ls} arrive at their minimum values, respectively.

Mode 1 [t_0, t_1]: At t_0, S_1 is turned off. Then, the energy stored in the magnetic components starts to charge and discharge the parasitic output capacitances of S_1 and S_2. Therefore, the voltage v_{S1} starts to rise from zero and the voltage v_{S2} starts to fall from V_{om}. With an assumption that the parasitic out capacitances of S_1 and S_2 are small and a sufficient energy is stored in the magnetic components, this time interval can be ignored.

Mode 2 [t_1, t_2]: At t_1, v_{S2} arrives at zero. Then, the body diode of S_2 is turned on. After that, the gate signal is applied to S_2. Since v_{S2} is zero before S_2 is turned on, zero-voltage turn-on of S_2 is achieved. In this mode, V_{in} is applied to L_B, i_{LB} increases linearly. Also, V_{om}+V_{oN}-V_{BN} is applied to L_s, i_{Ls} increases linearly.

Mode 3 [t_2, t_3]: At t_2, the current flowing through D_{Tx} arrives at zero and D_{Tx} is turned off. Then, i_{Ls} changes its direction and the lower side diode D_{Bx} ($1 \leq x \leq N$) is turned on. Since the changing rate of i_{DTx} flowing through D_{Tx} is controlled by L_s, the reverse-recovery is significantly alleviated. In this mode, i_{Ls} increases linearly. At the end of this mode, i_{LB} and i_{Ls} arrive at their maximum values, respectively.

Mode 4 [t_3, t_4]: At t_3, S_2 is turned off. Similar to mode 1, the voltage v_{S2} starts to rise and v_{S1} starts to fall. Also, this time interval can be ignored.

Mode 5 [t_4, t_5]: At t_4, v_{S1} arrives at zero and the body diode of S_1 is turned on. Then, the gate signal is applied to S_1. Since v_{S1} is zero before S_1 is turned on, zero-voltage turn-on of S_1 is achieved. With the turn-on of S_1, $-(V_{om}-V_{in})$ is applied to L_B and i_{LB} decreases linearly. Since D_{Bx} is still on, $-V_{BN}$ is applied to L_s and i_{Ls} decreases linearly.

Mode 6 [t_5, t_6]: Similar to mode 3, i_{DBx} arrives at zero and D_{Bx} is turned off. Then, D_{Tx} is turned on and its current increases linearly. Since the diode currents are controlled by L_s, the reverse-recovery problem of the output diodes is significantly alleviated. At the end of this mode, i_{LB} and i_{Ls} arrive at their minimum values, respectively.

By applying the volt-second balance law to the voltage across L_{LB}, V_{om} can be easily derived by $V_{in}/(1-D)$. As in a boost converter, the voltage stresses across the power switches are confined to V_{om}. Since V_{om} depends on D, the voltage stresses of S_1 and S_2 can be varied according to D and V_{in}. The voltage gain M of the proposed converter is given by

$$M = \frac{V_o}{V_{in}} = \frac{1}{1-D} + \frac{N(1-2k)D}{(D+(1-2D)k)(1-D-(1-2D)k)}, \tag{1}$$

$$k = \frac{1}{2}\left(1-\sqrt{1-\frac{8L_s I_o f_s}{NDV_{in}}}\right). \tag{2}$$

Due to the N-stage voltage multiplier cells, the voltages across all the output diodes are confined to $(V_o-V_{om})/N$.

3 Experimental Results

The performance of the proposed converter was verified on a 90W prototype. The prototype was designed to operate from a 24V input voltage and provide 200V output voltage. Its operating frequency was 75kHz. The proposed converter provides the voltage gain of 8.3 only with N=2. As N increases, the voltage gain can be raised. The inductor L_B was selected as 100uH. The inductor L_s was 14uH. The blocking capacitor C_{B1} and C_{B2} were chosen as 6.6uFx2. The output capacitors C_{om}, C_{o1}, and C_{o2} were selected as 100uF. Fig. 2 shows the measured key waveforms of the proposed converter and efficiency curve. The measured waveforms agree with the theoretical analysis. The measured maximum voltage stresses of S_1 and S_2 are around 80V, which agrees with the theoretical analysis. The ZVS operations of S_1 and S_2 are also shown in Fig. 2. Since the voltages across the switches go to zero before the switch currents change their directions, the ZVS turn-on of the switches is achieved. The proposed converter exhibits the maximum efficiency of 92.8% at full load. Due to its soft-switching characteristic and alleviated reverse-recovery problem, it shows a higher efficiency than other high step-up DC-DC converters.

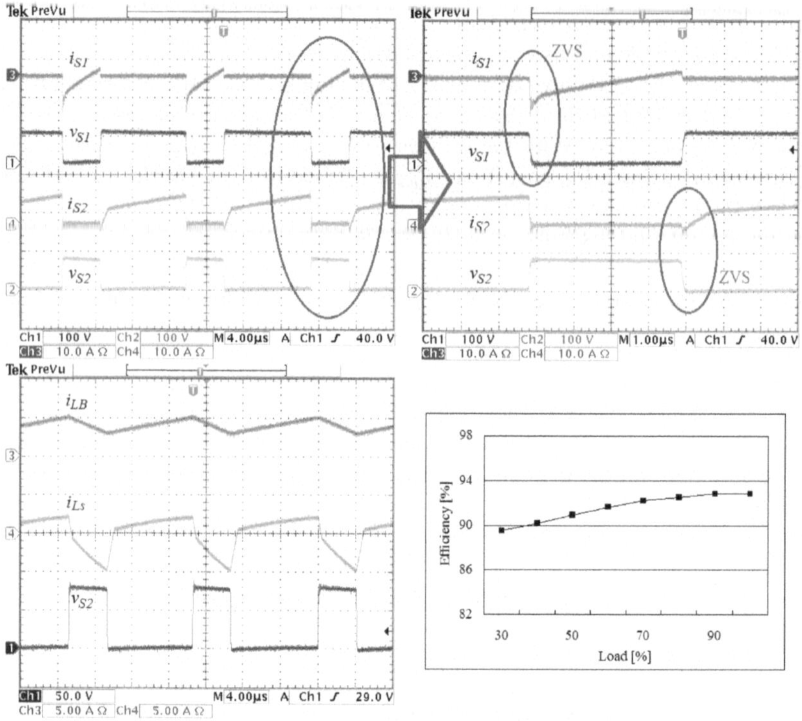

Fig. 2. Experimental waveforms and measured efficiency curve

4 Conclusion

A non-isolated high step-up ZVS DC-DC converter with voltage multiplier cells has been proposed. Both power switches operate with ZVS and the reverse-recovery problem of output diodes is dramatically alleviated by the serial inductor. Without any coupled inductors and transformers, the voltage gain of 8.3 has been achieved only with 2-stage voltage multiplier cell. Due to its soft commutation of power semiconductor devices, a high efficiency is achieved.

References

1. Wai, R.J., et al.: High-performance stand-alone photovoltaic generation system. IEEE Trans. Ind. Elec. 55, 240–250 (2008)
2. Yang, L.S., et al.: Transformerless DC-DC converters with high step-up voltage gain. IEEE Trans. Ind. Elec. 56, 3144–3152 (2009)
3. Grant, D.A., et al.: Synthesis of tapped-inductor switched-mode converters. IEEE Trans. Power Elec. 22, 1964–1969 (1997)

Energy Recovery Sustain Driver with Discharge Current Compensation

Hyun-Lark Do

Department of Electronic & Information Engineering,
Seoul National University of Science and Technology, Seoul, South Korea
hldo@seoultech.ac.kr

Abstract. An energy recovery sustain driver with discharge current compensation is proposed to drive a plasma display panel (PDP). It provides zero-voltage switching (ZVS) of main sustain switches. Therefore, the surge currents associated with hard switching are avoided. Moreover, a voltage notch across the panel is significantly reduced. Since the voltage notch in the sustain pulses reduce the accumulated amount of wall charge, it should be reduced to obtain the enough operating margin. Theoretical analysis and performance of the proposed sustain driver were verified on an experimental prototype operating at 200 kHz switching frequency.

Keywords: Energy recovery, sustain driver, resonance, plasma display.

1 Introduction

A plasma display panel (PDP) is a device for displaying characters or images by utilizing a gas discharge. The PDP provides a large-scale screen as well as an improved image quality owing to the recent development. Most of the ac PDPs utilize the address-display-separation (ADS) driving scheme. One TV field is divided into several subfields, each consisting of a reset period, an address period, and a sustain period. Since a PDP has its intrinsic capacitance, a considerable energy stored in the panel is consumed across the non-ideal resistance of the circuit and PDP without energy recovery circuits. Furthermore, the surge current which occurs when the panel is charged or discharged could cause serious resonance and electromagnetic interference (EMI) noise. To solve these problems, several energy-recovery sustain drivers have been proposed. Among them, the sustain driver proposed in [1] utilizes the series LC resonance between C_p and external inductors. It is widely used but still has several drawbacks. Inevitable conduction losses which occur in the semiconductors, the wires and the other parasitic components lead to a damped oscillation in the circuit. As a result, the panel voltage cannot rise perfectly from zero to a sustain voltage V_s and vice versa which induces surge currents and increases switching stresses of main sustain switches. Moreover, when plasma discharge occurs, a large discharge current flowing through sustain switches causes a voltage notch which reduces the accumulated amount of wall charge [2], [3].

D. Jin and S. Lin (Eds.): Advances in EECM Vol. 1, LNEE 139, pp. 555–558.
springerlink.com

To remedy these problems, a new energy recovery sustain driver is proposed. Since the inductor current compensates for the large portion of a discharge current, the voltage notch across the panel is significantly reduced. Moreover, sustain switches are turned on with ZVS and the surge currents associated with hard switching are avoided.

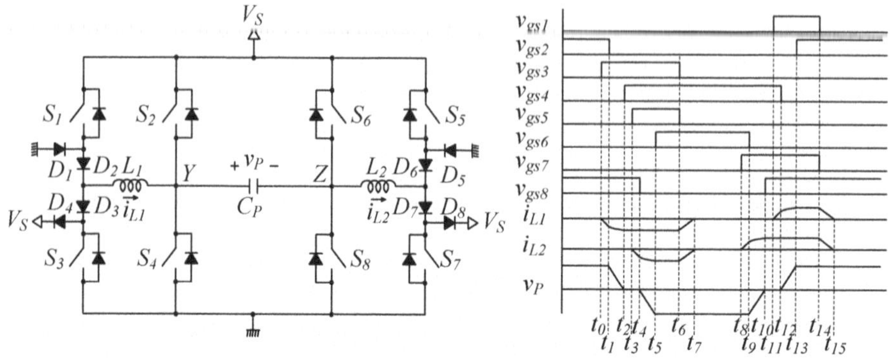

Fig. 1. Circuit diagram and key waveforms of proposed sustain driver

2 Analysis of the Proposed Sustain Driver

Fig. 1 shows the circuit diagram and key waveforms of the proposed sustain driver. Before t_0, the panel voltage v_p is maintained as $-V_s$ with S_4 and S_6 conducting. At t_0, the switch S_7 is turned on and the current i_{L2} increases linearly. At t_1, the switch S_6 is turned off and the resonance between L_2 and C_p occurs. At t_2, the panel voltage v_p arrives at zero. Since the voltage across S_8 is zero at t_2, the zero-voltage switching of S_8 is obtained. After the turn-on of S_8 at t_2, the inductor current i_{L2} freewheels through D_7, S_7, and S_8. At t_3, the switch S_1 is turned on and the inductor current i_{L1} increases linearly. At t_4, the switch S_4 is turned off and the resonance between L_1 and C_p occurs. At t_5, the panel voltage v_p arrives at V_s. Since the voltage across S_2 is zero at t_5, the zero-voltage switching of S_2 is obtained. After the turn-on of S_2 at t_5, the inductor current i_{L1} freewheels through D_2, S_1, and S_2. The freewheeling inductor currents i_{L1} and i_{L2} act as current sources and compensate for the large portion of a plasma discharge current during the time interval t_5 - t_6, the voltage notch across the panel can be significantly reduced. After the discharge current compensation, S_1 and S_7 are turned off and the inductor currents i_{L1} and i_{L2} decrease linearly and approach to zero. The circuit operations of t_0 - t_7 is similar to that of t_8 - t_{15}. For symmetric operation, the values of L_1 and L_2 are identical (L_1=L_2=L). All sustain switches S_2, S_4, S_6, and S_8 are turned on with ZVS. The surge currents, which are inevitable due to the parasitic components in the conventional series-resonant sustain driving method, are avoided. Moreover, the voltage notch across the panel can be significantly reduced because the inductor currents i_{L1} and i_{L2} compensate for a discharge current and reduce the voltage drops across the sustain switches. Fig. 2 shows the inductor current and the voltage waveforms. Equivalent resonant inductance L_{eq} includes the parasitic line inductance

L_l (L_{eq}=L+L_l). Also, the equivalent resonant capacitance C_{eq} includes the output capacitance C_{oss} of the sustain switch (C_{eq}=C_p+2C_{oss}). For a given transition time t_t, the required build-up current I_b and the required build-up time t_b are determined as follows:

$$I_b = \frac{V_s}{Z_o}\cot \omega_o t_t \, , \tag{1}$$

$$t_b = \frac{1}{\omega_o}\cot \omega_o t_t \, , \tag{2}$$

where ω_o=1/($L_{eq}C_{eq}$)$^{1/2}$ and Z_o=(L_{eq}/C_{eq})$^{1/2}$. The freewheeling current I_f is given by

$$I_f = I_b \cot \omega_o t_t + \frac{V_s}{Z_o}\sin \omega_o t_t \, . \tag{3}$$

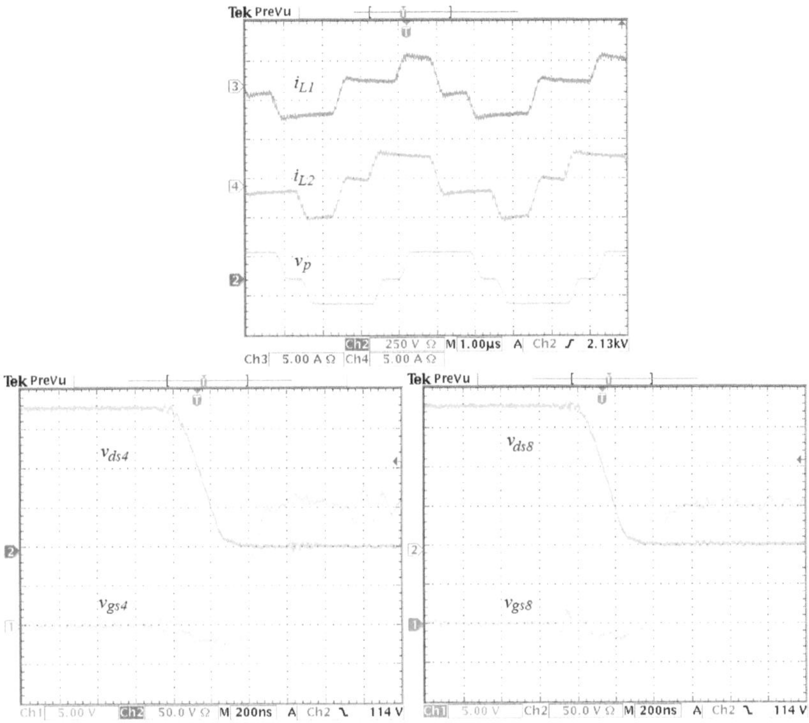

Fig. 2. Experimental waveforms

3 Experimental Results

The prototype sustain driver for a PDP has been implemented with specifications of V_s =180V, switching frequency=200kHz, C_p=3nF (for 7.5 inch PDP), and

L_1=L_2=12uH. Fig. 2 shows the experimental waveforms of the proposed sustain driver, when the white image is displayed. The build-up time and the transition time is 100ns and 250ns, respectively. With the discharge current compensation from i_{L1} and i_{L2}, the voltage notch across the panel disappears. It increases the sustain margin of the PDP. The panel voltage v_p can commutate perfectly from zero to V_S and vice versa. Therefore, ZVS of all sustain switches S_2, S_4, S_6, and S_8 can be achieved and EMI noise can be reduced. It is shown in Fig. 2 that S_4 and S_8 are turned on with ZVS. The gate-source voltages are applied after the drain-source voltages drop to zero. S_2 and S_6 are also turned on with ZVS.

4 Conclusion

A new sustain driver for a PDP is proposed to overcome the drawbacks of the prior method. The ZVS of main sustain switches is achieved and the surge current can be avoided. Due to the ZVS operation, the switching loss is reduced. Moreover, the external inductors act as current sources and compensate for the large portion of a discharge current. As a result, a voltage notch across the panel disappears. By eliminating the voltage notch in the sustain pulses, the operating margin is enlarged.

References

1. Weber, L.F., Wood, M.B.: Energy recovery sustain circuit for the AC plasma display. In: Proc. Symp. Society for Information Display, San Jose, CA, USA, pp. 92–95 (1987)
2. Hsu, H.B., et al.: Regenerative power electronics driver for plasma display panel in sustain-mode operation. IEEE Trans. Ind. Elec. 47, 1118–1124 (2000)
3. Liu, C.C., et al.: An energy-recovery sustaining driver with discharge current compensation for AC plasma display panel. IEEE. Ind. Elec. 48, 344–351 (2001)

Improved Harmony Search Algorithm in Strip Mine Vehicle Route Research

Ting-dong Hu

Shenzhen Kunyiziyuan Electronic Co. Ltd., Shenzhen, China
Td.hu@kunyiziyuan.com

Abstract. In order to solve the strip mine vehicles way computation complexity high problem, the harmony searching algorithm has introduced in this paper to carries on the solution of the question. Basic harmony searches algorithm can obtain the strip mine vehicles way plan question the solution, but basic harmony searches for the law algorithm has very big dependence .This article unify the genetic algorithm and basic harmony searching algorithm makes the algorithm improvement, overcome basic harmony searching algorithm to initial harmony memory storehouse dependence strong shortcoming, Through the example analysis, the improvement a harmony searching algorithm can solve the strip mine vehicles way plan question well, and has the very big enhancement in the algorithm stability and the solution accurate aspect.

Keywords: Strip mine, Vehicles way optimization, harmony searching algorithm, Genetic algorithm.

1 Introduction

Along with truck transportation distance's enlargement, the transportation cost and the work load can large scale increase, moreover, because transports in the strip mine production the influential role and the status is very big, however the efficiency is not high. Depending on artificially carried on the vehicles way plan and the management way already more and more could not satisfy the enhancement transportation efficiency the need, To raise enterprise's production efficiency and the whole management level unceasingly, reduces the transportation cost, thc promotion enterprise's development, appears very essential to the strip mine vehicles way plan. Therefore, how to effective and organization vehicles travel route raise the transportation efficiency as far as possible to reduce the transportation cost, realizes the mine haulage management, dispatch to computerize highly effective, the modernization into key.

2 Vehicle Path Program

The vehicles way plan question becomes the key target which the related domain study, uses to solve in the physical distribution allocation process transit route's definite question. Regardless in urban physical distribution allocation, in the open-air

D. Jin and S. Lin (Eds.): Advances in EECM Vol. 1, LNEE 139, pp. 559–564.
springerlink.com

mineral substance's transportation process, how to organization best route to cause install unifies effectively with the transit route, has become the transportation domain the key point, Therefore uses reasonable, scientific, effective method to choice transit route become raises the vehicular traffic efficiency and the economic efficiency, avoids in the practical work depending upon the experience blind direction the limitation, realizes the physical distribution allocation scientific style important action.

The vehicles route optimization problem (VRP) is one kind of special TSP question. The transportation process is mainly according to each user's different demand, carries on the vehicle using vehicles' capacity or the carrying capacity to carry the mineral the transportation to realize from the strip mine installs the loading point to various debarkation points spatial shift, thus determination allocation vehicles and route's optimization plan.

3 Strip Mine Vehicle Path Program

In the open-air mine, usually the charge of transport dissipation of link is take up 60% ~70% of the mine total cost. Therefore, excellent exaltation of path to mine economic benefit for turning mine carriage has to full important economic meaning. The mine transports path of excellent what main to turn is the carriage of minerals , attend to know together as the mine physical distribution dispatching route of excellent turn, it is a type of NP-HARD question. The solution vehicles way optimization question, usually uses the classics the genetic algorithm, the ant group algorithm and so on heuristic algorithm solution, This article innovation will lie apply the mine physical distribution allocation way optimization question for the first time with harmony searching algorithm in the research.

4 The Harmony Searches Algorithm

The Harmony searching algorithm is one kind of heuristic overall situation searching algorithm which the new century most innovates, with the ant group algorithm, the genetic algorithm and the taboo searching algorithm and so on heuristic algorithm compares, it has demonstrated a better performance in the processing question optimization aspect. Uses the harmony searching algorithm to solve the optimization problems, in the optimized environment needs to establish the question the objective function and the constraints [1].

Object function: $\max f(x)$ or $\min f(x)$ (1)

Constraint condition: $h_i(x) = 0;\ i = 1,\ldots,p;$ (2)

$$g_i(x) \geq 0;\ i = 1,\ldots,q; \quad (3)$$

$$x_i \in X_i = \{x_i(1),\ldots,x_i(k),\ldots,x_i(K_i)\} \text{ or } x_i^L \leq x_i \leq x_i^U \quad (4)$$

Explained: in the feasible solution space $x = (x_1,\ldots,x_n)$ search the globally optimal solution $f(x)$ vector Satisfies the objective function optimum value; If the optimized question has equal or not the equal other constraints, may use the function type (2), (3) to express. If make policy variable x_i is discrete value, the candidate election solves is $x_i \in X_i = \{x_i(1),...,x_i(k),...,x_i(K_i)\}$; If the decision variable is successive value, $x_i^L \leq x_i \leq x_i^U$ is the feasible solution space of x_i [2].

5 Strip Mine Vehicles Based on Harmony Searching Algorithm

The harmony searches algorithm according to giving musical performance a kind of intelligence that look for a wonderful harmony principle design algorithm. With the traditional optimized algorithm (for example genetic algorithm, ant group algorithm and so on) compares, this kind of optimized algorithm's superiority lies from has in filial generation's process to be able to consider that to the entire population in the hereditary property, is more advantageous in seeks to the question space optimal solution. In the solution of open pit mine vehicle routing problem, a harmony on behalf of a client or a vehicle identification code [3].

5.1 Encoding Rule

This paper adopts the natural number coding rules, assuming there are n clients, m vehicle path planning problem. The code is the n customers to use the natural number 1 to n logo, and n+1, n+2...n+m-1 identifies the m car, encoded in the n bits customers, m-1 said the vehicle identification code.

5.2 Illegal the Codes Processing Strategy

Harmony search algorithm in new generation process, accordance with the new generation of generation rules will inevitably generate repeat harmony, such as code 1, 11, 2, 3, 2, 5, 6, 7, 3, 8, 9, 10, 2 and 3 are repeated, and then the code is not legitimate coding. In order to ensure that the generation of coding $x_i^{'}$ are legitimate coding. This paper in the programming treatment refusal strategies. Set to generate a new solution vector of $X^{'} = (x_1^{'}, x_2^{'}, \cdots, x_i^{'}, \cdots x_N^{'})$, $X = \Phi$, x_i is lately solve according to the harmony the regulations generates of number solution quantity. The Refuse process is as follows:

(1) Judgment $i > N$ If is, then stops, otherwise transfer to (2);

(2) Judgment $i = 1$ If is, $x_i^{'} = x_i$, $X = X \cup \{x_i^{'}\}$, $X = X \cup \{x_i^{'}\}$, $i{=}i+1$ otherwise transfer to (3);

(3) Judgment $x_i \in X$ If is, then re-generate x_i until $x_i \notin X$, otherwise $x_i^{'} = x_i$, $X = X \cup \{x_i^{'}\}, i = i + 1$;

(4) transfer to (1);

6 Strip Mine Vehicles Way Plan Based on Improvement Harmony Searching Algorithm

In view of with harmony searching algorithm to initial and sound memory storehouse dependence strong characteristic, makes the improvement with the genetic algorithm to it, forms one kind of new improvement algorithm, solves the strip mine vehicles way plan question.

6.1 Harmony Searching Algorithm Improvement Strategy

This article to makes the improvement with the harmony searching algorithm, It is mainly revolves the enhancement and the sound searching algorithm stability and the search efficiency. In view of with harmony searching algorithm to initial and sound memory storehouse dependence strong characteristic, introduction genetic algorithm, through genetic algorithm high speed search ability and overall situation search ability, union the both finally through the example analysis, confirmed the algorithm is validity [4].

6.2 Genetic Algorithm for Improved Harmony Search Algorithm

In view of the harmony searching algorithm to initial harmony memory storehouse dependence very strong shortcoming, this article uses the genetic algorithm high parallelism and the earlier period can improve the solution fast the ability characteristic, produces and the sound searching algorithm with the genetic algorithm initial and the sound memory storehouse[5].

Code: For open-pit vehicle path planning problem, this paper uses natural number coding rules, That is, for n clients, m car vehicle path planning, Encoding is used in this customer using the n natural numbers 1 to n to identify, but with n +1, n +2, ..., n + m-1 to identify the m vehicles, said n-bit code in the client, m-1 bit is a vehicle identification code.

Cross: Cross is an important process in genetic algorithms, by cross progeny to inherit the parent's good traits. In this genetic algorithm, using the following cross-ways:

Variation: Genetic mutation operation is another important operation in the operation, is designed to increase population diversity to prevent the algorithm's "premature convergence."

Mutation operation is accordance with a certain probability of chromosomal genes in some of the other gene into the process. Mutation rate is generally below 5%. In this genetic algorithm, using the following mutation:

6.3 Case Analysis

An open pit with 3 ore of 8 tons cars of on the eight-point demand for transportation of ore, the single car largest transport distance is 20 km, The load and unload point, and each distance of as show table 1.

Table 1. Distance between the points

i \ j	0	1	2	3	4	5	6	7	8
0	0	2.7	4	5	6	13.3	6.7	10.7	5.3
1	2.7	0	4.3	2.7	6.7	3.3	5	7.3	6.7
2	4	4.3	0	5	6.7	6.7	5	5	5
3	5	2.7	5	0	6.7	3.3	6	6	10
4	6	6.7	6.7	6.7	0	6.7	5	5	6.7
5	13.3	3.3	6.7	3.3	6.7	0	4.7	6	5
6	6.7	5	5	6	5	4.7	0	4.7	6.7
7	10.7	7.3	5	6	5	6	4.7	0	6.7
8	5.3	6.7	5	10	6.7	5	6.7	6.7	0

The tipping point of demand as shown in Table 2:

Table 2. Quantity demanded

Tip point	1	2	3	4	5	6	7	8
Demand	1	2	1	2	1	4	2	2

Algorithm parameters are as follows: GA pop size = 80, = 0.9, = 0.02, NG = 50, HMS), HMCR = 0.8, PAR = 0.05, NI = 50. Basic harmony search algorithm and improved harmony search algorithm for the operation 10 times. The results are shown in Tables 3 and 4.

Table 3. The result of the basic HSA

Evaluation order	total path length	Path plan
1	55.4	0→6→5→1→0;0→3→7→2→0;0→4→8→0
2	55.4	0→6→5→1→0;0→3→7→2→0;0→4→8→0
3	57.4	0→2→7→3→0;0→4→5→1→0; 0→8→6→0
4	53	0→2→7→4→0;0→1→5→3→0;0→8→6→0
5	55.4	0→2→7→3→0;0→4→8→0;0→1→5→6→0
6	53	0→2→7→4→0;0→1→5→3→0;0→8→6→0
7	55.4	0→8→4→0;0→2→7→3→0;0→1→5→6→0
8	54	0→8→5→1→0;0→3→7→2→0;0→6→4→0
9	54	0→2→7→3→0;0→8→5→1→0; 0→4→6→0
10	54	0→3→7→2→0;0→8→5→1→0;0→4→6→0
Average total path:54.7		

Table 4. The result of the improved HSA

Evaluation order	total path length	Path plan
1	53	0→3→5→1→0;0→4→7→2→0;0→6→8→0
2	53	0→3→5→1→0;0→8→6→0;0→4→7→2→0
3	53	0→2→7→4→0;0→3→5→1→0; 0→8→6→0
4	53	0→1→5→3→0;0→8→6→0;0→2→7→4→0
5	53	0→1→5→3→0;0→6→8→0;0→4→7→2→0
6	53	0→2→7→4→0; 0→8→6→00→3→5→1→0;
7	57	0→8→5→1→0;0→2→7→3→0;0→4→6→0
8	53	0→3→5→1→0;0→4→7→2→0;0→6→8→0
9	53	0→2→7→4→0;0→3→5→1→0; 0→8→6→0
10	53	0→3→5→1→0;0→2→7→4→0;0→8→6→0
Average total path: 53.4		

From the above results can be seen, the basic harmony search algorithm in 10 times get second operation to the optimal solution, obtained the average total path length is 54.7; The improved algorithm in the calculation 10 times, nine times to get the optimal solution 53, there is an time calculated approximate optimal solution is 57, the total path average is 53.4. The improved algorithm not only can be good obtain open-pit vehicle path planning problem, but also has greatly improved in the algorithm stability and accuracy than the basic harmony search algorithm.

7 Conclusion

This paper use harmony search algorithm to solve the open-pit vehicle path planning, for basic harmony search algorithm in solving the open pit mine the car vehicle path planning problem the lack, Combined with the characteristics of genetic algorithm, with genetic algorithm to improve the basic harmony search algorithm, harmony search algorithm to improve the stability and accuracy.

References

1. Saka, M.P.: Optimum design of steel sway frames to BS5950 using harmony search algorithm. Journal of Constructional Steel Research 65(1), 36–43 (2009)
2. Cel Yan, H., Haldenb Len, H.S., Baskan, O.: Transport energy modeling with meta-heuristic harmony search algorithm, an application to Turkey. Energy Policy 36(7), 2527–2535 (2008)
3. Lee, K.S., Geem, Z.W., Lee, S.-H., Bae, K.-W.: The Harmony Search Heuristic Algorithm for Discrete Structural Optimization. Engineering Optimization 37(7), 663–684 (2005)
4. Geem, Z.W., Kim, J.H., Loganathan, G.V.: A new heuristic optimization algorithm: harmony earch. Simulation 76(2), 60–68 (2001)
5. Mahdavi, M., Fesanghary, M., Damangir, E.: An improved harmony search algotithm for solving optimization problems. Appl. Math. Comput 188, 1567–1579 (2007)

Computer-Aided Exploration for the Gracefulness of Digraph $n-\vec{c}_4$

Yun Xu

College of Mathematics and Computing Sciences, Huanggang Normal University, Huanggang, Hubei, China

Abstract. The purpose of the paper is to obtain a class new graceful digraph with n circles. In this paper, we defined digraph $n-\vec{c}_4$, and designed a algorithm to seek the graceful labeling of $n-\vec{c}_4$. With the program running, all the graceful labelings of $2k-\vec{c}_4$ corresponding to different values of k were displayed on the screen. By analyzing the labeling datas, we obtain the law of graceful labeling, and prove mathematically the digraph $2k-\vec{c}_4$ is graceful. In additional, we show a property.

Keywords: Algorithm, Digraph, Directed circle, Graceful labeling.

1 Introduction

Graceful graph was raised by S .W .Golomb[1] in 1972. Graceful graph has wide applications in military and science fields, such as error correcting code design, communication network, measuring atomic position in a crystal structure, radar pulse, missile guidance code design, etc, so the study on graceful graph becomes a hot topic in graph theory, and many mathematics workers had made a lot of research in this field. Now lots of research results on undirected graceful graph were obtained[2], but few research results on directed graph were presented. Researches on gracefulness of directed graph are mainly done in directed circles by far, among which, the research results on n directed circles which have a common vertex are more. But result on n directed circles that have a common edge only was presented as follows: when $n \equiv 0(\bmod 2)$, then $n-\vec{c}_3$ is graceful[3]. In this paper, we will explore the gracefulness of digraph $n-\vec{c}_4$ with the help of computer.

Definition 1[3]. Let $G=(V,E)$ be a simple digraph with p vertices and q arcs, mapping $f: V(G) \to \{0,1,2\cdots,q\}$, let $f^*(u,v)=[f(v)-f(u)](\bmod(q+1))$, where $(u,v)\in E$, if f is an injection and f^*: $E(G) \to \{1,2,\cdots,q\}$ is a bijection, then f is called a graceful labeling of digraph G, and digraph G is graceful.

D. Jin and S. Lin (Eds.): Advances in EECM Vol. 1, LNEE 139, pp. 565–569.
springerlink.com

Definition 2. Digraph which composed by 2k directed circles $\vec{C_4}$ which have a common edge is written as $2k-\vec{c_4}$.

2 Algorithm Design for Graceful Labelings of $2k-\vec{c_4}$

Computer provides a powerful tool for exploring the graceful labeling of graph, this problems related to selecting m integers from n integers, its order of complexity is $n!$, so running the program need to spend a lot of time, thus we should optimize the algorithm by the characteristics of the graph to shorten the time. Now to give the algorithm for graceful labelings of $2k-\vec{c_4}$ as follows:

begin
input n $(n \geq 2)$

1) Execute cycle 1 : for $t=0$ to $\left[\frac{3n+1}{2}\right]+1$. (where t is the value for one vertex of the common edge, considered the symmetry of the vertices, t only need to value to half of $3n+1$)
2) if $t=0$, then generate a set $A_1=\{1,2,\cdots,3n+1\}$, select $2n+1$ numbers from A_1, and assign a sort of them in turn to $x_0, x_1, x_2, \cdots, x_n, y_1, y_2, \cdots, y_n$, dim a array $C[3n+1]$(where $C[i]=[f(v)-f(u)](\bmod(q+1))$ $(u,v)\in E$, $q=3n+1$), and let $C[0]=(x_0-t) \bmod(3n+2)$,

Execute cycle 2: for $k=1$ to n, Let
$C[3k-2]=(t-x_k) \bmod(3n+2)$;
$C[3k-1]=(y_k-x_0) \bmod(3n+2)$
$C[3k]=(x_k-y_k) \bmod(3n+2)$
End cycle 2
Judge $\{C[0], C[1],\cdots,C[3n]\}=\{1,2,\cdots,3n+1\}$?, if true, then print $t, x_0, x_1, x_2, \cdots, x_n, y_1, y_2, \cdots, y_n$.
Select another sort of $2n+1$ numbers above to reassign to $x_0, x_1, x_2, \cdots, x_n, y_1, y_2, \cdots, y_n$; repeat cycle 2, till all sorts of $2n+1$ from A_1 have been assigned.
3) if $t \neq 0$, then generate a set $A_2=\{0,1,2,\cdots,t-1,t+1,\cdots,3n+1\}$, select $2n+1$ numbers from A_2, next steps are similar to 2).

End all the cycles above, print all the graceful labelings of $n-\vec{c_4}$ with corresponding n.
End.

After the program running, When $n=2$, $2-\vec{c}_4$ has nearly two hundred kind of graceful labelings; When $n=3$, no solution, so $3-\vec{c}_4$ is not graceful; When $n=4$, tens of thousands of solutions were obtained. When $n=5$, the final datas can not be waited for it is too long. By classifying and comparing the datas from $n=2$ and $n=4$, we find that they all have the solutions with 0 on one vertex of the common edge , so we select all the datas with 0 to analyze (*here datas is too many, so this is not showed*) . At last , we obtained the law of labeling.

3 Results and Proofs

Theorem. Digraph $2k-\vec{c}_4$ is graceful (where k is an arbitrary natural number)

Proof. Let the vertices set of digraph $2k-\vec{c}_4$ be $V=\{x_i \mid i=0,1,2,\cdots,2k\} \cup \{y_i \mid i=0,1,2,\cdots,2k\}$, edges set be E , then $q=|E|=6k+1$. The vertices of the common edge of $2k-\vec{c}_4$ were written as x_0 , y_0 , the vertices of $\vec{C}_4^{(i)}$ were written in turn as $x_0\ y_0\ y_i\ x_i$. Now give a graceful labeling of $2k-\vec{c}_4$ as follows:

Let

$f(x_0)=0$, $f(y_0)=3k+1$.

when $1\le i\le k$, $f(x_i)=3k+1-i$, $f(y_i)=3k+1+i$.

when $k+1\le i\le 2k$, $f(x_i)=3k+1+i$, $f(y_i)=7k+2-i$.

It is easy to prove that the labelings of vertex above are different each other, and $\max\{f(v)\mid v\in V\}=q$, so mapping $f:V\to\{0,1,2\cdots,q\}$ is an injection.

Now let $f^*(u,v)=[\,f(v)-f(u)\,]\,(\mathrm{mod}(q+1))$,where $(u,v)\in E$. It is going to proved mapping f^* is a bijection.

According to the labeling above,

$f^*(x_0,y_0)=[\,f(y_0)-f(x_0)\,](\mathrm{mod}(q+1))=3k+1$

For $1\le i\le k$:

$f^*(y_0,y_i)=[\,f(y_i)-f(y_0)\,](\mathrm{mod}(q+1))=i$

$f^*(y_i,x_i)=[\,f(x_i)-f(y_i)\,](\mathrm{mod}(q+1))=6k+2-2i$

$f^*(x_i,x_0)=[\,f(x_0)-f(x_i)\,](\mathrm{mod}(q+1))=3\mathrm{k}+1+\mathrm{i}$

For $k+1\le i\le 2k$:

$f^*(y_0,y_i)=[\,f(y_i)-f(y_0)\,](\mathrm{mod}(q+1))=4\mathrm{k}+1\text{-}i$

$$f^*(y_i,x_i)=[f(x_i)-f(y_i)](\mathrm{mod}(q+1)) = 2k+1+2i$$

$$f^*(x_i,x_0)=[f(x_0)-f(x_i)](\mathrm{mod}(q+1)) = 3k+1-i$$

Obviously :

$$A_1=\{f^*(y_0,y_i)\mid 1\le i\le 2k\}=\{1,2,\cdots,k\}\cup\{2k+1,2k+2,\cdots,3k\}$$

$$A_2=\{f^*(x_i,x_0)\mid k+1\le i\le 2k\}=\{k+1,k+2,\cdots,2k\}$$

$$A_3=\{f^*(x_i,x_0)\mid 1\le i\le k\}=\{3k+2,3k+3,\cdots,4k+1\}$$

$$A_4=\{f^*(y_i,x_i)\mid 1\le i\le k\}=\{4k+2,4k+4,\cdots,6k\}$$

$$A_5=\{f^*(y_i,x_i)\mid k+1\le i\le 2k\}=\{4k+3,4k+5,\cdots,6k+1\}$$

So $A=A_1\cup A_2\cup A_3\cup A_4\cup A_5\cup f^*(x_0,y_0\} =\{1,2,\cdots,6k+1\}$

It is obvious that mapping $f^*: E\to\{1,2,\cdots,q\}$ is a bijection. therefore, f is a graceful labeling of $2k-\vec{c}_4$, namely the theorem is true.

Next we show an example to illustrate the labeling for $4-\vec{c}_4$ as follows :

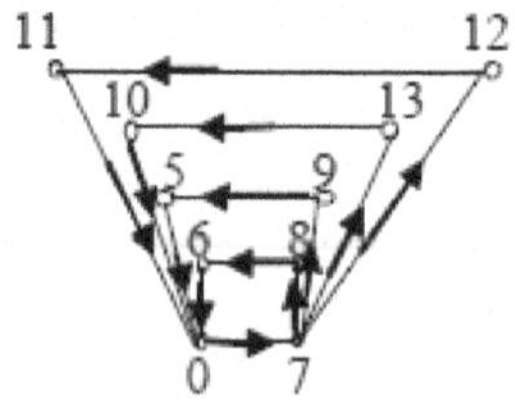

Fig. 1. Graceful labeling of digraph $4-\vec{c}_4$

According to the expressions above, $2k-\vec{c}_4$ has:

Property. for arbitrary natural number k, the sum of graceful value of all the edges of any directed circle $\vec{C}_4^{(i)}$ is $12k+4$.

Proof: According to the labeling mapping f above, for $1\le i\le k$, sum of graceful value of all the edges of any directed circle $\vec{C}_4^{(i)}$ is:

$$f^*(x_0,y_0)+f^*(y_0,y_i)+f^*(y_i,x_i)+f^*(x_i,x_0)$$
$$=[f(y_0)-f(x_0)](\mathrm{mod}(q+1)) +[f(y_i)-f(y_0)](\mathrm{mod}(q+1))$$
$$+[f(x_i)-f(y_i)](\mathrm{mod}(q+1)) +[f(x_0)-f(x_i)](\mathrm{mod}(q+1))$$
$$=3k+1+i+6k+2-2i+3k+1+i=12k+4$$

similarly, for $k+1\le i\le 2k$, it is also $12k+4$.

Proof end.

References

1. Golomb, S.W.: How to number a graph. In: Graph Theory and Computing, Academic Press, New York (1972)
2. Gallian, J.A.: A Dynamic Survey of Graph Labeling. The Electronic Journal of Combinatorics, Mol. Biol. 5, 1–148 (2005)
3. Graceful, M.J.: Graceful graph. Peking University Press, Beijing (1991)

Analysis of Nonlinear Circuit Subnet Tearing and Measurability

Yan Liu[1], Kai Wang[1], LiQiang Liu[2], and Li Cheng[1]

[1] Xi'an High-Tech Institute, Xi'an, 710025, China
[2] Inner Mongolia University of Technology, Hohhot, 010062, China

Abstract. Network torn is a key subject in circuit fault diagnose. This paper pays attention to partition condition and analyzing measurability of circuit containing nonlinear component. Study of subnet measurability laid the groundwork of further fault location.

Keywords: nonlinear, subnet tearing, measurability.

1 Introduction

Along with the development of circuit scale, fault diagnoses, based on subnet tearing, became a crucial subject gradually. More and more researchers worked on it and made some achievements [1-10]' among them, torn terminals accessibility are demanded [1-2]. Fault diagnose of subnet[4], breaks through restraint that all torn terminals should be accessible but is confined to the method of the consistent match and intercalibration between calculated value and measured ones. Partition conditions of solvable subnet those partial torn terminals inaccessible in nonlinear time invariant circuit are shown and on measurability of subnet is discussed [5]· The method of dummy accessible measurable nodes raised in [9] lowers the standards of nodes numbers in circuit diagnosing. [10] introduces the associating of tearing and BP net algorithm that reduces workload simplified test.

Based on [5], this paper presents partition conditions and discusses measurability of circuit subnet containing nonlinear components.

2 Partition Conditions of Nonlinear Solvable Subnet

Analog nonlinear time invariant circuit N is torn into some subnet, among which, solvability of subnet not containing nonlinear component is judged in [5]. Solvability of subnet containing nonlinear component is shown as follows:

Take for instance two-port component. Under the postulate of the connection between nonlinear ports and node j&k, torn nodes of connectivity subnet that contains nonlinear components include two parts: measurable set MT($mt \in MT$) and non-measurable set LT($lt \in LT$). Also, non-torn node include two parts: measurable set

D. Jin and S. Lin (Eds.): Advances in EECM Vol. 1, LNEE 139, pp. 571–576.
springerlink.com

UM(um∈UM) non-measurable set UU(uu∈UU). Obviously, m1=mt+um, m2=mt+lt. In analysis, first remove nonlinear components and list node voltage equation for linear part, then fill current supply i=f(u)(current direction is from node j to node k) according to substitution theorem. The analysis of nonlinear subnet solvability is as follows:

2.1 j, k Are Accessible and Torn

Therefore, subnet node voltage equation is:

$$\begin{bmatrix} Y_{LTLT} & Y_{LTMT} & Y_{LTUM} & Y_{LTUU} \\ Y_{MTLT} & Y_{MTMT} & Y_{MTUM} & Y_{MTUU} \\ Y_{UMLT} & Y_{UMMT} & Y_{UMUM} & Y_{UMUU} \\ Y_{UULT} & Y_{UUMT} & Y_{UUUM} & Y_{UUUU} \end{bmatrix} \begin{bmatrix} V_{LT} \\ V_{MT} \\ V_{UM} \\ V_{UU} \end{bmatrix} = \begin{bmatrix} I_{LT} \\ I_{XM} \\ 0 \\ 0 \end{bmatrix} + \begin{bmatrix} 0 \\ I_{MT} \\ I_{UM} \\ 0 \end{bmatrix} + \begin{bmatrix} 0 \\ F \\ 0 \\ 0 \end{bmatrix}$$

There into, F – nonlinear current supply added into linear subnet according to substitution theorem.

$$F = [0 \ \cdots \ 0 \ f(u) \ 0 \ \cdots \ 0 \ -f(u) \ 0 \ \cdots \ 0]^T \qquad (mt \times 1) \tag{1}$$

Eliminated voltage column vector Vuu, Measurable terminals equation is obtained while VLT、ILT、IXM are unknown variables:

$$HX = AZ + D = B + D \tag{2}$$

Expansion is:

$$\begin{bmatrix} Y_{LTLT} + Y_{LTUU}C_1 & -E & 0 \\ Y_{MTLT} + Y_{MTUU}C_1 & 0 & -E \\ Y_{UMLT} + Y_{UMUU}C_1 & 0 & 0 \end{bmatrix} \begin{bmatrix} V_{LT} \\ I_{LT} \\ I_{XM} \end{bmatrix} = \begin{bmatrix} -(Y_{LTMT} + Y_{LTUU}C_2) & 0 & -(Y_{LTUM} + Y_{LTUU}C_3) & 0 \\ -(Y_{MTMT} + Y_{MTUU}C_2) & E & -(Y_{MTUM} + Y_{MTUU}C_3) & 0 \\ -(Y_{UMMT} + Y_{UMUU}C_2) & 0 & -(Y_{UMUM} + Y_{UMUU}C_3) & E \end{bmatrix} \begin{bmatrix} V_{MT} \\ I_{MT} \\ V_{UM} \\ I_{UM} \end{bmatrix} + \begin{bmatrix} 0 \\ F \\ 0 \end{bmatrix}$$

There into: H- mixing coefficient matrix, matrix order is (p×q), p=m1+lt, q=m2+lt; B- known column vector,

$$C_1 = Y_{UUUU}^{-1} Y_{UULT} \ ; \ C_2 = Y_{UUUU}^{-1} Y_{UUMT} \ ; \ C_3 = Y_{UUUU}^{-1} Y_{UUUM}$$

Because j, k are accessible, F is known, right side of the equal sign is constant in equation above. Then, equation (1) is solvable as equation (1) that draw the conclusion 1.

Conclusion 1 Time-invariance analog network N, connecting to subnet N1 which contains one nonlinear component, is solvable when the followings conditions needs are met:

(1) m1≥m2
(2) rank($Y_{UMLT}+Y_{UMUU}C_1$)=lt (Namely, mixing matrix H is column filled).

In short, when j, k are accessible, or j, k are inaccessible and torn, N1 is solvable on condition that conclusion 1 is meet.

2.2 j, k Are Inaccessible and Not Torn

Equations obtained is as follows:

$$\begin{bmatrix} Y_{LTLT} & Y_{LTMT} & Y_{LTUM} & Y_{LTUU} \\ Y_{MTLT} & Y_{MTMT} & Y_{MTUM} & Y_{MTUU} \\ Y_{UMLT} & Y_{UMMT} & Y_{UMUM} & Y_{UMUU} \\ Y_{UULT} & Y_{UUMT} & Y_{UUUM} & Y_{UUUU} \end{bmatrix} \begin{bmatrix} V_{LT} \\ V_{MT} \\ V_{UM} \\ V_{UU} \end{bmatrix} = \begin{bmatrix} I_{LT} \\ I_{XM} \\ 0 \\ 0 \end{bmatrix} + \begin{bmatrix} 0 \\ I_{MT} \\ I_{UM} \\ 0 \end{bmatrix} + \begin{bmatrix} 0 \\ 0 \\ 0 \\ F \end{bmatrix}$$

There into, $F = [0 \ \cdots \ 0 \ f(u) \ 0 \ \cdots \ 0 \ -f(u) \ 0 \ \cdots \ 0]^T \ (uu \times 1)$

Eliminated part of variables in inner node voltage column vector V_{uu}, then:

$$H_l X_l = AZ = B \tag{3}$$

Expansion of equation is:

$$\begin{bmatrix} Y_{LTUU}Y_{UUUU}^{-1} & Y_{LTLT}+Y_{LTUU}C_1 & -E & 0 \\ Y_{MTUU}Y_{UUUU}^{-1} & Y_{MTLT}+Y_{MTUU}C_1 & 0 & -E \\ Y_{UMUU}Y_{UUUU}^{-1} & Y_{UMLT}+Y_{UMUU}C_1 & 0 & 0 \end{bmatrix} \begin{bmatrix} F \\ V_{LT} \\ I_{LT} \\ I_{XM} \end{bmatrix} = \begin{bmatrix} -(Y_{LTMT}+Y_{LTUU}C_2) & 0 & -(Y_{LTUM}+Y_{LTUU}C_3) & 0 \\ -(Y_{MTMT}+Y_{MTUU}C_2) & E & -(Y_{MTUM}+Y_{MTUU}C_3) & 0 \\ -(Y_{UMMT}+Y_{UMUU}C_2) & 0 & -(Y_{UMUM}+Y_{UMUU}C_3) & E \end{bmatrix} \begin{bmatrix} V_{MT} \\ I_{MT} \\ V_{UM} \\ I_{UM} \end{bmatrix}$$

Regard f(u) as two new variables z1, z2 while f(u) contains two unknown variables V_j、V_k , and eliminate column corresponding to coefficient matrix zero element in F, so, indirect accessible port equation is:

$$HX = AZ = B \tag{4}$$

Expansion of equation is:

$$\begin{bmatrix} D_1 & Y_{LTLT}+Y_{LTUU}C_1 & -E & 0 \\ D_2 & Y_{MTLT}+Y_{MTUU}C_1 & 0 & -E \\ D_3 & Y_{UMLT}+Y_{UMUU}C_1 & 0 & 0 \end{bmatrix} \begin{bmatrix} z_1 \\ z_2 \\ V_{LT} \\ I_{LT} \\ I_{XM} \end{bmatrix} = \begin{bmatrix} -(Y_{LTMT}+Y_{LTUU}C_2) & 0 & -(Y_{LTUM}+Y_{LTUU}C_3) & 0 \\ -(Y_{MTMT}+Y_{MTUU}C_2) & E & -(Y_{MTUM}+Y_{MTUU}C_3) & 0 \\ -(Y_{UMMT}+Y_{UMUU}C_2) & 0 & -(Y_{UMUM}+Y_{UMUU}C_3) & E \end{bmatrix} \begin{bmatrix} V_{MT} \\ I_{MT} \\ V_{UM} \\ I_{UM} \end{bmatrix}$$

New matrixes D_1, D_2, D_3 are obtained by eliminating zero elements column corresponding to F in equation (3) coefficient matrix respectively. The matrixes have 2 columns. The necessary and sufficient condition of equation (4) being solvable is rank(H)=rank(HB)=(q+2).

Conclusion 2 Time-invariance analog network N, connecting to subnet N1 which contains one nonlinear component when both of nonlinear component ports are not accessible and not torn, is solvable as followings conditions needs are met:

(1) $m_1 \geq (m_2+2)$;

(2) rank($[D_3 \ Y_{UMLT}+Y_{UMUU}C_1]$)=(lt+2) (Namely, mixing matrix H is column filled).

Sufficiency: When conditions (1), (2) are satisfied, matrix H expansion shows that all the columns of matrix H are linearly independent. Namely, rank(H)=(q+2). As shown by literature [6], when N1 is connectivity subnet, there must be rank(H)=rank(HB). To sum up the two points above, equation (4) has unique solution so that subnet N1 is solvable.

Necessity: Once N1 is solvable, equation (4) has unique solution, so, rank(H)=(q+2) and $m_1 \geq m_2+2$.When matrix H is column filled, all the columns of matrix H are linearly independent, namely, the first (lt+2) columns of H are linearly independent, so, rank($[D_3\ Y_{UMLT}+Y_{UMUU}C_1]$)=(lt+2)

Therefore, partition condition of solvable nonlinear subnet:

(1) Numbers of measurable nodes are not less than that of torn nodes in each linear subnet when being torn.

(2) In torn subnet containing nonlinear components, it is necessary to insure the numbers of accessible nodes not less than that of torn nodes when the both terminals of nonlinear components are accessible or torn. When the both terminals of nonlinear components are accessible or torn, numbers of accessible should be equal or greater than that of torn nodes plus 2.

(3) For each subnet, judge $[Y_{UMLT}+Y_{UMUU}C_1]$ or $[D_3\ Y_{UMLT}+Y_{UMUU}C_1]$ that whether it is column filled: if yes, the subnet is solvable; otherwise, not solvable.

3 Nonlinear Subnet Mensurability Analysis

Subnet is solvable and measurability of each subnet is analyzed when partition conditions above are met. Literature [5] represents measurability of linear subnet while this paper focuses on measurability of subnet contains nonlinear components. There must be left inverse of Matrix H when the subnet containing nonlinear components is solvable, equation (2), (4) can be simplified as:

$$X=H^{-L}AZ \tag{5}$$

Supposing that subnet as a multi input-out system, network function is $G=H^{-L}A$, then subnet measurability can be obtained by rank solving of Jacobi matrix for subnet component.

To sum up the points above, calculation summarization of partition and measurability analysis for subnet are as follows:

(1) Partition current into K subnets according to the structure. Each partitioned subnet should satisfy the partition conditions conclusion mentioned in literature [5]. Subnet containing nonlinear components should meet the needs of condition (1) in conclusion 1 according to both terminals of nonlinear component.

(2) For subnet J (J=1,2,…k), node voltage equation of notation format should be created according to improved method of nodes and substitution theorem.

(3) Judge $[Y_{UMLT}+Y_{UMUU}C_1]$ or $[D_3\ Y_{UMLT}+Y_{UMUU}C_1]$ that whether it is column filled: if yes, continue; if not, go to 1 redo partition, or add testing point inside subnet j, redo judgment.

(3) Calculate $G_j=H_j^{-L}A_j$ in each subnet.

(4) Calculate measurability T_j of each subnet.

4 Examples

Analog time invariant circuit N: nodes 1, 3, 5, 6 accessible, reference point 0. G_7 is nonlinear component whose characteristic equation is $i=f(u_2)$, $G_i=0.001(i=1,2,\ldots,14;\ i\neq 7)$. N1 is torn into linear subnet N1 (Fig.2) and subnet N1 containing nonlinear components G_7 (Fig.3). Judge the measurability of nonlinear subnet N2.

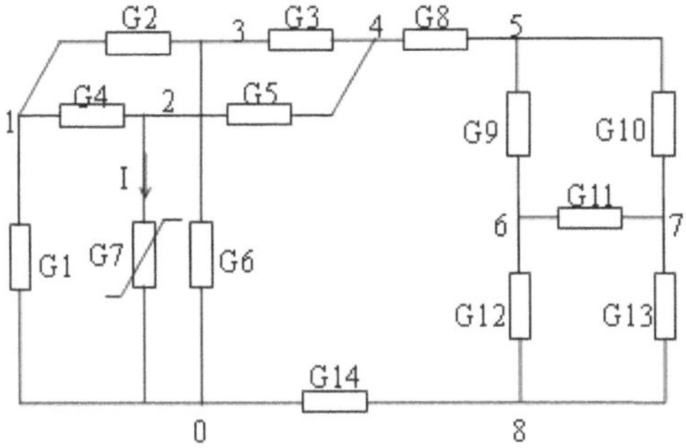

Fig. 1. Analog time invariant circuit N

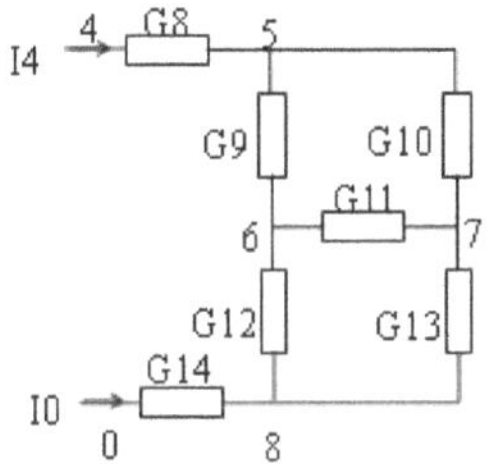

Fig. 2. Linear subnet N1

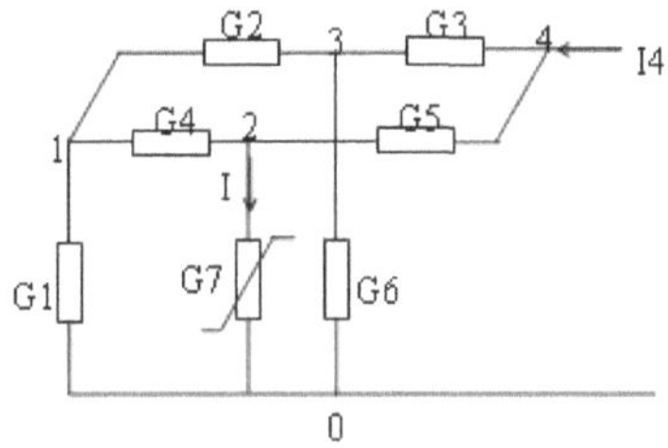

Fig. 3. Nonlinear subnet N1

shown in Fig.3, nodes 1 and 3 are accessible and not torn, node 4 is not accessible and torn. Characteristic equation is $i=f(u_2)=f(u_2)^2$ and current direction is from node 2 to 0. Take node 0 as reference point, subnet node voltage equation is:

$$\begin{bmatrix} G_3+G_5 & 0 & -G_3 & -G_5 \\ 0 & G_1+G_2+G_4 & -G_2 & -G_4 \\ -G_3 & -G_2 & G_3+G_4+G_6 & 0 \\ -G_5 & -G_4 & 0 & G_4+G_5 \end{bmatrix} \begin{bmatrix} V_4 \\ V_1 \\ V_3 \\ V_2 \end{bmatrix} = \begin{bmatrix} I_4 \\ 0 \\ 0 \\ 0 \end{bmatrix} + \begin{bmatrix} 0 \\ 0 \\ 0 \\ -f \end{bmatrix}$$

According equation (4), therefore: $H \cdot X = A \cdot Z$

Expansion:

$$\begin{bmatrix} G_4 & -G_1 G_5 & 0 \\ 0 & G_3+G_5 & -1 \\ 0 & G_3 & 0 \end{bmatrix} \begin{bmatrix} f \\ V_4 \\ I_4 \end{bmatrix} = \begin{bmatrix} -(G_1+G_2)(G_4+G_5)-G_4G_5 & G_2(G_4+G_5) \\ \dfrac{G_5(G_1+G_2+G_4)}{G_4} & G_3-\dfrac{G_2G_5}{G_4} \\ G_2 & -(G_3+G_4+G_6) \end{bmatrix} \begin{bmatrix} V_1 \\ V_3 \end{bmatrix}$$

Obviously, matrix H is column filled. So, subnet N2 is solvable:

$$X=H^{-L}AZ=GZ$$

Expansion:

$$\begin{bmatrix} f \\ V_4 \\ I_4 \end{bmatrix} = \begin{bmatrix} \frac{1}{G_4} & 0 & \frac{G_5}{G_3} \\ 0 & 0 & \frac{1}{G_3} \\ 0 & -1 & \frac{G_3+G_5}{G_3} \end{bmatrix} \begin{bmatrix} -(G_1+G_2)(G_4+G_5)-G_4G_5 & G_2(G_4+G_5) \\ \frac{G_5(G_1+G_2+G_4)}{G_4} & G_3-\frac{G_2G_5}{G_4} \\ G_2 & -(G_3+G_4+G_6) \end{bmatrix} \begin{bmatrix} V_1 \\ V_3 \end{bmatrix}$$

$$G = \begin{bmatrix} \frac{-G_3(G_1+G_2)(G_4+G_5)-G_3G_4G_5+G_2G_4G_5}{G_3G_4} & \frac{G_2G_3(G_4+G_5)-G_4G_5(G_3+G_4+G_6)}{G_3G_4} \\ \frac{G_2}{G_3} & -\frac{G_3+G_4+G_6}{G_3} \\ \frac{-G_3G_5(G_1+G_2+G_4)+G_2G_4(G_3+G_5)}{G_3G_4} & \frac{G_2G_3G_5-G_3^2G_4-G_4(G_3+G_5)(G_3+G_4+G_6)}{G_3G_4} \end{bmatrix}$$

rank G Jacobi matrix: rank(T)=6, so N2 measurable value T_2=6.

5 Concluding Remarks

Subnet fault diagnose, with engineering significance, is an accurate and rapid method in fault location for large scale analog circuit. Based on literature [5], this paper analyzes subnet measurability containing nonlinear components and proves the algorithm.

References

1. Tan, Y., He, Y., Chen, H., Wu, J.: Neural Network Method for Fault Diagnosis of Large-scale Analogue Circuits. Journal of Circuits And Systems 6(4), 25–28 (2001)
2. Zhang, Y., Huang, Y.: Fault Diagnosis of Single Sub-network for Black-box. Journal of Circuits and Systems 5(2), 28–32 (2000)
3. Zhang, C., Yang, J., Wang, Y.: The Research of the Fault Diagnose of the Electric Circuit Based on the Mixed Neural- network. Control & Automation 24(6-1), 169–170 (2008)
4. Yang, S.: Fault Diagnosis and Reliability Design of Analog System, pp. 132–137. Tsinghua University Press, Beijing (1993)
5. Wang, K., Liu, L., Cheng, L., Tian, Q.: Analysis of Analog Circuit Subnet Tearing and Measurability. Control & Automation 25, 125–126 (2009)
6. Huang, D.: Tearing Method for Fault Diagnosis in Analog Circuits. Acta Electronica Sinica 12(4), 28–34 (1984)
7. Cao, T., Peng, M., He, Y., Lv, J.: Neural Method of Fault Diagnosis at Subnetwork Level in Large Scale Analog Circuit. Microelectronics & Computer 21(12), 9–12 (2004)
8. Cai, J., Ma, X., Huang, D.: Fault Diagnosis For Networks By Neural Networks And Intersected-Torn. Mini-Micro Systems 21(7), 700–702 (2000)
9. Liu, B., Hu, C.: Fault Diagnosis in Analog Circuits Based on Virtual Accessible Testing Nodes. Electronics Optics & Control 16(4), 65–68 (2009)
10. Liu, M., Shi, J., Wu, Z.: Research on Fault Diagnosis of Large-scale Analog Circuit. Instrumentation Technology 3, 39–41 (2010)

Research on Data Integration of Smart Grid Based on IEC61970 and Cloud Computing*

ShaoMin Zhang, JingYan Wang, and BaoYi Wang

School of control and computer engineering, North China Electric Power University, Baoding, Hebei, 071003, China
{zhangshaomin,wangjingyan1013,wangbaoyi}@126.com

Abstract. At present, the demand of information integration in smart grid is very urgent and the phenomenon that various information systems of the power grid scattered and isolated from each other exists generally. A large number of information resources are difficult to share. Through the research on IEC61970 standards and data integration technology based on common information model CIM, we design the CIM model database based on HBase and develop the query processing engine on Hadoop which is an open-source platform based on cloud computing technology. Experiments proved that Hadoop based on cloud computing technology could integrate idle resources of power system and provide 'super-computing power' for the data integration platform of smart grid.

Keywords: Smart grid, Data integration, IEC61970, Cloud computing, Hadoop.

1 Introduction

Smart grid with accurate, fast, open, sharing information system as the foundation, this is the biggest difference with the traditional grid [1]. According to the specific needs of each business department, electric power enterprises respectively installed and developed different applications to achieve different functions in different period. The 'information island' is widespread [2]. China's smart grid construction is not to overthrow the current power grid operation mode, but on the basis of the traditional power grid to improve, develop and innovate.

CIM is a major component of IEC61970 and provide a standardized definition of power system. In literature [3], the East China Power Grid used techniques of CIM / XML model file exchange and model merging to achieve the sharing of information resources and a graded responsibility of maintenance between the superior and subordinate dispatching automation systems. In the literature [4], mainly for the

* This work is supported by Hebei higher education science research plan funding subject of research of information integration based on IEC61970 and the security techniques of smart grid. (Z2010290)

D. Jin and S. Lin (Eds.): Advances in EECM Vol. 1, LNEE 139, pp. 577–582.
springerlink.com

requirements of power grid dispatching automatic, the application of IEC61970 standard realized the integrated platform of dispatching automation.

In this paper, by using the CIM data model defined by IEC61970 standard established a global model for data integration platform. With the combination of cloud computing technology, it completed the mapping between conventional relational database of power system and the CIM model and designed the data integration model based on cloud computing.

2 Related Technologies

2.1 Cloud Computing

In view of the problems that the power system is facing, the introduction of 'cloud computing' to the power system will solve the problem of information and resource sharing in power dispatching system. Under the condition of wide area network hardware unchanged within the power system, the cloud computing system can maximize the integration of data and the processor resources of the current system to provide effective technical support for data integration of power dispatching center[5].

2.2 Open-Source Hadoop Platform and HBase

Hadoop is a distributed system infrastructure and open source platform based on cloud computing technology, developed by the Apache Foundation. Hadoop consists of two parts: a distributed file system HDFS (Hadoop Distributed File System) and MapReduce [6].

HBase database is based on the Hadoop project, is open-source implementation of Google's Bigtable. It is similar to Google's Bigtable [7].

3 Data Integration Platform for Smart Grid Based on CIM and Hadoop

3.1 Design of Data Integration Platform Based on CIM

Creating a database to storage and manage the CIM model is the core work of the platform's design, HBase is used to establish the model database, mainly to complete the following content.

(1) According to the criteria defined by CIM, to design structure of the database and map the CIM class package into database structure.

(2)To realize data platform's the bidirectional interface to various heterogeneous database.

(3) To realize mapping heterogeneous database schemas into the CIM model.

(4) To achieve global database's updates synchronized with the CIM.

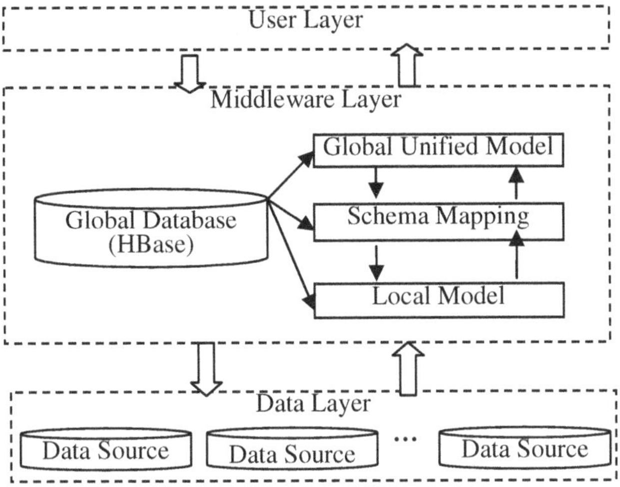

Fig. 1. Framework of Data Integration

3.2 Design of Global Database Based on CIM and HBase

The CIM is an abstract model which is based on the object-oriented idea and uses the object-oriented method to describe the complex information model in power system. It can be perfectly combined with object-oriented software [8-10]. In order to realize fast query and mass storage, in this paper, HBase is used to realize CIM.

The global database based on HBase is mainly responsible for the data locating and mapping when users interact with various heterogeneous database. In the global database, there are some tables that their design is very important.

Table 1. CIM Class

RowKey	Timetamp	Column 'Name'	Column 'Relation'	Column 'Attributes'
'Class_ID'		'ClassName'	' Relation :ParentClass ' 'Relation:Aggregation'	'Attributes :A0' … 'Attributes :An'

In order to follow CIM, each package's classes are realized with a table, naming table is consistent with the name of package. Table 1 is the implementation of CIM package in HBase which carries on the description to package's classes. The row key 'Class_ID' which is as a globally unique identifier identifies each class in CIM. Columns in column family 'Attributes' lists as the inherent attribute, the contents of these columns are strings which are formed by attribute name and the data type of attribute connected with the character '+', its inherited attributes can be obtained through the inquiry to its father class. When a new version of CIM appears, we only need to modify its father class. The synchronous update of subclasses' attributes is realized by inheritance. Moreover, such design can avoid redundant attributes' record. Column 'Relation:ParentClass' in 'Relation' indicates the father class of 'Class_ID'. Column 'Relation:Association' indicates the classes associated with 'Class_ID'. Column 'Relation:Aggregation' indicates partial classes that 'Class_ID' as the integral class aggregates . Such design solved the relations of inheritance, association and aggregation between CIM classes.

Table 2. Attributes Mapping

RowKey	TimeStamp	Column DB_ID('SQL_ID')		Column DB_ID ('Orcale_ID')		...
'Class_ID'+'+ Attr_Name'		'DB_ID :tb0'	'Ci'	'DB_ID :tb0'	'Ci'	
		...		...		
		'DB_ID :tbn'	'Ci'	'DB_ID :tbm'	'Ci'	

Table 2 mainly complete the mapping between attributes of CIM classes and the table fields of various databases. Columns in 'DB_ID' indicate the tables of database 'DB_ID' corresponding to the attributes of CIM's classes, the content of the column is the field of the tables which correspond to this attribute. This table's row key is the string that is composed of 'Class_ID' and 'Atrr_Name' by character '+'.

4 The Design of New Integration Platform's Query Engine

The data integration framework is shown in figure 1, the query engine in middleware layer is the most important parts of the model, Users' query is decomposed to generate executable queries, and then the engine will pass the data extracted according to custom format to users, the structure is shown in figure 2.

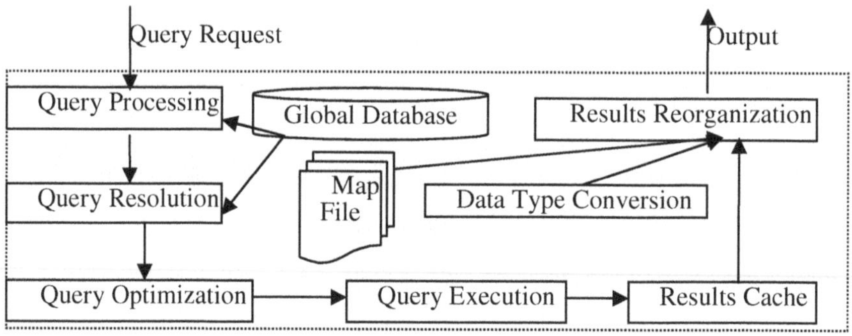

Fig. 2. Query Engine of Integration Platform

The processing engine is developed with the use of object-oriented technology in Hadoop platform. Data can be extracted from multiple data sources according to users' queries. Users' access to information and operations do not directly act on the data source, users' transparent access can be realized through the global database based on HBase. The major function of the middleware is described as follows.

(1) Query resolution module analyzes the information which processed by query processing module and returns the relevant data information and query sentence.

(2) Query execution module is responsible for the decomposed sub-queries assigned to each data source, and accepts the query results from each data source and caches query results.

(3) Result reorganization module is responsible for the integration of the sub-query results, and then passes the results in accordance with the users' custom data format.

5 Experiment and Analysis

In order to verify the HBase's efficient management on large data sets, the experiment has built a Hadoop cluster. In this experiment, Hadoop cluster has three computers that their configuration is the same, each computer's memory is 2G, hard disk space is 160G, and the operating system is Ubuntu 10.10. These three computers have installed hadoop-0.20.2, jdk-6u22-linux-i586 and hbase-0.20.6.

In order to inspect the performance of HBase and SQL Server 2005, Data import operation under the different scale is made on the cluster system. The experiment results are shown as Figure 3.The Y-coordinate of the line chart said time, unit MS. The horizontal coordinate of it shows the data size.

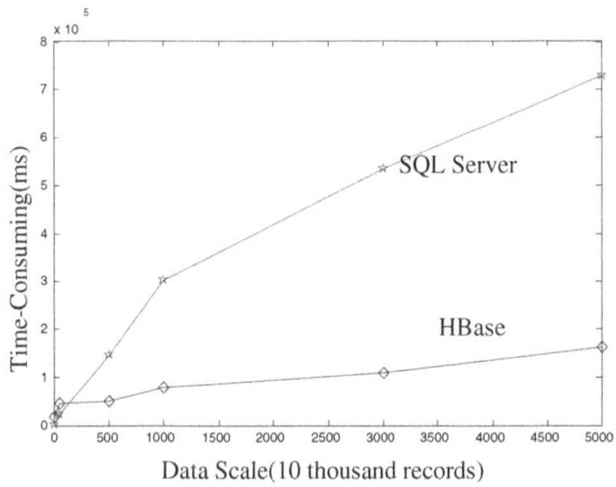

Fig. 3. Graphic about comparison between experimental results of different data scale

Through the contrast, we can see that with the increase in the amount of data, data import efficiency based on the HBase data management system is far ahead of traditional database system. Moreover along with the amount of data rise, the traditional single-machine way will even have the possibility of downtime, but Hadoop platform will be a good way to solve the problem. If the quantity of server nodes increases to hundreds, even thousands more, the use of distributed way to process mass data will have greater speed promotion.

6 Conclusion

This paper studied the architecture of data integration platform based on CIM, combined with Hadoop which is the implement of cloud computing technology, the CIM model is realized on HBase and use HBase to manage the metadata. The cloud computing technology can maximize the integration of calculation and storage capacity, reduce the power grid's redundant investment and increase the performance of current system. The experiment carried on the test to the performance of small Hadoop cluster, has verified the superiority of the platform in processing large data sets. Considering large quantity and diversity of data source in the actual grid, the implementation of this project will encounter more problems, such as safety problems. Subsequent work is to study the safety of the data integration platform based on cloud computing.

References

1. Jiang, D., Shen, T., Gang, U., Li, H., et al.: Significance and Roles of Standardized Basic Information in Developing Smart Grid. Automation of Electric Power Systems 33(20), 1–5 (2009) (in Chinese)
2. Lin, F., Hu, M., Jiang, Y., et al.: Architecture and Related Techniques of a Power Dispatching Data Platform. Automation of Electric Power Systems 31(1), 61–64 (2007) (in Chinese)
3. Qian, J., Ge, M., Mi, W., et al.: Management of East China Power Grid Hierarchy Model Based on IEC61970. Power System Technology 31, 295–298 (2007) (in Chinese)
4. Lu, Z.-Q., Wang, H.-C., Zhang, Z.-W., et al.: Open Dispatching Automation System Integration Platform Based on IEC 61970. Power System Technology 30, 89–92 (2006) (in Chinese)
5. Baker, S.: Google and Wisdom of Clouds(EB/OL) (December 13, 2007), `http://www.businessweek.com/magazine/content/07_52/b4064048925836.htm`
6. `http://hadoop.apache.org/common/docs/current/hdfs_design.html`
7. `http://hadoop.apache.org/hbase/`
8. Ding, Y., Ding, M., Bi, R., et al.: Research on CIM/XML model of microgrid system. Power System Protection and Control 38(9), 37–41 (2010) (in Chinese)
9. Ding, M., Yang, W., Zhang, Y., et al.: IEC61970 based MicroGrid energy management system. Electric Power Automation Equipment 29(10), 16–19 (2009) (in Chinese)
10. Xing, J., Yang, H.: A Novel CIM Data Processing Method for Application Systems. Automation of Electric Power Systems 34(18), 46–49 (2010) (in Chinese)

Realization Distributed Access Control Based on Ontology and Attribute with OWL*

ShaoMin Zhang, HongBian Yang, and BaoYi Wang

School of Control and Computer Engineering, North China Electric Power University, Baoding, Hebei, 071003, China
{Zhangshaomin,wangbaoyi}@126.com, yanghongbian2008@163.com

Abstract. For the problems of attribute elements maintenance difficulty and heterogeneous attribute of multi-domain in ABAC model, we propose the method of using ontology to maintain access control elements and distributed attribute management, which describe the logical relationships among attributes with OWL and introduce attribute mapping technology in access control decision-making. It can reduce the complexity of attribute management and raise the security of the cross-domain access. It makes up the defects in original ABAC model, and has a good reference to research the cross-domain access control.

Keywords: Ontology, OWL, attribute management, ABAC, attribute mapping.

1 Introduction

Attribute-based access control (ABAC) model is an authorization access control mechanism that makes decision based on the attributes of related entities. ABAC can make fine-grained access control, it is sufficient flexibility and scalability, can well adapt to an open, dynamic network environment, and it is a commonly method [1] of authentication and authorization in the distribution systems.

The research of ABAC model is mainly based on attribute certificates[1] (AC) recently, it binds the user name with one or more attributes by AC. Literature[1,2] combination attribute certificate with the policy set to achieve authorization management, which makes access control more flexible. However, maintain and management the attribute certificates are very complex for large-scale systems and authority is not accurate for heterogeneous attributes in cross-domain.

Literature [3] described resource metadata and user attributes with OWL, but it was only to the access control of a single domain. Literature[4] proposed the access control model that improved the cross-domain access security, but it took heavy workloads to maintain access control elements of attribute certificates forms. Literature[5] proposed a method to solve the problem of attribute heterogeneous in multi-domain by extracting the common public attributes to construct general attribute ontology, although for non-public attribute exists defects, but for the distributed property management has certain reference significance.

* This work is supported by Hebei higher education science research plan funding subject of research of information integration based on IEC61970 and the security techniques of smart grid. (Z2010290)

D. Jin and S. Lin (Eds.): Advances in EECM Vol. 1, LNEE 139, pp. 583–588.
springerlink.com

In response to these problems, we propose a method of using ontology to maintain access control elements and distributed attribute management in this paper. It introduces attribute mapping technology in decision-making to solve the problem of multi-domain attributes heterogeneous through establishment the association of cross-domain attributes.

2 Ontology and Description Language OWL

Web Ontology Language [6] (OWL) is the Semantic Web ontology description language standard which is recommended by the World Wide Web Consortium (W3C). OWL can declare classes, attributes and their hierarchy and define a class through attribute constraint. OWL also can declare the attribute is transitive, symmetric, function and so on. OWL has three subsets, OWL Lite, OWL DL and OWL Full. Considering the knowledge expression and the efficiency of inference we use OWL DL to describe ontology.

At present, the most popular OWL ontology editor tool is protégé of the Stanford University. It provides not only a graphical editor interface, but also includes much plug-ins to visual display the various relations of the ontology described. Syntax symbols in the OWL expressions as shown in Table 1.

Table 1. Protégé of symbolic logic

OWL element	symbol	OWL element	symbol
Owl:allValuesFrom	$\forall$	Owl:cardinality	$=$
Owl:someValuesFrom	$\exists$	Owl:intersectionOf	$\cap$
Owl:hasValue	$\ni$	Owl:unionOf	$\cup$
Owl:minCardinality	$\geq$	Owl:oneOf	$\{\ \}$
Owl:maxCardinality	$\leq$		

3 Realization of Distributed Access Control Management Based on Ontology and Attribute with OWL

3.1 Structure Attributes Library with Protégé

We establish attribute library for the entities with protégé. ABAC considers attribute as the basic elements of access control, which has flexible expression. Attribute is the characteristic of entity which includes subject, resource, environment and behavior. And correspond to the 4 core classes in the ontology, as shown in figure 1.

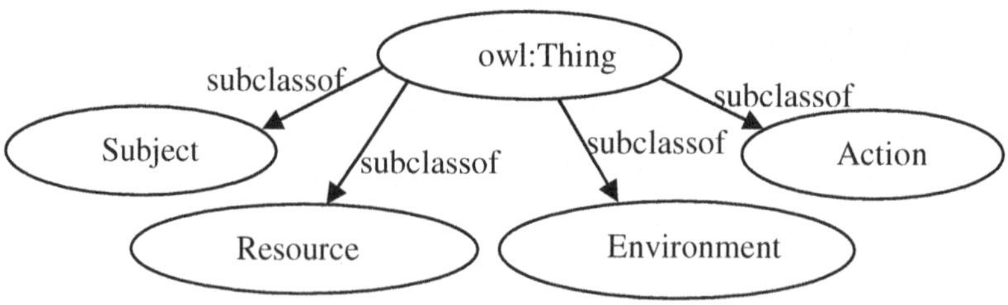

Fig. 1. Core classes of model

Experts of domain and safety predefine entities' attributes according to the actual demand. Attributes (A) are composed by attributes type (A_T) and attributes value (A_V.) A_V can be determined by the application environment. Figure2 describes a simply define of subject attributes, including domain, roles, jobs, duties, IP address, and so on, role as one of subject attributes.

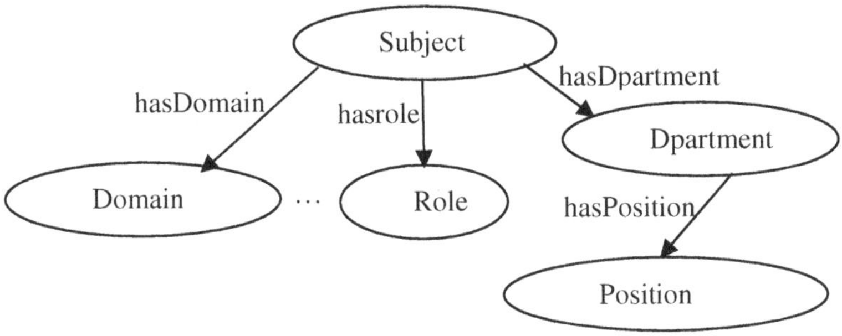

Fig. 2. Subjects Attribute Ontology

We can also restrain resources, environment and other attributes of the entity. Specific users and resources of the domain are the set of instances in ontology. It can automatically generate OWL format file through the protégé, which can solve the problem of ontology storage.

3.2 Realization of the Ontology Attribute Query

Jena provides a programming environment for parsing OWL ontology, which is a Java-based open source developed by HP. Jena API can directly fulfill ontology query function in ontology engineering. The steps of query entities attributes are as follows.

Step1: Using the functions of ModelFactory.createOntologyModel(), read() in Jena API to achieve the read of owl file. Codes are as follows: OntModel=ModelFactory.createOntologyModel(OntModelSpec.OWL_MEM); OntModel.read("file: data // AC.owl"); the ontology mode in OWL file is loaded into the ontModel, after the queries are done on the ontModel.

Step2: According to the ontology access information and the requested resources function, the entity class of access information can be got by calling the listClasses().

Step3: Calling listInstances() method, all individuals of the class can be got. Using getLocalName() to read the current class of instance name function and recycle method to judge the individual existence.

Step4: Calling listDeclaredProperties() method to obtain the individual all attributes and getPropertyValue (P) function to obtain the attribute value.

3.3 Design of Attribute Mapping Algorithm in Cross-Domain

Attribute mapping method means to calculate the similarity degree between attributes. Literature[7] used the similarity matrix to calculate attributes similarity degree, but it did not consider some semantic relations between attributes. Making full use of the semantic information among attributes, Literature[8] calculated similarity degree from the data type attributes and object attributes respectively, as follow the formula (1) and (2).

(1) Data type attribute comprehensive similarity degree is:

$$\text{dataTypeSim} = \frac{S_n + S_d + S_r + S_t}{n} \tag{1}$$

Among them, considering 4 aspects of characteristics: name (S_n), domain (S_d), range (S_r) and rdftype (S_t). n ($1 \leq n \leq 4$) is two data type attribute number of smaller value.
(2) The object attribute comprehensive similarity degree is:

$$\text{objectSim} = \frac{S_e + S_i + S_{dif} + S_d + S_r + S_t + S_s}{m} \tag{2}$$

Among them, considering the characteristics in 7 aspects: equal attribute (S_e), inverse attribute (S_i), different attribute (S_{dif}), domain (S_d), range (S_r), rdftype(S_t) and sameAs (S_s). m ($1 \leq m \leq 7$) is the smaller number of two objects attributes.

Formula (1) and (2) used equal weight method to set the weight in the calculation of attributes similarity, which ignored the effect of different factors, and can not reflect the difference of different factors. This paper introduces the sigmoid function to calculate the weight which can avoid the defects that bring by equal weight method or artificial set of weights. The following is the form of the sigmoid function[9]:

$$f_{(x)} = \frac{1}{1 + e^{-8(x-0.5)}} \tag{3}$$

Among them, $0 \leq x \leq 1$, x is similarity value of each factor, $f_{(x)}$ is the initial weight of each factor.

Obtained the each factor weight through the sigmoid function, then weighted and combined. The formula (1) and (2) can be expressed by the following formula (4):

$$\text{Sim}_{(A,B)} = \sum_{i=1}^{m} \left(\frac{f_{(i)}}{\sum_{i=1}^{m} f_{(i)}} \text{Sim}_{i(A,B)} \right) \tag{4}$$

Among them, $1 \leq i \leq m$, m is the number of influence factors.

3.4 Process of Cross-Domain Attribute Mapping

We use ontology library to provide the entity attribute information for the policy decision point (PDP) based on Literature[4]. PDP queries attributes from the ontology library according to the request, queries policies from the Policy library and makes decision. Attribute mapping only for the cross-domain access control decision-making. There are three main steps: Attribute extraction, Similarity computation, Attribute-mapping. Existing domain-A and domain-B are two security domains,

Figure 3 shows the mapping process of the requests information of domain-A's user and the entity attribute information of domain B. PDP-B on behalf of the policy decision point of domain B.

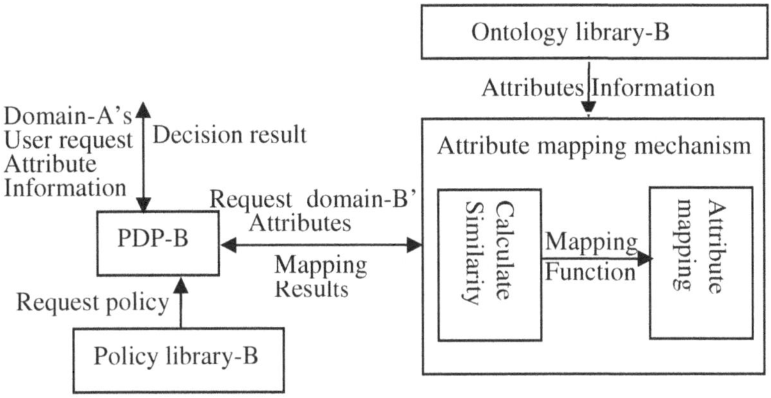

Fig. 3. The process of attribute-mapping in cross-domain

(1) Attribute extraction
The domain-A user's request information to the PDP-B, PDP-B will query attributes related with subject and resource by the function provided with Jena API from ontology library-B according to this request information.
(2) Similarity computation
Calculate the similarity value of each factor of the each attribute of access request information respectively, then through introduction of sigmoid function to set the weights of different factors to get overall similarity value of the attribute.
(3) Attribute mapping
General attribute similarity value is compared to the threshold value that attribute mapping function set, if the similarity value is greater than the threshold value is mapped and passed the results to PDP-B. If no mapping, that is the requested user of domain-A has not yet been authorized in domain-B then access to suspension; else, PDP-B will further queries security policy from policy library and to determine whether allows this cross-domain access.

4 Analysis of Performance

In summary, the method of using ontology to maintain access control elements and distributed attribute management has the following advantages:

(1) OWL can express the logical relationship among attributes, which meet the requirements of the semantics and simplified the management work of the attribute elements of the original model.

(2) The design of distributed attribute management can take advantage of attribute information of the each entity, and avoid the defect of using public attribute set.

(3) It not only simplifies the authorization policy management of heterogeneous attributes, but also raises the security of the cross-domain access, through the attribute mapping mechanism in cross-domain access control decision-making.

5 Conclusion

Multi-domain security interoperability is one of the key problems when distributed system fully and safely play roles. We propose the method of using ontology to maintain access control elements and distributed management. It simplifies the management of access control elements, and raises the security of the cross-domain access. In future work, there are still some issues to be further analysis and improvement, such as improved attribute similarity algorithm of the cross-domain to improve the accuracy of the attribute mapping, etc.

References

1. Chen, K., Lang, B.: Attribute Based Access Control Method Oriented to Hierarchical Resource. Computer Engineering 36(7) (2010) (in Chinese)
2. Zheng, Q.-Y., Shen, J.-J.: Access control model for web services security based on XACML. Computer Engineering and Design 28(16) (2007) (in Chinese)
3. Shen, H.: A semantic-aware attribute-based access control model for web services. In: Proceedings of the 9th International Conference on Algorithms and Architectures for Parallel Processing, pp. 693–703 (2009)
4. Zou, X., Jin, B., Ni, L.-S.: Research on cross-domain access control and border protection technique. Applications Research of Computers 27(4) (2010)
5. Hu, D.-Y., Zhang, B.: Attribute research of ABAC based on semantic in grid environment. Computer Engineering and Design 31(14) (2010)
6. Patel-Schneider, P.F., Hayes, P., Horrocks, I.O.: Web ontology language semantics and abstract syntax (EB/OL) (2004), http://www.w3.org//TR//owl-semantics
7. Nie, G.-H., Zuo, X.-R., Chen, D.-L.: Improved concept similarity computing approach in Ontology mapping. Computer Applications. 28 (6) (2008) (in Chinese)
8. Zhang, J.-X., Fang, L.-Y., Yan, J.-Z., Wang, P.: Ontology Mapping Approach Based 0n Property Structure. Computer Engineering 36(7) (2010) (in Chinese)
9. Xu, Q., Peng, J.-Y., Li, Z.: Integrated concept similarity computing method in ontology mapping. Computer Engineering and Applications 46(24), 34–36 (2010)

Attribute of SMB and HML as Non-systemic Risk Factors: An Empirical Study on CAPM Residuals*

XiaoGuang Lu, Tao Cai, and QingChun Lu

Business School, HoHai University, Nanjing 210098, China

Abstract. Based on basic assumptions and related corollaries about attribute of SMB and HML as non-systemic risk factors, considering CAPM residuals as non-systemic risk premium, proceeding cointegration ananlysis on CAPM residuals, SMB and HML, this paper sets up a long-term equilibrium model on them, revealing that SMB and HML are non-systemic risk premium factors, because SMB and HML can well explain the information about non-systemic risk premium contained in CAPM residuals.

Keywords: scale risk premium, value risk premium, SMB, HML, non-systemic risk factor.

1 Introduction

William Sharpe(1964)、John Lintner(1965) and Jan Mossin(1966) presented Capital Asset Pricing Model(CAPM) respectively.The theory regards systematic risk as the only risk factor, which is also named Market Risk Premium (MRP). MRP is the difference between portfolio expected return and riskless rate. The theory states that non-systemic risk can be eliminated by diversification. So we can made portfolio containing enough different kinds of stocks in order to eliminate non-systemic risk, which could be neglected and contained in residuals.

According to the study on stock return of American listed companies by Fama and French (1992), cross-sectional datas about stock return of different companies are very different. Share capital market value and book-to-market can well expain the marked difference in stock return of different companies, so they are regarded as risk premium, which is omitted in CAPM.

Fama and French(1993) used multifactor model combined with cross-section regression and analyzed all the listed non-financial companies in New York Stock Exchange, American Stock Exchange and NASDAQ from1963 to1990. After careful empirical research, they found that besides CAPM and MRP, scale and book-to-market can also expain the stock return differences.So SMB (small minus big), which means scale risk premium, is defined as the average return difference between big company portfolio and the small company portfolio. Moreover, HML (high minus low), which

* This Paper is supported by the Fundamental Research Funds for the Central Universities (2009B23314), and Funds for Philosophy and Social Science Research Projects of Jiangsu Province（2010SJD630039).

D. Jin and S. Lin (Eds.): Advances in EECM Vol. 1, LNEE 139, pp. 589–593.
springerlink.com

means value risk premium, is defined as the average return difference between high book-to-market portfolio and low book-to-market portfolio.The three-factor model of Fama-French cantains three risk factors(MRP, SMB and HML) as the follwing equation:

$$R_{it} - R_{it} = \alpha_i + b_i\left(R_{mt} - R_{ft}\right) + s_i SMB_t + h_i HML_t + \varepsilon_{it} \qquad (1)$$

As shown, i stands for the ith portfolio; t stands for time; R_f stands for risk free rate; R_m stands for portfolio return; ε_i stands for residual. The three factor model of Fama-French mades pricing model statistically significant based on cross-sectional data. Fama and French did not analyze the essential distinction among SMB, HML and MRP from the viewpoint of non-systematical risk, so they did not clearly indicate the non-systemic risk attribute of SMB and HML.

This paper argues that in order to make clear the non-systematical risk attribute of SMB and HML, it is necessary to analyze the essential distinction among SMB, HML and MRP. So based on the three-factor model of Fama-French, viewing CAPM residuals as non-systematical risk premium, presenting basic assumptions and related corollaries about attribute of SMB and HML as non-systemic risk factors, this paper chooses reports and data of listed companies from Shanghai Stock Exchange for empirical study. In analysis, make 6 portfolios according to scale of corporation and book-to-market, calculate the aggregate price index using modified DVI (dynamic volume index) method and then figure out portfolio return, risk premium, SMB and HML. This paper analyzes attribute of SMB and HML as non-systemic risk factors by proceeding cointegration ananlysis and regression analysis on CAPM residuals, SMB and HML.

2 The Basic Assumptions about Attribute of SMB and HML as Non-systemic Risk Factors

Assumption 1: SMB and HML are non-systemic risk factors which help achieve equilibrium regression.

Corollary 1: There is significant cointegration relationship among SMB, HML and CAPM residuals.

Assumption 2: SMB and HML of portfolio can expain CAPM residuals well.

Corollary 2: The regression model of SMB, HML and CAPM residuals is significantly effective.

Assumption 3: non-systemic risk premium of SMB and HML of different portfolios are very different.

Corollary 3: In CAPM residual model, SMB and HML of different profolios are different in regression coefficient and significance level.

3 Cointegration Analysis on CAPM, Its Residuals, SMB and HML

Based on the three-factor model of Fama-French, according to the difference between systemic risk factors and non-systemic risk factors, divide the three-factor model of

Fama-French into two models. The first one is CAPM, which reflects the regression relation between systemic risk factor MRP and portfolio risk premium $PRP_{SMB,HML,t}$. The other regression model, taking CAPM residuals as dependent variable, is to analyze if SMB and HML can well explain CAPM residuals and study the differences of explanatory capacity of different portfolios . As the following:

$$PRP_{SMB,HML,t} = C_{1,SMB,HML} + b_{SMB,HML} MRP_t + \xi_{SMB,HML,t} \tag{2}$$

$$\xi_{SMB,HML,t} = C_{2,SMB,HML} + s_{SMB,HML} SMB_t + h_{SMB,HML} HML_t + \varepsilon_{SMB,HML,t} \tag{3}$$

In the equation (2), SMB stands for scale risk premium, which takes B (big scale) and S (small scale); HML stands for value risk premium, which takes H (high book-to-market), M (moderate book-to-market) and L (low book-to-market). $MRP = R_m - R_f$ stands for market risk premium. R_f stands for risk free rate, which is monthly return calculated on compound interest from three-month deposit rate published by People's Bank of China. R_m stands for market portfolio return, which is first-order differential of natural logarithm of Shanghai composite index. $PRP_{SMB,HML,t}$ stands for a portfolio risk premium.

Take shanghai stock exchange A share listed companies as samples, not including companies with negative equity and financial companies. Their reports and exchange data from 2002 May to 2010 Jun were used. Regard equity minus deferred tax debits plus deferred tax credits as book value. Regard stock market value as circulation market value.Calculate BH、BM、BL、SH、SM and SL of six portfolios during the 97 months using modified momentum index method .Then, we can figure out the time series data of R_{BH}、R_{BM}、R_{BL}、R_{SH}、R_{SM} and R_{SL} during the 96 months by calculating the first-order differential of natural logarithm of the price index as follows:

$$R_{SMB,HML,t} = Ln\left(I_{SMB,HML,t}\right) - Ln\left(I_{SMB,HML,t-1}\right) \tag{4}$$

In equation (4), $I_{SMB,HML}$ stands for monthly price index of the portfolio. $R_{SMB,HML}$ stands for monthly return， t stands for month.

Using the rate of retrturn of the six portfolio: R_{BH}、R_{BM}、R_{BL}、R_{SH}、R_{SM} and R_{SL}, figure out non-systemic risk factors SMB and HML according to $SMB_t = \left[\left(R_{SH,t} + R_{SM,t} + R_{SL,t}\right) - \left(R_{BH,t} + R_{BM,t} + R_{BL,t}\right)\right]/3$ and $HML_t = \left[\left(R_{SH,t} + R_{BH,t}\right) - \left(R_{SL,t} + R_{BL,t}\right)\right]/2$.

Using Johansen multiple cointegration test, process cointegration test on CAPM residuals of six portfolios(ξ_{BH}、ξ_{BM}、ξ_{BL}、ξ_{SH}、ξ_{SM}、ξ_{SM} and ξ_{SL}) and non-systemic risk factors (SMB and HML). Trace Statistic and Max-Eigen Statistic are included in table 1.

Table 1. Cointegration test on SMB, HML and CAPM residuals of six portfolios

Number of cointegration	ξ_{BH}	ξ_{BM}	ξ_{BL}	ξ_{SH}	ξ_{SM}	ξ_{SL}
Trace Statistic						
None	101.849***	97.255***	96.836***	97.553***	96.705***	101.595***
At most 1	57.975***	55.418***	56.617***	56.985***	55.741***	58.509***
At most 2	25.818***	23.202***	23.920***	24.285***	23.169***	25.995***
Max-Eigen Statistic						
None	43.874***	41.838***	40.219***	40.568***	40.965***	43.086***
At most 1	32.158 ***	32.216***	32.697***	32.699***	32.572***	32.514***
At most 2	25.818***	23.202***	23.920***	24.285***	23.169***	25.995***

***, ** and * respectively stand for significant level 0.01, 0.05 and 0.1.

In table 1, for six portfolios, all the test statistic values of Trace Statistic and Max-Eigen Statistic are significant at level 0.01. Besides, there are three significant cointegration relationships aming SMB, HML and CAPM residuals. Thus, Corollary 1 can be concluded: there is significant cointegration relationship among CAPM residuals, SMB and HML .The conclusion provides a basis for analysis on the attribute of SMB and HML as non-systemic risk factors.

4 The Long-Term Equilibrium Analysis on CAPM Residuals, *SMB and HML*

In table 2, according to equation (3), use software to estimate the parameters of CAPM residual model about six portfolios. Modified R^2 and F-statistic are also included in table 2.

Table 2. Regression model on CAPM residuals, SMB and HML

CAPM residuals	ξ_{BH}	ξ_{BM}	ξ_{BL}	ξ_{SH}	ξ_{SM}	ξ_{SL}
C_2	0.000	0.003	0.005	0.001	0.003	0.004
s	0.267	0.366*	0.380*	1.464***	1.194***	1.329***
h	0.047	-0.712***	-1.170***	-0.199	-0.651***	-0.962***
Modified R^2	0.004	0.277	0.440	0.416	0.527	0.641
F-statistic	1.170	19.162***	38.261***	34.798***	53.998***	85.756***

***, ** and * respectively stand for significant level 0.01, 0.05 and 0.1.

In table 2, only thc residual model of PRP_{BH} is not significant. The rest are all significant. The modified R^2 vulue of the CAPM residual model of PRP_{BH} is just 0.004. The regression coefficient s of SMB, the regression coefficient h of HML and their F-statistic are not significant, meaning that the SMB and HML of the portfolio of big company and high book-to-market can not explain CAPM residuals well, not having significant scale effect and value effect.

Corollary 2 and corollary 3 can be concluded fron the regression analysis. The portfolio CAPM residual model is significant; in model, some of regression coefficient of SMB and HML are significant, under certain conditions about portfolio, SMB and HML can explain CAPM residuals well. Thus, for portfolio of different company scale and book-to-market, explanatory capacity of SMB and HML for CAPM residuals is different in terms of value and significance.

5 Conclusions

First, SMB and HML are non-systemic risk factors, which help achieve equilibrium regression.

Second, SMB and HML are non-systemic risk factors differing from MRP essentially.

Last, as non-systemic risk factors, SMB and HML of different portfolio are different.

References

1. Fama, Eugene, F., French, K.R.: The cross-section of expected stock returns. Journal of Finance 47, 427–465 (1992)
2. Fama, Eugene, F., French, K.R.: Common risk factors in the returns on stocks and bonds. Journal of Financial Economics 25, 23–49 (1993)

Structuring of Incentive Mechanism of Human Resource of Scientific and Technical Periodical

Ling Shen

Journal Editorial Department of Changchun University, Changchun, China
linglong1968@126.com

Abstract. In order to realize sustainable development, the scientific and technical periodical press shall enhance the optimum usage of human resources and structure the scientific and effective incentive mechanism. In structuring the incentive mechanism, the periodical press shall integrate the substance incentive with spiritual drive, and the positive incentive with negative incentive, and follow the principle of fairness, moderation, dynamism and differentiation, it shall set out the incentive strategy mainly including the salary incentive, promotion incentive, training incentive and emotional encouraging.

Keywords: Scientific and Technical Periodical, Human Resource, Incentive Mechanism, Principle, Strategy.

1 Introduction

As far as the operation of the scientific and technical periodical press is concerned, the human resource, information resource and material resource are the most essential production factors, among them, the most important one is human resource which has a dominant function. The development and utilization of information resource, material resource will ultimately depend on the quality level and ability development of human resource. Therefore, the optimum usage of human resource is very important in the resource development and optimum allocation of the periodical press, the scientific and technical periodical shall enhance the development and management of human resource, especially shall set up the scientific and effective human resource incentive mechanism, so as to achieve sustainable development.

2 Summary of Theories of Motivation of Human Resource

The establishment of incentive mechanism needs the support of scientific theory, some relatively typical motivation theories are:

Abraham Maslow's "Hierarchy of Needs Theory".Abraham Maslow thinks that people's basic need can be divided into physiological needs, safety needs, social needs, esteem needs, and actualization needs, which form and develop gradually from low hierarchy to high hierarchy. People's needs give rise to their purpose and motivation of work, if you want to motivate somebody, you shall know his/her current hierarchy of needs, then focus on meeting the needs at this hierarchy or above.

D. Jin and S. Lin (Eds.): Advances in EECM Vol. 1, LNEE 139, pp. 595–600.
springerlink.com

Richard Froom's "The Expectancy Theory". This theory thinks that incentive power is equal to the product of titer and expected value, that is to say, the power to drive people to achieve their goals is the product of two variables, if either variable is zero, the utility of motivation is zero. The titer and expected value is people's judgment of the value and possibility of achieving the goal, the higher the titer and expected value are, the higher the degree of motivation, on the contrary, it is low, down to zero.

Fredrick Herzberg's "Two-factor Theory". Through the survey research, Fredrick Herzberg finds that there are two factors affecting people's behaviors: One is hygienic factor, such as the provision of policy, the pay level, working environment, and human relations etc.; the other is incentive factor, such as achievement, appreciation, challenging work and opportunities for advancement etc. The hygienic factor is necessary, but it can only eliminate dissatisfaction, and can not produce larger incentive function; while the incentive factor can make people produce larger initiative and creativity.

3 Principles on Which Incentive Mechanism of Scientific and Technical Periodical Is Structured

3.1 Integration of Substance Incentive with Spiritual Drive

Substance incentive is to motivate the employees by means of material incentive such as bonus, allowance, welfare and fine etc.; spiritual drive is to achieve the goal of motivation by means of spiritual incentive such as authorization, promotion, training and punishment etc. The material requisite must be met, but excessive substance incentive can bring some side effect, and it can not have larger incentive function. While in the practical work, people have not only physical needs, but also spiritual needs, especially the employees of the scientific and technical periodical---the high knowledge hierarchy pay more attention to the satisfaction of the reputation and spirit. Therefore, in structuring the incentive mechanism of the scientific and technical periodical, only by the organic integration of substance incentive with spiritual drive, and laying particular emphasis on spiritual drive can the incentive effect double.

3.2 Integration of Positive Incentive with Negative Incentive

"Reinforcement Theory" (suggested by James Skinner---American psychology professor) thinks that: Reinforcement is to praise the employees with good achievements and punish those with bad performance by the special design of continuously changing the environment, so as to produce the inverse results to make people motivated. Positive reinforcement is to give the credit by material or spiritual encouragement such as salary raise, praise, promotion etc. immediately after the occurrence of some behavior, so that such behavior can be consolidated and maintained; negative reinforcement is to give negative or punishment to some behavior, such as bonus deduction, criticism, dismission etc., so that such behavior can be completely eradicated. Therefore, in structuring the incentive mechanism of the scientific and technical periodical, only by "Reinforcement Theory", giving consideration to positive incentive and negative incentive, and organically integrating the two can the incentive effect double.

3.3 Fairness Principle

Fairness principle is an important principle on which the incentive mechanism is structured. According to Bryan Adams' Equity Theory, whether an employee is satisfied with his pay or not is generally judged by traverse and longitudinal comparisons. Traverse comparison is to compare the pay-input ratio of oneself with that of the others in the organization; longitudinal comparison is to compare the present pay-input ratio of oneself with that of the past, so as to comprehend whether it is fair by comparison. Therefore, in structuring the incentive mechanism of the scientific and technical periodical, only by drawing up fair and reasonable motivation strategy to make the employees have a sense of fairness can the proper incentive effect be achieved. However, the fairness principle is not averaging, the equal distribution is equal to no incentive.

3.4 Moderation Principle

Only by moderate incentive can good incentive effect be achieved. The cost—— income principle applied in the business administration is also suitable for the structuring of the incentive mechanism of human resource. If the incentive power is too large, the cost of the incentive measures exceeds the income brought in by the incentive effect, it will have no continuous incentive function; if the incentive power is too small---although the incentive cost is reduced, the incentive income will also be reduced, it can not achieve the goal of incentive either. Therefore, in structuring the incentive mechanism of the scientific and technical periodical, only by grasping the scale of the incentive strategy and the magnitude of the incentive power according to the actual condition of the periodical press can the expected incentive effect be achieved.

3.5 Dynamics Principle

On the one hand, because human needs are always changing---when the needs at low levels are met, it will go up to the needs at higher levels; on the other hand, due to the factors of knowledge update and career development, the ability qualifications for different jobs are changing, while the individual ability and knowledge structure are also changing. These changes make the contradictions between individual pursuit and post value, individual ability and job requirements emerge, if we can not make timely adjustment, it will cause brain waste, lead to negative affect, or some posts will lose their proper functions. Therefore, in structuring the incentive mechanism of the scientific and technical periodical, only by grasping the dynamic balance between the periodical development and individual development, and continuously adjusting the ability-job congruent relationship and incentive strategy can every member of the periodical press be fully utilized, so as to produce more positive attitude.

3.6 Differentiation Principle

Because the needs of different individuals are different, the needs of the same individual are also different in different stages, this differentiation requires the incentive strategy fit for it. For example, seen from the gender, female employees pay more attention to the incentives in the environment, emotion, while male employees pay more attention to the incentives in the promotions, pay; seen from the age, young employee think more of the incentives in training and individual development, while middle-age employees think more of the incentives in pay and welfare; seen from the education background, high-education employees pursue the realization of self-value, and pay more attention to spiritual drive, while low-education employees pay more attention to substance incentive meeting basic needs. Therefore, in structuring the incentive mechanism of the scientific and technical periodical, only by drawing up the incentive measures varying with each individual can good incentive effect be achieved.

4 Strategy for Structuring Incentive Mechanism of Scientific and Technical Periodical

4.1 Pay Incentive Strategy

Pay incentive is to achieve the goal of motivating the employees by the rational design of employee salary system. According to its own actual condition, the scientific and technical periodical shall draw up a set of flexible payment systems and effective welfare systems, and can generally adopt the mode of"post wage+allowance+welfare". "post wage" is the fixed salary of every operating post; "allowance" is determined by the result of the performance evaluation carried out by different standards and modes based on different posts; "welfare" not only includes statutory welfare, but also includes various subsidies such as housing, transportation, communication, and education etc. As for the employees who are engaged in management work, the mode of "management by objective" shall be adopted, and the amount of bonus shall be determined by the implementation degree and effect of the goal. The pay model is "post wage+targeted performance bonus+welfare"; as for the personnel who are engaged in topic development, the mode of "project management" shall be adopted, the rewards are given by the mode of project commission according to the benefit for the periodical press brought in by the topic project. The pay model is "post wage+project performance commission+welfare"; as for the personnel who are engaged in marketing management, the mode of "performance management" shall be adopted, and the potential consumer market created by them and the promotion of the popularity of the periodical are included in the amount of work. The pay model is "post wage+marketing performance commission+welfare".

4.2 Promotion Incentive Strategy

Promotion incentive is a motivation pattern that can meet the employees' desire for individual development, and is also an effective measure of the scientific and technical periodical press for attracting more talents, keeping more talents. Due to the effect of traditional culture of our country, thc thought of official rank standard still roots in people's heart, while promotion itself is also a kind of credit for the employees' achievement, so promotion naturally becomes the goal and task motivation of most employees, its incentive function for the employees is also tacit. In general, the periodical press provides two equal promotion channels, one is the management channel; the other is the technical channel, the same rank of the two channels enjoys equal status and pay. Due to the individual differences between the employees, the periodical press shall provide suitable promotion channels for them according to their ability characteristics and likes and dislikes. For example, impel the employees with leadership consciousness and strong organizing ability to develop through the management channel; impel the employees with intensive-study spirit and strong professional ability to develop through the technical channel; while impel the employees with outstanding organizing ability and professional ability as well as vigorous energy to develop through the management and technical channels.

4.3 Training Incentive Strategy

Training incentive is to achieve the goal of motivating the employees by providing the employees with the opportunities of training and continuing education. On the one hand, only by continuous update of the knowledge and information can the editors of the scientific and technical periodical put forward the unique and valuable topics, and edit high quality manuscripts; on the other hand, the editors of the scientific and technical periodical is a knowledge-type group, they pay more attention to self-improvement, self-enriching, self-perfection. Therefore, training is not only the means of improving the overall quality of the employees, but also a route to realize self-development of the editors. Training incentive shall avoid becoming a mere formality, the employees at different posts with different characteristics shall be provided with different training opportunities, for example, provide the business training opportunities for young editors, and create the condition for receiving high education; provide further education opportunities for middle-age and old editors to urge the knowledge update; provide the opportunities for taking part in all kinds of academic meetings for scholar-type editors to expand the scope of academic exchange. Only by implementing this distinctive training incentive can it really obtain actual effect.

4.4 Emotional Encouraging Strategy

Emotional encouraging is a typical spiritual drive pattern. Psychological study shows that positive emotion can improve people's activity ability, make people produce positive mental attitude and initiative, and improve work efficiency; while negative emotion can weaken people's activity ability, make people produce negative affect, and lower work efficiency. Therefore, the purpose of emotional encouraging is to promote positive emotion, and restrain negative emotion. In people-oriented times, humanistic concern is the core of emotional encouraging. The leaders of the periodical press shall sincerely pay attention to the individual development of the employees, eliminate the misunderstandings between the employees, and solve the actual difficulties of the employees, so that all the employees can happily go into work with high morale. To set up the positive culture of the periodical press is an effective route of emotional encouraging. The concrete way of emotional encouraging includes regular public lectures, forums, cultural activities and social clubs etc., in addition, the leaders of the periodical press shall grasp the personal information of every employee and his/her family members, and give him/ her meticulous care at the right time, so as to make him/her feel the warmth of the organization.

5 Conclusion

To structure the incentive mechanism of the human resource of the scientific and technical periodical is a complicated system engineering. In the process of structuring the incentive mechanism of the human resource, the scientific and technical periodical press shall base on certain incentive theory, and draw up the scientific, reasonable, effective incentive strategy according to its own actual condition and following necessary principle, so that the incentive mechanism of human resource plays a positive role in human resource development and management of the scientific and technical periodical press.

References

1. Wang, X.: Research of Multi-incentive Mechanism in Human Resource Management of Periodical Press. Journal of Luoyang University 1, 42–45 (2006)
2. Sun, X., Chu, J.: On Incentive of Innovative Edition Talents. Pioneer 5, 190–194 (2004)
3. Jiang, M., Yang, J.: Optimized Allocation of Human Resource of Modern Small and Medium-sized Presses. Journal of Northwestern Polytechnical University 3, 94–98 (2008)
4. Hu, H.: On 7 General Principles for Human Resource Incentive. Contemporary Finance and Economics 4, 65–67 (2002)

Quantitative Analysis of Web Citations in Book Information Periodicals

Ling Shen

Journal Editorial Department of Changchun University, Changchun, China
linglong1968@126.com

Abstract. In order to understand the utilization of network information resources in the field of the library and information science of our country, this article makes an quantitative analysis of the variation trend of the quantity of web citations, citations type, and the distribution of the top-level domain of web citations by taking the papers in *Researches in Library Science* from 2002 to 2008 as samples, the result shows: The utilization degree of the network resources by the scholars in this field presents a feature of wave-mode uplift, and more resources from edu website are utilized, but the utilization of printed document resources is still the main stream.

Keywords: Book Information Periodicals, *Researches in Library Science,* Web Citations, Quantitative Analysis.

1 Introduction

With the rapid development of web information technology and the enrichment of web information resources of our country, people's awareness and ability of using web resources have been raised. In particular, the web information resources have been widely used in the field of academic research, which can be verified by the plenty of web quotations in the scholarly literatures. At the same time, the research on web quotations also raises the attention and interest of many scholars including Zhang Cuiying, Wang Jianfang (2004) [1], Wang Kairong (2004) [2], Qu Weiqun, Yao Xiaojiao, Wei dan, Gu Chaoran (2008) [3] and Hu Dehua, Jin Jianbin (2009) [4].They research the phenomenon of web quotations in academic periodicals from different perspectives. This paper sets *Researches in Library Science* as its research object to explore the law of development of the net quotations and to provide first-hand information of the situation on using web resources in the field of library information science.

2 Research Object and Statistical Method

Researches in Library Science is a kind of library research periodical sponsored by Jilin Provincial Library, which mainly publishes the academic research papers on library science, information science, information management science, literature

D. Jin and S. Lin (Eds.): Advances in EECM Vol. 1, LNEE 139, pp. 601–606.
springerlink.com

science which are related to information management and transmission. It started publication in 1979, since 1992, it has long been a Chinese core periodical in China and also an excellent periodical in the field of library and information science of our country. The paper makes an statistical analysis of the web quotations and its related factors through the index samples in 2002-2008 *Researches in Library Science*.

Statistical index shows: The percentages refer to the proportion of the number of the papers with web quotations/the number of the papers, the proportion of the number of the web quotations/the number of the quotations, the proportion of the number of the quotations of periodicals, network, books, newspapers and conference papers/the number of quotations, the proportion of the distribution number of the locations of first authors of the papers with web quotations/the number of quotations, and the proportion of the number of the quotations with top-level domains of com, edu, org, gov, net/the number of web quotations.

It is necessary to explain that the number of the quotations used in this paper includes bibliographic references (indirect quotations) and the entries marking the quotations source in the motes (direct quotations), without the entries for "further explanation or supplemental instructions for certain specific content of the text" (Retrieval and Evaluation Data Standards of China Academic Periodical (CD)).

As the current periodical data base contains a large number of non-academic data, to ensure that the data is complete and accurate, this paper mainly collects data by hand. Nevertheless, statistical process still has inevitable omissions, but there is no significant effect on the result.

3 Analysis of Statistical Results

3.1 Analysis of Quantity of Web Citations

The quantity of quotations is the basic clue that reflects the subject trends in the thesis's theory sources, and is the data basis of information metrology. [2] So the quantity of web quotations is the important index for investigating academic research's using the web information resources. From 2002 to 2008, *Researches in Library Science* had published 2443 papers with 16874 quotations, of which 694 papers used the web resources, including 2594 web citations. Table 1 lists the quantity of web quotations and related data of the papers published on *Researches in Library Science* from 2002 to 2008.

Table 1. Quantity of Quotations on 2002—2008 *Researches in Library Science*

Year	Papers	Papers with web quotations	Per.	Quotations	Web quotations	Per.
2002	228	28	12.3%	1165	120	10.3%
2003	366	67	18.3%	1985	201	10.1%
2004	394	69	17.5%	2327	230	9.9%
2005	371	98	26.4%	2605	353	13.6%
2006	376	128	34.0%	2908	549	18.9%
2007	361	153	42.4%	2953	535	18.1%
2008	347	151	43.5%	2931	606	20.7%
Total	2443	694	28.4%	16874	2594	15.4%

First of all, during 2002-2007, the proportion of the papers with web quotations to all the papers had increased 6%-9% every year, except 2004, which dropped 0.8 percent, and this increase didn't slow down until 2008. Through this, we can see that the group of scholars who use web information resource to do scientific researches in the field of library and information science is enlarging rapidly, which shows that the scholars' awareness and ability of using web information is improving.

Secondly, the proportion of web quotations to quotations had risen from 10.3% in 2002 to 20.7% in 2008, which had doubled and showed a trend of wave-like uplift, thus, we can divide the web quotations development of *Researches in Library Science* (2002-2008) into four stages: Steady and balanced development stage (2002-2004); fast rising development stage (2005-2006); steady development stage again (2007); another fast rising stage(2008). This variation tendency is different from the previous one, it proves that the web information resource utilization may be related to the development of web information technology and the supply of web resource, thus featuring intermittence.

Again, from the perspective of the comparison between the average quotations in each paper and the average web quotations in each paper, the variation tendency of the average web quotations in each paper resembles that of the quantity of web quotations, but varies from that of the average quantity of quotations in each paper. Therefore, it shows that scholars still tend to choose printed resources, which results from the subject limits of web information resource, low traceability and the low degree of recognition.

3.2 Analysis of Web Citations' Source

In order to understand the web citations' development condition in *Researches in Library Science* more clearly, the author analyses the 16874 quotations in *Researches in Library Science*(2002 to 2008)classified in periodicals, network, books, newspapers, conference papers and dissertations etc. The results are in Table 2.

Table 2. Quotation Type Distribution of 2002—2008 *Researches in Library Science*

Year	Periodicals	Per.	Net	Per.	Books	Per.	Conference	Per.	Newspapers	Per.
2002	792	68.0%	120	10.3%	223	19.1%	4	0.3%	8	0.7%
2003	1306	65.8%	201	10.1%	405	20.4%	12	0.6%	32	1.6%
2004	1742	74.9%	230	9.9%	318	13.7%	9	0.4%	19	0.8%
2005	1820	69.9%	353	13.6%	359	13.8%	13	0.5%	11	0.4%
2006	1982	68.2%	549	18.9%	283	9.7%	32	1.1%	22	0.8%
2007	1958	66.3%	535	18.1%	342	11.6%	30	1.0%	17	0.6%
2008	1911	65.2%	606	20.7%	328	11.2%	23	0.8%	8	0.3%
Total	11511	68.2%	2594	15.4%	2258	13.4%	123	0.73%	117	0.70%

From Table 2, we can see that the quotations in *Researches in Library Science*, from 2002 to 2008, can be classified into 6 in proper order as follows: Periodicals 68.2%, web15.4%, books 13.4%, meeting papers 0.73%, newspapers 0.70%, dissertations 0.4%. Among them, the quantity of quotations of the periodicals, books, and newspapers presented downtrend during 7 years, respectively from 68.0%,

19.1% and 0.7%(in 2002) to 65.2%, 11.2% and 0.3%(in 2008). Especially the quotations quantity of books dropped more quickly, its downtrend generally corresponds to the uptrend of the quotations quantity of network. While the quotations quantity of the network, meeting papers and dissertations quotations quantity showed an increase in these 7 years, respectively from 10.3%, 0.3% and 0.1% (in 2002) to 20.7%, 0.8% and 0.9 %(in 2008). Especially the web quotations quantity rose more quickly, which in 2006 surpassed the books quotations to become the second largest quotations type. Therefore, it further shows that the scholars in the field of book information science of our country are unceasingly strengthening the utilization degree of web information resources, and are enhancing the utilization capacity.

3.3 Analysis of Distribution of First Authors Whose Theses Have Web Citations

Through the statistical analysis of the distribution of the first authors whose papers have web citations, we can not only understand the scope of periodical's impact and the strength of its influence, but also understand the use of web resources in various regions. [5] Table 3 lists the distribution of the first authors (top 15) of the papers with web citations in 2002-2008 *Researches in Library Science*.

Table 3. Distribution of First Authors of Papers with Web Quotations in 2002 – 2008 *Researches in Library Science*

Year	GD	JL	HB	ZJ	JS	LN	HN	HuN	SD	BJ	HLJ	FJ	SH	SC
2002	2	5	2	1	3	1	1		2	3	2	2		
2003	10	10	4	5	7	5	1	4	2	2	2	3	1	
2004	15	5	11	6	2	4	2	4	2	2	2	5	2	1
2005	17	19	11	8	11	3	5	4	2		2		1	2
2006	17	15	12	8	5	10	6	5	7	5	7	5	7	3
2007	25	24	15	14	9	10	8	5	2	7	5	8	6	6
2008	34	10	12	5	7	10	12	10	10	7	6	2	3	7
Total	120	88	67	47	44	43	35	32	27	26	26	25	20	19
Percent	17.3%	12.7%	9.7%	6.8%	6.3%	6.2%	5.0%	4.6%	3.9%	3.7%	3.7%	3.6%	2.9%	2.7%

GD=Guangdong JL=Jilin HB=Hubei ZJ=Zhejiang JS=Jiangsu LN=Liaoning HN=Henan HuN=Hunan SD=Shandong BJ=Beijing HLJ=Heilongjiang FJ=Fujian SH=Shanghai SC=Sichuan

The statistical result indicates that there were 694 first authors of the papers with web citations published on 2002-2008 *Researches in Library Science*. They were distributed in 27 provinces, autonomous regions and municipalities, it is thus clear that this publication had the very big influence in our country. There were 120 authors from Guangdong, which occupies 17.3% standing by the highest. The following 5 provinces were Jilin, Hubei, Zhejiang, Jiangsu and Liaoning, the proportions were respectively 12.7%, 9.7%, 6.8%, 6.3% and 6.2%. The reason for Jilin province's ranking second was that it holds "the favorable terrain" (this publication's locus). The proportions of the other areas were lower than 6%. Obviously, during these 7 years, the authors' locus distribution was extremely imbalanced, which reflects the disproportionality of our country 's Internet popularity and the regional disproportionality of the development of the book information science.

3.4 Analysis of Distribution of Top-Level Domains of Web Citations

The study on the distribution of web citations domain name is helpful to understand the institution type of the information concerned and used by the scholars, such as commercial organizations, organizations, educational institutions, government agencies etc., so as to provide a scientific basis for better exploitation and utilization of web information resources. This article selects 5 top-level domain name as the statistical objects, including com (commercial organizations), edu (educational institutions), org (non-profit organizations), gov (government agencies) and net (computer networks), and investigates the distribution of the web citations of the papers published on *Researches in Library Science* (2002 - 2008) on these five domain names.

Table 4. Distribution of Web Citations Domain Name of 2002-2008 *Researches in Library Science*

Year	com	edu	org	gov	net	Total
2002	58	23	22	5	6	114
2003	76	41	26	15	18	176
2004	67	53	32	28	18	198
2005	109	84	37	27	35	292
2006	146	147	122	32	24	471
2007	142	129	102	49	38	460
2008	164	138	112	51	34	49
Total	762	615	453	207	173	2210
Percentage	29.4%	23.7%	17.5%	8.0%	6.7%	85.2%

From Table 4, we can see that there were 2210 web quotations containing the above 5 top domain names in *Researches in Library Science* (2002 - 2008), accounting for 85.2% of the total web quotations. The distribution was com> edu > org >gov >net, the proportions were respectively 29.4%, 23.7%, 17.5%, 8.0%, 6.7%. Obviously, the websites with these 5 top-level domain names were the main soruce where the authors of *Researches in Library Science* gained the web information resources. The com website is a commercial website, however, the registered domain names of many academic websites, especially individual or the commercial academic websites were com or com.cn, so the quantity of the web quotations with the domain name of com is large. In the aspect of distribution characteristics, the difference between *Researches in Library Science* and the information periodicals was that the quotations quantity from edu website was relatively high. The reason was that the authors of *Researches in Library Science* mostly come from universities' libraries, it's convenient for them to use the resources from universities' websites. In brief, the scholars gained lots of information resources from the above 5 domains because of the high technicality, stability and reliability of the information resources of these websites.

4 Conclusions

Through the statistical analysis of the web quotations of *Researches in Library Science* (2002-2008), and the correlative factors, we draw the following conclusion:

(1) The analysis of web quotations quantity shows that: In the field of our country's library information science, there are more and more scholars using web information resources; the extent of scholars' utilizing web information resources raises quickly, and presents a rising wave along with the development of web information technology. Even so, at present, the scholars' using the resources from printed documents is still the mainstream.

(2) The analysis of quotations source shows that: The quantity of the quotations of these three traditional types of quotations such as periodicals, books, and newspapers presents downtrend at varying degrees, while that of the web quotations shows a strong uptrend, which reflects that the scholars' capacity and degree of using web information resource are increasing.

(3) The analysis of the distribution of the first authors whose theses have web citations shows that: The disproportionalities of our country's Internet popularity and the subject development of book information science.

(4) The analysis of distribution of top-level domain names of the web citations shows that: The websites whose domain names are com, edu, org, gov, net are still the main source where books information researchers gain the web resources.

References

1. Zhang, C., Wang, J.: Research on Web Resources Utilization in Academic Research. Periodical of Information (4), 113–114 (2004)
2. Wang, K.: Qualitative Research of Web Citations in Information Science Core Periodical. Library and Information Service 48(10), 17–19 (2004)
3. Qu, W., Yao, X., Weidan, et al.: Research and Analysis of Quantity of Web Citations and Its Availability. Periodical of Information (12), 112–114 (2008)
4. Hu, D., Jin, J.: Research on Usage Efficiency of Web Academic Resources Based on the Web Citations. Information Science 27(3), 379–383 (2009)
5. Ju, L.: Quantitative Analysis of Web Citations During 1998 to 2008. Library Theory and Practice (7), 25–28 (1998)

Several of Improved Algorithms for Wavelet De-noising

Zhen-xian Lin

Department of Applied Mathematics, College of Science,
Xi'an University of Posts and Telecommunications, Xi'an, China
lzhx163@163.com

Abstract. With the development of Wavelet Theory in Image Processing, persons developed all sorts of wavelet de-noising algorithms. Especially, Lifting scheme which is the basis constructed tool for the second generation wavelet and it makes us interpret the basis theory in a simple method is commended by JPEG 2000, persons put more attention to use it to image processing. This paper first described the principle of several wavelet de-noising algorithms and compared with these algorithms. Based on these, the paper gives three kinds of improved methods at last. The simulation experiment shows that it is effective for image de-noising.

Keywords: Mallat algorithm, modulus maximum, lifting scheme, threshold.

1 Introduction

According to the practical image characteristic, the statistical property of noising and the principle of frequency spectral distributing, persons developed all sorts of de-noising manners: In 1989, Mallat brought forward Mallat algorithm, and could be forcibly filtered de-noising by Mallat algorithm[1]. In 1992, he put forward the theory of singularity verifying, thus, could utlize the wavelet transform modulus maximum method [2] to de-noise. In 1994, Xu's brought forward a kind of de-noising approach based on the relativity of space field [3]. Donoho and Johnstone brought forward the nonlinear wavelet threshold de-noising method in 1995 [4]. In the same age, Coifman and Donoho brought forward the translation invariance wavelet de-noising method [5]. Furthermore, different kinds of improved methods and some new approaches of methodological incorporation are continually emerged.

In 1995, Sweldens brought forward a constructed method of wavelet in space domain --Lifting scheme [6], it is the basis constructed tool for the second generation wavelet and it makes us interpret the basis theory in a simple method. Compared with the traditional wavelet transform, the more fast the computing rate is, the more simply the computing method is. However, it fits for self-suit, nonlinear, non-oddity sample and integer-to-integer transform. It is commended by JPEG 2000[7].

This paper compares with the Mallat forced de-noising algorithm, the wavelet transform modulus maximum method, the nonlinear wavelet threshold de-noising method and the de-noising based on the interrelation of wavelet domain, and presented three kinds of improved algorithm. The results show that it is practicable and effective for de-noised image.

D. Jin and S. Lin (Eds.): Advances in EECM Vol. 1, LNEE 139, pp. 607–612.
springerlink.com

2 Lifting Scheme

Lifting scheme gives a simple effective method to construct biorthogonal wavelet. The high-frequency of signal is obtained by polynomial interpolation, and the low – frequency of signal is obtained by constructing scale function. The main procedure includes three steps: split and prediction and update.

The lifting steps can be occurred repeatedly and the inverse transform of the lifting is that above three steps are operated in inverse order. The maximal virtue of wavelet transform by lifting scheme is that the course of wavelet filters is separated into the simple basic steps, and every step is reversible.

3 Several of Algorithms for Wavelet De-noising

3.1 The Mallat Forced De-noising Algorithm

In 1989, Mallat brought forward the fast algorithm for wavelet transform ---Mallat algorithm, and then could be forcibly filtered de-noising by wavelet decomposition and reconstruction.

The de-noising step by wavelet decomposition and reconstruction is given as follows: according to require, by the Mallat decomposition formula

$$\begin{cases} V_{j+1} = H \cdot V_j \\ W_{j+1} = G \cdot W_j \end{cases} \quad (j = 0,1,\cdots,J-1) \tag{1}$$

where H and G are the filters coefficient matrices, V_0 is the sampling value of the originality image , V_j and W_j are respectively approximation coefficient and wavelet coefficient for the scale j, the sampling value with noising at the certain scale are decomposed into different frequency bands, and then the frequency bands with noising are put zeros, and by the corresponding reconstruction formula

$$V_j = H^{\bullet} V_{j+1} + G^{\bullet} W_{j+1}, \quad (j = J-1,\cdots,1,0) \tag{2}$$

where $H^{\bullet}$ and $G^{\bullet}$ are the syntheses filters, and satisfied with: $H^{\bullet}H + G^{\bullet}G = 1$ proceeded the wavelet reconstruction, thereby, achieved the aim of de-noising.

The Mallat forced de-noising is also named as the approach of wavelet decomposition and reconstruction. This approach is basically able to wipe off noising, its computing velocity is fast, and it can restore image information in a great measure. But its de-noising results are not fine to the numerous available white-noising toward the practical application.

3.2 The Wavelet Transform Modulus Maximum Algorithm

The broken point is one important characteristic described a twinkling signal. The singularity point of signal is the broken point in signal. It is practical significance how to check up the signal broken point. Mallat's established the relation between wavelet transform and Lipschitz exponent described the signal broken characteristic: Suppose $0 \le \alpha \le 1$, presence constant $k > 0$, make Lipschitz exponent of the signal $f(x)$ and wavelet transform modulus maximum with

$$\log_2 \left| W_{2^j} f(x) \right| \le \log_2 k + \alpha j \tag{3}$$

The wavelet transform modulus maximum method is mostly applied to signal with white-noising, and have much more singularity points in signal. This manner is able to effectively reserve the signal singularity point information within de-noising. The signal after de-noising is the extreme best estimate to the originality signal, without redundant oscillation, and it possess the fine image quality.

3.3 The Nonlinear Wavelet Threshold De-noising Algorithm

Steps of the nonlinear wavelet threshold de-noising algorithm are: (1)Perform wavelet transform to the noising-signal $f(t)$, and obtain a group of wavelet decomposition coefficient $w_{j,k}$;(2)Handle the $w_{j,k}$ decomposed with threshold, and get the estimative wavelet coefficient $\hat{w}_{j,k}$, make $\left\| \hat{w}_{j,k} - u_{j,k} \right\|$ as possible as small. ($u_{j,k}$ is the wavelet transform coefficient without noising signal.) (3)Proceed to the wavelet reconstruction processing with the estimative wavelet coefficient $\hat{w}_{j,k}$, and achieve the estimative signal $\hat{f}(t)$, these are namely the de-noising signal.

The advantage of the nonlinear wavelet threshold de-noising method is that noising is almost completely gotten restrain, and the characteristic peak points of original signal are gotten commendably reserve.

3.4 The De-noising Based on Coefficient Interrelation of Wavelet Domain

The de-noising based on the coefficient interrelation of wavelet domain is, according to different status characteristic of signal and noising at different scales, construct the corresponding rule and handle the wavelet transform coefficient for signal and noising. The essentials of handling are that decrease up to completely wipe off the coefficient of noising, and that furthest reserve the wavelet coefficient of useful signal. The wavelet coefficient possess the strong interrelation at each scale, especially, at the signal edge neighborhood, its interrelation is much more distinctness; whereas, the wavelet coefficient of noising has not this obvious interrelation at each scale; therefore, utilize the corresponding relationship of wavelet coefficient at different scales to distinguish the sort of coefficient, proceed to accept or reject, and achieve the aim of de-noising.

The de-noising based on the coefficient interrelation of wavelet domain constructs interrelated images with the interrelated quantity of the coefficient which lies in the same position at the image neighboring scales with wavelet transform, then compares with the original wavelet image after the proper grey flexed, the farther large interrelated quantity is regarded as image properties for the corresponding image edge, and that be taken out from it, it is acted as the estimation of wavelet transform for original signal, then the de-noising signal is obtained with inverse transform. At the de-noising based on the coefficient interrelation of wavelet domain, the tiny offset of wavelet coefficient at each scale would lead to the interrelated coefficient inexact, and infinitely impact the algorithm's property.

4 Experiment Result and Improved Algorithm

Aimed at the upper four approaches, we join the Gaussian white-noising into "woman" image, and divide it into three layers by wavelet transform, result in Table 1:

Table 1. The signal-noise ratio (SNR) of image which contain different noising degree by the four de-noising handling

Original noising image	Mallat forced denoising	wavelet transform modulus max -imum method	denoising based on coefficient interrelation of wavelet domain	nonlinear wavelet threshold denoising method
32..24	29..06	33..16	31..22	32..97
31..09	28..87	32..07	29..34	32..15
28..76	28..53	30..23	29..04	30.48
26..88	27..96	28..12	27..87	27..98
23..95	26..11	25..36	25..76	26..06
22..54	25..69	23..99	24.47	25..12

Through the above table, we can make out: The Mallat forced de-noising is not satisfactory with the de-noising effect of the detail abundant image, but it is satisfactory with the serious noising. The de-noising based on coefficient interrelation of the wavelet domain could lead the de-noising effect to imperfect owing to the tiny excursion of wavelet coefficient at each scale. The de-noising effect for the wavelet transform modulus maximum method and the nonlinear wavelet threshold de-noising method are fine at the further high of the SNR of image, but the de-noising effect is drop at the lower of the SNR. This is described that the de-noising is insufficiency and could be improved retained.

4.1 Improved Algorithm 1

After the de-noising with the wavelet transform modulus maximum method, the wavelet transform coefficient only left the values of the modulus maximum point, and the left are all made zeros; its mistaken would be further large to directly reconstruct image with only the limit modulus maximum points; the alternation projection algorithm which Mallat brought forward although gives a kind of method for reconstructing wavelet coefficient, but it usually need several decade repeat for assuring the accuracy of reconstructed signal and improving the signal-noise ratio, its velocity is very slowly. Therefore, aiming at the de-noising signal with the modulus maximum principle, according to their properties of wavelet coefficient and modulus maximum, the subsection spline function is used instead, and can be rapidly and effectively reconstructed wavelet coefficient, then be restored the satisfied image combined with Mallat reconstructed algorithm.

4.2 Improved Algorithm 2

The most of de-noising algorithm are based on the extent property of the wavelet coefficient, as the extent of signal and noise possess different spread characteristic with wavelet transform, but, for the lower SNR image, at the wavelet field of small scale, the real information of the image are often covered by noise. So we utilize wavelet phase to filter de-noising.

We call the wavelet transform phase at this scale is the angle of the horizontal component and the vertical component at the corresponding scales.

Steps for the second improved method are: (1)Proceed to the noising image, get the wavelet coefficient $w_{j,k}$;(2)These coefficients are different the image's and the noise's with the relation of wavelet coefficient at each scale: the phase value of image possess the strong interrelation at the neighboring two scales, the phase information of the upper scale is perfectly transferred to the next scale, but the noise have not this performance. We can compare with the two neighboring phase and filter the wavelet coefficient of noise;(3) The extent and phase of each situation has the interrelation with the average of each neighbor around, select a window width, in this window, compare with phase value, can also filter the wavelet coefficient of noise, then proceed the wavelet reconstruction, get the de-noising image.

4.3 Improved Algorithm 3

2D DWT is carried out as a separable transform by cascading two 1D transforms in the horizontal and vertical direction. Now, each 1D wavelet transform can be factored into one or more lifting stages. Through these lifting stages, we can obtain the value signals by the poly-phase factorisation of the wavelet transforms. These values differ for wavelets of different vanishing and preserving moments.

We tested the "Woman" image which contain Gaussian noise with the upper three methods: with the first approach, the SNR of the "Woman" image which contain Gaussian noise is21.9764 dB, N=4, the SNR of de-noising image which we got is 25.3948dB. With the second approach, the SNR of the "Woman" image which contain Gaussian noise is 15.4398dB, the window width is 2, the threshold is the angle of 30, the SNR of de-noising image which we got is 22.2537dB. With the third approach, the SNR of the "Woman" image which contain Gaussian noise is 20.5565dB, the lifting stages is twice, the SNR of de-noising image which we got is 26.1235dB.Effect as Figure 1.

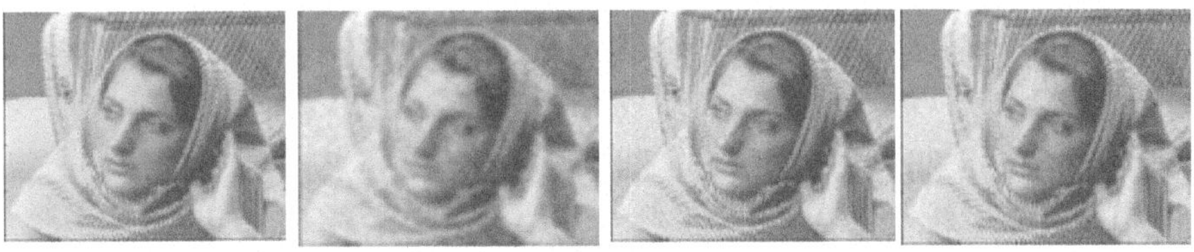

Fig. 1. Effect of these improved method

5 Conclusion

We compared and improved the above de-noising method. In general, we can see: the de-noising effect of the third improved algorithm is very fine when noise is Gaussian noise, but it is remained to research the other noise. The good or bad of image de-noising, in the certain meaning, could be restricted with the similitude degree of signal and noise sometime. If very similitude, the effect of de-noising is not very fine. Therefore, it is remained to research the better algorithm of wavelet image de-noising via lifting scheme in farther.

Acknowledgment. This work is supported by Shaanxi Provincial Department of Education Foundation 2009JK723.

References

1. Mallat, S., Zhong, S.: Characterization of signals from multiscale edges. IEEE Trans. on PAMI 14(7), 710–732 (1992)
2. Mallat, S., Hwang, W.L.: Singularity detection and processing with wavelets. IEEE Trans. on IT 38(2), 612–643 (1992)
3. Xu, Y., et al.: Wavelet transform domain filters: A spatially selective noise filtration technique. IEEE Trans. on IP 3(6), 217–237 (1994)
4. Donoho, D.L., Johnstone, I.: Ideal spatial adaptation via wavelet shrinkage. Biometrika 12(81), 425–455 (1994)
5. Coifman, R.R., Donoho, D.L.: Translation-invariant de-noising. In: Wavelets and Statistics. Springer Lecture Notes in Statistics, vol. 103, pp. 125–150. Springer, New York (1994)
6. Sweldens, W.: The lifting scheme: A construction of second generation wavelets. SIAM J. Math. Anal. 29(2), 511–546 (1997)
7. Cheng, L.-Z., Wang, H.-X., Luo, Y.: Theory and application of wavelet. Publishing Company of Science, Beijing (2004); Yu, Z. X., Chen, H. T., Wang, Y. J.: Research on Markov Delay Characteristic-Based Closed Loop Network Control System. Control Theory and Applications, 19(2), 263–267 (2002)

A Successive Genetic Algorithm for Solving the Job Shop Scheduling Problem

Rui Zhang

School of Economics and Management, Nanchang University
Nanchang 330031, P.R. China
zhangrui05@gmail.com

Abstract. A successive genetic algorithm is proposed for solving job shop scheduling problems in which the total weighted tardiness should be minimized. In each iteration, the following three steps are performed. First, a new subproblem is defined by extracting a subset of operations from the entire operation set. Then, the jobs' bottleneck characteristic values are introduced to depict the criticality of each operation in the current subproblem. Finally, a genetic algorithm is applied to optimize the production sequence of these operations based on the bottleneck information. Numeric computations show that the proposed algorithm is effective for solving the job shop scheduling problem.

Keywords: Job shop scheduling problem, Genetic algorithm, Computational experiments.

1 Introduction

The job shop scheduling problem has attracted a lot of research attention since the 1950s. Most types of job shop scheduling problems have been shown to be strongly $\mathscr{NP}$-hard [6]. Therefore, it is considerably difficult to obtain the optimal solution even for small-scale job shop instances. In recent years, local search strategies, such as genetic algorithm (GA) [4,7], have played a significant role in solving the problem.

In practical manufacturing environments, the scale of job shop scheduling problems is usually very large. For example, the number of operations may be up to 10,000 in some mechanical workshops. In this case, the $\mathscr{NP}$-hard property of job shop problems means that the performance of ordinary methods will not be satisfactory. To overcome the difficulty, several decomposition based algorithms have been proposed [9,1,11,10,12,2], which decompose the original large-scale problem into a series of smaller problems (subproblems), and obtain a final solution after solving these subproblems respectively. But there are some obvious drawbacks with some of these approaches. The main issue is that these algorithms are not sufficiently robust in the computational experiments. That is to say, it is hard to guarantee high-quality solutions consistently in a number of executions of the algorithm.

In order to promote the optimization robustness, we propose a successive genetic algorithm (SGA) to solve large-scale job shop problems with the objective of minimizing total weighted tardiness. In SGA, the original scheduling problem is successively decomposed and optimized by GA. The number of subproblems is predetermined and

D. Jin and S. Lin (Eds.): Advances in EECM Vol. 1, LNEE 139, pp. 613–619.
springerlink.com

each subproblem corresponds to a subset of operations. Before solving each subproblem, we calculate the bottleneck characteristic values (*JBN*) and then use this information to guide GA in order to improve the overall optimization efficiency.

2 Problem Formulation

In a job shop scheduling problem (JSSP), a set of n jobs $\{J_i\}_{i=1}^{n}$ are to be processed on a set of machines $\{M_k\}_{k=1}^{m}$ under some basic assumptions. Each job has a fixed processing route which traverses all the machines in a predetermined order. The problem discussed in this paper is noted as $J||\sum w_i T_i$ according to the three-field notation [3].

JSSP can also be described by a disjunctive graph $G(O,A,E)$ [12], in which $O = \{0,1,\cdots,*\}$ represents the set of nodes (including two dummy nodes, 0 and $*$); A is the set of conjunctive arcs and $E = \bigcup_{k=1}^{m} E_k$ is the set of disjunctive arcs if we denote by E_k the disjunctive arcs that correspond to machine M_k. Then the discussed JSSP can be formulated as follows:

$$\begin{cases} \min \; WT = \sum_{j \in C(J)} w_j (t_j + p_j - d_j)^+ \\ \text{s.t.} \\ \qquad t_i + p_i \le t_j, \;\; (i,j) \in A, \\ \qquad t_i + p_i \le t_j \vee t_j + p_j \le t_i, \;\; (i,j) \in E, \\ \qquad t_i \ge 0, \;\; i \in O. \end{cases}$$

In this formulation, $C(J)$ is the set of the final operations of all the jobs; w_j and d_j are respectively the tardiness weight and the due date of the job which operation j belongs to; p_j and t_j are the processing time and the starting time of operation j, respectively; $(x)^+ = \max\{x,0\}$.

3 The Proposed Algorithm

3.1 Algorithm Framework

In SGA, since the subproblems are solved sequentially, a decomposition policy will suggest their priority levels. To be precise, the decomposition method in our algorithm satisfies: (1) If the original problem is divided into p subsets $\{S_k\}_{k=1}^{p}$, then $\bigcup_{k=1}^{p} S_k = N$, and $\forall i \neq j$, $S_i \cap S_j = \emptyset$; (2) The operations in different subsets have different priority levels, which can be noted as $S_1 \prec S_2 \prec \cdots \prec S_p$ ("$X \prec Y$" suggests that X has a higher scheduling priority than Y); and (3) For any two operations O_i and O_j, if $(O_i, O_j) \in A$ and $O_i \in S_1$, $O_j \in S_2$, then $S_1 \prec S_2$ or $S_1 = S_2$.

The major steps of SGA are presented as follows.

Step 1: Input the number of subsets (subproblems) p and the number of operations in each subproblem $\{u_k\}_{k=1}^{p}$.

Step 2: Let $N \leftarrow O$, where O is the set of operations in the original problem, and let the set of scheduled operations $N^S \leftarrow \emptyset$.

Step 3: Let $k = 1$.

Step 4: If $k = p$ or $|N| \leq u_k$, then the decomposition procedure should stop; $N^S \leftarrow N^S \cup N$; use GA to solve subproblem N; $N \leftarrow \emptyset$; return the final schedule of the original problem. Otherwise, continue.
Step 5: Divide set N into two subsets, N_1 and N_2, where $|N_1| = u_k$.
Step 6: Use GA to solve subproblem N_1.
Step 7: $N^S \leftarrow N^S \cup N_1$; $N \leftarrow N_2$.
Step 8: $k \leftarrow k+1$. Go to Step 4.

Furthermore, to effectively utilize the characteristic information in different scheduling instances, we calculate the characteristic values concerning bottleneck jobs, which will be used in the next stage of subproblem solving. Fig. 1 offers an illustration of the overall SGA algorithm framework.

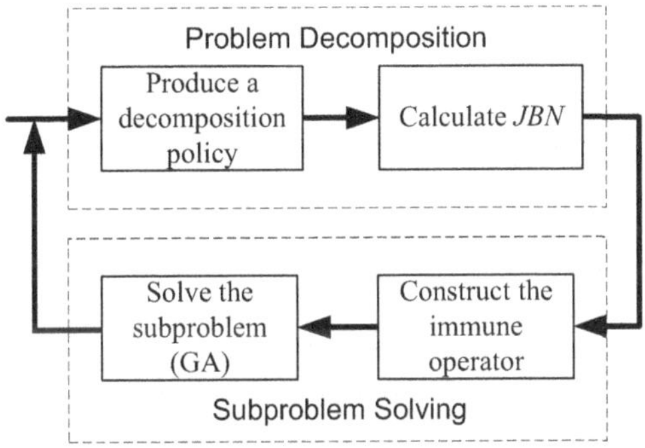

Fig. 1. The SGA framework

3.2 The Decomposition Policy

A decomposition policy determines the structure of the current subproblem and can obviously affect the subsequent optimization process, so in each iteration of SGA, we hope to find a policy that is beneficial for improving the scheduling performance. On the other hand, computational results show that the quality of the final solution is not very sensitive to slight changes of the decomposition policies. Here we use a heuristic method to produce the decomposition policy.

For each operation to be scheduled, the SLACK index is defined as

$$\theta_i = d_i - \left(\max_{j \in \mathrm{pre_s}(i)} \{C_j\} + \sum_{j \in \mathrm{pre_u}(i)} p_j \right) - \sum_{j \in \mathrm{suc}(i)} p_j,$$

Where C_j refers to the completion time of operation j; set $\mathrm{pre_s}(i)$ consists of operation i's preceding operations that have already been scheduled in previous subproblems; $\mathrm{pre_u}(i)$ includes operation i's preceding operations that have not been scheduled; $\mathrm{suc}(i)$ is the set of succeeding operations of operation i.

Then, we define a priority index α for each operation in N as

$$\alpha_i = \left(2-\bar{\theta}_i\right)\left(1+\bar{w}_i\right),\ \bar{\theta}_i, \bar{w}_i \in [0,1].$$

Where $\bar{\theta}$ and $\bar{w}$ denote normalized values, i.e., $\bar{\theta}_i = (\theta_i - \theta_{\min})/(\theta_{\max} - \theta_{\min})$, $\bar{w}_i = (w_i - w_{\min})/(w_{\max} - w_{\min})$.

Finally, we sort all the operations in N in a non-increasing order of α, and then assign the first u_k operations to N_1 and the remaining operations are assigned to N_2. In this way, we obtain an effective decomposition policy.

3.3 The Bottleneck Characteristic Values

Once the decomposition policy has been evaluated through simulation[1], we can obtain the estimated completion time of each job. With this information, here we define: (1) the predicted tardiness of job i: $\widehat{T}_i = (\widehat{F}_i - d_i)^+$, where $\widehat{F}_i$ and d_i respectively denote the predicted completion time and due date of job i; (2) the slack time of job i: $S_i = d_i - R_i - \sum_{j \in J_i \cap N} p_j$, where $R_i = \max_{j \in J_i \cap N^S} C_j$ equals the completion time of job i's scheduled operations and also the release time of its current operations in set N. $J_i \cap N$ ($J_i \cap N^S$) refers to those operations that belong to both job i and set N (N^S).

The bottleneck characteristic values for each job is defined as

$$JBN_i = w_i \cdot \widehat{T}_i \cdot \exp\{-S_i\}.$$

It is notable that this definition of bottleneck substantially differs from the ordinary approaches in the existing literature which usually aim at the idle time or utilization rate of machines. Our definition focuses on the final objective function and depicts the critical factors in different optimization stages.

3.4 A Genetic Algorithm for Subproblem Solving

When the subset N_1 is determined, we use a genetic algorithm to schedule these operations based on the partial schedule obtained in previous subproblems. We devise an immune genetic algorithm [5] that makes use of the characteristic information (*JBN* values) acquired in the previous step.

Encoding. The encoding scheme for the GA is based on operation priority lists. A solution relates each machine with a priority list of operations to be processed on this machine. Note that after the decoding procedure, the actual processing order of operations in the final feasible schedule may differ from these priority lists.

Initialization. We initialize the population using SLACK rule and random permutation of operations.

---

[1] We adopt a priority-rule-based simulation procedure to evaluate the decomposition policy. Since the objective function for the original problem is the total weighted tardiness, we choose ATC (Apparent Tardiness Cost) [8] as the priority rule in simulation for the situations where the decomposition policy does not impose any priority requirements.

Mutation & Crossover. The SWAP operator is applied to randomly selected individuals where two of the operations will be exchanged in the priority list.

Each crossover is performed using the LOX operator to a randomly selected individual and the best individual in the current population.

The Immune Operator

Step 1: Under the immune probability p_i, choose an individual from the current population, randomly select a machine, and then calculate the JBN values for all the operations to be scheduled on this machine.
Step 2: Randomly select an operation O_k from the priority list of the selected machine.
Step 3: Evaluate $i_1 = \arg\max_{i \in N^\mathrm{p}(k)} \{(JBN_k - JBN_i)^+\}$, where $N^\mathrm{p}(k)$ denotes the set of operations that are before O_k in the priority list.
Step 4: Evaluate $i_2 = \arg\max_{i \in N^\mathrm{s}(k)} \{(JBN_i - JBN_k)^+\}$, where $N^\mathrm{s}(k)$ denotes the set of operations that are after O_k in the priority list.
Step 5: Let $i^* = \arg\max_{i \in \{i_1, i_2\}} \{|JBN_k - JBN_i|\}$, and then swap O_k and O_{i^*} in the priority list.

After performing crossover and mutation to each generation of individuals, the immune operator is applied to a selected portion of the population. When vaccination is finished, we evaluate the fitness of the new individual, and if its fitness is improved, the new individual will be kept down to the next generation, otherwise this vaccination is discarded and the individual will be restored.

Decoding & Evaluation. The decoding process concerns iteratively scheduling the ready operations (whose preceding operations have been scheduled) in the current subproblem according to their priority order and as early as possible, based on the partial schedule obtained when solving the previous subproblems.

We try to obtain a complete schedule by scheduling the remaining operations reasonably. The sequence information on the unscheduled operations is extracted from the simulation result of the decomposition policy and thus we obtain the relative priority order among these operations. Then, we schedule the remaining operations according to these precedence relations to form a complete schedule for which we could calculate the objective value. It is used as an evaluation for the individuals in GA.

4 Numerical Computational Results

To test the performance of the SGA, we randomly generate different-scale job shop scheduling problem instances in which the route of each job is a random permutation of m machines and the (integral) processing time of each operation follows a uniform distribution $\mathscr{U}(1,99)$. The due date of each job is obtained by a series of simulation runs which apply different priority rules (such as SPT, SRPT, etc.) to the machines and finally we take the average completion time of each job among these simulations as its due date. The (integral) tardiness weight of each job is generated from a uniform distribution $\mathscr{U}(1,10)$.

In the numerical computations, the mutation probability (p_m) is 0.3; the crossover probability (p_c) is 0.8; the immune probability (p_i) is 0.5; the population size (PS) is 20 and the number of generations (GN) is 50.

In the following we will list the computational results with respect to different p values (the number of subproblems). Once p is fixed, the number of operations in each subproblem is

$$u_k = \begin{cases} \left\lfloor \frac{|O|}{p} \right\rfloor & k = 1, \cdots, p-1, \\ |O| - (p-1) \cdot \left\lfloor \frac{|O|}{p} \right\rfloor & k = p. \end{cases}$$

where $|O| = n \times m$ is the total number of operations in the original problem.

We compare the proposed SGA with a canonical genetic algorithm using operation-based encoding scheme (CGA) and the heuristic decomposition algorithm (BWS) presented in [2]. The results for eight different-scale problems are shown in Table 1 and Table 2.

Table 1. Performance of SGA on Smaller-scale Scheduling Instances

No.	Size	SGA								CGA		BWS	
		$p=1$		$p=2$		$p=3$		$p=4$					
		WT[a]	Time[b]	WT	Time	WT	Time	WT	Time	WT	Time	WT	Time
1	10×10	197*	1.7	214	5.8	260	8.1	289	9.7	193	3.4	317	9.5
2	20×10	2272	4.7	2197*	16.2	2293	24.1	2640	29.4	2216	8.6	2579	26.8
3	100×5	81592	16.8	80946	52.8	80871*	81.5	81129	104.9	81507	33.1	80931	90.4

[a] WT: the average objective value of the solutions obtained by the algorithm in 10 consecutive runs.

[b] Time (in seconds): the average running time of the program.

* The best objective value for different p.

Table 2. Performance of SGA on Larger-scale Scheduling Instances

No.	Size	SGA								CGA		BWS	
		$p=4$		$p=6$		$p=8$		$p=10$					
		WT	Time	WT	Time	WT	Time	WT	Time	WT	Time	WT	Time
4	60×10	60375*	114.3	60682	165.7	61258	190.4	62463	211.7	62351	42.2	61715	135.8
5	200×5	264318	193.4	258648	264.8	258341*	314.0	269437	360.1	300462	80.7	292723	290.7
6	100×20	10126	274.3	10003	310.2	9816*	351.7	9882	385.4	11052	113.6	9903	349.2
7	500×20	734615	946.1	719754	1252.5	706425*	1416.9	710552	1595.2	845862	673.0	763677	1318.0
8	400×30	695673	1103.7	646403	1427.9	641545	1522.1	606404*	1694.5	736791	948.1	695100	1583.7

* The best objective value for different p.

Table 1 suggests that for smaller-scale problem, SGA achieves better result when p is relatively small ($p < 4$). This is because, if the number of subproblems is unduly large compared with the problem size, the additional precedence constraints imposed by the decomposition policy will hinder the search process of GA. In addition, when the problem size is small, the difference between the computational results obtained by SGA and CGA is also trivial. It is concluded from from Table 2 that when the problem size grows, the optimal number of subproblems for SGA also increases and our algorithm substantially outperform CGA and BWS for larger-scale problems.

5 Conclusion

In this paper, a successive genetic algorithm for solving large-scale job shop scheduling problems is proposed. In the algorithm, the original large-scale problem is tackled by successively defining and solving a series of subproblems. Characteristic values are evaluated and used to accelerate the converging speed of the optimization process. Computational results show that the proposed algorithm is effective.

Acknowledgment. This work is partially supported by the National Natural Science Foundation of China under Grant No. 61104176.

References

1. Bassett, M.H., Pekny, J.F., Reklaitis, G.V.: Decomposition techniques for the solution of large-scale scheduling problems. AICHE Journal 42(12), 3373–3387 (1996)
2. Byeon, E.S., Wu, S.D., Storer, R.H.: Decomposition heuristics for robust job-shop scheduling. IEEE Transactions on Robotics and Automation 14(2), 303–313 (1998)
3. Jain, A.S., Meeran, S.: Deterministic job-shop scheduling: Past, present and future. European Journal of Operational Research 113(2), 390–434 (1999)
4. Jia, Z., Lu, X., Yang, J., Jia, D.: Research on job-shop scheduling problem based on genetic algorithm. International Journal of Production Research 49(12), 3585–3604 (2011)
5. Jiao, L., Wang, L.: A novel genetic algorithm based on immunity. IEEE Transactions on Systems, Man and Cybernetics, Part A 30(5), 552–561 (2000)
6. Lenstra, J.K., Kan, A.H.G.R., Brucker, P.: Complexity of machine scheduling problems. Annals of Discrete Mathematics 7, 343–362 (1977)
7. Liu, M., Sun, Z., Yan, J., Kang, J.: An adaptive annealing genetic algorithm for the job-shop planning and scheduling problem. Expert Systems with Applications 38(8), 9248–9255 (2011)
8. Pinedo, M.L.: Scheduling: Theory, Algorithms and Systems, 3rd edn. Springer, New York (2008)
9. Sidney, J.B.: Decomposition algorithms for single machine sequencing with precedence relations and deferral costs. Operations Research 23(2), 283–298 (1975)
10. Singer, M.: Decomposition methods for large job shops. Computers and Operations Research 28(3), 193–207 (2001)
11. Sun, D., Batta, R.: Scheduling larger job shops: a decomposition approach. International Journal of Production Research 34(7), 2019–2033 (1996)
12. Wu, S.D., Byeon, E.S., Storer, R.H.: A graph-theoretic decomposition of the job shop scheduling problem to achieve scheduling robustness. Operations Research 47(1), 113–124 (1999)

Application of Farmercurve Method in Construction Project Risk Management

Guo Zhanglin[1] and Shi Ying[2]

[1] Department of civil engineering, North China Institute of Science and Technology, East yanjiao, Beijing, China, 101601
[2] School of Economics and Management HeBei University of Engineering Handan, 056038 China
gg_zz_11@163.com, felixshi@163.com

Abstract. This paper introduces the construction project of risk management, and expounds the principle of Farmercurve, analysis the risk of the construction project with Farmercurve, and then to make the objective evaluation.

Keywords: Construction project, Farmercurve, Risk assessment.

1 Introduction

Construction project is a complex process. Due to the large investment, long time, this will need a lot of different professional staff, The influence of the external and internal conditions by many factors, and its facing risk than other industry. Construction risk will be a direct and indirect losses to the owner and the contractor. So to strengthen the construction of the building project risk management, raise the risk management consciousness, master risk identification technology. Carry out risk assessment and analysis, timely to guard against and dissolve the project risk.

2 Farmercurve principle

This is a relatively new project risk management method of quantitative analysis. It contains two factors: The probability of failure and the consequences of failure. This method has the identification of the risk is high, middle and low three categories. Low risk has slightly adverse effect to the the project target, it has small probability(Generally<0.3); medium risk is refers to the probability is bigger(Generally in 0.3 to 0.7); high risk is refers to the probability is very large(Generally>0.7), Farmercurve is risk assessment project risk level coefficient. If the project risk coefficient is R, so:

$$R=1-P_sC_s=1-(1-P_f)(1-C_f)=P_f+C_f-P_f\,C_f \tag{1}$$

Obviously, 0< R < 1. P_f and P_s mean project's failure and success probability.

So $P_s=1-P_f$.

D. Jin and S. Lin (Eds.): Advances in EECM Vol. 1, LNEE 139, pp. 621–624.
springerlink.com

C_f and C_s mean that project the consequences of failure the utility values and successful consequences utility values, according to utility theory, $C_f + C_s = 1, 0 < C_f < 1, 0 < C_s < 1$.

Farmercurve's paining as follows:

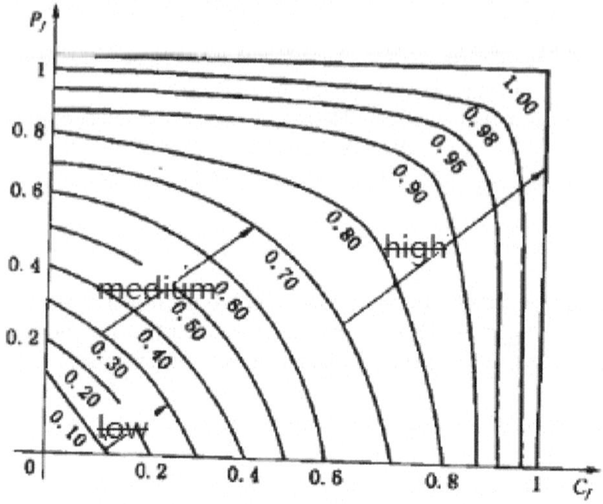

Project risk factors according to the calculation formula (1). Formula (1) of the calculation formula and visible (2) and (3) formula.

$$P_f = \frac{P_{f_1} + P_{f_2} + \cdots + P_{f_n}}{n} \tag{2}$$

$P_{f_1} + P_{f_2} + \cdots P_{f_n}$ respectively means each risk the probability of occurrence in the project, n is the number of the risk.

$$C_f = \frac{C_{f_1} + C_{f_2} + \cdots C_{f_n}}{n} \tag{3}$$

$C_{f_1} + C_{f_2+} \cdots C_{f_n}$ respectively means the consequences of the project risk utility values, n is the number of the consequence of the risk.

2 Example

A construction company in the construction engineering project construction. In order to make the construction project reasonable avoid risk, reduce loss, improve efficiency, it by many uncertain factors. According to investigation, influences of the main risk are: construction technology risk, construction organization risk, political and economic risk, natural risk and social risk. Some of these risk bring consequences, such as: material quality is not good, the construction organization is not reasonable, seasonal delay, shortage of funds, environmental protection and so on a series of problems.

2.1 The Results of the Survcy on Experts Statistics

The probability of occurrence in a major risk (Table 1)

Table 1. Experts on the main risk probability opposes summary form

Respondent	Political and economic risk	Construction organization risk	Construction technology risk	Natural risk	Social risk
Experts 1	0.2	0.5	0.6	0.4	0.1
Experts2	0.1	0.7	0.7	0.5	0.3
Experts3	0.3	0.8	0.6	0.4	0.1
Experts4	0.2	0.6	0.9	0.6	0.2
Experts5	0.4	0.5	0.8	0.7	0.1
Experts6	0.3	0.7	0.7	0.6	0.4
Experts7	0.1	0.6	0.6	0.8	0.2
Experts8	0.2	0.8	0.7	0.7	0.2
Experts9	0.3	0.5	0.8	0.5	0.1
Experts10	0.1	0.4	0.8	0.3	0.2
Total	2.2	6.1	7.2	5.5	1.9
Average	0.22	0.61	0.72	0.55	0.19

From Table 1, it is known that P_{f_1}=0.22, P_{f_2}=0.61, P_{f_3}=0.72, P_{f_4}=0.55, P_{f_5}=0.19. And then can get $P_f = \sum_{i=1}^{5} P_{f_i} / 5$ =0.458.

(2) The consequences of risk (Table 2)

Table 2. Experts to risk score consequences of summary

Respondent	The construction organization is not reasonable	The material quality not qualified	Shortage of funds
Experts 1	0.2	0.3	0.3
Experts2	0.1	0.5	0.4
Experts3	0.3	0.4	0.4
Experts4	0.2	0.5	0.5
Experts5	0.1	0.5	0.3
Experts6	0.1	0.4	0.4
Experts7	0.3	0.6	0.3
Experts8	0.2	0.5	0.3
Experts9	0.1	0.7	0.2
Experts10	0.3	0.4	0.5
Total	1.9	4.8	3.6
Average	0.19	0.48	0.36

From Table 2 can know that C_{f_1} =0.19, C_{f_2} =0.48, C_{f_3} =0.36.And then can get

$$C_f = \sum_{i=1}^{3} C_{f_i} / 3 = 0.34.$$

2.2 Calculate Risk Factor R

$$R=1-P_s C_s=1-(1-P_f)(1-C_f)=P_f+C_f-P_f C_f$$
$$=0.458+0.34-0.458\times 0.34=0.64$$

From the Farmercurve, 0.64 between 0.6 and 0.7,it is belong to the medium risk, which should manage strictly according to medium risk.

References

1. Zhang, J.: Application of Farmercurve method in mine project risk management. China Coal 3(2), 24–25 (2005)
2. Liu, G.: Modern Project management. Xi 'an Jiaotong university press (2007)
3. Shi, Z.: Project management for construction. Science press (2008)
4. Wang, Z.: Engineering project risk management. China's water conservancy and hydropower press (2009)

Single-Stage AC-DC Converter with a Synchronous Rectifier

Hyun-Lark Do

Department of Electronic & Information Engineering,
Seoul National University of Science and Technology, Seoul, South Korea
hldo@seoultech.ac.kr

Abstract. A single-stage AC-DC converter with a synchronous rectifier is proposed. It is based on a boost-flyback topology. The secondary winding is added in the boost inductor at power factor correction (PFC) stage to provide an additional discharging path and suppress DC-bus voltage. A flyback transformer with a center tap is utilized. Therefore, the direct energy transfer of a part of the input energy to the output improves the efficiency. Experimental results for a 60W converter at a constant switching frequency of 70*kHz* are obtained to show the performance of the proposed converter.

Keywords: Boost converter, flyback converter, power factor correction.

1 Introduction

Single-stage approach is especially attractive in low power applications due to its simple power stage and control circuit. However, it still has several drawbacks such as high current stress in power switch, high DC-bus voltage stress etc. Its major drawback is a high voltage stress on the dc-bus capacitor. Many single-stage PFC AC-DC converters suffer from high DC-bus voltage stress at light load and high line, which makes these converters impractical [1]. The voltage across the DC-bus capacitor varies with the variation of the input voltage and the load, especially while PFC stage operates in discontinuous conduction mode (DCM) and DC-DC stage is in continuous conduction mode (CCM). When the DC-DC stage operates in CCM, the duty cycle does not change according to load. When load becomes light, the duty cycle does not change immediately due to the CCM operation in the DC/DC stage. Then, the input power still keeps the same level as at heavy load. This unbalanced power causes the DC-bus voltage to rise. However, there is no DC-bus voltage stress problem in combinations of same operating mode, either in DCM or in CCM for the two parts. Since DCM boost converter has an inherent PFC property, PFC part is preferred to operate in DCM [1]. If both the PFC part and the DC/DC part are operated in DCM, then the DC-bus capacitor voltage is independent of load [2]. However, the high root-mean-square (RMS) value of the switch current requires a high current-rated switch and reduces the efficiency, as compared to CCM operation.

D. Jin and S. Lin (Eds.): Advances in EECM Vol. 1, LNEE 139, pp. 625–628.
springerlink.com

A single-stage AC-DC converter with a synchronous rectifier is proposed. Since the secondary winding is added in the boost inductor, DC-bus voltage is effectively suppressed. Also, a synchronous rectifier is used instead of output diode to increase the efficiency.

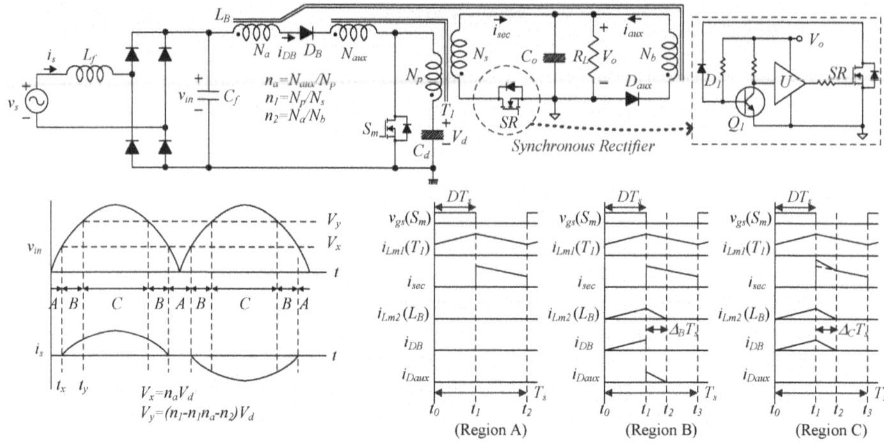

Fig. 1. Circuit diagram and key waveforms of the proposed converter

2 Analysis of the Proposed Converter

Fig. 1 shows the circuit diagram and key waveforms of the proposed converter. The flyback converter consisting of T_1, S_m, and *SR* can be operated either in CCM or DCM according to load. In Fig. 1, only its CCM operation is shown to simplify the analysis. The boost converter consisting of L_B, D_B, S_m, and C_d operates in DCM to provide PFC function. The secondary winding N_b of L_B provides an additional discharging path and it can suppress the DC-bus voltage. As shown in Fig. 1, the proposed converter has three operation regions. When v_{in} is lower than V_x, the converter operates in the region A where D_B and D_{aux} are reversely biased during turn-off of S_m. When v_{in} is between V_x and V_y, the converter operates in the region B where D_B is reversely biased and D_{aux} is conducting for a while during turn-off of S_m. When v_{in} is higher than V_y, the converter operates in the region C where D_{aux} is reversely biased and D_B is conducting for a while during turn-off of S_m.

2.1 Region A, B, and C Operation of the Proposed Converter

In the region A, the PFC stage is disabled and only the DC-DC stage operates. The DC-DC stage delivers power from C_d to the load through T_1. When S_m is turned on at t_0, the energy is stored in the magnetizing inductance of T_1. The magnetizing current i_{Lm1} increases linearly with the slope of V_d/L_{m1}. When S_m is turned off at t_1, the energy is transferred to the secondary side. The current i_{Lm1} decreases linearly with the slope of n_1V_o/L_{m1}.

In the region B, the PFC stage is enabled and it operates like a DCM flyback converter. At t_0, S_m is turned on. Then, v_{in}-n_aV_d is applied to the magnetizing inductance L_{m2} of L_B and i_{DB} increases linearly. At t_1, S_m is turned off and D_B is turned off and D_{aux} is turned on. At t_2, i_{aux} arrives at zero and D_{aux} is turned off. The operation of DC-DC stage is the same as in the region A.

In the region C, the PFC stage operates like a DCM boost converter. At t_0, S_m is turned on. Then, v_{in}-n_aV_d is applied to L_{m2} and i_{DB} increases linearly. At t_1, S_m is turned off and D_B is still on. Therefore, the energy stored in L_B is transferred to the secondary side through T_1. Therefore, both energies stored in L_{m1} and L_{m2} are transferred through T_1. At t_2, i_{DB} arrives at zero and D_B is turned off. Then, only the energy stored in Lm1 is transferred to the secondary side.

When the proposed converter operates in the region C, the DC-bus capacitor C_d can be charged. In the regions A and B, C_d is only discharged. In the region A, all the output power is provided from C_d. In the region B, part of the output power is provided from C_d. When V_d goes up, the region A is widened. Then, V_d goes down.

2.2 Synchronous Rectifier

The proposed synchronous rectifier is a voltage sensing type. By sensing the drain-source voltage of *SR*, the driving signal for *SR* is determined. When S_m is turned off, the body diode of *SR* is turned on at first. Then, the voltage of drain terminal of *SR* is lower than the source terminal. Therefore, the base charge of Q_1 is removed by turn-on of d_1 and Q_1 is turned off quickly. Then, "High" signal is applied to a gate driver *U* and the gate signal is applied to *SR*. Now, *SR* is fully turned on and all the current is flowing through the channel of *SR*. Under a CCM mode, i_{sec} decreases rapidly with turn-on of S_m. At the moment when i_{sec} changes its direction, the voltage of drain terminal of *SR* is higher than the source terminal. Then, d_1 is turned off and Q_1 is turned on. Then, "Low" signal is applied to the gate driver and *SR* is turned off rapidly. Under a DCM, *SR* is turned off when i_{sec} arrives at zero as under a CCM.

3 Experimental Results

A prototype is implemented and tested with specifications of v_s=90~265Vrms, V_o=12V, f_s=70kHz, and $P_{o,max}$=60W. The circuit parameters are L_f=1mH, C_f=0.1uH, L_{m1}=750uH, L_{m2}=72uH, n_a=0.5, n_1=6, n_2=6, C_d=220uF (450V), and C_o=4700uF (25V).

The experimental waveforms of the proposed converter are shown in Fig.2. The measured waveforms in three regions agree with the theoretical analysis. Also, the input current waveform at v_s=90Vrms and full load condition is shown. Measured power factor is 0.89 with the efficiency of 89 % at 110*V* input and 0.92 with 90.1% efficiency at 220*V* input. Maximum efficiency of the converter is 90.5%. The measured line current harmonics are also shown in Fig.2. All line current harmonics are below the IEC 61000-3-2 Class D requirements. In Fig. 2, the DC-bus voltage V_d according to load at the highest line input voltage of 265Vrms. The maximum DC-bus voltage is 401*V*, so the 450*V* rated bulk capacitor can be used.

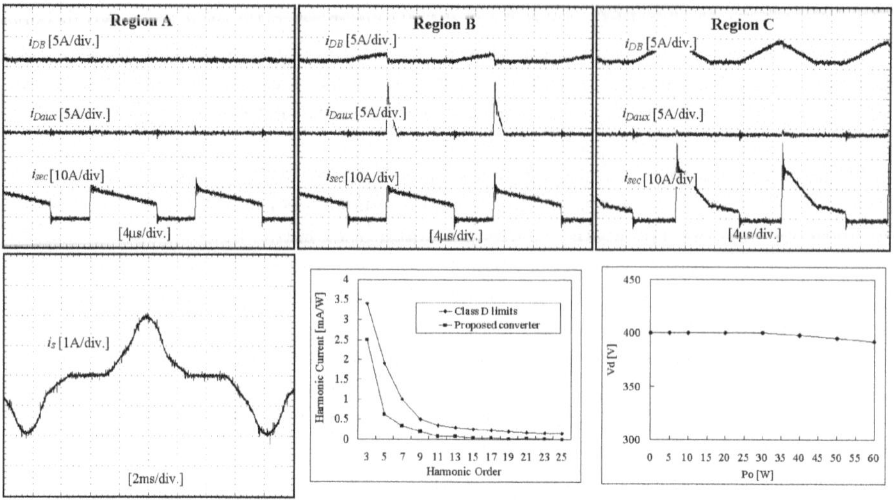

Fig. 2. Experimental waveforms, line current harmonics, and DC-bus voltage

4 Conclusion

A single-stage AC-DC converter with a synchronous rectifier has been proposed. Experimental results for a prototype at a constant switching frequency of 70kHz have been given to show the performance of the proposed converter. The voltage across the DC-bus capacitor was held below 401V even though the converter operates in a wide range of universal input. Due to direct energy transfer and a synchronous rectifier at the secondary, the proposed converter shows higher efficiency than conventional single-stage boost-flyback converters.

References

1. Madigan, M., Erickson, R., Ismail, E.: Integrated high quality rectifier-regulators. In: Proc. of IEEE PESC, pp. 1043–1051 (1992)
2. Daniele, M., Jain, P.K., Joos, J.: A single-stage power-factor-corrected AC/DC converter. IEEE Trans. Power Electronics 14, 1046–1055 (1999)

IGBT Based Cost-Effective Energy Recovery Sustain Driver for Plasma Display

Hyun-Lark Do

Department of Electronic & Information Engineering,
Seoul National University of Science and Technology, Seoul, South Korea
hldo@seoultech.ac.kr

Abstract. An insulated gate bipolar transistor (IGBT) based cost-effective energy recovery sustain driver for plasma display is proposed in this paper. In the proposed sustain driver, the number of switching devices is reduced and IGBTs are used as power switches to reduce the overall cost. Since all power switches are turned off under zero-current switching (ZCS) operation, the tail current problem associated with IGBT does not occur. Moreover, due to a simpler structure compared with the conventional sustain driver, higher efficiency can be achieved. Theoretical analysis and performance of the proposed sustain driver were verified on an experimental prototype operating at 200 kHz switching frequency.

Keywords: Energy recovery, sustain driver, IGBT, plasma display.

1 Introduction

Plasma display panel (PDP) has its intrinsic capacitance. So, a considerable energy stored in a panel is consumed across non-ideal resistance of its driving circuits and PDP without energy recovery circuits [1]. Furthermore, the surge current occurring while the panel is charged or discharged could cause serious resonance and electromagnetic interference noises. To solve these problems, several energy-recovery sustain drivers have been proposed in [1]-[6]. Among them, Weber's sustain driver in [2], [3] is widely used. It is utilizing the LC resonance between panel capacitance and external inductors. It is very widely used and known to be a very effective method due to its features such as high efficiency and high flexibility. High efficiency is one of the most important characteristics of sustain drivers. It is because large power consumption is a major drawback of PDP. Almost every PDP company adopts Weber's sustain driver. However, there are some drawbacks in Weber's sustain driver such as its complexity and high cost.

To overcome these problems, a cost-effective sustain driver with fewer components is proposed. Its circuit diagram and key waveforms are shown in Fig. 1. The energy storage capacitors are removed and there are just two power semiconductor devices on the resonant current path. It can provide both higher efficiency and lower cost.

D. Jin and S. Lin (Eds.): Advances in EECM Vol. 1, LNEE 139, pp. 629–632.
springerlink.com

Moreover, IGBTs are employed as power switches. Since all IGBTs are turned off under ZCS operation, the tail current problem occurring with turn-off IGBT does not occur. Since an IGBT is cheaper than a power MOSFET with the same rated current, cost-effectiveness can be achieved by adopting IGBTs as switches.

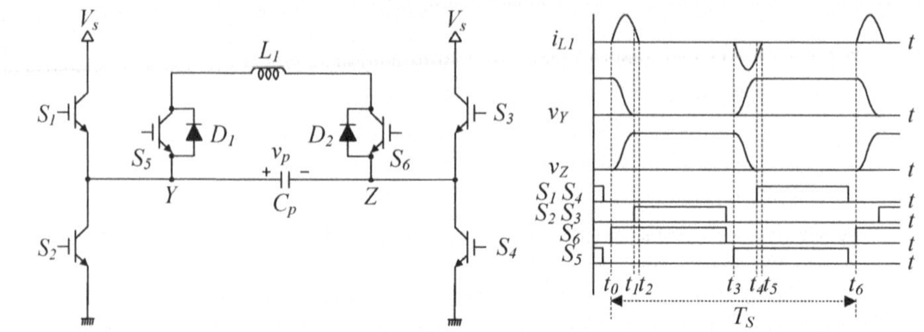

Fig. 1. Circuit diagram and key waveforms of proposed sustain driver

2 Analysis of the Proposed Sustain Driver

Mode analysis is carried out for the time interval $t_0 - t_6$ which is divided into six modes. Before t_0, the panel voltage v_p is maintained and there is no current flow. Gate signals for all switches are removed. At t_0, the switch S_6 is turned on. Then, the resonance between the inductor L_1 and the panel capacitor C_p occurs. The energy stored in the panel capacitance C_p is transferred to the external inductor L_1 with a resonant manner through D_1, L_1, and S_6. When v_p arrives at zero, the energy stored in L_1 starts to charge C_p in the negative direction. The inductor current i_{L1} and the panel voltage v_p are given by

$$i_L(t) = \frac{V_s}{\omega L} e^{-\frac{R}{2L}(t-t_0)} \sin \omega(t-t_0) \ , \tag{1}$$

$$v_p(t) = V_s e^{-\frac{R}{2L_1}(t-t_0)} \left(\cos \omega(t-t_0) - \frac{R}{2\omega L} \sin \omega(t-t_0) \right) , \tag{2}$$

where

$$\omega = \sqrt{\frac{1}{LC_p} - \left(\frac{R}{2L}\right)^2} \tag{3}$$

and R represents the parasitic resistances of the circuits such as the resistance of the printed circuit board and the parasitic resistances of the semiconductor devices.

At t_1, S_2 and S_3 are turned on and the panel voltage v_p is clamped as $-V_s$. Typically, since the main switches should be turned on before plasma discharge occurs to obtain operating margin, S_2 and S_3 is turned on before the resonance between the panel capacitance and the external inductor ends. In this mode, with an assumption that a voltage drop caused by the parasitic resistances of the circuits is constant as V_1, the voltage $-(V_s-V_1)$ is applied to L_1 and the inductor current i_{L1} decreases linearly as follows:

$$i_{L1}(t) = i_{L1}(t_1) - \frac{V_s - V_1}{L_1}(t - t_1) \quad . \tag{4}$$

At t_2, the inductor current i_{L1} arrives at zero. The gate signal for S_6 is removed just before t_3. At that time, the current flowing through S_6 is zero and S_6 is turned off with ZCS. Also, S_2 and S_3 are turned off just before t_3. Since the plasma discharge ends at that time, S_2 and S_3 are also turned off with ZCS. Although the gate signals for S_2 and S_3 are removed, the panel voltage v_p is maintained as $-V_s$ until S_5 is turned on at t_3.

The circuit operations of t_3 - t_6 is similar to that of t_0 - t_3. At t_3, the switch S_5 is turned on. Then, the resonance between L_1 and C_p occurs. The resonant current flows through D_2, L_1, and S_5. When v_p arrives at zero, the energy stored in L_1 starts to charge C_p in the positive direction. Similarly, to obtain operating margin, S_1 and S_4 is turned on before the resonance between the panel capacitance and the external inductor ends. At t_4, S_1 and S_4 are turned on and the panel voltage v_p is clamped as V_s. In this mode, with the same assumption, the voltage V_s-V_1 is applied to L_1 and the inductor current i_{L1} increases linearly as follows:

$$i_{L1}(t) = i_{L1}(t_4) + \frac{V_s - V_1}{L_1}(t - t_4) \quad . \tag{5}$$

At t_5, the inductor current i_{L1} arrives at zero. The gate signal for S_5 is removed just before t_6. At that time, the current flowing through S_5 is zero and S_5 is turned off with ZCS. Also, S_1 and S_4 are turned off just before t_6. Since the plasma discharge ends at that time, S_1 and S_4 are also turned off with ZCS. Although the gate signals for S_2 and S_3 are removed, the panel voltage v_p is maintained as V_s until S_6 is turned on at the next period.

3 Experimental Results

The prototype sustain driver for a 42-in PDP, which has a capacitance of about 100nF, has been implemented with specifications of V_s =190V, L_1=450nH, and the switching frequency f_s= 200kHz. Fig. 2 shows the experimental waveforms of the proposed sustain driver, when the white image is displayed. To compare the circuit performance, the power loss in the circuit is measured. In this case, no image is displayed. The power consumption according to the input voltage source V_s at the maximum sustain pulses is compared with the conventional sustain driver in Fig. 2. The proposed method has lower power consumption than the conventional method. It is mainly because there are fewer components in current paths and the conduction loss is reduced.

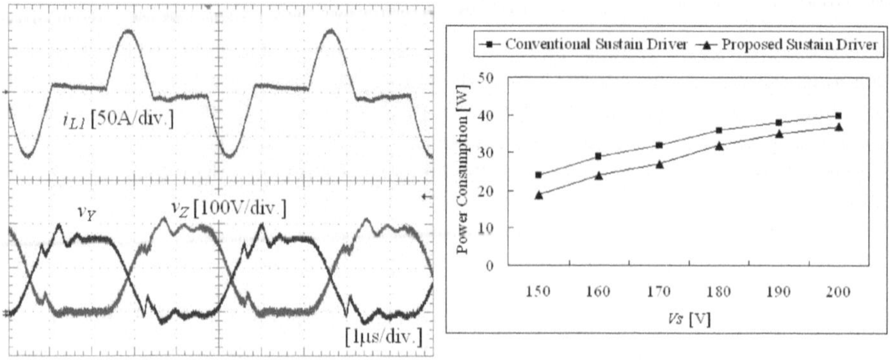

Fig. 2. Experimental waveforms and measured power consumption

4 Conclusion

A new sustain driver for a PDP is proposed to overcome the drawbacks of the prior method. In the proposed sustain driver, to obtain the cost-effectiveness, the number of switching devices is reduced compared to the conventional sustain driver and IGBTs are used as power switches. Since all power switches are turned off under zero-current switching (ZCS) operation, the tail current problem associated with IGBT does not occur. Moreover, due to a simpler structure compared with the conventional sustain driver, higher efficiency can be achieved.

References

1. Inanaga, Y., Iwata, A., Tanaka, M.: Measurement of wall charges in a surface discharge AC-PDP. In: Proceedings of the Fourth International Display Workshops, pp. 527–530 (1997)
2. Weber, L.F., Wood, M.B.: Energy recovery sustain circuit for the AC plasma display. In: Proc. Symp. Society for Information Display, San Jose, CA, USA, pp. 92–95 (1987)
3. Weber, L.F., Wood, M.B.: Power efficient sustain drivers and address drivers for plasma panel. U.S. Patent 5081400 (1992)
4. Ohba, M., Sano, Y.: Energy recovery driver for a dot matrix ac plasma display panel with a parallel resonant circuit allowing power reduction. U.S. Patent 5642018 (1997)
5. Hsu, H.B., et al.: Regenerative power electronics driver for plasma display panel in sustain-mode operation. IEEE Trans. Ind. Elec. 47, 1118–1124 (2000)
6. van der Broeck, H., Wendt, M.: Alternative sustain driver concepts for plasma display panels. In: IEEE Power Electronics Specialists Conf., pp. 267–2677 (2004)

Improved De-noising Algorithm on Directed Diffusion

Lixia Chen[1] and Zhaoyu Shou[2]

[1] School of Mathematics and Computing Science,
Guilin University of Electronic Technology, Guangxi, Guilin, P.R. China
[2] Science and technology department,
Guilin University of Electronic Technology, Guangxi, Guilin, P.R. China
clx_2001@126.com, guilinshou@guet.edu.cn

Abstract. The traditional directional diffusion model detected edges by the gradients of images, and they were susceptible to noise. Aiming at avoiding the weakness, an improved directional diffusion model is proposed. In this proposed model, a diffusion function is introduced and the modulus of gradient is substituted by the modulus of a wavelet transform, which gives the results that the newly introduced model could preserve edges better and has stronger ability to resist noise. The experimental results show improvements of the proposed model.

Keyword: Image de-noising, Directed diffusion, Wavelet transform.

1 Introduction

De-noising is currently one of the most interesting investigated topics in image processing from both a theoretical and an empirical point of view[1-3]. Noising commonly occurs in image acquisition and transmission in a variety of fields such as astronomy, remote sensing and so on. The important information, say edges, can be smoothed inevitably while de-noising, so a lot of studies focus on the algorithms which can remove noise and preserve characteristic information at the same time. In the past two decades, PDE models based on variation were widely used in image processing. Well known examples are Rudin- Osher-Fatemi's total variation[4], Perona-Malik's diffusion[5], Weichert's tensor diffusion[6], Chambolle-Lions's total variation analysis[7].

The earliest equation to process image comes from the heat equation, which describes the distribution of heat (or variation in temperature) in a given region over time. As a classical PDE applied in image processing, the heat equation is

$$u_t = c\Delta u \tag{1}$$

where Δ is a laplacian; c is a constant and the denotes diffusive coefficient.

D. Jin and S. Lin (Eds.): Advances in EECM Vol. 1, LNEE 139, pp. 633–637.
springerlink.com

Let ξ is a unit vector in the gradient direction, η is an other unit vector which is orthogonal to ξ, By the coordinate translation, the heat equation can be expressed as follows

$$u_t = c(u_{\xi\xi} + u_{\eta\eta}) \tag{2}$$

From (2), we can see that there are equal diffusions both in the gradient and the tangent directions. But actually, in order to well preserve the characteristic of an image while de-noising, the above formula (6) can be improved as

$$\frac{\partial u}{\partial t} = cu_{\eta\eta} \tag{3}$$

The above formula (3) is called directed diffusion model.

2 PDE and Wavelet Based De-noising Algorithm

From the above formula (3), we can see that there are equal diffusions both in the smooth regions or on the edges, which is unreasonable. So, a diffusion function is introduced to the model to preserve edges. In addition, the gradient magnitude which is easy to be affected by noise, in order to overcome this drawback, the magnitude of wavelet transform is used in [8]. Based on the above two points, in order to better preserve edges, an improved directed algorithm is presented.

Let $\psi(x, y)$ is a two-dimensional basic wavelet, the wavelet transform of the function $u(x, y)$ is defined as

$$W_a u(x, y) = \langle u(s,t), \psi_{a,x,y}(s,t) \rangle,$$

where $\psi_{a,x,y}(s,t) = \frac{1}{a}\psi(\frac{s-x}{a}, \frac{t-y}{a})$, $a \in R^+$ is the scale parameter. If $\psi(x, y)$ is separable, then $W_a u(x, y)$ has three components: horizontal detail $W_a^1 u(x, y)$, vertical detail $W_a^2 u(x, y)$ and diagonal detail $W_a^3 u(x, y)$. The magnitude of the wavelet transform of the function $u(x, y)$ is defined as

$$M_a u(x, y) = (|W_a^1 u(x, y)|^2 + |W_a^2 u(x, y)|^2 + |W_a^3 u(x, y)|^2)^{\frac{1}{2}}$$

which represents the intensity variation of the image $u(x, y)$ at point (x, y), $M_a u$ takes small values inside the region, and is relatively near the edges. Moreover, the stronger the singularity of image is, the larger $M_a u$ becomes. Therefore, $u_{\eta\eta}$ in formula (3) is replaced by $g(Mu)\frac{\partial^2 u}{\partial \eta^2}$, where $g(Mu)$ is a non-increase function of Mu, and satisfies $g(0) = 1$ and $\lim_{s\to\infty} g(s) = 0$. In this paper, we set

$g(Mu)=1/(1+|Mu|^2/k^2)$, where k is a threshold parameter of gradient. then the proposed model is

$$\frac{\partial u}{\partial t}=g(Mu)u_{\eta\eta} \tag{4}$$

from above, we can see that the diffusion only takes place in the η direction, which can better preserve or even enhance the edges of an image. In the process of filtering, $g(|\nabla u|)u_{\xi\xi}$ is a diffusion in ξ direction, and the diffusion function is $g(|\nabla u|)$. The filter process can be explained as: when (x,y) is in the smooth regions, the value of $M_a u(x,y,t)$ is relatively small. We know that $g(s)$ is a non-increase function, resulted in a large $g(M_a u)$ and a strong diffusion in η directions to remove noise; when (x,y) is on the edges, $M_a u(x,y,t)$ reaches the local maximum, resulted in a small $c(M_a u)$ and a weak diffusion to preserve edges; when (x,y) is the isolated noise, from the characteristics of the wavelet, we know that the value of $M_a u(x,y,t)$ is relatively small and a strong diffusion in η directions to remove noise.

In order to discrete formula (4), the forward difference is used in the time dimension, the forward and backward combined difference is used in the space dimension. Let the time step is Δt, the space step is h, Mu denotes the magnitude of the stationary wavelet transform of u .then the numerical scheme of (8) is:

$$\frac{u_{i,j}^{n+1}-u_{i,j}^{n}}{\Delta t}=g(Mu_{i.j})u_{\eta\eta}^{\alpha} \tag{5}$$

Where $(u_x)_{i,j}^n=(u_{i+1,j}^n-u_{i-1,j}^n)/2h$, $(u_y)_{i,j}^n=(u_{i,j+1}^n-u_{i,j-1}^n)/2h$, $(u_{xx})_{i,j}^n=(u_{i+1,j}^n-2u_{i,j}^n+u_{i-1,j}^n)/h^2$, $(u_{yy})_{i,j}^n=(u_{i,j+1}^n-2u_{i,j}^n+u_{i,j-1}^n)/h^2$,

$$(u_{xy})_{i,j}^n=(u_{i+1,j+1}^n-u_{i-1,j+1}^n-u_{i+1,j-1}^n+u_{i-1,j-1}^n)/4h^2,$$

$$u_{\eta\eta}^{\alpha}=\frac{u_y^2u_{xx}-2u_xu_yu_{xy}+u_x^2u_{yy}}{u_x^2+u_y^2+\alpha},$$

This discretization satisfies the stability condition $0\le\frac{\Delta t}{h^2}\le\frac{1}{4}$.

3 Experiments

In all experiments, $\Delta t=0.1$, $h=1$, and we take the stationary wavelet transform with the basic wavelet ‘db3’, the gradient threshold parameter k ranges from 10 to 40, and the performance evaluation of the proposed methods are measured by the peak signal to noise ratio (PSNR)

$$PSNR = 10\log_{10}(255^2 / MSE), MSE = \left[\sum_{i}^{M} \sum_{j}^{N} \frac{(u_0(i,j) - u(i,j)^2}{M.N} \right]^{1/2}$$

where $u_0(i, j)$ and $u(i, j)$ are the noised and the de-noised image, respectively. The test image is part of "Barbara". Table1 shows the de-noising effects of the proposed method on PSNR at different noise levels. The de-noised results by the improved algorithm and the directional diffusion model are listed in Fig.1. We can see that the trousers and necktie in (c) are clearer in details than those in (b).

4 Conclusions

The results of the present study indicated that the traditional directional diffusion model were susceptible to noise. Aiming at avoiding this weakness, an improved directional diffusion model is proposed. By introducing a new diffusion function to the traditional diffusion model and replacing the modulus of gradient by the modulus of a wavelet transform, the improved model smoothes the images according to it's characteristic, which gives the results that the new model could preserve edges better and has stronger ability to resist noise. The experimental results prove the conclusion.

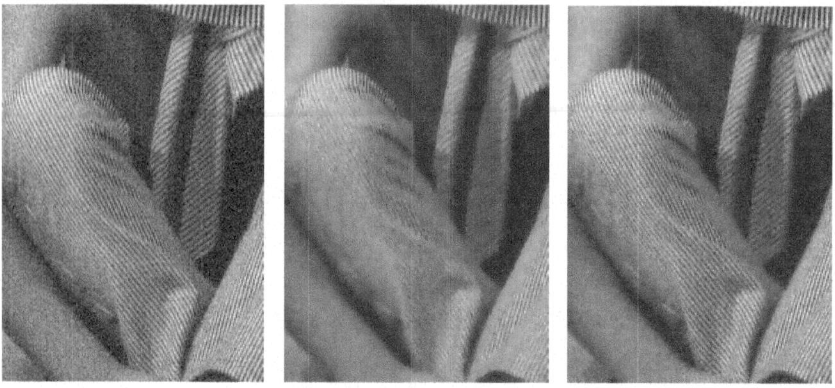

Fig. 1. From left to right, there is the noised image with the standard deviation 15, the denoised result of the standard directional diffusion model and the denoised result of the improved model

Table 1. PSNR of the proposed model and the directional diffusion model

Original image	Barbara		Lena		Plane	
σ	15	20	15	20	15	20
Noised image	24.5313	21.0024	24.5061	21.2116	25.1103	21.8842
Directionla diffusion	27.3529	26.1237	29.1012	28.1035	29.1436	28.7996
Improved model	28.0718	27.2015	29.5847	28.9185	29.8012	29.2236

Acknowledgments. This research is supported by the National Natural Science Foundation of China (10861005 &10871217) and the Guangxi Natural Science Foundation (2011GXNSFA018158) and the Guangxi Scientific Research and Technological Development Project (11107006-45).

References

1. Yin, S.F., Cao, L.G., Ling, Y.S., Jin, G.F.: Image denoising with anisotropic bivariate shrinkage. Signal Processing 91(8), 2078–2090 (2011)
2. Wu, J.H., Xu, J.T., Xia, R.B.: Surface mesh denoising via diffusion gradient field. Optics and Lasers in Engineering 49(1), 104–109 (2011)
3. Artur, L., David, B., Nishan, C., Alin, A.: Non-Gaussian model-based fusion of noisy images in the wavelet domain. Computer Vision and Image Understanding 114(1), 54–65 (2010)
4. Rudin, L.I., Osher, S., Fatemi, E.: Nonlinear total variation based noise removal algorithms. Physica D 60, 259–268 (1992)
5. Perona, P., Malik, J.: Scale-space and edge detection using anisotropic diffusion. IEEE Trans. on Pattern Anal. Machine. Intell. 12(7), 629–639 (1990)
6. Weickert, J.: Anisotropic Diffusion in image processing. ECMI Series, Teubner, Stuttgar (1998)
7. Chambolle, A., Lilons, P.L.: Image recovery via total variation minimization and related problems. Numerische Mathematic 79(2), 167–188 (1997)
8. Liu, F.: Diffusion filtering in image processing based on wavelet transform. Science in China Series F: Information Sciences 49(4), 494–503 (2006)

Exploring Differences of Consumers' Perceived Factors in Shopping Online: The Effects of Shopping Experience and Gender

Lingying Zhang[1,2], Yingcong Xu[2], Bin Ye[2], and Qingpeng Wang[2]

[1] Harbin Institute of Technology Shenzhen Graduate School, Shenzhen 518000, China
[2] College of Management, Shenzhen University, Shenzhen 518060, China
zly2009@gmail.com, yingcongxu@126.com

Abstract. There are two empirical studies in this paper. In the first study we investigated what were the important factors which would impact on the perceived factors of consumers' shopping online in China through interviewing, email and internet investigation, and obtained three important factors which were products, websites and personal factors based on 1075 valid respondents. In the second study we designed a questionnaire with 29 items used to measure the three kinds of influence factors verified in the first study and 792 valid data were collected. The differences between consumers with different shopping experiences and genders were compared and tested according to statistical and ANOVA analysis. The results indicate that: shopping experiences are important influent factors which will cause the differences of consumers' perceived factors in the process of shopping online; consumers' attitudes for online shopping would gradually be changed as the times of their shopping experience add up; there are significant differences about products factors between consumers both with different shopping experiences and with different genders.

Keywords: shopping experience, gender, differences, shopping online, perceived factors.

1 Introduction

More and more consumers are trying to add their shopping times with the rapid development of E-commerce in recent years. Most of them think that shopping online is an important choice of shopping channels when they have purchasing decision -making[1-3]. According to recent statistical data about the development of shopping online , we know that the number of Internet users in China goes up to the top 1 in the world, more and more experienced products are sold in Internet, and consumers become more caring about delivery and after-sale services of vendors when they decide to shop online [4,5].

Having reviewed the studies about the behavior of consumers in the Internet, we find that many of them focused on whether there were any factors which would impact on the behavior of consumers when they want to shop online[6-13]. Very little research carried out in the E-commerce has conducted a separate analysis of comparison of the difference of consumers' perceived factors in shopping online in considering the

D. Jin and S. Lin (Eds.): Advances in EECM Vol. 1, LNEE 139, pp. 639–646.
springerlink.com

characteristics of consumers, and indeed in their empirical studies, there is little study about the differences analysis of Chinese consumers[16].

In this paper, we have conducted two empirical studies. In the first study, we investigated what were the important factors which would impact on the perceived factors of Chinese consumers when they want to purchase online. In the second study, we designed a questionnaire with 29 items used to investigate the three kinds of influence factors verified in the first study. Based on the statistical and ANOVA analysis, the differences between consumers with different shopping experiences and genders were compared and tested.

2 Theoretical Background and Literature Review

2.1 Perceived Factors

Perceived factors are usually used to describe the perception of consumers in measuring their individual behaviours[1]. Previous researchers have tested different variables to measure individual behavior in the environment of E-commerce[1-16]. Jarvenpaa and Todd [1] analyzed the influences of products, shopping experience, and customer service on the behavior of consumers when they use the Internet as a shopping channel and purchasing online. Schaupp and Belanger [2] verified that convenience, trust to the venders, and delivery after sale were the three important characters of websites which would impact on the perception of consumers for shopping online. Crespo & Bosque[3] reviewed many related studies before and concluded that that there were four kinds of factors ,including product, technology, amusements, and information service, would influence the perception of consumers purchasing online.

Many researchers on E-commerce have proposed the important factors include technology, service, and product which have impacts on shopping online according their theoretical and empirical studies [2,6,7]. But they usually focused on the evaluation of these perceived factors, there is a little study about how these factors influence consumers before their decision-making for shopping online.

2.2 Shopping Experience and Gender

Shopping experience and gender are the two statistic variables related to consumers. Venkatesh et al[8], Porter and Donthu[9] studied their impacts on behavior of consumers by using them as adjusted variables based on TAM (Technology Accepted Model), verified that social statistics of consumers have significant relationship with their attitudes for new technology, and proposed that they are determining factors and blocks of consumers in using internet.

Garbarino and Strahilevitz[10] found that attitude and purchasing intension of consumers were different between different groups , females cared more about the loses caused by privacy and safety than male. Bassam Hasan [11] proposed that social statistics could be used as adjusted variables in analyzing the behavior online and new technology use. Sonia San Martín Gutiérrez et al[12] analyzed the correlation between product ,channel risk and participate of consumers. Considering gender, age and education as controlling variables, they analyzed the effects of these variables on perceived risks and shopping intensions of consumers and the results indicated that gender and education of consumers have significant influence on their perceived factors and participation in shopping online.

From the past literatures we find that the knowledge structure for consumers' purchasing online are very different from each other depending on different consumer's personal shopping experience, and the difference of knowledge structure will cause the difference of their personal behaviors such as searching information, diagnosing, analyzing, adjusting and applying[13].There are some empirical studies tested the impact of shopping experience on behavior, perceived risk and intension of consumers[13-15]. Cheolho Yoon[16] dedicated that the influence analysis about the antecedent and consumer satisfaction in condition that shopping experience as a potential adjusting variable.

Although there are some researches explored the effect of gender and shopping experience of consumers on their purchasing intensions[7,11,15,16], there are still little studies in-depth about the difference comparison between different groups of consumers with different gender and shopping experience.

3 Analysis of Important Influence Factors

3.1 Data Analysis

In this study, firstly we listed a series of possible questions related with shopping online on our investigation paper in order to collect the applicable data to analyze which are important influence factors that impact on the behavior of consumer purchasing online. Secondly, we conducted our investigation through interview, e-mail, and an investigation website(1diaocha.com) in China. The period of our investigation was from July 1st, 2010 to January 10th, 2011. We got 1075 valid data by checking up all the investigation data and the statistics are listed in Table 1. Finally, We picked up the top three answers of each factor and put them together to calculate the proportion of important influence factors as shown in Table2.

Most online Chinese consumers are young people with age between 20 to 40, just as the respondents in our study, in which the proportion of the age 20 to 40 is 72.19% and about 63.53% consumers have experience of shopping online . Analysis for the data collected from experienced consumers could make us get correct knowledge about the influence factors. But it left a clue needed to be studied in-depth as there are only 39.35% consumers who have definite attitude to shopping online and meanwhile many consumers, including experienced ones, still haven't definite attitude to shopping online.

The terms and proportion in table 1 indicate that there may be some relationship between the terms of social statistics and their attitudes to shopping online, which means in the second study we should consider the statistical variables of consumers as a kind of independent variables to analyze in-depth the differences of perceived factors for consumers shopping online.

Table 1. Statistics of consumers

Terms		Number	Proportion(%)
Gender	male	457	42.51
	female	618	57.49
Shopping Experience	yes	683	63.53
	no	392	36.47
Age	<20	108	10.05
	20-30	507	47.16
	30-40	269	25.03
	40-50	96	8.93
	>50	95	8.83
Education	middle	372	34.6
	college	627	627
	graduated	76	76
Attitude	Abs. no	159	14.79
	no	130	12.09
	Not sure	363	33.77
	agree	371	34.51
	Abs. agr.	52	4.84
Total		1075	100

3.2 Important Influence Factors

Besides the characteristics of consumers, there are another two kinds of influence factors: products and websites. For products, consumers in China usually worry about their quality, as whether they could receive real product with reliable and truth description as sales-web. According to the description in Table 2, there are many contents related with websites such as safety of payment, information, and economy; service of technology, delivery, and after-sale; trust to websites, products, and information. Sequence of these influence factors from higher to lower proportions are products, 33.02% ; service of website, 26.05%; safety, 22.79%; trust, 18.14%; as they are shown in Table 2.

Websites, as the shopping channel, play an important role in research on the important influence factor to the perception of consumers shopping online. Firstly, it is how to make sure the information of websites really and reliably as that information of product is real, the price of product clear, the credit rating of Agents reliable, and the introduction of websites truthful, etc. Secondly, it is about the safety of shopping online. The safety problems that make consumers worry about include the safety of payment system, personal information, and commitment for safety from websites. Thirdly, it is about the convenience and efficiency of the services such as quickly searching information before buying, easily proceeding shopping online, multiple kinds of payment choices, fast delivering of product and perfect service after-sale.

Products, as the target of shopping online, are always the core of consumers care about. The quality of products has become more important than before because it is difficult for consumers to gain the experience from internet. From the choices of our respondents we find that the kinds of products that consumers purchasing from websites have gradually from some numerical standard products changed into a widely kinds of products such as clothes, shoes, furniture, and many experience-needed products.

Table 2. Influence factors for shopping online

Factors	Description	Number	Proportion (%)
Products	Quality(reliable, real, truth description)	355	33.02
Websites	Safety(payment, information, economy)	245	22.79
	Service(technology, delivery, after-sail)	280	26.05
	Trust(websites, products, information)	195	18.14
Total		1075	100

Therefore, the results from Table 1 and 2 indicate that the former three important factors which influence the perception of Chinese consumers shopping online are websites, products, and consumer themselves.

4 Differences of Consumers' Perceived Factors—Effects of Shopping Experience and Gender

4.1 Questionnaire Design and Data Collection

We design a questionnaire with 32 items to investigate the three kinds of important perceived factors we obtained in above section based on our first study and related references. A pretest for the questionnaire was conducted in college students and graduated students in Shenzhen University. Then, the questionnaire was modified and adjusted based on the opinion and suggestion of our professional colleagues. Finally, we got our last questionnaire with 29 items about all the perceived factors. There are 3 items for product quality; 7 for safety of websites, 11 for services of shopping vendors in which 4 for technology, 4 for delivery and 3 for after-sale service; 3 items for trust, another 3 for attitude to shopping online; besides, there are 2 items about gender and shopping experience. We take likert 7 scale to measure the differences from minimum 1 means "definitely disagree" to maximum 7 means "definitely agree".

We did our data gathering from the random respondents coming from Beijing, Shanghai, Sichuan, Hunan, Guangdong and Shenzhen through interviewing, e-mail, BBS of some famous websites in China such as Taobao and Tesent, and internet investigation. We discriminate all the respondents and delete the answers with missing data and at last obtained 792 valid data, in which there are 435 male, the proportion is 54.9% and female 357, 45.1%; those who have shopping experience are 479, its proportion is 60.5%, there are 313 respondents who didn't shopping online before and its proportion is 39.5%.

4.2 Measures and Results

We use Cronbach'sαto analyze the reliability and validity for the questionnaire. Every α value of each factor is greater than 0.75 and the results indicate that the survey data in our research are true and reliable. In addition, the composite reliability value CR of each factor is greater than 0.8, which indicates that our survey data have high reliability. Besides, We use SPSS17 to perform the KMO and Bartlett's test of sphericity. The KMO of our survey is 0.821, which is more than 0.8, Bartlett's test chi-square p value is 0.000. These values indicate that the data from our questionnaire are acceptable and have a good validity.

Mean value comparison is used to analyze the average level and differences of data. All the factors are sorted in their mean value from higher to lower as follows: product, attitude, trust, safety, and service, as shown in Table 3. The mean values of product and attitude are more than 5 which indicate that all the respondents average agree that product and attitude are more important factors. The mean values of trust, safety, and service are not more than 5 but more than 4, which implicate that there are certain differences between different consumers.

Table 3. Statistics

Factor	Min	Max	Mean	Sta.D
Product	1	7	5.1488	0.98465
Attitude	1	7	5.06	1.11914
Trust	1	7	4.8267	1.02129
Safety	1	7	4.6723	1.011
Service	1	7	4.3159	0.83494

One-way ANOVA analysis is usually used to test the differences of dependent variables under a certain one independent factor. We use this method to explore and verify the differences of consumers' perceived factors in considering the effects of shopping experience and gender and the comparison results are shown in Table 4.

Table 4. One-way ANOVA analysis

<table>
<tr><th colspan="2" rowspan="2">Influence factors</th><th rowspan="2">Group variables</th><th colspan="2">One-way ANOVA</th></tr>
<tr><th>F</th><th>P</th></tr>
<tr><td colspan="2" rowspan="2">Product</td><td>shopping experience</td><td>13.86</td><td>0.000***</td></tr>
<tr><td>gender</td><td>3.303</td><td>0.039**</td></tr>
<tr><td rowspan="6">Website</td><td rowspan="2">trust</td><td>shopping experience</td><td>1.184</td><td>0.166</td></tr>
<tr><td>gender</td><td>1.574</td><td>0.183</td></tr>
<tr><td rowspan="2">safety</td><td>shopping experience</td><td>2.909</td><td>0.057*</td></tr>
<tr><td>gender</td><td>1.698</td><td>0.153</td></tr>
<tr><td rowspan="2">Convenience</td><td>shopping experience</td><td>4.128</td><td>0.018**</td></tr>
<tr><td>gender</td><td>0.909</td><td>0.46</td></tr>
<tr><td colspan="2" rowspan="2">attitude</td><td>shopping experience</td><td>14.633</td><td>0.000***</td></tr>
<tr><td>gender</td><td>0.036</td><td>0.556</td></tr>
<tr><td colspan="5">*Sig.<0.1, **Sig.<0.05, ***Sig.<0.01</td></tr>
</table>

For the factor product the results indicate that there are significant differences both for groups divided by shopping experience and by gender. There are not significant differences for the factor of trust between the two groups. Factor safety has a significant difference between the groups with different shopping experience, but not significant between the groups with different gender. For the perceived factor service, there is a significant difference between the groups with different shopping experience, but not significant for the groups with different gender.

5 Discussion and Implication

The results in Table 4 exhibit that there are significant differences for all the perceived factors between different groups with shopping experience. The effects of shopping experience will change the perceived factors of consumers including perceived knowledge for the quality of products; perceived safety, service, and trust of websites. It is worth to study the changes and related differences in-depth in the future. Meanwhile, the results also indicate that shopping experience should be used as an important independent variable in the research on the purchasing decision -making of consumers in Internet.

The attitudes of consumers to purchasing online will be changed as they have obtained more shopping experience. The results in Table 4 indicate that there is a significant difference between the groups with different shopping experience. It is worth for us to further study its influence of difference on shopping intension and the other behavior of consumers.

For the factor products, we find that there are significant differences not only between groups with different shopping experience but also with different gender. In one way it shows that consumers have been promoting their ability to distinguish the quality of products in websites; In another way it dedicates that the differences coming from their preference for products between groups of consumers would bring about significant differences for the factor products between groups with different gender.

For the factor trust, the results indicate that there are not significant differences between groups of consumers whatever with different shopping experience or gender. Together with the mean value analysis in Table 3, we find that Chinese consumers always think trust as an important factor of vendors in Internet and they would pay more attention to the trust of websites whatever they have many shopping experiences or not in Internet.

The conclusions obtained from our two studies have important implications for the business sector and for academic research, derived principally from the important influence factors of perceptions of customers and the differences analysis of effects of shopping experience and gender. Regarding the academic implications, our results contribute to the study in the field of E-commerce with that purchasing experience influences the evolution of customers' perceptions, attitude and behavior. Consequently, research on E-commerce must distinguish between customers who have no previous shopping experience in Internet and those who base their perceptions on their experiences, establishing behavior patterns for each sample. For the business sector contribution, the results suggest that the vendors cannot control the level of experience of customers but they can develop websites that include options for quality of products, safety and service, then, guide customers through a shopping experience that is confidence-building.

References

1. Jarvenpaa, S.L., Todd, P.A.: Consumer reactions to electronic shopping on the World Wide Web. Intern. J. Electronic Commerce 1(2), 59–88 (1997)
2. Schaupp, Belanger, F.: A conjoint analysis of online consumer satisfaction. Journal of Electronic Commerce Research 6(2), 95–111 (2005)
3. Crespo, A.H., del Bosque, I.R.: The influence of the commercial features of the Internet on the adoption of e-commerce by consumers. Electronic Commerce Research and Applications 9(6), 562–575 (2010)
4. CNNIC:The 26th Statistical Report about the Situation of Internet Development in China (in Chinese), http://research.cnnic.cn/html/1279173730d2350.html (July 15, 2010)
5. Wang, Z.: Consultation Corporation: Investigation Report about Consumers Shopping, Beijing (2010), http://www.iaskchina.cn/Report/view/id/88 (May 10, 2010) (in Chinese)
6. Chen, Y.-H., Hsu, I.-C., Lin, C.-C.: Website attributes that increase consumer purchase intention:A conjoint analysis. Journal of Business Research 63(9-10), 1007–1014 (2010)
7. Hernandez, B., Jimenez, J., Jose Martin, M.: Customer behavior in electronic commerce: The moderating effect of e-purchasing experience. Journal of Business Research 76, 964–971 (2010)
8. Venkatesh, V., Morris, M.G., Davis, G.B., Davis, F.D.: User acceptance of information technology: Toward a unified view. MIS Quarterly 27(3), 425–478 (2003)
9. Porter, C.E., Donthu, N.: Using the technology acceptance model to explain how attitudes determine Internet usage:the role of perceived access barriers and demographics. Journal of Business Research 59(9), 999–1007 (2006)
10. Garbarino, E., Strahilevitz, M.: Gender differences in the perceived risk of buying online and the effects of receiving a site recommendation. Journal of Business Research (57), 768–775 (2004)
11. Hasan, B.: Exploring gender differences in online shopping attitude. Computers in Human Behavior 26(4), 597–601 (2010)
12. Gutiérrez, S.S.M., Izquierdo, C.C., Cabezudo, R.S.J.: Product and channel-related risk and involvement in online contexts. Electronic Commerce Research and Applications 9(3), 263–273 (2010)
13. Castaneda, J.A., Munoz-Leiva, F., Luque, T.: Web acceptance model (WAM): moderating effects of user experience. Information & Management 44(4), 384–396 (2007)
14. Taylor, S., Todd, P.: Assessing IT usage: The role of prior experience. MIS Quarterly 19(4), 561–570 (1995)
15. Doolin, S., Thompson, F.: Perceived risk, the Internet shopping experience and online purchasing behavior:a new zealand perspective. Journal of Global Information Management 13(2), 66–88 (2005)
16. Yoon, C.: Antecedents of customer satisfaction with online banking in China: The effects of experience. Computers in Human Behavior 26(6), 1296–1304 (2010)

Self Tuning of PID Controller Based on Simultaneous Perturbation Stochastic Approximation

Ping Xu, Geng Li, and Kai Wang

Xi'an High-Tech Institute, Xi'an, 710025, China

Abstract. A model need to be identified for the controlled plant when applying advanced strategies to determine the parameters of a PID controller. An online, model-free parameter tuning method is proposed based on simultaneous perturbation stochastic approximation in order to free the control system from plant model. The main idea for this tuning method is to use the highly efficient simultaneous perturbation approximation to the gradient of performance index function for PID controller. All parameters of PID controller can be tuned online by stochastic approximation. Simulation results show that satisfactory performances can be obtained using the proposed tuning method.

Keywords: PID controller, two degree-of-freedom (2-DOF), tuning method, simultaneous perturbation stochastic approximation (SPSA).

1 Introduction

Although there is much artificial intelligence algorithm, proportional integral derivative controller (PID) holds a leading post by such advantages as simplifying structure, field debugging, being of good robustness, etc. Achievement has been made on PID parameter tuning during the past dozens of years by many researchers after Ziegler and Nichols raising the question of parameter tuning, such as Ziegler-Nichols formula, Cohen-Coon method, IAE、ITAE optimality index method, IMC method, relay self-tuning based on iteration feedback tuning. A proper system model should be selected by special experiment when these tuning strategy is adopted, excellent control effect is hard to be obtained for the difficulties in building system models or models identification.

PID parameter tuning method and tuning algorithm steps based on simultaneous perturbation stochastic approximation are proposed in this paper that realizes PID parameters' online, model-free self-tuning without knowing the control object model.

2 Problem Description

Thereinto:
r-reference, u-control, y-output, v- perturbation, G(s) -unknown control object, C(s) -standard PID controller:

$$C(s) = K\left(1 + \frac{1}{T_i s} + T_d s\right) \tag{1}$$

D. Jin and S. Lin (Eds.): Advances in EECM Vol. 1, LNEE 139, pp. 647–652.
springerlink.com

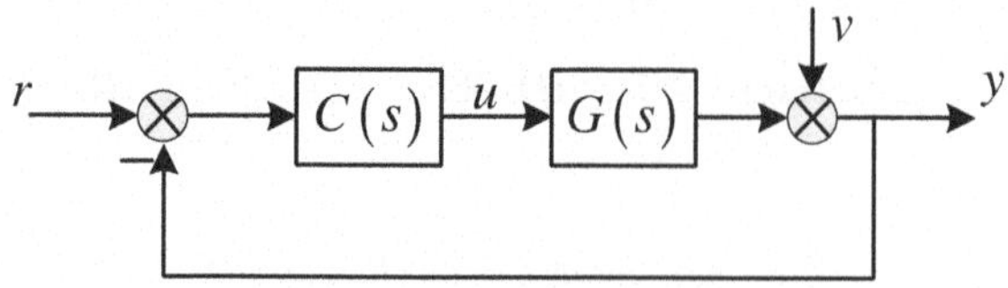

Fig. 1. PID controller closed-loop feedback control system

3 Parameter Tuning

3.1 Principles of Simultaneous Perturbation Stochastic Approximation (SPSA)

SPSA algorithm was first proposed in 1987 and fully discussed in 1992 by J.C. Spall, who proposes adaptive stochastic approximation by the simultaneous perturbation method. It is extensive used in many fields such as statistic parameter estimation, traffic control, feedback control, signal and image processing, etc.

SPSA algorithm is mainly used on optimization, such as minimization function, namely $\min(L(\theta))$. $L(\theta)$- differentiable and measurable variables, θ-continuous P-dimension adjustable vector parameters. Grads information or grads value of $L(\theta)$ is not necessary during optimization, namely, $\partial L(\theta)/\partial\theta$ could not be obtained directly. Only measured value of function, mostly influenced by noise, is needed during iteration. Adjustment of parameter θ in SPSA:

$$\hat{\theta}_{k+1} = \hat{\theta}_k - a_k \hat{g}_k(\hat{\theta}_k) \tag{2}$$

$\hat{g}_k(\hat{\theta}_k)$-the estimated value in step k of grad $\partial L(\theta)/\partial\theta$,

$a_k = a/(A+k+1)^\alpha$ -step factor

So, key of algorithm is to find optimal value θ^* that assures corresponding grad zero:

$$g(\theta^*) = \frac{\partial L(\theta)}{\partial\theta}\Big|_{\theta=\theta^*} = 0 \tag{3}$$

SPSA algorithm applies to parameter optimization of higher dimension because estimated value of grad is obtained by simultaneous perturbation and only two measured values with noise.

Grad estimated value is as follows:

$$\hat{g}_k(\hat{\theta}_k) = \frac{q(\hat{\theta}_k + c_k\Delta_k) - q(\hat{\theta}_k - c_k\Delta_k)}{2c_k}\begin{bmatrix}\Delta_{k1}^{-1}\\ \Delta_{k2}^{-1}\\ \vdots\\ \Delta_{kp}^{-1}\end{bmatrix} \tag{4}$$

$c_k = c/(k+1)^\gamma$ -- marginal perturbation value

$q(\theta)$ --measured value of function with noise

$\Delta_k = \left[\Delta_{k1}, \Delta_{k2}, \cdots, \Delta_{kp}\right]^T$ -- stochastic perturbation vector

3.2 Setting Algorithm

Purpose of controller parameter setting is to optimize system. Performance index is show in equation (6). Parameter setting of controller is to find a parameter group θ to minimize $J(\theta)$.

$$J(\theta) = \int_0^\infty e^2(t,\theta)dt \tag{5}$$

$e(t,\theta) = r(t) - y(t,\theta)$ is the difference between reference signal and output signal of closed control system.

$\theta \triangleq [K, T_i, T_d]^T$ includes 3 parameters of PID controller. Self-tuning of PID controller based on simultaneous perturbation stochastic approximation block flow diagram is shown in Fig.2.

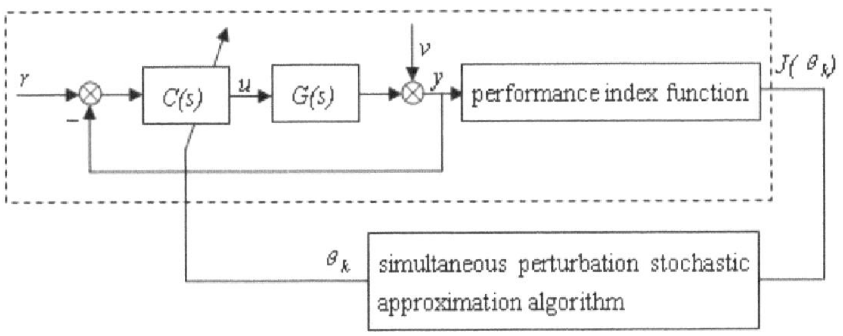

Fig. 2. PID parameter self-tuing

$q(\theta) = J(\theta)$ is supposed in this paper, so the self-tuning algorithm is as follows:

Step1 Algorithm initialization., $\hat{\theta}_0$, initial value of controller parameter, is calculated by Z-N method while $k = 0$. Parameter a, c, A, α and γ in gain sequence a_k and c_k are set.

Step2 build simultaneous perturbation stochastic vector. Build 3-dimension stochastic perturbation vector Δ_k by Monte Carlo. $E\{\Delta_{ki}\} = 0$, $i = 1,2,3$. Simultaneous perturbation vector is build in this paper by 1/2 probability of Bernoulli's ±1 distribution.

Step3 Calculate performance index. Calculate $J\left(\hat{\theta}_k + c_k\Delta_k\right)$ and $J\left(\hat{\theta}_k - c_k\Delta_k\right)$ corresponding to Δ_k and c_k in the former steps. $\hat{\theta}_k$ is controller parameter vector in step k.

Step4 Gradient estimation. Estimate gradient value in the next step by equation (4).

Step 5 adjustment. Use equation (2) to adjust controller parameter.

Step 6 Iterating or terminating. Turn to the next step when equation (3) is satisfied. Otherwise, $k = k + 1$, back to step (2). The principle of termination is that there is not much difference of performance index in continuous iteration.

Step 7 Output. Put optimized controller parameters in 2 freedom PID and output system response curve.

4 Emulation Research

$G_1(s)$ -object of non-minimum phase

$G_2(s)$ -object by first order inertia delay link

Use $G_1(s)$ and $G_2(s)$ make verification experiments for the methods in this paper, Z-N and paper [4].

$$G_1(s) = \frac{1-5s}{(1+10s)(1+20s)} \tag{6}$$

$$G_2(s) = \frac{1}{1+20s}e^{-20s} \tag{7}$$

Response result is in Fig.3 and Fig.4. The table 1 shows the controller parameter tuning results by 3 different tuning methods for different control object.

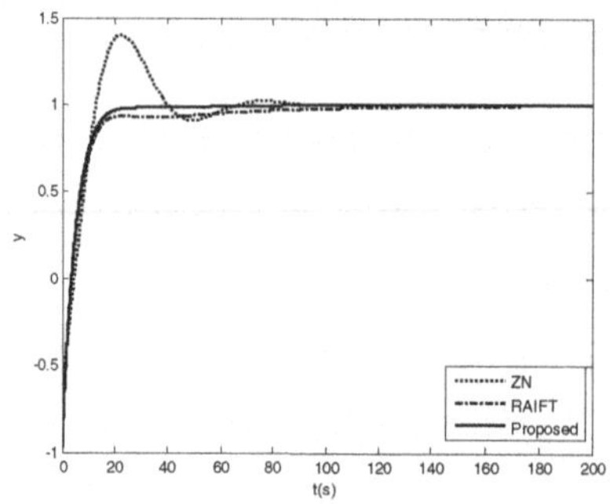

Fig. 3. Dynamic response curve of $G_1(s)$

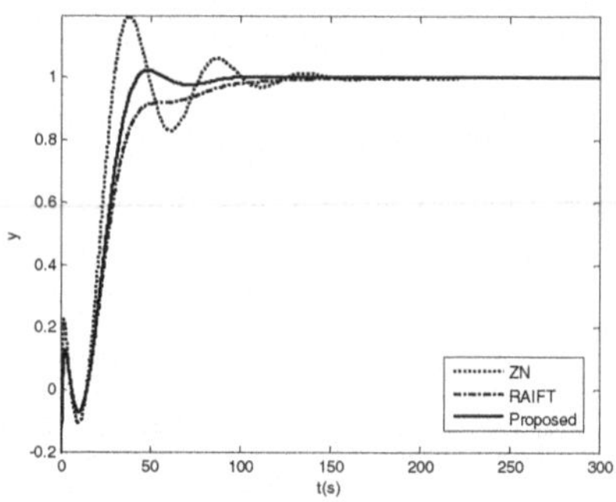

Fig. 4. Dynamic response curve of $G_2(s)$

Table 1. Controller parameter by 3 different tuning methods

object	Tuning method	K	T_i	T_d
$G_1(s)$	ZN	3.622	16.711	4.178
	RAIFT	3.028	46.318	6.079
	SPSA	3.0772	33.0591	6.4144
$G_2(s)$	ZN	1.363	30.800	7.70
	RAIFT	0.930	30.106	6.075
	SPSA	1.1097	27.6749	6.4048

Compared with traditional tuning method and to certify robustness of algorithm model error, made perturbation for $G_1(s)$ as follows:

A: add 50% steady-state gain, change extended time constant from 20 to 25.
B: lead in 3 seconds delay.
Then, after perturbation:

$$G_{1a}(s) = \frac{1.5(1-5s)}{(1+10s)(1+25s)} \tag{8}$$

$$G_{1b}(s) = \frac{1-5s}{(1+10s)(1+20s)} e^{-3s} \tag{9}$$

Fig. 5and Fig.6 shows response curve of $G_1(s)$.

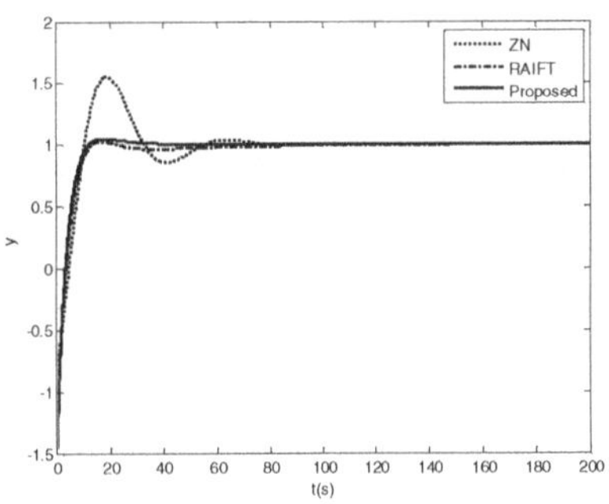

Fig. 5. Dynamic response curve of $G_{1a}(s)$

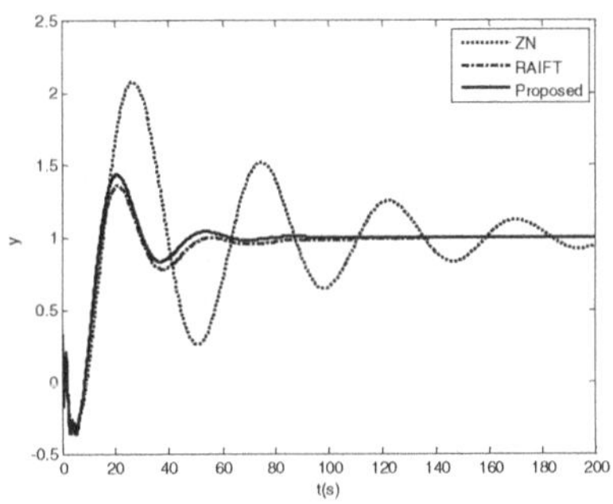

Fig. 6. Dynamic response curve of $G_{1b}(s)$

Emulation shows: Self tuning of PID controller based on simultaneous perturbation stochastic approximation can meet the needs of time-lag and non-minimum phase objects control. Compared with traditional methods, it has smaller overshoot and spends less adjustment time.

Robustness of model error: controller of all tuning methods get good robustness. The controller of traditional tuning need more adjustment time and suffers oscillation when there is delay error in system. Obviously, the tuning method in this paper gets better robustness.

5 Conclusions

This paper offers self tuning of PID controller based on simultaneous perturbation stochastic approximation and the specific steps that enable controller parameter reflect demands of performance index, making closed loop performance index meet the needs of control system. Meanwhile, this method does not need mathematical model or prior knowledge, especially applied to nondeterministic system for the advantages of simple, easy and convenient. Emulation shows: the method gains good performance in online PID parameter setting and be independent of precise system model.

References

1. Astrom, K.J., Hagglund, T.: The future of PID control. Control Engineering Practice 9, 1163–1175 (2001)
2. Wang, W., Zhang, J., Chai, T.: A Survey of Advanced Pid Parameter Tuning Methods. Acta Automatica Sinica 26(3), 347–355 (2000)
3. Astrom, K.J.: Automatic tuning of PID regulators. Instrument Society of America, North Carolina (1988)
4. Liu, X., Liu, H., Zhang, Q., An, J.: Relay Auto-tuning of PID Controller Based on Iterative Feedback Tuning. Journal of Northwestern Polytechnical University 22(3), 380–383 (2004)
5. Spall, J.C.: A stochastic approximation technique for generating maximum likelihood parameter estimates. In: Proceedings of the American Control Conference, pp. 1161–1167 (1987)
6. Spall, J.C.: Multivariate stochastic approximation using a simultaneous perturbation gradient approximation. IEEE Transactions on Automatic Control 37(3), 332–441 (1992)
7. Spall, J.C.: Adaptive Stochastic Approximation by the Simultaneous Perturbation Method. IEEE Transactions on Automatic Control 45(10), 1839–1853 (2000)
8. Spall, J.C.: Feedback and weighting mechanisms for improving jacobian estimates in the adaptive simultaneous perturbation algorithm. IEEE Transactions on Automatic Control 54(6), 1216–1229 (2009)
9. Song, Q., Spall, J.C., Soh, Y.C., Ni, J.: Robust neural network tracking controller using simultaneous perturbation stochastic approximation. IEEE Transactions on Neural Networks 19(5), 817–835 (2008)
10. Guo, C.Y., Song, Q., Cai, W.J.: A neural network assisted cascade control system for air handling unit. IEEE Transactions on Industrial Electronics 54(1), 620–628 (2007)

A Study on the Benefit Distribution of Mobile Publishing Industrial Chain Based on the Cooperative Game Theory

Jing Su and Lianjia Ren

Beijing University of Posts and Telecommunications (BUPT)
School of Economics and Management
No.10 Xi Tucheng Road, Haidian District, 100876 Beijing, P.R.C
bysj@sina.com, renlianjia_8608@126.com

Abstract. In recent years, more and more digital publishing products appeared. Among the digital publishing products, mobile publishing is the fastest developing one. However, one of the biggest bottlenecks that restrict the development of mobile publishing is the copyright issues. The uneven distribution of benefits is the core problem. In order to promote the development of mobile publishing and the formation of industry alliance, this paper builds the theoretical model based on the cooperative game theory to research the rational distribution of benefit in the mobile publishing industrial chain.

Keywords: Mobile Publishing, Cooperative Game Theory, Distribution of Benefit, Industry Alliance.

1 Introduction

The protection target of copyright system is works; however, mordern copyright is born with the invention and widely application of printing, not works [1].The widely application of printing changes the benefit pattern of publishing industry deeply. The creation of modern copyright is to solve the rearrangement of benefit pattern. So, modern copyright is called as the son of print publishing [2].With the progress of technology, the copyright law and publishing induatry has changed a lot.

In recent years, more and more digital publishing products appeared which has become an important force and changed the development direction of publishing industry. Mobile publishing is the most important one in the digital products. During this article, we will use the cooperative game theory to research the rational distribution of benefit in the mobile publishing industrial chain.

2 Present Status of Publishing Industry in Digital Age

2.1 The Development of Digital Publishing Industry

According to the 2010 annual report of digital publication of China, which is posted by the National News Publication Researching Institute of China, we find that, the total

D. Jin and S. Lin (Eds.): Advances in EECM Vol. 1, LNEE 139, pp. 653–659.
springerlink.com

value of digital publishing industry is 105,179,000,000￥, which has expanded by 31.97 percents than last year. Among them, mobile publishing is responsible for 33.26 %(online game and internet advertising for 30.78% and 30.54% respectively). The total of them has exceeded 90%. Correspondingly, electronic book and digital newspaper only share 2.36% and 0.57% respectively (as figure 1 and figure 2 shows).

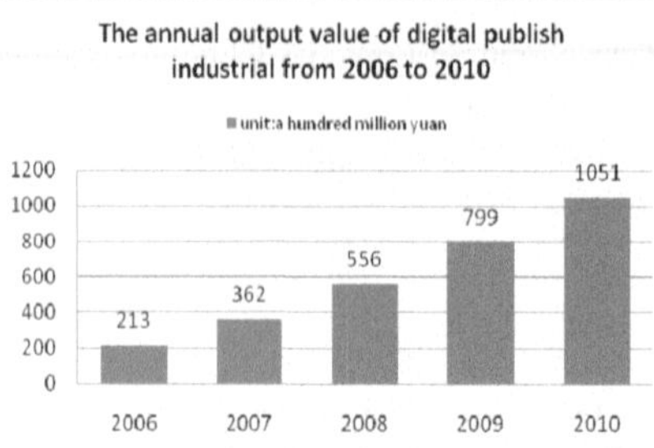

Fig. 1.

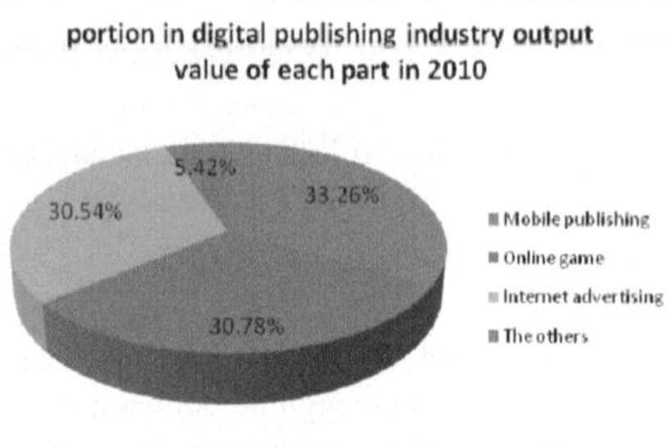

Fig. 2.

Data sources: The annual report of Chinese digital publishing

2.2 The Blowout Type Development of Mobile Publishing

With the coming of 3G age, mobile phone has converted into portable media. Among digital publishing products, mobile publishing is the fastest developing type and shares the biggest portion

All over the world, the flourishing of mobile publishing has become an inevitable trend. Based on the mature charge mode, mobile publishing has taken the biggest portion in the digital publishing industry, although its starting is late in China. Mobile reading is the major type of online reading now. The number of mobile internet users in China is 303,000,000 in the December of 2010, which increased by 69,300,000 from the end of 2009. Mobile phone users have taken a bigger and bigger portion among the netizens.It increased to 66.2% from 60.8% in the 2010(as Figure 3 shows). It has become the major force which promotes the number of netizen in China. So, we can find that mobile publishing has a prosperous prospect.

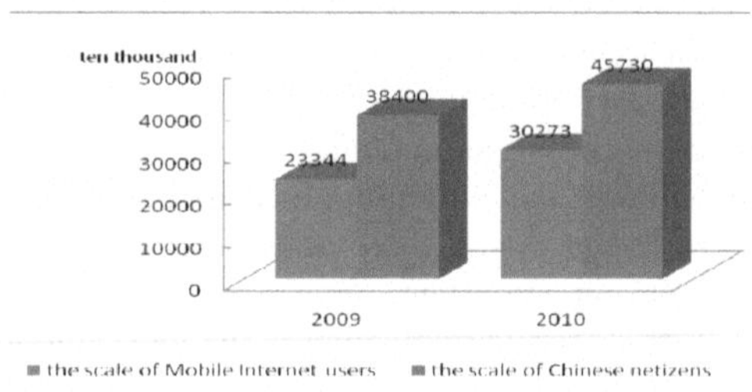

Fig. 3. Data sources: The 27 times Chinese Internet network development statistics report

3 The Problems Existed in the Development of Mobile Publishing

Although mobile phone is developing rapidly as a new publishing media, and has shown obvious advantages, the development of mobile publishing is still facing many problems.

First, the profit of mobile publishing industrial chain is unevenly distributed. At present, relatively complete industrial chain of mobile publishing has been established (as Figure 4 shows) , which includes content provider (CP, mainly includes traditional publishing house, traditional recording industry, film industry and individual), service provider (SP, mainly includes hardware manufacturer, software provider, and platform operator), and telecom operator. Operators play a dominant role in the whole industry chain; however, CP is in the weakness position, and gets the least profit. In the past developments, channel and operation take the most important place. The content had the least weight. Most of the profit which created by mobile publishing was shared by several major telecom operators and middlemen. The traditional publishers probably got less than 10%. Traditional publishers and authors get too little from the digital publishing. So, their initiative is low. It does harm to the development of the industry if win-win situation cannot be achieved.

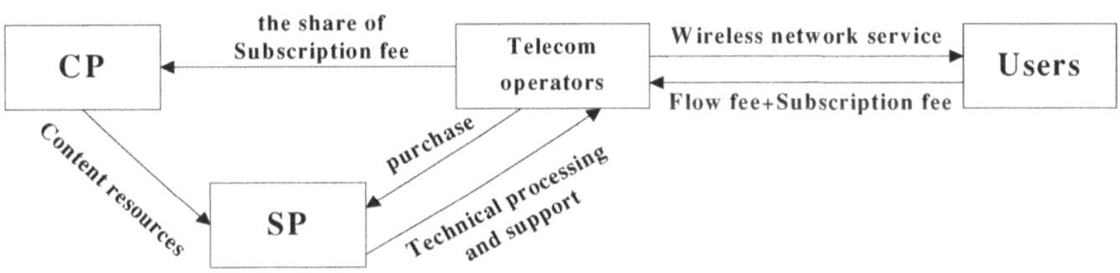

Fig. 4. The structure of mobile publishing industrial chain

Second, the content of mobile publishing is simplex. No matter how the new technique changes, how the new media develops, quality content resources are always the core of competition of publishing industry. However, at present the common problems of many mobile phone publications are content problems. The content resources are scarce, homogeneity serious, lack of initiative. These problems cause the content has no sufficient attraction. Also, because the profit of the industrial chain is unevenly distributed, mobile publishing in China is actively motivated by telecom operators and SP. The passion of CP is not very high.

Moreover, the profit model is unclear. Although current mobile publishing has mobile value-added services fees and advertising benefit by inserting advertisement into publications, the profit mainly comes from the subscription. For this reason, the profit model is simplex. The profit space is small. The motivation of users, development of subscription, and the profit are all decreased, since the subscription fee of mobile publishing is higher than that of traditional publication. The current market of mobile publishing is in the period of cultivation, and needs plenty of money for market expending. Totally speaking, the profit space of mobile publishing is limited.

4 Solving thc Problem of Mobile Publishing with the Cooperative Game Theory

4.1 The Analysis of Benefit Distribution of Each Participant in Mobile Publishing Industrial Chain by Cooperative Game Theory

The problem existing in mobile publishing is due to the uneven benefit distribution of every participant as the preceding part shows, and we will discuss this problem and give particular solution.

Game theory is divided into cooperative game theory and non-cooperative game theory due to different assumption. The former one highlights the collective rationality and focus on efficiency, just and equitable[3]; while the latter one mainly study how people choose strategy to maximize their interest in a situation in which people's interest influences each other and focus on the individual rationality and individual optimal decision. In the real world, cooperation exists definitely as long as it can bring more profit [4]. Thus, this article believes that the benefit distribution behavior of every participant in the industrial chain of mobile publishing is suitable for the suit condition of the cooperative game theory, and lies in two aspects:

Firstly, the key point of identifying cooperative game and non-cooperative game is whether every principle part can negotiation with each other efficiently [5]. In the whole industrial chain of mobile publishing, the publishing activity is mainly formed by content provider (CP), service provider (SP) and telecom operator. It can be seen that every participant of the mobile publishing industrial chain has its own 'core superiority', and thus, if the core superiority complement is achieved, the mobile publishing will have a larger space for development and will develop towards a healthy direction.

Secondly, the alliance of mobile publishing industry needs to consider both the collective rationality and individual rationality. This is the basement of cooperative game, i.e., the profit of the whole is larger than the sum of profit made by individual operation [6]. To each corporation, the profit created by the cooperation is at least larger than that created by individual operation [7], otherwise they would not cooperate.

4.2 Cooperative Game Model of the Distribution of Interests

First of all, let me introduce the mathematical expressions related to the cooperative game:

Any subset (S) of N—referring to all the participants (N= {1, 2……n})—is called a union.

The characteristic function: if given a game including n persons and S is a union, V (S) refers to the maximum effect of S in the game between S and $N\text{-}S=\{ i \mid i \in N, i \notin S \}$ and V (S) is called the characteristic function of union S. Generally, we use (N, V) to stand for the cooperative game in which there are N participants and the characteristic function is V [8].

Second, we should set a model. We can reach two corollaries in the following basic on the hypothesis of 'rational economic man 'in the game theory:

Corollary 1: At least, the improvement of income should equal to the direct loss of all the participants caused by the cooperation after all the partners in the mobile phone industry form the alliance.
Corollary 2: if $S \subseteq S'$ and i can't participate in union S and S ' in the meantime, the improvement of the income of S' should be larger than the one of S after i participate in them.

Because the participants have different positions in the mobile phone industry currently, expected earnings will not be identical. If the members whose expected return is less cannot get reasonable compensation, the cooperation is difficult to form [9]. From the above, we know that Shapley distribution vector is one core of the group cooperative game. There must contain interest transfer between members to ensure the cooperation.

In the game (N, V), in which the effect can be transferred, if there exists a fixed number h_j^i, which makes $S \subseteq N$ and $i \notin S$, ,

$$V(S \cup \{i\}) = V(S) + V(\mathrm{i}) + \sum_{j \in S} (h_i^j + h_j^i)$$ then the value of Shapley is:

$$\varphi_i(N,V) = V(i) + \frac{1}{2} \sum_{j \in N} (h_i^j + h_j^i)$$

In order to analyze conveniently, we assume that the cooperative behavior of all the members taking part in the cooperation have no influence on the non-members and the policy is consistent, we can define the following characteristic functions:

$$V(S) = \sum_{i \in S} \left(R_0^i + \sum_{j \in N} K_j^i - \sum_{j \notin S} \theta_j^i + \sum_{j \notin S} \pi_j^i \right) \tag{1}$$

In this form, the first term is the reserved utility respectively before their cooperation; The second one is the total benefit of all the members in the union S when the alliance N was formed; The third one is benefit of the member of union S if it cooperates with the other members (among union N); The fourth one is the loss of earnings that the members not attending the alliance S (but still in the N) caused.

For, $g \notin S$ there is $$V(S \cup \{g\}) = V(S) + V(g) + \sum_{j \in S} \left[(\theta_g^j - \pi_g^j) + (\theta_j^g - \pi_j^g) \right]$$

If g = i and $h_i^j = \theta_i^j - \pi_i^j$; $h_j^i = \theta_j^i - \pi_j^i$ then

$$\varphi_{\mathrm{i}}(N,V) = R_0^i + \sum_{j \in N} K_j^i - \sum_{j \neq \mathrm{i}} \theta_j^i + \sum_{j \neq \mathrm{i}} \pi_j^i + \frac{1}{2} \sum_{j \neq i} \left[(\theta_i^j - \pi_i^j) + (\theta_j^i - \pi_j^i) \right]$$

$$= R_0^i + \sum_{j=i}^{n} k_j^i + \frac{1}{2} \sum_{j=i}^{n} \left[(\pi_j^i - \theta_j^i) - (\pi_i^j - \theta_i^j) \right]$$

As for the cooperative game (N, V), the value of Shapley can be got from the following formulas:

$$\varphi_i(N,V) = R_0^i + \sum_{j=1}^{n} K_j^i + \frac{1}{2}\sum_{j=1}^{n}\left[\left(\pi_j^i - \theta_j^i\right) - \left(\pi_i^j - \theta_i^j\right)\right] \quad \text{i=1, 2,..., n} \tag{2}$$

If alliance N is formed but there is no utility transferred, the benefit of the member i

$$R_0^i + \sum_{j=1}^{n} K_j^i \qquad \text{i=1,2,...,n}$$

According (2), we can see that the benefit of each member attending the cooperation equals to the benefit that get from the cooperation but have no utility transferred adds half of the benefit that get from cooperation, and the difference is the income distribution when the Shapley value was executed. And from the above, we can get several results as follows:

When all the members participate in the cooperation, Shapley interest distribution vector will give every member the compensation of interests.

$$T_i = \frac{1}{2}\sum_{j\neq i}^{n}\left[\left(\pi_j^i - \theta_j^i\right) - \left(\pi_i^j - \theta_i^j\right)\right] \qquad \text{i=1,2,...,n} \tag{3}$$

For any member i and j, π_j^i is the benefit that i get from the cooperation with others not with j and θ_j^i is the loss that i get without cooperation with j. Therefore, $\pi_j^i - \theta_j^i$ is the net income that i get from their non-cooperation between i and j; similarly, $\pi_i^j - \theta_i^j$ is the net income that j get from the non-cooperation between j and i.

Therefore, the total net income compensated to i or taken from i to compensate other members (i.e. the value of T_i can be positive or negative) is the sum of all the differences that i gets from working with its partners. For any bilateral compensation quantity, type (3) means: in the cooperation, members who get less benefit should get compensation from the members who get more benefits. Therefore, the latter should give the former some compensation. Then, the agreement may be formed. Meanwhile, the benefit should be more than the one before participating in the cooperation.

5 Conclusion

By means of establishing a industry alliance, participants involved in mobile publishing industry chain could obtain various genuine products with high quality and copyright protection. In the cooperation, members who get fewer benefits should get compensation from the members who get more benefits. In a particular situation, this compensation is set, as formula (3).

The establishment of mobile publishing association, will solve the unbalanced beneficial problem for publications, and push CP to participate willingly. With more contents, SP could provide even more personal, customized services, thus lift up user experiences, and increase users' degree of satisfaction. With fruitful resources, personal services, high satisfaction, the profit model for mobile publishing could evolve clearly. Moreover, industry alliance can also provide supervision, to prevent the distribution of pirate contents.

This model of cooperation and distribution may reduce the earnings of the dominant operators temporally in industrial chain of mobile publishing; however, by establishing industry alliance, participants in industrial chain of mobile publishing can make use of their advantages respectively without worries of benefit infringement or loss of existing core competitive advantages. Cooperation can create optimum complementation. It will benefit long term development of mobile publishing. In addition, cooperation can bring participants more benefits and any breach of collaboration will lead to profits decrement.

Although the research here mainly focuses on mobile publishing, it also can be a good reference for other digital publishing models. Due to limited time and energy, the empirical analysis here is not sufficient. Because of so many affecting factors, revenue transfer in formula 3 should be used together with other compensation mode in order to make better effect in practice.

References

1. Zheng, S.: Copyright law, p. 8. Chinese People's University Press (1990)
2. Duan, R.: Overview of Intellectual Property Law, p. 28. Guangming Daily Press (1988)
3. Yang, B., Gao, Z.: Non-empty Property of the Core in Cooperative Games of the Supply Chain Value Network. Systems Engineering 27(11), 77–81 (2009)
4. Krus, L., Bronisz, P.: Cooperative Came Solution Concepts to A Cost Allocation Problem. European Journal of Operational Research 122(2), 258–271 (2000)
5. Belenky, A.S.: Cooperative Games of Choosing Partners and Forming Coalitions in the Marketplace. Mathematical and Computer Modelling 36(11-13), 1279–1291 (2002)
6. Zang, W.: The Theory of Games and Information Economics, pp. 78–93. Shanghai People's Press (1996)
7. Mccain, R.A.: Cooperative Games and Cooperative Organizations. Journal of Socio-Economics 37(6), 2155–2167 (2008)
8. Li, B., Wang, Y.: Game Theory and its Application, p. 124. China Machine Press (2010)
9. Zhang, P.: Non-complete Common Interest Group Cooperation & Game Theory and Application, p. 41. Shanghai jiaotong University Press (2006)

The Gas Pipeline Risk Assessment Base on Principal Component Analysis and BP Neural Network

Guo Zhanglin[1] and Zhang Huanjun[2]

[1] Department of civil engineering, North China Institute of Science and Technology, East yanjiao, Beijing, China, 101601
gg_zz_11@163.com
[2] School of Economics and Management HeBei University of Engineering Handan, 056038 China
qiuriliange2006@26.com

Abstract. In this article ,we combine the status of Handan City gas pipeline, then proposed comprehensive evaluation ideas based on principal component analysis and BP neural network, The BP neural network has the advantage of parallel processing, adaptive learning and fault tolerance, so a BP neural network model based on principal component analysis will be created ,thus we can make the effectively risk assessment on gas pipeline.

Keywords: Principal component analysis, BP neural network, Risk model.

1 Introduction

The rapid development of city gas is an important symbol of the modern city, with the progress of the laying of Handan municipal pipe network, in order to speed up the rate of the gas pipeline, to further improve the pipe network; there are some measures which ensure the safety of users:

(1) Encryption pipeline project: 2011 will improve the supporting pipe network 10 kilometers, and Handan gas pipeline was most developed before 1997, corrosion is more and more serious, In order to ensure the safety of users of gas, this year will complete the five gardens gas pipeline reconstruction tasks.
(2) Universal gas project: According to the development plans of Handan City, the gas penetration will reach 99% after three years, thus making the transformation should also be focusing on the development of new gas customers, to ensure gas users in 2011 more than in 10000.
(3) Security Engineering: Gas safety as a great social and political responsibility, Handan city put forward a series of measures to ensure that the normal aging long-term dress pipe work, so as to improve the safety operation of the whole.
(4) County ventilation engineering : In order to better use of fuel gas, the popularity of Handan city will be in urban areas under the condition of basic saturated, positive development each county, accomplish truly popularization, better service to the people ,and at the same time also the development of fuel gas projects for Handan city laid a stable foundation.

D. Jin and S. Lin (Eds.): Advances in EECM Vol. 1, LNEE 139, pp. 661–665.
springerlink.com

2 Security Risk Assessment (Based on the Principal Component Analysis and the BP Neural Network of Safety Evaluation Model) and Case Analysis

In this paper we select the gas pipeline of Handan as the research object and then make simulation analysis .This pipeline is located in the city's busy district, the relatively large flow of people. Through the detection of the corresponding index factors to obtain the corresponding data collection, we can make use of principal component extraction and get the corresponding data input software calculation related coefficient matrix, get the Eigen value and eigenvector and variance contribution .According to the principles of the main ingredients to determine the (cumulative contribution rate of over 85% of the smallest integer m the number of main components),the selected principal components as the BP neural network input layer node index, instead of the above indicators, through principal component analysis of the input layer node index significantly cut, reducing the complexity of the network model, finally input the index data to BP neural network, get the learning sample.

(1) Evaluation index system.General effect of the gas pipeline risk index and hierarchical structure such as table 1.

Table 1. Gas pipeline safety evaluation index

First-class indexes	second-class indexes
Gas Users A_1	Residents of unauthorized alterations or renovation Unauthorized use of gas facilities residents X_1 Improper use (long-term change is not timely, not closed stove, stepped on, etc.) X_2
Gas Company A_2	No intention of destruction pipeline when testing X_3
Construction Company A_3	Illegal construction X_4 Design defect X_5 Weld defects X_6 Material defects X_7 External Corrosion X_8
Corrosion A_4	Internal Corrosion X_9 Stress Corrosion X_{10} Lightning X_{11}
Other natural conditions A_5	Rainstorms and floods X_{12} Frost Crack Pipe X_{13}
Design A_6	Design personnel level X_{14} Design personnel negligence X_{15}

(2) Sample data collected. This paper collected 10 sections of the pipeline sample data, and the specific sample data value such as table 2.

Table 2. Sample data value

	X_1	X_2	X_3	X_4	X_5	X_6	X_7	X_8	X_9	X_{10}	X_{11}	X_{12}	X_{13}	X_{14}	X_{15}
1	0.10	0.05	0.10	0.12	0.01	0.03	0.03	0.15	0.16	0.05	0.05	0.09	0.13	0.10	0.05
2	0.20	0.06	0.13	0.10	0.02	0.01	0.02	0.14	0.19	0.09	0.08	0.08	0.12	0.13	0.02
3	0.13	0.04	0.15	0.13	0.02	0.02	0.01	0.18	0.15	0.04	0.04	0.05	0.10	0.05	0.05
4	0.14	0.12	0.08	0.08	0.01	0.03	0.03	0.15	0.18	0.08	0.09	0.04	0.07	0.12	0.03
5	0.16	0.13	0.12	0.11	0.03	0.03	0.01	0.19	0.15	0.05	0.05	0.05	0.04	0.09	0.07
6	0.17	0.08	0.13	0.07	0.01	0.01	0.01	0.16	0.14	0.05	0.02	0.06	0.10	0.10	0.05
7	0.15	0.14	0.14	0.10	0.03	0.02	0.02	0.15	0.11	0.01	0.06	0.05	0.10	0.11	0.08
8	0.20	0.16	0.15	0.06	0.02	0.02	0.02	0.15	0.12	0.05	0.05	0.02	0.08	0.15	0.02
9	0.10	0.10	0.18	0.09	0.01	0.01	0.01	0.12	0.15	0.06	0.01	0.01	0.16	0.07	0.05
10	0.11	0.10	0.10	0.10	0.03	0.01	0.03	0.11	0.15	0.02	0.05	0.05	0.10	0.08	0.08

(3) The extraction of the principal component. Using the principal component analysis of SPSS software for original data processing, the variables of the complex relationship between interrelated simplified analyses, high dimension variables space will be reduced processing, getting the corresponding Eigen values and eigenvectors and variance contribution rate, such as shown in table 3.

Table 3. Principal component characteristic value and the cumulative contribution

PC	characteristic value	variance contribution	cumulative contribution
Y_1	3.883	25.887	25.887
Y_2	3.261	21.740	47.627
Y_3	3.029	20.196	67.823
Y_4	2.039	13.590	81.414
Y_5	1.204	8.028	89.441

According to the judgment principle of principal component, we will select the first five principal components as theinput layer nodes indicators of the integrated the BP nerve network model,replacing the original 15 indicators,in this way,reduced the complexity of the network model effectively.

Table 4. The principal component factor loading matrix

Load factor Evaluation index	Y_1	Y_2	Y_3	Y_4	Y_5
X_1	0.485	-0.596	-0.202	0.350	0.490
X_2	0.081	-0.891	0.134	-0.270	-0.229
X_3	-0.638	-0.282	-0.525	0.293	0.129
X_4	-0.257	0.665	0.530	0.266	0.132
X_5	-0.201	-0.384	0.651	-0.081	0.500
X_6	0.412	0.004	0.589	0.250	-0.570
X_7	0.536	0.230	0.270	-0.733	-0.075
X_8	0.116	-0.145	0.401	0.872	-0.115
X_9	0.565	0.665	-0.194	0.098	0.022
X_{10}	0.660	0.241	-0.562	0.298	-0.103
X_{11}	0.829	0.010	0.340	-0.153	0.161
X_{12}	0.351	0.590	0.332	0.110	0.457
X_{13}	-0.385	0.513	-0.596	0.310	0.100
X_{14}	0.730	-0.506	-0.191	-0.201	0.162
X_{15}	-0.639	0.024	0.662	-0.224	0.052

3 The BP Neural Network Evaluation Method

Generally use the three layer structure of BP neural network model which including input node layer,the output node layer and a hidden layer. In this three-layer perception, input vector is $X=(x_1,x_2,\ldots x_I,\ldots x_n)_T$,In addition to $x_0=-1$ is set for the threshold introduction of hidden layer neuron, hidden layer output vector is $Y=(y_1,y_2,\ldots y_j,\ldots y_m)_T$, In addition to $y_0=-1$ is set for the threshold introduction of output layer neuron ,the output vector of output layer is $O=(o_1,o_2,\ldots o_k,\ldots o_n)_T$; the expected output vector is $d=(d_1,d_2\ldots d_k,\ldots,d_l)_T$. the weight matrix between the input layer and the hidden layer is express with V,$V=(V_1, V_2\ldots V_j,\ldots Vm)_T$, The BP neural network training process is as follows: we can get N samples about N typical projects .network was trained under the premise that the principal component analysis of indicator data as the network's input vector matrix and the reviews set as the output vector. Then the stable training results will be offered as knowledge base after training, so that we can use it to evaluate the risk management of gas pipeline.

4 Maintenance Management Measures

(1)Establishing the gas pipeline information database in whole country. First, we should establish a reasonable data acquisition system .Before we evaluate the gas pipeline risk what we must do is collecting and screening data. Secondly we should ensure the data correct and effective and the data that is good or not directly concern the efficiency and effectiveness of evaluation.

(2)Establishing and perfecting the system of safety management. The system is combined with modern technology, adopted new technology and new methods. we can establish an operational and supervisory mechanism which is scientific, perfect and reasonable.

(3)Popularizing the safety knowledge of gas. we should set up a supervisory net for the gas security by the masses version.

5 Concluding Remarks

The detecting method based on neural network is a kind of automatic, smart and new-type leakage detecting method. And this method possesses the function of self-learning, and it can constantly improve the leakage detecting performance by its own ability .Its development depends on high-speed computer, mechanics of communication and the progress of mathematics. It is the developmental direction of future detective technology. Furthermore this method make for improving the economic benefits of gas industry and ensuring the safe operation of gas pipeline throughout the country, enhancing the technological capability of management for gas pipeline network in our country city ,extending the service life of gas pipeline in city and avoiding or reducing the unpreventable accidents of the gas network.

References

1. Zhang, W.: Study on the road safety evaluation based on BP neural network. Chanan University, Xi'an (2006)
2. Shen, X.Y.: A renovated safety evaluation model based on the principal component analysis and BP neural network. Journal of Safety and Environment 9(1), 180–183 (2009)
3. Cai, L.J.: Analysis on present risk assessment situation of city gas pipelines. City gas 402(8), 15–18 (2008)
4. Wen, S.Q.: Design and implementation of gas conduits evaluation expert system based on BP neural network. Computer Engineering and Design 28(2), 425–439 (2007)

Author Index

The manufacturer's authorised representative in the EU is Springer Nature Customer Service Centre GmbH, Europaplatz 3, 69115 Heidelberg, Germany. If you have any concerns regarding our products, please contact ProductSafety@springernature.com

Printed and bound by CPI Group (UK) Ltd, Croydon, CR0 4YY
25/09/2026
02222839-0015